P9-DGF-921

ALGEBRA

A... ZEJ WODZINSKI

ALGEBRA

SAUNDERS MacLANE

Max Mason Distinguished Service Professor of Mathematics
The University of Chicago

GARRETT BIRKHOFF

Putnam Professor of Pure and Applied Mathematics
Harvard University

CHELSEA PUBLISHING COMPANY
NEW YORK, N. Y.

THIRD EDITION

Copyright ©, 1993, by Saunders Mac Lane
and Garrett Birkhoff

Copyright ©, 1988 by Saunders Mac Lane
and Garrett Birkhoff

Copyright ©, 1979, by Saunders Mac Lane
and Garrett Birkhoff

Copyright ©, 1967, by Saunders Mac Lane
and Garrett Birkhoff

Library of Congress Catalog Card Number 87-71728
International Standard Book Number 0-8284-0330-9

512′.02

Printed in the United States of America

Preface to the Third Edition

THIS book aims to present modern algebra from first principles, so as to be accessible to undergraduates or graduates, and this by combining standard materials and the needed algebraic manipulations with the general concepts which clarify their meaning and importance.

The modern conceptual approach to algebra starts with the description of algebraic structures by means of axioms chosen to suit the examples at hand; as for instance with the axioms for groups, rings, fields, lattices, and vector spaces. This axiomatic approach, emphasized by David Hilbert and developed in Germany in the 1920's by Emmy Noether, Emil Artin, B. L. Van der Waerden, and others became available on the graduate level in the 1930's, and was then popularized on the undergraduate level in the 1940's and 1950's in part by our own *A Survey of Modern Algebra*. Since that time, algebra has expanded vigorously; this book is designed to present also the basic new concepts which have appeared.

In linear algebra, the notions of module and tensor product are emphasized, because of their use in topology, differential geometry, and elsewhere. Here the standard axioms used to describe a vector space with scalars from a field also apply when the scalars are elements of a ring; the axioms then define a module over that ring. By means of examples, we aim to make this notion easy to handle.

Homomorphisms are the means of comparing two algebraic structures of the same type. Such homomorphisms can be composed, and under this composition constitute the morphisms of a category, while many constructions on algebraic structures apply also to these morphisms, and so can be understood as functors (= morphisms) on the category at issue.

The construction of a new algebraic object will often solve a specific problem in a universal way, in the sense that every other solution of the given problem is obtained from this one by a unique homomorphism. The basic idea of an adjoint functor arises from the analysis of such universals.

Our presentation starts with integers, groups, and rings. In Chapter I the integers are constructed from the natural numbers so as to provide the universal way to make subtraction possible, thus providing a fundamental explanation of this operation. Chapter II begins with the origin of groups in the study of symmetry. Given a group G with a normal subgroup N, there is a universal construction of a homomorphism to another group G/N, so that the given N is mapped to the identity element; this description of the quotient group is decisive, because it provides all the needed information about the behavior of this G/N. Rings (Chapter III) involve both addition and multiplication; here the extension of a ring to a ring of polynomials in an indeterminate x is described as the universal way of adding one new element (to wit, x) to the original ring. With these and other examples of universal constructions at hand, Chapter IV gives the general description of a universal element, plus the axiomatic description of categories (of morphisms) and of lattices (of subobjects).

The next three chapters treat linear algebra: modules, vector spaces, and matrices, with emphasis on the idea that matrices are tools used to represent and compute with linear transformations between vector spaces. The proof of the existence and properties of eigenvalues and eigenvectors for linear transformations in Chapter X is based on the special properties of the real and complex fields, as developed in Chapter VIII. There, other examples of fields (of power series and of p-adic numbers) are also given. The intermediate Chapter IX constructs the tensor product of two vector spaces conceptually in terms of universal bilinear functions, and includes related notions such as that of determinant and exact sequence—an idea important for topology.

Up to this point, the chapters fall in a natural sequence; thereafter, the chapters are largely independent of each other. Chapter XI uses the concept of module over a principle ideal domain to provide a unified treatment of both the rational canonical form for matrices under similarity and the structure theorem for finitely generated groups—a fine illustration of the use of conceptual unification. A second chapter on group theory develops notions such as that of a composition series, to be used in the subsequent chapter presenting Galois Theory. There are also chapters on Lattices, on Categories, and on Multilinear Algebra, including tensor and exterior algebra. These chapters were also present in our second edition, but we have also included the chapter on Affine and Projective Geometry, with an appropriate axiomatization—a chapter from the first edition of our book, but omitted from the second. Moreover, a number of misprints from the second edition have now been corrected.

As in the prior editions, we extend our thanks to the many friends who have helped and commented on our work, together with our thanks to the Chelsea Publishing Company for making this new edition available.

SAUNDERS MAC LANE
GARRETT BIRKHOFF

Dune Acres, Indiana
Cambridge, Massachusetts

From the Preface to the First Edition

RECENT years have seen striking developments in the conceptual organization of Mathematics. These developments use certain new concepts such as "module", "category", and "morphism" which are algebraic in character and which indeed can be introduced naturally on the basis of elementary materials. The efficiency of these ideas suggests a fresh presentation of algebra.

The starting point of this whole development is the systematic use of abstract and axiomatic methods, as in the modern algebra of the decade of the 1920's. At that time it became clear that algebra deals not primarily with the manipulation of sums and products of numbers (such as rationals, reals, or complexes) but with sums and products of elements of any sort—under the assumption that the sum and product for the elements considered satisfy the appropriate basic laws or "axioms"; more explicitly, the axioms for a "ring" (addition, subtraction, multiplication) or a "field" (these three, plus division). There was a similar transformation in the treatment of vectors. Initially, a vector in three-space was given in terms of its components relative to a given system of axes, so a vector was described as a triple of numbers. Emphasis on the operations of vector addition and multiplication of a vector by a real number (a scalar) showed that the vectors could be better treated, independently of any choice of axes, as the elements of a real "vector space" in which these operations are defined and are required to satisfy suitable axioms. The same axioms (and most of the theorems) still apply when the scalars are not real numbers but elements of any field. Thereby the algebra of matrices appeared in clearer and invariant guise as the algebra of linear transformations. Other branches of algebra were clarified by similar reformulations. For example, Galois theory was seen to deal not with substitutions on the roots of a polynomial, but with the group of automorphisms of the field generated by these roots. All these ideas of modern algebra made their way into instruction in the following decade (that of the 1930's), on the graduate level with van der Waerden's influential *Moderne Algebra* and later on the undergraduate level with various books such as our own *A Survey of Modern Algebra*. By now the use of these ideas in instruction is generally accepted.

In the meantime, modern algebra itself has continued to develop vigorously. For example, a vector space over a field is defined by axioms on the addition of vectors and the multiplication of a vector by a scalar from that field. If the field of scalars is replaced by a ring of scalars, the same axioms make sense and define the notion of a module over that ring. This more general notion of module has proved to have widespread utility in topology and geometry—and even in the study of vector spaces. Similarly, linear algebra (i.e., the study of linear transformations) has been supplemented by multilinear algebra (the study of multilinear functions), with the resulting concepts of tensor product and exterior product. At the same time the understanding of algebra has deepened; it is now clear that we study not just a single algebraic system (a group or a ring) by itself, but that we also study the homomorphisms of these systems; that is, the functions mapping one system into another so as to preserve the operations (of addition and/or multiplication).

All the systems of a given type together with the homomorphisms between them are said to form a "category" consisting of "objects" (the systems) and "morphisms" (the homomorphisms between them). The essential operation is that of composing two morphisms to form a third one, and the axioms for this composition define the abstract notion of a category. A homomorphism from one category to another is called a "functor", and certain pairs of such functors are said to be "adjoint". For example, the construction of the free group on a given set of generators determines a functor (on the category of sets to that of groups); this functor is adjoint to the "forgetful" functor (from groups to sets) which sends each group G to the set of all elements of G. Such adjoint pairs of functors occur very frequently; their properties can be used systematically to organize many parts of geometry, analysis, and algebra.

This book proposes to present algebra for undergraduates on the basis of these new insights. In order to combine the standard material with the new, it seemed best to make a wholly new start. At the same time, just as in our *Survey*, we hold that the general and abstract ideas needed should grow naturally from concrete instances. With this in view, it is fortunate that we do not need to begin with the general notion of a category. The most basic category is the category whose objects are all sets and whose morphisms are all functions (from one set to another); hence we can start Chapter I with sets—more accurately, with sets, functions, and the composition of functions—as the fundamental materials. On this background, Chapter II introduces the integers as the most basic example of an algebraic system. All the other categories which we need are quite "concrete" ones—each object A in the category is a set (with some structure), and each morphism from an object A to an object B in the category is a function (one which preserves the structure) on the set A to the set B. Hence we can give in

Chapter II an easy, explicit definition of a "concrete" category, leaving the full treatment of the more general notion of a category to Chapter XV. In the same spirit, the idea fundamental to the notion of adjoint functor turns out to be the simple one of a "universal" construction. This idea, introduced in Chapter I for sets and in Chapter II for other concrete categories (such as monoids and lattices), is then developed with successive examples throughout the subsequent chapters. In a similar way, the notions of least upper and greatest lower bounds in a partially ordered set are introduced (with examples) in Chapter II to be ready for use in the many subsequent illustrations.

The effective completion of any book depends on the help of many people; it is a pleasure for us to acknowledge that help for our book. That superb critic, Arthur Mattuck, read the whole manuscript and made many pertinent suggestions, from most of which we have profited greatly (probably we could have profited even more from the rest). We have, likewise, drawn great help from the incisive comments of M. F. Smiley and L. J. Ratliff, Jr. O. F. G. Schilling suggested a number of exercises and other topics for inclusion. Paul Palmquist and Richard H. Ivan read several chapters of the manuscript with eagle-eyed attention. Students in classes at the University of California (Riverside) and at The University of Chicago worked over a preliminary edition thoroughly; in particular, Louis Crane made vital comments. Successive drafts of the book were carefully typed by Dorothy Mac Lane, Gretchen Mac Lane, and Merilee Benson. Dorothy Mac Lane prepared the index, especially necessary here to organize the terminology needed for the new concepts in this book. Some of the ideas presented here depend on current research and so owe a great deal to the generous and staunch support which the Air Force Office of Scientific Research has provided for a number of years for the research of one of the authors (S. M.). To all these—and others—we express our sincere thanks and appreciation.

SAUNDERS MAC LANE
GARRETT BIRKHOFF

Dune Acres
Cambridge

From the Preface to the Second Edition

THE treatment of many of these topics in the first edition has been simplified—and clarified—in this second edition. The material on universal constructions, formerly introduced at the end of the first chapter, has now been assembled in Chapter IV, at a point where there are at hand many more effective examples of these constructions. A great many points in the exposition have been clarified, for instance in a simpler construction of the integers, a more elementary description of polynomials, and a more direct treatment of dual spaces. The chapter on special fields now includes power series fields and a treatment of the p-adic numbers. There is a wholly new chapter on Galois theory; in exchange, the chapter on affine geometry has been dropped. New exercises have been added and some old slips have been excised.

The effective completion of any book depends on the help of many people; it is a pleasure to acknowledge that help for this second edition. A number of readers pointed out to us errors and possible improvements in the published version. They include Dominique Bernaroli, J. L. Brenner, R. E. Johnson, T. Karenakaran, E. Klemperer, Ronald Nunke, L. E. Pursell, S. Segal, Jacques Weil, and Charles Wells. We are notably indebted to Frank Gerrish, who thoughtfully provided us with an especially large number of astute comments. Neal Koblitz helped with material on Chapter VIII. Kathy Edwards, Joel Fingerman, Leo Katzenstein, Gaunce Lewis, Miguel LaPlaza, and others examined this revision with care and attention. For typing assistance, we are indebted to Karen McKeown and Janet Mezgolits. Dorothy Mac Lane prepared the index. To all these—and to many others—we express our sincere thanks and appreciation.

SAUNDERS MAC LANE
GARRETT BIRKHOFF

Dune Acres, Indiana
Cambridge, Massachusetts

Contents

List of Symbols

Symbol	*Usage*	*Meaning*	*Reference*
$\in$	$x \in S$	x is an element of S	I.1
$\notin$	$x \notin S$	x is not an element of S	I.1
$\subset$	$S \subset X$	S is a subset of X	I.1
	$S \subset G$	S is a subgroup of G	II.4
$\varnothing$	$\varnothing$	The empty set	I.1
$\{\ \}$	$\{x \mid —\}$	All x such that	I.1
$\cap$	$S \cap T$	Intersection of S and T	I.1
$\cup$	$S \cup T$	Union of S and T	I.1
$(\ )$	(k, m)	Binomial coefficient $(k + m)!/(k!)(m!)$	III
$(\ )$	(x, y)	Ordered pair of x and y	I.3
$\times$	$X \times Y$	Product of X and Y	I.3
$\Rightarrow$	$\cdots \Rightarrow —$	$\cdots$ implies —	I.1
$\Leftrightarrow$	$\cdots \Leftrightarrow —$	$\cdots$ if and only if —	I.1
$\circ$	$f \circ g$	Composite, f following g	I.2
$\rightarrow$	$X \rightarrow Y$	Function on X to Y	I,2
$\cdots\blacktriangleright$	$X \cdots\blacktriangleright T$	Function to be constructed	IV.1
$\longmapsto$	$x \longmapsto x^2$	Function assigning x^2 to x	I.2
$\cong$	$X \cong Y$	Bijection (isomorphism) X to Y	I.3
	Y^X	All functions on X to Y	IV.8
$\square$	$x \square y$	Binary operation on x, y	I.3
$\vee$	$x \vee y$	Join of x and y	IV.6
$\wedge$	$x \wedge y$	Meet of x and y	IV.6
	$u \wedge v$	Exterior product of u and v	XVI.6
$\leqslant$	$x \leqslant y$	x less than or equal to y	I.6
		x contained in y (in a lattice)	IV.6
$\mid$	$m \mid n$	m divides n	III.8
$\equiv$	$k \equiv m (\text{mod } n)$	k congruent to m, modulo n	I.8
‘	f‘	Function given by polynomial f	III.7

Symbol	*Usage*	*Meaning*	*Reference*
$\triangleleft$	$N \triangleleft G$	N normal subgroup of G	II.9
$:$	$[G : S]$	Index of S in G	II.8
$/$	X/E	Quotient set of X by E	I.9
	G/N	Quotient group of G by N	II.10
	R/A	Quotient ring of R by ideal A	III.3
	A/D	Quotient module	V.4
$*$	z^*	Conjugate complex number of z	VIII.7
	$f_* S$	Image of S under f	IV.8
	$f^* T$	Inverse image of T under f	IV.8
	V^*	Dual vector space to V	V.7
	t^*	Dual map to t	V.7
	t^*	Adjoint map to t	X.6
$\oplus$	$A \oplus B$	Biproduct of modules A, B	V.6
$\otimes$	$A \otimes B$	Tensor product of A, B	IX.8
$\vert\ \vert$	$\vert a\vert$	Absolute value of a	VIII.1
	$\vert u\vert$	Length of vector u	X.5
	$\vert A\vert$	Determinant of matrix A	IX.2
$\langle\ \rangle$	$\langle u, v\rangle$	Inner product of u, v	X.5
$\perp$	$u \perp v$	u orthogonal to v	X.5
∞	∞	Infinite (characteristic)	III.1
op	G^{op}	Opposite group	II.2
	R^{op}	Opposite ring	III.2
	$\mathbf{X}^{\mathrm{op}}$	Opposite category	XV.1

Standard Abbreviations	*Meaning*	*Reference*
Aut (G)	Automorphisms of G	II.1
Bilin $(A, B; C)$	Bilinear functions $A \times B \to C$	IX.10
End (A)	Endomorphisms of A	II.2, V.2
hom (X, Y)	Morphisms of X to Y	IV.2, XV.1
Hom (A, B)	Group of morphisms, A to B	V.2
$GL(n, F)$	General linear group	VII.6, IX.3
O_n	Orthogonal group	X.7
$SL(n, K)$	Special linear group	IX.3
dim V	Dimension of V	VI.2
rank A	Rank of A	VI.3, VII.5

Letters with Fixed Meanings	*Meaning*	*Reference*
A_n	Alternating group	II.6
$\mathbf{C}$	Field of complex numbers	VIII.7
$\mathbf{N}$	Set of natural numbers	I.4
$\mathbf{Q}$	Field of rational numbers	III.5
$\mathbf{R}$	Field of real numbers	VIII.5
S_n	Symmetric group	II.6
$\mathbf{Z}$	Ring of integers	I.7
$\mathbf{n}$	$\{1, 2, \cdots, n\}$	I.7
Δ_n	Dihedral group	II.5
δ_{ij}	Kronecker delta	VI.4
$\boldsymbol{\varepsilon}_i$	Unit vectors	V.5

CHAPTER I

Sets, Functions, and Integers

ALGEBRA starts as the art of manipulating sums, products, and powers of numbers. The rules for these manipulations hold for all numbers, so the manipulations may be carried out with letters standing for the numbers. It then appears that the same rules hold for various different sorts of numbers, rational, real, or complex, and that the rules for multiplication even apply to things such as transformations which are not numbers at all. An algebraic system, as we will study it, is thus a set of elements of any sort on which functions such as addition and multiplication operate, provided only that these operations satisfy certain basic rules. The rules for multiplication and inverse are the axioms for a "group", those for addition, subtraction, and multiplication are the axioms for a "ring", and the functions mapping one system to another are the "morphisms". This chapter starts with the necessary ideas about sets, functions, and relations. Then the natural numbers are used to construct the integers and the integers modulo n, with their addition and multiplication. This serves as an introduction to the notion of a morphism from one algebraic system to another.

Many developments in algebra depend vitally upon defining the right concept. When our presentation reaches any definition, the term being defined is put in italics, as *group, ring, field,* and so on. However, terms little used in the sequel as well as terminology alternative to that selected here are put in quotation marks; thus "range" stands for *codomain* and "onto" for *surjective* (see §2 below).

A reference such as Theorem 3 is to Theorem 3 of the current chapter, while Theorem II.3 is to Theorem 3 of Chapter II. In like manner, Corollary IV.5.2 refers to Corollary 2 of Theorem 5 of Chapter IV, and Equation (VI.11) to Equation (11) of Chapter VI. Within each Chapter, Theorems and Propositions are numbered in a single series. More difficult exercises and sections which may be omitted on first reading are denoted by an asterisk, *.

1. Sets

Intuitively, a "set" is any collection of elements, and a "function" is any rule which assigns to each element of one set a corresponding element of a second set.

Examples of sets abound: The set of all lines in the plane, the set $\mathbf{Q}$ of all rational numbers, the set $\mathbf{C}$ of all complex numbers, the set $\mathbf{Z}$ of all integers (positive, negative, or zero). Sets with only a finite number of different elements may be described by listing all their elements, often indicated by writing these elements between braces. Thus the set of all even integers between 0 and 8, inclusive, may be exhibited as $\{0, 2, 4, 6, 8\}$, while the set of all positive divisors of 6 is the set $\{1, 2, 3, 6\}$. The order in which the elements of a set are listed is irrelevant: $\{1, 3, 6, 2\} = \{1, 2, 3, 6\}$.

More formally, "$x \in S$" stands for "x is an *element* of the set S" or equivalently, "x is a member of the set S" or "x belongs to S". Also, $x \notin S$ means that x is *not* an element of S. Since a set is completely determined by giving its elements, two sets S and T are *equal* if and only if they have the same elements; in symbols:

$$S = T \quad \Leftrightarrow \quad \text{For all } x,\ x \in S \text{ if and only if } x \in T. \tag{1}$$

(Here the two-pointed double arrow "$\Leftrightarrow$" stands for "if and only if".) Also, S is a *subset* of T (or, is *included* in T) when every element of S is an element of T, so that, if $x \in S$, then $x \in T$; in symbols:

$$S \subset T \quad \Leftrightarrow \quad \text{For all } x,\ x \in S \Rightarrow x \in T.$$

(Here, on the right, the one-pointed double arrow "$\Rightarrow$" stands for "implies".) By this definition, $S \subset T$ and $T \subset U$ imply $S \subset U$, while the equality of sets, as defined above, may be rewritten as

$$S = T \quad \Leftrightarrow \quad S \subset T \text{ and } T \subset S.$$

A set S is *empty* if it has no elements. By the equality rule (1), any two empty sets are equal. Hence, we speak of *the* empty set, written $\varnothing$. It is also called the *null set* or the *void set;* it is a subset of every set. Also, S is a *proper subset* of a set U when $S \subset U$ but $S \neq \varnothing$ and $S \neq U$.

A particular subset of a given set U is often described as the set of all those elements x in U which have a specified property. Thus the subset of those complex numbers z such that $z^2 = -1$ is written $\{z | z \in \mathbf{C}$ and $z^2 = -1\}$, while the formulas

$$E = \{x | x \in \mathbf{Z} \text{ and } x = 2y \text{ for some } y \in \mathbf{Z}\}, \qquad N = \{x | x \in \mathbf{Z} \text{ and } x \geqslant 0\}$$

describe the set E of all even integers and the set N of all nonnegative integers, respectively. Different properties may describe the same subset; thus

$$\{n | n \in \mathbf{Z} \text{ and } 0 < n < 1\} \qquad \text{and} \qquad \{n | n \in \mathbf{Z} \text{ and } n^2 = -1\}$$

both describe the empty set $\varnothing$.

Next we consider the operations of intersection and union on sets. If R and S are given sets, their *intersection* $R \cap S$ is the set of all elements common to R and S:

$$R \cap S = \{x | x \in R \text{ and } x \in S\},$$

while their *union* $R \cup S$ is the set of all elements which belong either to R or to S (or to both):

$$R \cup S = \{x | x \in R \text{ or } x \in S\}.$$

These definitions may be stated thus:

$$x \in (R \cap S) \quad \Leftrightarrow \quad x \in R \text{ and } x \in S,$$

$$x \in (R \cup S) \quad \Leftrightarrow \quad x \in R \quad \text{or} \quad x \in S.$$

This display correlates the operations of intersection and union with the logical connectives "and" and "or". The corresponding correlate of "not" is the operation of "complement": If S is a subset of U, the *complement* S' of S *in* U is the set of all those elements of U which do not belong to S:

$$S' = \{x | x \in U \text{ and } x \notin S\}.$$

For example, for the sets E and N above, $E \cap N$ is the set of even nonnegative integers, $E \cup N$ the set of all integers except the negative odd ones, while the complement E' of E in $\mathbf{Z}$ is the set of all odd integers.

The operations of intersection, union, and complement satisfy various "identities", valid for arbitrary sets. A sample such identity is

$$R \cap (S \cup T) = (R \cap S) \cup (R \cap T), \tag{2}$$

valid for any three sets R, S, and T. (This equation states that the operation "intersection" is distributive over the operation "union".) To prove this statement, consider any element x. By the definitions of $\cap$ and $\cup$ above,

$$\begin{aligned} x \in [R \cap (S \cup T)] \quad &\Leftrightarrow \quad x \in R \text{ and } x \in S \cup T \\ &\Leftrightarrow \quad x \in R \text{ and } (x \in S \text{ or } x \in T). \end{aligned}$$

For similar reasons,

$$x \in [(R \cap S) \cup (R \cap T)] \Leftrightarrow (x \in R \text{ and } x \in S) \text{ or } (x \in R \text{ and } x \in T).$$

Now, in view of familiar properties of "and" and "or", the two different statements made about x at the right of the two displays above are logically equivalent. Hence, the two sets in question have the same elements and therefore are equal. In other words, this proof reduces property (2) of intersection and union to an exactly corresponding property of the logical connectives "and" and "or".

A similar argument gives another distributive law,

$$R \cup (S \cap T) = (R \cup S) \cap (R \cup T). \tag{3}$$

Other algebraic properties of intersection, union, and complement will be considered in the exercises in §3 below.

Two sets, R and S, are called *disjoint* when $R \cap S = \varnothing$.

Given a set U, the set $P(U)$ of all subsets S of U is called the *power set* of U; thus $P(U) = \{S \mid S \subset U\}$. For example, if U has two elements, it has four different subsets which are the four elements of $P(U)$. Explicitly, $P(\{1, 2\}) = \{\{1, 2\}, \{1\}, \{2\}, \varnothing\}$. Here $\varnothing$ is the empty set (a subset of every set, as above).

EXERCISES

1. For subsets R, S, and T of a set U, establish the following identities:
(a) $R \cap S = S \cap R$, $\quad R \cap (S \cap T) = (R \cap S) \cap T$.
(b) $R \cup S = S \cup R$, $\quad R \cup (S \cup T) = (R \cup S) \cup T$.
(c) $(R \cap S)' = R' \cup S'$, $\quad (R \cup S)' = R' \cap S'$.
(d) $S \cap (S \cup T) = S$, $\quad S \cup (S \cap T) = S$.

2. Show that any one of the three conditions $S \subset T$, $S \cap T = S$, and $S \cup T = T$ on the sets S and T implies both of the others.

3. For $S \subset U$, show that $S \cap S' = \varnothing$ and $S \cup S' = U$.

4. List the elements of the sets $P(P(\{1\}))$ and $P(P(P(\{1\})))$.

5. Show that a set of n elements has 2^n different subsets.

6. If $m < n$, show that a set of n elements has $(n!)/(n-m)!(m!)$ different subsets of m elements each, where $m! = 1 \cdot 2 \cdots m$.

2. Functions

A *function* f *on* a set S *to* a set T *assigns* to each element s of S an element $f(s) \in T$, as indicated by the notation

$$s \mapsto f(s), \qquad s \in S.$$

The element $f(s)$ may also be written as fs or f_s, without parentheses; it is the *value* of f at the *argument* s. The set S is called the *domain* of f, while T is the *codomain*. The *arrow* notation

$$f : S \to T \qquad \text{or} \qquad S \xrightarrow{f} T$$

indicates that f is a function with domain S and codomain T. A function is often called a "map" or a "transformation".

To describe a particular function, one must specify its domain and its codomain, and write down its effect upon a typical ("variable") element of its domain. Thus the squaring function $f: \mathbf{R} \to \mathbf{R}$ for the set $\mathbf{R}$ of real numbers may be described in any of the following ways: As the function f with $f(x) = x^2$ for any real number x, or as the function $(\text{—})^2$, where — stands for the argument, or as the function which sends each $x \in \mathbf{R}$ to x^2, or as the function given by the assignment $x \mapsto x^2$ for $x \in \mathbf{R}$. We systematically use the *barred arrow* to go from argument to value of a function and the *straight arrow* $S \to T$ to go from domain to codomain.

Note that a letter such as f or g stands for a function, while an expression such as $f(x)$ or $g(x)$ stands for a value of that function for an element x of its domain. For example, in trigonometry the expression "sin x" stands for a number, so we speak not of "the function sin x" but of the function $\sin:\mathbf{R} \to \mathbf{R}$. By using a barred arrow, we can describe particular functions without naming them as f or g; for example, $x \mapsto x^2 + 3x + 2$ for x real describes a function $\mathbf{R} \to \mathbf{R}$.

Two functions f and g are *equal* (in symbols, $f = g$) when they have the same domain, the same codomain, and the same value $f(s) = g(s)$ for each element s of this common domain. For example, the assignment $x \mapsto x + 2$ (add two) defines on the integers $\mathbf{Z}$ a function $f:\mathbf{Z} \to \mathbf{Z}$; on the set $\mathbf{R}$ of real numbers it also defines a function $g:\mathbf{R} \to \mathbf{R}$; these are *different* functions because they have different domains.

The *image* of a function $f:S \to T$ is the set $f(s)$ of all values $f(s)$ for $s \in S$; it is always a subset of the codomain of f.

For any set S, the *identity function* $1_S:S \to S$ is that function $s \mapsto s$ which maps each element s of S onto itself. Different sets have different identity functions. If S is a subset of U, the *insertion* $i:S \to U$ is that function on S to U which assigns to each element of S the same element, now in U. Note that "insertion" is a function $S \to U$ and "inclusion" a relation $S \subset U$; every inclusion relation gives rise to an insertion function.

The *composite* $f \circ g = fg$ of two functions is the function obtained by applying them in succession; first g, then f—provided this makes sense; that is, provided the domain of f is the codomain of g. More formally, given the functions

$$g : R \to S, \qquad f : S \to T,$$

their *composite* is the function $f \circ g:R \to T$ with values given by

$$(f \circ g)(r) = f(g(r)), \qquad \text{all } r \in R. \tag{4}$$

This definition may be visualized by the "mapping diagram" displayed below:

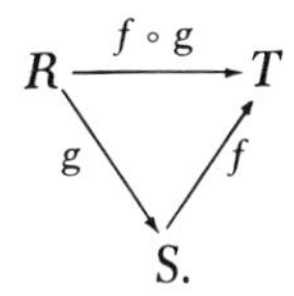

To go from the set R directly to the set T by the composite $f \circ g$ is the same as going through S in two steps, the first by g and the second by f. We also express this fact by saying: This triangular diagram "commutes".

Composition of functions obeys the

Associative law: $(f \circ g) \circ h = f \circ (g \circ h),$

whenever the composites involved are defined. This is obvious intuitively; both $(f \circ g) \circ h$ and $f \circ (g \circ h)$ have the effect of applying first h, then g, and finally f, in that order. Formally, given $h:P \to R$, $g:R \to S$, and $f:S \to T$, both triple composites $(f \circ g) \circ h$ and $f \circ (g \circ h)$ are functions on P to T, while the first composite assigns to each $p \in P$ the value

$$[(fg)h]p \underset{(fg)h}{=} (fg)(hp) \underset{fg}{=} f(g(hp)) \underset{gh}{=} f((gh)p) \underset{f(gh)}{=} [f(gh)]p;$$

here each step depends on applying the definition (4) of composition to the composite indicated below the equality symbol for that step. By the definition of equality for functions, this proves the associative law $(fg)h = f(gh): P \to T$. Note that here (and often later) it is convenient to omit the symbol "$\circ$" in $f \circ g$ and the parentheses in $h(p)$ or $(gh)p$.

Under composition each function $f:S \to T$ obeys the

Identity law: $f \circ 1_S = f = 1_T \circ f : S \to T.$

To prove the first equality, note by (4) that $(f1_S)s = f(1_S s) = fs$ for all $s \in S$; hence, $f1_S = f$, by the definition of equality for functions. The second equality is proved similarly.

A function $r:S \to T$ is said to be a *restriction* of a function $f:U \to V$ when $S \subset U$, $T \subset V$, and $r(s) = f(s)$ for each $s \in S$. (One also then says that f is an *extension* of the function r.) For example, given a subset $S \subset U$, the insertion $i:S \to U$ is a restriction of the identity $1_U:U \to U$.

Certain useful special types of functions will now be defined.

A function $f:S \to T$ is *injective* or an *injection* when $s_1 \neq s_2$ in S implies $fs_1 \neq fs_2$ in T; that is, when f carries distinct elements of its domain to distinct elements of its codomain. For example, every insertion is an injection. A function $h:S \to T$ is *surjective* or a *surjection* when its image is the whole codomain T; that is, when to each $t \in T$ there exists at least one $s \in S$ with $hs = t$. Finally, a *bijection* $b:S \to T$ is a function which is both an injection and a surjection; thus b is *bijective* if and only if to each $t \in T$ there is exactly one element $s \in S$ with $bs = t$. The notation $\cong$, as in $b:S \cong T$, indicates that b is a bijection of S to T.

For example, among the functions $\mathbf{Z} \to \mathbf{Z}$, the function $n \mapsto (-n)$ is a bijection, the function $n \mapsto 2n$ is an injection but not a surjection, and the function $n \mapsto n^2$ is neither an injection nor a surjection.

Again, for example, if $\mathbf{R}^+$ is the set of all nonnegative real numbers, the squaring function $g:\mathbf{R} \to \mathbf{R}^+$ given by $g(x) = x^2$ is surjective, because every nonnegative real is the square of some real number. However, the squaring function $f:\mathbf{R} \to \mathbf{R}$ with $f(x) = x^2$ and codomain *all* the real numbers is *not* surjective. These two squaring functions (though they have the same values) count as different functions because they have different codomains: Whether or not a function is a surjection *depends on its codomain.*

There is another parallel terminology for these ideas:

Injection $S \to T$ =	"one-one"	map of S "into" T;
Surjection $S \to T$ =		map of S *onto* T;
Bijection $S \to T$ =	one-one	map of S onto T,

or, in the last case, a "one-one correspondence" of S to T. The older terminology (that to the right) will not be used in this book.

Any function f can be written as a composite $f = g \circ h$, where g is injective and h surjective. Indeed, if $f: S \to T$ has image $U \subset T$, its restriction $r: S \to U$ is a surjection, the insertion $i: U \to T$ is an injection, and f itself is the composite $f = i \circ r$.

Certain functions have "inverses". Suppose that $g: T \to S$ and $f: S \to T$, so that the composite $f \circ g$ is defined. If this composite is the identity $1_T = f \circ g$, call f a *left inverse* of g and g a *right inverse* of f. When the composites in both orders are identities, so that $f \circ g = 1_T$ and $g \circ f = 1_S$, call f a *two-sided inverse* of g (and hence g a two-sided inverse of f).

Theorem 1. *A function with non-empty domain is an injection if and only if it has a left inverse. A function is a surjection if and only if it has a right inverse.*

Proof: Suppose first that $g: T \to S$ has a left inverse $f: S \to T$. Then $fg = 1_T$, so $g(t_1) = g(t_2)$ implies $t_1 = fg(t_1) = fg(t_2) = t_2$. Therefore, g is injective. Conversely, suppose that $g: T \to S$ is injective with domain $T \neq \varnothing$; pick some $t_0 \in T$. Since g is injective, there is to each $s \in S$ at most one t with $s = g(t)$; hence, a function $f: S \to T$ is defined by

$$\begin{aligned} f(s) &= \text{that } t \text{ with } g(t) = s, && \text{when } s \in \text{image } (g), \\ &= t_0, && \text{otherwise.} \end{aligned}$$

This function f sends each $g(t)$ "back where it came from"; so $f(g(t)) = t$ for every t. This states that $f \circ g = 1_T$, so f is the desired left inverse for g.

Note, however, that an injection g which is not a bijection will have in general many left inverses; for example, one for each choice of the t_0 above.

It remains to prove the second half of the theorem, concerning surjections. Suppose first that a function $f: S \to T$ has a right inverse g. Now $1_T = f \circ g$ means that $t = f(gt)$ for all t, so each $t \in T$ is in the image of f, and f is surjective, as required. Conversely, suppose that $f: S \to T$ is surjective; this means that to each $t \in T$ there is at least one $s \in S$ with $f(s) = t$. Choose one such s for each t and define $g: T \to S$ by letting $g(t)$ be the chosen s. Then $f(g(t)) = t$, so $f \circ g = 1_T$, and g is the desired right inverse. This completes the proof of the theorem.

Note: This proof depends on making a (possibly) infinite number of choices (one $s \in S$ with $f(s) = t$ for each $t \in T$). In an axiomatic treatment of set theory, when all the operations on sets are derived from a complete list of formal axioms on the membership relation $x \in S$, one of the axioms states that such a set of choices can be made. This axiom, called the "axiom of choice", states that to each set $\mathfrak{F}$ whose elements are disjoint nonvoid sets, there exists a set C such that each $C \cap S$, for $S \in \mathfrak{F}$, has exactly one element. This axiom is equivalent to the assumption that every surjection has a right inverse (Exercise 11).

COROLLARY. *The following properties of a function $g: T \to S$ are equivalent:*

(*i*) *g is a bijection.*
(*ii*) *g has both a left inverse f and a right inverse h.*
(*iii*) *g has a two-sided inverse.*

When this is the case, any two inverses (left, right, or two-sided) of g are equal. This unique inverse of g (written g^{-1}) is bijective, and satisfies

$$(g^{-1})^{-1} = g. \tag{5}$$

Proof: First suppose that $T \neq \varnothing$ and $S \neq \varnothing$, so the theorem can be used. Since a bijection is both surjective and injective, the theorem at once gives the equivalence of (i) and (ii). As for (iii), any two-sided inverse is trivially both a left and a right inverse; thus (iii) implies (ii). Conversely, (ii) implies

$$f = f \circ 1_S = f \circ (g \circ h) = (f \circ g) \circ h = 1_T \circ h = h,$$

which means that $f = h$ is a two-sided inverse for g; hence (ii) gives (iii). This argument also shows that any left inverse f of g must equal any right inverse h; this is the next clause of the corollary. Finally, the inverse $f = h$ of g has g for a two-sided inverse, hence, it is also bijective and has g as its inverse. This is the conclusion (5) of the corollary.

Only the (uninteresting) case of the corollary when T or S is empty remains: Now a function $g: \varnothing \to S$ with empty domain must assign to each element of $\varnothing$ an element of S. But there are no elements of the empty set $\varnothing$, so there is exactly one function $\varnothing \to S$ (namely, the one which involves *no* assignments). If $S \neq \varnothing$, this function g is not a bijection; on the other hand, there can be no function $S \to \varnothing$, so g has no inverses of any sort. Thus the corollary holds in this case. If $S = \varnothing$ it holds trivially, and hence in all cases.

Note incidentally that the function $g:\varnothing \to S$ with $S \neq \varnothing$ is injective but has no left inverse.

If $g:T \to S$ and $k:S \to R$ are both bijections, so is their composite $k \circ g$; its inverse is given by

$$(k \circ g)^{-1} = g^{-1} \circ k^{-1} \qquad \text{(reverse the order).} \tag{6}$$

Indeed, $k^{-1} \circ k = 1_S$ and $g^{-1} \circ g = 1_T$, so, by the associative law,

$$g^{-1}k^{-1}kg = g^{-1}1_S g = g^{-1}g = 1_T,$$

and $g^{-1}k^{-1}$ is a left inverse for kg. A similar calculation shows that it is also a right inverse, hence the conclusion and (6).

Conventions on functions differ. In this discussion (as elsewhere in this book), we have written each function to the left of its argument, as in $f(s)$—and as is customary in analysis and topology. In consequence, the composite $f \circ g$ has meant *first* apply g, *then* apply f. Functions may also be written to the right of their arguments; then a composite has the opposite meaning.

EXERCISES

1. If $S = \{0, 1\}$ is a set with exactly two elements, exhibit all functions $S \to S$ and classify them as injective, surjective, bijective, or none of these.

2. If $f \circ g$ is defined and both f and g have left inverses, show that $f \circ g$ has a left inverse.

3. Show that the composite of two surjections is a surjection, and similarly for injections.

4. If f is a bijection and $f \circ g$ is defined, show that g is an injection if and only if $f \circ g$ is, and a surjection if and only if $f \circ g$ is one.

5. With $\mathbf{N}$ the set of nonnegative integers, show that the function $f:\mathbf{N} \to \mathbf{N}$ given by $n \mapsto n^2$ has no right inverse, and exhibit explicitly two left inverses.

6. If f is injective, while both $f \circ g$ and $f \circ g'$ are defined, show that $f \circ g = f \circ g' \Rightarrow g = g'$.

7. Find an analog of Exercise 6 for surjections.

8. For any function $f:S \to T$ with $S \neq \varnothing$, construct a function $h:T \to S$ with $fhf = f$. Deduce from this the results of Theorem 1.

9. Show that a function which has a unique right inverse is necessarily bijective.

10. Prove the Corollary of Theorem 1 without using the axiom of choice.

***11.** Assuming that every surjection has a right inverse, prove the axiom of choice as stated in the text. (*Hint:* Let U be the union of all $S \in \mathfrak{F}$, define $f:U \to \mathfrak{F}$ by $f(u) = S$ when $u \in S$, and show f surjective.)

3. Relations and Binary Operations

To treat functions of two variables or relations between two variables we use "ordered pairs". The *ordered pair* consisting of two elements, s and t, in that order, is written (s, t). The equality of two ordered pairs is defined by the rule

$$(s, t) = (s', t') \quad \Leftrightarrow \quad s = s' \quad \text{and} \quad t = t'.$$

The *cartesian product* $S \times T$ of two sets S and T is defined to be the set of all ordered pairs (s, t) of elements from S and T, respectively. Thus

$$S \times T = \{(s, t) | s \in S, \quad t \in T\}.$$

Thus, if $\mathbf{R}$ is the set of all real numbers, $\mathbf{R} \times \mathbf{R}$ is the set of all ordered pairs (x, y) of real numbers; in other words, $\mathbf{R} \times \mathbf{R}$ is just the set of all cartesian coordinates of points in the plane (relative to given coordinate axes).

Any cartesian product $S \times T$ may be "projected" onto its "axes", S and T:

$$S \stackrel{p}{\leftarrow} S \times T \stackrel{q}{\rightarrow} T.$$

These *projections* are the functions p and q defined by $p(s, t) = s$ and $q(s, t) = t$, as in the diagram

$$\begin{array}{ccc} T & \xleftarrow{q} & S \times T \\ & & \downarrow p \\ & & S. \end{array}$$

We call this the *cartesian-product diagram.*

Note the bijection $S \times T \cong T \times S$ given by $(s, t) \mapsto (t, s)$.

Ordered triples may be described in terms of ordered pairs. Given r, s, and t, define the *ordered triple* (r, s, t) to be $(r, (s, t))$. Write $R \times S \times T$ for the set $R \times (S \times T)$ of all triples $(r, (s, t))$ for $r \in R$, $s \in S$, and $t \in T$, and note that the assignment $(r, (s, t)) \mapsto ((r, s), t)$ is a bijection $R \times (S \times T) \cong (R \times S) \times T$. Ordered "quadruples" (r, s, t, u) and the like are defined similarly.

One may also form the cartesian product of functions. Given two functions $u: S \rightarrow S'$ and $v: T \rightarrow T'$, their *cartesian product* is the function $u \times v$: $S \times T \rightarrow S' \times T'$ defined by $(u \times v)(s, t) = (us, vt)$.

The cartesian product is useful in describing the usual functions of two or more arguments; a function F with two arguments $s \in S$ and $t \in T$ and with values in a set W is a function

$$F : S \times T \rightarrow W$$

on the cartesian product $S \times T$ to the set W. Such a function F assigns to each ordered pair $(s, t) \in S \times T$ a value $F(s, t) \in W$.

For any two sets X and Y, a subset $R \subset X \times Y$ is called a *binary relation* on X to Y (or, a relation "between" X and Y). Then $(x, y) \in R$ is usually written as xRy and is read: "x stands in the relation R to y", as in "$x \leqslant y$" for numbers or in "x divides y" for integers x and y. For example, if $X = \{1, 3, 5\}$ and $Y = \{0, 2, 4\}$, then the set $R = \{(1, 2), (1, 4), (3, 4)\}$ consists of all pairs (x, y) with $x < y$, $x \in X$ and $y \in Y$, so is the relation $<$ between X and Y. Again, for example, set inclusion $S \subset T$ is a binary relation on the power set $P(U)$ to itself. The *equality relation* $I = I_X$ on any set X to itself is given by xIy if and only if $x = y$.

The *converse* of a relation $R \subset X \times Y$ is the relation $R^{\smile} \subset Y \times X$ defined by $yR^{\smile} x$ if and only if xRy. If R is a relation on X to Y, its converse is one on Y to X, and $(R^{\smile})^{\smile} = R$.

Thus, for numbers, $\geqslant$ is the converse of $\leqslant$.

The "composite" $R \circ S$ of a relation R with a relation S, in that order, is defined when $R \subset X \times Y$ and $S \subset Y \times Z$; it is the relation $R \circ S \subset X \times Z$ described by

$$x(R \circ S)z \quad \Leftrightarrow \quad \text{For some } y \in Y,\ xRy \text{ and } ySz.$$

In analytic geometry one constructs to each function $\mathbf{R} \to \mathbf{R}$ its "graph" as a subset of $\mathbf{R} \times \mathbf{R}$. The same construction works generally. If $f: X \to Y$ is a function, its *graph* is the subset

$$G(f) = \{(x, y) | x \in X \text{ and } y = f(x)\}$$

of $X \times Y$. Thus the graph of a function on X to Y is a relation on X to Y. However, not every relation R is the graph of a function; this is the case only when the "vertical line" through each point of X meets the set $R \subset X \times Y$ in exactly one point (see Figure I-1). This fact, as formulated in the following theorem, gives another way of constructing functions (this time, from relations).

Theorem 2. *A relation $R \subset X \times Y$ is the graph of a function $X \to Y$ if and only if there is to each $x \in X$ exactly one $y \in Y$ with xRy.*

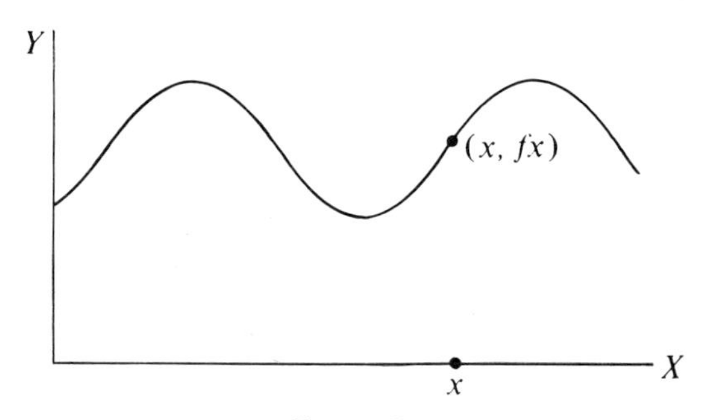

Figure I-1

Proof: Each graph $G(f)$ has the given property. Conversely, if a relation $R \subset X \times Y$ has this property, one may assign to each x in X the unique y in Y with xRy. The resulting function $f: X \to Y$ has R as its graph.

Now consider relations on a set X to itself. A relation R on X to X is called an *equivalence relation* on X when it has each of the following three properties:

Reflexive: xRx for all $x \in X$.
Symmetric: xRy implies yRx for all x, y in X.
Transitive: xRy and yRz imply xRz for all x, y, z in X.

For example, let X be a set whose elements x, y, z are themselves sets, while "xRy" means "there is a bijection on the set x to the set y". This defines an equivalence relation on X: For each x, the identity 1_x is a bijection $x \cong x$; if $x \cong y$ is a bijection, so is its inverse $y \cong x$; bijections $x \cong y$ and $y \cong z$ have composite a bijection $x \cong z$. Again, if T is the set of all triangles in the plane, the relation, "triangle s is congruent to triangle t", and the relation "triangle s is similar to triangle t", are both equivalence relations on T to T. On the other hand, the inclusion relation $S \subset T$ on subsets of a set U is reflexive and transitive, but not symmetric.

In §9 we shall return to equivalence relations.

A *linear order* of a set X is a binary relation on X to X, written $x < y$, which is transitive and satisfies the *trichotomy law:* For all x and y in X, exactly one of $x < y$, $x = y$, or $x > y$ will hold. For example, the usual inequality gives a linear order for $\mathbf{N}$, for $\mathbf{Z}$, or for the set $\mathbf{R}$ of real numbers. For any linear order, $x \leqslant y$ stands for the alternative $x < y$ or $x = y$. This binary relation $\leqslant$ is both reflexive and transitive.

The sum $(m, n) \mapsto m + n$ and the product $(m, n) \mapsto mn$ of two integers m, n give the functions

$$+ : \mathbf{Z} \times \mathbf{Z} \to \mathbf{Z}, \qquad \cdot : \mathbf{Z} \times \mathbf{Z} \to \mathbf{Z}$$

of addition and multiplication for the set $\mathbf{Z}$ of all integers. There are similar functions for the set $\mathbf{R}$ of real numbers. Also, the intersection operation $(S, T) \mapsto S \cap T$ and the union $(S, T) \mapsto S \cup T$ provide functions

$$\cap : P(U) \times P(U) \to P(U), \qquad \cup : P(U) \times P(U) \to P(U)$$

for the power set $P(U)$ of all subsets $S, T, \cdots$ of U. It is convenient to have a common terminology for the properties of such functions as these.

A function $X \times X \to X$ is called a *binary operation* on the set X. Thus addition and multiplication are binary operations on $\mathbf{Z}$. The fact that the symbol $+$ for the operation of addition is usually written between its arguments, as in $m + n$, rather than in front, as in $+(x, y)$ for a function $+$, is just a notational convenience.

More generally, functions such as

$$u : X \to X, \qquad b : X \times X \to X, \qquad t : X \times X \times X \to X$$

are called, respectively, *unary, binary,* and *ternary operations* on the set X. If n is any natural number, while $X^n = X \times \cdots \times X$ (n factors) is the n-fold cartesian product of X with itself, then an *n-ary operation* on X is defined to be a function $f: X^n \to X$. In particular, if $n = 0$, define X^0 to be the standard set $\mathbf{1} = \{1\}$ with just one element. Then a *nullary operation* on X is a function $c: \mathbf{1} \to X$. Since this function has only one value, giving the function c amounts to giving the element $c(1) \in X$. In other words, each nullary operation on a set X amounts to "selecting" an element in X.

Let $\square$ denote an arbitrary binary operation $(x, y) \mapsto x \square y$ on a set X. This operation is said to be

Commutative: when $\quad x \square y = y \square x,$
Associative: when $x \square (y \square z) = (x \square y) \square z,$

in both cases for *all* elements $x, y, z \in X$. Moreover, an element $u \in X$ is a (two-sided) *unit* for the operation $\square$ when for all x

$$(u \text{ a unit}); \qquad x \square u = x = u \square x.$$

A unit element for $\square$ is often called an *identity element* for $\square$. If only $x = u \square x$ for all x, we call u a *left unit* for $\square$. A second binary operation $\triangledown$ on X is said to be (left) *distributive* over $\square$ when

$$x \triangledown (y \square z) = (x \triangledown y) \square (x \triangledown z),$$

for all elements $x, y, z \in X$.

The *opposite* of a binary operation $\square$ on X is the operation $\square^{\text{op}}$ on the same set X defined by $x(\square^{\text{op}})y = y \square x$. Hence, $\square$ is commutative exactly when $\square^{\text{op}} = \square$.

For a fixed set U (U for "universe of discourse") we now consider properties of the binary operations $\cap$ and $\cup$ on $P(U)$. In §1 we already proved, in (2) and (3), that intersection is distributive over union and union distributive over intersection. Similar proofs show that both intersection and union are commutative and associative operations. Moreover, the definitions show that $\varnothing \cup S = S$ and $U \cap S = S$, so that the empty set $\varnothing$ is a unit for union and the universe U a unit for intersection. Still another pair of properties is given by the "idempotency" laws:

$$S \cap S = S, \qquad S \cup S = S.$$

If S' is the complement of the set S in U (S' = all elements of U not in S), then $S \mapsto S'$ is a unary operation on $P(U)$. This operation satisfies with $\cap$ and $\cup$ the identities

$$(S \cap T)' = S' \cup T', \qquad (S \cup T)' = S' \cap T', \qquad (S')' = S.$$

These identities for the algebra of sets will be further studied in Chapter XIV.

EXERCISES

1. (a) Determine which of the three properties "reflexive", "symmetric", and "transitive" apply to each of the following relations on the set $\mathbf{Z}$ of integers:

"m divides n", "$m \leqslant n$", "$m < n$", "$m^2 + m = n^2 + n$".

(b) Do the same for each of the following relations on the set $\mathbf{R}$ of reals:

"$|x| \leqslant |y|$", "$x - y$ is an integral multiple of 2π", "$x^2 + y^2 = 1$".

2. A relation R on a set X to itself is called "circular" if xRy and yRz imply zRx. Show that a relation is reflexive and circular if and only if it is reflexive, symmetric, and transitive.

3. What is wrong with the following "proof" that the symmetric and transitive laws for a relation R imply the reflexive law? "By the symmetric law, xRy implies yRx; by the transitive law, xRy and yRx imply xRx".

4. Let R be any binary relation on a set X to itself. Describe the smallest transitive relation T on X with $R \subset T$. (This relation is called the "ancestral" of R.)

5. Show that the composite of three relations, when defined, is associative.

6. If the composite $R \circ S$ is defined, prove that $(R \circ S)^{\smile} = S^{\smile} \circ R^{\smile}$.

7. Show that the composite of the graphs of functions corresponds to the graph of their composite in the following sense (note the reversal of order): $X \xrightarrow{f} Y \xrightarrow{g} Z$ gives $G(g \circ f) = G(f) \circ G(g)$.

8. Let R be a relation on X to X. Show that R is reflexive if and only if $R \supset I_X$, symmetric if and only if $R^{\smile} \subset R$, and transitive if and only if $R \circ R \subset R$.

9. Show that a relation R on X to Y is the graph of some function on X to Y if and only if $R \circ R^{\smile} \supset I_X$ and $R^{\smile} \circ R \subset I_Y$. Deduce that R is the graph of a bijection if and only if $R \circ R^{\smile} = I_X$ and $R^{\smile} \circ R = I_Y$.

10. Assume that a binary operation $\square$ on a set X has a left unit and satisfies the identity $x \square (y \square z) = (x \square z) \square y$. Prove that $\square$ is associative and commutative.

11. The "symmetric difference" $S \Delta T$ of subsets $S, T \subset U$ is

$$S \Delta T = (S \cap T') \cup (S' \cap T).$$

Prove that Δ is associative, commutative, and that $\cap$ is distributive over Δ.

12. Which of the following binary operations $(m, n) \mapsto m \square n$ on integers m and n are associative and which are commutative?

$$m \square n = m - n, \qquad m^2 + n^2, \qquad 2(m + n), \qquad -m - n.$$

4. The Natural Numbers

Intuitively, the set $\mathbf{N} = \{0, 1, 2, \cdots\}$ of all *natural numbers* may be described as follows: $\mathbf{N}$ contains an "initial" number 0; there is a *successor function* $\sigma:\mathbf{N} \to \mathbf{N}$, as in the picture

$$0 \overset{\sigma}{\mapsto} 1 \overset{\sigma}{\mapsto} 2 \overset{\sigma}{\mapsto} 3 \overset{\sigma}{\mapsto} 4 \mapsto \cdots,$$

and $\mathbf{N}$ is "generated" from 0 by σ. Formally, we shall describe $\mathbf{N}$ by axioms essentially due to G. Peano (1858–1932).

PEANO POSTULATES. *The system* $\mathbf{N}$ *of natural numbers is a set* $\mathbf{N}$ *with a function* $\sigma:\mathbf{N} \to \mathbf{N}$ *and a selected element* $0 \in \mathbf{N}$ *such that: (i)* σ *is injective; (ii)* 0 *is not in the image of* σ*; (iii) any subset* $U \subset \mathbf{N}$ *with the two properties*

$$(a)\quad 0 \in U; \qquad (b)\quad \textit{For all } n \in \mathbf{N},\ n \in U \Rightarrow \sigma(n) \in U$$

must be the whole set $\mathbf{N}$.

Selecting the element $0 \in \mathbf{N}$ amounts to a nullary operation on $\mathbf{N}$, while σ is a unary operation.

Postulate (iii) is called the *Principle of Mathematical Induction*. Instead of "$n \in U$" one might say, "n has the property U". In this language, the Peano Postulates read thus: The set $\mathbf{N}$ of natural numbers has one unary operation σ and one nullary operation, "select 0", such that: (i) $\sigma n = \sigma m$ implies $n = m$; (ii) σn is never zero; and (iii) any property of natural numbers holding for 0 and holding for σn whenever it holds for n is true for all natural numbers n.

From these axioms all the properties of natural numbers may be derived; in particular, properties of addition and multiplication. After addition has been defined, it will turn out that the successor function σ is the function defined by $\sigma(n) = n + 1$ for all n.

Many useful facts about natural numbers can be proved by induction (cf. the Exercises). As a first simple use of mathematical induction, we will prove that if $n \neq 0$, then n is σk for some $k \in \mathbf{N}$. For let $U \subset \mathbf{N}$ be the set

$$U = \{0\} \cup \{m \mid m = \sigma k \text{ for some } k \in \mathbf{N}\}.$$

Then $0 \in U$. Also if $n \in U$, then $\sigma n \in U$, taking k to be n. Hence, by the postulate of mathematical induction, $U = \mathbf{N}$. This is the desired result.

The natural numbers enter into the description of the iterates of any unary operation f. Each such operation $f:X \to X$ has iterates $f^2 = f \circ f$, $f^3 = f \circ f \circ f$, and so on, with $f^1 = f$ and f^0 the identity $1:X \to X$. For $n > 1$, the nth *iterate* is $f^n = f \circ \cdots \circ f$, the composite of n factors f. Instead of indicating these n factors by the customary three dots we can write $f^{n+1} = f \circ f^n$. This suggests the following description of the nth iterate in terms of the successor function σ:

$$f^0 = 1_X, \qquad f^{\sigma n} = f \circ f^n : X \to X \qquad \text{for all } n \in \mathbf{N}. \tag{7}$$

This is called a "recursive" definition, since it gives each iterate of f in terms of the previous iterates (and f).

Two unary operations f and g on X *commute* when $f \circ g = g \circ f$. This may be visualized by the square diagram, as displayed. This is called *commutative diagram* because the composites along both paths from upper left to lower right are equal.

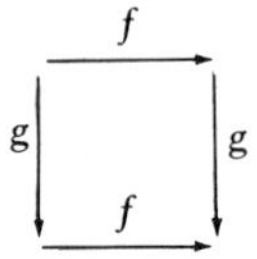

$$\begin{array}{ccc} & \xrightarrow{\ f\ } & \\ g\big\downarrow & & \big\downarrow g \\ & \xrightarrow{\ f\ } & \end{array}$$

Proposition 3. *If $f, g : X \to X$ commute, then, for all $m, n \in \mathbf{N}$,*

$$f^n \circ g^m = g^m \circ f^n, \qquad (f \circ g)^n = f^n \circ g^n. \tag{8}$$

The proof of these formulas will illustrate proofs by mathematical induction. First, assuming that f and g commute, we will show that

$$f^n \circ g = g \circ f^n \tag{9}$$

for all $n \in \mathbf{N}$. To prove this, consider the set U of all those $n \in \mathbf{N}$ for which (9) holds. Since $f^0 = 1_X$ is the identity, $0 \in U$. If $n \in U$, then the definition of the iterate plus (9)—which holds because $n \in U$—give

$$f^{\sigma n} \circ g = f \circ f^n \circ g = f \circ g \circ f^n.$$

But the result is $g \circ f \circ f^n = g \circ f^{\sigma n}$, since f and g commute. This gives $f^{\sigma n} \circ g = g \circ f^{\sigma n}$, which states that $\sigma n \in U$. By the postulate (iii) of mathematical induction, $U = \mathbf{N}$. This is (9) for all $n \in \mathbf{N}$. It states that when g commutes with f, it also commutes with each iterate f^n. Since f^n then commutes with g, it also commutes with each iterate g^m. In other words, $f^n \circ g^m = g^m \circ f^n$, as in the first equation of (8).

The second equation $(f \circ g)^n = f^n \circ g^n$ is also to be proved by induction on n. Since $f^0 = 1_X = g^0$ and $(f \circ g)^0 = 1_X$, the equation holds for $n = 0$. Assume that this equation holds for some n. (This is the "induction assumption".) Then by the definition of an iterate and this assumption,

$$(f \circ g)^{\sigma n} = (f \circ g) \circ (f \circ g)^n = f \circ g \circ f^n \circ g^n.$$

Since g commutes with f, it commutes with f^n; hence, using again the definitions of $f^{\sigma n}$ and $g^{\sigma n}$,

$$(f \circ g)^{\sigma n} = \cdots = f \circ f^n \circ g \circ g^n = f^{\sigma n} \circ g^{\sigma n}.$$

This is the desired equation with n replaced by its successor σn; the induction is complete.

Observe that in this second proof by induction, as often in the future, the set U at issue (all those $n \in \mathbf{N}$ for which $(f \circ g)^n = f^n \circ g^n$) is not explicitly named.

Since any operation f commutes with itself, this proposition implies (for all m and n) that $f^m \circ f^n = f^n \circ f^m$. In particular this gives $f^{\sigma n} = f \circ f^n = f^n \circ f$, an alternate to our original definition of the iterate.

Proposition 4. *For $f:X \to X$ and all $m, n \in \mathbf{N}$, $(f^m)^n = (f^n)^m$.*

The proof will establish the property, "For all m, $(f^m)^n = (f^n)^m$", by induction on n. This means that the induction axiom will be applied to the set $U \subset \mathbf{N}$ with $n \in U$ if and only if $(f^m)^n = (f^n)^m$ for every $m \in \mathbf{N}$. But $f^0 = 1$, so $0 \in U$. If $n \in U$, then, for every m, $(f^m)^{\sigma n} = f^m \circ (f^m)^n = f^m \circ (f^n)^m$, by the induction assumption ($n \in U$). Since f and f^n commute, the last expression becomes $(f \circ f^n)^m = (f^{\sigma n})^m$ by (7); this shows that $\sigma n \in U$ and so completes the induction.

EXERCISES

1. Using the fact that the composite of injections is an injection, prove by induction that $f:X \to X$ injective implies f^n injective for all $n \in \mathbf{N}$.
2. Prove $f:X \to X$ surjective implies f^n surjective for all $n \in \mathbf{N}$.
3. By induction, prove that $\sigma^n(0) = n$ for all $n \in \mathbf{N}$.
4. Prove by induction that $1 + 2 + \cdots + n = n(n + 1)/2$. (You may use familiar properties of addition and multiplication.)
5. Prove similarly by induction the summation formulas:
   (a) $1 + 4 + 9 + \cdots + n^2 = n(n + 1)(2n + 1)/6$.
   (b) $1 + 8 + 27 + \cdots + n^3 = [n(n + 1)/2]^2$.
   (c) $(1 + 2^5 + \cdots + n^5) + (1 + 2^7 + \cdots + n^7) = 2[n(n + 1)/2]^4$.
6. Construct three different left inverses for the successor function σ.
7. By induction, prove that $n^2 - n$ is always even.
8. What ails the following "proof" that all the elements of a finite set are equal? All elements of a set with no elements are equal, so make the induction assumption that any set with n elements has all its elements equal. In a set with $n + 1$ elements, the first n and the last n are equal by the induction assumption. They overlap at n, so all are equal, completing the induction.

9. For any two of the three Peano Postulates find a set X with an element 0 and a unary operation σ which satisfy thse two axioms but not the third. (*Hint:* In two cases out of three, finite sets will do.)

5. Addition and Multiplication

Sums and *products* of natural numbers m and n are defined by

$$m + n = \sigma^n(m), \qquad mn = (\sigma^m)^n(0). \tag{10}$$

These are the usual elementary definitions: To add n, make an n-fold addition of 1; to multiply m by n, iterate n times the operation σ^m of adding m. Or, put the recursive description of the iterate $\sigma^{\sigma n} = \sigma \circ \sigma^n$ in the above definition of the sum. It becomes

$$m + 0 = m, \qquad m + \sigma n = \sigma(m + n). \tag{11}$$

These "recursive" formulas give each sum $m + \sigma n$ in terms of a previous sum. For example, define $1 = \sigma(0)$, $2 = \sigma(1)$, $3 = \sigma(2)$, and $4 = \sigma(3)$; then, by these equations,

$$2 + 2 = 2 + \sigma(1) = \sigma(2 + 1) = \sigma[2 + \sigma(0)] = \sigma[\sigma(2 + 0)] = \sigma[\sigma(2)] = 4,$$

just as we knew it should.

Theorem 5. *Addition is a commutative and associative binary operation on* $\mathbf{N}$ *with* 0 *as unit. It has the "cancellation" properties*

$$k + n = m + n \Rightarrow k = m, \qquad m + n = 0 \Rightarrow m = n = 0, \tag{12}$$

for all $k, m, n \in \mathbf{N}$. *For any function* $f: X \to X$ *and for all* $m, n \in \mathbf{N}$,

$$f^m \circ f^n = f^{m+n}. \tag{13}$$

First, we prove by induction on n that (13) holds for all m. It holds for $n = 0$; assuming it for n, this assumption, the recursion for $f^{\sigma n}$ (twice), and the recursion (11) for $m + \sigma n$ give

$$f^m \circ f^{\sigma n} = f^m \circ f \circ f^n = f \circ f^m \circ f^n = f \circ f^{m+n} = f^{\sigma(m+n)} = f^{m+\sigma n}.$$

This is (13), for all m, with n replaced by σn; the induction is complete.

Next, an induction on n proves that $\sigma^n(0) = n$.

To prove addition commutative, use the definition of $+$ and (8) to get

$$m + n = \sigma^n(m) = \sigma^n(\sigma^m(0)) = (\sigma^n \circ \sigma^m)(0) = (\sigma^m \circ \sigma^n)(0)$$
$$= \sigma^m(n) = n + m.$$

The argument for associativity uses the definition of $+$, commutativity, and (13):

$$k + (m + n) = \sigma^{m+n}(k) = \sigma^{n+m}(k) = (\sigma^n \circ \sigma^m)(k) = \sigma^n(\sigma^m(k))$$
$$= (k + m) + n.$$

Since $\sigma^0 = 1_{\mathbf{N}}$, $m + 0 = \sigma^0(m) = m$, and 0 is a unit for addition.

The first cancellation law of (12) is proved for all k and m by induction on n. For $n = 0$, $k + 0 = m + 0$ trivially gives $k = m$. If cancellation holds for n, then $k + \sigma(n) = m + \sigma(n)$ implies $\sigma(k + n) = \sigma(m + n)$, hence $k + n = m + n$, since σ is an injection (Peano Postulates); hence $k = m$ by the induction assumption.

Finally, suppose $m + n = 0$. If $n \neq 0$, then $n = \sigma(k)$ for some $k \in \mathbf{N}$. Then $0 = m + n = m + \sigma(k) = \sigma(m + k)$ makes 0 the successor of $m + k$, a contradiction to the postulate that 0 is not a successor. This contradiction shows $n = 0$. By commutativity, we also get $m = 0$, completing the proof.

Now turn to multiplication. From the definitions for $m(\sigma n)$ and $f^{\sigma n}$,

$$m(\sigma n) = (\sigma^m)^{\sigma n}(0) = \left[\sigma^m \circ (\sigma^m)^n\right](0) = \sigma^m(mn) = mn + m.$$

Hence, the definition (10) of the product may be written as a "recursion"

$$m0 = 0, \qquad m(\sigma n) = mn + m. \tag{14}$$

The definition of the product also gives, for all $k, m, n \in \mathbf{N}$,

$$(\sigma^m)^n(k) = (\sigma^m)^n\left[\sigma^k(0)\right] = \sigma^k\left[(\sigma^m)^n(0)\right] = mn + k.$$

With these equations, we leave the reader to prove

Theorem 6. *Multiplication is a commutative and associative binary operation on* $\mathbf{N}$ *with* 1 *as unit. It is distributive over addition, and* $mn = 0$ *only when* $m = 0$ *or* $n = 0$. *Also, for any function* $f: X \to X$ *and all* $m, n \in \mathbf{N}$,

$$(f^m)^n = f^{mn}. \tag{15}$$

Observe that our theorems included the "laws of exponents"

$$f^n \circ f^m = f^{n+m}, \qquad (f^n)^m = f^{nm}$$

of (13) and (15). From Proposition 3, we know that $f^n \circ f^m = f^m \circ f^n$; both are now expressed as f^{n+m}. Similarly, Proposition 4 gave $(f^n)^m = (f^m)^n$; both are now f^{nm}.

EXERCISES

1. Prove formula (15) of Theorem 6.

2. (a) Prove that multiplication in $\mathbf{N}$ is commutative and associative, using formula (15).

(b) Prove that $m(n + n') = mn + mn'$ for all $m, n, n' \in \mathbf{N}$.

3. (a) Construct a function $\tau:\mathbf{N} \to \mathbf{N}$ other than σ which satisfies Peano's Postulates.

(b) Show that, if $\tau:\mathbf{N} \to \mathbf{N}$ satisfies Peano's Postulates, then $\tau\beta = \beta\sigma$ for some bijection $\beta:\mathbf{N} \to \mathbf{N}$.

4. (a) For σ the successor function, find all functions $\phi:\mathbf{N} \to \mathbf{N}$ such that $\phi\sigma = \sigma\phi$.

(b) Show that, if $\tau:\mathbf{N} \to \mathbf{N}$ satisfies Peano's Postulates and $\sigma\tau = \tau\sigma$, then $\tau = \sigma$.

5. Define $P_m:\mathbf{N} \to \mathbf{N}$ by $P_m(n) = mn$, and m^n as $(P_m)^n(1)$.

(a) Show that m^n may also be defined by the recursion $m^0 = 1$, $m^{\sigma n} = m(m^n)$.

(b) Prove that $k^{m+n} = k^m k^n$, $k^{mn} = (k^m)^n$, and $(k^n)(m^n) = (km)^n$, for all $k, m, n \in \mathbf{N}$.

6. Derive the first cancellation law for addition from the result of Exercise 4.1.

6. Inequalities

The usual binary relation $m < n$, which states that the natural number m is smaller than the natural number n, can be defined in terms of the addition of natural numbers:

$$m < n \quad \Leftrightarrow \quad \text{there is an } x \neq 0 \text{ in } \mathbf{N} \text{ with } n = m + x. \tag{16}$$

(By the cancellation law for addition, such an x is unique if it exists.) As is customary, we shall write

$$m \leqslant n \qquad \text{for either } m < n \text{ or } m = n,$$
$$n > m \qquad \text{for } m < n.$$

Thus "$>$" is the converse of "$<$"; similarly, let "$\geqslant$" be the converse of "$\leqslant$"; all are binary relations on $\mathbf{N}$ to $\mathbf{N}$.

From these definitions we deduce the usual properties of inequalities. For example, the transitive law for the relation $<$ asserts that

$$k < m \quad \text{and} \quad m < n \Rightarrow k < n.$$

For a proof, note that $k + x = m$ and $m + y = n$ give $k + (x + y) = n$; here $x + y$ can be zero only if both x and y are zero.

Proposition 7 (*The Trichotomy Law*). *For given natural numbers m and n, exactly one of the following alternatives holds*:

$$m < n, \qquad m = n, \qquad m > n.$$

Proof: First note that at most one of these can hold. For example, if both $m < n$ and $m = n$, then $n = m + x$ for some $x \neq 0$, but the cancellation law would give $x = 0$, a contradiction. Were both $m < n$ and $m > n$, there would be a similar contradiction.

It remains to show by induction on n that for each m at least one of these alternatives holds. For $n = 0$ this is clear (each m is either 0 or not 0). Hence, make the induction assumption that the alternatives hold for all $m \in \mathbf{N}$ and for some one n, then show for $\sigma(n)$ and for every $k \in \mathbf{N}$ that one of

$$k < \sigma n, \qquad k = \sigma n, \qquad k > \sigma n$$

will hold. If $k = 0$, the first holds. Otherwise $k \neq 0$, so $k = \sigma m$ for some $m \in \mathbf{N}$; by our induction assumption we have one of $m < n$, $m = n$, or $m > n$. Apply σ to both sides of these inequalities; this gives the three alternatives listed above, for clearly $m < n$ implies $\sigma(m) < \sigma(n)$.

Proposition 8. *Addition and multiplication of natural numbers are isotonic; that is, for all k, m, $n \in \mathbf{N}$,*

$$m < n \quad \textit{and} \quad k \in \mathbf{N} \quad \textit{imply} \quad m + k < n + k;$$

$$m < n \quad \textit{and} \quad k \neq 0 \quad \textit{imply} \quad km < kn.$$

The proof is left to the reader.

Corollary. *A positive factor k may be canceled; that is,*

$$k > 0 \quad \textit{and} \quad km = kn \quad \textit{imply} \quad m = n.$$

Otherwise, by trichotomy, $m < n$ or $m > n$; by the proposition, $km < kn$ or $km > kn$, in contradiction to the hypothesis $km = kn$.

Proposition 9. *For natural numbers n and m, $n > m$ implies $n \geqslant \sigma m$.*

This seems intuitively clear, but we wish to prove it from the Peano Postulates. The hypothesis $n > m$ means that $n = m + x$ for some natural number $x \neq 0$. As already noted, this means that $x = \sigma y$ from some natural number y. Hence, $n = m + x = m + \sigma y = \sigma(m + y)$ by (11). However, $m + y \geqslant m$, so $\sigma(m + y) \geqslant \sigma m$ and $n \geqslant \sigma m$, as required.

Corollary. *There is no natural number between 0 and 1.*

Proposition 10 ($\mathbf{N}$ *is "well-ordered"*). *Every nonvoid set V of natural numbers contains a first element f; that is, an element $f \in V$ such that $x \in V$ implies $x \geqslant f$.*

Proof: Suppose that some $V \neq \varnothing$ has no first element. For this V we shall prove by induction on n that

$$x \in V \Rightarrow x \geqslant n.$$

For $n = 0$ this property is immediate. Make the induction assumption that it holds for some n. Then we cannot have $n \in V$, for if so, n would be a first element of V. Therefore, $n \notin V$, so every $x \in V$ has $x > n$; that is, by Proposition 9, $x \geqslant \sigma n$. This is the statement displayed with n replaced by σn; hence, this completes the inductive proof of that statement.

Now V is not empty, so there is some element $k \in V$. The displayed statement holds for $n = k + 1$, so gives $k \geqslant k + 1$, a contradiction. The proposition is established.

Remark: An *ordered set* is a set S together with a linear order (cf. §3), usually written $<$, on S to S. An ordered set S is *well-ordered* if *every* nonvoid subset of S has a first element. In this language, Proposition 10 states that the set of natural numbers is well-ordered. This statement gives the following "Second Principle of Mathematical Induction". The first Principle inducts from $n - 1$ to n; the second, from all $m < n$ to n.

Theorem 11. *If a set $U \subset \mathbf{N}$ has for every $n \in \mathbf{N}$ the property*

$$P_n: \quad \textit{if every } m < n \textit{ is in } U, \textit{ then } n \in U,$$

then $U = \mathbf{N}$.

Proof: Suppose $U \neq \mathbf{N}$. By the preceding proposition, there is a first natural number f not in U. Since f is first, every $m < f$ belongs to U. From this assumption, the property P_f implies that $f \in U$, a contradiction. Hence $U = \mathbf{N}$.

Caution: In applying this principle one must establish the property P_n in particular for $n = 0$; since the set of all natural numbers $m < 0$ is the empty set $\varnothing$, this case of the hypothesis reads "The assumption that every $m \in \varnothing$ belongs to U implies that $0 \in U$"; this is just "$0 \in U$".

Mathematical induction is often applied just to *positive* natural numbers n (those $n \in \mathbf{N}$ with $n > 0$). The first form of the principle reads: If W is a set of positive natural numbers such that $1 \in W$ and such that, for each positive n, $n \in W$ implies $(n + 1) \in W$, then W contains all the positive natural numbers. This is a direct consequence of the induction axiom: Just take $U = \{0\} \cup W$.

Inequalities may be used to define various subsets of **N**. In particular, we will systematically denote by **n** the set

$$\mathbf{n} = \{m|m \in \mathbf{N} \text{ and } 1 \leqslant m \leqslant n\}.$$

This will be our "standard" finite set with n elements; $\mathbf{n} = \{1, 2, \cdots, n\}$. Note in particular that **0** is the empty set $\varnothing$ and $\mathbf{1} = \{1\}$ a one-element set.

A function $l:\mathbf{n} \to X$ on this set **n** is determined by listing its n values, say as $l_1, \cdots, l_n$. Hence, we call such a function l a *list*, or "a list of n elements of X". Similarly, a function $A:\mathbf{m} \times \mathbf{n} \to X$ is called an $m \times n$ *matrix* with "entries" in the set X.

Finally, a function $s:\mathbf{N} \to X$ with domain the set **N** of all natural numbers is called a *sequence* or a "sequence of elements" of X. Its values will be written as $s_0, s_1, s_2, \cdots$.

EXERCISES

1. Prove that the relation $\leqslant$ on **N** to **N** is transitive and reflexive.

2. Give a proof of Proposition 8.

3. Deduce the (first) Principle of Mathematical Induction from the fact that **N** is well-ordered.

4. Prove that any subset of a well-ordered set is well-ordered.

5. (a) If S and T are ordered sets (with $s, s' \in S$, $t, t' \in T$), show that the definition $(s, t) < (s', t') \Leftrightarrow (s < s'$ or both $s = s'$ and $t < t')$ makes $S \times T$ an ordered set. (This is called the "lexicographic" order of $S \times T$.)

(b) If S and T are both well-ordered, prove that $S \times T$ is also well-ordered.

6. Show that a well-ordered set S can contain no infinite descending sequence s; that is, there is no $s:\mathbf{N} \to S$ with

$$s_1 > s_2 > s_3 > \cdots.$$

7. The Integers

This section will show how the algebraic system **Z** of all integers with the usual binary operations of addition and multiplication may be explicitly constructed from the algebraic system **N** of natural numbers. From this construction the familiar properties of addition and multiplication in **Z** will be derived from those of **N**. These properties are formulated completely in Theorems 12, 13, and 14 below. A reader who wishes to accept these familiar results as known may at first reading skip the following explicit construction of **Z**.

In **N**, subtraction is not always possible; that is, an equation $m + x = n$ for given m, n in **N** need not have a solution x in **N**. To make subtraction always possible, **N** will be enlarged to **Z**. Informally, **Z** is to be the set

$$\mathbf{Z} = \{\cdots, -3, -2, -1, 0, 1, 2, 3, \cdots\}.$$

More formally, $\mathbf{Z}$ is constructed from $\mathbf{N}$ and the set $\mathbf{P} = \{p \mid p \in \mathbf{N}$ and $p > 0\}$ of all positive natural numbers as follows: Take a set $\{-\}$ with just one element "$-$"; in the cartesian product $\{-\} \times \mathbf{P}$ write an ordered pair $(-, p)$ as $-p$. Then $\{-\} \times \mathbf{P}$ and $\mathbf{N}$ are disjoint; let $\mathbf{Z}$ be the union $(\{-\} \times \mathbf{P}) \cup \mathbf{N}$. This ensures that every element $a \in \mathbf{Z}$ is either a natural number n or the (formal) negative $-p$ of a positive natural number, and never both.

In this section, elements of $\mathbf{Z}$ will be written as a, b, c, while elements of $\mathbf{N}$ are k, m, n. We call elements of $\mathbf{Z}$ integers.

We now wish to define addition and multiplication for these integers, using only the corresponding operations for natural numbers. It is easy to guess how this might be done. First, any integer a can be written (in many ways) as a difference $a = m - n$ of two natural numbers m and n. To get the sum $a + b$ we can then write $b = k - l$ and

$$a + b = (m - n) + (k - l) = (m + k) - (n + l).$$

This amounts to taking pairs (m, n) and (k, l), adding these pairs to get $(m + k, n + l)$, and then taking the difference of the result. This type of construction first defines the binary operation of addition on a new set—the set of pairs—and then uses the difference function $(m, n) \mapsto m - n$ to get a surjection from the pairs onto the desired integers. Similar constructions occur often, so we will carry this out quite formally without assuming that we "know" the answer.

If we write $u = (m, n)$ for a pair of natural numbers m and n, then the operation d for "take their difference" is a function $d: \mathbf{N} \times \mathbf{N} \mapsto \mathbf{Z}$, which may be defined for all m, n, and k in $\mathbf{N}$ with $k \neq 0$ by the rules

$$d(n + k, n) = k, \qquad d(n, n) = 0, \qquad d(m, m + k) = -k.$$

These rules do define the difference $d(m, n)$ for all pairs (m, n), because the law of trichotomy for the inequality in $\mathbf{N}$ states that either $m > n$, in which case $m = n + k$, or that $m = n$, or that $m < n$, in which case $n = m + k$ for some positive natural number k. The difference function $d: \mathbf{N} \times \mathbf{N} \to \mathbf{Z}$ is a surjection because $d(k, 0) = k$ and $d(0, k) = -k$ for all k. The equations mean that a right inverse $s: \mathbf{Z} \to \mathbf{N} \times \mathbf{N}$ for d may be defined by

$$s(m) = (m, 0), \qquad s(-k) = (0, k), \qquad m \geqslant 0, \quad k > 0. \tag{17}$$

Then the composite $ds = 1_{\mathbf{Z}}$, so s is a right inverse. However, the composite sd is not the identity on $\mathbf{N} \times \mathbf{N}$, because

$$sd(n + k, n) = (k, 0), \qquad sd(m, m + k) = (0, k). \tag{18}$$

Addition can be defined for pairs of natural numbers as

$$(m, n) + (m', n') = (m + m', n + n'), \qquad m, m', n, n' \in \mathbf{N}. \tag{19}$$

With this definition, the binary operation $+$ on $\mathbf{N} \times \mathbf{N}$ is commutative, associative, and has the pair $(0, 0)$ as unit. Moreover, the way in which sd in (18) fails to be the identity function can now be expressed, via addition, as

$$(n, n) + sd(n + k, n) = (n + k, n), \ (m, m) + sd(m, m + k) = (m, m + k).$$

In other words, to each pair u there is another pair u_0, either (n, n) or (m, m), such that

$$u = sd(u) + u_0, \qquad du_0 = 0 \in \mathbf{Z}; \tag{20}$$

thus sd misses being the identity only by a pair u_0 of difference zero (a *diagonal* pair). If u_0 is any diagonal pair, while w is any pair whatever, a calculation from the definitions of d and addition shows that

$$d(w + u_0) = dw. \tag{21}$$

An operation $\oplus$ of addition for integers a and b can now be defined by

$$a \oplus b = d(sa + sb). \tag{22}$$

In words, represent each integer a as the difference of some pair, say of the pair sa, add these pairs, and take the difference of the resulting pair. The operation $\oplus$ so-defined is really just the old familiar process of adding integers. Specifically, if a and b are both positive integers m and n, our formula (22) gives $m \oplus n = m + n$, while if a and b are both negative, so that $a = -p$ and $b = -q$ for positive integers p and q, then $sa = (0, p)$, $sb = (0, q)$, so the formula gives

$$(-p) \oplus (-q) = d[(0, p) + (0, q)] = d(0, p + q) = -(p + q).$$

The formula (22) for $\oplus$ also works in the expected way in the remaining cases, when just one of a or b is negative.

Lemma. *Taking the difference preserves addition, in the sense that*

$$du \oplus dv = d(u + v), \qquad u, v \in \mathbf{N} \times \mathbf{N}. \tag{23}$$

Proof: Write $u = sd(u) + u_0$ and $v = sd(v) + v_0$, as in (20). Then

$$\begin{aligned} u + v &= sdu + sdv + u_0 + v_0, \\ d(u + v) &= d(sdu + sdv + u_0 + v_0) \\ &= d(sdu + sdv + u_0) = d(sdu + sdv) \qquad \text{by (21)}, \\ &= du \oplus dv, \qquad \text{by the definition of } \oplus. \end{aligned}$$

This proves (23), and hence the lemma.

Equation (23) can now be used to prove various properties for the addition of integers. For example, to prove that $a \oplus b$ is commutative, we can write $a = du$ and $b = dv$, for pairs u and v, because d is a surjection. Then, since addition of pairs is commutative,

$$a \oplus b = du \oplus dv = d(u + v) = d(v + u) = dv \oplus du = b \oplus a,$$

as required. The associative law and the fact that 0 is a unit for the addition in $\mathbf{Z}$ are proved similarly.

The relation of the addition of the integers to that of the natural numbers can now be stated (without using pairs) as follows:

THEOREM 12. *The set $\mathbf{Z} \supset \mathbf{N}$ has an associative and commutative binary operation $\oplus$ with 0 as unit and such that:*

(*i*) (*Subtraction is possible.*) *For every $a, b \in \mathbf{Z}$ there is an x in $\mathbf{Z}$ with $a \oplus x = b$.*

(*ii*) *The sum of two elements m and n of $\mathbf{N}$ is the same in $\mathbf{N}$ as in $\mathbf{Z}$; that is, $m + n = m \oplus n$.*

(*iii*) *To every element $a \in \mathbf{Z}$ there exist $m, n \in \mathbf{N}$ with $n \oplus a = m$.*

The first condition gives $b - a = x \in \mathbf{Z}$; the second condition states that the addition in $\mathbf{Z}$ is obtained by extending the given addition in $\mathbf{N}$, while the last condition states that all the new elements added to $\mathbf{N}$ are really necessary to make subtraction possible.

To show subtraction possible in $\mathbf{Z}$, write $a \in \mathbf{Z}$ as $a = d(m, n)$ and set $a' = d(n, m)$. Since $(m, n) + (n, m) = (m + n, n + m) = w_0$ is a diagonal pair,

$$a \oplus a' = d(m, n) \oplus d(n, m) = d[(m, n) + (n, m)] = f(w_0) = 0.$$

This states that a' is an "additive inverse" of a in $\mathbf{Z}$. For any b in $\mathbf{Z}$, $a \oplus (a' \oplus b) = (a \oplus a') \oplus b = 0 \oplus b = b$, so the equation $a \oplus x = b$ has a solution $x = a' \oplus b$, as required for (i). Finally, each $n \in \mathbf{N}$ satisfies $0 + n = n$, while each new element $-p$ in $\mathbf{Z}$ is the solution of the equation $p \oplus (-p) = 0$. This proves property (iii) and so completes the proof of the theorem.

From this theorem it is easy to prove that subtraction is unique in $\mathbf{Z}$; in other words, for $a, b \in \mathbf{Z}$ there is exactly one $x \in \mathbf{Z}$ with $a \oplus x = b$. As usual, we write $x = b - a$, and we stop using the special symbol $\oplus$ for addition in $\mathbf{Z}$.

A similar process will yield the multiplication of integers.

Theorem 13. *The set* $\mathbf{Z} \supset \mathbf{N}$ *has a commutative and associative binary operation of multiplication, distributive over the addition of Theorem 12, with* $1 = \sigma 0$ *as unit, and such that the product of two elements of* $\mathbf{N}$ *is the same in* $\mathbf{N}$ *as in* $\mathbf{Z}$.

Proof: Were the differences $m - n$ and $m' - n'$ present in $\mathbf{N}$, their product would be given by the usual formula

$$(m - n)(m' - n') = (mm' + nn') - (mn' + nm').$$

Define a binary operation of multiplication in $\mathbf{N} \times \mathbf{N}$ by the corresponding formula

$$(m, n)(m', n') = (mm' + nn', mn' + nm'). \tag{24}$$

A calculation shows that this operation is commutative and associative with $(1, 0)$ as unit, and that it is distributive over the addition of pairs previously defined in (19). For example, to prove the distributive law, put in the definitions and calculate that

$$\begin{aligned}(r, s)[(m, n) + (m', n')] &= (r(m + m') + s(n + n'), r(n + n') + s(m + m')) \\ &= (rm + rm' + sn + sn', rn + rn' + sm + sm').\end{aligned}$$

Calculating $(r, s)(m, n) + (r, s)(m', n')$ gives the same result; hence, the distributive law holds for pairs.

Now define the product of two integers $a, b \in \mathbf{Z}$ to be

$$ab = d[(sa)(sb)]. \tag{25}$$

This does define the usual product of integers; for example, if $a = m$ is positive and $b = -n$ negative, then $sa = (m, 0)$, $sb = (0, n)$ and the definition becomes

$$m(-n) = d[(m, 0)(0, n)] = d(0, mn) = -(mn),$$

as expected. Moreover, the difference $d: \mathbf{N} \times \mathbf{N} \to \mathbf{Z}$ satisfies

$$d(uv) = (du)(dv), \qquad u, v \in \mathbf{N} \times \mathbf{N}; \tag{26}$$

that is, d carries products (in $\mathbf{N} \times \mathbf{N}$) to products in $\mathbf{Z}$. The proof is like that for the lemma above, if one observes also that $du_0 = 0$ gives $d(u_0 v) = 0$. With this property (26) and the lemma, the distributive, associative, and other laws now carry over from $\mathbf{N} \times \mathbf{N}$ to $\mathbf{Z}$, as stated.

Since $\mathbf{Z} \supset \mathbf{N}$, $\mathbf{Z}$ does contain the set $\mathbf{P}$ of all *positive* integers m (all $m \in \mathbf{N}$ with $0 < m$). For these one readily proves the following:

THEOREM 14. *Sums and products of positive integers are positive, while for each $a \in \mathbf{Z}$ exactly one of the following alternatives holds: $a \in \mathbf{P}$, $a = 0$, or $(-a) \in \mathbf{P}$.*

As for the natural numbers, an order relation "$<$" for $\mathbf{Z}$ is now defined by

$$a < b \quad \Leftrightarrow \quad \text{there exists a positive } m \text{ with } a + m = b.$$

The usual properties of "$<$" now follow; they will be treated in more detail in §VIII.1.

Now that $\mathbf{Z} \supset \mathbf{N}$ with all the expected properties is at hand, there is no longer any need to refer to the difference function $d: \mathbf{N} \times \mathbf{N} \to \mathbf{Z}$ or to its right inverse s, as used in the construction above.

EXERCISES

1. Complete the proof of (26).
2. Complete the proof of the distributive law in $\mathbf{Z}$.
3. Show that there is only one binary operation of addition in $\mathbf{Z}$ which satisfies (23), relative to the given addition in $\mathbf{N} \times \mathbf{N}$.
4. Prove that subtraction in $\mathbf{Z}$ is unique. (*Hint:* Use cancellation for addition in $\mathbf{N}$.)
5. Show that there is only one binary operation of multiplication in $\mathbf{Z}$ satisfying (26) relative to the given multiplication in $\mathbf{N} \times \mathbf{N}$.
6. State and prove the usual properties of the binary relation $<$ on $\mathbf{Z}$.

8. The Integers Modulo n

A simple *finite* algebraic system arises from the familiar observation that one can add and multiply "even" and "odd"—meaning even integers and odd integers—according to the rules

$$\text{even} \oplus \text{even} = \text{even} = \text{odd} \oplus \text{odd}, \qquad \text{even} \oplus \text{odd} = \text{odd},$$
$$\text{even} \otimes \text{even} = \text{even} = \text{even} \otimes \text{odd}, \qquad \text{odd} \otimes \text{odd} = \text{odd}.$$

This amounts to giving binary operations of addition $\oplus$ and multiplication $\otimes$ on a set of just two elements 0 (even) and 1 (odd), by the rules

$$0 \oplus 0 = 0 = 1 \oplus 1, \qquad 0 \oplus 1 = 1 = 1 \oplus 0,$$
$$0 \otimes 0 = 0 = 0 \otimes 1 = 1 \otimes 0, \qquad 1 \otimes 1 = 1.$$

This system is called the *integers modulo* 2. In telling time, there are similar operations on the set $\{0, 1, \cdots, 11\}$ of twelve elements standing for the hours; the resulting algebraic system is the *integers modulo* 12: To add two hours, take their sum and then its "remainder" after possibly subtracting 12, as in 7 hours after 8 o'clock is 3 o'clock. We will now show that there is a similar system formed with remainders modulo n, for any natural number n.

The ordinary process of dividing an integer k by $n \neq 0$ yields a *quotient* q and a *remainder* r, as in $k/n = q + r/n$. The result may be expressed without any "division" in the following way.

Theorem 15 (*The Division Algorithm for* **Z**). *To given integers k and n with $n > 0$ there exist integers q and r such that*

$$k = qn + r, \qquad 0 \leqslant r < n; \tag{27}$$

these properties determine q and r uniquely from k and n.

Proof: The desired remainder r has the form $r = k - qn$; it is a non-negative difference, k less a multiple of n. Consider the set

$$M = \{m | m \in \mathbf{N} \text{ and } m = k - qn \text{ for some integer } q\}$$

of all such differences. This set M is not empty: If k is nonnegative, then $k = k - 0n \in M$; while if k is negative, $(-k)(n - 1) = k - kn \in M$.

Since M is non-empty and **N** is well-ordered, this set M must have a first element. Call this element r, so that $k = qn + r$ for some q. This element will indeed satisfy the condition (27). First $r \geqslant 0$, since r is in M, hence nonnegative. Second, if $r \geqslant n$, then $r = n + x$ for some $x \in \mathbf{N}$; then $x = r - n = k - (q + 1)n$, so that $x \in M$ is smaller than the intended first element r of M, a contradiction. Therefore, (27) has at least one solution.

It remains to show that (27) has at most one solution. Suppose instead that there were two different solutions $k = qn + r$ and $k = q'n + r'$ with both $0 \leqslant r < n$ and $0 \leqslant r' < n$. The notation may be chosen so that $r \geqslant r'$; then $n > r - r' \geqslant 0$, while $r - r' = (k - qn) - (k - q'n) = (q' - q)n$ is a multiple of n between 0 and $n - 1$ inclusive. But by the corollary to Proposition 9, there are no such multiples. Hence, $r = r'$, and also $qn = q'n$, so $q = q'$ and any two solutions of (27) are equal, as asserted in the theorem.

For n fixed, this division algorithm can be used to map each integer k onto its remainder "modulo n". These remainders modulo n can be added and multiplied: Take the usual sum or product; if the result exceeds n, replace by its remainder on division by n. For the remainders $\mathbf{Z}_5 = \{0, 1, 2, 3, 4\}$ this sum $\oplus$ and the product $\otimes$ are those given by the following tables:

$\oplus$	0	1	2	3	4
0	0	1	2	3	4
1	1	2	3	4	0
2	2	3	4	0	1
3	3	4	0	1	2
4	4	0	1	2	3,

$\otimes$	0	1	2	3	4
0	0	0	0	0	0
1	0	1	2	3	4
2	0	2	4	1	3
3	0	3	1	4	2
4	0	4	3	2	1.

(28)

This is called the algebra $\mathbf{Z}_5$ of integers "modulo 5".

For each positive integer n, let $\mathbf{Z}_n = \{0, 1, \cdots, n-1\}$ be the set of all r with $0 \leqslant r < n$, while $\rho: \mathbf{Z} \to \mathbf{Z}_n$ is the function which assigns to each integer k its remainder $\rho(k) = r$ on division by n. This function ρ is a surjection; with it the addition of remainders modulo n may be described as follows.

Theorem 16. *If $\rho: \mathbf{Z} \to \mathbf{Z}_n$ is the function assigning to each integer its remainder modulo n, there is exactly one binary operation $\oplus$ on $\mathbf{Z}_n = \{0, \cdots, n-1\}$ such that*

$$\rho(k+m) = (\rho k) \oplus (\rho m) \tag{29}$$

for all integers k and m. This operation $\oplus$ is commutative, associative, and has zero as unit. To each $r \in \mathbf{Z}_n$ there is an $r' \in \mathbf{Z}_n$ with $r \oplus r' = 0$.

Proof: Let r and s be remainders modulo n; that is, integers with $\rho r = r$ and $\rho s = s$. Then Equation (29) states that $r \oplus s$ must be $\rho(r+s)$. In words, the only possible operation $\oplus$ is this: To add two remainders in $\mathbf{Z}_n$, take their ordinary sum in $\mathbf{Z}$ and then its remainder modulo n.

This addition does satisfy (29) for all k and m. For, let k and m have remainders r and s, so that $k = r + qn$ and $m = s + q'n$ for integers q and q'. Then $k + m = r + s + (q + q')n$, so the remainder of $k + m$ is $\rho(k+m) = \rho(r+s)$, which is defined to be $r \oplus s$. Therefore, $\rho(k+m) = r \oplus s = (\rho k) \oplus (\rho m)$, just as required.

Now consider the commutative law for $\oplus$. By the Equation (29) just proved,

$$(\rho k) \oplus (\rho m) = \rho(k+m) = \rho(m+k) = (\rho m) \oplus (\rho k).$$

This shows that any two remainders commute in $\mathbf{Z}_n$. But every element of $\mathbf{Z}_n$ is a remainder, so the operation $\oplus$ in $\mathbf{Z}_n$ is commutative. Put differently: $\rho: \mathbf{Z} \to \mathbf{Z}_n$ is a surjection; by (29) it carries $+$ to $\oplus$, hence it carries the commutative law for $+$ to the commutative law for $\oplus$. A parallel argument shows that $+$ associative in $\mathbf{Z}$ makes $\oplus$ associative in $\mathbf{Z}_n$, and 0 a unit for $+$ in $\mathbf{Z}$ with $\rho(0) = 0$ makes 0 a unit for $\oplus$. For any non-zero remainder r, the integer $r + (n - r)$ has remainder zero, so $r \oplus (n - r) = 0$. This completes the proof.

Here is the corresponding result for multiplication.

Theorem 17. *There is exactly one binary operation $\otimes$ on $\mathbf{Z}_n$ with*

$$\rho(km) = (\rho k) \otimes (\rho m) \tag{30}$$

for all integers k and m. This operation is commutative, associative, distributive over $\oplus$, and has 1 as unit.

Proof: Take remainders r and s. The assumption (30) shows that their product $r \otimes s$ must be $\rho(rs)$. In words: To multiply two remainders in $\mathbf{Z}_n$, take their product in $\mathbf{Z}$ and then its remainder modulo n.

This multiplication does satisfy (30). For any integers $k = r + qn$ and $m = s + q'n$, the remainder of $km = rs + (qs + rq' + qq'n)n$ is $\rho(rs)$; hence (30) holds. Since ρ is a surjection, *identities* such as the distributive law valid in $\mathbf{Z}$ are then valid in $\mathbf{Z}_n$, Q.E.D.

These arguments show that *identities* valid for addition and multiplication in $\mathbf{Z}$ imply corresponding identities for the new addition and multiplication in $\mathbf{Z}_n$. They do *not* show that other properties valid in $\mathbf{Z}$ carry over to $\mathbf{Z}_n$; for example, non-zero factors may be canceled in $\mathbf{Z}$ but not necessarily in $\mathbf{Z}_n$. For instance, $2 \neq 0$ in $\mathbf{Z}_4$ but $2 \otimes 2 = 2 \otimes 0 = 0$ there.

In this construction of the finite algebraic system $\mathbf{Z}_n$, the fact that $\rho:\mathbf{Z} \to \mathbf{Z}_n$ is a surjection satisfying (29) and (30) for addition and multiplication is important, while the choice of the remainders $\{0, 1, \cdots, n - 1\}$ as the elements of $\mathbf{Z}_n$ is incidental. For example, we might have used instead the set $\{-n, -n + 1, \cdots, -1\}$ of largest negative remainders. Alternatively, one frequently replaces each remainder r by the set of *all* integers with the remainder r. This set, called the "residue class" or the "coset" of an integer k, modulo n, is

$$n\mathbf{Z} + k = \{nq + k \mid \text{all } q \in \mathbf{Z}\}. \tag{31}$$

For example, $n\mathbf{Z} = n\mathbf{Z} + 0$ is just the set of all integral multiples qn of n. In particular, if $n = 2$, there then are two cosets: The set $2\mathbf{Z}$ of all even integers and the set $2\mathbf{Z} + 1$ of all odd integers. For any n, there are just n different cosets.

$$n\mathbf{Z},\ n\mathbf{Z} + 1, \cdots, n\mathbf{Z} + (n - 1).$$

Let $\mathbf{Z}/n$ denote the *set* of all these cosets, while $p:\mathbf{Z} \to \mathbf{Z}/n$ is the function which assigns to each integer k its coset $pk = n\mathbf{Z} + k$. Comparing remainders modulo n to cosets modulo n, there is a bijection $r \mapsto n\mathbf{Z} + r$ from the set $\mathbf{Z}_n$ of remainders to the set $\mathbf{Z}/n$ of cosets. Its inverse θ is the unique function $\theta:\mathbf{Z}/n \to \mathbf{Z}_n$ which makes the following diagram commute:

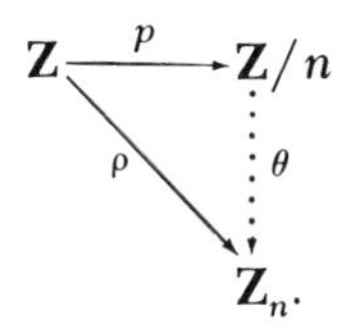

Explicitly, θ maps each coset $n\mathbf{Z} + k$ to the remainder of k, modulo n.

This means that the set of remainders can be replaced by the set of cosets, so that Theorems 16 and 17 hold with $\rho:\mathbf{Z} \to \mathbf{Z}_n$ replaced by $p:\mathbf{Z} \to \mathbf{Z}/n$ to give unique sums and products of cosets, as in

$$(n\mathbf{Z} + h) \oplus (n\mathbf{Z} + k) = n\mathbf{Z} + (h + k): \tag{32}$$

To add (or multiply) two cosets, pick an integer in each coset, add (or multiply) these integers and take the coset of the result; moreover, the final result does not depend on the choice of the integers h and k in their respective cosets; indeed, $h' = nq + h$ and $k' = nq' + k$ give $h' + k' \in n\mathbf{Z} + (h + k)$.

In the future, these binary operations $\oplus$ and $\otimes$ on cosets or remainders will be written (undecorated) as $+$ and $\cdot$, respectively, while $p:\mathbf{Z} \to \mathbf{Z}/n$ and $\rho:\mathbf{Z} \to \mathbf{Z}_n$ will be called the *projection* of the integers onto the integers modulo n.

The remainders modulo n can also be described by a binary relation of congruence modulo n. Given n, this relation is defined for integers h and k by

$$h \equiv k \pmod{n} \quad \Leftrightarrow \quad h - k \in n\mathbf{Z};$$

Thus $h \equiv k \pmod{n}$ if and only if $h - k$ is divisible by n, or, if and only if h and k leave the same remainder upon division by n. This relation $\equiv$ of *congruence* modulo n is clearly transitive, as well as reflexive and symmetric, so it is an equivalence relation in the sense defined at the end of §3. Moreover, the cosets modulo n can be described directly in terms of this congruence relation as

$$n\mathbf{Z} + k = \{h \mid h \equiv k \pmod{n}\}. \tag{33}$$

The surjection $p:\mathbf{Z} \to \mathbf{Z}/n$ of (31) to the set of cosets has the property that

$$ph = pk \quad \Leftrightarrow \quad h \equiv k \pmod{n}. \tag{34}$$

Furthermore, the fact that cosets (or remainders) can be added and multiplied, as in Theorems 16 and 17, could also be expressed by observing that congruences can be added and multiplied: $h \equiv k$ and $r \equiv s \pmod{n}$ imply

$$h + r \equiv k + s \pmod{n}, \qquad hr \equiv ks \pmod{n}. \tag{35}$$

EXERCISES

1. Construct addition and multiplication tables for $\mathbf{Z}_6$, and show that multiplication in $\mathbf{Z}_6$ does not satisfy the cancellation law.
2. For each n, show that the relation of congruence modulo n is reflexive, symmetric, and transitive.
3. Prove that congruences can be added and multiplied, as in (35).
4. If m is an integer, show that $m^2 \equiv 0$, 1, or 4 (mod 8).
5. Use Exercise 4 to prove that no integer $k \equiv 7$ (mod 8) can be expressed as the sum of three squares.

6. Show that the sum of the cubes of any three consecutive integers is divisible by 9.

7. In the usual decimal notation, the addition and multiplication of large numbers may be checked by the rule of "casting out 9's": Replace each number to be added by the sum of its digits, add these, and compare with the sum of the digits of the proposed answer (in comparing, use the sum of the digits of the sum of the digits . . .). Explain why this rule works.

9. Equivalence Relations and Quotient Sets

We have just seen in (33) that cosets of integers modulo n can be constructed directly from the relation of congruence modulo n. Much the same can be done with *any* equivalence relation on a set X, to yield a "quotient set" of X.

First note that any function yields an equivalence relation on its domain. For, given $f:X \to S$, define a binary relation E_f on X to X by

$$zE_fx \Leftrightarrow fz = fx$$

for $x, z \in X$. Then E_f is reflexive ($fx = fx$), symmetric ($fz = fx \Rightarrow fx = fz$), and transitive ($fz = fx$ and $fx = fy \Rightarrow fz = fy$). Hence, E_f is an equivalence relation on X, called the *equivalence kernel* of the function f.

We wish to show that every equivalence relation E on a set X is the equivalence kernel of some function. Given E, first define the *equivalence class* under E of any element $x \in X$ to be the set

$$p_Ex = \{\, y \mid y \in X \text{ and } yEx\}$$

of all those elements y of X which bear the relation E to x. Any subset C of X which has the form $C = p_Ex$ for some x is called an *equivalence class* for E (or an *E-class* for short). The set of all possible E-classes will be written as

$$X/E = \{C \mid C \subset X \text{ and } C = p_Ex \text{ for some } x \in X\}$$

and called the *quotient set* of X by E. Since $x \mapsto p_Ex$ assigns to each x in X an E-class, it is a function $p_E:X \to X/E$, called the *projection* of X on its quotient by E.

Theorem 18. *If E is an equivalence relation on a set X, the projection $p_E:X \to X/E$ is a surjection with equivalence kernel E.*

Proof: Since the elements of X/E are the equivalence classes p_Ex, the function p_E is necessary surjective. By the definition of an equivalence kernel, it remains only to show that xEz if and only if $p_Ex = p_Ez$.

First suppose that $p_Ex = p_Ez$. Since E is reflexive, xEx. Therefore, $x \in p_Ex = p_Ez$, hence, xEz.

Conversely, suppose that xEz. Then $y \in p_E x$ means that yEx and hence, by transitivity, yEz. This means that $y \in p_E z$. This argument shows that xEz implies $p_E x \subset p_E z$. Since E is symmetric, the same argument will give $p_E z \subset p_E x$, hence, $p_E x = p_E z$.

This completes the proof that p_E has equivalence kernel E.

COROLLARY. *Each element of X belongs to exactly one equivalence class for E.*

This corollary means that the function $p_E: X \to X/E$ may be described as that function assigning to each x the unique E-class C with $x \in C$.

Proof: An element x belongs to an equivalence class $p_E z$ if and only if xEz and hence if and only if $p_E x = p_E z$. Thus $p_E x$ is the only equivalence class containing x, Q.E.D.

It is suggestive and customary (but in no way necessary for our purposes) to describe quotient sets as partitions. A "partition" π of a set X is a set π whose elements are subsets of X such that each $x \in X$ is an element of exactly one $S \in \pi$. The corollary states that each equivalence relation E on a set X yields a partition X/E of X, namely, the partition $\pi = X/E$ into the equivalence classes for E.

To illustrate, take X to be a cartesian product $S \times T$, while E is the relation with $(s, t)E(s', t')$ if and only if $s = s'$; in other words, two points of $S \times T$ stand in the relation E if and only if they have the same first coordinates. This relation E is reflexive, symmetric, and transitive; an equivalence class is a vertical line $\{s\} \times T$ for some $s \in S$, as displayed in Figure I-2. Thus X/E is the set of all vertical lines in $S \times T$, while $p_E: X \to X/E$ assigns to each point (s, t) the vertical line containing that point.

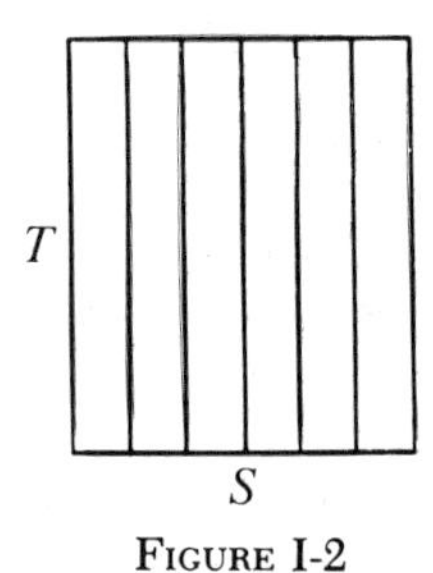

FIGURE I-2

Another useful equivalence relation comes from plane geometry. Let L be the set of all lines $k, l, \cdots$ in the (usual) Euclidean plane, while $k \| l$ has its usual meaning: "The line k is parallel to or coincides with the line l". Thus $\|$ is a binary relation on L to L; it is clearly reflexive, symmetric, and transitive, hence an equivalence relation. The equivalence class $p(k)$ then consists of all the lines l which are parallel to or coincide with k, so that $p(k) = p(l)$ if and only if k and l are parallel or $k = l$; that is, if and only if k and l have the same "direction". The quotient set $L/\|$ can thus be regarded as the set of all possible directions of lines, so that the construction of a quotient set in this case gives a precise meaning to "direction" of a line.

We now state the fundamental property of the quotient set X/E.

Theorem 19. *Given an equivalence relation E on the set X, let $f:X \to S$ be any function such that xEy implies $fx = fy$. Then there is exactly one function $g:X/E \to S$ for which $f = g \circ p_E$. If f is a surjection and $fx = fy$ implies xEy, then g is a bijection.*

This property may be visualized by the diagram displayed below. Given are f and p_E (solid arrows); there exists a unique g (dotted arrow) which makes diagram (36) commute.

Proof: Since yEx implies $fy = fx$, the function f carries all the elements y of an equivalence class $p_E x$ into a single element s of S. The assignment $p_E x \mapsto s$ is then a function $g:X/E \to S$ with $g \circ p_E = f$. This is clearly the only such function, exactly as required.

This proof is often put thus: By hypothesis, the function f is constant on each equivalence class; hence, it can be regarded as a function g defined on these equivalence classes. The value of g on any class C is just the value of f on an arbitrary element $x \in C$.

For example, on $\mathbf{Z}$ the unary operation $-:\mathbf{Z} \to \mathbf{Z}$ given by $k \mapsto -k$ (take the negative) has the property for each n that $h \equiv k \pmod{n}$ implies $-h \equiv -k \pmod{n}$. Therefore, there is a corresponding operation g on the quotient set which makes the following diagram commute,

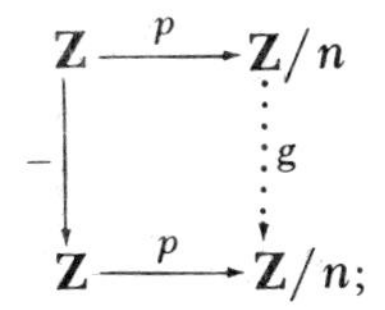

explicitly, this is Theorem 19 applied to $f = p\,-\,:\mathbf{Z} \to \mathbf{Z}/n$. Clearly, g is "take the negative" of a coset $n\mathbf{Z} + h \mapsto n\mathbf{Z} + (-h)$. The addition of cosets as in §8 can also be described in this way (Exercise 2 below).

Another example already occurred in our construction in §7 of the integers from the natural numbers. The difference function $d{:}\mathbf{N} \times \mathbf{N} \to \mathbf{Z}$ is a surjection; this means that $\mathbf{Z}$ could be described as a quotient set of $\mathbf{N} \times \mathbf{N}$ by an equivalence relation. (Two pairs, (m, n) and (m', n'), are equivalent under E when they have the same difference.) The crucial part of the construction was the definition of addition $\oplus$ in $\mathbf{Z}$ from the addition $+$ of pairs by the equation (22): $a \oplus b = d(sa + sb)$. This states that the sum of two integers a and b is the class of the sum in $\mathbf{N} \times \mathbf{N}$ of elements sa and sb in the classes a and b. This states that $\oplus$ is the g of Theorem 19 for two variables.

Theorem 19 can also be used to prove the quite familiar fact that any function f can be expressed as a surjection followed by an injection.

Corollary. *Any function $f{:}X \to S$ can be written as a composite $f = g \circ p$,*

$$X \xrightarrow{p} X/E_f \xrightarrow{g} S$$

with g injective and p the projection to a quotient set; namely, to the quotient set of X by the equivalence kernel E_f of f.

Proof: Given f, construct the equivalence kernel E_f and the corresponding projection $p = p_E$. By the theorem, $f = g \circ p$ for some g; it remains only to verify that g is an injection. Suppose that $g(pz) = g(px)$. This means that $fz = fx$, hence $z(E_f)x$, hence $pz = px$. Therefore, g is indeed injective.

EXERCISES

1. For the set T of all triangles t in the plane, let a binary relation S be defined by $t\ S\ t'$ if and only if t and t' are similar. Show informally that the quotient set T/S may be regarded as the set of all possible "shapes" of triangles.

2. Show that the addition of cosets of integers modulo n can be described according to Theorem 19 as the unique function $\oplus$ which makes the following diagram commute:

$$\begin{array}{ccc} \mathbf{Z} \times \mathbf{Z} & \xrightarrow{p \times p} & \mathbf{Z}/n \times \mathbf{Z}/n \\ \downarrow + & & \vdots\ \oplus \\ \mathbf{Z} & \xrightarrow{p} & \mathbf{Z}/n. \end{array}$$

3. For points (x, y) in the real coordinate plane $\mathbf{R}^2$, define $(x, y)E(x', y')$ to mean that $x - x'$ and $y - y'$ are both integers. Prove that E is an equivalence relation and show that the quotient set $\mathbf{R}^2/E$ may be described as the set of points on a torus (= surface of a doughnut).

10. Morphisms

Sums $\oplus$ and products $\otimes$ in $\mathbf{Z}_n$ were *defined* to make the projection $p:\mathbf{Z} \to \mathbf{Z}_n$ satisfy, as in (29) and (30),

$$p(k + m) = (pk) \oplus (pm), \qquad p(km) = (pk) \otimes (pm).$$

In other words, p "carries" the sum and the product in $\mathbf{Z}$ to the corresponding operations in $\mathbf{Z}_n$. We also say that the function p is a "morphism" of addition and of multiplication. This notion of morphism has far-reaching consequences. Our discussion will start with the definition of morphism for any binary operation.

Let $\square$ be a binary operation on a set X, while $\square'$ is another such operation on a set X'. A *morphism* $f:(X, \square) \to (X', \square')$ is defined to be a function on X to X' which "carries" the operation $\square$ on X into the operation $\square'$ on X', in the sense that

$$f(x \square y) = (fx)\square'(fy) \tag{37}$$

for all $x, y \in X$. On the left, one applies to an element $(x, y) \in X \times X$ first the operation $\square$, then the function f; on the right one applies first $f \times f$ and then $\square'$. In other words, f is a morphism if and only if the diagram below is commutative ($f \circ \square = \square' \circ (f \times f):X \times X \to X'$):

$$\begin{array}{ccc} X \times X & \xrightarrow{\ \square\ } & X \\ {\scriptstyle f \times f}\downarrow & & \downarrow{\scriptstyle f} \\ X' \times X' & \xrightarrow{\ \square'\ } & X'. \end{array}$$

For any binary operation $\square$ on a set X, the identity map 1_X of X is a morphism $1_X:(X, \square) \to (X, \square)$. If

$$f : (X, \square) \to (X', \square'), \qquad g : (X', \square') \to (X'', \square'')$$

are two morphisms, their composite $g \circ f$ is also a morphism. Indeed,

$$(g \circ f)(x \square y) = g[f(x \square y)] = g[(fx)\square'(fy)] = [(g \circ f)x]\square''[(g \circ f)y].$$

For example, the usual logarithm satisfies

$$\log (xy) = \log x + \log y$$

for all real $x, y > 0$. In more detail, write $(\mathbf{P}^*, \cdot)$ for the set of all positive real numbers with the binary operation of multiplication and $(\mathbf{R}, +)$ for the set of *all* real numbers under the binary operation of addition; then the logarithm (to any base) is a morphism log: $(\mathbf{P}^*, \cdot) \to (\mathbf{R}, +)$.

When both the binary operations involved are operations of addition (or both of multiplication), we speak of a *morphism of addition* (or, of a morphism of multiplication, as the case may be). Thus the addition in $\mathbf{Z}$ was defined in (23) so as to make the difference function $d:\mathbf{N} \times \mathbf{N} \to \mathbf{Z}$ a morphism of addition: $d(u + v) = d(u) + d(v)$. By (26), it is also a morphism of multiplication. Functions such as this which are simultaneously morphisms for several operations occur frequently.

Morphisms occur more generally. If (X, h) and (X', h') are sets with unary operations $h:X \to X$, $h':X' \to X'$, a *morphism* $f:(X, h) \to (X', h')$ of unary operations is a function $f:X \to X'$ with $(f \circ h)x = (h' \circ f)x$ for all $x \in X$.

$$\begin{array}{ccc} X & \xrightarrow{h} & X \\ {\scriptstyle f}\downarrow & & \downarrow{\scriptstyle f} \\ X' & \xrightarrow{h'} & X'. \end{array}$$

This states that the diagram above commutes. If $t:X^3 \to X$ is a ternary operation on X, and t' a similar operation on X', a morphism $f:(X, t) \to (X', t')$ of ternary operations is a function $f:X \to X'$ with $(f \circ t)(x, y, z) = t'(fx, fy, fz)$ for all $x, y, z \in X$. (Draw the diagram.) An analogous definition applies to n-ary operations for any natural number n. In particular, if $u:\mathbf{1} \to X$ and $u':\mathbf{1} \to X'$ are nullary operations, a morphism $f:(X, u) \to (X', u')$ is a function $f:X \to X'$ with $f \circ u = u'$. Now the nullary operation u amounts to selecting an element $u(1) = u_1 \in X$, so a morphism f of nullary operations is just a function $X \to X'$ with $f(u_1) = u_1'$.

A morphism $f:(X, \square) \to (X', \square')$ is said to be a

monomorphism	if the function f is an injection;
epimorphism	if the function f is a surjection;
isomorphism	if the function f is a bijection;

often a morphism itself is called a "homomorphism". A morphism $(X, \square) \to (X, \square)$ of $(X, \square)$ to itself is called an *endomorphism*, or, if bijective, an *automorphism*. This terminology will be used for each type of morphism (for example, for morphisms of several operations, binary or otherwise). In any such case the composite of two monomorphisms, when defined, is again a monomorphism; the same holds for the composites of epimorphisms or of isomorphisms. Moreover, one has

Theorem 20. *The inverse of an isomorphism is an isomorphism.*

Proof (*for one binary operation* $\square$): We must show that if a bijection $f:X \cong X'$ is a morphism, so is the function $f^{-1}:X' \to X$ inverse to f. Thus we must prove that

$$f^{-1}(u \,\square'\, v) = (f^{-1}u) \,\square\, (f^{-1}v) \tag{38}$$

for all $u, v \in X'$. Since f is a bijection, it is enough to show that both sides of this equation become equal when f is applied: $u \,\Box'\, v = f[(f^{-1}u) \,\Box\, (f^{-1}v)]$. But f is a morphism, so the right side here is $(ff^{-1}u) \,\Box'\, (ff^{-1}v) = u \,\Box'\, v$, as required.

Two sets $(X, \Box)$ and $(X', \Box')$, each with a binary operation, are called *isomorphic* when there exists some function f which is an isomorphism $f:(X, \Box) \cong (X', \Box')$. Since the identity function 1_X is always an isomorphism and since the inverse or the composite of isomorphism(s) is again such, it follows that the relation <'isomorphic'' is reflexive, symmetric, and transitive. For many purposes, however, it is not enough to know just that two objects are isomor*phic*: One needs to specify the isomorph*ism*.

EXERCISES

1. Find all additive endomorphisms of $\mathbf{Z}$.
2. In each of the following cases list all morphisms of addition:

$$\mathbf{Z}_6 \to \mathbf{Z}_3, \qquad \mathbf{Z}_3 \to \mathbf{Z}_9, \qquad \mathbf{Z}_5 \to \mathbf{Z}_6, \qquad \mathbf{Z} \to \mathbf{Z}_4.$$

3. List all endomorphisms (of addition) of $\mathbf{Z}_6$, of $\mathbf{Z}_8$.
4. List all isomorphisms $(\mathbf{Z}_4, +) \to (Q, \circ)$, where Q is the set of all rotations of the square into itself with composition as binary operation.
5. (a) If $m: Y \to Z$ is a monomorphism, prove that $m \circ f = m \circ f'$ for any two morphisms $f, f': X \to Y$ implies $f = f'$.
   (b) If $e: T \to X$ is an epimorphism, prove that $f \circ e = f' \circ e$ implies $f = f'$.
6. If $f:(X, t) \to (X', t')$ and $g:(X', t') \to (X'', t'')$ are morphisms of ternary operations, show explicitly that $g \circ f$ is also such a morphism, and draw a diagram.
7. Show that the set of nonzero real numbers under multiplication is not isomorphic to the set of all real numbers under addition.

11. Semigroups and Monoids

An *algebraic system* is a set A together with one or more n-ary operations on A which satisfy specified axioms (identities or other conditions). In this book we are concerned principally with certain types of algebraic systems such as groups (Chapter II), rings (Chapter III), fields (Chapter VIII), modules (Chapter V), and vector spaces (Chapter VI). A first example of an algebraic system is a "monoid"; its chief importance for us is that it will be useful in helping to describe several of the later systems.

A "semigroup" $(S, \Box)$ is a set S together with an associative binary operation $\Box: S \times S \to S$. A *monoid* $(M, \Box, u)$ is a semigroup $(M, \Box)$ with an element $u \in M$ which is a unit for $\Box$; that is, $u \,\Box\, x = x \,\Box\, u = x$ for all $x \in M$. As we have noted in §3, the selected element u may be regarded as a nullary operation on M, so a monoid is an algebraic system with two operations, $\square$ and u. For any set X, the set consisting of all functions $f:X \to X$ is a monoid with composition as the binary operation and with the identity function 1_X as unit. Many other monoids arise in arithmetic; we now mention a few.

Call a monoid "additive" when the binary operation is written as + and the unit as 0. Thus the definition of an additive monoid A reads: $0 \in A$ and $+ :A \times A \to A$ satisfy the axioms

$$a + (b + c) = (a + b) + c, \qquad 0 + a = a = a + 0 \tag{39}$$

for all elements $a, b, c \in A$. For example, $\mathbf{N}$, $\mathbf{Z}$, and $\mathbf{Z}_n$ are all additive monoids; they are also *commutative monoids*, since $a + b = b + a$ in all three examples.

Likewise a "multiplicative" monoid is one in which the binary operation is written as multiplication and the unit (= "identity" element) as 1. Thus the axioms for such a monoid M are

$$a(bc) = (ab)c, \qquad 1a = a = a1 \tag{40}$$

for all elements $a, b, c \in M$. For example, $\mathbf{N}$, $\mathbf{Z}$, and all $\mathbf{Z}_n$ are multiplicative monoids under the multiplications defined earlier.

One can define "powers" of an element in any multiplicative monoid M. Thus a^n is $aa \ldots a$ (n factors), so that each $n \in \mathbf{N}$ gives a unary operation $a \mapsto a^n$ on M, "take the nth power". Formally, a^n is defined, by recursion on n, by the equations

$$a^0 = 1, \qquad a^{n+1} = aa^n. \tag{41}$$

Moreover, one has for all $m, n \in \mathbf{N}$ the rules

$$a^{m+n} = a^m a^n, \qquad a^{mn} = (a^m)^n. \tag{42}$$

These rules may be proved by induction on n, just as for (13) and (15) in the special case of the iterated functions f^n defined in (7). Similar formulas hold in additive monoids, where they assume the form

$$0 \cdot a = 0, \qquad (n + 1)a = na + a, \tag{43}$$

$$(m + n)a = ma + na, \qquad (mn)a = m(na). \tag{44}$$

Any *list* $a_1, \cdots, a_n$ of $n \geqslant 2$ elements in a multiplicative monoid M has an n-fold iterated product, written as

$$a_1 a_2 \cdots a_n = \prod_{i=1}^{n} a_i.$$

Expressed explicitly with parentheses, the left-hand side should be

$((\cdots(a_1a_2)\cdots)a_n)$, but by the associative law for multiplication the actual arrangement of parentheses does not change the resulting iterated product. (This fact is often called the "general associative law".) Formally, the n-fold iterated product is defined by recursion on n, starting with $n = 2$:

$$\prod_{i=1}^{2} a_i = a_1a_2, \qquad \prod_{i=1}^{n+1} a_i = \left(\prod_{i=1}^{n} a_i\right)a_{n+1}.$$

This n-fold product is a function $M^n \to M$; the index i conventionally used in the product formula can be replaced by any other index without altering the value of the product. Thus we write variously

$$\prod_{i=1}^{n} a_i = \prod_{j=1}^{n} a_j = \prod_{k\in \mathbf{n}} a_k = \textstyle\prod_i a_i = \prod a_i,$$

the latter two when the range of $i = 1, \cdots, n$ is clear from the context.

If the monoid A is additive, iterated products become iterated sums, usually written

$$((a_1 + a_2) + \cdots) + a_n = \sum_{i=1}^{n} a_i,$$

or more simply as $\Sigma_i\, a_i$.

In a *commutative* monoid, a permutation of the order of summands (or factors, in the multiplicative case) does not change the sum. For example, if $b{:}\mathbf{m} \times \mathbf{n} \to A$ is any rectangular array or "matrix" of elements of a commutative additive monoid, then one has for "double sums" the equality

$$\sum_{i=1}^{m}\left(\sum_{j=1}^{n} b_{ij}\right) = \sum_{j=1}^{n}\left(\sum_{i=1}^{m} b_{ij}\right) = \sum_{\mathbf{m}\times\mathbf{n}} b_{ij}. \tag{45}$$

Next consider morphisms of monoids. Since a monoid $(M, \square, u)$ is a set equipped with two operations (binary and nullary), a *morphism* of monoids $f{:}(M, \square, u) \to (M', \square', u')$ is a function $f{:}M \to M'$ which is a morphism for both pairs of operations. For example, if M and M' are multiplicative, a morphism $f{:}M \to M'$ of monoids is a function f satisfying

$$f(ab) = (fa)(fb), \qquad f(1) = 1' \tag{46}$$

for all $a, b \in M$ and for 1 and $1'$ the identities of M and M'.

For any fixed $n \in \mathbf{N}$, the operation $a \mapsto a^n$ (raise a to the nth power) is a unary operation on M. Any morphism $f{:}M \to M'$ of monoids is also a morphism for this unary operation, for one can prove that $f(a^n) = (fa)^n$ by induction on n, using the recursive definitions $a^0 = 1$, $a^{n+1} = aa^n$ of (41).

An induction also shows that any morphism of multiplicative monoids is also a morphism for the n-ary operation $(a_1, \cdots, a_n) \mapsto \prod_i a_i$.

If M is a monoid, the identity function 1_M of M is a morphism $1_M: M \to M$ of monoids. If $f: M \to M'$ and $g: M' \to M''$ are two morphisms of monoids, so is their composite $g \circ f: M \to M''$.

We now show that the additive monoid defined by **N** has a "universal" element 1, in the following sense.

Proposition 21. *If a is an element in the monoid M, there is exactly one morphism $f: \mathbf{N} \to M$ of monoids with $f(1) = a$.*

Proof: The properties $a^0 = 1$ and $a^{m+n} = a^m a^n$ state that $n \mapsto a^n$ carries sums to products, so defines a morphism of the additive monoid **N** to the multiplicative monoid M. On the other hand, if $f: \mathbf{N} \to M$ is any such morphism with $f(1) = a$, then $f(n+1) = f(n)f(1) = f(n)a$, so $f(n) = a^n$ by induction on n.

EXERCISES

1. Prove: A semigroup can have at most one unit.
2. If a is a list of $m + n$ elements in a multiplicative monoid, prove that

$$\left(\prod_{i=1}^{m} a_i\right)\left(\prod_{j=1}^{n} a_{m+j}\right) = \prod_{k=1}^{m+n} a_k.$$

3. In a multiplicative monoid, prove the rules (42).
4. Prove the first equality (45) for double sums in a commutative monoid.
5. For monoids $(M, \square, u)$ and $(M', \square', u')$ show by an example that a morphism $f:(M, \square) \to (M', \square')$ of binary operations need not be a morphism of units.

CHAPTER II

Groups

THIS CHAPTER will introduce one of the simplest yet most useful types of algebraic systems: the group.

1. Groups and Symmetry

A *group* G is a set G together with a binary operation $G \times G \to G$, written $(a, b) \mapsto ab$, such that:

(i) This operation is associative.
(ii) There is an element $u \in G$ with $ua = a = au$ for all $a \in G$.
(iii) For this element u, there is to each element $a \in G$ an element $a' \in G$ with $aa' = u = a'a$.

In other words, a group is a monoid in which every element is invertible.

The element ab is called the *product* of a and b in G, while u is the *unit* or the *identity* of G, and a' the *inverse* of a in G. Here the binary operation in G has been written as a product; often it may be written as a sum $(a, b) \mapsto a + b$; we say accordingly that the group is *multiplicative* or *additive*. The letter G will stand both for the set of elements of the group and for this set together with its binary operation.

If G and G' are groups, a *morphism* $\phi: G \to G'$ of *groups* is a function on G to G' which is a morphism for the binary operations involved. In case G and G' are both multiplicative groups, this requirement means that $\phi(ab) = (\phi a)(\phi b)$ for all $a, b \in G$.

A morphism $\phi: G \to G'$ of groups which is a bijection is called an *isomorphism*. For example, the additive group $\mathbf{Z}_2$ of integers modulo 2 is isomorphic to the multiplicative group $\{-1, +1\}$, since $0 \mapsto +1$, $1 \mapsto -1$ is a morphism $\mathbf{Z}_2 \to \{-1, +1\}$.

We have already explicitly constructed several groups of numbers. For example, the set $\mathbf{Z}$ of integers is a group under addition; indeed, Theorem I.12 states that the binary operation $+$ on $\mathbf{Z}$ is associative and has zero as unit, and that to each element $a \in \mathbf{Z}$ there is an x with $a + x = 0$ (and hence $x + a = 0$). Similarly, Theorem I.16 asserts that the set $\mathbf{Z}_n$ of integers modulo n is a group under addition and that the projection $\mathbf{Z} \to \mathbf{Z}_n$ is a morphism of groups. In particular, the group $\mathbf{Z}_5$ has five elements, with the addition table as displayed in (I.28).

Any finite group G may be explicitly described by such a table. If G is a multiplicative group with n elements, then its *multiplication table* is an $n \times n$ square array headed both to the left and above by a list of the n elements of G. In this table the entry in the row headed by a and the column headed by b is the product ab in G. Not every such $n \times n$ table defines a group; for example, the multiplication table (I.28) does not make $\mathbf{Z}_5$ a group under multiplication because 1 is the (only) unit while $0 \otimes r = 0$ for every r, so the number 0 has no inverse in $\mathbf{Z}_5$. On the other hand, if 0 is omitted the remaining elements of $\mathbf{Z}_5$ have the multiplication table

×	1	2	3	4
1	1	2	3	4
2	2	4	1	3
3	3	1	4	2
4	4	3	2	1;

from this it is evident that 1 is a unit and each element has an inverse, while we already know that this multiplication is associative. Hence, these four elements form a multiplicative group G. In this group G, $2^2 = 4$, $2^3 = 3$, and $2^4 = 1$, so all elements are powers of 2. This suggests that $\mathbf{Z}_4$ be compared with this group G by the function indicated below:

$$\mathbf{Z}_4 \to G \ (\text{mod } 5)$$

$$0 \mapsto 1 = 2^0$$

$$0 \mapsto 2 = 2^1$$

$$2 \mapsto 4 = 2^2$$

$$3 \mapsto 3 = 2^3;$$

it turns out that this function is an isomorphism of the additive group $\mathbf{Z}_4$ to the multiplicative group G.

Other familiar groups of numbers will be constructed in later chapters. For example, the set P^* of positive real numbers is a group under the usual binary operation of multiplication, while the set $\mathbf{R}$ of *all* real numbers is a group under addition. In this case the function $\log_e : \mathrm{P}^* \to \mathbf{R}$ is a morphism of groups, because $\log_e(xy) = \log_e x + \log_e y$. For that matter, this function is an isomorphism; its inverse is the function $x \mapsto e^x$, which is also a morphism of groups (by the familiar property $e^{x+y} = e^x e^y$).

A *permutation* on a set X is a bijection $X \to X$. Many important groups consist of permutations with product the composition of permutations.

Indeed, from our study (§I.2) of functions we know that the composite of two permutations on X is also a permutation, that this operation of composition is associative, that 1_X is a permutation, and that the inverse of any permutation is again such. These properties state that the set $S(X)$ of all permutations on X is a group under the operation of composition. In particular, the group consisting of all the permutations $\sigma:\mathbf{n} \to \mathbf{n}$ on the finite set $\{1, 2, \cdots, n\} = \mathbf{n}$ is called the *symmetric group* S_n, of "degree" n. It clearly has $n! = n(n-1) \cdots 2 \cdot 1$ different permutations σ. For, the image $\sigma(1)$ of the first element may be chosen in n different ways, that of the second element can then be chosen in $n - 1$ ways from those elements not $\sigma(1)$, and so on.

As another example, consider permutations on the set F of all points of some geometric figure in the plane. Call a permutation $f:F \to F$ a "symmetry" of F when it preserves distances; that is, when the distance between any two points p and q of F equals the distance between their images $f(p)$ and $f(q)$ under f. Then the inverse function f^{-1} is also a symmetry of F. Moreover, the composite of two symmetries of F is also a symmetry of F, as is the identity function $F \to F$. It follows that the set of all symmetries of F under composition is a group, called the *group of symmetries* of F.

For instance, let F be the set T of points on the perimeter of an equilateral triangle T (Figure II-1). Three obvious symmetries are the counterclockwise rotations through 120°, 240°, and 360°; write them as R, $R \circ R = R^2$, and $R \circ R \circ R = 1_T$. Three other symmetries are the reflections D_1, D_2, D_3 in the altitudes through the three vertices 1, 2, 3. This gives a set Δ_3 of six symmetries of T:

$$\Delta_3 = \{1, R, R^2, D_1, D_2, D_3\};$$

they give every possible rearrangement of the three vertices 1, 2, 3. Since any symmetry of the equilateral triangle T is determined by its effect on the three vertices, the set Δ_3 contains *all* symmetries of T. It follows as above that Δ_3 is a group under composition.

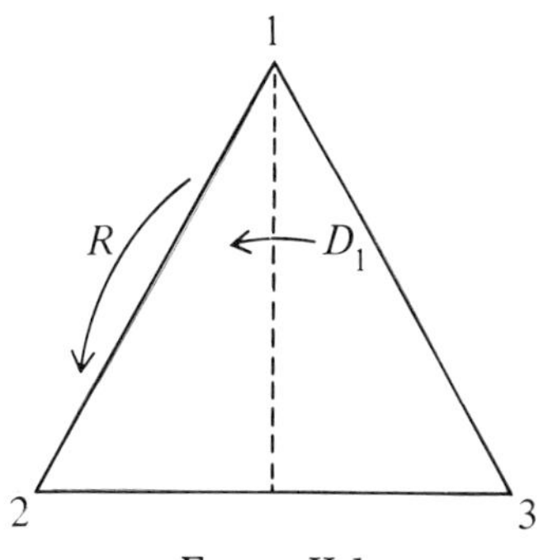

FIGURE II-1

One can easily compute a group table for Δ_3, either by keeping track of where each vertex goes, or by experimentation with a triangular piece of cardboard. For instance, to find $D_1 \circ R$, first rotate the cardboard through $120°$, then reflect in the axis which is *now* vertical; the final effect is that of D_2, so $D_1 \circ R = D_2$. A similar calculation gives $R \circ D_1 = D_3$; hence in this group $D_1 \circ R \neq R \circ D_1$: Group multiplication need not be commutative.

Analogous methods apply to a regular polygon of n sides; such a polygon has exactly $2n$ different symmetries, n of rotation and n of reflection. They form a group Δ_n of $2n$ elements which will be studied in §5. To each symmetry s of such a polygon assign the permutation $f(s) = \sigma$ of the vertices which s effects; we obtain for each n a monomorphism $f:\Delta_n \to S_n$ of Δ_n to the symmetric group. For $n = 3$, f is a surjection, hence an isomorphism $\Delta_3 \cong S_3$.

There are analogous groups of symmetries for algebraic systems. For instance, let G be a group, and consider the set of all automorphisms $G \to G$. The composite of two such automorphisms is again an automorphism, and this composition is associative. Moreover, the identity 1_G is an automorphism, and any automorphism has a two-sided inverse which is, by Theorem I.20, an automorphism. Hence, the set of all automorphisms of any group G is itself a group, which we shall write as $\mathrm{Aut}(G)$. For example, the additive group $\mathbf{Z}_4$ has two automorphisms, the identity and $a \mapsto -a$; the square of the second automorphism is the identity, so $\mathrm{Aut}(\mathbf{Z}_4) \cong \mathbf{Z}_2$.

To groups of numbers and groups of symmetries, we add a third source of groups: The construction of new groups from given ones. Thus, if G and G' are given (multiplicative) groups, their *product* $G \times G'$ is the product set consisting of all pairs (a, a') for $a \in G$, $a' \in G'$ with the binary operation

$$(a, a')(b, b') = (ab, a'b') \tag{1}$$

of "termwise multiplication" of these pairs. This does make $G \times G'$ a group, for termwise multiplication is manifestly associative, has as identity the pair (u, u') for u, u' the identities of G and G', respectively, and has for each pair an inverse. Moreover, the projections $(a, a') \mapsto a$ and $(a, a') \mapsto a'$ of the product set on its "axes" are morphisms of groups

$$G \leftarrow G \times G' \to G'.$$

For example, the product $\mathbf{Z}_2 \times \mathbf{Z}_2$ of the additive group $\mathbf{Z}_2$ with itself is an additive group with four elements $0 = (0, 0)$, $a = (1, 0)$, $b = (0, 1)$, and $c = (1, 1)$, with addition table $a + a = 0$, $b + b = 0$, $c + c = 0$, $a + b = c$, $b + c = a$, $c + a = b$. This group is known as the *four group*.

EXERCISES

1. Give multiplication tables for S_3 and for Δ_3.
2. Exhibit an isomorphism of the additive groups $\mathbf{Z}_6$ and $\mathbf{Z}_2 \times \mathbf{Z}_3$.

3. Show that the four-group $\mathbf{Z}_2 \times \mathbf{Z}_2$ is not isomorphic to $\mathbf{Z}_4$, but is isomorphic to the group of all symmetries of a rectangle.

4. Describe the six symmetries of a cube with one vertex held fixed, and show that they form under composition a group isomorphic to S_3.

5. Show that a cube has exactly 48 different symmetries.

6. Show that the group of all symmetries of a regular tetrahedron is isomorphic to S_4.

7. Show that $\mathbf{Z}_6$ with zero deleted is not a multiplicative group, but that $\mathbf{Z}_7$ with zero deleted is a multiplicative group, isomorphic to the additive group $\mathbf{Z}_6$.

8. State exactly how a proposed multiplication table can be tested for: (i) presence of an identity, and (ii) presence of a two-sided inverse for each element.

9. Establish isomorphisms $G \times G' \cong G' \times G$ and $G \times (G' \times G'') \cong (G \times G') \times G''$.

10. Let $\mathbf{R}$ be the set of all real numbers. For a and b real with $a \neq 0$, the assignment $x \mapsto ax + b$ is a permutation $\mathbf{R} \to \mathbf{R}$. Show that the set of all these permutations is a group under composition.

11. Let a, b, c, d be real numbers with $ad - bc = 1$. (a) Show that $x \mapsto (ax + b)/(cx + d)$ defines a permutation on the set $\mathbf{R}$ of all real numbers plus a symbol ∞; (b) show that the set of all these permutations is a group under composition.

12. Let H be the set consisting of the following permutations on the plane: All translations and all reflections in any line of the plane. Show that H is not a group under the operation of composition.

13. (a) Show that $\mathbf{Z}_3$ has two automorphisms and that $\mathrm{Aut}(\mathbf{Z}_3) \cong \mathbf{Z}_2$.
(b) Show that $\mathbf{Z}_2 \times \mathbf{Z}_2$ has six automorphisms with $\mathrm{Aut}(\mathbf{Z}_2 \times \mathbf{Z}_2) \cong S_3$.
*(c) Show that $\mathrm{Aut}(\Delta_3) \cong \Delta_3$.

2. Rules of Calculation

The axioms for a group require that its multiplication:

(i) Be associative.
(ii) Have a two-sided unit u, with $au = a = ua$ for all a.
(iii) Have to each element a a two-sided inverse a', with $aa' = u = a'a$.

These axioms have a number of simple but important consequences. We shall now establish them in detail, to illustrate the notion of a formal proof from axioms.

RULE 1. In any group G there is exactly one element u with

$$au = a \text{ for all } a \in G \qquad \text{and} \qquad ua = a \text{ for all } a \in G.$$

Either one of these two properties uniquely determines u.

By axiom (ii) there is at least one such element u. It remains to prove (say) that any "right unit" r is equal to u; that is, that $ar = a$ for all $a \in G$ implies $r = u$. Now

1. $r = ur$ (u is a unit by axiom (ii)).
2. $ur = u$ (by hypothesis, r is a right unit).
3. $r = u$ (steps 1, 2, transitive law for equality), Q.E.D.

The unique unit element u for a multiplicative group G will usually be written as 1, and called the *identity* element of G.

RULE 2. Any group G satisfies the right and left cancellation laws: For all $a, b, c \in G$,

$$ab = ac \quad \Rightarrow \quad b = c; \qquad ba = ca \quad \Rightarrow \quad b = c.$$

We prove the first of these laws ("left cancellation"). Given a, take the element a' given by axiom (iii). Then

1. $a'(ab) = a'(ac)$ (by hypothesis, $ab = ac$).
2. $(a'a)b = (a'a)c$ (associative axiom).
3. $ub = uc$ (axiom (iii) for a').
4. $b = c$ (u is a left unit).

The right cancellation law can be proved similarly.

RULE 3. To each element a in a group G there is exactly one element $a^{-1} \in G$, called the *inverse* of a in G, with $aa^{-1} = u$ and $a^{-1}a = u$. Either property uniquely determines a^{-1}.

Proof: Axiom (iii) gives to each a one such element a'. We must show, say, that any "right inverse" b for the element a necessarily equals this a'; that is, that $ab = u$ implies $b = a'$. But:

1. $ab = u$ (hypothesis).
2. $ab = aa'$ (axiom (iii), $u = aa'$).
3. $b = a'$ (step 2, cancellation by Rule 2).

RULE 4. For u the identity of G and for all $a, b \in G$,

$$u^{-1} = u, \qquad (a^{-1})^{-1} = a, \qquad (ab)^{-1} = b^{-1}a^{-1}.$$

Proof: Since u is a left identity, $uu = u$; this equation states that u is the (unique) left inverse u^{-1} of u. Similarly, $a^{-1}a = u$ states that a is the right inverse of a^{-1}, hence equal to $(a^{-1})^{-1}$. By the associative law,

$$(b^{-1}a^{-1})(ab) = b^{-1}(a^{-1}(ab)) = b^{-1}((a^{-1}a)b) = b^{-1}(ub) = u.$$

Hence, $b^{-1}a^{-1}$ is the (unique) left inverse of ab—and also the right inverse.

This gives all three conclusions; let the reader reduce this argument to a step-by-step form.

RULE 5. For given elements a and b in a group G, the equations $xa = b$ and $ay = b$ have in G unique solutions $x = ba^{-1}$ and $y = a^{-1}b$, respectively.

Proof: If $xa = b$ has any solution x, this solution is unique, for, by the right cancellation law, $xa = b = x_1a$ implies $x = x_1$. On the other hand, $(ba^{-1})a = b(a^{-1}a) = bu = b$, so $x = ba^{-1}$ is indeed a solution. That $ay = b$ has the unique solution $y = a^{-1}b$ is proved similarly.

Above, it has sufficed to prove one of a pair of left- and right-hand laws because the axioms of a group are left–right symmetric. For the same reason, if G is a group under a binary operation $\square:(a, b) \mapsto a \square b$, then it is also a group under the "opposite" operation $\square'$ given by $(a, b) \mapsto b \square a$. The *opposite group* of G so defined is written G^{op} (same set, opposite operation). We have proved the following rule:

RULE 6. The opposite of a group G is a group.

For example, in the symmetric group S_n of all permutations on $\mathbf{n}$, the product $\sigma \circ \tau$ means first apply τ, then σ. In the opposite group S_n^{op}, the product $\sigma\tau$ is first σ, then τ.

For any group G the identity function 1_G is a morphism of groups; also, if $f:G \to G'$ and $f':G' \to G''$ are morphisms of groups, so is their composite $f' \circ f:G \to G''$. As always, a morphism $f:G \to G'$ of groups is called a monomorphism, an epimorphism, or an isomorphism when the function f is injective, surjective, or bijective, respectively. The general properties of morphisms (Theorem I.20) hold in particular for morphisms of groups. In addition, we can now prove the following special result.

PROPOSITION 1. *Any morphism $\phi:G \to G'$ of groups is also a morphism for the identity and for the inverse.*

Proof: If u and u' are the identities of G and G', we must first prove that $\phi(u) = u'$. Since $uu = u$ in G and ϕ is a morphism, $(\phi u)(\phi u) = \phi u = (\phi u)u'$; by cancellation, $\phi u = u'$, as required.

Second, $a \mapsto a^{-1}$ is a unary operation on G; we must prove that ϕ is a morphism for this operation; that is, that $\phi(a^{-1}) = (\phi a)^{-1}$ for all $a \in G$. But $aa^{-1} = u$ and ϕ a morphism give $(\phi a)(\phi a^{-1}) = \phi u = u'$, and this equation implies that $(\phi a)^{-1} = \phi(a^{-1})$. Hence, any morphism carries inverses to inverses, as required.

Alternative axiom systems for groups are discussed in Exercises 5–7.

A group G is called *abelian* when its binary operation is commutative. Thus a multiplicative group G is abelian when $ab = ba$ for all $a, b \in G$. However, the binary operation in an abelian group is frequently written as addition; the unit is then written as 0, and the inverse a^{-1} as $-a$. Hence an additive abelian group is a set A with a binary operation $(a, b) \mapsto a + b$ with the following properties: For all a, b, c,

$$a + (b + c) = (a + b) + c \qquad \text{(associative law);} \tag{2}$$

$$a + b = b + a \qquad \text{(commutative law);} \tag{3}$$

there exists an element $0 \in A$ such that, for all $a \in A$,

$$a + 0 = a \qquad \text{(0 an "additive unit");} \tag{4}$$

and such that to each $a \in A$ there exists an element $-a \in A$ with

$$a + (-a) = 0 \qquad (-a \text{ an "additive inverse"}). \tag{5}$$

Rules 1–5 above still apply; in the additive notation they read: The property (4) uniquely determines the *zero element* 0, $a + b = a + c$ implies $b = c$, property (5) uniquely determines the *negative* $-a$ of each a, and

$$-0 = 0, \qquad -(-a) = a, \qquad -(a + b) = (-a) + (-b).$$

Moreover, for given a and b, $a + x = b$ has a unique solution which is $x = b + (-a)$, usually called the *difference* $b - a$.

EXERCISES

1. Prove each of the following rules in a group:
(a) The right cancellation law.
(b) $a(b(cd)) = ((ab)c)d$.
(c) $((ab^{-1})c)^{-1} = (c^{-1}b)a^{-1}$.

2. Prove the following extension of Rule 1 in any group G: If $ar = a$ for even *one* element $a \in G$, then $r = u$.

3. In a group of $2n$ elements, prove that there is an element not the unit which is its own inverse.

4. Prove that any two groups with three elements each are isomorphic.

5. Let S be a nonvoid set with an associative binary operation such that both cancellation laws hold (that is, Rule 2 holds). If S is finite, prove that S is a group. If S is infinite, show that S need not be a group.

6. If a binary operation of multiplication on a set X is associative, has a left identity u (so that $ua = a$ for all $a \in X$), and if each element a has a left inverse a' (so that $a'a = u$), prove that X is a group. (*Hint:* First show that cancellation on the left is possible, and thence that u is a right identity and a' is a right inverse.)

7. Prove: If a binary operation of multiplication on a nonvoid set G is associative, and if all equations $xa = b$ and $ay = b$ have solutions x and y in G, then G is a group. (*Hint:* Pick any $a_0 \in G$, solve $ua_0 = a_0$ for u, prove that this u is a left identity, and apply Exercise 6.)

8. (a) Show that the permutations $(x, y) \mapsto (\varepsilon x + b, y)$ on the (x, y)-plane, with $\varepsilon = \pm 1$ and b real, form a nonabelian group under composition.

(b) Show that the permutations $(x, y) \mapsto (x + a, \varepsilon y)$ on the (x, y)-plane, with $\varepsilon = \pm 1$ and a real, form an abelian group under composition.

(c) Interpret geometrically the permutations defined by the preceding formulas.

9. Prove that the following postulates define an (additive) abelian group: (i) $(a + b) + c = a + (c + b)$ for all a, b, and c; (ii) there exists an element 0 with $0 + a = a$ for all a; (iii) there exists to each a an element a' with $a' + a = 0$.

10. For any group G, show that $a \mapsto a^{-1}$ is an isomorphism $G \cong G^{\mathrm{op}}$.

11. If $(G, \cdot)$ is a group, $\square$ a binary operation on a set X, and $p:(G, \cdot) \to (X, \square)$ an epimorphism of binary operations ($\cdot$ to $\square$), prove that X is a group under the operation $\square$.

12. If an element b of a monoid has both a left inverse and a right inverse, prove these inverses equal.

3. Cyclic Groups

For a regular hexagon, let R be the operation of counterclockwise rotation by 60°. Then the group of all rotational symmetries of the hexagon consists of the six rotations

$$\{1, R, R^2, R^3, R^4, R^5\}$$

with $R^6 = 1$. This is called a "cyclic" group of "order" 6, generated by R.

If L is the usual infinite line with all integer points marked, then one symmetry of this line L is a translation T to the right by one unit. All the translations of this marked line form a group

$$\cdots, T^{-3}, T^{-2}, T^{-1}, 1, T, T^2, T^3, \cdots,$$

an "infinite" cyclic group. It is isomorphic to the additive group of all integers.

The general description of such cyclic groups uses exponents (powers of elements).

In a multiplicative group G with identity 1 the nonnegative *powers* of an element g are $g^0 = 1$ and $g^k = g \cdots g$, with k factors, just as in a monoid. In a group one can also define the negative powers g^{-k} of g as $g^{-k} = (g^k)^{-1}$.

The "exponents" so defined have the property that

$$g^m g^n = g^{m+n} \tag{6}$$

for all integers m and n. We have already proved this property for m and n both nonnegative (by induction; see (I.42)). If both exponents are negative, the definitions give

$$g^{-h}g^{-k} = (g^h)^{-1}(g^k)^{-1} = (g^k g^h)^{-1} = (g^{h+k})^{-1} = g^{(-h)+(-k)},$$

as desired. If $m = -h$ and $0 \leqslant h \leqslant n$, set $k = n - h \geqslant 0$; then

$$g^{-h}g^n = g^{-h}g^{h+k} = (g^h)^{-1}g^h g^k = g^k = g^{n+(-h)},$$

again as desired. The remaining cases of (6) are treated similarly. A corresponding case subdivision will prove that

$$(g^m)^n = g^{mn} \tag{7}$$

for all integers m and n.

The formula (6) states for each g that the function $n \mapsto g^n$ is a morphism $\mathbf{Z} \to G$ from the additive group $\mathbf{Z}$ to the multiplicative group G. Since *any* morphism of groups carries inverses to inverses,

$$g^{-n} = (g^n)^{-1}, \qquad n \in \mathbf{Z}. \tag{8}$$

Moreover, any morphism $\mathbf{Z} \to G$ of groups has the form $n \mapsto g^n$ for some g:

Theorem 2. *For a fixed element g of a group G, the assignment $n \mapsto g^n$ is a morphism $\mathbf{Z} \to G$ of groups, and the only morphism $\mathbf{Z} \to G$ with $1 \mapsto g$.*

Proof: Any morphism $\phi: \mathbf{Z} \to G$ has

$$\phi(0) = 1, \qquad \phi(n+1) = \phi(n)\phi(1), \qquad \phi(-k) = \phi(k)^{-1}, \tag{9}$$

for a morphism must take the additive unit 0 to the multiplicative unit, sums to products, and additive to multiplicative inverses. Now suppose that $\phi(1) = g$. The first two equations become the recursive definition of $\phi(n) = g^n$ for $n \geqslant 0$, while the last equation is the one used above to define g^{-k}. Hence $\phi(n) = g^n$ for all integers n, Q.E.D.

In an additive group A, an exponent is usually written in front as a "multiple", so that the definition of these multiples for $a \in A$ reads

$$0a = 0, \qquad (n+1)a = na + a, \qquad (-k)a = -(ka). \tag{10}$$

If we regard the first two equations as a recursion in n with parameter a, they define a function $\mathbf{Z} \times A \to A$. Upon translation to this "multiple" notation, the properties (6), (8), and (7) for exponents read

$$(m + n)a = ma + na, \qquad (-m)a = -(ma), \qquad m(na) = (mn)a \quad (11)$$

for all $m, n \in \mathbf{Z}$. In case the additive group A is abelian, the multiples have a further property

$$m(a + b) = ma + mb, \qquad m \in \mathbf{Z}, \quad a, b \in A. \qquad (12)$$

In the special case of the additive group $\mathbf{Z}$ of integers, the multiple na is the same as the product na—indeed, both functions are defined by the same recursion on n.

A (multiplicative) group is *cyclic* when all its elements are powers c^n of some one element c. For any element g in a group G, the set $\{g^n | n \in \mathbf{Z}\}$ of all powers of g is clearly a group under multiplication and indeed a cyclic group; we call it the *cyclic subgroup of* G *generated by* g. The additive group $\mathbf{Z}$ is cyclic with generator 1, while each additive group $\mathbf{Z}_n$ is cyclic, with generator the remainder of 1, modulo n. We shall now show that every cyclic group is isomorphic either to $\mathbf{Z}$ or to some $\mathbf{Z}_n$.

To do this, consider again the morphism $\phi: \mathbf{Z} \to G$ with $\phi 1 = g$, thus

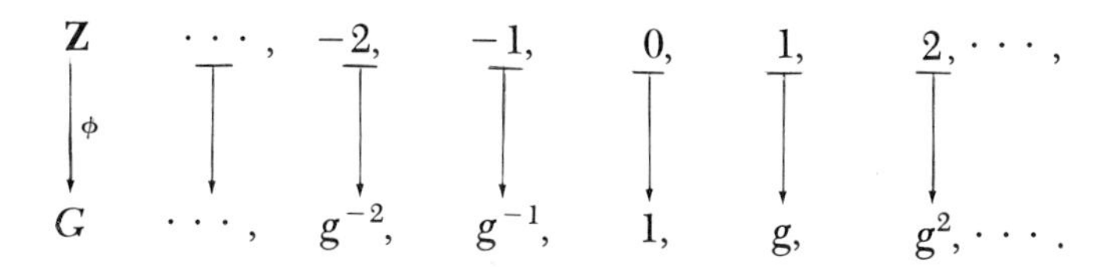

The image of the function ϕ is the cyclic subgroup generated by g. Ask when two of the listed powers of g are equal. If these powers are all different, we say that g has *infinite order* in G. If not, $g^k = g^m$ for some $k < m$. Then $g^{m-k} = 1$, so some positive power of g is the identity element. Take the *least* positive integer n such that $g^n = 1$, and call this integer the *order* of the element g. Thus every element of a group has some positive integer or infinity as its order. Note in particular that the unit element of G is the unique element of G with order 1.

THEOREM 3. *Any cyclic group with generator of infinite order is isomorphic to* $\mathbf{Z}$; *more generally, if an element* g *has order* ∞ *in a group* G, *the morphism* $\mathbf{Z} \to G$ *with* $1 \mapsto g$ *is a monomorphism with image the cyclic subgroup generated by* g.

Proof: The isomorphism is that described above. In particular, we defined "g has infinite order" to mean that the corresponding morphism $\mathbf{Z} \to G$ is injective.

Theorem 4. *Any cyclic group with generator of order n is isomorphic to $\mathbf{Z}_n$; more generally, if any element g has order n in a group G, a monomorphism $\mathbf{Z}_n \to G$ is given by assigning to each remainder r (modulo n) the power g^r.*

Proof: Let g have order n, so $g^n = 1$. For any power g^k, write $k = qn + r$ with $0 \leqslant r < n$ by the division algorithm, so that $g^k = g^{qn}g^r = g^r$. Thus the subgroup generated by g is $\{1, g, \cdots, g^{n-1}\}$. These elements are all different, for $g^r = g^s$ implies $g^{r-s} = 1$. Hence, $r \mapsto g^r$ is a bijection from the set $\mathbf{Z}_n = \{0, 1, \cdots, n-1\}$ of remainders modulo n to the subgroup $\{1, g, \cdots, g^{n-1}\}$ generated by g. Now, as in (I.29), $r \oplus s = t$ in $\mathbf{Z}_n$ when t is the remainder of $r + s = qn + t$; then $g^r g^s = g^{qn+t} = (g^n)^q g^t = g^t$ in G. This states that the injection $r \mapsto g^r$ carries sums in $\mathbf{Z}_n$ to products in G. Hence, it is a monomorphism, as asserted.

This result also shows when two powers of an element g of order n are equal:

$$g^k = g^m \quad \Leftrightarrow \quad k \equiv m \pmod{n} \qquad (g \text{ of order } n). \tag{13}$$

In particular, $g^k = 1$ if and only if n divides k. The condition (13) states that the equivalence kernel (§I.9) of the morphism $\phi: \mathbf{Z} \to G$ with $\phi(1) = g$ is the relation $\equiv_n$ of congruence modulo n.

Proposition 5. *Any epimorphic image of a cyclic group is cyclic.*

Proof: Let the multiplicative group C be cyclic with generator c. If $h: C \to G$ is an epimorphism, then every element of G has the form $h(c^k) = (hc)^k$ for some $k \in \mathbf{Z}$. Hence, G is generated by the image hc of the generator of C. Also, $c^n = 1$ implies $(hc)^n = 1$, so the order of the generator hc of G must divide the order of c.

As we have already suggested, a "subgroup" S of a given group G means a subset of G which is itself a group—under the same binary operation. (Subgroups will be formally defined in the next section.)

Proposition 6. *Any subgroup of a cyclic group is cyclic.*

Proof: For the infinite cyclic group $\mathbf{Z}$, we show more explicitly that any subgroup of $\mathbf{Z}$ is the set $n\mathbf{Z}$ of all integral multiples of some $n \geqslant 1$. Indeed, we easily see that these sets $n\mathbf{Z}$ are subgroups. Conversely, let S be any proper subgroup of the additive group $\mathbf{Z}$. Pick the least positive $n \in$ S; every integral multiple of n is then in S. Conversely, if $k \in$ S, the division algorithm gives $k = qn + r$ with $0 \leqslant r < n$, and $k \in$ S, $qn \in$ S imply

$r \in S$. Since n is the least positive element of S, this remainder r must be 0, and $k = qn$. The elements of S are then exactly the multiples of n.

For a finite cyclic group C with generator c of order n, we show more explicitly that any subgroup of C is cyclic with generator c^k of order n/k, where k is a positive divisor of n. Indeed, given k with $n = km$, the distinct powers of c^k are $1, c^k, c^{2k}, \cdots, c^{(m-1)k}$, with $c^{mk} = 1$; they form a subgroup of C which is cyclic of order m; that is, with m elements. Conversely, if S is any subgroup of C, there exists a least positive integer k for which $c^k \in S$. Since $1 = c^n \in S$, the division algorithm shows that k is a divisor of n and that S consists exactly of the $m = n/k$ distinct powers of c^k, just as claimed.

Theorem 3 provides a more conceptual interpretation for our construction of the integers $\mathbf{Z}$ from the natural numbers $\mathbf{N}$. The aim of that construction was to so enlarge $\mathbf{N}$ that subtraction would be possible; that is, to embed the additive monoid $\mathbf{N}$ in an additive group. The result of that construction may be reformulated as the following statement that the insertion i is "universal" among morphisms from $\mathbf{N}$:

Theorem 7. *The insertion $i:\mathbf{N} \to \mathbf{Z}$ is a morphism of additive monoids. If A is any group and $f:\mathbf{N} \to A$ any morphism of monoids, there is a unique morphism $f':\mathbf{Z} \to A$ of groups with $f = f' \circ i:\mathbf{N} \to A$.*

Proof: If $f(1) = a$, then $n \mapsto na$ is the only possible map f'; it is a morphism, as required.

EXERCISES

1. List all possible generators of the cyclic group $\mathbf{Z}_6$.
2. Show that $\mathbf{Z}_n$ has exactly n endomorphisms.
3. Show that $\mathbf{Z}_5$ is generated by any element not the identity.
4. Show that the cyclic group $\mathbf{Z}_{14}$ is generated by any one of six generators.
5. Show that the additive group $\mathbf{Z}$ of integers has exactly two generators.
6. For exponents in any group, prove that $(g^m)^n = g^{mn}$.
7. In a group G, prove that $(g_1 g_2)^m = g_1{}^m g_2{}^m$ for all integers m if and only if $g_1 g_2 = g_2 g_1$.
8. If $g^2 = 1$ for all elements g of a group G, prove G abelian.
9. With Aut (G) the group of all automorphisms of the group G prove

$$\text{Aut}\,(\mathbf{Z}) \cong \mathbf{Z}_2, \quad \text{Aut}\,(\mathbf{Z}_5) \cong \mathbf{Z}_4, \quad \text{Aut}\,(\mathbf{Z}_6) \cong \mathbf{Z}_2, \quad \text{Aut}\,\mathbf{Z}_8 \cong \mathbf{Z}_2 \times \mathbf{Z}_2.$$

*10. If M is a commutative monoid satisfying the cancellation law, show that there exists an abelian group A and a morphism $u:M \to A$ of monoids with the following "universal property": To any group B and any morphism

$f:M \to B$ of monoids, there is a unique morphism $f':A \to B$ of groups with $f = f' \circ u:M \to B$.

4. Subgroups

Much information about the structure of a group can be obtained from a knowledge of its subgroups.

DEFINITION. A group S is a subgroup *of a group G, in symbols $S \subset G$, when S is a* subset *of G and the function $S \to G$ which inserts the set S in the set G is a morphism of groups.*

In particular, the set G itself is a subgroup of G, as is the set $\{1\}$ consisting only of the identity of G; this latter subgroup is usually denoted simply as 1. Subgroups of G other than these two "improper" subgroups G and 1 are called *proper subgroups* of G. If S is a subgroup of G, we also say that G is an *extension* of S.

DEFINITION. A subset T of a multiplicative group G is closed *under the operation of multiplication in G when $t, t' \in T$ imply that $tt' \in T$.*

When is a subset T of a group G a subgroup? The requirement that the insertion $T \to G$ be a morphism of multiplication means exactly that any two elements $t, t' \in T$ have a product in T equal to their product in G. Therefore, T must be closed under the product in G, and this product, a restriction of that in G, must make T a group. The conclusion is as follows:

PROPOSITION 8. *A subset T of a multiplicative group G is a subgroup of G if and only if T is closed under the multiplication in G and is a group under the binary operation $T \times T \to T$ given by restriction of the multiplication in G.*

Usually, the distinction between the two binary operations $T \times T \to T$ and $G \times G \to G$ is dropped, so a subgroup T of G is *defined* as a subset of G which is a group under "the" multiplication of G. Our definition in terms of the insertion $S \to G$ has the merit that it also defines a "subobject" in other cases. For example, a monoid N is a *submonoid* of M when N is a subset of M and the insertion $N \to M$ is a morphism of monoids.

In any group G, $g \mapsto g^{-1}$ is a unary operation, while selecting the identity is a nullary operation. "Closure" of a subset of G under these operations is defined as before.

Theorem 9. *A subset T of a multiplicative group G is a subgroup of G if and only if T is closed under the following three operations of G: Identity, inverse, and multiplication.*

In other words, a subset T is a subgroup if and only if

$$1 \in T, \qquad t \in T \;\Rightarrow\; t^{-1} \in T, \qquad t, t' \in T \;\Rightarrow\; tt' \in T. \tag{14}$$

Proof: First suppose T a subgroup as in Proposition 8. Then T is closed under multiplication. As a group, T must have some unit u, with $uu = u$ in T. Then $u = u1$ in G gives $uu = u1$ and so, by cancellation, $u = 1$; the unit u of T is the identity 1 of G. Also, each $t \in T$ has an inverse in T; since the inverse in a group G is unique, this can only be its inverse in G. This proves the closure conditions (14) necessary.

Conversely, if T satisfies these closure conditions, it has a product with identity 1 and with inverses, and this product is automatically associative because the product in the larger set G is associative. Hence, T satisfies all the group axioms and so is a subgroup of G, by Proposition 8.

Note that any subgroup of an abelian group is abelian.

Corollary. *A nonvoid subset T of a finite multiplicative group G is a subgroup if and only if it is closed under multiplication.*

Proof: For any element t in a finite group the powers $t, t^2, \cdots$ cannot all be different, so t has some finite order m. Thus $t^m = 1$ and $t^{-1} = t^{m-1}$; closure under product implies closure under identity and inverse.

For any group G, the inclusion relation $S \subset T$ on subgroups is reflexive, transitive, and antisymmetric (the last means that $S \subset T$ and $T \subset S$ imply $S = T$). This relation is illustrated in the inclusion diagram for all subgroups of the additive group $\mathbf{Z}_{12}$, as displayed below:

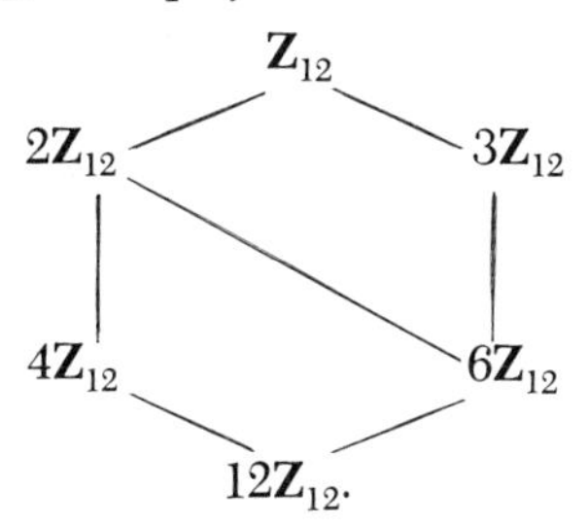

As in Proposition 6, every subgroup of $\mathbf{Z}_{12}$ is cyclic and generated by some divisor k of 12, so is the set $k\mathbf{Z}_{12}$ of all multiples of k in $\mathbf{Z}_{12}$; for instance, $4\mathbf{Z}_{12} = \{0, 4, 8\}$.

PROPOSITION 10. *If S and T are subgroups of G, the intersection $S \cap T$ of the sets S and T is a subgroup of G which contains every subgroup contained in both S and T.*

Proof: Since both S and T contain the identity of G, so does their intersection $S \cap T$. Given group elements $u, v \in S \cap T$, their product uv is then both in the subgroup S and in the subgroup T, hence in the intersection $S \cap T$. Similarly, u^{-1} is in $S \cap T$. By Theorem 9, these three closure properties of $S \cap T$ show it a subgroup. Clearly, any subgroup R with $R \subset S$ and $R \subset T$ has $R \subset S \cap T$.

More is true. Let U be any set whose elements $S \in U$ are themselves subgroups of G. The *intersection* $\bigcap U$ of the elements of U is the subset

$$T = \bigcap_{S \in U} S = \{\text{all } g \mid g \in G \text{ and } S \in U \Rightarrow g \in S\} \tag{15}$$

of G. By the arguments above, this subset T is a subgroup of G. If U is empty, then $T = G$.

This construction allows us to define "generators" of a group G. Let X be any subset of the group G. Take U to be the set of all those subgroups S of G each of which contains X. Then the intersection $T = \bigcap U$ is a subgroup of G; indeed, the least subgroup of G containing X. We call it the group *generated* by X.

PROPOSITION 11. *The subgroup of G generated by the subset X is the set T consisting of 1 and all products $y_1 \cdots y_n$ of any $n > 0$ elements, each of which is either an element of X or the inverse of an element of X.*

Proof: This set T is closed under identity, product, and inverse, hence is a subgroup by Theorem 9. On the other hand, every subgroup S containing all $x \in X$ must contain all their inverses and, hence, all products $y_1 \cdots y_n$. Thus $S \supset T$, so T is the intersection of all these subgroups S, as claimed.

PROPOSITION 12. *To any two subgroups S and T of a group G there is a least subgroup containing them both; that is, a subgroup L of G with*

$$L \supset S; \qquad L \supset T; \qquad R \supset S \text{ and } R \supset T \quad \text{imply} \quad R \supset L, \tag{16}$$

the latter for any subgroup R.

Proof: Let U be the set of all subgroups R of G which contain both S and T. Then the intersection L of all sets $R \in U$ is a subgroup with the desired property (16).

This subgroup L is called the *join* of S and T, written $L = S \vee T$. It is usually much larger than the union of the sets S and T. Indeed, $S \vee T$ is the set of all those elements of G which can be written for some k as a product $g = s_1 t_1 \cdots s_k t_k$ of $2k$ factors $s_i \in S$, $t_i \in T$. (*Proof:* The set of all these products is a subgroup, and the least such containing S and T.)

In the inclusion diagram of all subgroups of G, the join $S \vee T$ is the lowest subgroup situated (along rising lines) above both S and T. Thus, among the subgroups of $\mathbf{Z}_{12}$ in the diagram on page 57,

$$4\mathbf{Z}_{12} \vee 6\mathbf{Z}_{12} = 2\mathbf{Z}_{12}, \qquad \text{but} \qquad 4\mathbf{Z}_{12} \vee 3\mathbf{Z}_{12} = \mathbf{Z}_{12}.$$

EXERCISES

1. In the group of symmetries of a regular hexagon, show that the subgroup mapping a given diagonal onto itself is isomorphic to the four group $\mathbf{Z}_2 \times \mathbf{Z}_2$.

2. Make an inclusion diagram for all subgroups of (a) S_3; (b) $\mathbf{Z}_{18}$; (c) Δ_3.

3. Determine the join and the meet of every pair of subgroups of $\mathbf{Z}_8$.

4. Show that a nonvoid subset of G closed under product and inverse is a subgroup of G.

5. Show that a nonvoid subset S of G closed under $(s, t) \mapsto st^{-1}$ is a subgroup of G.

6. (a) Show that the elements of finite order in an abelian group A form a subgroup of A.

(b) Show by an example that the corresponding statement for non-abelian groups is false.

7. Show that a group with no proper subgroups must be cyclic of order 1 or a prime.

8. (a) In a multiplicative abelian group A, show that for each integer n the elements a in A which satisfy $a^n = 1$ constitute a subgroup. Show also that this result does not hold for nonabelian groups, for example, in S_3.

(b) For each n show that the set of all nth powers in A is a subgroup.

9. Show that a subset T of a multiplicative monoid S is a submonoid if and only if T is closed under identity and product.

5. Defining Relations

If an element c of order n generates a (multiplicative) cyclic group C, this element satisfies the relation

$$c^n = 1.$$

From this one relation (and the group axioms) we can derive the whole multiplication table of the group C, for the table is $c^k c^m = c^{k+m}$, where the exponent $k + m$ is to be reduced modulo n. This also means that c is a "universal" element of order n, in the sense that it can be compared (uniquely) to any other such element g: If an element g in any group G satisfies the relation $g^n = 1$ (so that the order of g is a divisor of n), there is exactly one group homomorphism $\theta: C \to G$ with $\theta c = g$. Indeed, θ is the map $c^k \mapsto g^k$ for all k.

Noncyclic groups have analogous sets of defining relations on their generators. Thus for $n \geqslant 3$ the *dihedral group* Δ_n is defined as the group of symmetries of a regular polygon P_n of n sides. One element of Δ_n is the rotation R of the polygon in its plane through $360°/n$ about the center; this rotation clearly has order n (cf. the triangle of §1, where $n = 3$). Another element of Δ_n is the reflection D of the polygon in an axis through one of its vertices, say the vertex numbered 1; this reflection has order 2. Together R and D generate $2n$ different symmetries

$$1, R, R^2, \cdots, R^{n-1}, D, RD, \cdots, R^{n-1}D.$$

This list contains all the symmetries of the polygon P_n, because any symmetry is determined by its action on the vertices $1, 2, \cdots, n$ of P_n; and if a symmetry takes vertex 1 to vertex i, it must either keep the vertices in the same cyclic order, as does R^{i-1}, or it must reverse that order, as does $R^{i-1}D$. Therefore, the list above is a complete list of symmetries of P_n. Thus the group Δ_n contains $2n$ elements, and it is generated by the two elements R and D.

The composite symmetry DR is equal to $R^{n-1}D$, for both these composites reverse the order of the vertices and carry vertex 1 to vertex n. Hence, D and R satisfy the defining relations

$$R^n = 1, \qquad D^2 = 1, \qquad DR = R^{n-1}D. \tag{17}$$

These three equations do completely determine the multiplication table of the dihedral group Δ_n. For, write any one of the $2n$ listed elements of the group in the standard form $R^i D^j$, with exponents i, j in the ranges $0 \leqslant i < n$, $0 \leqslant j < 2$. To compute the product of two elements both in this form, we need only use the third equation of (17) to move factors D to the right past factors R, and then use the first two equations, giving the orders of R and D, to reduce the new exponents to the desired range.

This can be stated as a "universal" property as follows. If G is any group with elements R' and D' which satisfy equations like (17), then the product of two elements of the form $R'^i D'^j$ can be calculated just like the product of two elements $R^i D^j$ in Δ_n. Therefore, the function $R^i D^j \mapsto R'^i D'^j$ from Δ_n to G preserves products, so is a morphism of groups $\Delta_n \to G$. It is the unique morphism carrying R to R' and D to D'. Because there is a unique such morphism for any two elements satisfying the relations (17), we call these relations the *defining relations* for the generators R and D of the dihedral group Δ_n.

As another example, consider the product $E = B \times C$ of two finite cyclic groups B and C with generators b and c of orders m and n, respectively. Every element in E has the form of a product $b^i c^j$ with integral exponents i and j. Since $b^m = 1$ and $c^n = 1$, one can reduce i modulo m and j modulo n, and E has just mn elements. Thus E is generated by two elements b and c and its multiplication table may be computed directly from the following three defining relations:

$$b^m = 1, \qquad c^n = 1, \qquad bc = cb. \tag{18}$$

Indeed, from these relations one may calculate any product of two elements of E as $(b^i c^j)(b^k c^l) = b^{i+k} c^{j+l}$, where the exponents are to be reduced modulo m or modulo n, as the case may be. Thus the "defining relations" (18) on the generators b and c of E again describe this group up to isomorphism. (This may also be stated as a universality property, as in Exercise 10.)

The product $G \times H$ of two groups G and H is generated by subgroups

$$G' = \{(g, 1) |\ \text{all } g \in G\}, \qquad H' = \{(1, h) |\ \text{all } h \in H\}$$

isomorphic to G and H, respectively. These subgroups of $G \times H$ have intersection $G' \cap H' = 1$ and join $G' \vee H' = G \times H$; moreover, every element of G' commutes with every element of H'. These properties of the subgroups characterize $G \times H$ in the following way.

Proposition 13. *If a group D has subgroups G and H with*

$$G \cap H = 1, \qquad G \vee H = D, \qquad gh = hg, \tag{19}$$

for all $g \in G$ and $h \in H$, then there is an isomorphism $\psi: G \times H \cong D$ of groups with $\psi(g, 1) = g$ and $\psi(1, h) = h$ for all $g \in G$ and $h \in H$.

Proof: We try the definition of $\psi: G \times H \to D$ by $\psi(g, h) = gh$. Then for a product we have

$$\psi[(g, h)(g', h')] = (gg')(hh'), \qquad \psi(g, h)\psi(g', h') = ghg'h'.$$

Since h commutes with g', the results are equal, so ψ is a morphism of groups. It remains to prove ψ an isomorphism.

First, $\psi(g, h) = \psi(g', h')$ means that $gh = g'h'$, hence that $g(g')^{-1} = h'h^{-1} \in G \cap H$. Since $G \cap H = 1$ by hypothesis, this implies that $g = g'$ and $h = h'$. Hence, ψ is a monomorphism.

The hypothesis $G \vee H = D$ means that every element $d \in D$ is an iterated product of elements of G and H. Since always $gh = hg$, this implies that d is a product $g'h' = \psi(g', h')$. Hence, ψ is an epimorphism, and therefore an isomorphism, as required.

A group D satisfying the hypotheses (19) of Proposition 13 may be called a "direct product" of its subgroups G and H. From this definition it is immediate that the direct product of two abelian groups is abelian. Since any cyclic group is abelian, any iterated direct product of cyclic groups is abelian. Later, in Chapter XI, we shall prove a partial converse of this result by showing that any *finite* abelian group is isomorphic to a direct product of cyclic groups, and even more, to a direct product of cyclic groups of prime power order.

EXERCISES

1. (a) Use the relations (17) to calculate the complete multiplication table for the group Δ_3.

(b) Show that Δ_3 is also generated by the elements denoted D_1 and D_2 (in §1), and write defining relations for these generators of Δ_3.

2. Determine the order of every element in the group Δ_4 of the square.

3. In Δ_5, the group of the pentagon, show that every element except the identity has order 2 or order 5.

4. Construct an inclusion diagram for all subgroups of Δ_5.

5. In the dihedral group Δ_n, express as a product of the generators R and D the operation of reflecting the polygon P_n in the axis through vertex i.

***6.** Prove that Δ_6 is isomorphic to $S_3 \times S_2$.

7. Find generators and defining relations for the group of all symmetries of each of the following (infinite) ornamental patterns:

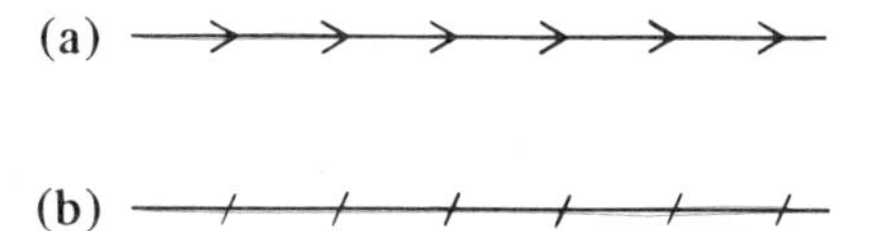

8. (a) Show that the relations $a^4 = 1$, $b^2 = a^2$ and $b^{-1}ab = a^{-1}$ define a group of order 8. (It is called the "quaternion group".)

(b) Show that this group is not isomorphic to Δ_4 (also of order 8).

9. If a group K has subgroups G and H with $gh = hg$ for all $g \in G$ and $h \in H$:

(a) Show that there is a morphism $\psi: G \times H \to K$ of groups with $\psi(g, 1) = g$ for all $g \in G$ and $\psi(1, h) = h$ for all $h \in H$.

(b) Show that ψ is a monomorphism if and only if $G \cap H = 1$ and an epimorphism if and only if $G \vee H = K$.

10. Let E be the direct product of two cyclic groups G and H with generators b and c of orders m and n, respectively. If a group K has elements u, v with $u^m = 1$, $v^n = 1$, and $uv = vu$, prove that there exists a unique morphism $\psi: E \to K$ of groups with $\psi(b) = u$ and $\psi(c) = v$.

6. Symmetric and Alternating Groups

As in §1, a permutation σ on the finite set $\mathbf{n} = \{1, \cdots, n\}$ is a bijection $\sigma : \mathbf{n} \to \mathbf{n}$, and the *symmetric group* S_n of "degree" n is the group of all permutations on $\mathbf{n}$ under the operation of composition. For example, if $n = 5$, the assignments

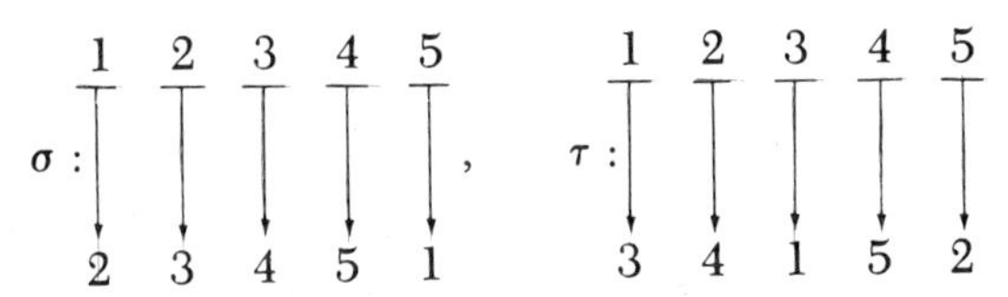

define two permutations σ, τ, usually exhibited as

$$\sigma = \begin{pmatrix} 1 & 2 & 3 & 4 & 5 \\ 2 & 3 & 4 & 5 & 1 \end{pmatrix}, \qquad \tau = \begin{pmatrix} 1 & 2 & 3 & 4 & 5 \\ 3 & 4 & 1 & 5 & 2 \end{pmatrix}.$$

In this "two-row" notation for a permutation, the digits (or "letters") 1, 2, 3, 4, 5 to be permuted are listed in order in the first row; below each digit is its image under the function in question. Such a two-row symbol represents a bijection precisely when the second row contains all the digits in some order and (hence) with no repetitions. In the present case the composite $\sigma \circ \tau$ is $1 \mapsto 3 \mapsto 4$, $2 \mapsto 4 \mapsto 5, \cdots,$ while the composite $\tau \circ \sigma$ is $1 \mapsto 2 \mapsto 4$, $2 \mapsto 3 \mapsto 1, \cdots$. In the two-row notation,

$$\sigma \circ \tau = \begin{pmatrix} 1 & 2 & 3 & 4 & 5 \\ 4 & 5 & 2 & 1 & 3 \end{pmatrix}, \qquad \tau \circ \sigma = \begin{pmatrix} 1 & 2 & 3 & 4 & 5 \\ 4 & 1 & 5 & 2 & 3 \end{pmatrix};$$

thus $\sigma \circ \tau \neq \tau \circ \sigma$.

The permutation σ above amounts to a circular rearrangement of the symbols permuted, as indicated in Figure II-2. Such a permutation is called a *cyclic permutation* or a *cycle*. For such a cycle there is a briefer "one-row" notation: Write first any digit involved, then its image, and so on until the cycle closes. In this notation the permutation σ above thus appears in any one of the equivalent forms (1 2 3 4 5), (2 3 4 5 1), (3 4 5 1 2), (4 5 1 2 3), or (5 1 2 3 4). In this case $\sigma^5 = 1$, so that σ has order 5.

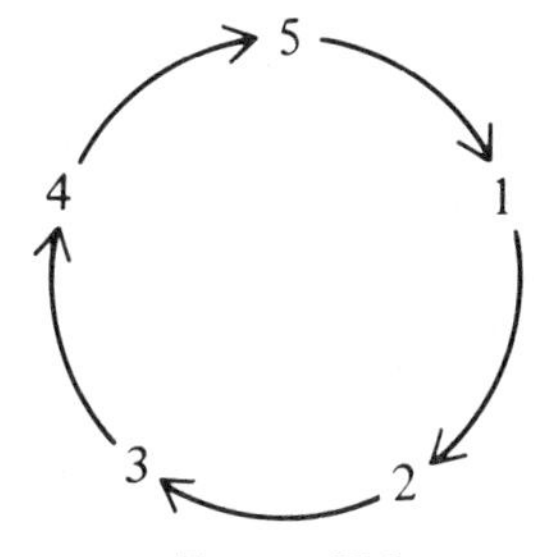

FIGURE II-2

Similar considerations prove that any cyclic permutation of k letters has order k. The number of letters in a cycle is also called its *length*; a cycle of length 2 is called a *transposition*.

This notation for cycles can be adapted to more general permutations. Thus the permutation τ displayed above interchanges the digits 1 and 3, and permutes 2, 4, and 5 cyclically, and so is the composite of these two cycles, in either order, as

$$\tau = (1\,3) \circ (2\ 4\ 5) = (2\ 4\ 5) \circ (1\,3).$$

There is a similar decomposition for any permutation. First, note that the two cycles above—each regarded as a permutation on $\{1, 2, 3, 4, 5\}$—have no digits in common. In general, two permutations α, β on $\mathbf{n}$ are called "disjoint" when there is no digit moved by both (that is, when $\alpha(i) \neq i$ and $i \in \mathbf{n}$ imply $\beta(i) = i$). For disjoint permutations, $\alpha \circ \beta = \beta \circ \alpha$, for both $(\alpha \circ \beta)(i)$ and $(\beta \circ \alpha)(i)$ are $\alpha(i)$, $\beta(i)$, or i according as α, β, or neither displaces i.

This preparation leads to the following result:

Theorem 14. *Any permutation $\sigma \neq 1$ on a finite set X is a composite $\gamma_1 \cdots \gamma_k$ of disjoint cyclic permutations γ_i, each of length 2 or more. Except for changes in the order of the cyclic factors, σ has only one such decomposition.*

Proof: The *orbit* C of a point $x \in X$ under the permutation σ is defined to be the set $\{x, \sigma x, \sigma^2 x, \cdots\}$ of all images of x under the powers σ^i of σ. Since the orbit is finite, one must have some equality $\sigma^n x = \sigma^{n+k} x$ for positive n and k; hence, applying σ^{-n}, one has $\sigma^k x = x$. If m is the least positive integer with $\sigma^m x = x$, the orbit C must consist of exactly the m different points $C = \{x, \sigma x, \cdots, \sigma^{m-1} x\}$; also, each of the points $\sigma^i x$ in this set C has the same orbit (just the same points in different cyclic order). The permutation σ restricted to this subset $C \subset X$ is just a cyclic permutation $\gamma = (x\ \sigma x \cdots \sigma^{m-1} x)$ of length m.

Every point $x \in X$ belongs to exactly one such orbit of σ. All told there are, say, k orbits $C_1, \cdots, C_k$ for σ, and σ restricted to each C_i is a cyclic permutation γ_i. Moreover, if $i \neq j$, the cycles γ_i and γ_j are disjoint. The composite $\gamma_1 \cdots \gamma_k$ of these disjoint cycles is thus a permutation on X which has the same effect on each point $x \in X$ as does the original σ, because σx is $\gamma_i x$ if x belongs to the ith orbit C_i. Therefore, σ is the composite $\gamma_1 \cdots \gamma_k$ of disjoint cycles. In this composite, any cycle of length 1, that is, any fixed point, may be omitted.

On the other hand, for any decomposition $\sigma = \beta_1 \cdots \beta_l$ of σ into disjoint cycles β_j, the letters moved by a cycle β_j form one of the orbits C_i of σ, and hence β_j is the corresponding cycle γ_i of the previous decomposition

$\gamma_1 \cdots \gamma_k$. Therefore, the two decompositions differ only in the arrangement of their factors, as asserted.

Corollary. *The order of a permutation is the least common multiple of the lengths of its disjoint cycles.*

Proof: In the cyclic representation $\sigma = \gamma_1 \cdots \gamma_k$ the γ's are disjoint, so $\gamma_i \circ \gamma_j = \gamma_j \circ \gamma_i$, and hence, for any integer m, $\sigma^m = \gamma_1^m \cdots \gamma_k^m$. Therefore, $\sigma^m = 1$ if and only if each $(\gamma_i)^m = 1$, hence if and only if m is a common multiple of the lengths of all these cycles γ_i. The order of σ is the least such m, hence the conclusion. (For a discussion of least common multiples, see §III.10.)

If τ is a fixed permutation in the symmetric group S_n, the assignment $\sigma \mapsto \tau\sigma\tau^{-1}$ for variable $\sigma \in S_n$ is an automorphism of S_n, for

$$\tau(\sigma_1\sigma_2)\tau^{-1} = \left(\tau\sigma_1\tau^{-1}\right)\left(\tau\sigma_2\tau^{-1}\right).$$

This automorphism is called *conjugation* by τ. We now compute its effect.

Proposition 15. *If $\gamma \in S_n$ is a cycle of length m, so is any conjugate $\tau\gamma\tau^{-1}$ of γ.*

Proof: If γ is the cycle $(x_1 \cdots x_m)$, we show that $\tau\gamma\tau^{-1}$ is the cycle

$$\tau(x_1\ x_2 \cdots x_m)\tau^{-1} = \left(\tau(x_1)\ \tau(x_2) \cdots \tau(x_m)\right). \tag{20}$$

Indeed, let $\tau\gamma\tau^{-1}$ act on any letter y. Now $y = \tau(\tau^{-1}y)$. If $x = \tau^{-1}y$ is not one of the x_i, the action of $\tau\gamma\tau^{-1}$ on y is $\tau(x) \mapsto x \mapsto x \mapsto \tau(x)$, while if $x = x_i$, it is $\tau(x_i) \mapsto x_i \mapsto x_{i+1} \mapsto \tau(x_{i+1})$. But this is exactly the effect of the cycle written on the right of our formula (20).

The conjugate $\tau\sigma\tau^{-1}$ of any permutation σ may now be calculated. Write σ as the product $\gamma_1 \cdots \gamma_k$ of disjoint cycles. Since conjugation is an automorphism, $\tau\sigma\tau^{-1} = (\tau\gamma_1\tau^{-1}) \cdots (\tau\gamma_k\tau^{-1})$, and each cycle on the right is expressed as in (20). In words: To conjugate σ by τ, apply the function τ to each letter in the disjoint cycle representation of σ.

Proposition 16. *Any permutation σ on* **n** *is a composite of transpositions.*

Since σ is a composite $\gamma_1 \cdots \gamma_k$ of cycles γ_i, it is enough to prove this for a cycle. But this is easy:

$$(1\ 2 \cdots m) = (1\ m) \cdots (1\ 3)(1\ 2).$$

Next we divide the permutations on $\mathbf{n}$ into two classes, "even" and "odd". Consider the set D of all those ordered pairs $(i, j) \in \mathbf{n} \times \mathbf{n}$ with $i < j$, and say that the permutation $\sigma : \mathbf{n} \to \mathbf{n}$ inverts the pair $(i, j) \in D$ when $\sigma i > \sigma j$. Let $\operatorname{sgn}(\sigma)$ denote the total number of such inversions for σ, and call $(-1)^{\operatorname{sgn} \sigma}$ the *parity* of σ. Finally, call σ *even* when $(-1)^{\operatorname{sgn} \sigma} = +1$ and *odd* when $(-1)^{\operatorname{sgn} \sigma} = -1$; in other words, an odd permutation is one that inverts an odd number of ordered pairs. For example, the transposition $(3\ 6)$ sends $3\ 4\ 5\ 6$ to $6\ 4\ 5\ 3$, so inverts the five pairs $(3, 4)$, $(3, 5)$, $(3, 6)$, $(4, 6)$, and $(5, 6)$. In general, the transposition $(h\ k)$ with $h < k$ inverts the pair (h, k) and all the pairs (h, i) and (i, k) with $h < i < k$, but no other pairs. Hence, *any transposition is odd.*

What matters is not $\operatorname{sgn} \sigma$, but just the parity $(-1)^{\operatorname{sgn} \sigma} = \pm 1$, an element of the multiplicative group $\{+1, -1\}$ (cyclic of order 2).

Theorem 17. *The function $\sigma \mapsto (-1)^{\operatorname{sgn} \sigma}$ assigning to each permutation σ its parity is a morphism $S_n \to \{+1, -1\}$ of groups.*

In other words, even times odd is odd, and odd times odd is even.

Proof: The number $\operatorname{sgn} \sigma$ specifies how many pairs (i, j) the permutation σ will invert in the set D of all pairs (i, j) in $\mathbf{n}$ with $i < j$. This involves applying σ to the set D to get the set σD of all pairs $(\sigma i, \sigma j)$ with $i < j$. Also, for each $k < l$ in $\mathbf{n}$ this set σD must contain exactly one of the two possible pairs (k, l) and (l, k). Now apply another permutation τ to σD, giving the set $\tau(\sigma D)$ which contains either $(\tau k, \tau l)$ or $(\tau l, \tau k)$. In either case, this pair will be inverted (from its order in σD) precisely when the pair (k, l) is inverted by τ. Thus τ inverts $\operatorname{sgn} \tau$ pairs of σD, so that the passage $D \to \sigma D \to \tau(\sigma D)$ has, all told, inverted $\operatorname{sgn} \sigma + \operatorname{sgn} \tau$ pairs. Some of them may have been inverted twice, and thus returned to the original state. On the other hand, the direct passage $D \to (\tau \circ \sigma) D$ inverts $\operatorname{sgn}(\tau\sigma)$ pairs. Hence (writing this modulo 2 to account for pairs inverted twice)

$$\operatorname{sgn}(\tau \circ \sigma) \equiv \operatorname{sgn} \tau + \operatorname{sgn} \sigma \qquad (\text{mod } 2).$$

Written with exponents, this states that

$$(-1)^{\operatorname{sgn}(\tau \circ \sigma)} = (-1)^{\operatorname{sgn} \tau}(-1)^{\operatorname{sgn} \sigma} \tag{21}$$

and hence that $\sigma \mapsto (-1)^{\operatorname{sgn} \sigma}$ is a morphism, as required.

This result has several striking consequences. For example, since any transposition is odd, the product of two transpositions must be even; moreover:

Corollary 1. *A product of k transpositions is odd or even according as k is odd or even.*

A permutation σ can be written in many ways as a product of transpositions; this shows that if one such factorization has an even number of transpositions, so does every other such factorization.

The parity of any permutation may be calculated from its disjoint cycle representation once we know the parity of a cycle. But this is easy.

Corollary 2. *A cycle γ of length m has parity $(-1)^{\operatorname{sgn}\gamma} = (-1)^{m-1}$.*

Proof: As in Proposition 16, the cycle $(1\ 2 \cdots m)$ is a product of $m-1$ transpositions $(1\ m) \cdots (1\ 3)(1\ 2)$, each of which is odd.

A very important consequence of Theorem 17 is

Corollary 3. *For $n > 1$, the set A_n of all even permutations on* **n** *is a subgroup of S_n with $(n!)/2$ elements.*

This subgroup A_n is called the *alternating group* of degree n.

Proof: Since $\sigma \mapsto (-1)^{\operatorname{sgn}\sigma}$ is a morphism, $\sigma \mapsto 1$ and $\tau \mapsto 1$ imply that $\sigma \circ \tau \mapsto 1$. Thus $A_n \subset S_n$ is closed under product (and identity), hence by the corollary of Theorem 9 is a subgroup. List its elements as $\sigma_1, \cdots, \sigma_t$. Multiply each by some convenient odd permutation, say $(1\ 2)$. This gives a list $\sigma_1(1\ 2), \cdots, \sigma_t(1\ 2)$ of odd permutations, also all different. But *any* odd permutation ρ has the product $\rho(1\ 2) = \sigma_i$ even, so $\rho = \sigma_i(1\ 2)$, and ρ is in this list. We conclude that the number of odd permutations is the same as the number of even permutations, so each is half the total number $n!$ of permutations in S_n.

This style of "counting" argument for the subgroup $A_n \subset S_n$ will soon be applied to other subgroups.

EXERCISES

1. Express each of the following permutations as a product of disjoint cycles:

$$\begin{pmatrix} 1 & 2 & 3 & 4 & 5 & 6 \\ 3 & 5 & 6 & 1 & 2 & 4 \end{pmatrix}, \begin{pmatrix} 1 & 2 & 3 & 4 & 5 & 6 \\ 6 & 1 & 5 & 3 & 4 & 2 \end{pmatrix}, \begin{pmatrix} 1 & 2 & 3 & 4 & 5 & 6 & 7 \\ 6 & 4 & 5 & 7 & 3 & 1 & 2 \end{pmatrix};$$

find the order and the inverse of each of these permutations.

2. Represent the following composites as products of disjoint cycles:

$$(1\ 2\ 3\ 4\ 5)(1\ 5\ 6)(2\ 4\ 6), \quad (1\ 2\ 3\ 4)(2\ 3\ 4\ 5)(3\ 4\ 5\ 1),$$
$$(1\ 2\ 3)(3\ 4\ 5)(1\ 3\ 5), \quad (1\ 2)(2\ 3)(3\ 4)(4\ 5)(5\ 1).$$

3. Which symmetric groups and which alternating groups are abelian?

4. Describe the following subgroups of S_4 and determine the number of elements in each subgroup:

(a) All permutations carrying the set $\{1, 2\}$ into the set $\{1, 2\}$.

(b) All permutations carrying $\{1, 2\}$ into either $\{1, 2\}$ or $\{3, 4\}$.

5. Find four different subgroups of S_4 isomorphic to S_3 and nine isomorphic to S_2.

6. Show that there are at least 30 different subgroups of S_6 isomorphic to S_3.

7. Show that two conjugate permutations σ and $\tau\sigma\tau^{-1}$ have the same parity but not necessarily the same number of inversions.

8. Show that a product of not necessarily disjoint cycles is even if and only if it contains an even number of cycles of even length.

9. Prove that every permutation of order 14 on 10 letters is odd.

***10.** Show that every even permutation can be written as a product of cycles of length 3.

11. Prove that S_n is generated by the cycles $(1\ 2 \cdots n-1)$ and $(n-1\ n)$.

***12.** In Exercise 11 determine defining relations on these two generators.

13. Prove that S_n is generated by the transpositions $(1\ 2), (2, 3), \cdots, (n-1\ n)$.

***14.** In Exercise 13 show that the generators $\tau_i = (i\ i+1)$, $i = 1, \cdots, n-1$, satisfy the defining relations $(\tau_i)^2 = 1$, $\tau_i\tau_j = \tau_j\tau_i$ if $i - j \neq \pm 1$, and $(\tau_i\tau_{i+1})^3 = 1$ and that this is a complete list of relations.

***15.** A flat square box is filled with 16 flat metal squares, numbered in order as shown.

1	2	3	4
5	6	7	8
9	10	11	12
13	14	15	

The last square is removed, making it possible for other squares to move by sliding. Consider any sequence of such moves ending with the lower right corner again vacant. Prove that the permutations possible by such sequences of moves are exactly those in A_{15}.

7. Transformation Groups

A bijection $X \to X$ is often called a *transformation* on the set X, so a *transformation group* T on X is a set of bijections on X which form a group under composition. In more detail, a transformation group T is a set of bijections $t: X \to X$ which contains the identity bijection 1_X, to each pair of bijections their composite, and to each bijection its inverse. In other words, a transformation group (cf. Theorem 9) is just a subgroup of the group $S(X)$ of all permutations on X. Typically, groups of symmetries are transformation groups. Thus the group of all symmetries of the triangle is a transformation group on the set of points of the triangle. Analogously, the group Aut (G) of all automorphisms of a group G is a group of transformations on the set G.

A transformation group on a finite set is called a *permutation* group.

Theorem 18. *Any group G is isomorphic to a transformation group.*

This theorem, due to Cayley, shows that the axioms for the operation of multiplication in a group imply all formal properties valid for the operation "composition of transformations".

Proof: Each fixed $a \in G$ defines a function $f_a : G \to G$ by the formula $f_a(x) = ax$ for each $x \in G$. Let T be the set consisting of all functions on G to G of the form f_a. Since

$$(f_a f_b)x = f_a(f_b x) = f_a(bx) = a(bx) = (ab)x = f_{ab}x$$

for all $x \in G$, the functions $f_a f_b$ and f_{ab} are equal; hence, in particular, the composite of two functions in T is again in T. On the other hand, $f_1(x) = 1x = x$, so $f_1 = 1_G$ is the identity function on the set G; it follows that $f_a f_{(a^{-1})} = 1 = f_{(a^{-1})} f_a$, so $(f_a)^{-1} = f_{(a^{-1})}$ is in T. Therefore, every element f_a of T is a bijection, and T is a group of transformations on the set G. Now the assignment $a \mapsto f_a$ is a function $f : G \to T$. Since $f_{ab} = f_a f_b$, this function is a morphism of groups. Since T was defined to consist only of the f_a, this morphism f is surjective. Also, $f_a 1 = a$, so $a \neq b$ implies $f_a \neq f_b$. Hence, f is a monomorphism, and therefore is an isomorphism $f : G \cong T'$ of G to a transformation group T, Q.E.D.

This isomorphism f can be explicitly visualized if G is finite, in which case T is a group of permutations on the finite set G. Observe that f_a is given by left multiplication (="left translation") by a. If $x_1, x_2, \cdots, x_n$ is the list of all the distinct elements of G, then f_a is the permutation

$$f_a = \begin{pmatrix} x_1 & x_2 & \cdots & x_n \\ ax_1 & ax_2 & \cdots & ax_n \end{pmatrix}.$$

In other words, f_a is the permutation given by the ath row in the multiplication table for G.

A morphism of a group G to a transformation group is often called a "representation" of G (or, a "linear representation" when the transformations are linear in the sense of Chapter V). The particular representation f constructed in the proof of this theorem is known as the "left regular representation" of G.

A given group may have many representations. Thus the dihedral group Δ_n is by definition the group of all symmetries of a regular n-gon P_n, hence is defined as a transformation group on the set of *all* points of P_n. The function which assigns to each symmetry of P_n the induced permutation on the n

(numbered) vertices is clearly an isomorphism of Δ_n to a subgroup of S_n; for example, the generators R and D of Δ_n are mapped into the permutations

$$R \mapsto \begin{pmatrix} 1 & 2 & 3 & \cdots & n \\ 2 & 3 & 4 & \cdots & 1 \end{pmatrix}, \quad D \mapsto \begin{pmatrix} 1 & 2 & 3 & \cdots & n \\ 1 & n & n-1 & \cdots & 2 \end{pmatrix}. \tag{22}$$

This is a second representation of Δ_n. On the other hand, Cayley's theorem represents Δ_n as a subgroup of S_{2n} (namely, as permutations on the set of all $2n$ elements in the group Δ_n).

Many properties of transformation groups hold under the more general circumstances to be described now.

DEFINITION. A group G acts *on a set* X *when there is given a function* $G \times X \to X$, *written* $(g, x) \mapsto gx$ *and called the "action" of* $g \in G$ *on* $x \in X$, *and such that always (for all* $x \in X$ *and* $g \in G$)

$$1x = x, \qquad (g_1 g_2)x = g_1(g_2 x). \tag{23}$$

For example, if T is a transformation group consisting of permutations t on X, the assignment $(t, x) \mapsto t(x)$ defines an action of T on X. More generally, any representation $\theta: G \to T$ of G gives by $(g, x) \mapsto (\theta g)x$ an "action" of G on X. In particular, the symmetric group S_n acts on the set $\mathbf{n}$. This action induces others; for instance, if P is the set of all pairs (i, j) in $\mathbf{n}$, then $(i, j) \mapsto (\sigma i, \sigma j)$ is an action of S_n on P.

Take an element g in a multiplicative group G. For $x \in G$, $x \mapsto gxg^{-1}$ defines an automorphism of G called *conjugation* by g, as discussed in §6 for the case of symmetric groups G. Since $x \mapsto gxg^{-1}$ is an automorphism of G, $(g, x) \mapsto gxg^{-1}$ defines an action of G on itself; every group G *acts on itself by conjugation*. If S is any subgroup of G, the set

$$gSg^{-1} = \{ gsg^{-1} | s \in S \} \qquad \text{(the "conjugate of } S \text{ by } g\text{")}$$

is also a subgroup of G; indeed, it is the image of S under $x \mapsto gxg^{-1}$. Thus $(g, S) \mapsto gSg^{-1}$ defines an action of G on the set of all subgroups of G; one says that G *acts on its subgroups* by *conjugation*.

If a group G acts on a set X, two points x and $x' \in X$ are called *equivalent* under G if there is a $g \in G$ with $gx = x'$. Then $g^{-1}x' = x$ by (23), so this relation of equivalence is symmetric. Similarly, by the conditions (23), it is reflexive and transitive; hence, it partitions the set X into disjoint equivalence classes. In view of geometric examples, these equivalence classes may be called the *orbits* of G on X. The orbit of a point $x_0 \in X$ under the action of G (that is, the equivalence class containing that point) evidently is the set $\{ gx_0 | g \in G \}$, often written as Gx_0.

Orbits have already been used. Given a permutation σ on $\mathbf{n}$, the cyclic group generated by σ acts on the set $\mathbf{n}$; the orbits under this action are the sets appearing in the decomposition of σ into disjoint cycles (Theorem 14). When G acts by conjugation, an orbit is called a *conjugate class*, and two elements of G (or two subgroups of G) are said to be *conjugate* in G when they lie in the same orbit. Since conjugation is an automorphism, any two conjugate elements have the same order—as we have already seen in (20) for conjugate permutations.

Given an action $G \times X \to X$, a point $x \in X$ is *fixed* under $g \in G$ when $gx = x$. Given x, the set F_x of all group elements h fixing x is a subgroup of G, known variously as the *subgroup fixing* x, the "isotropy group" of x and the "stabilizer" of x.

An action $G \times X \to X$ is *transitive* on X when to each pair of points $x, y \in X$ there is at least one $g \in G$ with $gx = y$. This amounts exactly to saying that an action is transitive on X when the only orbit of any $x \in X$ under this action is the whole set X. (Under any action, any orbit of G on X is often called a "set of transitivity".)

A group H is said to *act on the right* on a set Y when there is given a function $Y \times H \to Y$, written $(y, h) \mapsto yh$, with always $y1 = y$ and $y(h_1h_2) = (yh_1)h_2$. Given such a right action of H, $(h, y) \mapsto yh$ defines an action of the opposite group H^{op} (see §2) on the left of Y. For instance, in any group G, conjugation, written as $x \mapsto g^{-1}xg$, defines a right action of G on itself.

A Note on Notation: Algebra includes many formal calculations drawing consequences from axioms, so the notation should be chosen to make these calculations efficient. The device of juxtaposing two letters u and v, as in uv, is so efficient that it is used in many different senses, according to the meaning previously announced for the letters. Thus, if u is a function and v an element of its domain, uv denotes the value of u for the argument v. If u and v are both functions with the domain of u equal to the codomain of v, uv denotes the composite function. If u is an element of a multiplicative group acting on a set which has v as element, uv denotes the result of this action. But if u and v are both elements of a multiplicative group, uv will always denote their product, since the result of acting on v by conjugation with u can be written as a triple product uvu^{-1}.

EXERCISES

1. Verify that the permutations appearing in the representation (22) of the dihedral group Δ_n satisfy the defining relations (17) of that group.
2. Exhibit the left regular representation of the symmetric group S_3.
3. Exhibit the left regular representation of the dihedral group Δ_4.
4. Let T be the group of all symmetries of the cube. Represent the isotropy subgroup of a vertex as a group of permutations (on the remaining vertices).

5. For the transformation group S_n, what are the isotropy subgroups? Show that they are all conjugate in S_n.

6. Show that there are three classes of conjugate elements in S_3 and five in S_4.

7. Show that the left regular representation of the additive group $\mathbf{R}$ of real numbers makes $\mathbf{R}$ isomorphic to a group of translations of the real line. Give a similar description of the product group $\mathbf{R} \times \mathbf{R}$.

8. Let the group G act transitively on the set X. Given x and y in X, prove that the subgroup fixing x is conjugate to that fixing y. (*Hint:* Use g with $gx = y$.)

9. Regard Δ_4 as the group of the square (that is, as a group acting on the square). Show how *every* subgroup of Δ_4 can be regarded as an isotropy group for a suitable action of Δ_4. (*Suggestion:* Use actions on the diagonals, sides, etc., of the square.)

10. If G acts on X, call a subset $I \subset X$ "invariant" under G when $g \in G$ and $x \in I$ imply $gx \in I$. Prove that every invariant subset of X is a union of disjoint orbits, and that the orbit of each $x \in X$ is the least invariant subset of X containing x.

11. The following statements show that every action on X can be described by a representation in the group $S(X)$ of all permutations on the set X:

(a) If G acts on X while, for each $g \in G$, ϕg is the permutation $x \mapsto gx$ on X, show that $\phi: G \to S(X)$ is a morphism of groups (and hence a representation).

(b) If $\phi: G \to S(X)$ is a morphism of groups, show that $(g, x) \mapsto (\phi g)x$ defines an action $\alpha: G \times X \to X$ of G on X.

(c) Show that $\phi \mapsto \alpha$ is a bijection (from representations to actions).

12. For a fixed element b in a group G the assignment $x \mapsto xb$ gives a bijection $g_b: G \to G$ called *right multiplication* (or, "right translation") by b. The function $g: G \to S(G)$ is called the *right regular representation* of G. Using the opposite group G^{op} of §2, show that it is a monomorphism $G^{\mathrm{op}} \to S(G)$ of groups.

8. Cosets

Let S be a subgroup of a multiplicative group G. Restriction of the multiplication of G defines a function $S \times G \to G$ which is an action $(s, g) \mapsto sg$ of S on G, called "left multiplication" by S. The orbits of S under this action are called *left*† *cosets* of S in G. Thus two elements $a, b \in G$ belong to the same left coset of S if and only if $b = sa$ for some $s \in S$; that is, if and only if $ba^{-1} \in S$. This is an equivalence relation between a and b; the standard properties of equivalence relations (§I.9) establish:

†Some authors call such an orbit a "right" coset. Our terminology is chosen because it brings out the analogy with left ideals (§III.3) and left modules (§V.1).

Proposition 19. *If S is a subgroup of G, each $a \in G$ belongs to exactly one left coset of S in G; namely, the coset*

$$Sa = \{\text{all } sa \mid s \in S\}. \tag{24}$$

If G/S denotes the set of all these left cosets, $a \mapsto Sa$ is a surjection $G \to G/S$.

Put differently: Distinct left cosets of S are disjoint, and G is the union of all left cosets of S. Note that two left cosets Sa and Sb are equal—that is, are equal sets—precisely when $a = s_0 b$ for some $s_0 \in S$. For example, for G cyclic with generator c of order 12, the cyclic subgroup $S = \{1, c^4, c^8\}$ generated by c^4 has four different left cosets:

$$S = \{1, c^4, c^8\}, \quad Sc = \{c^1, c^5, c^9\}, \quad Sc^2 = \{c^2, c^6, c^{10}\}, \quad Sc^3 = \{c^3, c^7, c^{11}\}.$$

Again, in the additive group $\mathbf{Z}$ the cosets of the subgroup $n\mathbf{Z}$, written additively, are the sets $n\mathbf{Z} + k$ already used in §I.8 as the elements of $\mathbf{Z}_n$.

The *order* of a finite group G is defined to be the number of elements in the set G, while the *index* of a subgroup $S \subset G$ is the number of distinct left cosets of S in G. Now right multiplication by a is a bijection $S \to Sa$, so any left coset of S has just as many elements as does S. Since G is the union of disjoint left cosets, the number of elements in the whole G is just the number of cosets times the number of elements in S:

$$\text{order } G = (\text{index } S \text{ in } G)\,(\text{order } S). \tag{25}$$

Often, the index of S in G is written $[G:S]$; in particular, $[G:1]$ denotes the order of G, so this equation takes the form $[G:1] = [G:S][S:1]$.

This formula includes the following classical result:

Theorem 20 (*Lagrange*). *The order of a finite group is an integral multiple of the order of each of its subgroups.*

Lagrange's theorem sharply restricts the possible orders of subgroups. For example, the order m of an element g of G was so defined as to be the order of the cyclic subgroup $\{1, g, g^2, \cdots, g^{m-1}\}$ generated by g. Hence the

Corollary. *Every element of a finite group G has order a divisor of the order of G.*

This allows us to find *all* groups of a given order n, if n has very few prime factors. For instance, if G has order 5, all elements have orders dividing 5, hence orders 1 or 5. The identity is the only element of order 1, so there must be at least one element of order 5, so G is cyclic. Similarly, any group of prime order is cyclic.

A similar argument shows that any group G of order 4 is isomorphic to $\mathbf{Z}_4$ or to $\mathbf{Z}_2 \times \mathbf{Z}_2$. Indeed, if G contains an element of order 4, it is cyclic, so isomorphic to $\mathbf{Z}_4$. Otherwise, all elements $g \neq 1$ must have order 2. Call these elements a, b, and c. By the cancellation law ab cannot be $a = a1$, $b = 1b$, or $1 = aa$; hence $ab = c$. Symmetrically, $ba = c$, $ac = ca = b$, and $bc = cb = a$. These, together with $a^2 = b^2 = c^2 = 1$, give the whole multiplication table for G and show $G \cong \mathbf{Z}_2 \times \mathbf{Z}_2$.

For any subgroup S a *right coset* $aS = \{as \mid s \in S\}$ is the set of the right multiples of a by all $s \in S$. The arguments above apply, *mutatis mutandis*, to these right cosets, and, if S is finite, the number of elements in a right coset is the same as the order of S, hence the same as the number of elements in a left coset. However, a right coset need not be a left coset. For example, in the symmetric group S_4 take the subgroup fixing 4. This subgroup is (isomorphic to) S_3. The left coset $S_3(1\ 4)$ consists of all those permutations which carry 1 to 4, while the right coset $(1\ 4)S_3$ of the same transposition $(1\ 4)$ consists of all those permutations which carry 4 to 1.

Since each coset Sa has as many elements as S, we know the size of the orbits Sa under action by left multiplication by S. Here is the corresponding result for any action by a finite group.

Theorem 21. *If a finite group G acts on a set X, then the number of points in the orbit of a point $y \in X$ is the index $[G{:}F_y]$ in G of the subgroup F_y fixing y.*

Proof: The orbit consists of the points gy, while $g_1 y = g_2 y$ if and only if $g_2^{-1}g_1 \in F_y$; that is, if and only if the cosets g_2F_y and g_1F_y are equal. Thus $gF_y \mapsto gy$ is a bijection from the set of all right cosets to the orbit of y. When G is finite, both sets have the same number of elements, hence the theorem.

EXERCISES

1. In any group, show that the bijection $a \mapsto a^{-1}$ carries right cosets to left cosets, and hence that the number of right cosets (when finite) of any subgroup is the number of left cosets.

2. In Δ_6, let S be the subgroup leaving vertex number 1 fixed. Describe (in terms of the effect on vertex 1) both the right cosets and the left cosets of S in Δ_6.

3. In S_4, let T be the subgroup which consists of all permutations carrying the set $\{1, 2\}$ into itself. Describe the right and left cosets of T (in terms of their effects upon sets like $\{1, 2\}$).

4. (a) If G is the group of all permutations $x \mapsto ax + b$ on the set $\mathbf{R}$ of real numbers, with $a \neq 0$ and b real, while S is the subgroup of all such permutations with $a = 1$, describe the right and left cosets of S in G.

(b) Do the same for the subgroup T of all the permutations above with $b = 0$.

5. If S and T are subgroups of a finite group G, prove that

$$[S:1][T:1] \leqslant [S \cap T:1][S \vee T:1].$$

6. Prove that any group of order 6 is isomorphic to either $\mathbf{Z}_6$ or S_3.

***7.** Show that any group of order 10 is isomorphic to either $\mathbf{Z}_{10}$ or Δ_5.

***8.** Determine all groups of order $2p$, for p an arbitrary odd prime.

9. (a) Given subgroups $S, T \subset G$, each element a of G determines the S-T *double coset* SaT consisting of all products sat for $s \in S$ and $t \in T$. Prove that two S-T double cosets are disjoint or equal, and that the union of all such double cosets is G.

(b) If G acts transitively on a set X, then the subgroup F fixing a point $x_0 \in X$ also acts on X. Show that there is a bijection from the set of all double cosets FgF to the set of orbits of X under F.

9. Kernel and Image

Let $\phi:G \to H$ be any morphism of groups. Its *image* is the set

$$\text{Im}\,(\phi) = \phi_* G = \{\text{all } \phi(g) | \, g \in G\}. \tag{26}$$

This subset of H is closed under product, inverse, and identity, hence is a subgroup of H. On the other hand, the set

$$\text{Ker}\,(\phi) = \{\text{all } g | \, g \in G, \phi(g) = 1 \in H\} \tag{27}$$

is a subgroup of G called the *kernel* of the morphism ϕ. Furthermore:

$$\begin{aligned}
&\phi : G \to H \text{ is an epimorphism} &&\Leftrightarrow \quad \text{Im}\,(\phi) = H;\\
&\phi : G \to H \text{ is a monomorphism} &&\Leftrightarrow \quad \text{Ker}\,(\phi) = 1;\\
&\phi : G \to H \text{ is an isomorphism} &&\Leftrightarrow \quad \text{Ker}\,(\phi) = 1 \text{ and Im}\,(\phi) = H.
\end{aligned}$$

The first statement just reformulates the definition, for ϕ is an epimorphism precisely when the function ϕ is surjective. As for the second, ϕ a monomorphism means that the function ϕ is injective; hence it implies that Ker $(\phi) = 1$. Conversely, suppose that Ker $(\phi) = 1$. If $\phi(g) = \phi(g')$, then $\phi(g'g^{-1}) = \phi(g')\phi(g)^{-1} = 1$, so $g'g^{-1} \in$ Ker (ϕ) is 1 and $g = g'$; hence, ϕ is indeed injective. The third statement follows from the first two.

The kernel K of $\phi:G \to H$ tells how much this morphism ϕ "collapses" G. For let h be any element of Im (ϕ), so that $h = \phi(g)$ for some $g \in G$. Then also $h = \phi(g')$ for all $g' = kg$ in the coset Kg—and only for these elements, since $\phi(g') = \phi(g)$ implies $\phi(g'g^{-1}) = 1$, and so $g' = kg$ where k has $\phi(k) = 1$. In other words, ϕ sends all the kernel K to 1 in H and all of each coset Kg of K in G to the element ϕg in H. Thus $Kg \mapsto \phi g$ is a bijection $G/K \to \text{Im}\,\phi$. In other words, the image "consists" of cosets of the kernel.

For $\phi:G \to H$, the insertion $j:\text{Im}(\phi) \to H$ is a morphism of groups, and ϕ "factors through" Im (ϕ) as the composite

$$G \xrightarrow{\phi'} \text{Im}(\phi) \xrightarrow{j} H, \qquad \phi = j \circ \phi',$$

where the restriction ϕ' of ϕ is an epimorphism. Hence, any morphism of groups can be written as an epimorphism followed by a monomorphism (Cf. Cor. I.19).

Next consider the behavior of subgroups under the morphism ϕ. For each subgroup $S \subset G$ the set

$$\phi_* S = \{h | h \in H \text{ and } h = \phi(s) \text{ for some } s \in S\} \tag{28}$$

is a subgroup of H, called the *image* of S under ϕ. The assignment $S \mapsto \phi_* S$ defines a function ϕ_* on the set of subgroups of G to that of H. When $S \subset S'$ in G, then $\phi_* S \subset \phi_* S'$ in H; in other words, ϕ_* is a morphism of inclusion.

There is a reverse process. For each subgroup $T \subset H$, the set

$$\phi^* T = \{g \,|\, g \in G \text{ and } \phi g \in T\} \tag{29}$$

is a subgroup of G, called the *counterimage* of T under ϕ; it is sometimes called the *inverse image* and written $\phi^{-1}T$, even though there may be no function ϕ^{-1} which is an inverse of ϕ. For example, the inverse image of the trivial subgroup 1 of H is just the kernel of ϕ, and the inverse image of any subgroup always contains the kernel. Again, ϕ^* is a morphism of inclusion.

Theorem 22. *Any morphism $\phi:G \to H$ of groups yields a bijection*

$$\{S | \text{Ker}\,\phi \subset S \subset G\} \cong \{T | 1 \subset T \subset \text{Im}\,\phi\}$$

which assigns to each subgroup S containing Ker ϕ *its image $\phi_* S$, and whose inverse assigns to each subgroup T contained in* Im ϕ *its counterimage $\phi^* T$. Each of these sets of subgroups is closed under intersection and join, and this bijection is an isomorphism of these operations.*

Proof: If Ker $\phi \subset S$, then $\phi^*(\phi_* S) = S$. Similarly, if $T \subset \text{Im}\,\phi$, then $\phi_*(\phi^* T) = T$. The assignment $S \mapsto \phi_* S$ is therefore a bijection, as claimed. This bijection may be visualized by the figure

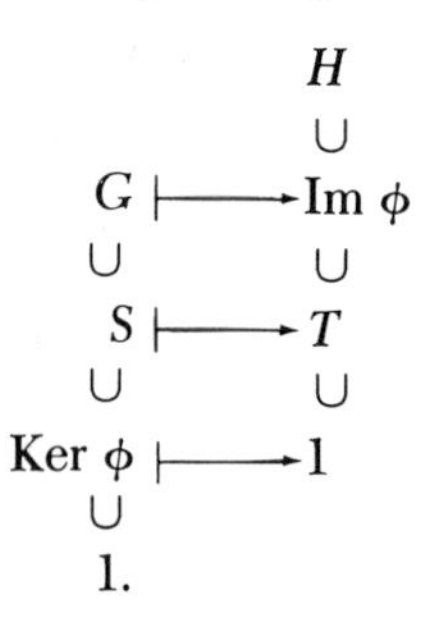

The set of all subgroups of G containing Ker ϕ goes by ϕ_* to the set of all subgroups of H contained in Im ϕ, and ϕ_* is a morphism of inclusion. For two subgroups $S_1, S_2 \supset$ Ker ϕ, the definitions (or, alternatively, Proposition 10) show that $\phi_*(S_1 \cap S_2) = \phi_* S_1 \cap \phi_* S_2$. Similarly, $\phi_*(S_1 \vee S_2) = \phi_* S_1 \vee \phi_* S_2$. Therefore ϕ_* is an isomorphism of intersection and join.

We now discuss a necessary condition on a subgroup S of a group G in order that S be the kernel of some morphism of groups. In the next section, we shall show that this necessary condition is also sufficient.

DEFINITION. A subgroup N of G is said to be normal *in G (in symbols $N \lhd G$) when $n \in N$ and $a \in G$ imply $ana^{-1} \in N$.*

Thus a normal subgroup (a "self-conjugate" or "invariant" subgroup) must contain with any one of its elements n all conjugates of that element. Any subgroup of an abelian group is trivially normal (because any conjugation is the identity automorphism). On the other hand, many subgroups of non-abelian groups are not normal; for example, in $S_3 \cong \Delta_3$ the three cyclic subgroups of order 2 are not normal.

Theorem 23. *The kernel of a morphism $\phi: G \to H$ is a normal subgroup of its domain G.*

Proof: If K is the kernel of ϕ, then $k \in K$ if and only if $\phi(k) = 1$. For any $g \in G$,

$$\phi(gkg^{-1}) = \phi(g)\phi(k)[\phi(g)]^{-1} = \phi(g)[\phi(g)]^{-1} = 1,$$

so that K contains with each k all its conjugates gkg^{-1}.

In particular $A_n \lhd S_n$, for the alternating group A_n is the kernel of the "parity" morphism $S_n \to \{\pm 1\}$. In any product $G \times H$ the subgroups $G' = \{(g, 1) | g \in G\}$ and $H' = \{(1, h) | h \in H\}$ are normal because they are the kernels of the projections $G \times H \to H$ and $G \times H \to G$, respectively.

Proposition 24. *A subgroup N of G is normal in G if and only if every right coset of N in G is a left coset (and hence also if and only if every left coset is a right coset).*

Proof: First assume $N \lhd G$. For each g, $n \mapsto gng^{-1}$ is then a bijection $N \to N$; thus $gn = (gng^{-1})g$ with $gng^{-1} \in N$ shows that $gN = Ng$. In other words, each right coset gN is a left coset. Conversely, if the right coset gN is some left coset, this left coset contains g, so must be the left coset Ng, and $gN = Ng$ means that each gn is $n'g$ for some $n' \in N$, so each conjugate $gng^{-1} = n'$ is in N, as required for $N \lhd G$.

It is a corollary that any subgroup S of index 2 in G must be a normal subgroup. For an S of index 2 has exactly two right cosets, say S and gS; trivially, $S1 = 1S$, hence $gS = Sg$ must be the complement of S in G.

EXERCISES

1. Let $\phi: G \to H$ be a morphism of groups.

(a) For S and S' subgroups of G, prove that $\phi^*(\phi_* S) \supset S$ and that $\phi_*(S \cap S') \subset (\phi_* S) \cap (\phi_* S')$, $\phi_*(S \vee S') = (\phi_* S) \vee (\phi_* S')$.

(b) For T and T' subgroups of H, prove that $\phi_*(\phi^* T) \subset T$ and that $\phi^*(T \cap T') = (\phi^* T) \cap (\phi^* T')$, $\phi^*(T \vee T') \supset (\phi^* T) \vee (\phi^* T')$.

2. Show that A_3 is the only proper normal subgroup of S_3.

3. In the dihedral group Δ_p, p an odd prime, show that the only proper normal subgroup is the cyclic group of order p, consisting of rotations.

4. List all normal subgroups of Δ_4; of Δ_6.

5. The "center" of a group G is the set $Z(G)$ of all those elements $a \in G$ with $ax = xa$ for every $x \in G$.

(a) Show that the center of any group G is a normal subgroup of G.

(b) Show that the center of Δ_n is 1 or $\mathbf{Z}_2$ according as n is odd or even.

(c) Show that the center of S_n for $n > 2$ is always 1.

6. Prove that the join and the intersection of any two normal subgroups of G are normal in G.

7. For $g \in G$, call $x \mapsto gxg^{-1}$ an "inner automorphism" of G.

(a) Show that the set In (G) of all inner automorphisms of G is a group under composition.

(b) Prove that In $(G) \triangleleft$ Aut (G).

8. If a group K has normal subgroups G and H with $G \cap H = 1$ and $G \vee H = K$, show that there is an isomorphism $\theta: G \times H \cong K$ defined by $\theta(g, h) = gh$ for all $g \in G$ and all $h \in H$.

9. If S is any subgroup of G, show that the intersection of all conjugates of S is a normal subgroup of G.

***10.** If $S \subset G$ with G finite, while $\phi: G \to H$ is a morphism with kernel N, prove that $[G:S] = [\phi_* G : \phi_* S][N : S \cap N]$.

11. For dihedral groups show that $\Delta_3 \subset \Delta_6$, as suggested by the diagram below (equilateral triangle inside regular hexagon). Determine all conjugate subgroups to Δ_3 in Δ_6 and describe the cosets of Δ_3.

12. Generalize the regular representation of G (Cayley's Theorem 18), as follows. For $T \subset G$ and fixed $a \in G$ show that the assignment $gT \mapsto agT$ is a permutation $h_a: G/T \to G/T$ on the set G/T of right cosets of T in G.

Hence (varying a) show that $h:G \to S(G/T)$ a morphism of groups. Describe the kernel of h; if $T = 1$, show that h is the left regular representation of G.

10. Quotient Groups

By Theorem 23 any morphism $G \to H$ has a kernel which is a normal subgroup of G. Conversely, given any normal subgroup N of G, we can construct a morphism (indeed, an epimorphism) with exactly this subgroup N as kernel. Indeed, we can construct this morphism by using the previous observation (§9) that the image consists (up to isomorphism) of the cosets of the kernel.

Theorem 25. *If $N \lhd G$, then the set G/N of left cosets of N in G can be made into a group in such a way that the function*

$$p : G \to G/N, \qquad g \mapsto Ng, \tag{30}$$

is an epimorphism of groups with kernel N.

Proof: We wish to define a product of cosets. Given two left cosets C and D of N in G, pick an element in each, so that (say) $C = Nc$ and $D = Nd$. If the function p of (30) is to be a morphism of multiplication, then

$$N(cd) = p(cd) = (pc)(pd) = (Nc)(Nd).$$

Therefore, the product of the cosets C and D must be the coset $N(cd)$.

To define now the product of $C = Nc$ and $D = Nd$, consider any two elements $c' \in C$ and $d' \in D$. Thus $c' = mc$ and $d' = nd$ for some elements $m, n \in N$, and

$$c'd' = (mc)(nd) = m(cnc^{-1})cd = k(cd), \qquad k = m(cnc^{-1}).$$

But cnc^{-1} is a conjugate of n, hence in the normal subgroup N, and thus $k \in N$. Therefore, the product $c'd'$ is in the same coset $N(cd)$ as is the product cd: Any product of an element in C by an element in D lies in the same coset, so we may define this coset to be the product

$$(Nc)(Nd) = N(cd). \tag{31}$$

This multiplication on G/N satisfies the group axioms. For, the associative law in G clearly makes the product (31) associative. The coset $N = N1$ containing the identity of G is an identity for the product (31), because $(N1)(Nd) = N(1d) = Nd$. With this identity, the coset Nh^{-1} is an inverse for Nh. Hence, G/N is a group. Finally, the function p clearly is an epimorphism and has kernel N.

Call G/N the *quotient group* (or "factor group") of G by N and p the *projection* of G onto its quotient G/N. The essential property of this quotient is that *every* morphism with domain G and kernel N has its image isomorphic to G/N (see Corollary 2 below). This property has a more general formulation: The projection p on G/N has $p_*N = 1$, and every morphism ϕ on G with $\phi_*N = 1$ factors uniquely through the projection p (Theorem 26 below). This property, sometimes stated as "p is universal among morphisms killing N", is the basic result from which all properties of quotient groups derive:

THEOREM 26 (*Main Theorem on Quotient Groups*). *Let $N \lhd G$. For each morphism $\phi: G \to L$ of groups with $\phi_*N = 1$ there is a unique morphism $\phi': G/N \to L$ such that $\phi = \phi' \circ p$.*

Proof: The situation is as indicated in the diagram below:

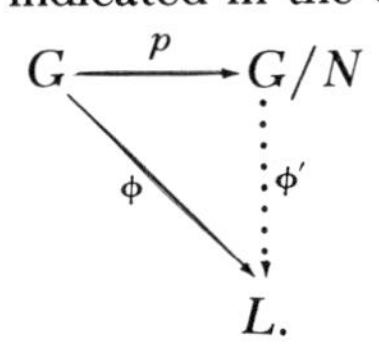

Given are the morphisms p and ϕ, required is a morphism ϕ' to make the diagram commute.

By hypothesis, $\phi(n) = 1$ for each $n \in N$. Hence, ϕ carries all the elements of any one coset Nc to a single element $\phi(nc) = \phi(c)$ of L. This means that $Nc \mapsto \phi c$ is the (unique) *function* ϕ' on G/N to L with $\phi' \circ p = \phi$.

This function ϕ' is a morphism. For, the product (31) of cosets satisfies

$$\phi'(Nc)\phi'(Nd) = (\phi c)(\phi d) = \phi(cd) = \phi'[N(cd)],$$

so that ϕ' carries products to products. This completes the proof.

COROLLARY 1. *For the morphism $\phi': G/N \to L$ with $\phi' \circ p = \phi$ above,*

$$\text{Im } \phi' = \text{Im } \phi, \qquad \text{Ker } \phi' = (\text{Ker } \phi)/N.$$

Proof: As for the first conclusion, $\phi'(pg) = \phi(g)$ and p a surjection imply that ϕ' and ϕ have the same image. As for the second, let $K = \text{Ker } \phi$. By assumption, $K \supset N$. Since N is normal in G, it is certainly normal in the smaller group K, so that the quotient group K/N is defined—and consists of some of the cosets of N in G, hence is indeed a subgroup of G/N. Now,

$\phi'(Ng) = 1$ if and only if $\phi(g) = 1$; that is, if and only if $g \in K$, so if and only if $Ng \in (K/N)$. Therefore, $\operatorname{Ker} \phi' = K/N$.

Corollary 2. *If $\phi : G \to L$ is any epimorphism of groups with kernel N, there is a unique isomorphism $\phi' : G/N \cong L$ such that ϕ is the composite $\phi' \circ p$.*

Proof: Since ϕ is an epimorphism, $\operatorname{Im} \phi' = \operatorname{Im} \phi = L$; hence, ϕ' is also an epimorphism. Since $\operatorname{Ker} \phi = N$, $\operatorname{Ker} \phi' = N/N = 1$, and ϕ' is a monomorphism. Altogether, then, ϕ' is an isomorphism, as desired.

This corollary states that any epimorphism $\phi : G \to L$ has codomain L a (copy of a) quotient group:

$$\text{Codomain } \phi \cong \text{Domain } \phi / \operatorname{Ker} \phi, \qquad \phi \text{ an epimorphism.}$$

(The corollary states more than this; it specifies *which* isomorphism.)

Corollary 3. *If $\phi : G \to L$ is any morphism of groups with kernel $N \lhd G$, there is a unique monomorphism $\phi' : G/N \to L$ such that $\phi = \phi' \circ p$. Hence, ϕ and ϕ' have the same image.*

Proof: $\operatorname{Ker} \phi = N$ gives $\operatorname{Ker} \phi' = N/N = 1$, hence ϕ' a monomorphism. This corollary states (in part) that

$$\operatorname{Im} \phi \cong \text{Domain } \phi / \operatorname{Ker} \phi \tag{32}$$

for ϕ any morphism of groups. It also implies the following "standard" factorization.

Corollary 4. *Any morphism $\phi : G \to L$ with kernel $N \lhd G$ and image $T \subset L$ can be written as a composite $i \circ \theta \circ p$,*

$$G \xrightarrow{p} G/N \xrightarrow{\theta} T \xrightarrow{i} L$$

with the projection p an epimorphism, θ an isomorphism, and the insertion i a monomorphism.

Here are some applications. Let $\mathbf{R}^*$ be the multiplicative group of nonzero real numbers, and let $\{\pm 1\}$ be the subgroup consisting of ± 1. What is the quotient group $\mathbf{R}^*/\{\pm 1\}$? It is isomorphic to the multiplicative group $\mathbf{P}^*$of positive reals, for $x \mapsto |x|$ is an epimorphism $\mathbf{R}^* \to \mathbf{P}^*$ with kernel $\{\pm 1\}$; apply Corollary 2. Again, let $\mathbf{R}$ be the additive group of real numbers, $\mathbf{Z} \subset \mathbf{R}$ the subgroup of integers. What is the quotient group $\mathbf{R}/\mathbf{Z}$? *Claim:* It is isomorphic to the multiplicative group $\mathbf{C}_1$ of complex numbers of absolute value 1, for the function $x \mapsto e^{2\pi i x}$ is an epimorphism $\mathbf{R} \to \mathbf{C}_1$ with kernel $\mathbf{Z}$.

Here—and in many similar cases—one can *determine* a quotient group such as $\mathbf{R}/\mathbf{Z}$ up to isomorphism without ever bothering to look at the cosets which are purportedly the elements of the quotient group. Here are two more examples.

Proposition 27. *If $N \lhd G$ and $S \subset G$, the join $N \vee S$ consists of all products ns for $n \in N$ and $s \in S$. If both $N \cap S = 1$ and $N \vee S = G$, then $G/N \cong S$.*

Proof: Consider first the set of all products ns; it contains 1. Since N is normal in G, this set is closed under multiplication ($nsn's' = [n(sn's^{-1})]ss'$) and under inverse: $(ns)^{-1} = s^{-1}n^{-1} = (s^{-1}n^{-1}s)s^{-1}$. Hence this set is a subgroup of G. It contains both N and S and is contained in any subgroup containing them both. Hence, it is the join $N \vee S$.

Now let θ be the composite $S \to G \to G/N$ (insertion followed by projection). Since $N \cap S = 1$, Ker $\theta = 1$. Since $G = N \vee S$ is the join just described, Im $\theta = G/N$. Hence, θ is an isomorphism $S \cong G/N$, as required.

For applications of this result, see Exercise 5b.

In such cases, when $N \cap S = 1$ and $N \vee S = G$, the subgroup S is called a "complement" of the normal subgroup N. A normal subgroup may not have a complement; for example, in $\mathbf{Z}_4$ the cyclic subgroup of order 2 has none.

Proposition 28. *If $N_1 \lhd G_1$ and $N_2 \lhd G_2$ are normal subgroups, then the product $N_1 \times N_2$ is a normal subgroup of the product $G_1 \times G_2$, and*

$$(G_1 \times G_2)/(N_1 \times N_2) \cong (G_1/N_1) \times (G_2/N_2). \tag{33}$$

Proof: For $i = 1, 2$, let $p_i: G_i \to G_i/N_i$ be the projection. Then setting $\psi(g_1, g_2) = (p_1 g_1, p_2 g_2)$ for $g_i \in G_i$ defines an epimorphism $\psi: G_1 \times G_2 \to (G_1/N_1) \times (G_2/N_2)$. Its kernel is $N_1 \times N_2$, so must be normal, while its codomain is then isomorphic to $(G_1 \times G_2)/(N_1 \times N_2)$.

To illustrate: The additive group $\mathbf{Z}$ is a subgroup of the additive group $\mathbf{R}$ of real numbers. The quotient group $(\mathbf{R} \times \mathbf{R})/(\mathbf{Z} \times \mathbf{Z}) \cong (\mathbf{R}/\mathbf{Z}) \times (\mathbf{R}/\mathbf{Z})$ is called, for geometric reasons, the group of the torus (cf. Exercise I.9.3).

The quotient group provides an effective means for building new groups from given ones. Here are some instances.

To each $g \in G$, conjugation by g is an automorphism $\gamma_g: G \to G$, $\gamma_g(x) = gxg^{-1}$, called an "inner automorphism". In the group Aut (G) of *all* automorphisms of G (under composition), two automorphisms α and β are said to be in the same "automorphism class" when $\alpha = \gamma \circ \beta$ for some inner automorphism γ. One may prove (Exercise 9.7) that the inner automorphisms form a subgroup In (G) normal in Aut (G). Therefore, the automorphism classes are the elements of a group Aut $(G)/$In (G), called the group of "outer automorphisms" of G.

Consider morphisms of a given group G to abelian groups. With quotient groups one can construct a universal such morphism (Exercise 7).

Any action of a group G on a set X leads to a transformation group on X as follows. Let F be the set of all $g \in G$ with $gx = x$ for every x; thus F is the intersection of all isotropy subgroups of points $x \in X$. Then $F \lhd G$, and the given action of G on X defines an action of G/F on X which makes G/F isomorphic to a transformation group on X (Exercise 11).

Let a, b, c, d be four real numbers with $ad - bc \neq 0$. Then $L(x, y) = (ax + cy, bx + dy)$ defines a bijection $L: \mathbf{R}^2 \to \mathbf{R}^2$; all such L form a group, the "general linear group" $GL(2, \mathbf{R})$. Each bijection L carries lines through the origin of $\mathbf{R}^2$ into lines, so $GL(2, \mathbf{R})$ acts on the set X of all such lines. The normal subgroup F fixing all these lines is the group of all bijections $(x, y) \mapsto (kx, ky)$ for $k \neq 0$. The quotient group $GL(2, \mathbf{R})/F$ is called the real projective group in dimension 1. Similar groups in higher dimensions will appear in the later chapters on linear algebra.

EXERCISES

1. If S is the cyclic subgroup of order n in the dihedral group Δ_n, show that $\Delta_n/S \cong \mathbf{Z}_2$.

2. List all quotient groups of the group Δ_4 of the square.

3. (a) For $4\mathbf{Z}$ all integral multiples of 4, prove that $4\mathbf{Z}/20\mathbf{Z} \cong \mathbf{Z}_5$ and $\mathbf{Z}_6/3\mathbf{Z}_6 \cong \mathbf{Z}_3$.

   (b) For positive integers m and k, determine $k\mathbf{Z}/mk\mathbf{Z}$ and $(\mathbf{Z}/mk\mathbf{Z})/(k\mathbf{Z}/mk\mathbf{Z})$.

4. If $\mathbf{Q}^*$ is the multiplicative group of all non-zero rational numbers and $\{\pm 1\}$ the subgroup consisting of the two elements $+1$ and -1, to what familiar group is $\mathbf{Q}^*/\{\pm 1\}$ isomorphic?

5. (a) In the symmetric group S_4, show that the following subgroup is normal:

$$V = \{1, (1\ 2)(3\ 4), (1\ 3)(2\ 4), (1\ 4)(2\ 3)\}.$$

   (b) Prove that $S_4/V \cong S_3$ and $A_4/V \cong \mathbf{Z}_3$.

6. If Z is the center of G (Exercise 9.5), prove that G/Z is isomorphic to the group of inner automorphisms of G.

7. (a) A *commutator* in a group G is an element of the form $ghg^{-1}h^{-1}$. Prove that the set $[G, G]$ of all products $C_1 \cdots C_n$ of commutators C_i is a normal subgroup of G (called the "commutator subgroup" of G).

   (b) Prove that $G/[G, G]$ is abelian and that any morphism $G \to A$ to an *abelian* group A factors uniquely through the projection $G \to G/[G, G]$.

8. If $N \lhd G$, $G/N \cong \mathbf{Z}_5$ and $N \cong \mathbf{Z}_2$, prove that G is abelian.

9. For subgroups $N \subset M$ of G, both normal in G, prove that $(G/N)/(M/N) \cong G/M$.

10. For $\phi : G \to H$ an epimorphism of groups and T a subgroup of H, prove that $T \lhd H$ if and only if $\phi^* T \lhd G$; when this is the case, prove that $G/(\phi^* T) \cong H/T$.

11. If G acts on X and F consists of those $g \in G$ fixing every $x \in X$, as in the text, prove that $F \lhd G$. If $p : G \to G/F$ is the projection, prove that there is a unique action of G/F on X with $(pg)x = gx$. If ϕ maps pg to the permutation $x \mapsto gx$ on X, prove that $\phi : G/F \to S(X)$ is a monomorphism (to a transformation group).

CHAPTER III

Rings

1. Axioms for Rings

A *ring* $R = (R, +, \cdot, 1)$ is a set R with two binary operations, addition and multiplication, and a nullary operation, "select 1", such that:

(i) $(R, +)$ is an abelian group.
(ii) $(R, \cdot, 1)$ is a monoid.
(iii) Multiplication is distributive (on both sides) over addition.

The last requirement means that all triples of elements a, b, c in R satisfy the identities

$$a(b + c) = ab + ac, \qquad (a + b)c = ac + bc. \tag{1}$$

A *commutative ring* is one in which the multiplication is commutative. Familiar systems of numbers are commutative rings under the usual operations of sum and product; examples are

$\mathbf{Z}$,	the ring of all integers,
$\mathbf{Q}$,	the ring of all rational numbers,
$\mathbf{R}$,	the ring of all real numbers,
$\mathbf{C}$,	the ring of all complex numbers.

The ring $\mathbf{Q}$ will be constructed from $\mathbf{Z}$ in §5 of this chapter, while the last two systems of numbers will be studied in Chapter VIII. There are also finite rings; for example, the set $\mathbf{Z}_n$ of integers modulo n, with addition and multiplication as constructed in §I.8, is a ring. For that matter, the set consisting of just one element 0, with addition and multiplication given (in the only possible way!) by $0 + 0 = 0$, $00 = 0$, is a ring; it will be called the "trivial ring".

The definition of a ring amounts to the statement that a ring is a set R with a selected element $1 \in R$ and two binary operations $(a, b) \mapsto a + b$ and $(a, b) \mapsto ab$ which are both associative, so that

$$a + (b + c) = (a + b) + c, \qquad a(bc) = (ab)c \tag{2}$$

for all $a, b, c \in R$, which have addition commutative, so that

$$a + b = b + a \qquad \text{for all } a, b \in R, \tag{3}$$

which contains a *unit* 1 and a *zero* 0 such that

$$a + 0 = a, \qquad a1 = a = 1a, \qquad \text{for all } a \in R, \tag{4}$$

which contains to each element a an *additive inverse* $(-a)$ with

$$a + (-a) = 0, \tag{5}$$

and in which both distributive laws (1) hold.

Some elementary properties of rings follow from known facts about additive groups or multiplicative monoids. For instance, since a ring R is a group under addition, its zero is unique, as is the additive inverse (the "negative") of each element a; moreover, for all $a, b \in R$,

$$-(-a) = a, \qquad -(a + b) = (-a) + (-b), \qquad -0 = 0.$$

As in any additive group, the *multiples* na of a ring element a by an intege n are defined and have the familiar formal properties (II.11) and (II.12). Since R is a monoid under multiplication, its unit 1 is unique. As in any monoid, the *powers* a^n of an element a can be defined for any natural number n as exponent so as to satisfy the identities of (I.42)

$$a^{m+n} = a^m a^n, \qquad a^{mn} = (a^m)^n.$$

Rings enjoy other properties whose validity depends in part on the distributive law, and which are thus not consequences of properties of abelian groups or monoids. Among these properties are various familiar rules of calculation, such as the following.

RULE 1. For all $a \in R$, $a0 = 0 = 0a$.

Proof: Since 0 is the zero for addition, $a + 0 = a$. Multiply both sides by a, to get $a(a + 0) = aa$. Now the distributive law and the definition of 0 give

$$aa + a0 = aa = aa + 0.$$

Subtract aa from both sides to get $a0 = 0$. A symmetrical proof yields $0 = 0a$.

RULE 2. If R is not the trivial ring, $0 \neq 1$.

Proof: If $0 = 1$, $a = a1 = a0 = 0$ for all a, so R contains only one element 0; hence it is the trivial ring.

RULE 3. For all a, b in R, $(-a)b = -(ab) = a(-b)$.

Proof: By Rule 1 and distributivity,

$$0 = 0b = [a + (-a)]b = ab + (-a)b.$$

But the additive inverse of ab is defined to be the unique solution x for the equation $0 = ab + x$, hence $-(ab) = (-a)b$; the case of $a(-b)$ is analogous.

Since $-(-a) = a$, a corollary is $(-a)(-b) = ab$; in particular, one has $(-1)(-1) = 1$. Another corollary is $(-a) = (-1)a$.

RULE 4 (*General Distributive Law*). For all elements $a_1, \cdots, a_m$ and $b_1, \cdots, b_n$ in R,

$$(a_1 + \cdots + a_m)(b_1 + \cdots + b_n) = \sum_{i=1}^{m} \sum_{j=1}^{n} a_i b_j.$$

For $m = 1$ the proof is by induction on n, the case $n = 2$ being the first half of the distributive law (1). Given this rule for $m = 1$ and all n, a second induction, this time on m, gives the whole rule, now using the other half of the distributive law (1).

RULE 5. For all integers n and all a, b in R,

$$n(ab) = (na)b = a(nb).$$

Proof: For n positive, the first equation is $ab + \cdots + ab = (a + \cdots + a)b$, to n summands, hence is a case of the general distributive law. For n negative, use Rule 3.

For commutative rings, another important rule is the binomial theorem (Exercise 9).

A *morphism of rings*, $\alpha: R \to R'$, is a function α on one ring R to a second ring R', which is a morphism of addition, of multiplication, and of unit In other words, α must satisfy the conditions

$$\alpha(a + b) = \alpha(a) + \alpha(b), \qquad \alpha(ab) = (\alpha a)(\alpha b), \qquad \alpha(1) = 1' \qquad (6)$$

for all elements $a, b \in R$ and for 1 the unit of R, $1'$ that of R'. The first condition of (6) implies, as in Proposition II.1, that $\alpha(0) = 0$. However, the condition $\alpha(1) = 1'$ does not follow from the other conditions of (6); this is why we have included the nullary operation "select 1" as part of the structure in the definition of a ring.

As in the case of groups, a morphism α of rings is called a *mono-*, *epi-*, or *isomorphism* when the function α is in-, sur-, or bijective, respectively. For example, the insertions $\mathbf{Z} \to \mathbf{Q} \to \mathbf{C}$ are monomorphisms of rings. For the integers modulo n, the projection $\rho: \mathbf{Z} \to \mathbf{Z}_n$ which sends each integer to its remainder (mod n) is an epimorphism of rings (see Theorems I.16 and I.17).

For any ring R, the identity function $1_R: R \to R$ is a morphism of rings. If $\beta: R'' \to R$ and $\alpha: R \to R'$ are two morphisms of rings, so is their composite $\alpha \circ \beta: R'' \to R'$.

By induction on n, each morphism $\alpha : R \to R'$ of rings has $\alpha(n1) = n1'$. This determines the effect of α on all multiples of the unit element 1. For example, every element in the ring $\mathbf{Z}$ of integers is a multiple of 1, so, given R', there is only one possible morphism $\mathbf{Z} \to R'$.

Proposition 1. *For each ring R' there is exactly one morphism $\mu : \mathbf{Z} \to R'$.*

Proof: We have just shown that the only possible choice for μ is $\mu(n) = n1'$, where $1'$ is the unit of R'. The function μ so defined is clearly a morphism of addition and of units. To show that it is a morphism of multiplication, we need only show

$$(m1')(n1') = (mn)1', \qquad m, n \in \mathbf{Z}. \tag{7}$$

If m is nonnegative, this may be proved by induction. Indeed, (7) is immediate for $m = 0$, so make the induction assumption that (7) holds for some $m \geqslant 0$ and all n. Then

$$\begin{aligned}((m + 1)1')\,(n1') &= (m1' + 1')\,(n1') = (m1')\,(n1') + n1' \\ &= (mn)1' + n1' = (mn + n)1' = ((m + 1)n)1';\end{aligned}$$

this is (7) for $m + 1$, so the induction is complete. Finally, if m is negative, (7) follows from the case when m is positive by Rule 3 above.

This morphism $\mu : \mathbf{Z} \to R'$ with $\mu(n) = n1'$ depends essentially upon the unit elements, so will be called the *unital morphism* for the ring R'.

The *characteristic* of a ring R' is defined to be the (additive) order of its unit $1'$. Thus a nontrivial ring R' has a finite characteristic m if m is the least positive integer with $m1' = 0$ and has characteristic ∞ if no such multiple $m1'$ is zero. (*Note:* Some authors say "characteristic 0" for our "characteristic ∞".) For example, $\mathbf{Z}$ has characteristic ∞, while each ring $\mathbf{Z}_n$ has characteristic n.

Proposition 2. *If a ring R' has characteristic ∞, the unital morphism $\mu : \mathbf{Z} \to R'$ is a monomorphism of rings. If a ring R' has finite characteristic m, there is a monomorphism $\mathbf{Z}_m \to R'$ of rings, with $m\mathbf{Z} + k \mapsto \mu(k) = k1'$, and the order of every element of the additive group of R' is a divisor of m.*

Proof: Consider the unital morphism $\mu : \mathbf{Z} \to R'$ for the moment just as a morphism of additive groups. Its kernel, an additive subgroup of $\mathbf{Z}$, is determined by the characteristic of R'. If R' has characteristic ∞, this kernel is zero, so that μ is a monomorphism (of additive groups, and hence also of rings). If R' has characteristic m, this kernel is the set $m\mathbf{Z}$ of all integral multiples of m, so that μ factors as $\mu = \mu' \circ \rho$, where $\rho:\mathbf{Z} \to \mathbf{Z}_m$ is the projection sending each $k \in \mathbf{Z}$ to its coset $m\mathbf{Z} + k$ and $\mu'(m\mathbf{Z} + k) = k1'$ is a monomorphism of additive groups. By the rule for multiplying cosets, μ' is also a morphism of multiplication and hence a monomorphism $\mu':\mathbf{Z}_m \to R'$ of rings, as asserted. Finally, for each element a, Rule 5 gives $ma = m(1'a) = (m1')a = 0a = 0$, so the additive order of a does indeed divide the characteristic m.

Now reconsider the construction of the ring $\mathbf{Z}$ of integers from the system $\mathbf{N}$ of natural numbers. The operations of addition and multiplication in $\mathbf{Z}$ are, as in Theorems I.12 and I.13, extensions of these operations in $\mathbf{N}$, while both $\mathbf{N}$ and $\mathbf{Z}$ have the same unit for multiplication. These facts amount to the statement that the insertion $i:\mathbf{N} \to \mathbf{Z}$ is a morphism of addition, multiplication, and multiplicative unit. This gives the following characterization of the embedding $i:\mathbf{N} \to \mathbf{Z}$ among all other possible morphisms of $\mathbf{N}$ into a ring.

Proposition 3. *If R' is any ring and $\alpha:\mathbf{N} \to R'$ any morphism of addition, multiplication, and multiplicative unit, there is a unique morphism $\alpha':\mathbf{Z} \to R'$ of rings with $\alpha' \circ i = \alpha$, where $i:\mathbf{N} \to \mathbf{Z}$ is the insertion.*

Proof: The only morphism $\alpha':\mathbf{Z} \to R'$ of rings is the unital morphism $\alpha' = \mu$. On the other hand, the unital morphism $\alpha' = \mu$ does have $\alpha'(n) = n1' = \alpha(n)$, so $\alpha' \circ i = \alpha$, as required. (See the commutative diagram below.)

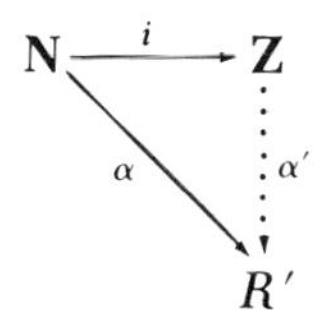

Our definition of a ring requires the presence of a multiplicative unit; our definition of morphism of rings regards this unit—along with the operations of addition and multiplication—as an essential part of the "structure" of a ring. Some authors, in defining a ring, do not require the presence of a unit for multiplication; with such a definition, both the set of all even integers and the set of all continuous functions $f:\mathbf{R} \to \mathbf{R}$ with $\int_{-\infty}^{\infty} |f(x)|\, dx < \infty$ are rings. With our definition, neither is.

EXERCISES

1. Prove that the following rules hold in any ring:

$$(a + b)(c + d) = ac + ad + bc + bd, \qquad a(b + c)d = abd + acd.$$

2. Prove that the following rules hold for the operation $(a, b) \mapsto a - b$

in any ring:

$$(a - b)(c - d) = ac + bd - (ad + bc),$$
$$(a - b)c = ac - bc, \qquad a(b - c) = ab - ac.$$

3. Are the following sets of rational numbers rings under the usual operations of addition and multiplication? If so, why; if not, why not?

(a) All rational numbers with denominator 1, 2, or 4.

(b) All rational numbers with denominator 1 or some power of 2.

4. In each of the following cases, show that there is exactly one morphism between the rings displayed:

$$\mathbf{Z}_4 \to \mathbf{Z}_2, \qquad \mathbf{Z}_{12} \to \mathbf{Z}_3, \qquad \mathbf{Z}_{12} \to \mathbf{Z}_4.$$

5. Prove that there can be no morphism $\mathbf{Z}_3 \to \mathbf{Z}_4$ of rings.

6. For m and n positive integers, find conditions that there be a morphism $\mathbf{Z}_m \to \mathbf{Z}_n$ of rings.

7. Prove: Any ring with two elements is isomorphic to $\mathbf{Z}_2$.

8. Prove: Any ring with seven elements is isomorphic to $\mathbf{Z}_7$.

9. The *binomial coefficient* (k, m) is defined for all $k, m \in \mathbf{N}$ by $(k, m) = (m + k)!/[(m!)(k!)]$. (The notation $(k, m) = \binom{k+m}{k}$ is also used for this coefficient.)

(a) Show that $(k - 1, m) + (k, m - 1) = (k, m)$.

(b) For natural numbers k and m, prove (k, m) an integer.

(c) In any commutative ring K, prove by induction the *binomial formula* for all $a, b \in K$:

$$(a + b)^n = \sum_{k+m=n} (k, m)a^k b^m.$$

10. Prove that $(n, 0) + (n - 1, 1) + \cdots + (0, n) = 2^n$.

11. Obtain a formula for the nth derivative of the product of two n-times differentiable real functions and prove the formula by induction on n.

2. Constructions for Rings

We present here several ways of getting new rings from old ones.

DEFINITION. A ring S is a subring *of a ring R, in symbols $S \subset R$, when S is a subset of R and the insertion $S \to R$ is a morphism of rings.*

Just as in the case of subgroups (Theorem II.9) one has

THEOREM 4. *A subset T of a ring R is a subring if and only if it is closed under the following operations of R: Multiplicative unit, subtraction, and multiplication.*

Proof: These three closure conditions are:

(i) $1 \in T$ (1 the unit of R).
(ii) $s, t \in T$ implies $s - t \in T$.
(iii) $s, t \in T$ implies $st \in T$.

The first two make T an additive subgroup of R, while the third and first ensure that it is also a submonoid under multiplication. The distributive laws—like the other identities—then automatically hold in T, so T is a ring and the insertion $T \to R$ is a (mono-)morphism, as required.

Any morphism $\alpha: R \to R'$ of rings, when considered just as a function, has as image a subset Im (α) of R'. By Theorem 4, this subset is a subring of R'. For example, the image of the unital morphism $\mathbf{Z} \to R'$ is the subring of R' consisting of all multiples of its unit $1'$. By Proposition 2, this subring is isomorphic to $\mathbf{Z}$ (if R has characteristic ∞) or to $\mathbf{Z}_m$ (if R has characteristic m). If S and T are subrings of a ring R, their intersection $S \cap T$ as sets is, by Theorem 4, a subring, while the intersection of all subrings containing both S and T is a subring. The latter subring is called the *join* of S and T.

To exhibit another example of subrings, assume that the usual complex numbers $a + bi$, with a, b real and $i = \sqrt{-1}$, form a ring $\mathbf{C}$. A *gaussian integer* is then defined to be a complex number of the form $m + ni$ for $m, n \in \mathbf{Z}$. In particular, the usual rules for subtraction and multiplication of complex numbers give the following rules for gaussian integers:

$$(m + ni) - (m' + n'i) = (m - m') + (n - n')i,$$
$$(m + ni)(m' + n'i) = (mm' - nn') + (mn' + nm')i.$$

Since $1 \in \mathbf{Z}$ also, it follows that the set G of all gaussian integers is a subring of the ring $\mathbf{C}$. In §13 we shall see that it is possible to construct this ring G of gaussian integers directly from the ring $\mathbf{Z}$ without assuming the complex numbers themselves already at hand.

Next we construct a product $R \times R'$ of two given rings R and R' with units 1 and $1'$. Take the set $R \times R'$ of all ordered pairs (r, r') of elements $r \in R$, $r' \in R'$, take a unit $(1, 1')$ in $R \times R'$ and define addition and multiplication in $R \times R'$ by

$$(r, r') + (s, s') = (r + s, r' + s'), \qquad (r, r')(s, s') = (rs, r's').$$

(These are "termwise" operations, just as for groups in (II.1).) It is easy to verify that $R \times R'$ is a ring under these operations; its additive group is just the product of the additive groups of R and R', and its zero is $(0, 0)$. Moreover, the projections p and p' of the product set on its factors yield ring epimorphisms

$$R \xleftarrow{p} R \times R' \xrightarrow{p'} R', \qquad p(r, r') = r, \quad p'(r, r') = r'. \tag{8}$$

This ring $R \times R'$ is called the *product* of R and R' (sometimes the "direct sum" of R and R').

The assignment $r \mapsto (r, 0)$ defines a function $R \to R \times R'$ which is a morphism of addition and multiplication, but not of rings, because it does not preserve the units. Put differently, all the elements $(r, 0)$ in $R \times R'$ form a ring, but *not* a subring of $R \times R'$.

The product of two commutative rings is evidently commutative. For instance, the product $\mathbf{Z}_2 \times \mathbf{Z}_2$ of two copies of the ring $\mathbf{Z}_2$ of integers modulo 2 is a commutative ring with four elements. Now $\mathbf{Z}_4$ is another commutative ring with four elements; the two are not isomorphic, since the additive group of $\mathbf{Z}_4$ is cyclic, that of $\mathbf{Z}_2 \times \mathbf{Z}_2$ is not (it is the four group).

We next construct a "function ring" R^X from any set X and any ring R. Given two functions $f, g: X \to R$, their *pointwise sum* $f + g$ and *pointwise product* fg are the functions defined for each $x \in X$ by

$$(f + g)(x) = f(x) + g(x), \qquad (fg)(x) = f(x)g(x). \tag{9}$$

If $X = \mathbf{R}$ and $R = \mathbf{R}$, these are exactly the familiar definitions of the sum and product of two real functions, as used in calculus. (The functions cos, sin$:\mathbf{R} \to \mathbf{R}$ have as product the function cos $\cdot$ sin, usually written $x \mapsto \cos x \sin x$.) In general, the definitions (9) give binary operations of addition and multiplication in the set R^X of all functions $f: X \to R$, while $1'(x) = 1$ for all $x \in X$ selects a (pointwise) unit element $1' \in R^X$.

Proposition 5. *If R is a ring, so is R^X under pointwise unit, sum, and product. If R is commutative, R^X is also, both for any set X.*

Proof: We must verify the axioms (1)–(5) defining a ring. Consider, for instance, the first distributive law of (1) for three functions f, g, and $h: X \to R$. Now $f(x)[g(x) + h(x)] = f(x)g(x) + f(x)h(x)$ holds for each $x \in X$ because the distributive law holds in R. Hence, by the pointwise definitions (9), the two functions $f[g + h]$ and $fg + fh$ are equal; this proves the first distributive law in R^X. Similar pointwise arguments establish the second distributive law and each of the other axioms required for a ring. The zero $0'$, unit $1'$, and additive inverse $-f$ in R^X are all defined pointwise, as

$$0'(x) = 0, \qquad 1'(x) = 1, \qquad (-f)(x) = -(fx), \qquad x \in X.$$

All told, R^X is a ring, commutative if R is commutative.

The particular functions $0': X \to R$ and $1': X \to R$ just defined are called "constant" functions. Each element $r \in R$ similarly yields a *constant func-*

tion $r':X \to R$ defined by $r'(x) = r$ for all $x \in X$. The assignment $r \mapsto r'$ is a monomorphism $R \to R^X$ of rings; often, the element r is identified with the corresponding "constant" function r'; this makes R a subring of R^X.

A useful example arises when X is the closed unit interval $[0, 1]$ on the real axis, while $R = \mathbf{R}$ is the ring of real numbers; in this case $\mathbf{R}^{[0,1]}$ is the ring of *all* real-valued functions on the unit interval. Many of its subrings are useful in analysis; they include the rings of all bounded, all continuous, all continuously differentiable, or all Lebesgue-measurable functions on $[0, 1]$ to $\mathbf{R}$.

Endomorphisms of abelian groups provide an important class of rings; usually, these rings are noncommutative. If f and g are two endomorphisms of an additive abelian group A, their pointwise sum $f + g$ and their composite $f \circ g$ are defined for all $a \in A$ by the equations

$$(f + g)(a) = fa + ga, \qquad (f \circ g)a = f(ga). \tag{10}$$

We already know that the composite $f \circ g$ is still an endomorphism of addition in A. The same is true for the pointwise sum because, for all $a, b \in A$,

$$\begin{aligned}
(f + g)(a + b) &= f(a + b) + g(a + b) && \text{(definition of pointwise sum)},\\
&= fa + fb + ga + gb && (f, g \text{ morphisms of addition}),\\
&= fa + ga + fb + gb && (\text{addition commutative in } A),\\
&= (f + g)(a) + (f + g)(b) && \text{(definition of pointwise sum)}.
\end{aligned}$$

For any abelian group A the set End (A) of all endomorphisms of A is thus an algebraic system with two binary operations, (pointwise) addition and multiplication (by composition), and a nullary operation, select the identity automorphism.

Theorem 6. *If A is an abelian group* End (A) *is a ring under the operations of pointwise sum, composition, and selection of* 1_A *as unit.*

Proof: The commutative law $f + g = g + f$ holds for pointwise addition because $(f + g)a = fa + ga$ is $ga + fa = (g + f)a$ for each $a \in A$. This and similar arguments show that End (A) is an abelian group under addition. Also, composition is associative and has the identity endomorphism 1_A as unit. It remains only to verify the two distributive laws by the separate calculations, for each $a \in A$, that

$$\begin{aligned}
[(f + g) \circ h]a &= (f + g)(ha) = f(ha) + g(ha) = (f \circ h + g \circ h)a,\\
[h \circ (f + g)]a &= h[(f + g)a] = h(fa + ga)\\
&= hfa + hga = (h \circ f + h \circ g)a.
\end{aligned}$$

Note especially that the second calculation uses the fact that h is a morphism of addition.

For example, it is easy to exhibit End ($\mathbf{Z}$). Recall that an endomorphism $f:\mathbf{Z} \to \mathbf{Z}$ of the additive group $\mathbf{Z}$ is completely determined by the image $f(1) = n$ of the generator $1 \in \mathbf{Z}$; moreover, pointwise addition of those endomorphisms adds these images and composition of endomorphisms multiplies them. Hence, the assignment $f \mapsto f(1)$ is an isomorphism End $(\mathbf{Z}) \cong \mathbf{Z}$ of rings (the first $\mathbf{Z}$ is the *additive group* of integers, the second the *ring* of integers).

Endomorphisms of the additive group $\mathbf{Z} \times \mathbf{Z}$ give something new. For instance, $f(r, s) = (s, r)$ and $g(r, s) = (r, 0)$ define particular endomorphisms f, g of $\mathbf{Z} \times \mathbf{Z}$ with $(f \circ g)(r, s) = (0, r)$ and $(g \circ f)(r, s) = (s, 0)$, and hence with $f \circ g \neq g \circ f$. In other words, the ring End $(\mathbf{Z} \times \mathbf{Z})$ is not commutative —our first such example.

Our final construction is that of the opposite ring. The *opposite* R^{op} of a ring R is defined to be the same set R with the same addition and the opposite binary operation of multiplication. Since the ring axioms are left-right symmetric (for example, *both* distributive laws are required) it follows, much as for opposite groups in Rule 6 of §II.2, that the opposite of a ring is a ring.

We have now constructed subrings, products of rings, function rings, endomorphism rings, and opposite rings. Later in the chapter we shall construct polynomial rings and quotient rings; in Chapter XVI we shall define tensor rings.

EXERCISES

1. Show that $\mathbf{Z}$ and $\mathbf{Z}_n$ have no proper subrings.
2. Show that $\mathbf{Z}_4 \times \mathbf{Z}_4$ has exactly three subrings.
3. Establish an isomorphism $\mathbf{Z}_6 \cong \mathbf{Z}_2 \times \mathbf{Z}_3$ of rings.
4. In the ring $\mathbf{C}$ of all complex numbers, determine which of the following subsets are subrings:
   (a) $\{m + n\sqrt{5} \mid m, n \in \mathbf{Z}\}$.
   (b) $\{m + n\sqrt[3]{5} \mid m, n \in \mathbf{Z}\}$.
   (c) $\{k + m\sqrt[3]{2} + n\sqrt[3]{4} \mid k, m, n \in \mathbf{Z}\}$.
5. In the ring $\mathbf{Z}_{10}$, show that the multiples of 5 form a ring isomorphic to $\mathbf{Z}_2$ which is not a subring of $\mathbf{Z}_{10}$ (different unit!).
6. For $\mathbf{2} = \{1, 2\}$, show that $f \mapsto (f(1), f(2))$ is an isomorphism $R^{\mathbf{2}} \cong R \times R$ of rings.
7. Obtain a ring isomorphism $(R \times R')^X \cong R^X \times R'^X$.
8. If X is a set and M a multiplicative monoid, show that the function set M^X of all $f:X \to M$ is a monoid under pointwise multiplication and unit.
9. (a) If X is a set and G a multiplicative group, show that the function set G^X of all functions $f:X \to G$ is a group under pointwise multiplication, abelian when G is abelian.

(b) Prove the group isomorphisms $G^1 \cong G$ and $G^2 \cong G \times G$.

10. Let S, T be sets with binary operations $\square$ and $\square'$. On the product set $S \times T$, define a binary operation $\square''$ by the formula $(s_1, t_1) \square'' (s_2, t_2) = (s_1 \square s_2, t_1 \square' t_2)$ for any $s_1, s_2 \in S$ and $t_1, t_2 \in T$, and assume $S, T \neq \varnothing$.

(a) Show that $\square''$ is commutative if and only if $\square$ and $\square'$ are both commutative.

(b) Show that $\square''$ is associative if and only if both $\square$ and $\square'$ are associative.

(c) Show that $\square''$ has a unit if and only if both $\square$ and $\square'$ have units.

3. Quotient Rings

For groups, as we have seen, certain subgroups are called "normal", the kernel of a morphism $G \to G'$ is a normal subgroup N of the domain G, and the quotient group G/N is constructed so that the projection $G \to G/N$ is a "universal" morphism on G with kernel N. Similarly, for rings, certain subsets will be called "ideals", the kernel of a morphism $R \to R'$ will be an ideal A in the domain R, and the quotient ring R/A will be so constructed that the projection $R \to R/A$ is a universal morphism of rings from R with kernel A.

The *kernel* of a morphism $a{:}R \to R'$ of rings is the set

$$\operatorname{Ker} \alpha = \{a \mid a \in R \text{ and } \alpha(a) = 0 \text{ in } R'\}; \tag{11}$$

it is just the kernel of α as a morphism of additive groups. Hence, $\alpha r = \alpha s$ for $r, s \in R$ if and only if $r - s \in \operatorname{Ker} \alpha$; the kernel tells which pairs of elements of R become equal under α. It is thus useful to notice what sorts of subsets of R can be kernels. Now Ker α is closed under subtraction and multiplication in R, but not under the operation "select the unit", so it is *not* a subring of R. However, $a \in \operatorname{Ker} \alpha$ and $r \in R$ imply that both products ra and ar are in Ker α. Hence, every kernel is an ideal in R, in the following sense:

DEFINITION. A (two-sided) ideal *A in a ring R is a nonvoid subset A of R with:*

(*i*) a_1 *and* $a_2 \in A \quad \Rightarrow \quad a_1 - a_2 \in A$.
(*ii*) $r \in R$ *and* $a \in A \quad \Rightarrow \quad ra \in A$ *and* $ar \in A$.

In other words, an ideal is an additive subgroup of R closed under all the operations "multiply by $r \in R$ on the left" or "on the right". (An additive subgroup closed only under all operations $a \mapsto ra$ is known as a "left ideal" in R.)

In a commutative ring K, the set $\{kb \mid k \in K\}$ of all "multiples" of a fixed element b in K is an ideal in K, denoted Kb or simply (b) and called a "principal" ideal. For example, in $\mathbf{Z}$ the kernel of the projection $\mathbf{Z} \to \mathbf{Z}_n$ is the ideal $n\mathbf{Z}$ consisting of all multiples of n.

Consider the set of all ideals of a ring R. If A and B are two ideals of R, so is their intersection $A \cap B$ as sets. The set

$$A + B = \{\text{all } a + b \mid a \in A \text{ and } b \in B\}$$

is also an ideal of R, since it is closed under subtraction and multiplication (left or right) by any $r \in R$. This set $A + B$ is called the *sum* of the ideals A and B. It is contained in any ideal of R which contains both A and B. In particular, the subset $\{0\} = 0$ consisting of 0 alone and the ring R itself are ideals in R. These two ideals 0 and R are called the *improper* ideals of R; all other ideals are called proper.

Now take a fixed ideal A in R. Since A is an additive subgroup of R, each $a \in A$ acts on R as $r \mapsto a + r$. The orbit of each element $r \in R$ under this action is the coset

$$A + r = \{a + r \mid \text{all } a \in A\}$$

of the additive subgroup A in R. The set of all these cosets is the quotient set R/A; the projection $p: R \to R/A$ sends each $r \in R$ to its coset $A + r$.

THEOREM 7. *For each ideal A in a ring R there are unique binary operations of addition and multiplication on the quotient set R/A which make the projection $p: R \to R/A$ a morphism of both addition and multiplication. Under these operations R/A is a ring with unit $p(1)$ and p is an epimorphism of rings with kernel A.*

Proof: For addition this result is known, from the construction in §II.10 of the additive quotient group R/A; the sum of two cosets C and D is the set of all sums $c + d$ with $c \in C$ and $d \in D$. Explicitly,

$$(A + r) + (A + s) = A + (r + s); \tag{12}$$

this formula shows that the projection $r \mapsto A + r$ is a morphism of addition.

Now consider the product. Given two cosets C and D, we claim that there is a unique coset, to be called CD, which contains (perhaps properly) all products cd of an element $c \in C$ by a $d \in D$. For, if $C = A + r$ and $D = A + s$, such a product cd is

$$(a + r)(a' + s) = (aa' + as + ra') + rs, \qquad a, a' \in A.$$

Because A is an ideal, each of the first three terms on the right is an element of A, so the whole sum does lie in the coset of A containing rs, as claimed. Explicitly, this product of two cosets is

$$(A + r)(A + s) = A + rs, \tag{13}$$

for any choice of r in the coset $C = A + r$ and any choice of s in $D = A + s$. With this product, $r \mapsto A + r$ is a morphism p of multiplication, and this is the only definition of a product of cosets which makes p such a morphism.

Finally, with these operations R/A is a ring. We already know that R/A is an abelian group under addition. The distributive law $r[s + t] = rs + rt$ holds in R; apply the morphism p of addition and multiplication to get

$$(A + r)[(A + s) + (A + t)] = (A + r)(A + s) + (A + r)(A + t);$$

this is the same distributive law, valid for any cosets in R/A. The same easy argument proves the other distributive law and the associativity of multiplication, and shows the coset $A + 1$ to be a multiplicative unit of R/A. Thus $p: R \to R/A$ is indeed a morphism of rings. Since the zero of R/A is the coset $A + 0$ consisting of the elements of the given ideal A, the kernel of this morphism is precisely A, as asserted.

In case B is another ideal in R with $B \supset A$, the formulas (12) and (13) for the addition and multiplication of cosets show that the set

$$B/A = \{\text{all cosets } A + b \mid b \in B\} \tag{14}$$

is a two-sided ideal in R/A. It is the image p_*B of B under the projection p. Every ideal of the quotient ring R/A has the form B/A for some such B.

The *quotient ring* R/A (often called the "residue class ring") can be characterized by the following property of the projection p.

Theorem 8 (*Main Theorem on Quotient Rings*). *Let A be any ideal in the ring R. To each morphism $\alpha: R \to S$ of rings with $A \subset \operatorname{Ker} \alpha$ there exists a unique morphism $\alpha': R/A \to S$ of rings such that the composite $\alpha' \circ p$ is the given morphism α.*

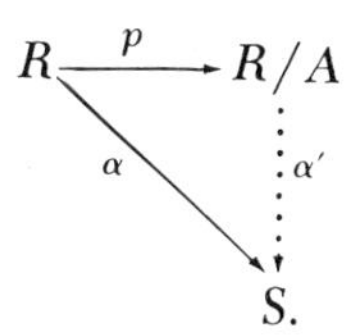

For this morphism α',

$$\operatorname{Im} \alpha' = \operatorname{Im} \alpha, \qquad \operatorname{Ker} \alpha' = (\operatorname{Ker} \alpha)/A. \tag{15}$$

This theorem says in effect that the projection $p: R \to R/A$ on the quotient ring is "universal" among all morphisms from R which kill the ideal A.

Proof: The requirement that $\alpha' \circ p = \alpha$ already determines α' as the function with $\alpha'(A + r) = \alpha(r)$, this for any representative r of the coset $A + r$. This formula immediately shows α' a morphism for unit, for addition, and for multiplication, and shows that its image is that of α. Its kernel is the set of all those cosets $A + r$ with $\alpha(r) = 0$, so is the set of all cosets $A + b$ with $b \in \text{Ker } \alpha$. In the notation (14), this kernel is exactly the ideal $(\text{Ker } \alpha)/A$ of R/A. The proof is complete.

In particular, if $\text{Ker } \alpha = A$, then α' is a monomorphism. A quicker but less complete statement of this case is

$$\text{Im}(\alpha) \cong \text{Domain } \alpha/\text{Ker } \alpha, \qquad \alpha \text{ a morphism of rings.} \tag{16}$$

In particular, the *codomain of any epimorphism is isomorphic to a quotient ring of its domain*. For example, consider the projection $R \times R' \to R$ of the product of two rings R, R' onto the first factor. The kernel, call it $0 \times R'$, is the set of all pairs $(0, r')$ for $r' \in R'$; hence, $R \cong (R \times R')/(0 \times R')$. As another example, take the morphism $\mathbf{Z}_{12} \to \mathbf{Z}_3$ given for each $n \in \mathbf{Z}$ as $12\mathbf{Z} + n \mapsto 3\mathbf{Z} + n$. This is an epimorphism with kernel the set $3\mathbf{Z}_{12}$ of all multiples of 3 in $\mathbf{Z}_{12}$. This kernel is an ideal in $\mathbf{Z}_{12}$ and, by (16), $\mathbf{Z}_{12}/3\mathbf{Z}_{12} \cong \mathbf{Z}_3$.

EXERCISES

1. Display the inclusion diagram of all ideals of the ring $\mathbf{Z}_{48}$.

2. If a ring R contains two ideals B and C with $B + C = R$ and $B \cap C = 0$, prove that B and C are rings and $R \cong B \times C$.

3. Show that every ideal in $\mathbf{Z}$ is the set of all multiples of some integer n.

4. Find all ideals in the product $\mathbf{Z} \times \mathbf{Z}$.

5. Show that the binary operations of intersection and sum of ideals are each associative and each commutative.

6. Prove both of the following isomorphisms of rings: $\mathbf{Z}_{96}/6\mathbf{Z}_{96} \cong \mathbf{Z}_6$; $\mathbf{Z}_{48}/4\mathbf{Z}_{48} \cong \mathbf{Z}_4$.

7. Prove: A nonvoid subset of a commutative ring K is an ideal if and only if it contains with each pair of elements b_1 and b_2 all elements of the form $\kappa_1 b_1 + \kappa_2 b_2$, for $\kappa_1, \kappa_2 \in K$.

8. Let $\alpha: R \to R'$ be a morphism of rings.

(a) If S is a subring of R, prove its image $\alpha_*(S)$ a subring of R'.

(b) Show that $S \mapsto \alpha_* S$ is a bijection from the set of all subrings of R containing $\text{Ker } \alpha$ to the set of all subrings of R' contained in $\text{Im } \alpha$.

9. Let $f:(S, \square) \to (T, \square')$ be an epimorphism of binary operations $\square$, $\square'$.

(a) Show that $\square$ commutative implies $\square'$ commutative.

(b) Show that $\square$ associative implies $\square'$ associative.

(c) Show by an example that the analogous result is not true for the cancellation law.

4. Integral Domains and Fields

The rest of this chapter will deal with commutative rings K.

A familiar property of the ring $\mathbf{Z}$ of integers is

$$\text{if } mn = 0, \text{ then either } m = 0 \text{ or } n = 0. \tag{17}$$

This property fails in many other rings. Thus in the product ring $\mathbf{Z} \times \mathbf{Z}$, the non-zero elements $r = (1, 0)$ and $s = (0, 1)$ have product zero.

DEFINITION. *Two elements a, b of a commutative ring K are* zero divisors *in K when $a \neq 0$, $b \neq 0$, and $ab = 0$. An* integral domain *is a nontrivial commutative ring with no zero divisors.*

In other words, a commutative ring is an integral domain when $0 \neq 1$ and the implication (17) holds.

Proposition 9. *A nontrivial commutative ring K is an integral domain if and only if it satisfies the* cancellation law: *For all a, b, c in K,*

$$ab = ac \quad \text{and} \quad a \neq 0 \quad \Rightarrow \quad b = c. \tag{18}$$

Proof: First assume the cancellation law (18). If $ab = 0$, then $ab = a0$, so either $a = 0$ or a can be canceled to give $b = 0$, as in (17). Conversely, assume (17). Then $ab = ac$ and $a \neq 0$ give $a(b - c) = 0$, therefore by (17) $b - c = 0$ and hence $b = c$ as in (18).

DEFINITION. A field F *is a nontrivial commutative ring in which every non-zero element a has a multiplicative inverse a^{-1}.*

This implies that every $a \neq 0$ in a field can be canceled, so a field is an integral domain—but not necessarily conversely.

In a field F the set F^* of all non-zero elements is an abelian group under multiplication; in particular, the inverse a^{-1} of each element $a \neq 0$ is unique. The product ab^{-1} will be written in the usual way as the quotient a/b. This quotient $x = a/b$ is the unique solution x of the equation $bx = a$. The equality of two quotients (with a, $c \neq 0$) is governed by

$$a/b = c/d \quad \Leftrightarrow \quad ad = bc. \tag{19}$$

Moreover, sum, product, negative, and inverse of quotients are given by

$$(a/b) + (c/d) = (ad + bc)/bd, \qquad b, d \neq 0, \tag{20}$$

$$(a/b)(c/d) = (ac)/(bd), \qquad b, d \neq 0, \tag{21}$$

$$-(a/b) = (-a)/b = a/(-b), \qquad b \neq 0, \tag{22}$$

$$(a/b)^{-1} = b/a, \qquad a, b \neq 0. \tag{23}$$

For example, to prove (20), let $x = a/b$ and $y = c/d$ denote the solutions of the equations $bx = a$ and $dy = c$. These equations may be combined to give

$$dbx = da, \qquad bdy = bc, \qquad bd(x + y) = da + bc.$$

Thus $x + y$ is the unique solution $t = (da + bc)/(bd)$ of the equation $(bd)t = da + bc$; this is (20).

The rationals $\mathbf{Q}$, the reals $\mathbf{R}$, and the complex numbers $\mathbf{C}$ are fields; both $\mathbf{R}$ and $\mathbf{C}$ have many subfields. Here a *subfield* S of a field F is defined to be a subring of F which is itself a field. Theorem 4 for subrings has as a corollary the corresponding result for subfields:

Theorem 10. *A subset of a field F is a subfield if and only if it is closed under the operations multiplicative unit, subtraction, multiplication, and multiplicative inverse (of non-zero elements).*

Proposition 11. *A nontrivial commutative ring is a field if and only if it has no proper ideals.*

Proof: Suppose first that F is a field. If A is an ideal in F, $A \neq 0$ gives an element $a \neq 0$ in A, so each $r \in F$ is $r = (ra^{-1})a \in A$, and $F = A$: The only ideals of F are the improper ideals 0 and F.

Conversely, suppose that L is a commutative ring with no proper ideals. If $a \neq 0$ is an element of L, the ideal La of all multiples of a must be all of L. In particular, the unit element 1 is a multiple $1 = ba$, so each $a \neq 0$ has an inverse b in L, and therefore L is a field.

If F and F' are fields, a "morphism of fields" is a function $\alpha : F \to F'$ which is a morphism of rings. Then the kernel of α, as an ideal in F, must be improper. Now Ker $\alpha = F$ would imply $\alpha(1) = 0$ in F', but $\alpha(1)$ is also the unit $1'$ of F', a contradiction to $0 \neq 1'$ in the nontrivial ring F'. Hence, the kernel of α must be the ideal 0; in other words, any morphism of fields is a monomorphism.

A *division ring* D is defined to be a nontrivial ring (not necessarily commutative) in which every non-zero element has a two-sided multiplicative inverse. A commutative division ring is thus the same thing as a field. An example of a noncommutative division ring is furnished by the quaternions, to be discussed in Chapter VIII.

EXERCISES

1. In any field, prove (19), (21), (22), and (23).

2. For b, c, d all non-zero in a field, prove that $(a/b)/(c/d) = (ad)/(bc)$.

3. Which of $\mathbf{Z}_3$, $\mathbf{Z}_4$, $\mathbf{Z}_5$, and $\mathbf{Z}_6$ are fields, and why?

4. Show that any finite integral domain is a field.

5. Show that every subring of a field is an integral domain.

6. Show that a subset of a finite field is a subfield if and only if it is closed under addition and multiplication and contains more than one element.

7. Show that any morphism $\alpha: F \to F'$ of fields is a morphism of quotients, in the sense that $b \neq 0$ and a in F give $\alpha(a/b) = (\alpha a)/(\alpha b)$ in F'.

8. Assume that $\mathbf{Q} \subset \mathbf{R}$ are fields:

(a) Prove for $d = 7$ or 11 that the set $\mathbf{Q}(\sqrt{d}\,)$ of all real numbers $a + b\sqrt{d}$ for $a, b \in \mathbf{Q}$ is a field.

(b) Show that the function $a + b\sqrt{7} \mapsto a + b\sqrt{11}$ is *not* an isomorphism $\mathbf{Q}(\sqrt{7}\,) \cong \mathbf{Q}(\sqrt{11}\,)$.

*(c) Prove that there is no isomorphism $\mathbf{Q}(\sqrt{7}\,) \cong \mathbf{Q}(\sqrt{11}\,)$ of fields.

5. The Field of Quotients

The integral domain $\mathbf{Z}$ is not itself a field, but it is contained in the familiar field $\mathbf{Q}$ of all rational numbers. Now each rational number $x = m/n$, for $m, n \in \mathbf{Z}$ and $n \neq 0$, may be described as the solution x in $\mathbf{Q}$ of the equation $nx = m$ with coefficients m and n in $\mathbf{Z}$. This suggests that the field $\mathbf{Q}$ might be formally constructed from $\mathbf{Z}$ as the set of all solutions m/n to such equations. The field $\mathbf{Q}$ so constructed will turn out to be the "smallest" field containing $\mathbf{Z}$.

This construction of a field of quotients applies not just to the domain $\mathbf{Z}$ of integers, but to any integral domain D; it will embed that domain D in a field $Q(D)$, the *field of quotients* of D, which may be described as follows.

Theorem 12. *For each integral domain D there is a field $Q(D)$ and a monomorphism $j: D \to Q(D)$ of rings such that every element $x \in Q(D)$ is a quotient $(ja)/(jb)$, where a and $b \neq 0$ are elements of D. Moreover, any monomorphism $\alpha: D \to F$ on D to a field can be written as a composite $\alpha = \alpha' \circ j$ for a unique morphism $\alpha': Q(D) \to F$ of fields.*

Proof: Since the elements of the field $Q(D)$ are to be quotients a/b of elements a and $b \neq 0$ of D, we start with all such pairs (a, b), introduce for them an equality like the equality of quotients a/b, and define for them operations of addition and multiplication by the formulas (20) and (21) used for sums and products of actual quotients.

In detail, let D^* be the set of all non-zero elements of D. In the product set $D \times D^*$ define a relation of congruence by

$$(a, b) \equiv (a', b') \quad \Leftrightarrow \quad ab' = a'b. \tag{24}$$

A straightforward argument shows that this relation is reflexive, symmetric, and transitive, so it is an equivalence relation. Let $Q(D)$ be the quotient set $(D \times D^*)/(\equiv)$, so that each element of $Q(D)$ is the equivalence class of some ordered pair (a, b). Write $[a, b]$ for this equivalence class; hence, by the definition (24) of congruence,

$$[a, b] = [a', b'] \quad \Leftrightarrow \quad ab' = a'b. \tag{25}$$

Now define sums and products in $D \times D^*$ by the formulas

$$(a, b) + (c, d) = (ad + bc, bd), \qquad (a, b)(c, d) = (ac, bd);$$

note that D is an integral domain, so $b \neq 0$ and $d \neq 0$ imply $bd \neq 0$. We wish to also define sums and products of equivalence classes so that the projection

$$D \times D^* \to (D \times D^*)/(\equiv) \quad = \quad Q(D)$$

will be a morphism of both addition and multiplication. To do this, we must show that $(a, b) \equiv (a', b')$ implies both

$$(a, b) + (c, d) \equiv (a', b') + (c, d) \qquad \text{and} \qquad (a, b)(c, d) \equiv (a', b')(c, d).$$

For example, by the definitions of addition and of congruence, the first conclusion reads $(ad + bc)b'd = (a'd + b'c)bd$; this is a direct consequence of the hypothesis $ab' = a'b$. The second conclusion is proved similarly. We conclude that there are in $Q(D)$ binary operations of addition and multiplication, given on equivalence classes $[a, b]$ and $[c, d]$ by the formulas (compare Theorem I.19 and Exercise I.9.2)

$$[a, b] + [c, d] = [ad + bc, bd], \qquad [a, b][c, d] = [ac, bd]. \tag{26}$$

Next, we verify that $Q(D)$ is a field under these operations. Consider, for example, the distributive law. By the definition of sum and product we calculate that

$$([a, b] + [c, d])[r, s] = [adr + bcr, bds],$$
$$[a, b][r, s] + [c, d][r, s] = [ards + bscr, bsds];$$

the results are equal by (25). Similar arguments prove the other necessary identities (commutative and associative laws). Also, the class $[0, 1]$ is a zero for addition, the negative of the class $[a, b]$ is $-[a, b] = [-a, b]$, and $[1, 1]$ is the unit for multiplication. Hence, $Q(D)$ is a commutative ring. To show that it is a field, note that $[a, b] \neq [0, 1]$ means that $a \neq 0$ in D and hence that $[b, a]$ is a class in $Q(D)$, with $[b, a][a, b] = [ba, ab] = [1, 1]$; in other words, $[a, b] \neq 0$ has $[b, a]$ as its multiplicative inverse—as it should, since, for actual quotients, $(a/b)^{-1} = (b/a)$. This completes the proof that $Q(D)$ is a field.

The assignment $a \mapsto [a, 1]$ defines a function $j:D \to Q(D)$ which is injective, since $[a, 1] = [a', 1]$ if and only if $a = a'$ in D. The definitions (26) of addition and multiplication show that this function j is a morphism of both these operations. Since $j(1) = [1, 1]$ also, j is a morphism of rings. If $x = [a, b]$ is any element of $Q(D)$, then $[b, 1]x = [ab, b] = [a, 1]$, so x is a quotient $(ja)/(jb)$, as required in the first part of the theorem.

It remains to compare j with any morphism α on D to a field. If there is a morphism α' of fields with $\alpha' \circ j = \alpha$, as indicated in the following commutative diagram,

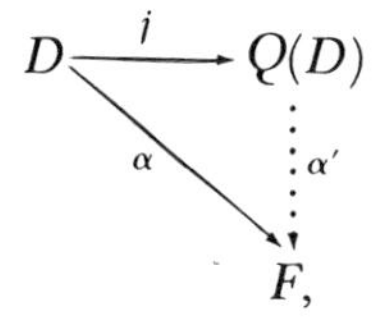

then α', j, and hence α must be injective and α' must satisfy

$$\alpha'[a, b] = \alpha'((ja)/(jb)) = (\alpha a)/(\alpha b), \qquad b \neq 0, \quad a \in D,$$

for any morphism of fields must carry quotients to quotients. Hence, α' is uniquely determined, if it exists. If α is injective, $[a, b] = [a', b']$ implies that $(\alpha a)/(\alpha b) = (\alpha a')/(\alpha b')$, so this formula does define a function $\alpha':Q(D) \to F$ which is a morphism of fields, because the rules for adding and multiplying equivalence classes $[a, b]$ are the same as those for quotients a/b. This shows that α' exists uniquely, as required in the theorem.

The property of $j:D \to Q(D)$ above is often expressed by stating that j is "universal" among monomorphisms $\alpha:D \to F$ of D to a field.

It is convenient to "identify" each element a of the original domain D with its image $j(a) = [a, 1] \in Q(D)$. Such an identification can be accomplished by modifying $Q(D)$ as follows. Replace each of the elements $[a, 1]$ of $Q(D)$ by the original element $a \in D$, leaving all the other elements of $Q(D)$ unaltered, and make the same replacements in each formula for a sum or a product, thus replacing the field $Q(D)$ by an isomorphic field. This identification has the effect of making the monomorphism $j:D \to Q(D)$ an insertion and hence D a subdomain of $Q(D)$. Hence, we have shown that *every integral domain D is a subdomain of a field $Q(D)$.*

In particular, the field $\mathbf{Q}$ of rational numbers is now defined to be the quotient field $Q(\mathbf{Z})$, containing $\mathbf{Z}$ as subdomain. It has all the familiar properties of the rational numbers.

The field $Q(D)$ of quotients may also be characterized (up to isomorphism, as usual) in the following terms:

Corollary. *If an integral domain D is a subring of a field F in which every element x is a quotient a/b of elements a and $b \neq 0$ of D, then $Q(D) \cong F$.*

Proof: The monomorphism $\alpha': Q(D) \to F$ of the theorem has image a subfield of F; this subfield must be all of F by the hypothesis on the elements $x \in F$.

EXERCISES

1. Prove in detail the commutative and the associative laws for multiplication in $Q(D)$.
2. Prove that the relation defined by (24) is reflexive, symmetric, and transitive.
3. If F is a field, prove that $Q(F) \cong F$.
4. For G the ring of gaussian integers of §2, construct an isomorphism of the quotient field $Q(G)$ to the ring of all complex numbers $r + si$ for r, s rational numbers.
5. Show that $D \cong D'$ implies $Q(D) \cong Q(D')$.
6. Show that any monomorphism $D \to D'$ of domains yields a corresponding monomorphism $Q(D) \to Q(D')$ of fields.

*7. Prove that any rational number $r/s \neq 0$ can be expressed uniquely in the form $r/s = b_1 + b_2/2! + b_3/3! + \cdots + b_n/n!$, where n is a suitable integer, and each b_k is an integer, with $0 \leqslant b_k < k$ if $k > 1$, and $b_n \neq 0$.

6. Polynomials

A polynomial with integral coefficients is an expression such as

$$f = 3 - 7x + 2x^2 + 5x^3 - 9x^4.$$

On the one hand, such an expression is a shorthand for a function: To find its values, substitute numbers in place of x. On the other hand, polynomials are formal expressions which may themselves be added and multiplied; thus, given two quadratic polynomials f and g, their sum is the quadratic polynomial

$$(f_0 + f_1x + f_2x^2) + (g_0 + g_1x + g_2x^2) = (f_0 + g_0) + (f_1 + g_1)x + (f_2 + g_2)x^2,$$

while their product is a polynomial of fourth degree:

$$f_0g_0 + (f_0g_1 + f_1g_0)x + (f_0g_2 + f_1g_1 + f_2g_0)x^2 + (f_1g_2 + f_2g_1)x^3 + f_2g_2x^4.$$

With these two operations, the polynomials themselves form a ring, as we shall show, and this when the "coefficients" f_i and g_j are taken from any commutative ring K.

In order to do this objectively, without assuming that we know what a polynomial is, we will replace the usual polynomial expression $f_0 + f_1 x + \cdots + f_k x^k$ by the list $(f_0, f_1, \cdots, f_k)$ of its coefficients. Instead of using lists like this with $k + 1$ terms for various degrees k, we might as well add zeros beyond f_k to get an infinite sequence $(f_0, f_1, \cdots, f_k, 0, 0, \cdots)$.

We consider only those sequences $f:\mathbf{N} \to K$ which are *ultimately zero*, in the sense that there is some k with $f_n = 0$ for all $n > k$. The sum of two such sequences f and g is then defined by the formula (pointwise sum!)

$$(f + g)_n = f_n + g_n, \qquad n = 0, 1, 2, \cdots . \tag{27}$$

The *product* (often called the "convolution" product) of two such sequences f and g is the sequence $f * g:\mathbf{N} \to K$ defined for all $n \in N$ as

$$(f * g)_n = f_0 g_n + f_1 g_{n-1} + f_2 g_{n-2} + \cdots + f_n g_0. \tag{28}$$

In briefer notation,

$$(f * g)_n = \sum_{k+m=n} f_k g_m = \sum_{k=0}^{n} f_k g_{n-k}.$$

(This does correspond exactly to the usual product of two polynomial expressions f and g.)

Theorem 13. *For any commutative ring K, the set S of all sequences $f:\mathbf{N} \to K$ which are ultimately zero is a commutative ring under pointwise sum and convolution product, with unit the sequence* $(1, 0, 0, \cdots)$.

Proof: It is straightforward to check that these sequences form an abelian group under pointwise addition. Since the product in K is commutative, so is the convolution product (28), while the special sequence $1':\mathbf{N} \to K$ with $1'_0 = 1$ and $1'_n = 0$ for $n > 0$ serves as a unit $(1, 0, 0, \cdots)$ for this product. To prove the convolution associative, use the definition (28) twice to calculate the value of a triple product $(f * g) * h$ at n to be

$$[(f * g) * h]_n = \sum_{k+m=n} (f * g)_m h_k = \sum_{k+m=n} \Big(\sum_{i+j=m} f_i g_j \Big) h_k = \sum_{i+j+k=n} f_i g_j h_k,$$

where in the last step we have used the "general" distributive law for a sum of $m + 1$ terms. A similar calculation of $[f * (g * h)]_n$ comes out with this same value in K; thus the two functions $(f * g) * h$ and $f * (g * h)$ are equal. An analogous argument establishes the distributive law.

This formally defined ring of sequences is really just the usual ring of polynomial expressions in a symbol x. To see this, we write x for the special sequence $x = (0, 1, 0, 0, \cdots)$. Also, for each element a in the original ring K we write a sequence $a' = (a, 0, 0, \cdots)$. Clearly $(a + b)' = a' + b'$ and $(ab)' = a' * b'$, while $1' = (1, 0, 0, \cdots)$ is the unit of S; this means that the assignment $a \mapsto a'$ is a monomorphism $K \to S$ of rings, which makes the original ring K isomorphic to its image, a subring of the ring S of sequences. Henceforth, we drop the prime, write $a' = a$, and so identify K with a subring of S.

The special sequence $x = (0, 1, 0, \cdots)$ is called the *indeterminate* x. For each $n \in N$ the (convolution) power $x^n = x * \cdots * x$ with n factors is the sequence

$$x^n = (0, \cdots, 0, 1, 0, \cdots) \qquad \text{(1 in position } n + 1),$$

as one may prove by induction on n. The convolution product $a * x^n$ is therefore the sequence $(0, \cdots, 0, a, 0, \cdots)$, with a in the position $n + 1$. Consequently, any sequence $(f_0, \cdots, f_n, \cdots)$, zero beyond $n = k$, can be rewritten as

$$\begin{aligned}(f_0, f_1, \cdots, f_n, \cdots) &= (f_0, 0, \cdots) + (0, f_1, 0, \cdots) + \cdots \\ &\qquad + (0, 0, \cdots, f_k, 0, \cdots) \\ &= f_0 + f_1 * x + \cdots + f_k * x^k,\end{aligned}$$

where x is the special sequence above, the sum is the pointwise sum, and the product (and powers) are formed by convolution. But this is then just the usual polynomial "expression". So we might as well write the convolution product as fg, without the $*$; and call the sequence $f = f_0 + \cdots + f_k x^k$ a polynomial. If $g = g_0 + \cdots + g_l x^l$ is another polynomial, the (convolution) product fg is

$$fg = f_0 g_0 + (f_0 g_1 + f_1 g_0)x + \cdots + f_k g_l x^{k+l}.$$

This is the "usual" formula. Theorem 13 becomes

Theorem 14. *For each commutative ring K, the polynomials form a commutative ring which contains the ring K (more exactly, a subring identified with K according to a natural isomorphism) and an element x such that any polynomial $f \neq 0$ can be expressed uniquely in the form*

$$f = f_0 + f_1 x + \cdots + f_k x^k, \tag{29}$$

with elements $f_n \in K$ as coefficients and $f_k \neq 0$.

We now write $K[x]$ for this ring of polynomials.

The *degree* of the polynomial $f \neq 0$ is this integer k; it can be described as the smallest natural number m such that $f_n = 0$ for all $n > m$. Polynomials of degrees 1, 2, 3, 4, 5, are said to be *linear*, *quadratic*, *cubic*, *quartic*, and *quintic*. The constants $a \in K$ are polynomials of degree zero, except for the constant 0, sometimes said to have degree $-\infty$. We write $\deg f$ for the degree of f.

The *leading coefficient* of a polynomial f of degree k is the constant f_k, while f is called *monic* if its leading coefficient is $f_k = 1$. The formula above for the product of two polynomials shows that the product of monic polynomials is monic and that

$$\deg(f + g) \leqslant \max(\deg f, \deg g), \qquad \deg(fg) \leqslant (\deg f) + (\deg g). \tag{30}$$

If K is an integral domain, the second inequality may be replaced by an equality. This shows that if K is an integral domain, so is $K[x]$.

Given K, the polynomial ring $K[x]$ is the most general possible commutative ring containing the given ring K and one new element x; we say that this is a "universal" way of adjoining a new element x to K. To express this formally, we use the process of substituting an arbitrary constant c for the indeterminate x; this process turns out to be a morphism E_c of rings.

Theorem 15. *Let a commutative ring L contain K as a subring. To each $c \in L$ there is a unique morphism $E_c : K[x] \to L$ of rings with*

$$E_c(a) = a \qquad \textit{for all } a \in K, \qquad\qquad E_c(x) = c. \tag{31}$$

Proof: Suppose first that there is such a morphism E_c. Write each polynomial $f \in K[x]$ in the standard form (29). By the assumptions (31), $E_c(f_i) = f_i$ for each coefficient. Since E_c is a morphism, $E_c(x^i) = c^i$ and

$$E_c(f_0 + f_1x + \cdots + f_kx^k) = f_0 + f_1c + \cdots + f_kc^k. \tag{32}$$

This proves that E_c, if it exists, is uniquely determined by this formula.

Conversely, given c, we define a function $E_c : K[x] \to L$ by the formula (32). This function E_c is evidently a morphism for addition; as for the (convolution) product, note that this product may be written as a double sum,

$$f * g = \left(\sum_i f_ix^i\right)\left(\sum_j g_jx^j\right) = \sum_i \sum_j f_ig_jx^{i+j}. \tag{33}$$

But the definition of E_c and the general distributive law give

$$E_c(f * g) = \sum_i \sum_j f_ig_jc^{i+j} = \left(\sum_i f_ic^i\right)\left(\sum_j g_jc^j\right) = (E_cf)(E_cg).$$

This shows E_c a morphism for multiplication and so completes the proof.

The morphism $E_c : K[x] \to L$ of (32) acts on the polynomial f by replacing x by c; hence, we call it *evaluation* of f at $x = c$, or *substitution* of c for x in f. In this language, the conditions (31) of the theorem require that any substitution in a "constant" a yields that constant, while substitution of c for x in x yields c and substitution is a morphism of rings.

Proposition 16. *To each morphism $\alpha : K \to K'$ of commutative rings there is a unique morphism $\alpha_\# : K[x] \to K'[x]$ with $\alpha_\#(x) = x$ and $\alpha_\#(a) = \alpha(a)$ for each $a \in K$.*

Note that we use the same letter x for an indeterminate over different rings K, K'.

Proof: Since any morphism $\alpha_\#$ must preserve both sums and products, the only possible choice for $\alpha_\#$ is

$$\alpha_\#(f_0 + f_1 x + \cdots + f_k x^k) = (\alpha f_0) + (\alpha f_1)x + \cdots + (\alpha f_k)x^k.$$

Thus $\alpha_\#$ simply replaces each coefficient f_i by its image αf_i. One then verifies that this $\alpha_\#$ is a morphism of addition, of multiplication, and of unit.

EXERCISES

1. Calculate the following products in $\mathbf{Z}_5[x]$:

$$(3x^2 + 2x - 4)(4x^3 - 2x + 3), \qquad (3x^6 - 2x^2 + 1)(7x^2 + 2x + 2).$$

2. The formal derivative of a polynomial $f = f_0 + f_1 x + \cdots + f_k x^k$ is defined to be the polynomial $f' = f_1 + 2f_2 x + \cdots + kf_k x^{k-1}$. Prove, for any commutative ring of coefficients, that:

(a) $(af)' = af'$. (b) $(f + g)' = f' + g'$.
(c) $(fg)' = fg' + f'g$. (d) $(f^n)' = nf^{n-1}f'$.

3. When does equality hold for the first inequality (30) on degrees?

***4.** Show that, if F is a field, the group of all those automorphisms of $F[x]$ which leave all elements of F fixed, consists of substitutions given by $x \mapsto ax + b$, $a \neq 0$ and b in F.

5. If a ring R is not commutative, an element c is called *central* in R if $cr = rc$ for every $r \in R$. Construct from R a ring $R[x]$ containing R as a subring and a central element x such that to any ring R' with a central element c and containing R as a subring there is a unique morphism $E_c : R[x] \to R'$ with $E_c(r) = r$ for all $r \in R$ and $E_c(x) = c$.

6. If S is an additive monoid, the *support* of a function $f : \mathrm{S} \to K$ is the set of all those $s \in \mathrm{S}$ with $f(s) \neq 0$. Show that the set of all functions f with finite support constitute a ring $K^{(\mathrm{S})}$ under pointwise sum and convolution product, with a suitable unit, and construct a monomorphism $K \to K^{(\mathrm{S})}$ of rings. (*Hint:* If S is the additive monoid $\mathbf{N}$, $K^{(\mathrm{S})}$ is the ring $K[x]$.) The ring $K^{(\mathrm{S})}$ is called the "monoid ring" for S over K.

7. Polynomials as Functions

Polynomials have been introduced formally, as sequences; that is, as expressions in an indeterminate x which can be added and multiplied so as to form a ring. The same polynomials can be regarded as functions. Indeed, the symbol $E_c f$ means "evaluate the polynomial f for $x = c$". For each polynomial f, the assignment $c \mapsto E_c f$ is thus a function $K \to K$, which we call the function $f^{\lq}: K \to K$. Thus $f^{\lq}c = E_c f$; in other words,

$$f = f_0 + f_1 x + \cdots + f_k x^k \quad \text{gives} \quad f^{\lq}c = f_0 + f_1 c + \cdots + f_k c^k. \tag{34}$$

Clearly, $f^{\lq}$ is exactly the function usually associated with the polynomial expression f. For example, the function $x^{\lq}$ is just the identity function $1: K \to K$. Any function on K to K which can be written as a function $f^{\lq}$ for some $f \in K[x]$ is called a *polynomial function*. Most functions $K \to K$ are not polynomial functions: For example, if K is the field of real numbers, the exponential function $x \mapsto e^x$ is not a polynomial function. Note again how we have analyzed the usual idea of a "polynomial" $f(x)$. Our f stands for the function $f: \mathbf{N} \to K$, which is the sequence of coefficients in the usual polynomial expression, while our $f^{\lq}: K \to K$ is the function had by evaluating this expression. We do not use the familiar notation $f(x)$, which ordinarily stands ambiguously both for the expression of a function and for the corresponding evaluation. After this section we will drop the special notation $f^{\lq}$ and write $f(c)$ for $f^{\lq}c$.

THEOREM 17. *Let K be a commutative ring. The assignment $f \mapsto f^{\lq}$ which sends each polynomial f to the associated polynomial function $f^{\lq}: K \to K$ is a morphism $K[x] \to K^K$ of rings, where K^K is the function ring with pointwise unit, pointwise sum, and pointwise product.*

The image of this morphism is called the *ring of polynomial functions*.

Proof: For each $c \in K$, $E_c: K[x] \to K$ is a morphism of rings, so

$$E_c(f + g) = E_c f + E_c g, \qquad E_c(fg) = (E_c f)(E_c g), \qquad E_c 1 = 1.$$

Since we now write $f^{\lq}c$ for $E_c f$, these equations become

$$(f + g)^{\lq}c = f^{\lq}c + g^{\lq}c, \qquad (fg)^{\lq}c = (f^{\lq}c)(g^{\lq}c), \qquad 1^{\lq}c = 1.$$

The first two state that the function $(f + g)^{\lq}$ or $(fg)^{\lq}$ is the pointwise sum or the pointwise product, respectively, of the functions $f^{\lq}$ and $g^{\lq}$, while the third states that the unit 1 of $K[x]$ yields the pointwise unit function. Therefore, $f \mapsto f^{\lq}$ is a morphism of rings, as required.

Observe also that each element $a \in K$, regarded as a polynomial $a' \in K[x]$, yields a function $(a')^{‘}$ which is a *constant function* $K \to K$; namely, the function which assigns to every $c \in K$ the value a.

This morphism $f \mapsto f^{‘}$ need not be a monomorphism; in other words, *different* polynomials may determine the *same* polynomial function. This necessarily occurs when the (nontrivial) commutative ring K is finite, for there are then infinitely many polynomials (at least one for each degree) and only a finite number of functions $K \to K$. Thus, if $K = \mathbf{Z}_2$, $1 + 1 = 0$, so $1^2 + 1 = 0$; the polynomial $f = x + x^2$ evaluates to zero at each $c \in \mathbf{Z}_2$, so determines the same polynomial function as does the zero polynomial. However, we shall soon see (Corollary 4 of Theorem 19) that K an infinite integral domain makes $f \mapsto f^{‘}$ a monomorphism.

The polynomial ring $K[x]$ has another "universal" property:

THEOREM 18. *If d is an element of a commutative ring L, each morphism $\alpha : K \to L$ of commutative rings extends to one and only one morphism $\alpha' : K[x] \to L$ of commutative rings with $\alpha'(x) = d$.*

Proof: Take the morphism $\alpha_{\#} : K[x] \to L[x]$ of Proposition 16 and the evaluation E_d; their composite

$$K[x] \xrightarrow{\alpha_{\#}} L[x] \xrightarrow{E_d} L$$

is a morphism $\alpha' = E_d \alpha_{\#}$ with the properties $\alpha'(a) = \alpha(a)$ for $a \in K$ and $\alpha'(x) = d$, as required. Explicitly, for any polynomial f,

$$\alpha'(f_0 + f_1 x + \cdots + f_k x^k) = (\alpha f_0) + (\alpha f_1) d + \cdots + (\alpha f_k) d^k;$$

α' simultaneously replaces x by d and acts on the coefficients by α. This formula shows α' unique, hence completes the proof.

Note that this theorem includes Theorem 15 and Proposition 16 as special cases. This will be used in §IV.4 for polynomials in two indeterminates.

The evaluation map $E_c : K[x] \to L$ with $c \in L$ was used with $L = K$ to represent each polynomial f as a function $f^{‘}$, with $f^{‘}(c) = E_c f$. More generally, take $L = K[x]$ and c an element of $K[x]$; that is, $c = g$ a polynomial. Then the morphism $E_g : K[x] \to K[x]$ sends x to g and any polynomial f in x to the corresponding polynomial with x replaced by g; that is,

$$E_g(f_0 + f_1 x + \cdots + f_k x^k) = f_0 + f_1 g + \cdots + f_k g^k.$$

Thus $E_g f$ acts like a composite function; see Exercise 7.

EXERCISES

1. For $a \in K$, show that E_{x+a} is an automorphism of $K[x]$.
2. For $a, b \in K$, show that E_{ax+b} is an endomorphism of $K[x]$, and find

conditions on a and b necessary and sufficient to make it an automorphism.

3. (a) Show that any endomorphism of $\mathbf{Q}[x]$ is E_g for some g.

(b) Find all automorphisms of $\mathbf{Q}[x]$.

4. Prove: Any endomorphism of $K[x]$ which is the identity on K is E_g for some $g \in K[x]$.

5. With the notation (b) for the ideal of all multiples of b in a commutative ring, as in §3, establish the following isomorphisms for quotient rings:

(a) $K[x]/(x + 1) \cong K$. (b) $\mathbf{Z}[x]/(3) \cong \mathbf{Z}_3[x]$.

6. For each natural number n, prove $E_{x^n}:K[x] \to K[x]$ a monomorphism of rings.

7. Use the uniqueness property of evaluation to prove that $(E_g f)`(c) = f`(g`c)$.

8. The Division Algorithm

We will soon prove from our assumptions on $\mathbf{N}$ that every natural number not 0 or 1 can be factored uniquely into positive primes, and will show that much the same argument applies also to polynomials in one indeterminate over a field. To obtain these results, we start with certain properties of divisibility which apply to the elements of any integral domain D.

In any commutative ring K the following definitions apply: b is *divisible* by a in K (in symbols, $a|b$) when there exists some $c \in K$ with $ac = b$, and b is *associate* to a in K when both $a|b$ and $b|a$. In particular, $u|1$ means that u has in K an inverse u^{-1}, hence is *invertible* in K; so the associates of the unit 1 are the invertible elements of K. The relation $a|b$ is reflexive and transitive, so that "a is associate to b" is an equivalence relation.

In an integral domain D, a and b are associates if and only if $a = bu$ for some invertible u. For $a|b$ and $b|a$ give elements c and d with $ac = b$ and $bd = a$; hence, $acd = a$. If $a \neq 0$, the cancellation law (18) gives $cd = 1$, so both c and d are invertible. If $a = 0$, then also $b = 0$, so that $a = b1$ with 1 invertible. Conversely, $a = bu$ with u invertible gives $b = au^{-1}$; hence, $a|b$ and $b|a$. An element $u \in D$ invertible in D is often called a "unit" of D; in this book, however, "unit" is reserved for *the* unit 1 for multiplication.

Every element b of the integral domain D is divisible by all its associates and by all invertibles (= all associates of 1). Together, these are called the *improper divisors* of b in D. If $b \neq 0$ has no proper divisors in D and is not invertible in D, it is called a *prime* in D.

In a field F, every non-zero element is invertible, and there are no prime elements.

In the ring $\mathbf{Z}$, the only invertibles are ± 1. Hence, two integers m and n are associates in $\mathbf{Z}$ if and only if $m = \pm n$. An integer $p \neq \pm 1$ is a prime in $\mathbf{Z}$ if and only if its only divisors are ± 1 and $\pm p$. Every prime is thus associate to a positive prime number; the first few positive primes in $\mathbf{Z}$ are 2, 3, 5, 7, 11, 13, 17, $\cdots$.

In the ring $D[x]$ of polynomials with coefficients in an integral domain D, the product fg of two polynomials f and g has degree the sum of $\deg f$ and $\deg g$ and leading coefficient the product of the leading coefficients of f and g. Hence a product fg can be 1 only when both factors are constants, each invertible in D. Therefore, the invertible polynomials of $D[x]$ are those constants c which are invertible in D, and two polynomials f and g are associates in $D[x]$ if and only if $g = cf$ for c an invertible constant of D. A non-constant polynomial which is prime in $D[x]$ is called *irreducible* over D. For example, all monic linear polynomials such as $x + b$ are irreducible as polynomials of $D[x]$. Over most domains D, there are irreducible polynomials of higher degree; for example, $x^2 + 1$ is irreducible in $\mathbf{Z}[x]$.

Consider now the ring $F[x]$ of polynomials with coefficients in a field F. In this ring the invertible elements are all the non-zero constants, and two polynomials f and f' are associates if and only if $f' = cf$ for a constant $c \neq 0$ in F. Every linear polynomial is irreducible, and every non-zero polynomial is associate to a monic polynomial of the same degree.

There is a division algorithm like that for integers (Theorem I.15) for polynomials over fields or commutative rings.

Theorem 19. *Let K be a commutative ring and g a polynomial in $K[x]$ with leading coefficient invertible in K. To each polynomial $f \in K[x]$ there is one and only one pair of polynomials q, r in $K[x]$ with*

$$f = qg + r \qquad \text{and} \qquad \deg r < \deg g. \tag{35}$$

Proof: The idea is to subtract from f successive multiples of g until the difference r has degree less than that of g. This idea can be turned into a proof by induction on the degree n of the polynomial f. As in the Second Principle of Mathematical Induction (Theorem I.11) we can assume that there is a representation (35) for any polynomial of degree less than n. Now take f of degree n. If n is less than $\deg g = m$, then $f = 0 \cdot g + f$ satisfies (35). Otherwise, $n = \deg f \geqslant m = \deg g$. Let f and g have the leading coefficients f_n and g_m, with g_m invertible in K by hypothesis. Then the product $(f_n g_m{}^{-1} x^{n-m})g$ has leading coefficient f_n and degree n. Hence, if we subtract from f this product by g, we get a difference f' of degree less than n, which by the induction assumption can be written as $f' = q'g + r'$. Then

$$f = \left(f_n g_m{}^{-1} x^{n-m} + q'\right)g + r', \qquad \deg r' < \deg g,$$

exactly as desired.

In (35), the "quotient" q and the "remainder" r are uniquely determined by f and g. For, $qg + r = f = \bar{q}g + \bar{r}$ implies that $r - \bar{r} = (\bar{q} - q)g$ with $\deg(r - \bar{r}) < \deg g$ and $g \neq 0$, hence $\bar{q} - q = 0$ and $r = \bar{r}$.

The remainder r can be zero (in which case its degree, by our convention, is $-\infty$); namely, precisely when g is a divisor of f in $K[x]$.

Corollary 1. *Let K be a commutative ring. If a polynomial $f \in K[x]$ is divided by a monic linear polynomial $g = x - c$, the remainder is $f^{\prime}c$.*

Proof: By the division algorithm,

$$f = q(x - c) + r,$$

where the degree of r is less than the degree 1 of $x - c$. Therefore, the polynomial r is a constant (an element of K). Evaluate both sides of this equation at c; since the evaluation E_c is a morphism and $x - c$ evaluates to zero, the result is $E_c(f) = f^{\prime}(c) = r$, exactly as desired.

In particular, this corollary gives

$$f^{\prime}(c) = 0 \quad \Leftrightarrow \quad r = 0 \quad \Leftrightarrow \quad f = (x - c)q \quad \Leftrightarrow \quad (x - c) | f \text{ in } K[x].$$

An element $c \in K$ with $f^{\prime}c = 0$ is called a *zero* or a *root* of f. We have proved

Corollary 2. *A polynomial f is divisible in $K[x]$ by the monic linear polynomial $x - c$ if and only if c is a zero of f.*

This is a basic principle: Zeros correspond to linear factors.

Corollary 3. *If D is an integral domain, a polynomial $f \in D[x]$ of degree k $(\geqslant 0)$ has at most k zeros in D.*

Proof: Let c be a zero of f. By the previous corollary, $f = (x - c)q$, where q has degree $k - 1$. By induction on k, q has at most $k - 1$ zeros in D; since D is an integral domain, a zero of $f = (x - c)q$ is either a zero of q or (the unique) zero c of $x - c$. Hence, f has at most k zeros.

Corollary 4. *If D is an infinite integral domain, then the morphism $D[x] \to D^D$ of Theorem 17 is a monomorphism.*

Proof: This is the morphism $f \mapsto f^{\prime}$ sending each polynomial f to the corresponding polynomial function $f^{\prime}$. Suppose it is not a monomorphism, so that there are polynomials $f \neq g$ with $f^\flat = g^\flat$. Then $f - g$ has some degree $k \geqslant 0$, so has at most k zeros in D. But $(f - g)^\flat = 0$, so $f - g$ has all the (infinitely many) elements of D as zeros, a contradiction.

EXERCISES

1. In any commutative ring, show that $a|b$ and $a|c$ imply $a|(b + c)$.

2. If the elements $a \neq 0$ and a' are associates in the integral domain D, prove $c|a \Leftrightarrow c|a'$; $a|b \Leftrightarrow a'|b$; a prime in $D \Leftrightarrow a'$ prime in D.

3. In any integral domain D, let E be the binary equivalence relation with aEb if and only if a is associate to b in D, while $p: D \to D/E$ is the projection of D to the corresponding quotient set D/E.

(a) Show that "$pa \leqslant pb$ if and only if $a|b$ in D" defines a binary relation $\leqslant$ on D/E which is reflexive and transitive.

(b) Show that there is a unique binary operation of multiplication in D/E which makes p a morphism of multiplicative monoids. (In particular this includes a proof that multiplication in D/E is associative and has a unit.)

4. Show that $x^2 - 1$ has four zeros in $\mathbf{Z}_{15}$. Why does not this result contradict Corollary 3?

5. Show that $x^2 + 1$ is irreducible as an element of $\mathbf{Q}[x]$ but reducible as an element of $\mathbf{Z}_5[x]$.

6. Show that $x^3 + x + 1$ is irreducible in $\mathbf{Z}_5[x]$.

7. List all the associates of $x^4 + 3x^2 - 2x + 4$ in $\mathbf{Z}_5[x]$.

8. If F is a field, show that a quadratic or a cubic polynomial f is irreducible in $F[x]$ if and only if it has no zero in F.

9. In the division algorithm for $\mathbf{Z}[x]$ with $f = 7x^4 + 2x^3 + 9x^2 + 5$, determine the quotient q and the remainder r in the following cases:

$$g = x^4, \qquad g = x + 4, \qquad g = x^2 + 2x, \qquad g = x^3 + x + 1.$$

***10.** Let D be a finite integral domain with exactly n distinct elements $c_1, \cdots, c_n$. Let d denote the polynomial $(x - c_1) \cdots (x - c_n)$.

(a) Prove: Two polynomials f and g have $f^\flat = g^\flat$ if and only if $d|(f - g)$.

(b) Compute the polynomial d for $D = \mathbf{Z}_3$ and $D = \mathbf{Z}_5$.

11. Let $c_0, \cdots, c_n$ be $n + 1$ distinct elements of a field F. For $i = 0, \cdots, n$, use the polynomial

$$q_i = (x - c_0) \cdots (x - c_{i-1})(x - c_{i+1}) \cdots (x - c_n)$$

to construct a polynomial p_i of degree n in $F[x]$ such that

$$p_i(c_i) = 1, \qquad p_i(c_j) = 0, \qquad i \neq j.$$

12. Given c_i as in Exercise 11 and elements $a_0, \cdots, a_n \in F$, prove that there is exactly one polynomial f of degree n in $F[x]$ with $f^\flat c_i = a_i$ for $i = 0, \cdots, n$. (This is the polynomial for *Lagrange interpolation*.)

13. If F is a finite field, prove that every function in F^F is a polynomial function.

14. If $f \in D[x]$ has $f^{\text{`}}a = 0 = f^{\text{`}}b$ for $a \neq b$ in D, show that f is divisible by $(x - a)(x - b)$.

9. Principal Ideal Domains

The division algorithm is closely connected with properties of ideals.

Recall that an ideal B in a commutative ring K is a subset of K closed under addition and under multiplication by any element of K. Thus if an ideal B contains elements $b_1, \cdots, b_m$, it must contain all the "linear combinations" $k_1b_1 + \cdots + k_mb_m$ of these elements, with coefficients $k_i \in K$. On the other hand, given any m elements $b_1, \cdots, b_m$ in K, the set

$$(b_1, \cdots, b_m) = \{\text{all } k_1b_1 + \cdots + k_mb_m | k_1, \cdots, k_m \in K\} \tag{36}$$

of all linear combinations of these elements is an ideal of K. It is called the ideal with "basis" $b_1, \cdots, b_m$ or, the ideal "generated" by $b_1, \cdots, b_m$; any ideal of K which contains $b_1, \cdots, b_m$ must contain this ideal.

In particular, the ideal (b) of all multiples kb of any one $b \in K$ is called a *principal ideal* of K.

The principal ideals in any integral domain D are closely associated with divisibility properties in D. Thus $(a) \subset (b)$ for principal ideals means that $a = bc$ for some c in D, hence that $b|a$ in D; conversely, $b|a$ implies $(a) \subset (b)$. Note that a *proper* divisor b of a determines a *larger* principal ideal than (a); thus in $\mathbf{Z}$, $(36) \subset (18)$. Similarly, $(a) = (b)$ if and only if a and b are associates in D; in particular, $(a) = D$ if and only if a is invertible in D.

A *principal ideal domain* D is an integral domain in which every ideal is principal. This is not often the case. However:

THEOREM 20. *If F is a field, the polynomial ring $F[x]$ is a principal ideal domain.*

Proof: Let B be any ideal in $F[x]$. Either $B = \{0\}$, in which case B is the principal ideal (0), or B contains a non-zero polynomial. Choose such a polynomial $g \in B$ of least degree d. For each polynomial $b \in B$ the division algorithm yields a remainder r of smaller degree with $r = b - qg$, hence certainly with r in the ideal. Since g was chosen with least degree, the remainder r must be zero, so $b = qg$ is a multiple of g. Therefore, $B = (g)$, as claimed.

A similar argument applies to the ring $\mathbf{Z}$. Any ideal $B \neq (0)$ in $\mathbf{Z}$ contains a least positive integer m; the division algorithm for $\mathbf{Z}$ then shows that $B = (m) = m\mathbf{Z}$. This proves

Theorem 21. *The ring* **Z** *of integers is a principal ideal domain.*

The following basic property is known as the "ascending chain condition" on ideals. (It will be needed in §10 and again in §XI.2.)

Proposition 22. *In a principal ideal domain D any ascending sequence*

$$C_1 \subset C_2 \subset \cdots \subset C_k \subset \cdots$$

of ideals is ultimately constant; that is, there is an index m for which $C_m = C_{m+1} = \cdots$.

Proof: Since each C_k is an ideal, the union C of all the ascending sequence of sets C_k is closed under sum and multiples, hence is an ideal of D. Thus $C = (c)$ for some element c. This element c must occur in some one of the ideals C_k of the union, say in C_m. Then $C_m \subset C$ is already all of C, so that $C_m = C_{m+1} = \cdots$, as required.

EXERCISES

1. In $\mathbf{Z}[x]$, show that the set of all polynomials with even constant term is an ideal, but not a principal ideal.

2. Find a nonprincipal ideal in $\mathbf{Z}_9[x]$.

3. Show that every ideal in $\mathbf{Z}_n$ is principal.

4. If F and F' are two fields, show that there are just two proper ideals in the ring $F \times F'$ and that every ideal in $F \times F'$ is principal.

5. Show that the ring of all rational numbers m/n with odd denominator n is a principal ideal domain.

6. Show that the set of all principal ideals in an integral domain D is a monoid under the binary operation $((a), (b)) \mapsto (ab)$.

7. In any commutative ring K, show that the sum of the principal ideals (b_1) and (b_2) is the ideal (b_1, b_2).

8. Prove the following rules for transforming a basis of an ideal in K into another basis for the same ideal:

$$(b_1, b_2, \cdots, b_m) = (b_2, b_1, b_3, \cdots, b_m),$$
$$(b_1, b_2, \cdots, b_m) = (ub_1, b_2, \cdots, b_m), \qquad u \text{ invertible in } K,$$
$$(b_1, b_2, \cdots, b_m) = (b_1 + kb_2, b_2, \cdots, b_m), \qquad k \in K.$$

***9.** Show that every ideal in the ring of gaussian integers (§2) is principal.

10. Unique Factorization

Every positive integer $n > 1$ can be written as a product of positive primes; thus $90 = 2 \cdot 5 \cdot 3^2$ or $154 = 2 \cdot 7 \cdot 11$; experience shows that there is just one such factorization for each n. On the other hand, every monic polynomial with coefficients in a field can be written as a product of irreducible monic polynomials, as for the polynomial $x^4 - 2x^3 + x^2 - 2x = x(x^2 + 1)(x - 2)$, with coefficients in $\mathbf{Q}$. Again in this case, the list of irreducible factors of a given polynomial f is unique except for order. Now both the domain $\mathbf{Z}$ of integers and the domain $F[x]$ of polynomials are principal ideal domains; here we shall show that this principal ideal property accounts for the unique factorization in these two domains. First, we consider common divisors.

An element d is called a *greatest common divisor* or g.c.d. of the elements a, b of a domain D when, for all c in D,

$$d|a, \quad d|b, \quad c|a \text{ and } c|b \quad \Rightarrow \quad c|d. \tag{37}$$

Note that the adjective "greatest" in this definition means not that d has a greater *magnitude* than any other common divisor c, but that d is a *multiple* of any such c. This definition also shows that any two different g.c.d. for given a and b are associates (multiples of each other).

Theorem 23. *In a principal ideal domain D any pair of non-zero elements a, b has a g.c.d. which can be written as a linear combination*

$$d = sa + tb \tag{38}$$

of a and b with suitable coefficients s, $t \in D$.

Proof: The set $C = (a, b)$ of all possible linear combinations $xa + yb$, $x, y \in D$, is an ideal of D, hence must be a principal ideal (d) for some d of the form $sa + tb$. Thus all elements of C, and in particular both the given elements $a = 1a + 0b$ and $b = 0a + 1b$, are multiples of this d. On the other hand, if c is any common divisor of a and b, so that $a = a'c$ and $b = b'c$ for some a' and b', then $d = sa + tb = (sa' + tb')c$ is a multiple of c. These conclusions show that d has the properties (37) requisite for a g.c.d.

If $(a, b) = (1)$ in this proof, then a and b have the g.c.d. 1; in this case we write $(a, b) = 1$ and call a and b *relatively prime* in D. In other words, two elements of D are relatively prime when their only common divisors are the invertibles of D. More generally, we sometimes write (a, b) for the element d —the unique positive g.c.d. of a and b in $\mathbf{Z}$, or the unique monic g.c.d. of the polynomials a and b in $F[x]$.

Corollary 1. *For a prime p in a principal ideal domain D,*

$$p|ab \quad \Rightarrow \quad p|a \quad \textit{or} \quad p|b, \qquad a, b \in D. \tag{39}$$

This property is often used as a definition of "prime".

Proof: By definition, a prime p has only invertibles or associates as divisors. Hence the g.c.d. of p and a is either an associate of p or of the identity 1. In the former case, $p|a$; in the latter case, use (38) to write the g.c.d. as $1 = sa + tp$ and so to write b as

$$b = b \cdot 1 = sab + tbp.$$

Here p divides the product ab by hypothesis, hence divides both terms on the right, and therefore divides b on the left, as required for the corollary.

The argument used to prove Corollary 1 will also prove

Corollary 2. *If $(a, c) = 1$ and $c|ab$, then $c|b$.*

An element $m \in D$ is called a *least common multiple* or l.c.m. of the elements a, b of D if it is a common multiple of a and b which divides all such common multiples; that is, if for all $e \in D$

$$a|m; \qquad b|m; \qquad a|e \text{ and } b|e \Rightarrow m|e. \tag{40}$$

If D is a principal ideal domain, such a l.c.m. always exists: The ideal of all common multiples e is a principal ideal (m) for some element m, and this element m has the required properties (40).

Consider an element a in D which can be written as a product

$$a = p_1 p_2 \cdots p_m \tag{41}$$

of primes p_i of D, not necessarily distinct. One may find other such representations of the same a by permuting the factors or by replacing each factor p_i by an associate $q_i = u_i p_i$, with u_i invertible and $u_1 u_2 \cdots u_m = 1$. The element a has *unique prime factorization* in D if it can be written as a product (41) of primes, unique up to such changes of order and associates.

Theorem 24 (*Unique Prime Factorization*). *In a principal ideal domain D, every element $a \neq 0$ is invertible or a product $a = p_1 \cdots p_m$ of primes p_i. If also $a = q_1 \cdots q_n$ with each q_i a prime, then $n = m$ and there is a permutation σ on $\{1, \cdots, n\}$ such that each $q_{\sigma i}$ is associate to p_i.*

Proof: First we show that such a factorization does exist. In case D is the domain $\mathbf{Z}$ of integers, we can prove by induction that every $n > 1$ has a prime factorization. For if not, the well-ordering of the positive integers gives a first integer $n > 1$ with no such factorization. This integer n is not itself prime, for otherwise $n = n$ is a prime factorization. Hence, n has two proper positive factors $n = km$. Both k and m are less than n, hence both are products of primes, and so is their product $n = km$, a contradiction. In case D is the principal ideal domain $F[x]$ of all polynomials in one indeterminate over a field, the existence of a prime factorization can be proved by a similar induction, this time on the degree.

For an arbitrary principal ideal domain, this argument by mathematical induction is replaced by a use of the ascending chain condition on ideals, as follows. Suppose that some element $a \neq 0$ in D is not invertible and has no prime factorization. Then a is not itself a prime of D; hence, it has some proper factorization $a = a_1b_1$. Now prime factorizations for a_1 and for b_1 would combine to give such a factorization for a; hence, one of a_1 and b_1, say the element a_1, has no prime factorization. Repetition of this argument yields $a_1 = a_2b_2$, where a_2 has no such factorization, and so on indefinitely. Since each a_{k+1} so obtained is a proper divisor of a_k, this gives an infinite properly ascending sequence $(a) \subset (a_1) \subset (a_2) \subset \cdots$ of principal ideals of D, in contradiction to Proposition 22.

To show the uniqueness of the factorization in the sense stated in the theorem, we will prove by induction on m that if $a \neq 0$ has one factorization with m prime factors, this factorization is unique. For $m = 1$, this means that $a = p$ is a prime, and this is then trivially the only factorization. So assume that uniqueness holds for any element which is a product of m prime factors and consider two factorizations,

$$p_1 p_2 \cdots p_m p_{m+1} = a = q_1 q_2 \cdots q_n. \tag{42}$$

The prime p_{m+1} divides the left side, hence the right side and hence (Corollary 24.1) one of the factors q_i on the right. Since both are primes, p_{m+1} and q_i are associates so $q_i = up_{m+1}$ for some unit u; canceling p_{m+1} from the equation displayed, we may apply the induction assumption to complete the proof.

The prime factorization may take special forms in particular domains.

In the domain **Z** of integers, each prime is associate to a positive prime. In the factorization of an integer n the same positive prime may occur several times. Collecting these occurrences, we write the factorization as

$$n = \pm p_1^{e_1} p_2^{e_2} \cdots p_k^{e_k}, \qquad 2 \leqslant p_1 < p_2 < \cdots < p_k \tag{43}$$

with natural numbers e_i as exponents. The uniqueness of the factorization now asserts that the exponent e_i to which each prime p_i occurs is uniquely determined by n.

THEOREM 25 (*Euclid*). *There are infinitely many prime numbers in* **Z**.

Proof: If not, let q be the last one. The number $n = 1 + q!$ then leaves remainder 1 upon division by any prime, but it also has a prime factorization, hence must be divisible by at least one prime—a contradiction.

In the domain $F[x]$ of polynomials over a field F, the invertibles are the non-zero constants of $F \subset F[x]$, and each polynomial is associate to a unique monic polynomial. Hence, each nonconstant polynomial $f \neq 0$ has a factorization

$$f = cp_1^{e_1}p_2^{e_2} \cdots p_k^{e_k}, \qquad c \in F, \tag{44}$$

with distinct monic irreducibles $p_1, \cdots, p_k$. Uniqueness asserts that f determines the exponent e_i of each monic p_i.

EXERCISES

All of the following exercises refer to elements in a given principal ideal domain.

1. Prove: If $(a, b) = 1$, $a|c$ and $b|c$, then $ab|c$.
2. Prove Corollary 2 of Theorem 23.
3. Show that $ab(a, b)^{-1}$ is a l.c.m. of a and b.
4. If $1 = sa + tb$, show that a and b are relatively prime.
5. Prove: $(a, c) = 1$ and $(b, c) = 1$ imply $(ab, c) = 1$.
6. Show that any three elements a, b, c have a g.c.d. which can be expressed in the form $sa + tb + uc$.
7. If $q \neq 0$ has the property that $q|ab$ always implies $q|a$ or $q|b$, prove that q is invertible or a prime.
8. If $q \neq 0$ has the property that for every a either $q|a$ or $(q, a) = 1$, prove that q is invertible or a prime.
9. If q is a power of a prime, show that $q|ab$ implies $q|a$ or $q|b^e$ for some power of b.
10. (a) If $a = p_1^{e_1} \cdots p_k^{e_k}$ and $b = p_1^{f_1} \cdots p_k^{f_k}$, show that the g.c.d. (a, b) is $p_1^{d_1} \cdots p_k^{d_k}$, with each d_i the minimum of e_i and f_i.
    (b) Obtain a similar formula for the l.c.m. of a and b.
11. Show that the poset of all principal ideals of D under inclusion is a lattice (§IV.6), with meet given by the l.c.m. and join by the g.c.d.

11. Prime Fields

In this section we make some elementary applications of prime numbers and unique prime factorization to the construction of fields.

Theorem 26. *The characteristic of an integral domain D is ∞ or a prime.*

Proof: The characteristic m of D was defined in §1 to be the additive order of the identity element $1'$ of D. If this m is finite but not a prime, it has proper positive factors $m = hk$. Then $0 = m1' = (hk)1' = (h1')(k1')$. Since D is an integral domain, either $h1' = 0$ or $k1' = 0$—but both h and k are less than m, contrary to the choice of m as the least positive integer with $m1' = 0$.

Theorem 27. *For each prime number p, $\mathbf{Z}_p$ is a field.*

The proof will illustrate the use of the expression (38) of the g.c.d. of two integers as a linear combination $sa + tb$. For, take a coset $p\mathbf{Z} + a$ in $\mathbf{Z}_p = \mathbf{Z}/p\mathbf{Z}$. Then $p\mathbf{Z} + a \neq 0$ and p prime means that $(a, p) = 1$, so that there are integers s, t with $sa + tp = 1$. The morphism $\mathbf{Z} \to \mathbf{Z}/p\mathbf{Z}$ applied to this equation carries the prime p to zero, hence s to an inverse $p\mathbf{Z} + s$ of $p\mathbf{Z} + a$, as required.

This result may also be stated in terms of congruences: If p is prime and $a \not\equiv 0 \pmod{p}$, each congruence $xa \equiv b \pmod{p}$ has an integral solution $x = sb$ which is unique, modulo p.

A similar argument shows that g irreducible in the polynomial ring $F[x]$ implies that the quotient ring $F[x]/(g)$ is a field.

Any subfield of a field F must contain the identity element $1 \in F$. Since the intersection of a set of subfields is again a subfield, the intersection of *all* the subfields of F is a subfield called the *prime subfield* P of F. It may also be described as the subfield generated by 1; like any subfield of F, it has the same characteristic (∞ or a prime number) as does F. Up to isomorphism, the only prime subfields are $\mathbf{Q}$ and $\mathbf{Z}_p$:

Theorem 28. *If F is a field of characteristic ∞, the unital morphism $\mu:\mathbf{Z} \to F$ extends to a unique morphism $\nu:\mathbf{Q} \to F$ of fields; the image of ν is the prime subfield of F. If F is a field of prime characteristic p, there is exactly one morphism $\nu:\mathbf{Z}_p \to F$ of fields; it is a monomorphism with image the prime subfield of F.*

Proof: In both cases, start with the unital morphism $\mu:\mathbf{Z} \to F$. If F has characteristic ∞, then since the insertion $\mathbf{Z} \to \mathbf{Q}$ of $\mathbf{Z}$ in its quotient field $\mathbf{Q}$ is universal as in Theorem 12, the monomorphism μ extends uniquely to $\nu:\mathbf{Q} \to F$ as in the diagram

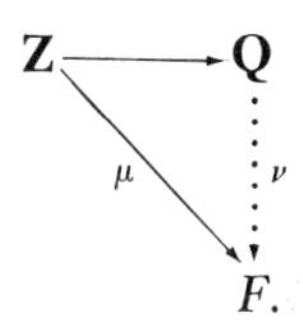

If F has characteristic p, the kernel of μ is the principal ideal (p), so yields by Theorem 8 a unique monomorphism $\nu:\mathbf{Z}_p \to F$ on the quotient ring $\mathbf{Z}_p = \mathbf{Z}/(p)$. This monomorphism takes $1 \in \mathbf{Z}_p$ to $1 \in F$, and has image the prime subfield of F

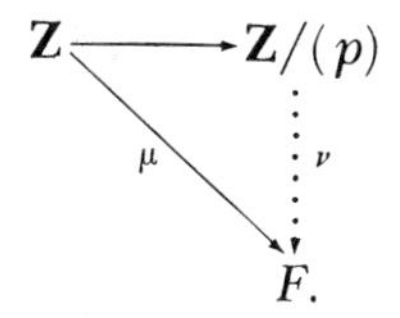

The unique decomposition of polynomials also enters in the description of the "rational functions" over a given field F. Since the polynomial ring $F[x]$ is an integral domain, it has a field of quotients, constructed as in §5. This field is written $F(x)$ and is called the *field of rational functions* over F. It is a field containing F and x in which each element $q \in F(x)$ can be written as a quotient $q = f/g$ of two polynomials f and g in x, with $g \neq 0$. Since any common factor of f and g in such a quotient may be canceled, we can write q as $q = f/g$ with f and g relatively prime and g monic. To interpret such a quotient as a function, one must avoid the zeros of the denominator g, if any. So let $c_1, \cdots, c_k$ be all the zeros of g in F, and let $F - \{c_1, \cdots, c_k\}$ denote the complement in F of the set of all these zeros. Then q defines a function $q^{\text{`}}:F - \{c_1, \cdots, c_k\} \to F$; namely, the function with values $q^{\text{`}}a = (f^{\text{`}}a)/(g^{\text{`}}a)$ for each $a \in F - \{c_1, \cdots, c_q\}$. In this way each element q of the field $F(x)$ of quotients can be interpreted as a function $q^{\text{`}}$.

EXERCISES

1. If Q is the field of quotients of the integral domain D, prove that the field $Q(x)$ of rational functions is (isomorphic to) the field of quotients of $D[x]$.

2. In what fields does the conventional formula for the solution of a quadratic equation hold?

3. Using the multiplicative group of the field $\mathbf{Z}_p$ for p a prime, prove for all integers x that $x^p \equiv x \pmod{p}$ (Fermat's "little" theorem).

*12. The Euclidean Algorithm

In the principal ideal domains $\mathbf{Z}$ and $F[x]$, there is a direct method of calculating both the greatest common divisor d of two elements a and b of the domain and the coefficients s and t in the expression $d = sa + tb$ of d as a linear combination of a and b.

To explicitly calculate the g.c.d. of two polynomials b_0 and b_1 in $F[x]$, first divide b_0 by b_1, as in the division algorithm, to get a remainder b_2 of smaller degree, then divide b_1 by the remainder b_2, getting a new remainder b_3, and so on. Since the degrees of the remainders decrease, some remainder, say b_{n+1}, is finally zero. With successive quotient polynomials q_i this process yields a table of equations

$$\begin{aligned} b_0 &= q_1 b_1 + b_2, & \deg b_2 &< \deg b_1, \\ b_1 &= q_2 b_2 + b_3, & \deg b_3 &< \deg b_2, \\ &\vdots & &\vdots \\ b_{n-2} &= q_{n-1} b_{n-1} + b_n, & \deg b_n &< \deg b_{n-1}, \\ b_{n-1} &= q_n b_n, \end{aligned} \tag{45}$$

called the *Euclidean Algorithm* for the polynomials b_0 and b_1.

Now take in $F[x]$ the ideal (b_0, b_1) of all linear combinations $fb_0 + gb_1$ for polynomials f and g. By the first equation of the algorithm, each such linear combination may be written as

$$fb_0 + gb_1 = f(q_1 b_1 + b_2) + gb_1 = (fq_1 + g)b_1 + fb_2,$$

so $(b_0, b_1) \subset (b_1, b_2)$ and similarly $(b_1, b_2) \subset (b_0, b_1)$, which proves that $(b_1, b_2) = (b_0, b_1)$. Continuing,

$$(b_0, b_1) = (b_1, b_2) = \cdots = (b_{n-1}, b_n) = (b_n, 0) = (b_n). \tag{46}$$

This expresses the ideal (b_0, b_1) as a principal ideal (b_n), so the polynomial b_n is the required g.c.d. of b_0 and b_1: In brief, this g.c.d. is the last non-zero remainder in the Euclidean algorithm.

To express this g.c.d. b_n explicitly as a linear combination of the original b_0 and b_1 we need only use the repeated equality (46). Namely, solve the equations of the algorithm backward to express b_n in terms of successive pairs of previous b's; thus

$$\begin{aligned} b_n = b_{n-2} - q_{n-1}b_{n-1} &= b_{n-2} - q_{n-1}(b_{n-3} - q_{n-2}b_{n-2}) \\ &= (1 + q_{n-1}q_{n-2})b_{n-2} - q_{n-1}b_{n-3} = \cdots \\ &= (- \; -)b_{n-3} - (- \; -)b_{n-4} = \cdots . \end{aligned}$$

The same algorithm applies to **Z**, with polynomials replaced by nonnegative integers; the remainders then decrease not in degree but in magnitude. For example, to get the g.c.d. (14, 9) in **Z**, write

$$14 = 9 + 5, \qquad 9 = 5 + 4, \qquad 5 = 4 + 1,$$

so the desired g.c.d. is 1 (as was clear). To express it as a linear combination, solve these equations to get

$$1 = 5 - 4 = 5 - (9 - 5) = 2 \cdot 5 - 9 = 2(14 - 9) - 9 = 2 \cdot 14 - 3 \cdot 9.$$

EXERCISES

1. (a) In each of the following cases, use the Euclidean algorithm in **Z** to express the indicated g.c.d. as a linear combination of the two given integers.

(i) (36, 15). (ii) (180, 252). (iii) (4148, 7684).

(b) Do the same exercises in $\mathbf{Q}[x]$ for:

(i) $(x^3 - 1, x^4 + x^3 + 2x^2 + x + 1)$. (ii) $(x^{18} - 1, x^{33} - 1)$.

2. Decompose the following polynomials into irreducibles in $\mathbf{Z}_3[x]$:
(a) $x^2 + x + 1$. (b) $x^3 + x + 2$. (c) $x^4 + x^3 + x + 1$.

3. List, to within associates, all divisors of $x^7 - x$ in $\mathbf{Q}[x]$.

4. Decompose $x^4 - 5x^2 + 6$ into irreducibles in $\mathbf{Q}[x]$ and in $\mathbf{R}[x]$.

5. Without using ideals, show directly from the equations (45) of the Euclidean algorithm that the last non-zero remainder b_n is a g.c.d. for b_0 and b_1.

***6.** For any positive integers m and n show that the set of all $sm + tn$, with s and t positive integers, includes all the multiples of (m, n) larger than mn.

7. If G and H are finite cyclic groups with relatively prime orders, prove that $G \times H$ is cyclic.

8. (*Eisenstein's irreducibility criterion.*) Prove: If p is a prime and the monic polynomial $f = f_0 + \cdots + f_{k-1}x^{k-1} + x^k$ has integral coefficients with $p | f_0, \cdots, p | f_{k-1}$ but f_0 not divisible by p^2, then f is irreducible in $\mathbf{Z}[x]$.

9. For positive integers m and n, show that $(m, n) = 1$ and mn a square implies m a square.

10. To find all triples of integers x, y, z with $x^2 + y^2 = z^2$ it suffices to assume that $(x, y) = 1$. Show that x and y cannot then both be odd. If y is even, apply Exercise 9 to show that $y = 2mn$ where m and n are integers with $x = m^2 - n^2$, $z = m^2 + n^2$. (*Hint:* Factor $z^2 - x^2$.)

13. Commutative Quotient Rings

In a commutative ring K, each ideal C determines, as in §3, a quotient ring K/C which is commutative. With properties of divisibility at hand, we can describe in more detail the structure of some of these rings K/C.

Proposition 29. *For any two relatively prime elements a and b in a principal ideal domain D there is an isomorphism of rings*

$$D/(ab) \cong [D/(a)] \times [D/(b)] \tag{47}$$

which sends each coset $(ab) + r$ to the pair of cosets $((a) + r, (b) + r)$.

Proof: Construct the diagram

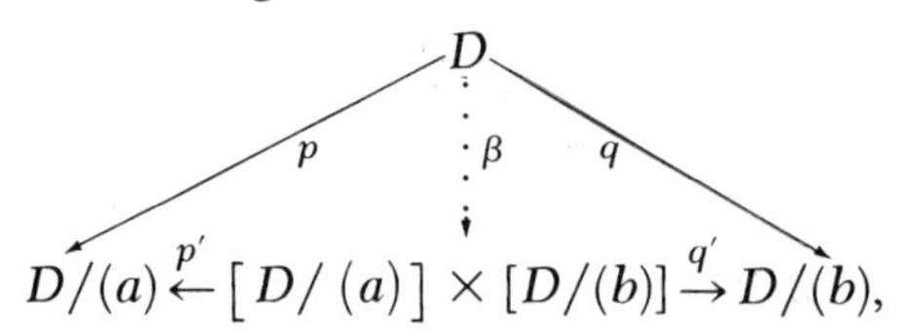

$$D/(a) \xleftarrow{p'} [D/(a)] \times [D/(b)] \xrightarrow{q'} D/(b),$$

where the morphisms p and q project D upon its quotient rings, while as in (8), p' and q' project the product ring upon its factors. We can fill in at β with a unique morphism of rings which makes the diagram commute; explicitly, β is the function with

$$\beta(x) = (px, qx) = ((a) + x, (b) + x), \qquad x \in D.$$

Now study the kernel and image of β. The kernel consists of all $x \in D$ divisible both by a and by b; since $(a, b) = 1$, these are exactly all x divisible by the product ab. The image consists of all elements in the product ring. For, consider any pair of cosets $((a) + c, (b) + d)$, for $c, d \in D$. As $(a, b) = 1$, the difference $c - d$, like any element of D, can be written as a linear combination $c - d = sa + tb$. Then the element $c - sa = x = d + tb$ in D has $\beta(x) = ((a) + c, (b) + d)$, as required.

Since $\beta: D \to [D/(a)] \times [D/(b)]$ is an epimorphism of rings with kernel the principal ideal (ab), Theorem 8 on quotient rings, in the form (16), gives the desired isomorphism (47).

Now take $D = \mathbf{Z}$ and recall that an element $(a) + x$ of the quotient ring $\mathbf{Z}/(a)$ is just the same thing as an equivalence class for the relation of congruence modulo a. In this language the proposition becomes the following

Corollary. *(Chinese Remainder Theorem). For any integers c, d, a, and b with $(a, b) = 1$, the pair of congruences*

$$x \equiv c \pmod{a}, \qquad x \equiv d \pmod{b} \tag{48}$$

has an integral solution x which is uniquely determined, modulo ab.

Next consider two special types of ideals. An ideal $M \neq K$ in a commutative ring K is *maximal* when any ideal C with $M \subset C \subset K$ is either $C = M$ or $C = K$; in other words, when there are no ideals in K properly between M and the whole ring K. An ideal $P \neq K$ in K is *prime* if it has for all elements $a, b \in K$ the property

$$ab \in P \quad \Rightarrow \quad a \in P \quad \text{or} \quad b \in P. \tag{49}$$

Recall by (39) that a prime element p in a principal ideal domain D has a similar property: $p|ab$ implies $p|a$ or $p|b$. In terms of the principal ideal (p) of D, this property reads: $ab \in (p)$ implies $a \in (p)$ or $b \in (p)$. Hence, it states that each prime *element* p of a principal ideal domain D determines an ideal (p) which is a prime *ideal*. More is true: (p) is maximal in D, for if an ideal C contains p and an element c not in (p), the g.c.d. of p and c must be $1 = sc + tp$, so $1 \in C$ and $C = D$.

Proposition 30. *For any ideal A in a nontrivial commutative ring K*

$$A \text{ is maximal} \quad \Leftrightarrow \quad K/A \text{ is a field}, \tag{50}$$

$$A \text{ is prime} \quad \Leftrightarrow \quad K/A \text{ is an integral domain}. \tag{51}$$

Proof: As for the first conclusion, recall by Proposition 11 that a nontrivial commutative ring L is a field if and only if L has no proper ideals. But by the discussion of (14), the ideals in the quotient ring K/A correspond to ideals in K between A and K. Hence, K/A is a field if and only if there are no ideals of K properly between A and K; that is, if and only if A is maximal in K.

As for the second conclusion (51), the statement that K/P is an integral domain amounts to the cancellation law

$$(P + a)(P + b) = P \quad \Rightarrow \quad P + a = P \quad \text{or} \quad P + b = P.$$

This is immediately the same as the statement (49) that P is a prime ideal.

Here are some direct consequences:

Corollary 1. *Every maximal ideal is prime.*

Corollary 2. *For each prime number p, $\mathbf{Z}_p = \mathbf{Z}/p\mathbf{Z}$ is a field.*

Corollary 3. *For F a field, each polynomial g irreducible in the polynomial ring $F[x]$ gives a quotient ring $F[x]/(g)$ which is a field.*

As an example of such a quotient field, consider the set $\mathbf{Q}(\sqrt{2})$ of all real numbers of the form $a + b\sqrt{2}$ with a and b rational. Sums and products of two such numbers are

$$(a + b\sqrt{2}) + (c + d\sqrt{2}) = (a + c) + (b + d)\sqrt{2},$$
$$(a + b\sqrt{2})(c + d\sqrt{2}) = (ac + 2bd) + (ad + bc)\sqrt{2};$$

from these formulas a tedious check shows that $\mathbf{Q}(\sqrt{2})$ is a ring (without assuming anything about real numbers). Here the formula for the product makes use of the fact that $\sqrt{2}$ is a root of $x^2 - 2 = 0$. Thus the evaluation $x \mapsto \sqrt{2}$ yields an epimorphism $\mathbf{Q}[x] \to \mathbf{Q}(\sqrt{2})$ of rings with kernel the principal ideal $(x^2 - 2)$. Therefore (by the main Theorem 8 on quotient rings), there is an isomorphism $\mathbf{Q}[x]/(x^2 - 2) \cong \mathbf{Q}(\sqrt{2})$; since $x^2 - 2$ is irreducible in $\mathbf{Q}[x]$, $\mathbf{Q}(\sqrt{2})$ is a field. In other words, the quotient ring may be used to construct a root $\sqrt{2}$ for the polynomial $x^2 - 2$, by making $\sqrt{2}$ the coset of x in $\mathbf{Q}[x]/(x^2 - 2)$. Moreover, the elements of this quotient ring appear as polynomials of degree at most 1 in this root $\sqrt{2}$.

A similar description applies to any quotient ring $F[x]/(g)$, where g is a polynomial in $F[x]$ of degree $n > 0$. By the division algorithm, every polynomial f in $F[x]$ can be written in the form $f = qg + r$ with a unique remainder r of degree less than n. Hence, every coset $(g) + f$ in the quotient ring $F[x]/(g)$ can be written, and indeed uniquely, in the form $(g) + r$. If we identify each coset $(g) + r$ with this remainder r, then the elements of $F[x]/(g)$ appear as remainders modulo g [polynomials of degree less than

$n = \deg(g)$]; to add or multiply two such remainders, take their sum or product as polynomials and then the remainder on division by g. This is just like the description of $\mathbf{Z}_n$ as the ring with elements the integral remainders modulo n (§I.8). If we let the symbol ζ denote the remainder x, we have proved that the quotient ring has the following form.

Proposition 31. *If F is a field and $g = g_0 + g_1 x + \cdots + g_n x^n$ a polynomial in $F[x]$ of degree $n > 0$, while ζ is the coset $(g) + x$, then each element of the quotient ring $F[x]/(g)$ can be written uniquely in the form*

$$r_0 + r_1\zeta + \cdots + r_{n-1}\zeta^{n-1}, \qquad r_0, \cdots, r_{n-1} \in F. \tag{52}$$

The sum of two such elements is their sum as polynomials in ζ; the product, their product as polynomials, reduced to this form (52) by using the relation

$$g_0 + g_1\zeta + \cdots + g_n\zeta^n = 0 \tag{53}$$

to replace ζ^n by lower powers of ζ.

With this in view, the quotient ring $F[x]/(g)$ is often described as a ring generated by F and one new element ζ subject only to the relation (53).

By the same argument, $\mathbf{Z}[x]/(x^2 + 1)$ consists of all elements $m + n\zeta$ for $m, n \in \mathbf{Z}$ with $\zeta^2 = -1$, so this ring is the ring G of Gaussian integers, just as described in §2, but with ζ and $\zeta^2 = -1$ in place of i and $i^2 = -1$.

Finally, we examine some subrings of the field $Q(D)$ of quotients of an integral domain D (§5).

Proposition 32. *If P is a prime ideal in an integral domain D, then the set D_P of all quotients a/b, with $a, b \in D$ and denominator $b \notin P$, is a subring of the field $Q(D)$ of quotients of D.*

Proof: To say that P is prime is to say that $b \notin P$ and $d \notin P$ imply $bd \notin P$. Hence, the product $(a/b)(c/d)$ of two quotients in D_P is in D_P. For the same reason, D_P is closed under addition. Since $1 \in D_P$, it is a subring of $Q(D)$.

This subring D_P is called the *localization* of the domain D at the prime ideal P. For example, the localization of $\mathbf{Z}$ at the prime ideal (2) is the ring of all rational numbers with odd denominators. The use of these localizations has the advantage of "separating" the effects of different prime ideals of D.

EXERCISES

1. Solve the following simultaneous congruences:
   (a) $x \equiv 2 \pmod 7$, $x \equiv 3 \pmod 6$.
   (b) $3x \equiv 2 \pmod 5$, $x \equiv 6 \pmod 7$, $x \equiv 1 \pmod 6$.

2. Show that two congruences in **Z** of the form

$$mx \equiv c \pmod{a}, \qquad nx \equiv d \pmod{b}$$

have a common solution $x \in \mathbf{Z}$ when the coefficients satisfy the conditions $(a, b) = 1$, $(m, a) = 1$ and $(n, b) = 1$.

3. (*The Chinese remainder theorem for k congruences.*) The integers $a_1, \cdots, a_k$ are called "relatively prime in pairs" when $i \neq j$ implies $(a_i, a_j) = 1$. If this is the case, show for each list of integers $c_1, \cdots, c_k$ that there is an integer x with $x \equiv c_i \pmod{a_i}$ for $i = 1, \cdots, k$, and that this x is unique, modulo the product $a_1 a_2 \cdots a_k$.

4. Show that a congruence $mx \equiv c \pmod{a}$ has a solution x in **Z** if and only if $(m, a) | c$; when this is the case, show that there are exactly (m, a) incongruent solutions, modulo a.

5. If A and B are ideals in a ring R with sum $A + B = R$, establish the isomorphism

$$R/(A \cap B) \cong [R/A] \times [R/B].$$

6. If $n = p_1^{e_1} \cdots p_k^{e_k}$ with $p_1, \cdots, p_k$ distinct primes in **Z**, show

$$\mathbf{Z}_n \cong \mathbf{Z}_{p_1^{e_1}} \times \cdots \times \mathbf{Z}_{p_k^{e_k}}.$$

7. (a) If p is a prime, show each binomial coefficient $(k, p - k)$ for $0 < k < p$ divisible by p.

(b) If a domain D has characteristic p, show that $(b + c)^p = b^p + c^p$ for all $b, c \in D$.

(c) Deduce that $b \mapsto b^p$ is an endomorphism of D.

8. If the rings R and R' have the finite characteristics m and m', respectively, show that the characteristic of the product ring $R \times R'$ is the least common multiple of m and m'.

9. For polynomial rings over a field F prove that

$$F[x]/(x^2 - 1) \cong F[x]/(x^2 - 4),$$
$$F[x]/(x^2 + 1) \cong F[x]/(x^2 + 2x + 2).$$

10. Establish the isomorphisms $\mathbf{Z}[x]/(3, x) \cong \mathbf{Z}_3$; $\mathbf{Z}[x]/(6) \cong \mathbf{Z}_6[x]$.

11. Show that the ideal (x) is prime but not maximal in $\mathbf{Z}[x]$.

12. If M is a maximal ideal of an integral domain D, show that the localization D_M has exactly one maximal ideal.

13. Find all subrings of **Q**.

***14.** Find all those subgroups of the additive group of **Q** which contain the subgroup **Z**.

CHAPTER IV
Universal Constructions

THE PRECEDING two chapters have described a number of examples and basic properties of groups and rings. In both cases, the morphisms (of groups or of rings) play an important part. This chapter will explain some ideas about morphisms of algebraic systems in general—ideas which are illustrated by and unify what we have proved so far. A collection of algebraic systems and their morphisms is called a "category", and a construction of new systems from given ones is a "functor". Finally, certain special constructions give properties which are "universal". These three closely related concepts of category, functor, and universality pervade all of algebra.

1. Examples of Universals

Several of our constructions of groups and rings can be characterized by the fact that they construct something with a desirable property and "universal" with that property. Thus one may construct an element which is universal with some characteristic, in the sense that this universal element maps uniquely into any other element with that characteristic. Or, one may construct a new algebraic system with some given property and a morphism (from a given system into this new system) which is universal, in the sense that any other morphism into a system with this property factors uniquely through the universal morphism. In each of these cases, the object which is universal in this way turns out to be unique, up to an isomorphism.

Here are some examples of universal elements.

A Universal Element of a Monoid. The number 1 in the additive monoid $\mathbf{N}$ of natural numbers has the following universal property: If a is any element in any monoid M, there is exactly one morphism $f:\mathbf{N} \to M$ of monoids which maps 1 into a (Proposition I.21).

A Universal Element of a Group. The number 1 in the additive group $\mathbf{Z}$ of integers has the following universal property: If g is any element in any group G, there is exactly one morphism $f:\mathbf{Z} \to G$ of groups which maps 1 into g. This is the morphism $n \mapsto g^n$ (Theorem II.2). This universal property is often described by saying that $\mathbf{Z}$ is the "free group" generated by one element.

Next we give some examples of a universal epimorphism from a given system to some "smaller" quotient system.

Quotient Sets. Given an equivalence relation E on a set X, there is a set X/E and a function $p_E:X \to X/E$ such that xEy implies $p_E x = p_E y$ and which has the following universal property: If $f:X \to S$ is any function to a set S such that xEy implies $fx = fy$ for all $x, y \in X$, there is a unique function $f':X/E \to S$ such that $f = f' \circ p_E$ (Theorem I.19). In this case, we say that p_E is "universal" among all functions f for which xEy implies $fx = fy$.

Quotient Groups. Given a normal subgroup N of a group G, there is a group G/N and a morphism $p:G \to G/N$ of groups with $p_*N = 1$ which has the following universal property: If $\phi:G \to L$ is any morphism of groups with $\phi_*N = 1$, there is a unique morphism $\phi':G/N \to L$ of groups with $\phi = \phi' \circ p$ (Theorem II.26). In other words, p is universal among morphisms killing N. As shown in detail in §II.10, this property of the projection p onto the quotient group G/N can be used to get all the useful properties of that quotient group, without referring to its construction via cosets.

Quotient Rings. Given an ideal A in a ring R, there is a ring R/A and a morphism $p:R \to R/A$ of rings with $p_*A = 0$ which has the following universal property: If $\alpha:R \to S$ is any morphism of rings with $\alpha_*A = 0$, there exists a unique morphism $\alpha':R/A \to S$ of rings with $\alpha = \alpha' \circ p$ (Theorem III.8). In other words, p is "universal" among morphisms which annihilate A.

In each of these examples, the epimorphism p represents a contraction of the given system. Next we will show how five of the methods we have used for constructing *extensions* of given algebraic systems can be viewed as morphisms with a universality property.

Field of Quotients. Given an integral domain D, there is a field $Q(D)$ and a monomorphism $j:D \to Q(D)$ of rings with the following universal property: If $\alpha:D \to F$ is any monomorphism of D to a field F, there is a unique morphism $\alpha':Q(D) \to F$ of fields such that $\alpha = \alpha' \circ j$ (Theorem III.12). As shown in §III.5, this property characterizes the construction of "fractions" in $Q(D)$; in particular, the construction of rational numbers from the integers.

Polynomial Rings. Given a commutative ring K, there is a ring $K[x]$, an element x in that ring, and a monomorphism $i:K \to K[x]$ of rings with the following universal property: If L is a commutative ring, d an element of L, and $\alpha:K \to L$ a morphism of rings, there is a unique morphism $\alpha':K[x] \to L$ of rings with $\alpha' \circ i = \alpha$ and $\alpha'(x) = d$ (Theorem III.18).

A Universal Ring. The ring $\mathbf{Z}$ of integers has the following universal property: For any ring R' (with unit $1'$) there is exactly one morphism $\mu:\mathbf{Z} \to R'$ of rings. This is the "unital" morphism μ which sends $n \in \mathbf{Z}$ into the multiple $n1'$ (Proposition III.1).

The Universal Embedding of $\mathbf{N}$ *in a Group.* The insertion $i:\mathbf{N} \to \mathbf{Z}$ of the natural numbers in the group of integers is a morphism of additive monoids with the following universal property: If A is any group and $f:\mathbf{N} \to A$ any morphism of monoids, there is a unique morphism $f':\mathbf{Z} \to A$ of groups with $f = f' \circ i$, as in the commutative diagram (Theorem II.7)

The Universal Way of Embedding the Natural Numbers in a Ring. The insertion $i:\mathbf{N} \to \mathbf{Z}$ of the natural numbers in the ring of integers is a morphism of addition, multiplication, and multiplicative unit with the following universal property: If R' is a ring and $\alpha:\mathbf{N} \to R'$ a morphism of addition, multiplication, and multiplicative unit, there is a unique morphism $\alpha':\mathbf{Z} \to R'$ of rings with $\alpha' \circ i = \alpha$ (Proposition III.3). The corresponding diagram is like (1), with f replaced by α and the group A by the ring R'. This "universal" property of $\mathbf{Z}$ describes what is meant by extending the system $\mathbf{N}$ of natural numbers to an essentially unique larger system in which subtraction is possible.

There are many more such "universal" constructions, with similar descriptions. We now turn to a general formulation and description of these kinds of universality.

2. Functors

Many constructions of a new algebraic system from a given one also construct suitable morphisms of the new algebraic systems from morphisms between the given ones. These constructions will be called "functors" when they preserve identity morphisms and composites of morphisms. Specifically, a *functor* $\mathfrak{F}$ from rings to groups assigns to each ring R a group $\mathfrak{F}(R)$ and to each morphism $f:R \to S$ of rings a morphism $\mathfrak{F}(f):\mathfrak{F}(R) \to \mathfrak{F}(S)$ of groups in such a way that

$$\mathfrak{F}(1_R) = 1_{\mathfrak{F}(R)} \qquad \text{and} \qquad \mathfrak{F}(g \circ f) = \mathfrak{F}(g) \circ \mathfrak{F}(f), \tag{2}$$

the latter condition holding whenever the composite morphism $g \circ f$ is defined. These conditions state that a functor takes identity morphisms to identities and composites to composites. For morphisms $f:R \to S$ and $g:S \to T$, these conditions may be visualized by the commutative diagrams:

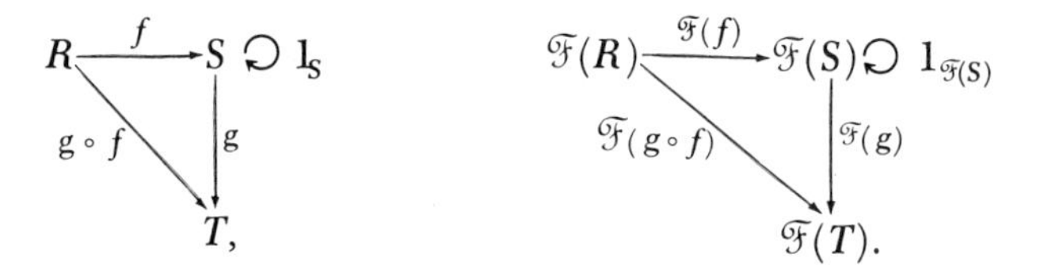

Indeed, the conditions (2) state that the functor $\mathfrak{F}$ must take each commutative diagram like that on the left into a commutative diagram (namely, that on the right).

One functor from rings to groups is the *forgetful functor* $\mathfrak{U}$, which assigns to each ring R the same set $\mathfrak{U}(R) = R$, but regarded just as an additive group, and to each morphism $f:R \to S$ of rings the same function $\mathfrak{U}(f) = f$, which is still a morphism of additive groups. The conditions (2) evidently hold. Another more interesting such functor $\mathfrak{V}$ assigns to each ring R the set $\mathfrak{V}(R)$ of all invertible elements of R, regarded as a multiplicative group. Since each morphism $f:R \to S$ of rings must carry invertible elements $x \in R$ into elements $f(x)$ invertible in S, the map f restricted to these elements is a function $\mathfrak{V}(f):\mathfrak{V}(R) \to \mathfrak{V}(S)$. This definition again satisfies (2).

We have defined the notion of a functor $\mathfrak{F}$ from rings to groups. Functors from rings to (say) monoids are given by the same definition, with the variation that each $\mathfrak{F}(R)$ is now required to be a monoid and each $\mathfrak{F}(f)$ a morphism of monoids, with conditions (2) still required. For example, one such functor $\mathfrak{M}$ is another "forgetful" functor assigning to each ring R the same set $\mathfrak{M}(R) = R$, regarded just as a monoid under multiplication, and to each morphism $f:R \to S$ of rings the same function $\mathfrak{M}(f) = f$, but regarded just as morphism of multiplication. There are many relevant functors from rings to rings. One is the operation $A \times -$ for A a fixed ring; it sends each ring R to the product ring $A \times R$ and each morphism $f:R \to S$ to the morphism $1 \times f:A \times R \to A \times S$; conditions (2) evidently hold. Another is the functor assigning to each ring R its opposite R^{op} (§III.2). The construction of the ring of polynomials in one indeterminate x is a functor $K \mapsto K[x]$ on commutative rings to commutative rings, since each morphism $\alpha:K \to K'$ of commutative rings gives a corresponding morphism $\alpha_{\#}:K[x] \to K'[x]$ of the associated polynomial rings, by Proposition III.16.

Next consider functors from groups to sets. One such is the forgetful functor $\mathfrak{U}$, which assigns to each group G the set $\mathfrak{U}(G)$ of its elements and to each morphism $f:G \to H$ of groups the same arrow, regarded just as a function between sets. For two groups L and G we regularly denote by

$$\text{hom}\,(L, G) = \{t \mid t:L \to G \text{ is a morphism of groups}\} \tag{3}$$

the *hom-set* consisting of all morphisms t of groups from L to G. Also, if $f:G \to H$ is a morphism with domain G, there is a corresponding function on hom-sets

$$\text{hom}\,(L, f):\text{hom}\,(L, G) \to \text{hom}\,(L, H), \tag{4}$$

which sends each $t \in \text{hom}(L, G)$ to the composite $f \circ t$, so $\text{hom}\,(L, f)t = f \circ t$. These functions compose, because $(g \circ f) \circ t = g \circ (f \circ t)$. Hence for each fixed group L, the construction $\text{hom}\,(L, -)$ is a functor on groups to sets, sending each group G to the set $\text{hom}\,(L, G)$ and each morphism

$f: G \to H$ of groups to the function hom (L, f) of (4). It is called a (covariant) *hom-functor*.

For example, take L to be the infinite cyclic group $\mathbf{Z}$ of all integers under addition. Then for any element $g \in G$ the assignment $n \mapsto g^n$ is a morphism $t: \mathbf{Z} \to G$ and, as stated in Theorem II.2, it is the only morphism $\mathbf{Z} \to G$ with $1 \mapsto g$. This means that the morphism t is completely determined by $t(1) = g$, so that the set hom $(\mathbf{Z}, G)$ of all such t "is" just the set $\mathfrak{U}(G)$ of all elements $g \in G$. This set $\mathfrak{U}(G)$ is called the "underlying set" of the group G; indeed, we usually do not distinguish in notation between the group G and its underlying set $\mathfrak{U}(G)$.

Similarly, if $L = \mathbf{Z}_k$ is the cyclic group of order k, the possible morphisms $\mathbf{Z}_k \to G$ all have the form $n \mapsto g^n$, where g is now an element of G with $g^k = 1$. Therefore, hom $(\mathbf{Z}_k, G)$ "is" just the subset $\{g \mid g \in G, g^k = 1\}$ of G consisting of those elements with order dividing k.

When L is a group, hom $(L, -)$ is a functor on groups to sets, as just noted. Again, if L is a fixed ring, there is a functor hom $(L, -)$ on rings to sets, which sends each ring R into the set hom (L, R) of all morphisms of rings from L to R. Much the same applies to other types of algebraic systems. We call all the systems of a given type a "category" (see §5), and we then speak of functors on a category to sets. More exactly, a category consists both of objects (the systems in question) and of morphisms between two such objects, while a functor $\mathfrak{F}$ from a category to sets consists of two functions: An object function $\mathfrak{F}_0$ sending each object to a set and a morphism function $\mathfrak{F}_M$ sending each morphism to a function between sets.

EXERCISES

1. If X is a fixed set, show that $S \mapsto X \times S$ and $f \mapsto 1_X \times f$ define a functor from sets to sets.

2. If $\mathfrak{G}$ and $\mathfrak{F}$ are given functors on sets to sets, show that the equations $(\mathfrak{G} \circ \mathfrak{F})(S) = \mathfrak{G}(\mathfrak{F}(S))$, $(\mathfrak{G} \circ \mathfrak{F})(f) = \mathfrak{G}(\mathfrak{F}(f))$ define a new functor $\mathfrak{G} \circ \mathfrak{F}$ on sets to sets. (It is called the "composite" of $\mathfrak{G}$ with $\mathfrak{F}$.)

3. Show that the construction Q of the field of quotients is (the object function of a) functor from integral domains to fields.

4. Each group G determines a subgroup $Z(G)$, the center of G, as in Exercise II.9.5. Show however that Z is *not* naturally a functor on groups to groups, because a morphism $f: G \to H$ of groups does not necessarily carry $Z(G)$ into $Z(H)$.

5. Each abelian group A determines a ring End A (Theorem III.6). Show, however, that End is not naturally a functor on abelian groups to rings, because a morphism $A \to B$ of groups does not necessarily induce a mapping of End A into End B.

6. If X is a fixed set and R a variable ring, show that $R \to R^X$ (the latter defined as in §III.2) can be regarded as (the object function of) a functor on rings to rings.

7. For X a fixed set and G a variable group, show that $G \to G^X$ (of Exercise III.2.9) can be regarded as a functor on groups to groups.

8. In an additive abelian group A, let nA, for a fixed integer n, denote the subgroup of all multiples na for $a \in A$. Show that $A \to A/nA$ can be regarded as a functor on abelian groups to abelian groups.

3. Universal Elements

We can now define a notion of "universal element" for certain functors in such a way as to subsume all the examples of "universals" given in §1 as well as many more to come in later chapters.

DEFINITION. Let $\mathcal{G}$ *be a functor on some category to sets.* A universal element *for* $\mathcal{G}$ *is an ordered pair* (u, R), *where* R *is an object of the category while* u *is an element of the set* $\mathcal{G}(R)$ *with the following property: To any object* S *of the category and any element* $s \in \mathcal{G}(S)$ *there is exactly one morphism* $h:R \to S$ *in the category for which* $\mathcal{G}(h)u = s$.

In other words, u is a universal element for $\mathcal{G}$ when u is in one of the sets $\mathcal{G}(R)$ and *any* element s of any one of the sets $\mathcal{G}(S)$ can be obtained from u by exactly one morphism $h:R \to S$ such that $\mathcal{G}(h)$ sends u into s. Thus to give a universal element for a functor $\mathcal{G}$ to sets is to give *both* an object R and an element u of the associated set $\mathcal{G}(R)$, this in such a way that each $s \in \mathcal{G}(S)$ has the form $\mathcal{G}(h)u$ for a unique h. A universal element (u, R) will often be written not as a pair (u, R) but as $u \in \mathcal{G}(R)$ or even just as u, when the object R is clear from the context.

This concept does include the examples of universals listed in §1:

The number 1 in the monoid $\mathbf{N}$ of natural numbers is universal for the forgetful functor $\mathcal{U}$, which maps each monoid M into the set $\mathcal{U}(M)$ of its elements, and each morphism $g:M \to M'$ of monoids into the same g, regarded as a function; the universal property quoted in §1 says exactly that each element $a \in \mathcal{U}(M)$ is $a = f(1)$ for a unique $f:\mathbf{N} \to M$.

The number $1 \in \mathbf{Z}$ is similarly a universal element for the forgetful functor $\mathcal{U}$ from groups to sets (as discussed in the second example of §1).

There is an evident functor from rings to sets which maps each ring R into the one-point set consisting only of the unit element $1 \in R$. Thus each unit 1 is an element of this functor. The uniqueness of the unital morphism $\mu:\mathbf{Z} \to R$ then amounts to the statement that $1 \in \mathbf{Z}$ is a universal element for this functor. In this case, what matters is not so much the universal element as the object $\mathbf{Z}$ in which it lies.

Using the additive monoid $\mathbf{N}$ we can construct to each abelian group A the set hom $(\mathbf{N}, A)$ of all morphisms $t:\mathbf{N} \to A$ of additive monoids. Much as in (3) and (4), this provides a functor hom $(\mathbf{N}, -)$ on abelian groups to sets.

The embedding $i:\mathbf{N} \to \mathbf{Z}$ of the natural numbers in the integers is then an element $i \in \hom(\mathbf{N}, \mathbf{Z})$ of this functor. Just as in (1) above, it is a universal element. As already noted at (1), it is this universality which characterizes the integers as the "universal" additive group containing the additive monoid $\mathbf{N}$.

In this case, the universal element $i \in \hom(\mathbf{N}, \mathbf{Z})$ of a hom-functor is also called a *universal arrow*. More generally, in considering arrows (that is, morphisms) $T \to X$ from a fixed algebraic system to a variable system X, we call a particular arrow $u:T \to R$ a *universal arrow* when (u, R) is universal as an element of the functor $\hom(T, -)$. The insertion $i:\mathbf{N} \to \mathbf{Z}$ is also a universal arrow when $\mathbf{Z}$ is considered not as a group but as a ring (as explained in §1).

For the projection $p:G \to G/N$ on the quotient of a group G by a normal subgroup N, we consider the set $\hom_N(G, L)$ of those morphisms $t:G \to L$ of groups with $t_*N = 1$. Much as in (3) and (4), this set becomes a functor $\hom_N(G, -)$ on groups L to sets. The projection $p:G \to G/N$ onto the quotient group is a universal element of this functor, by the main Theorem II.26 on quotient groups. For rings, the projection $R \to R/A$ on a quotient ring is a universal element of a similar functor. For each group L in the former case, there is an inclusion of sets

$$\hom_N(G, L) \subset \hom(G, L);$$

we say that $\hom_N(G, -)$ is a "subfunctor" of $\hom(G, -)$. In general, a functor $\mathcal{G}$ is said to be a *subfunctor* of a functor $\mathcal{F}$ to sets when every $\mathcal{G}(S)$ is a subset of $\mathcal{F}(S)$ and every $\mathcal{G}(f)$ is a restriction of $\mathcal{F}(f)$.

The remaining examples of §1 are similarly subsumed under the definition of universal element. Thus for an equivalence relation E on X, $p_E:X \to X/E$ is universal for the subfunctor of $\hom(X, -)$, which consists of all those functions t on X for which xEy implies $tx = ty$.

In all these examples, the universal element is unique, up to an isomorphism. Here an *isomorphism* (of groups, rings, or of objects in any category) can be described as a morphism $f:G \to H$ to which there is another morphism $t:H \to G$ which is a two-sided inverse of f, in the sense that $t \circ f = 1_G$ and $f \circ t = 1_H$. The uniqueness of universal elements can now be stated in general.

Theorem 1. *If $u \in \mathcal{G}(R)$ and $u' \in \mathcal{G}(R')$ are both universal elements for the same functor $\mathcal{G}$ to sets, there is an isomorphism $h:R \to R'$ with $\mathcal{G}(h)u = u'$.*

Proof: Since u is universal, there is a morphism $h:R \to R'$ with $u' = \mathcal{G}(h)u$. Since u' is universal, there is also a morphism $f:R' \to R$ with

$u = \mathcal{G}(f)u'$. Therefore,

$$u = \mathcal{G}(f)u' = \mathcal{G}(f)\mathcal{G}(h)u = \mathcal{G}(f \circ h)u.$$

But $\mathcal{G}$ is a functor, so $\mathcal{G}(1_R)$ is the identity; hence, u is also $\mathcal{G}(1_R)u$. Thus there are two morphisms 1_R and $f \circ h: R \to R$ both of which carry u to u. But u universal means there is only one such morphism; hence, $1_R = f \circ h$. Similarly, u' universal gives $1_{R'} = h \circ f$. Thus h has a two-sided inverse f, hence is an isomorphism $h: R \to R'$, just as required.

This useful result says that it does not matter how a universal object R is constructed, provided it *has* the universal property. Any other universal object will be isomorphic, so have all the same properties. For example (§I.8), the ring $\mathbf{Z}_n$ of integers modulo n may be constructed so that its elements are remainders modulo n or residue classes modulo n; in either way one gets a ring $\mathbf{Z}_n$ and a morphism $\mathbf{Z} \to \mathbf{Z}_n$ universal among the morphisms $f: \mathbf{Z} \to R$ for which $h \equiv k \pmod{n}$ implies $fh = fk$. Similarly, the quotient group G/N was constructed using cosets, but these do not really matter, since the properties of G/N all follow from the universality of the projection $G \to G/N$.

There are other examples of isomorphic universal elements. For instance, given the real line $\mathbf{R}$, consider for each S the set $\mathcal{G}(\mathrm{S})$ of all functions $f: \mathbf{R} \to \mathrm{S}$ with "period" 2π; these are the functions with $f(x + 2\pi) = f(x)$ for each $x \in \mathbf{R}$. Clearly, $\mathcal{G}$ is a functor on sets to sets; it is a subfunctor of hom $(R, -)$. It has several different universal elements. First, if E is the equivalence relation on $\mathbf{R}$ for which xEy means that $x - y$ is an integral multiple of 2π, then the projection $p_E: \mathbf{R} \to \mathbf{R}/E$ to the quotient set is, as noted above, a universal element for $\mathcal{G}$. Next, let S^1 be the unit circle in the plane and $w: \mathbf{R} \to \mathrm{S}^1$ the function with $w(x) = (\cos x, \sin x)$; thus w is the function which "wraps" the real line $\mathbf{R}$ uniformly around the circle. Any function f of period 2π can be written as a composite $f = g \circ w$ for a unique function $g: \mathrm{S}^1 \to \mathrm{S}$; this states that w is a universal element for $\mathcal{G}$ (a universal function of period 2π). Alternatively, if $J = \{y \mid y \in \mathbf{R} \text{ and } 0 \leqslant y < 2\pi\}$, the function $t: \mathbf{R} \to J$ which assigns to each real number x its "remainder" modulo 2π (that number y in J with $x - y$ an integral multiple of 2π) is also universal for $\mathcal{G}$. Finally, the usual trigonometric function tan x is infinite at $x = \pi/2 \pm n\pi$, so we may regard it as a function $\tan: \mathbf{R} \to \mathbf{R} \cup \{\infty\}$. Now tan x has period π. On consulting its graph, we see that $\tan(x/2)$ will be universal of period 2π. Thus we have at least four familiar functions, each universal for $\mathcal{G}$. By the uniqueness theorem, there are bijections $\mathbf{R}/E \cong \mathrm{S}^1 \cong J \cong \mathbf{R} \cup \{\infty\}$!

Now return to the definition of a universal element $u \in \mathcal{G}(R)$. It states that each element $s \in \mathcal{G}(\mathrm{S})$ has the form $\mathcal{G}(h)u$ for a unique morphism $h: R \to \mathrm{S}$, so that $h \mapsto \mathcal{G}(h)u$ is a bijection. This proves

Theorem 2. *If $u \in \mathcal{G}(R)$ is a universal element for a functor $\mathcal{G}$ to sets, then, for each S, $h \mapsto \mathcal{G}(h)u$ is a bijection* $\hom(R, S) \cong \mathcal{G}(S)$.

This is called the "representation theorem" because the bijection represents each set $\mathcal{G}(S)$ as a hom-set $\hom(R, S)$,

$$\hom(R, S) \cong \mathcal{G}(S). \tag{5}$$

EXERCISES

1. If Q denotes the field of quotients, show that $j: D \to Q(D)$ is a universal element of a functor, and specify the functor.
2. Show in any category that the identity morphism 1_A is universal for the functor $\hom(A, -)$.
3. For the defining relations (II.17) of the dihedral group, show that the generators R, D of Δ_n satisfying these relations do constitute a universal element. Specify the functor.
4. Show that the operation "cartesian square" is a functor $\mathcal{G}(S) = S \times S$ and $\mathcal{G}(f) = f \times f$ on sets to sets, and construct a universal element (*Hint:* Use a set with just two elements).
5. Construct a universal element for the functor $\mathcal{H}$ on sets to sets with $\mathcal{H}(S) = S \times S \times S$.
6. Given two functions $h, k: X \to Y$, consider for each set S the set $\mathcal{G}(S)$ of all functions $f: Y \to S$ with $f \circ h = f \circ k$. (These are the functions which "equalize" h and k.) Show that G is a subfunctor of $\hom(Y, -)$, and prove that it has a universal element (called a "coequalizer" of h and k).
7. Let $\mathcal{F}$ be a functor to sets and G_0 a function which assigns to each object S a subset $G_0(S) \subset \mathcal{F}(S)$. Prove that G_0 is the object function of a subfunctor of $\mathcal{F}$ if and only if $f: S \to T$ and $s \in G_0(S)$ imply that $F(f)s \in G_0(T)$.

4. Polynomials in Several Variables

Our construction of the polynomial ring $K[x]$ from the commutative ring K in Chapter III was really a "universal" way of adjoining one new element x to a ring K. This way of viewing polynomials turns out to be useful in constructing polynomials in several variables. The uniqueness of universal elements allows us to readily change the order in which these variables are added.

The construction of the polynomial ring $K[x]$ in one indeterminate x had the following form. From the given commutative ring K we constructed a new ring $K[x]$ containing a designated element x, called "the indeterminate", together with a monomorphism $i: K \to K[x]$ (which could be regarded as "the insertion"). This monomorphism i, by Theorem III.18, had the following universal property: Given any commutative ring L and a designated element $d \in L$ and given a morphism $\alpha: K \to L$ of rings, there is a unique morphism $\alpha': K[x] \to L$ of rings with $\alpha' \circ i = \alpha$ and $\alpha'(x) = d$ (this last equation states that α' carries the designated indeterminate x into the designated element $d \in L$). The commutative diagram is

$$\begin{array}{ccc} K & \xrightarrow{i} & K[x] \ni x \\ & \underset{\alpha}{\searrow} & \vdots \alpha' \quad \downarrow \\ & & L \ni d. \end{array} \tag{6}$$

To include this example formally under the definition of universal element, we define a *pointed* commutative ring (L, d) to be a commutative ring L together with a designated element $d \in L$. (The name "pointed ring" is chosen because a designated or selected element d can be considered as a "point" in the "space" L.) Such a pointed commutative ring (L, d) can be regarded simply as a commutative ring with one additional nullary operation "select d" (see the definition of nullary operations in §I.3). Thus a morphism $\beta: (L, d) \to (L, d')$ of such rings is just a morphism of rings which also carries d into d' (so that $\beta(d) = d'$).

Now (6) means that the insertion $i: K \to K[x]$ is universal among morphisms on K to pointed commutative rings. In more detail, the set hom (K, L) of all morphisms $\alpha: K \to L$ of rings can be regarded as a functor hom $(K, -)$ of the pointed ring (L, d): It sends (L, d) to the set hom (K, L) of all α's and sends β to the function $\alpha \mapsto \beta\alpha$ from hom (K, L) to hom (K, L). The universal property of Theorem III.18, as restated with the diagram above, is just the statement that $(i, (K[x], x))$ is a universal element of this functor hom $(K, -)$. Moreover, the uniqueness theorem for universals is now the statement that any different construction of a "polynomial ring" $K[y]$ in a different "indeterminate" y but with the same universal property would give an isomorphic ring $K[y] \cong K[x]$—where the isomorphism is an isomorphism of *pointed* rings (that is, carries y to x).

Now consider polynomials in two indeterminates x and y. Such a polynomial might be written as a polynomial in y with coefficients polynomials in x, say as

$$(3 + x + 7x^2) + (-2 + 9x - x^2)y + (1 - 4x)y^2; \tag{7}$$

the same expression may be rewritten as a "simultaneous" polynomial in x and y,

$$3 + (x - 2y) + (7x^2 + 9xy + y^2) - (x^2y + 4xy^2),$$

or as a polynomial in x with coefficients polynomials in y,

$$3 - 2y + y^2 + (1 + 9y - 4y^2)x + (7 - y)x^2. \tag{8}$$

With these three forms in mind, we may construct three nominally different but actually isomorphic polynomial rings $K[x, y]$. These may be handled quickly by the universality technique.

Theorem 3. *For each commutative ring K there exists a commutative ring $K[x, y]$ containing K as subring and two elements x and y with the following universal property: To each commutative ring L with two selected elements $d, e \in L$ and to each morphism $\alpha: K \to L$ of rings there exists a unique morphism $\alpha'': K[x, y] \to L$ of rings which extends α and has the property $\alpha''(x) = d$, $\alpha''(y) = e$.*

Proof: For $K[x, y]$ we will take the ring $(K[x])[y]$ which consists of polynomials in y with coefficients polynomials in an indeterminate x over K. The given morphism $\alpha: K \to L$ then extends by (6) to a morphism $\alpha': K[x] \to L$ with $\alpha'(x) = d$, as in the figure below. This α' in its turn extends by the universality (6) again,

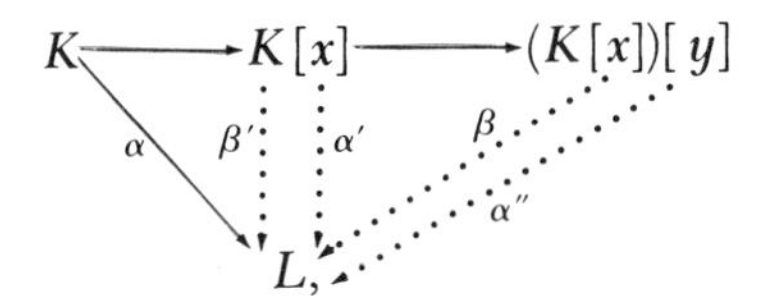

to a morphism α'' with $\alpha''(x) = d$ and $\alpha''(y) = e$. These conditions make α'' unique, for if $\beta: (K[x])[y] \to L$ were another morphism with $\beta(x) = d$ and $\beta(y) = e$, then β restricted to $K[x]$ is a morphism β' with $\beta'(x) = d$, so $\beta' = \alpha'$ by the uniqueness assertion of universality. Then β is an extension of $\beta' = \alpha'$ with $\beta(y) = e$, so the uniqueness assertion of universality gives $\beta = \alpha''$. This completes the proof.

This argument is symmetric in x and y. For instance, we could take L to be the ring $(K[y])[x]$ of polynomials in x with coefficients polynomials in y and $\alpha: K \to (K[y])[x]$ to be the insertion. Then the construction gives a unique morphism α'' as shown,

$$(K[x])[y] \underset{\gamma''}{\overset{\alpha''}{\rightleftarrows}} (K[y])[x],$$

with $\alpha''(x) = x$, $\alpha''(y) = y$. Reversing the roles of x and y yields a morphism γ'' as shown, again with $\gamma''(x) = x$ and $\gamma''(y) = y$. Then compare the composite $\gamma'' \circ \alpha''$ to the identity of $(K[x])[y]$. Both extend the insertion $K \to (K[x])[y]$ and map $x \mapsto x$, $y \mapsto y$. Hence, by the uniqueness part of Theorem 3, they must be equal: $\gamma''\alpha'' = 1$. Symmetrically, $\alpha''\gamma'' = 1$. In other words, $\alpha'' = \theta$ is an isomorphism. We have proved

Corollary. *For indeterminates x and y and any commutative ring K there is an isomorphism*

$$\theta\colon (K[x])[y] \cong (K[y])[x]$$

of rings with $\theta(x) = x$, $\theta(y) = y$ and $\theta(a) = a$ for each $a \in K$.

This is exactly the isomorphism exhibited above in (7) and (8), which takes a polynomial in y, with coefficients polynomials in x, and rewrites it as a polynomial in x, with coefficients in y. Hence, the uniqueness part of the definition of universals proves that this "rewriting" process is an isomorphism of rings (a direct proof of this fact would be much fussier).

The elements x, y in $K[x, y]$ are called *simultaneous indeterminates* over K. Each element $f \in K[x, y]$ is a polynomial in these indeterminates and defines a function $K \times K \to K$. Indeed, take α in the theorem to be the identity $K \to K$. Call the corresponding morphism $\alpha''\colon K[x, y] \to K$ the "evaluation" $E_{d,e}\colon K[x, y] \to K$. It is a morphism of rings with

$$E_{d,e}(x) = d, \qquad E_{d,e}(y) = e, \qquad E_{d,e}(a) = a, \qquad\qquad a \in K.$$

For each fixed $f \in K[x, y]$ the assignment $(d, e) \mapsto E_{d,e}(f)$ is a function $f^{\text{‘}}\colon K \times K \to K$ with the values $f^{\text{‘}}(d, e) = E_{d,e}f$, just as in the case of one indeterminate x.

EXERCISES

1. Describe the universality established in Theorem 3 as a universal element of a suitable functor on the category of bipointed rings (rings with *two* designated elements).

2. State and prove a universality property for the polynomial ring $K[x_1, \cdots, x_n]$ in n indeterminates x_i. Deduce that each permutation $\sigma\colon \mathbf{n} \to \mathbf{n}$ in the symmetric group S_n gives an automorphism of $K[x_1, \cdots, x_n]$ which is the identity on K and permutes the x_i as σ does.

3. For any set X, not necessarily finite, construct a polynomial ring $K[X]$ with an appropriate universal property. (*Hint:* Any one polynomial of the ring will involve only a finite number of the elements ("indeterminates") of X.)

4. Using the evaluation E, show that each polynomial in $K[x, y]$ can be interpreted as a function $K \times K \to K$, so as to yield a morphism of $K[x, y]$ into the function ring $K^{K \times K}$ (compare Theorem III.17).

5. For any ring R (possibly noncommutative) construct a polynomial ring $R[x]$ in which the indeterminate x commutes with all the elements of R.

Show that the insertion $i: R \to R[x]$ is universal among morphisms from R to those pointed rings (S, d) in which d commutes with every element of S.

6. If D is an integral domain, show that the set (x, y) in the ring $D[x, y]$ consisting of all polynomials with constant term 0 is an ideal in $D[x, y]$ but not a principal ideal.

7. In $D[x, y]$, with D an integral domain, show that (x^2, xy, y^2) is not a principal ideal. Does it have a basis of two elements?

8. If F is a field, find all ideals in the quotient ring $F[x, y]/(x^2, xy, y^2)$ and show that this ring has ideals which are not principal.

5. Categories

Our functors have been defined on the collection (or *class*) of all groups, or all rings, or all algebraic systems of some other fixed type. Classes like these are called "categories"; they consist of the objects (groups or rings) and the morphisms between objects. These morphisms can be described in terms of the set hom (P, Q) of all morphisms from the object P to the object Q. This suggests the following definition of a "concrete" category; there is a more general notion of an "abstract" category which will not be needed before Chapter XV.

A *concrete category* consists of:

(i) A class C of elements P, called the *objects* of the category.

(ii) A function $\mathfrak{U}$ which assigns to each object P a set $\mathfrak{U}(P)$, called the "underlying set" of P.

(iii) A function hom which assigns to each pair of objects P and Q a set hom (P, Q) whose elements are some of the functions f from the set $\mathfrak{U}(P)$ to the set $\mathfrak{U}(Q)$.

When f is an element of hom (P, Q), we write $f: P \to Q$ and say that f is a *morphism* of the category from P to Q—or, a morphism with *domain* P and *codomain* Q. The data for a category must satisfy two axioms:

(iv) For each object P, the identity function $1_{\mathfrak{U}(P)}$ is in hom (P, P).

(v) For any three objects P, Q, and R, if $f: P \to Q$ and $g: Q \to R$ are morphisms, then the composite $g \circ f$ is a morphism $g \circ f: P \to R$.

The second axiom asserts that suitable pairs of morphisms in a category can be composed. Since the composition of functions is associative, this composition of morphisms in any concrete category is also associative. Also, the identity functions $1_{\mathfrak{U}(P)}$ act as identities for this composition.

Here are some examples of categories we have already used.

The *category of sets* is the class of all sets, with $\mathfrak{U}$ the identity so that each set is its own "underlying" set. Also hom (P, Q) is just the set of *all* functions from P to Q.

Recall that a monoid $(M, \square, u)$ has been defined (§I.11) to be a set M with a binary operation $\square$ and an element $u \in M$ satisfying certain axioms ($\square$ associative, u a unit for $\square$). The notation in this definition has been chosen so as to distinguish between the set M and the monoid $(M, \square, u)$.

The *category of monoids* is the class of all monoids $(M, \square, u)$, with $\mathcal{U}$ given by $\mathcal{U}(M, \square, u) = M$, while hom $((L, \square', u'), (M, \square, u))$ is the set of all morphisms $h:L \to M$ of monoids.

The *category of groups* is the class of all groups G. The function $\mathcal{U}$ assigns to each group G the set of all elements in G (our usual notation does not distinguish between the set $\mathcal{U}(G)$ and the group G). Also hom (G, H) is the set of all morphisms of groups from G to H. The axioms (iv) and (v) above hold because the identity function is a morphism of groups and because the composite of any two morphisms of groups is itself a morphism of groups.

The *category of finite groups* is the class of all *finite* groups, with $\mathcal{U}$ and hom as before. It is clearly a "subcategory" of the category of groups.

The *category of rings* is the class of all rings R, with $\mathcal{U}(R)$ the set of elements in the ring R and hom (R, S) the set of all morphisms of rings $f:R \to S$. It has as subcategories the *category of commutative rings* and the *category of finite rings*.

These categories (and many others like them) are very large ones. There are also many "smaller" categories. For example, a single set P with its identity morphism $1:P \to P$ constitutes a category with only one morphism. The same set P with the whole set hom (P, P) of all functions $f:P \to P$ is also a concrete category. This particular category is really just the monoid $(P^P, \circ, 1_P)$ of all functions $f:P \to P$. On the other hand, in *any* category the set hom (P, P) of all endomorphisms of a fixed object P is closed under composition and hence is a monoid. Thus a category can be regarded as a "monoid with several objects".

Here is another, smaller example. Consider four different sets P, Q, R, and S and four functions f, g, f', and g' such that the diagram

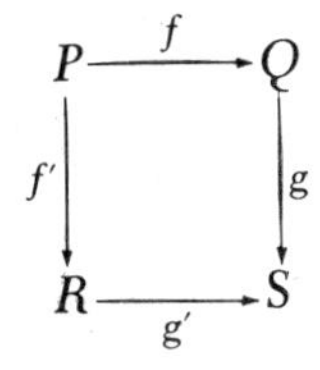

commutes (that is, such that $g' \circ f' = g \circ f$). This diagram determines a concrete category whose objects are the four sets P, Q, R, and S, with $\mathcal{U}(P) = P$, $\mathcal{U}(Q) = Q$, and so on. The morphisms are f, g, f', g', the composite h with $h = g \circ f = g' \circ f'$, and the four identity functions $1_P, 1_Q, \ldots$. Each hom-set is either empty or consists of a single morphism: thus hom $(P, Q) = \{f\}$, hom $(P, S) = \{h\}$ and hom $(P, P) = \{1_P\}$, while hom (Q, P) and hom (S, Q) are empty. This category is evidently a (very small) subcategory of the category of all sets. Other commutative diagrams of sets will lead to similar categories.

A functor (defined much as in §2) is essentially just a morphism of categories. Specifically, a *functor* $\mathfrak{F}: C \to C'$ on the category C to the category C' must assign to each object P of C an object $\mathfrak{F}(P)$ of C' and to each morphism $f: P \to Q$ of C a morphism $\mathfrak{F}(f): \mathfrak{F}(P) \to \mathfrak{F}(Q)$ of the second category C' in such a way that identities and composites are preserved, as in (2). This also means that the assignment $f \mapsto \mathfrak{F}(f)$ is a function

$$\hom(P, Q) \to \hom(\mathfrak{F}(P), \mathfrak{F}(Q))$$

between the corresponding hom-sets, but note, however, that the definition of a functor makes no reference to the "underlying set" function $\mathfrak{U}$. In general, the underlying sets are not of importance in the treatment of categories, and they will be dropped when we shift from concrete categories to categories in Chapter XV.

Examples of functors have already been described in §2. For any category C, the function $\mathfrak{U}$ defines a functor on C to the category of sets, called the *underlying set* functor. Also, if P is an object in C, then hom $(P, -)$ is a functor on C to sets, just as described in §2.

EXERCISES

1. Describe the concrete category of all abelian groups.
2. Describe the concrete category of all semigroups and give a functor from monoids to semigroups which "forgets" the unit element.
3. Given functors $\mathfrak{F}: C \to C'$ and $\mathfrak{G}: C' \to C''$, show that the composite $\mathfrak{G} \circ \mathfrak{F}$, suitably described, is a functor $\mathfrak{G} \circ \mathfrak{F}: C \to C''$.

6. Posets and Lattices

Algebraic systems have morphisms and subsystems. Their morphisms can be analyzed by the use of categories; to handle their subsystems we now consider lattices.

Inclusion relations between the subgroups of a group can be suggestively represented by an *inclusion diagram*, as in §II.4. There are similar diagrams for the inclusion relations between the ideals in a ring. Such inclusion relations are called "partial" orders, to distinguish them from orderings which are "total" or "linear", because they satisfy the trichotomy law (§I.3). These partial orders have some simple but important general properties, to which we now turn our attention.

A binary relation, written $x \leqslant y$, on a set X is said to be *antisymmetric* when, for all $x, y \in X$,

$$x \leqslant y \quad \text{and} \quad y \leqslant x \quad \Rightarrow \quad x = y.$$

A binary relation $\leqslant$ on a set X is called a *partial order* on X when it is reflexive, transitive, and antisymmetric. Read "$x \leqslant y$" as "x is contained in y". A set X with a partial order $\leqslant$ is called a *poset* $(X, \leqslant)$—for example, $\mathbf{N}$ and $\mathbf{Z}$ are posets under the usual linear order relation, while the power set $P(U)$ of all subsets of a set U is a poset under the inclusion relation $S \subset T$. Again, the set of all positive divisors k of a given integer n is partly ordered by the relation of divisibility: k *divides* m (in symbols, $k|m$) when $m = qk$ for some integer q. A finite poset may be displayed by an *inclusion diagram*, as with the diagrams

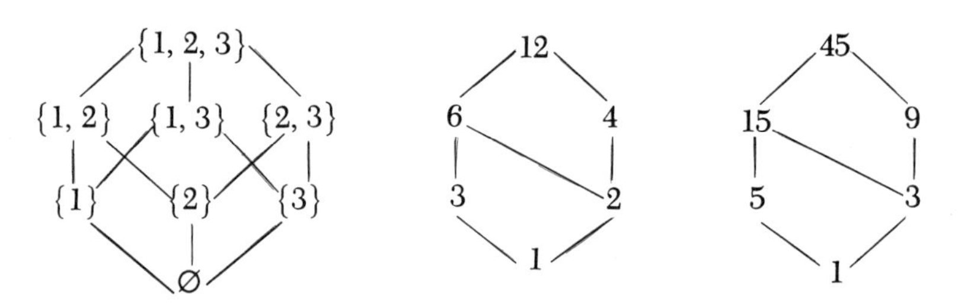

for the subsets of $\{1, 2, 3\}$, the divisors of 12, and the divisors of 45. In such a diagram, x is joined to y by a rising line precisely when $x < y$ and there is no z with $x < z < y$.

In many posets, one can define two operations somewhat resembling multiplication and addition. Thus two positive integers h and k have a greatest common divisor (g.c.d.) d; that is, a common divisor which is a multiple of any other common divisor. Likewise, any two subsets S and T of a set U have an intersection $D = S \cap T$, and this is a set which is included in both S and T and contains any other set so included. Note that these two descriptions of g.c.d. and intersection use only the relation $\leqslant$; as applied to posets in general, we say that they describe a *meet* (for example, a g.c.d. or an intersection). Similarly, the least common multiple (of positive integers) and the union (of subsets) have a common description, called a *join* in posets generally.

Let x and y be two elements of a poset X. An element $b \in X$ is a *lower bound* for x and y in X when $b \leqslant x$ and $b \leqslant y$. An element $m \in X$ is a *meet* of x and y when it is a lower bound of x and y which contains all lower bounds b. In other words, m is a meet of x and y when

$$m \leqslant x, m \leqslant y, \qquad \text{while} \quad b \leqslant x \text{ and } b \leqslant y \quad \Rightarrow \quad b \leqslant m.$$

If m and m' are both meets of x and y, this definition shows that $m \leqslant m'$ and $m' \leqslant m$, hence $m = m'$ by antisymmetry. Therefore, a meet, when it exists, is unique; we write it as $m = x \wedge y$. Such a meet is also called a "greatest lower bound" (g.l.b.) of x and y, because it is the "greatest" element in the set of all lower bounds of x and y.

An element $u \in X$ is an *upper bound* for x and y when both $x \leqslant u$ and $y \leqslant u$. An element $l \in X$ is a *join* (or, a "least upper bound") of x and y when it is an upper bound contained in all other upper bounds of x and y; that is, when

$$x \leqslant l, \; y \leqslant l, \qquad \text{while} \quad x \leqslant u \text{ and } y \leqslant u \quad \Rightarrow \quad l \leqslant u.$$

Because this definition may be obtained from the definition of meet simply by replacing the relation $x \leqslant y$ by its converse, we say that "join" is the "order dual" of "meet". From the corresponding "dual" result for meets it follows that the join l of two elements x, y in the poset X is unique, when it exists; we write it as $l = x \vee y$.

A *lattice* is a poset L in which any two elements x and y have both a meet and a join. Thus, if L is a lattice, meet and join are binary operations $\wedge : L \times L \to L$ and $\vee : L \times L \to L$. For example, the sets **N** and **Z** with the usual order relation are lattices, for the lesser of two numbers m and n is their meet, and the greater is their join. Again, the power set $P(U)$ of all subsets of U is a lattice under the inclusion relation; the meet of two subsets is their intersection; the join is their union.

The following proposition illustrates the way in which lattices describe subsystems of an algebraic system.

Proposition 4. *The subgroups of any group G form a lattice $L(G)$ under inclusion, join, and meet.*

Proof: By Proposition II.10, the intersection $S \cap T$ of two subgroups S and T (as sets) is a subgroup, and thus is their meet $S \wedge T$ in the poset of subgroups. Also, the least subgroup containing both S and T is their join $S \vee T$, by Proposition II.12. This join can also be described (as in §II.4) as the set of all products $s_1t_1s_2t_2 \cdots s_nt_n$ of a finite number of elements of S and T, in alternation.

Similarly, the set of all normal subgroups of G is a lattice. So is the set of all ideals $A, B, \cdots$ in a given ring R, partially ordered by inclusion, with meet the intersection $A \cap B$ and join (see §III.3) the sum $A + B$ of the ideals A and B.

Morphisms of Lattices. If L and L' are lattices, a *morphism* $f{:}L \to L'$ *of lattices* is a function $f{:}L \to L'$ which is a morphism both for the binary operation "meet" and for the operation "join". The composite of two morphisms of lattices is again such, while the identity function for any lattice L is a morphism $L \to L$. We shall later meet many examples of lattices and their morphisms; in Chapter XIV we shall then make a systematic study of them.

Likewise, given two posets, a *morphism* $f:(X, \leqslant) \to (X', \leqslant')$ of *posets* (or, a *morphism of partial order*) is a function $f:X \to X'$ such that

$$x \leqslant y \quad \Rightarrow \quad f(x) \leqslant' f(y) \qquad \text{for all } x, y \in X. \tag{9}$$

As usual, the composite of two morphisms of posets is again such a morphism, and for each poset $(X, \leqslant)$ the identity function 1_X is a morphism of posets. For instance, the last two diagrams above suggest an isomorphism from the poset of all divisors of 12 to that of all divisors of 45.

An *isomorphism* $f:(X, \leqslant) \to (X', \leqslant')$ of *posets* is a bijection $f:X \to X'$ such that both f and f^{-1} are morphisms of posets. An *isomorphism* $f:L \to L'$ of *lattices* is a bijection which is also a morphism of lattices; its inverse f^{-1} is then automatically also a morphism of lattices. For example, the bijection in Theorem II.22 is an isomorphism of lattices.

Although an isomorphism of posets which are lattices necessarily preserves meet and join, hence is an isomorphism of lattices, and *any* morphism of lattices is a morphism of posets, most morphisms of posets are not morphisms of lattices. Thus the mapping which carries each subset S of a group into the subgroup $\gamma(S)$ which it generates is a morphism of posets (even of joins), but not a morphism of lattices because it does not preserve meets.

EXERCISES

1. Find an isomorphism of posets from the set of all subsets of $\{1, 2, 3\}$ to the set of all positive divisors of 30. Find *all* such isomorphisms.

2. (a) If the posets X and X' are lattices, prove in detail that any isomorphism $X \cong X'$ of posets is also an isomorphism of lattices.

(b) Prove that any morphism $X \to X'$ of lattices is a morphism of posets. (*Hint:* $x \leqslant y$ in X if and only if $x \wedge y = x$.)

3. If R is a fixed subset of a set U, show that $S \mapsto S \cap R$ is an endomorphism of the lattice $P(U)$ of all subsets of U.

4. Show that the set of all positive natural numbers is a lattice under the relation of divisibility. If n is a fixed natural number, show that $k \mapsto nk$ is an endomorphism of this lattice.

5. In a monoid, show that the set of all submonoids, partially ordered by inclusion, is a lattice.

7. Contravariance and Duality

A "diagram", informally speaking, consists of vertices $p, q, \cdots$ together with arrows from one vertex to another, with each vertex p labeled by an object S_p and each arrow $p \to q$ labeled by a morphism f on S_p to S_q. A "path" in a diagram is a succession of arrows such as $p \xrightarrow{f} q \xrightarrow{g} r \xrightarrow{h} t$; each path determines the corresponding composite morphism $h \circ g \circ f$ from S_p to S_t. Finally, a diagram is *commutative* when any two paths from any one vertex p to any other vertex s in the diagram yield by composition the same morphism $S_p \to S_s$.

The "dual" of a diagram is the one obtained by reversing all arrows. If a concept is defined by means of a diagram, the concept defined by the dual diagram will be called (informally) the "dual" concept. For example, "X is the domain of the function f" means that "X is at the tail of the arrow f". Since reversing arrows changes tail to head, the dual concept is just "X is at the head of the arrow f", which means that X is the codomain of the function f. In these cases, once a concept is named, the dual concept may receive the same name with the prefix "co". But take heed; no terminological rules are absolute: "f has a left inverse" is dual to "f has a right inverse".

A functor is "contravariant" when it reverses arrows. More formally, a *contravariant functor* $\mathcal{C}$ on a category to sets assigns to each object S a set $\mathcal{C}(S)$ and to each morphism $f: S \to T$ a function $\mathcal{C}(f): \mathcal{C}(T) \to \mathcal{C}(S)$ (backward!) in such a way that

$$\mathcal{C}(1_S) = 1_{\mathcal{C}(S)} \qquad \text{and} \qquad \mathcal{C}(g \circ f) = (\mathcal{C}f) \circ (\mathcal{C}g), \tag{10}$$

the latter whenever the composite $g \circ f$ is defined. These conditions may be visualized by the following commutative diagrams:

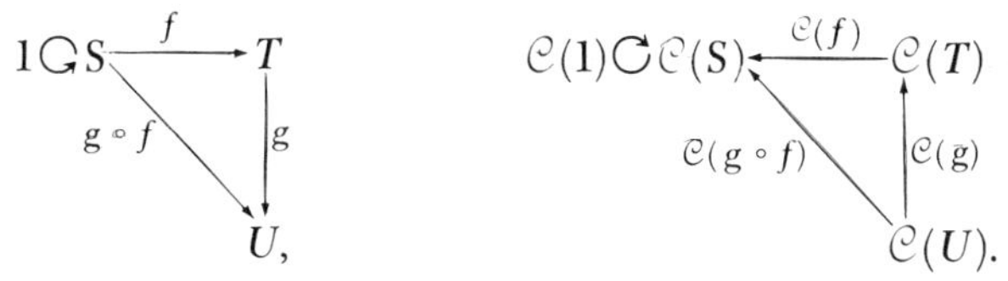

$$\begin{array}{ccc} 1 \circlearrowleft S & \xrightarrow{f} & T \\ & {\scriptstyle g \circ f} \searrow & \downarrow {\scriptstyle g} \\ & & U, \end{array} \qquad \begin{array}{ccc} \mathcal{C}(1) \circlearrowleft \mathcal{C}(S) & \xleftarrow{\mathcal{C}(f)} & \mathcal{C}(T) \\ & {\scriptstyle \mathcal{C}(g \circ f)} \nwarrow & \uparrow {\scriptstyle \mathcal{C}(g)} \\ & & \mathcal{C}(U). \end{array}$$

Note that $\mathcal{C}$ takes the left-hand commutative diagram into the dual diagram on the right, with all arrows reversed.

Functors, as previously defined, are sometimes said to be "covariant functors" in contrast to the present contravariant ones. The word "functor", without qualifying adjective, will always mean a covariant functor.

For example, let A be a fixed object of the category, while, for each object T, $\mathcal{C}(T) = \hom(T, A)$ is the set of all morphisms $T \to A$ in the category. If $f: S \to T$ is a morphism, there is a corresponding function

$$\mathcal{C}(f) : \mathcal{C}(T) = \hom(T, A) \to \mathcal{C}(S) = \hom(S, A),$$

which sends each $t: T \to A$ to the composite $tf: S \to A$. By associativity of composition, it follows readily that rules (10) hold, so that $\mathcal{C} = \hom(-, A)$ is a contravariant functor, on the category in question to sets. It is called the *contravariant hom-functor* (for the object A).

There are other such functors, say on the category of groups. In this category the set hom (A, B) is never empty, since it contains the trivial morphism $A \to B$ which sends every element of A to the unit element of B. Now consider a fixed morphism $\phi : A \to B$ of groups and for each group T let the set $N_\phi(T)$ consist of those morphisms $t : T \to A$ with ϕt trivial:

$$N_\phi(T) = \{t | t : T \to A, \phi t \text{ trivial}\}.$$

Then for any morphism $f : S \to T$, one has $\phi t f$ trivial when ϕt is, so that $t \mapsto tf$ is a function

$$N_\phi(f) : N_\phi(T) \to N_\phi(S), \qquad t \mapsto tf.$$

This makes N_ϕ a contravariant functor on groups to sets. Actually, each set $N_\phi(T)$ is a subset of hom (T, A) and $N_\phi(f)$ is a restriction of $\mathcal{C}(f)$ to this subset, so that N_ϕ is a subfunctor of the hom-functor $\mathcal{C} = \text{hom}(-, A)$.

The kernel $K = \text{Ker}\ \phi$ of a morphism $\phi : A \to B$ of groups was originally described (§II.9) as the set of those elements $a \in A$ with $\phi a = 1$, the unit of B. Alternatively, we can now give a universal property which describes the whole kernel K—or, more exactly, describes its insertion $\text{Ker}\ \phi \to A$.

Theorem 5. *If $\phi : A \to B$ is a morphism of groups with kernel* $\text{Ker}\ \phi \subset A$, *then for the insertion $i : \text{Ker}\ \phi \to A$ the composite ϕi is then trivial and is universal with this property, in the sense that any morphism $t : L \to A$ with ϕt trivial has $t = ih$ for a unique morphism $h : L \to \text{Ker}\ \phi$.*

This is shown in the following commutative diagram

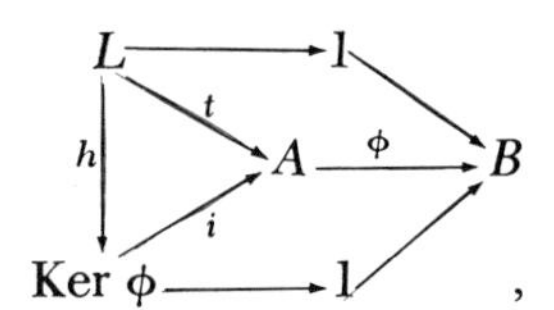

,

where a trivial morphism ϕt is equal to the composite $L \to 1 \to B$; it can be "factored through the group 1". If t has ϕt trivial, there is a unique h with $t = ih$, so the triangle shown will commute.

The proof of the theorem is immediate. If the composite ϕt is trivial, the image $t(x)$ of each $x \in L$ must be in $\text{Ker}\ \phi$, so $x \mapsto tx$ is a morphism $h : L \to \text{Ker}\ \phi$ with $ih = t$. It is clearly the only such.

This description uses the set of all t with ϕt trivial; this is just the set $N_\phi(L)$ for the functor N_ϕ considered above. In this language, the theorem states that $i : \text{Ker}\ \phi \to A$ is a "universal element" of the contravariant functor N_ϕ. Here is the general definition.

DEFINITION. If $\mathcal{C}$ is a contravariant functor on a concrete category to the category of sets, then a universal *element for $\mathcal{C}$ is a pair (v, R), where R is an object of the category while v is an element $v \in \mathcal{C}(R)$ with the following property: Given any object S and any element $s \in \mathcal{C}(S)$ there is exactly one morphism $f: S \to R$ with $\mathcal{C}(f)v = s$.*

Just as in the covariant case, this implies that the representing object R is unique, up to isomorphism. It also gives the representation theorem, stating that

$$\hom(S, R) \cong \mathcal{C}(S). \tag{11}$$

This description of kernels via universality has two effects. First, the same description of a kernel will (and does) work in any category with "trivial" morphisms. Second, it raises a question: If i is universal for $\phi \circ i$ trivial, what morphism j is universal for $j \circ \phi$ trivial (see Exercise 1b)?

There is also a categorical description of the product $G \times G'$ of two groups (§II.1) or, more exactly, for the projections

$$G \xleftarrow{p} G \times G' \xrightarrow{p'} G', \qquad g \leftarrow\!\shortmid (g, g') \mapsto g'$$

of that product, as follows.

Proposition 6. *For any group L and any pair $k: L \to G$, $k': L \to G'$ of morphisms of groups there is a unique morphism $h: L \to G \times G'$ with $p \circ h = k$ and $p' \circ h = k'$.*

Proof: Let the morphisms k and k' be given. Then for $x \in L$, $h(x) = (k(x), k'(x))$ does define a morphism $h: L \to G \times G'$ such that $ph = k$ and $p'h = k'$. This is the only way to define such a morphism h, for the condition $ph(x) = k(x)$ forces the first component of the pair $h(x)$ to be $k(x)$, and similarly for the second component, Q.E.D.

Exactly the same proof applies to the product of rings (§III.2) and to the product of sets (§I.3).

This proposition may be phrased in terms of a diagram:

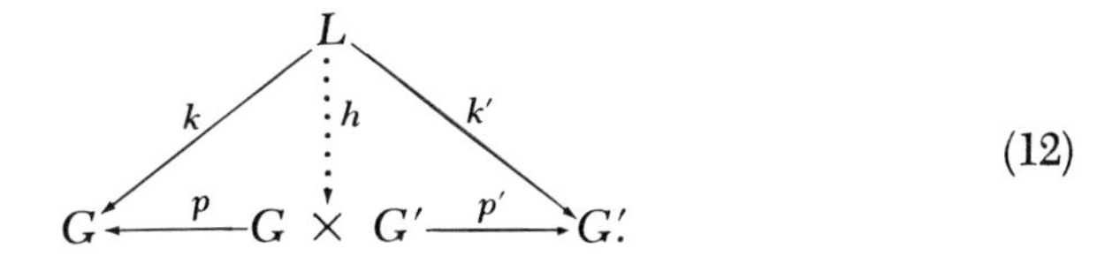

(12)

Given p, p' and k, k', there is a unique way to fill in at h to make both triangles commute. In other words, (p, p') is a universal element for the following contravariant functor on groups to sets: To each group L, assign the set $\hom(L, G) \times \hom(L, G')$ of all pairs (k, k'); to each morphism $\alpha: M \to L$ of groups assign the function $(k, k') \mapsto (k \circ \alpha, k' \circ \alpha)$. This functor could be described more briefly as the product $\hom(-, G) \times \hom(-, G')$ of two hom-functors. This description applies to *any* category and so suggests the following definition of the product object *with* its projections p, p'.

DEFINITION. A product of two objects G, G' in any category is a universal element for the contravariant functor $\hom(-, G) \times \hom(-, G')$. *If such an element exists, it is a pair of morphisms $p: D \to G$, $p': D \to G'$ of the category with a common domain D. This domain is called a* product object, *written $D = G \times G'$, while p and p' are called the* projections *from the product object.*

In this language, Proposition 6 above states that the product of two objects exists for any two objects in the category of all groups. For that matter, G and G' abelian imply $G \times G'$ abelian, so a product also exists in the category of all abelian groups.

In any category with product *objects* there are also product *morphisms*. Indeed, given two morphisms $\alpha: G \to H$ and $\alpha': G' \to H'$ of the category, form both products $G \times G'$ and $H \times H'$ and their projections as in the diagram

$$\begin{array}{ccccc} G & \xleftarrow{p} & G \times G' & \xrightarrow{p'} & G' \\ \downarrow{\scriptstyle\alpha} & & \vdots{\scriptstyle\alpha \times \alpha'} & & \downarrow{\scriptstyle\alpha'} \\ H & \xleftarrow{q} & H \times H' & \xrightarrow{q'} & H' \end{array} \tag{13}$$

(dotted arrow to be added later). Here the two composites $\alpha \circ p$, $\alpha' \circ p'$ have codomains H, H', while (q, q') is a universal pair of morphisms to these codomains. Therefore, there is a unique morphism $\alpha \times \alpha'$, as shown at the dotted arrow, which will make this diagram commute. Call this morphism $\alpha \times \alpha': G \times G' \to H \times H'$ the *product morphism*. In the category of sets it is just the cartesian-product function of §I.3; for groups it is the morphism $(g, g') \mapsto (\alpha(g), \alpha'(g'))$. For any category with products, one may prove that

$$1_G \times 1_{G'} = 1_{G \times G'}, \qquad (\beta \circ \alpha) \times (\beta' \circ \alpha') = (\beta \times \beta') \circ (\alpha \times \alpha'), \tag{14}$$

the latter for any pair of morphisms $\beta: H \to K$, $\beta': H' \to K'$ of the category. These equations state that $(G, G') \mapsto G \times G'$, $(\alpha, \alpha') \mapsto \alpha \times \alpha'$ form a functor (of two variables).

The notion of a product has a dual, called a "coproduct". The description is given by simply reversing all the arrows in the categorical description (12) of a product. Thus a coproduct of two objects G and G' in a category is an object C with a pair of morphisms $e:G \to C$ and $e':G' \to C$ which are universal among such pairs from G and G'. This means that to any object B with a pair of morphisms $f:G \to B$ and $f':G' \to B$ there is exactly one morphism $h:C \to B$ with $h \circ e = f$ and $h \circ e' = f'$, as in the following commutative diagram:

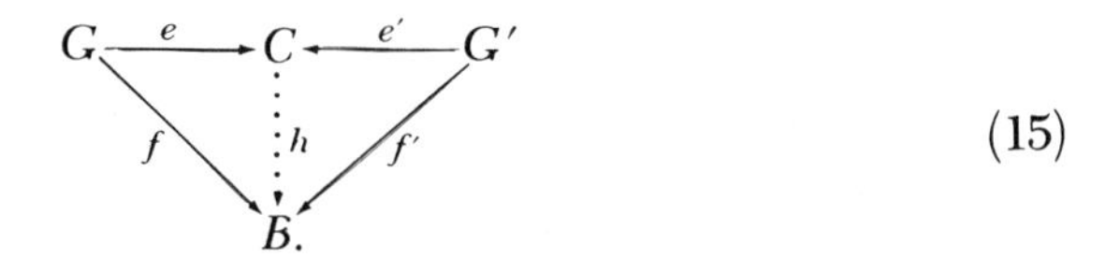

(This is the diagram (12) above for the product, with all arrows reversed.) This definition can be stated formally, as follows:

DEFINITION. A coproduct *of two objects* G, G' *in a category is a universal element for the functor* hom $(G, -) \times$ hom $(G', -)$. *If such an element exists, it is a pair of morphisms* $e:G \to C$, $e':G' \to C$ *of the category with the same codomain* C. *This codomain is called the* coproduct object; e *and* e' *are called the* injections *to the coproduct object.*

Coproducts may or may not exist in a given concrete category; when they do exist they have quite varied forms. This will be illustrated by describing the coproduct of two abelian groups and the coproduct of two sets.

For two (additive) abelian groups A and A' it turns out that the product $A \times A'$ is also a coproduct. In the product, the addition is given by the formula

$$(a_1, a_1') + (a_2, a_2') = (a_1 + a_2, a_1' + a_2').$$

Also there are two morphisms

$$A \xrightarrow{e} A \times A' \xleftarrow{e'} A'$$

defined by $a \mapsto (a, 0)$ and $a' \mapsto (0, a')$ for $a \in A$, $a' \in A'$. To show that these are the injections of a coproduct, consider any other pair of morphisms $f:A \to B$, $f':A' \to B$ into an abelian group B. Then we can define a function $h:A \times A' \to B$ by setting $h(a, a') = fa + f'a'$. If this function h is to be a morphism of abelian groups, the addition formula above shows that one must have

$$fa_1 + f'a_1' + fa_2 + f'a_2' = fa_1 + fa_2 + f'a_1' + f'a_2'.$$

This is indeed true because the two middle terms can be interchanged (the addition is commutative). The morphism h so constructed evidently is the only one with $h \circ e = f$ and $h \circ e' = f'$. Thus the product $A \times A'$, taken with the morphisms e and e', is a coproduct. Because this group $A \times A'$ is both a product (with projections p and p') and a coproduct in the category of abelian groups, it is called a *biproduct*.

In the category of *all* groups, this construction would fail because the function h as constructed above is not always a morphism. Hence the group $A \times A'$ is *not* a coproduct in this category. As Exercise 10 below indicates, there is a (much bigger) coproduct in the category of all groups.

In the category of sets, the construction of the coproduct C of two sets X and Y is quite different. The intention is that C will contain both X and Y as subsets, in such a way that each function h on C is completely determined by its values on X and Y. To make this work, X and Y need to be disjoint. It may be that the sets X and Y as given have common elements, so we take first a bijection $b:Y \to Y'$ to a new set Y' (a "copy" of Y) disjoint from X. Now form the union $X \cup Y'$ and the diagram

$$X \xrightarrow{i} X \cup Y' \xleftarrow{j} Y$$

of functions, where i sends each $x \in X$ to itself and j sends each $y \in Y$ to $by \in Y' \subset X \cup Y'$. Now if we are given functions $f:X \to Z$ and $g:Y \to Z$ to some third set Z, then since $X \cap Y' = \varnothing$, we can construct a unique function $h:X \cup Y' \to Z$ sending each $x \in X$ to fx and each $by \in Y'$ to gy. This function h has both $hi = f$ and $hj = g$; it is the only such function, since these two conditions determine h on X and on Y' and hence on all of $X \cup Y'$. We call this new set $X \cup Y'$ a *disjoint union* of X and Y, and sometimes write it as $X \dot{\cup} Y$.

This disjoint union is clearly not unique; in its construction we could have replaced Y' by any other copy of Y disjoint from X; for that matter, we could have replaced X by a suitable copy. However, like all universals, the disjoint union $X \dot{\cup} Y$ is unique up to an isomorphism of sets (that is, up to a bijection).

EXERCISES

1. (a) For any subgroup S of G construct a normal subgroup $N_h(S)$ of G with the properties: (i) $S \subset N_h(S)$; (ii) $S \subset N \lhd G$ implies $N_h(S) \subset N$. This subgroup $N_h(S)$ is called the "normal hull" of S in G.

   (b) The "cokernel" of a morphism $\phi:G \to H$ of groups is the quotient $H/N_h(\phi_* G)$ of H by the normal hull of the image. Prove that the set of all $\psi:H \to L$ with $\psi \circ \phi$ trivial defines a subfunctor of hom $(H, -)$, and that the projection $H \to H/N_h(\phi_* G)$ on the cokernel is a universal element for this functor.

***2.** Given morphisms $\phi: G \to H$ and $\phi': G' \to H$ of groups, define a contravariant functor $\mathcal{F}$ on groups to sets with $\mathcal{F}(L)$ the set of all those pairs (k, k') of morphisms $k: L \to G$, $k': L \to G'$ with $\phi \circ k = \phi' \circ k'$ (see the figure).

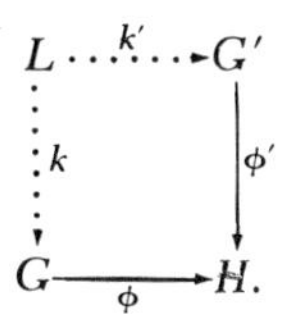

Prove that there exists a universal element for this functor. (It is called the "fibered product" of G and G', and it yields the "pull-back" diagram sketched above.)

3. Given two functions $f, g: X \to Y$, their *equalizer* is the set E of all $x \in X$ with $fx = gx$. Show that the insertion $E \to X$ is a universal.

4. In the category of all abelian groups, show that $\hom(A, -) \times \hom(A', -)$ has a universal element. (*Hint:* Use Proposition II.13.)

5. Show that products exist in the category of monoids.

6. Show that products exist in the category of posets.

7. Prove the rules (14) for the product of morphisms.

8. Let G, G', G'' be three objects in a category with products. Exhibit in $(G \times G') \times G''$ a universal element for the contravariant functor $\hom(-, G) \times \hom(-, G') \times \hom(-, G'')$.

9. In Exercise 8, show that $G \times (G' \times G'')$ has another universal element for the same functor. Deduce that the product is associative, up to an isomorphism.

10. Carry out the following construction of the coproduct $G * H$ (also called the "free product") of two groups G and H.

(a) Let a "word" be a formal product $w = g_1 h_1 \cdots g_n h_n$, that is, a list of $2n$ elements alternately from G and from H. As elements of $G * H$ take the "reduced words"—those with no g_i or $h_j = 1$ save perhaps g_1 or h_n. Show that every word can be brought to a unique reduced form by successive combinations.

(b) Show that the set $G * H$ of all reduced words is a group, with product given by juxtaposition and subsequent reduction.

(c) Show that the assignments $g \mapsto g1$ and $h \mapsto 1h$ give morphisms $e: G \to G * H$, $e': H \to G * H$.

(d) Prove that the pair (e, e') is universal, as required for a coproduct.

8. The Category of Sets

In the category of sets there are three fundamental functors: The power set $P(U)$, the product set $X \times Y$, and the hom-set $\hom(Y, X)$. We consider them here, noting by the way that the axiomatic properties of these functors can be used (in the theory of elementary topoi) to give a new foundation for set theory.

The power set $P(U)$ of a set U can be considered either as a covariant functor or as a contravariant one. For consider the effect of a function $g: U \to V$ on subsets of U. The *image* under g of $S \subset U$ is defined to be the subset

$$g_*(S) = \{v | v \in V \text{ and } v = gs \text{ for some } s \in S\} \tag{16}$$

of V. On the other hand, each subset $T \subset V$ has a *counterimage* under g which is the subset of U consisting of all those elements which go by g to the subset T; in symbols

$$g^*(T) = \{u | u \in U \text{ and } gu \in T\}. \tag{17}$$

This set is often written $g^{-1}(T)$ and is often called the *inverse image* of T under g, even though g as a function may have no inverse $g^{-1}: V \to U$.

Observe now that g_* is a function $S \mapsto g_* S$ from the power set $P(U)$ consisting of all subsets S of U to $P(V)$. One readily observes that $1_* = 1$, and for $h: V \to W$ that $h_* g_* = (hg)_*: PU \to PW$. Hence, $U \mapsto PU$ and $g \mapsto g_*$ define a covariant functor $\mathcal{P}$ on sets to sets.

On the other hand, $g: U \to V$ determines a function $g^*: PV \to PU$ sending each $T \subset V$ into its inverse image $g^* T$. Since one can prove $(hg)^* = g^* \circ h^*$ and $1^* = 1$, it follows that these assignments $U \mapsto PU$ and $g \mapsto g^*$ determine a contravariant functor on sets to sets, called the *contravariant power set functor* $\mathcal{P}^*$.

For sets X and Y the hom-set

$$\hom(Y, X) = \{t | t: Y \to X\} = X^Y \tag{18}$$

consists of all functions on Y to X, so is also called a *function set*. For example, if Y is the standard set $\mathbf{2} = \{1, 2\}$ of two elements, each function $t: \mathbf{2} \to X$ is determined by the two values $t(1)$, $t(2)$, both in X; hence, $t \mapsto (t(1), t(2))$ is a bijection $X^2 \cong X \times X$. Similarly, $X^1 \cong X$ and $X^3 \cong X \times X \times X$. On the other hand, a function $t: Y \to \mathbf{2}$ is completely determined by the subset $S \subset Y$ with $S = \{y | ty = 1\}$—the inverse image of 1. Conversely, every subset $S \subset Y$ arises in this way from its *characteristic* function t_S, defined by $t_S(y) = 1$ if $y \in S$ and $t_S(y) = 2$ otherwise. Thus the assignment $S \mapsto t_S$ is a bijection $P(Y) \cong \mathbf{2}^Y$ from the power set $P(Y)$ of all subsets of Y to the function set $\mathbf{2}^Y$.

The function set X^Y can be regarded as a functor of either variable set X or Y. If Y is fixed, it is the covariant functor $\hom(Y, -)$ which sends each set X to $X^Y = \hom(Y, X)$ and each function $f: X \to X'$ to the function

$$\hom(Y, f) = f^Y: X^Y \to X'^Y, \qquad t \mapsto ft. \tag{19}$$

If X is fixed, it is the contravariant functor hom $(-, X)$ which sends each set Y to X^Y and each function $g: Y' \to Y$ to

$$\text{hom}(g, X) = X^g : X^Y \to X^{Y'}, \qquad t \mapsto tg. \tag{20}$$

These two definitions can be combined to define a function

$$\text{hom}(g, f) = f^g : X^Y \to X'^{Y'}, \qquad t \mapsto ftg. \tag{21}$$

Moreover, since $f(tg) = (ft)g$, one has $f^g = f^{Y'} \circ X^g = X'^g \circ f^Y$ (draw a diagram). In such cases, one says that hom (Y, X) is a "bifunctor" of X and Y to sets, covariant in X and contravariant in Y.

The function sets X^Y are closely connected to products. On the one hand, the universality property of the product $S \times T$ gives to each pair of functions $f: Y \to S$ and $g: Y \to T$ a unique $h: Y \to S \times T$ and thus a bijection

$$S^Y \times T^Y \cong (S \times T)^Y. \tag{22}$$

This is like one of the laws of exponents for numbers. On the other hand, one has the following basic relation between products and exponents.

Theorem 7. *For given sets X and Y, the contravariant functor* hom $(- \times Y, X)$ *on sets to sets has a universal element.*

Proof: This is the functor which sends each set Z to hom $(Z \times Y, X)$ and each function $h: Z' \to Z$ to the function hom $(h \times 1_Y, X)$. We are asked to produce a representing object (a set) R and an arrow $R \times Y \to X$ which is universal among arrows $Z \times Y \to X$. For R we take the function set X^Y, for the arrow we choose the function

$$e: X^Y \times Y \to X, \qquad e(t, y) = t(y); \tag{23}$$

it is called *evaluation* because, when applied to a function $t: Y \to X$ and an element $y \in Y$, it takes the value $t(y)$ of that function t on that argument y. To prove e universal, consider any $f: Z \times Y \to X$; that is, an arbitrary function of two variables. From it we can define a function $F: Z \to X^Y$ as that function with each $F(z): Y \to X$ defined by $F(z)(y) = f(z, y)$; in other words, $F(z)$ is just f for fixed z, regarded as a function only of its second argument y. This means exactly that the evaluation $e(F(z), y)$ is $f(z, y)$. In other words, the diagram

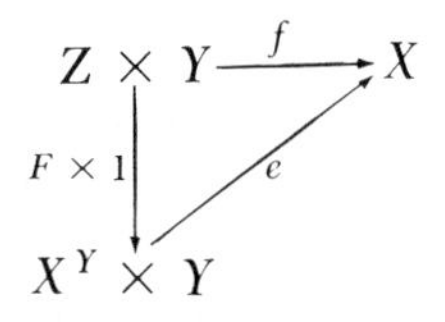

is commutative; moreover, F is the only function $Z \to X^Y$ which makes this diagram commute. Hence, every f factors uniquely through e, so e is a universal arrow from $- \times Y$ to X.

As in the representation theorem for universals, this result can also be formulated thus: There is a bijection from functions $f: Z \times Y \to X$ to functions $F: Z \to X^Y$; that is, $\hom(Z \times Y, X) \cong \hom(Z, X^Y)$, or

$$X^{Z \times Y} \cong (X^Y)^Z. \tag{24}$$

This bijection assigns to each function f of two variables (in Z and in Y) the corresponding function F of one variable (in Z) whose values are functions of one variable (in Y). In particular, if $Z = X^Y$ and if F is the identity function $1: Z = X^Y \to X^Y$, the corresponding function f is the evaluation function e. Moreover, the bijection (24) for finite sets implies the corresponding (and familiar) property of exponents of numbers.

If R is a ring, the function set R^X for any X can be made into a ring, called a *function ring*, as in §III.2.

EXERCISES

1. Show that $\varnothing^X = \varnothing$ for $X \neq \varnothing$, but that $X^\varnothing \cong \{1\}$ for all X.
2. For $f: X \to X'$, $g: Y' \to Y$, prove that $f^{Y'} \circ X^g = f^g = X'^g \circ f^Y$.

9. The Category of Finite Sets

Finite sets present several striking examples of concrete categories. For each natural number n we can take the standard finite set $\mathbf{n} = \{1, \cdots, n\}$ with n elements, as described in §I.6. If we think of this set $\mathbf{n}$ as an ordered set, then a morphism $f: \mathbf{n} \to \mathbf{m}$ is a function f on $\mathbf{n}$ to $\mathbf{m}$ which preserves the linear order, in the sense that $i \leqslant j$ implies $fi \leqslant fj$. The composite of two such order-preserving functions is again such a function. Hence, there is a concrete category with objects the sets $\mathbf{n}$ for all natural numbers n and morphisms the order-preserving functions f. This category, called Δ, plays an important role in algebraic topology.

Alternatively, consider $\mathbf{n}$ just as a set of elements with no order. The concrete category Σ with objects all these sets $\mathbf{n}$ and morphisms *all* functions $f: \mathbf{n} \to \mathbf{m}$ is called the "skeletal" category of finite sets—skeletal because it has just one set of each size.

Two sets X and Y are said to have the same "cardinal number" if there is a bijection $f: X \cong Y$; that is, if there is a function f which is a one-to-one correspondence between the elements of X and those of Y. In particular, a set Y is said to be *finite* if for some natural number n there is a bijection

$\mathbf{n} \to Y$. Clearly, the finite sets (with all functions between them) form a concrete category. An infinite set is one which is not finite. Finite sets have various familiar special properties not enjoyed by infinite sets; for example, any injection $g: Y \to Y$ is a surjection if Y is finite. These properties all follow from the Peano Postulates, as will be shown in this section. Readers willing to accept them as obvious can skip this section.

Proposition 8. *If $n < m$, there is no bijection $f: \mathbf{n} \to \mathbf{m}$.*

Proof: For $n = 0$ the proposition is clear, since no bijection can map the empty set $\mathbf{0} = \varnothing$ to a non-empty set $\mathbf{m}$. Suppose then by induction that the proposition holds for some one n and for all $m > n$. Take the next set $\mathbf{n+1} = \mathbf{n} \cup \{n+1\}$, and suppose, contrary to the conclusion of the proposition, that there is a bijection $f: \mathbf{n+1} \to \mathbf{m}$ for some $m > n + 1$. Now the set $\mathbf{n+1}$ contains the number $n + 1$; let $k = f(n+1)$ be the value of f at $n + 1$. Write

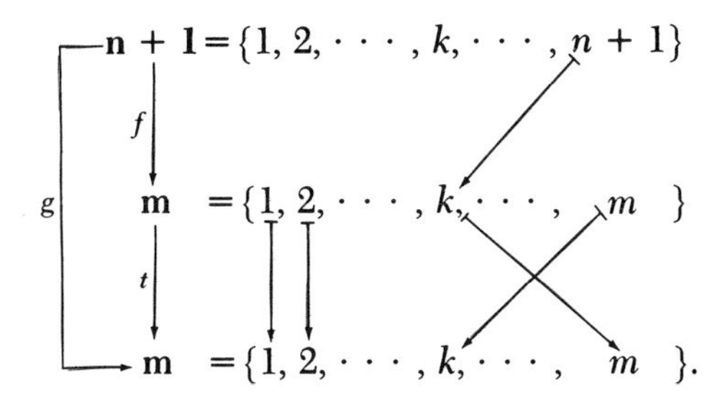

Here t denotes that function $t: \mathbf{m} \to \mathbf{m}$ which interchanges k and m and is otherwise the identity; it is defined for any argument $i \in \mathbf{m}$ by

$$
\begin{aligned}
t(i) &= k, && \text{if } i = m,\\
&= m, && \text{if } i = k,\\
&= i, && \text{if } i \neq m \text{ and } i \neq k.
\end{aligned}
$$

If it happens that $k = m$, this function t is just the identity on $\mathbf{m}$. Otherwise, t actually interchanges k and m, and is (cf. §II.6) the *transposition* of k and m. In either event $(t \circ t)(k) = t(m) = k$ and $(t \circ t)(m) = m$, so $t \circ t$ is the identity. Thus t is its own inverse, hence is a bijection. The composite $g = t \circ f: \mathbf{n+1} \to \mathbf{m}$, as displayed in the figure, is also a bijection and moreover has $g(n+1) = m$. Now restrict the domain of g to the subset $\mathbf{n} \subset \mathbf{n+1}$ and the codomain of g to $\mathbf{m-1} = \{1, \cdots, m-1\}$; that is, define $g': \mathbf{n} \to \mathbf{m-1}$ by $g'(i) = g(i)$ for each $i \in \mathbf{n}$. This function g' is a bijection $g': \mathbf{n} \to \mathbf{m-1}$ with $n < m - 1$, in contradiction to the induction assumption. This contradiction completes the proof.

The composite $t \circ f$ used in this proof can also be used to prove a familiar related result.

LEMMA. *For $n \in \mathbf{N}$, any injection $f:\mathbf{n} \to \mathbf{n}$ is a surjection.*

The proof is again by induction on n. The case $n = 0$ being trivial, assume the lemma true for some n and consider an injection $f:\mathbf{n+1} \to \mathbf{n+1}$; we must prove f surjective. Use the diagram above with $m = n + 1$ to define a composite $g = i \circ f:\mathbf{n} + \mathbf{1} \to \mathbf{n} + \mathbf{1}$, still injective, which has $g(n + 1) = n + 1$. The restriction $g':\mathbf{n} \to \mathbf{n}$ is then injective. The induction assumption states that it must be a bijection; this in turn shows g and therefore $f = t^{-1} \circ g$ a bijection. This completes the induction.

A set X is said to be *finite* when there is a natural number n and a bijection $h:\mathbf{n} \to X$. If there is also a bijection $g:\mathbf{m} \to X$ for the same set X and some natural number m, the composite $g^{-1} \circ h:\mathbf{n} \to \mathbf{m}$ is a bijection, so the proposition above shows that $n = m$. To each finite set X there is therefore a unique natural number n for which there is a bijection $\mathbf{n} \cong X$ (though if $n > 1$, there is more than one such bijection). Call this n the *cardinal number* of X; in symbols, $\#(X) = n$. Clearly, two finite sets X and Y have the same cardinal number if and only if there is a bijection $X \cong Y$.

THEOREM 9. *If the set X is finite, a function $h:X \to X$ is surjective if and only if it is injective.*

Proof: Since X is finite, there is a bijection $b:\mathbf{n} \cong X$. The composite $\mathbf{n} \xrightarrow{b} X \xrightarrow{h} X \xrightarrow{b^{-1}} \mathbf{n}$ is then injective (or surjective) when h is injective (or surjective, as the case may be). Hence, it suffices to prove the theorem when X is the special finite set $\mathbf{n}$. But the Lemma just above states that every injective $f:\mathbf{n} \to \mathbf{n}$ is surjective. Conversely, if $h:\mathbf{n} \to \mathbf{n}$ is surjective, it must (Theorem I.1) have a right inverse g. Now $h \circ g = 1$ implies that this right inverse is injective and therefore (by the Lemma again) bijective. Then $h \circ g = 1$ gives $h = g^{-1}$; hence, h is bijective, as desired.

A familiar principle of counting states that the addition and multiplication of cardinal numbers correspond, respectively, to the disjoint union $X \dot\cup Y$ (§7) and the product of finite sets X and Y; in symbols

$$\#(X \dot\cup Y) = (\# X) + (\# Y), \qquad \#(X \times Y) = (\# X)(\# Y). \quad (25)$$

To prove these rules, set $m = \#(X)$ and $n = \#(Y)$. This means that there are bijections $f:\mathbf{m} \cong X$ and $g:\mathbf{n} \cong Y$; hence, there is a bijection $f \times g:\mathbf{m} \times \mathbf{n} \cong X \times Y$ on the product sets. Therefore, $\#(X \times Y) = \#(\mathbf{m} \times \mathbf{n})$, and similarly $\#(X \dot\cup Y) = \#(\mathbf{m} \dot\cup \mathbf{n})$, where $\mathbf{m} \dot\cup \mathbf{n}$ is a disjoint union of the sets **m** and **n**. Equations (25) now become

$$\#(\mathbf{m} \dot{\cup} \mathbf{n}) = m + n, \qquad \#(\mathbf{m} \times \mathbf{n}) = mn. \tag{26}$$

To prove the first of these it suffices to show that the set $\{1, \cdots, m + n\}$ is a disjoint union (§7) of **m** and **n**; this may be seen in the following figure, where the arrows indicate an evident bijection

$$\{1, 2, \cdots, m+n\} = \{1, \cdots, m\} \cup \{m+1, m+2, \cdots, m+n\}$$

$$\qquad\qquad\qquad\qquad\qquad\qquad\qquad\qquad\downarrow \qquad \downarrow \qquad\qquad \downarrow$$

$$\qquad\qquad\qquad\qquad\qquad\qquad\qquad\quad\{\ 1, \qquad 2, \qquad \cdots, \qquad n\ \}.$$

To prove the second equation we need only find a bijection ϕ from $\mathbf{m} \times \mathbf{n}$ to the set $\{1, 2, \cdots, mn\}$. This can be given by the explicit formula $\phi(i, j) = n(i-1) + j$ for $i \in \mathbf{m}$ and $j \in \mathbf{n}$. This formula is a familiar one for counting the elements in an $m \times n$ matrix or array: Start counting the first row $(i = 1)$, then proceed with the second row, and so on. We leave the reader to give the explicit proof that ϕ is a bijection. (*Hint:* Use the division algorithm to construct its inverse.)

Cardinal numbers may also be defined for infinite sets. For example, a set X is said to be "denumerable" when there is a bijection $\mathbf{N} \cong X$. There are infinite sets which are not denumerable; a famous "diagonal" argument due to G. Cantor shows that the power set $P(\mathbf{N})$ is not denumerable (cf. Exercise 9 below).

EXERCISES

1. (a) Prove that there is an injection $\mathbf{m} \to \mathbf{n}$ if and only if $m \leqslant n$.
(b) Prove that there is a surjection $\mathbf{m} \to \mathbf{n}$ if and only if $m \geqslant n$.

2. For finite sets X and Y, show that $\#(X^Y) = (\# X)^{(\# Y)}$.

3. Show that the disjoint union of two denumerable sets is denumerable.

4. Show that the product of two denumerable sets is denumerable.

5. Prove that $\mathbf{Z}$ is denumerable.

6. (The paradox of Galileo.) If D is denumerable, prove that there is an injection $D \to D$ which is not a bijection.

7. (a) Using the axiom of choice, construct for every infinite set X an injection $\mathbf{N} \to X$.
(b) Prove: X is infinite if and only if there is a bijection $X \cong S$ with S a proper subset of X.

8. Prove that the category Δ of the text has coproducts and that the category Σ of the text has both products and coproducts.

***9.** Prove that $\mathbf{N}^{\mathbf{N}}$ is not denumerable as follows: If $s:\mathbf{N} \to \mathbf{N}^{\mathbf{N}}$ is any sequence of functions $s_n:\mathbf{N} \to \mathbf{N}$, define $g:\mathbf{N} \to \mathbf{N}$ by $g(n) = s_n(n) + 1$ and show that $g \in \mathbf{N}^{\mathbf{N}}$ is not in the image of s.

CHAPTER V

Modules

A MODULE is an additive abelian group whose elements can be suitably multiplied by the elements from some ring R of "scalars". Such modules exist for any ring R. Thus ordinary geometric vectors (in the plane or in space) form modules over the real field $\mathbf{R}$. The polynomials over any commutative ring K form a module over K. The elements of any additive abelian group form a module over the ring $\mathbf{Z}$ of integers. This chapter will be concerned with general properties of modules over arbitrary rings; special properties of modules over fields ("vector spaces") will be studied in the next chapter.

1. Sample Modules

Let R be any ring. We call it the "ring of scalars" and write its elements as lowercase Greek letters $\kappa, \lambda, \mu, \ldots$.

DEFINITION. An R-module A *is an additive abelian group together with a function* $R \times A \to A$, *written* $(\kappa, a) \mapsto \kappa a$, *and subject to the following axioms, for all elements* $\kappa, \lambda \in R$ *and* $a, b \in A$:

$$\kappa(a + b) = \kappa a + \kappa b, \tag{1}$$

$$(\kappa + \lambda)a = \kappa a + \lambda a, \tag{2}$$

$$(\kappa\lambda)a = \kappa(\lambda a), \tag{3}$$

$$1a = a. \tag{4}$$

More explicitly, such a module A is a *left* module, because in forming κa the scalar κ is written on the left of the module element a.

An R-module is thus an algebraic system A with one binary operation $+: A \times A \to A$ and with a set of unary operations $A \to A$, one for each scalar κ in the ring R. This operation $a \mapsto \kappa a$ is called *scalar multiple* or "scalar multiplication by κ". The axioms for a module, as stated above, require that A be an abelian group under $+$, that both distributive laws (1) and (2) hold, that the "mixed" associative law (3) hold, and that scalar multiplication $a \mapsto 1a$ by the unit 1 of the ring R be the identity function $A \to A$. (Without this last axiom, any abelian group A could be made into an R-module trivially, by defining each scalar multiple κa to be zero.)

Axiom (1) with κ fixed states that $a \mapsto \kappa a$ is a morphism $A \to A$ of addition. Since any morphism of addition is also (Proposition II.1) one of zero and one of additive inverse, (1) implies that

$$\kappa 0 = 0, \qquad \kappa(-a) = -(\kappa a). \tag{5}$$

Axiom (2) with a fixed states that $\kappa \mapsto \kappa a$ is a morphism $R \to A$ of addition. This implies as before that

$$0a = 0, \qquad (-\kappa)a = -(\kappa a). \tag{6}$$

We now examine some typical examples of modules.

EXAMPLE 1: Vectors from a fixed origin O in the plane form a module over the field of real numbers. When a vector is represented as a directed line segment OA from the origin O, the sum of two vectors is given by the usual parallelogram law. Given a vector u and a positive real number ρ, the scalar multiple ρu is the vector in the same direction as u with ρ times the length—or, if ρ is negative, in the opposite direction with $|\rho|$ times the length. Relative to x, y coordinate axes on the plane, each vector OA is represented by the coordinates (x, y) of its endpoint A; moreover, the operations of addition and scalar multiple become operations on these pairs:

$$(x, y) + (x', y') = (x + x', y + y'), \qquad \rho(x, y) = (\rho x, \rho y).$$

We note that the coordinates (x, y) of the vector depend on the choice of axes.

EXAMPLE 2: Let R be any ring. In the set $R \times R = R^2$ of all ordered pairs (ξ_1, ξ_2) of elements $\xi_i \in R$, the equations

$$\begin{aligned}(\xi_1, \xi_2) + (\eta_1, \eta_2) &= (\xi_1 + \eta_1, \xi_2 + \eta_2),\\ \kappa(\xi_1, \xi_2) &= (\kappa\xi_1, \kappa\xi_2)\end{aligned} \tag{7}$$

define a termwise sum and, for each $\kappa \in R$, a termwise scalar multiple. A check of the module axioms shows that this set $R \times R$ is an R-module under these two operations. In the plane, each choice of coordinates, as in the previous example, gives an isomorphism of the **R**-module of plane vectors to the module $\mathbf{R} \times \mathbf{R}$. By the same token, vectors from a fixed origin in three-space form a module isomorphic to the module $\mathbf{R} \times \mathbf{R} \times \mathbf{R}$ of all triples (x, y, z) of real numbers.

These constructions by pairs and triples can be generalized.

EXAMPLE 3: Given a ring R and a positive integer n, the set R^n of all lists $\xi = (\xi_1, \cdots, \xi_n)$ of n elements $\xi_i \in R$ is an R-module under the termwise operations defined by

$$\begin{aligned}(\xi_1, \cdots, \xi_n) + (\eta_1, \cdots, \eta_n) &= (\xi_1 + \eta_1, \cdots, \xi_n + \eta_n),\\ \kappa(\xi_1, \cdots, \xi_n) &= (\kappa\xi_1, \cdots, \kappa\xi_n).\end{aligned} \tag{8}$$

The module axioms may be readily verified; the zero for addition is the list $\mathbf{0} = (0, \cdots, 0)$, and the additive inverse of $(\xi_1, \cdots, \xi_n)$ is $(-\xi_1, \cdots, -\xi_n)$. In particular, R^2 is the module $R \times R$ of Example 2, while R^1 is just the ring R regarded as a module over itself. One can also interpret R^0 to be the *trivial R-module*; that is, the abelian group $\{0\}$ consisting of zero alone, with scalar multiples $\kappa 0 = 0$ for all $\kappa \in R$.

EXAMPLE 4: Any abelian group A is automatically a module over the ring $\mathbf{Z}$ of integers. For, in an additive abelian group A we have already defined to each integer n the multiples $a \mapsto na$, and the identities (II.10), (II.11), and (II.12) which these multiples satisfy include those required in (1)–(4) for a module over $\mathbf{Z}$. Vice versa, in any $\mathbf{Z}$-module, each "scalar" multiple na is just the "multiple" na already defined in (II.10) for an element a in an abelian group. For, by the module axiom (4), $1a = a$; by axiom (2), $(n + 1)a = na + a$; by property (6), $(-n)a = -(na)$ and $0a = 0$. These are just the equations (II.10).

In brief, a $\mathbf{Z}$-module *is* an abelian group.

EXAMPLE 5: Any ring T with a subring $R \subset T$ is an R-module. The operations are the addition in T and a restriction of the multiplication in T; namely, the function $(\kappa, t) \mapsto \kappa t$ which takes the product of an element κ in the subring by any t in the whole ring T. The case $T = R$ gives the module $R^1 = R$. Note especially the case when $T = K[x]$ is the ring of polynomials over a *commutative* ring K, considered only as a module over the subring K of coefficients.

EXERCISES

1. In any R-module, prove that $n(\kappa a) = \kappa(na)$ for all integers n.
2. Verify the module axioms (1)–(4) in the case of the module R^n.
3. Which of the following sets of polynomials in $K[x]$, K commutative, are K-modules: (a) all polynomials of degree exactly 4; (b) all polynomials of degree at most 4; (c) all monic polynomials; (d) all polynomials of even degree?
4. Prove that the K-module of all polynomials in $K[x]$ of degree less than n is isomorphic to the K-module K^n.
5. For any natural number $n > 1$, show that an additive abelian group A in which $na = 0$ for all $a \in A$ can be regarded as a $\mathbf{Z}_n$-module.
6. In the axioms for a module, show that the commutativity assumption for addition is redundant. (*Hint:* Expand $(1 + 1)(a + b)$ in two ways.)
7. Show that a left ideal A in a ring R is an R-module.
8. An $m \times n$ matrix A over R is a function $A : \mathbf{m} \times \mathbf{n} \to R$. Show that all such matrices form an R-module under termwise sum and scalar multiple, defined for all $i \in \mathbf{m}$ and $j \in \mathbf{n}$ by

$$(A + B)(i, j) = A(i, j) + B(i, j), \qquad (\lambda A)(i, j) = \lambda(A(i, j)).$$

2. Linear Transformations

Given two R-modules A and A', a *morphism of R-modules* is a function $t:A \to A'$ which is *additive* and *homogeneous*, in the sense that

$$t(a+b) = t(a) + t(b), \qquad t(\kappa a) = \kappa(ta) \tag{9}$$

for all $a, b \in A$ and all scalars $\kappa \in R$. These two conditions are equivalent to the single requirement that

$$t(\kappa a + \lambda b) = \kappa(ta) + \lambda(tb) \tag{10}$$

for all $a, b \in A$ and for all scalars $\kappa, \lambda \in R$. More generally, from scalars κ_i and module elements a_i for $i = 1, \cdots, n$ one can form in the module A the R-*linear combination* $\kappa_1 a_1 + \cdots + \kappa_n a_n$; any morphism of modules preserves such combinations:

$$t(\kappa_1 a_1 + \cdots + \kappa_n a_n) = \kappa_1(ta_1) + \cdots + \kappa_n(ta_n). \tag{11}$$

With this in view, a morphism of R-modules is called an R-*linear map* or a *linear transformation*. Monomorphisms, isomorphisms, and the like are defined as usual, as morphisms which are injections or bijections, as the case may be.

The composite $t' \circ t$ of two morphisms of modules, when defined, is a morphism. For any R-module A the identity function $1_A : A \to A$ is an endomorphism of A. These two facts mean that for each ring R we can introduce the concrete category of all R-modules. The objects of the category are the R-modules; to each object A we assign as underlying set the set $\mathfrak{U}(A)$ of all elements in the module A; to each pair of objects A, A' we assign the set of morphisms

$$\hom(A, A') = \{t \mid t:A \to A' \text{ is a morphism of } R\text{-modules}\}. \tag{12}$$

Then (as required for a concrete category) each morphism t is a function on the set $\mathfrak{U}(A)$ to the set $\mathfrak{U}(A')$. For A fixed, $\hom(A, -)$ gives a functor on R-modules to sets, while for A' fixed, $\hom(-, A')$ is a contravariant functor on R-modules to sets.

Linear transformations may be illustrated by the following examples, numbered to correspond to those of the preceding section.

EXAMPLE 1: In the $\mathbf{R}$-module V of all vectors from a fixed origin in the plane there is for each angle θ the function $r_\theta : V \to V$ which rotates the plane counterclockwise about the origin through the angle θ. Now a parallelogram rotates to a parallelogram, and for each real scalar ρ the ρ-fold of a vector u rotates to the ρ-fold of $r_\theta u$; hence, the operation r_θ of rotation is an endomorphism of the module V.

For each real σ the operation $s_\sigma : V \to V$ which stretches each vector v to σ times its length is a map which preserves parallelograms and scalar multiples, hence is an endomorphism of V. Many other geometric transformations are linear: The reflection in a line through the origin; the operation of stretching away from one of the axes by multiplying the (other) coordinate by a scalar, say by 2. All these linear transformations $t: V \to V$, when they are bijections, carry straight lines to straight lines, planes to planes, and so on. This is one expression of their importance in geometry.

EXAMPLE 2: For the R-module $R \times R$ of all ordered pairs (ξ_1, ξ_2) of elements of any ring R, each of the three assignments

$$(\xi_1, \xi_2) \mapsto (\xi_2, \xi_1), \qquad (\xi_1, \xi_2) \mapsto (\xi_1 + \xi_2, \xi_2), \qquad (\xi_1, \xi_2) \mapsto (\xi_1, 0)$$

respects the termwise module operations of $R \times R$, hence is an endomorphism of $R \times R$. In the geometrical case, when $R = \mathbf{R}$, the first transformation reflects the coordinate plane $\mathbf{R}^2$ in the 45° line through the origin in the first and third quadrants, the second "shears" the plane parallel to the x-axis (imagine a deck of cards!), while the third projects each point perpendicularly onto the x-axis (see Figure V-1).

EXAMPLE 3: For the module R^n of all lists $\boldsymbol{\xi}:\mathbf{n} \to R$ of n scalars there are many linear transformations $R^n \to R$, such as those given by the assignments $(\xi_1, \cdots, \xi_n) \mapsto \xi_1 + \xi_2$ or $(\xi_1, \cdots, \xi_n) \mapsto \xi_1$. Endomorphisms of R^n (to be studied systematically in §VII.1) include those given by permutation of the components and those which are given by many other assignments such as $(\xi_1, \cdots, \xi_n) \mapsto (\xi_1 + \xi_2, \xi_2 + \xi_3, \cdots, \xi_n + \xi_1)$.

EXAMPLE 4: Any morphism $A \to A'$ of abelian groups is a morphism of integral multiples; hence morphisms of abelian groups coincide with morphisms of $\mathbf{Z}$-modules.

EXAMPLE 5: If R is a subring of T, an endomorphism $T \to T$ of rings is an endomorphism of the R-module T provided it leaves the elements of the subring R fixed. On the other hand, an endomorphism $T \to T$ of modules need not be an endomorphism of rings. An example is the K-linear map $K[x] \to K[x]$ which multiplies each polynomial by x; another is the K-linear map $f \mapsto f'$ which sends each polynomial $f \in K[x]$ to its formal derivative (§III.6, Exercise 2). For each $c \in K$, evaluation at c is evidently a morphism $E_c: K[x] \to K$ of K-modules.

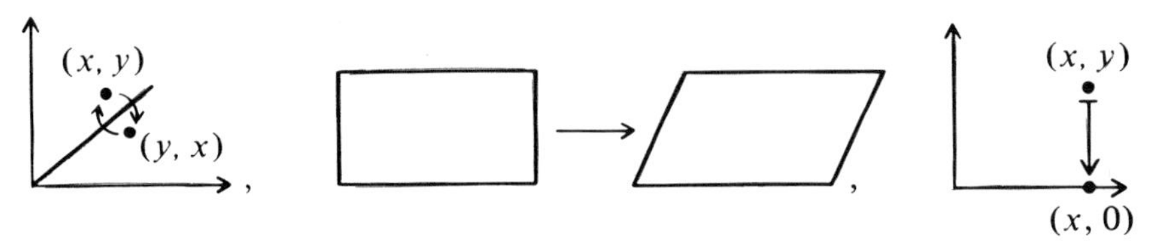

FIGURE V-1

Next we consider operations on morphisms themselves which will make the set hom (A, A') of all morphisms $t:A \to A'$ into an abelian group $\mathrm{Hom}_R\,(A, A')$.

Theorem 1. *If A and A' are two R-modules, the set*

$$\mathrm{Hom}_R\,(A, A') = \{t \mid t:A \to A' \text{ is a morphism of } R\text{-modules}\}$$

is an abelian group under pointwise addition of morphisms.

Hereafter "hom", with lower case "h" refers to the *set* of these morphisms t and "Hom", with capital "H", to the additive *group* of these t.

Proof: The pointwise sum of two morphisms $s, t:A \to A'$ is the function $s + t$ defined for all $a \in A$ by $(s + t)a = sa + ta$. For any scalar κ, $(s + t)(\kappa a) = \kappa(sa + ta) = \kappa[(s + t)a]$, so $s + t$ is a morphism for any scalar multiple $a \mapsto \kappa a$. It is also a morphism for addition because, for any two elements $a, b \in A$,

$$(s + t)(a + b) = s(a + b) + t(a + b) = sa + sb + ta + tb,$$

while

$$(s + t)a + (s + t)b = sa + ta + sb + tb;$$

the two results are equal because addition in A' is commutative. This sum gives a binary operation $(s, t) \mapsto s + t$ on $\mathrm{Hom}_R\,(A, A')$. As a pointwise sum, it is associative and commutative. The zero for this sum is that morphism 0 which sends every a to 0 (the *zero morphism* $0:A \to A'$). The additive inverse of t is the function $-t$ defined by $(-t)a = -(ta)$. Hence Hom_R is indeed an abelian group.

Theorem 2. *For R-module maps in the configuration*

$$B \xrightarrow{h} A \underset{t}{\overset{s}{\rightrightarrows}} A' \xrightarrow{k} C$$

both distributive laws hold:

$$k \circ (s + t) = k \circ s + k \circ t, \qquad (s + t) \circ h = s \circ h + t \circ h. \qquad (13)$$

To prove the first law, we take any $a \in A$ and write

$$[k(s + t)]a = k(sa + ta) = k(sa) + k(ta) = (k \circ s + k \circ t)a,$$

using the fact that k is linear. The proof of the second distributive law is similar, but does not use the linearity of h.

COROLLARY. *For each R-module A the set*

$$\mathrm{End}_R(A) = \{t \mid t : A \to A \text{ is an } R\text{-module endomorphism}\} \tag{14}$$

is a ring under pointwise addition and multiplication by composition.

Proof: By Theorem 1, $\mathrm{End}_R(A) = \mathrm{Hom}_R(A, A)$ is an abelian group. Composition of endomorphisms is known to be associative and to have $1_A : A \to A$ as unit. Since both distributive laws hold by Theorem 2, the set $\mathrm{End}_R(A)$ is a ring, as claimed.

In particular, for $R = \mathbf{Z}$ this again proves for any abelian group A that $\mathrm{End}_{\mathbf{Z}}(A)$ is a ring—as already noted in Theorem III.6, with essentially the same proof.

If R is a ring, any R-module A is automatically an abelian group, hence a $\mathbf{Z}$-module, and any morphism of R-modules is automatically a morphism of $\mathbf{Z}$-modules. Thus $\mathrm{End}_R(A)$ is a subring of $\mathrm{End}_{\mathbf{Z}}(A)$, under the same operations of pointwise addition and composition. When $R \neq \mathbf{Z}$, $\mathrm{End}_R(A)$ is usually a proper subring of $\mathrm{End}_{\mathbf{Z}}(A)$. However, in a context dealing only with R-modules for a fixed ring R, we write just $\mathrm{End}(A)$ for $\mathrm{End}_R(A)$ and $\mathrm{Hom}(A, A')$ for $\mathrm{Hom}_R(A, A')$, dropping the subscript R.

LEMMA. *If K is a commutative ring and $t : A \to A'$ a morphism of K-modules, then for each $\lambda \in K$ the pointwise multiple λt is also a morphism of K-modules.*

Proof: Here $\lambda t : A \to A'$ is the function defined by $(\lambda t)(a) = \lambda(ta)$ for each $a \in A$. Clearly, λt is a morphism of addition. Moreover, for any scalar κ and any $a \in A$,

$$(\lambda t)[\kappa a] = \lambda[t(\kappa a)] = (\lambda\kappa)(ta) = (\kappa\lambda)(ta) = \kappa[(\lambda t)a];$$

here we can write $\lambda\kappa = \kappa\lambda$ *because the ring K is commutative*. Hence, $\lambda t : A \to A'$ is a morphism of K-modules, as asserted.

This lemma (and hence Theorem 3) fails for R-modules when R is not commutative. As in Chapter III, we normally denote a commutative ring by K. Modules over such rings K have many useful properties not shared by modules over arbitrary rings. For example:

THEOREM 3. *For modules A and A' over a commutative ring K, the set $\mathrm{Hom}_K(A, A')$ is itself a K-module under pointwise addition and pointwise scalar multiplication.*

The proof is a straightforward verification of the module axioms for the multiples $(\lambda, t) \mapsto \lambda t$ defined by the lemma.

EXERCISES

1. Which of the following transformations of the plane are linear?
   (a) A translation.
   (b) Reflection in the origin.
   (c) Reflection in a line through the origin.
   (d) Reflection in a line not through the origin.

2. Show that the set of all infinitely differentiable functions $f:\mathbf{R} \to \mathbf{R}$ is an $\mathbf{R}$-module under termwise operations (for example, $(f + g)x = fx + gx$ for all $x \in \mathbf{R}$). Show that the operation D sending each f into its first derivative Df is linear.

3. Prove (11) by induction on n.

4. Show that any R-module map $R \times R \to R$ has the form $(\xi_1, \xi_2) \mapsto \lambda_1\xi_1 + \lambda_2\xi_2$ for suitable scalars λ_1 and $\lambda_2 \in R$.

5. For a commutative ring K regarded as a K-module, construct an isomorphism $\text{End}_K(K) \cong K$ of rings.

6. For the finite cyclic groups $\mathbf{Z}_m$, $\mathbf{Z}_n$, prove $\text{Hom}_{\mathbf{Z}}(\mathbf{Z}_m, \mathbf{Z}_n) \cong \mathbf{Z}_{(m, n)}$, where (m, n) denotes the greatest common divisor of m and n.

7. Let A and C be abelian groups while m and n are integers such that $ma = 0$ and $nc = 0$ for all elements $a \in A$ and $c \in C$. Prove that every element of $\text{Hom}_{\mathbf{Z}}(A, C)$ has order dividing the g.c.d. (m, n).

8. For K a commutative ring and $t:K \to A$ a morphism of K-modules, prove that the assignment $t \mapsto t(1)$ defines an isomorphism $\text{Hom}_K(K, A) \cong A$ of K-modules.

9. Using the result of Exercise 4 in the case of a commutative ring K, construct an isomorphism $\text{Hom}_K(K \times K, K) \cong K \times K$ of K-modules.

10. Give an example with $R = \mathbf{Z} \times \mathbf{Z}$ to show that $\text{Hom}_R(A, A') \neq \text{Hom}_{\mathbf{Z}}(A, A')$.

11. For a fixed R-module A, show that $A' \mapsto \text{Hom}_R(A, A')$ yields a functor on R-modules to abelian groups.

3. Submodules

An R-module D is a *submodule* of an R-module A when D is a subset of A and the insertion $D \to A$ is a morphism of R-modules. This implies that the operations of addition and scalar multiples in D are restrictions of the operations of addition and scalar multiples in A. Hence, a subset D of A is a submodule of A if and only if D is an R-module under restriction of the module operations in A.

Theorem 4. *A non-empty subset D of an R-module A is a submodule if and only if D is closed under addition and under multiplication by each scalar $\kappa \in R$.*

Proof: Closure of D under scalar multiplication $d \mapsto (-1)d$ by -1 implies closure of D under subtraction, since D is also closed under addition;

D is therefore an additive subgroup of A. As in the case of groups (Theorem II.9) and rings (Theorem III.4), closure of a subgroup D under all scalar multiples automatically implies that the module axioms (1)–(4) hold for these multiples in D. Hence, D is a submodule, as required.

Alternatively, a non-empty subset D of A is a submodule if and only if D contains all R-linear combinations of its elements.

Among the submodules of A are A itself and the set 0 consisting of the zero element alone. Any submodule of A different from these two is said to be a *proper submodule*. The set of all submodules D of A is partly ordered by the inclusion relation $\subset$, where $D_1 \subset D_2$ means that D_1 is a submodule of D_2. If D and E are any two submodules in this poset, so is their intersection $D \cap E$ as sets, and so is their *sum*, defined to be the subset

$$D + E = \{\text{all } d + e \mid d \in D \text{ and } e \in E\}. \tag{15}$$

(Since any submodule containing both D and E must contain all such sums $d + e$, the sum $D + E$ may also be described as the least submodule of A containing both D and E.) This means that the poset of all submodules of A, partly ordered by inclusion, is a lattice with $(D, E) \mapsto D \cap E$ as meet and $(D, E) \mapsto D + E$ as join, in the sense described in §IV.6. More generally, if F is any family of submodules D of A, the intersection of all $D \in F$ is a submodule of A; the sum $D + E$ of two modules therefore may be described as the intersection of the family F of all those submodules of A which contain both D and E.

Take any element $a \in A$. The subset

$$Ra = \{\text{all } \kappa a \mid \kappa \in R\} \tag{16}$$

of A is a submodule of A, called the *cyclic module* with generator a or the module *spanned* by the element a. If Ra is such a cyclic module, the assignment $\kappa \mapsto \kappa a$ is an epimorphism $R \to Ra$ of modules.

Take any list $a_1, \cdots, a_n$ of elements of A. The sum of the cyclic modules

$$Ra_1 + \cdots + Ra_n = \{\text{all } \kappa_1 a_1 + \cdots + \kappa_n a_n \mid \kappa_1, \cdots, \kappa_n \in R\}$$

consists of all R-linear combinations of the elements a_i of the given list and is a submodule of A, called the submodule *spanned* by $a_1, \cdots, a_n$. It is the intersection of all submodules of A containing these n elements. Similarly, the submodule of A *spanned* by any subset X of A is the intersection of all the submodules of A containing X. Since every submodule contains the element 0, the submodule of A spanned by the empty subset of A is the submodule 0 which consists only of the zero element of A. A module A is said to be *finitely spanned* (or, of *finite type*) if $A = Ra_1 + \cdots + Ra_n$ for some finite list $a_1, \cdots, a_n$ of elements of A.

For example, in the $\mathbf{R}$-module of all vectors in a three-dimensional space, each non-zero vector v spans a submodule which consists of all the vectors along a line through the origin—namely, the line containing the given vector v. Each pair of noncollinear vectors spans a submodule consisting of all the vectors in a plane through the origin—namely, the plane containing the two given noncollinear vectors. Any proper submodule of three-space is either such a line or such a plane. If D and E are submodules represented by two such planes, their intersection $D \cap E$ is the submodule represented by the intersection of these planes; it is a line except when $D = E$.

Under morphisms, submodules of a module behave as do subgroups of a group. Let $t:A \to A'$ be a morphism of modules. Its *image* is the set

$$\operatorname{Im}(t) = t_*A = \{t(a)|a \in A\}. \tag{17}$$

This subset is closed under addition and all scalar multiples, hence is a submodule of A'. The *kernel* of t is the submodule

$$\operatorname{Ker}(t) = t^*\{0\} = \{a|a \in A \text{ and } t(a) = 0\} \tag{18}$$

of A. As an additive subgroup of A, this is just the kernel of t considered as a morphism of additive groups. As in that case, t is a monomorphism if and only if $\operatorname{Ker}(t) = 0$ and an epimorphism if and only if $\operatorname{Im}(t) = A'$.

If $t:A \to A'$ and D is a submodule of A, the image t_*D of the set D is a submodule of A', called the *image* of D under t. For any two submodules D and E of A, $D \subset E$ implies $t_*D \subset t_*E$. In other words the function t_*, on submodules of A to those of A', is a morphism of inclusion.

Next consider inverse images under t. If D' is a submodule of A', the inverse image set t^*D', consisting of all $a \in A$ with $t(a) \in D'$, is a submodule of A, called the *inverse image* of D' under t. For two submodules D' and E' of A', $D' \subset E'$ implies $t^*D' \subset t^*E'$. Thus the function t^* on submodules of A' to those of A is a morphism of inclusion.

Theorem II.22 for subgroups has the following analog for submodules.

Theorem 5. *Any morphism $t:A \to A'$ of modules yields a bijection between the set of all those submodules of A containing* Ker t *and the set of all submodules of* Im (t). *This bijection assigns to each submodule D containing* Ker (t) *its image t_*D, while the inverse assigns to each submodule D' of* Im (t) *its inverse image under t. Each of these sets of submodules is a lattice under the partial order given by inclusion, and this bijection is an isomorphism of lattices.*

Proof: Any t_*D is contained in $\operatorname{Im}(t)$, while $\operatorname{Ker}(t) \subset D$ implies $t^*(t_*D) = D$ and $D' \subset \operatorname{Im}(t)$ implies $t_*(t^*D') = D'$. Hence, $D \mapsto t_*D$ is a bijection with inverse $D' \mapsto t^*D'$, as stated. Since intersection and sum of submodules containing Ker (t) again will contain Ker (t), this set of submodules is a lattice, as is the set of all $D' \subset \operatorname{Im}(t)$. Since $D \mapsto t_* D$ is an isomorphism of inclusion, it must carry meets and joins to meets and joins, respectively, so is a lattice isomorphism.

EXERCISES

1. For any ring R, prove that an R-submodule of the R-module R is exactly the same thing as a left ideal in R (§III.3).

2. Each of the following conditions describes a subset of the module R^3 of all triples (ξ_1, ξ_2, ξ_3) of scalars. Which of these subsets are submodules?

(a) $\xi_3 = 0$. (b) Either ξ_1 or $\xi_2 = 0$. (c) $\xi_1 = 1$.
(d) $\xi_1 + \xi_2 + \xi_3 = 0$. (e) $\xi_1 + \xi_2 + \xi_3 = 1$.

3. Show that intersection and sum of submodules are commutative and associative binary operations on the set of all submodules of A.

4. If C, D, and E are three submodules of A, prove the *modular law*: If $C \supset E$, then $C \cap (D + E) = (C \cap D) + E$.

5. In $\mathbf{Q}^3$ show that the $\mathbf{Q}$-submodule of all $(\xi_1, \xi_2, 0)$ is spanned by any of the following pairs of elements: $(0, 1, 0)$ and $(1, 1, 0)$; $(2, 2, 0)$ and $(1, 3, 0)$; $(3, 2, 0)$ and $(-3, 2, 0)$.

6. Let $D \subset \mathbf{Q}^3$ be the $\mathbf{Q}$-submodule of all $(0, \xi_2, \xi_3)$ and E the submodule spanned by $(1, 1, 0)$ and $(0, 1, 1)$. What are $D \cap E$ and $D + E$?

7. For $t: A \to A'$ linear and D, E submodules of A prove that

$$t_*(D \cap E) \subset (t_* D) \cap (t_* E), \qquad t_*(D + E) = (t_* D) + (t_* E).$$

8. For $t: A \to A'$ linear and D', E' submodules of A' prove that

$$t^*(D' \cap E') = (t^* D') \cap (t^* E'), \quad t^*(D' + E') \supset (t^* D') + (t^* E').$$

9. Show that $K[x]$, regarded as a K-module, is not finitely spanned.

10. Prove that an R-module A is finitely spanned if and only if there is for some natural number n an epimorphism $R^n \to A$ of modules.

***11.** For each left ideal C in the ring R, show that the (additive) quotient group R/C is an R-module; more precisely, show that there is exactly one way to make it an R-module so that the projection $R \to R/C$ becomes a morphism of R-modules.

***12.** Prove that any cyclic submodule $Ra \subset A$ is isomorphic to some module R/C, formed as in the previous exercise.

13. (a) Let $t: A \to A'$ be a morphism of modules with kernel D and $i: D \to A$ the insertion. Prove that every morphism $s: A'' \to A$ of modules with $t \circ s = 0$ can be written as a composite $s = i \circ s''$ for a unique morphism $s'': A'' \to D$. (Draw the diagram!)

(b) Show that this result means that i is a universal element of a suitable functor (cf. Theorem IV.5).

4. Quotient Modules

The construction of quotient modules is like that of quotient groups (§II.10). If D is a submodule of the R-module A, it is automatically a normal subgroup of the additive group of A. Hence, the quotient group A/D exists; it is an additive abelian group with the cosets $D + a = \{d + a \mid d \in D\}$ of D as elements, and the projection $p:A \to A/D$, which sends each element $a \in A$ to its coset $p(a) = D + a$ is a morphism of additive groups. This means that the sum of any two cosets in A/D is defined, as in (II.31), by

$$(D + a) + (D + b) = D + (a + b).$$

Note that a coset $D + a$ may be regarded as a "translate" of the submodule D by the element a; for example, if a is a vector in a three-dimensional vector space and D a two-dimensional subspace, the coset $D + a$ is the set of all vectors in a plane parallel to the plane D and passing through a.

Theorem 6. *Let D be a submodule of the R-module A, and A/D the (additive) quotient group. There is on A/D a unique set of operations of scalar multiplication by elements $\kappa \in R$ which make the projection $p:A \to A/D$ a morphism of scalar multiples. With these operations, A/D is an R-module and p an epimorphism of modules with kernel D.*

Proof: Scalar multiplication $a \mapsto \kappa a$ by κ is a morphism $\kappa:A \to A$ of groups which maps D to D. Its composite with the projection p is thus a morphism $p \circ \kappa:A \to A/D$ of groups which maps D to 0. As p is a universal morphism mapping D to 0, there is a unique morphism κ' of groups which makes the diagram

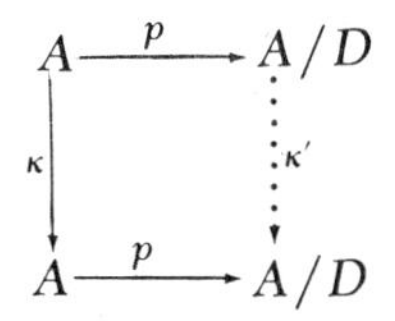

commute. For the coset $D + a = p(a)$, this commutativity states that

$$\kappa'(D + a) = D + \kappa a.$$

This is a definition of the scalar multiple $\kappa'(D + a)$ of a coset $D + a$ by κ, and is the only definition which makes p a morphism of scalar multiples. With this definition, the axioms (1)–(4) for the scalar multiples in the module A/D are immediately verified, and $p:A \to A/D$ is an epimorphism, as stated.

This module A/D is called the *quotient module* of A by its submodule D. The "main theorem" on these quotient modules states that $p:A \to A/D$ is universal among morphisms from A which kill D:

THEOREM 7. *Let D be a submodule of the R-module A. To each R-linear transformation $t:A \to A'$ with kernel containing D there is a unique R-linear transformation $t':A/D \to A'$ with $t' \circ p = t$:*

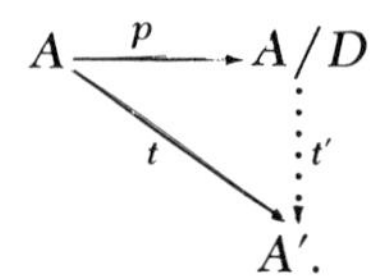

For this morphism t',

$$\text{Im}\,(t') = \text{Im}\,(t), \qquad \text{Ker}\ t' = (\text{Ker}\ t)/D. \tag{19}$$

Proof: By the main theorem for quotient groups (Theorem II.26) there is a unique morphism $t':A/D \to A'$ of additive groups with $t' \circ p = t$, just as in the commutative diagram above. For each coset $p(a) = D + a$, this commutativity condition states that $t'(D + a) = t(a)$; this formula and the definition of scalar multiples show that t' is also a morphism for these multiples. The properties (19) of t' follow at once from the corresponding results in the group case (Corollary II.26.1).

In particular, if $\text{Ker}\ t = D$, then t' is a monomorphism, so

$$\text{Im}\,(t) \cong \text{Domain}\ t/\text{Ker}\ t, \qquad t \text{ any linear map.} \tag{20}$$

As in the case of groups and the case of rings, the main theorem on quotient modules may be used to determine various quotient modules explicitly (Exercises 1–3 below).

EXERCISES

1. For any ring R, show that the assignment $(\xi_1, \xi_2, \xi_3) \mapsto (\xi_1, \xi_2)$ is an epimorphism $R^3 \to R^2$ of R-modules; if $D \subset R^3$ is the set of all triples $(0, 0, \xi_3)$, deduce that D is a submodule of R^3 with $R^3/D \cong R^2$ (a module isomorphism).

2. In the three-dimensional vector space V over the real number field, let D be the submodule of all vectors on a given line through the origin. Describe geometrically the quotient module V/D.

3. Consider the polynomial ring $K[x]$ as a K-module, with D the submodule of all polynomials $f_0 + f_2x^2 + \cdots + f_{2n}x^{2n}$ in *even* powers of x. Show that the quotient module $K[x]/D$ is isomorphic (as a K-module!) to $K[x]$.

4. Define the *cokernel* of a morphism $t:A \to A'$ of modules to be the quotient module Coker $(t) = A'/(\text{Im } t)$. Let $q:A' \to A'/(\text{Im } t)$ be the corresponding projection; note that $q \circ t = 0$.

(a) Show that to any morphism $s:A' \to A''$ of modules with $s \circ t = 0$ there is a unique morphism s':Coker $(t) \to A''$ with $s = s' \circ q$.

(b) Given t, show that the result of (a) describes q as a universal element of a suitable functor (a "universal left annihilator" of t).

(c) Show that this characterization of the cokernel of t is dual (in the sense of §IV.7) to the characterization of the kernel given in Exercise 3.13.

5. Using the cokernel described in Exercise 4, prove the following properties of the projection to the cokernel and the insertion of the kernel:

(a) For $t:A \to A'$ a monomorphism, $A \cong$ Ker $(A' \to \text{Coker }(t))$.

(b) For $t:A \to A'$ an epimorphism, $A' \cong$ Coker (Ker $(t) \to A$).

6. The "coimage" of a morphism $t:A \to A'$ is defined to be the quotient $A/(\text{Ker } t)$. Prove that this module is isomorphic to Im (t).

7. For submodules D_1 and D_2 of A show that the projection $p:A \to A/(D_1 + D_2)$ is universal among morphisms $t:A \to A'$ with $t_*D_1 = 0 = t_*D_2$.

5. Free Modules

The modules R^n, for R any ring, have some useful special properties. Recall that R^n is the set of all lists $\boldsymbol{\xi} = (\xi_1, \cdots, \xi_n)$ of n elements of R, with module operations defined termwise, as in (7) and (8); that is, by

$$(\boldsymbol{\xi} + \boldsymbol{\eta})_i = \xi_i + \eta_i, \qquad (\kappa\boldsymbol{\xi})_i = \kappa(\xi_i), \qquad i \in \mathbf{n}, \kappa \in R. \tag{21}$$

(Here and later, lists—functions on $\mathbf{n}$—will be written with boldface letters such as $\boldsymbol{\xi}$ and $\boldsymbol{\eta}$.)

For $n = 3$ and R the field $\mathbf{R}$ of real numbers, the module $\mathbf{R}^3$ is just the set of all vectors (ξ_1, ξ_2, ξ_3) in ordinary three-space. If we label the three *unit vectors* (vectors of length 1 along each of the three axes) as $i = (1, 0, 0)$, $j = (0, 1, 0)$, and $k = (0, 0, 1)$, then any vector in the space can be expressed as a linear combination of these unit vectors:

$$(\xi_1, \xi_2, \xi_3) = \xi_1 i + \xi_2 j + \xi_3 k. \tag{22}$$

We aim to describe the sense in which $\mathbf{R}^3$ is "freely generated" by i, j, and k.

More generally, for any ring R and any dimension n the R-module R^n contains n special elements

$$\boldsymbol{\varepsilon}_1 = (1, 0, \cdots, 0), \boldsymbol{\varepsilon}_2 = (0, 1, 0, \cdots, 0), \cdots, \boldsymbol{\varepsilon}_n = (0, \cdots, 0, 1), \tag{23}$$

called the n *unit elements* of R^n. For any scalar ξ_1 the multiple $\xi_1\boldsymbol{\varepsilon}_1$ is the list $(\xi_1, 0, \cdots, 0)$, and so on for $\xi_2\boldsymbol{\varepsilon}_2, \cdots$. Thus any element $\boldsymbol{\xi}$ of R^n can be expressed, just as in $\mathbf{R}^3$, as a linear combination

$$\boldsymbol{\xi} = (\xi_1, \cdots, \xi_n) = \xi_1\boldsymbol{\varepsilon}_1 + \cdots + \xi_n\boldsymbol{\varepsilon}_n \tag{24}$$

of the n unit elements with the ξ_i as coefficients. This expression can be used to calculate the effect of any linear transformation $t:R^n \to A$ on R^n to any other R-module A. It turns out that t is completely determined by giving the n elements $t(\boldsymbol{\varepsilon}_1), \cdots, t(\boldsymbol{\varepsilon}_n)$ in A; in other words, this list $\boldsymbol{\varepsilon}_1, \cdots, \boldsymbol{\varepsilon}_n$ of n vectors from R^n is universal among such lists, in the following sense.

Theorem 8. *For any n elements $c_1, \cdots, c_n$ in any R-module A, there is a unique morphism $t:R^n \to A$ of modules with $t(\boldsymbol{\varepsilon}_i) = c_i$ for each $i \in \mathbf{n}$. This morphism t is defined by the formula*

$$t\left(\sum_{i=1}^{n} \xi_i\boldsymbol{\varepsilon}_i\right) = \sum_{i=1}^{n} \xi_i c_i. \tag{25}$$

Proof: If there is any morphism t with $t(\boldsymbol{\varepsilon}_i) = c_i$, it must preserve linear combinations, so that

$$t\left(\sum \xi_i\boldsymbol{\varepsilon}_i\right) = \sum t(\xi_i\boldsymbol{\varepsilon}_i) = \sum \xi_i(t\boldsymbol{\varepsilon}_i) = \sum \xi_i c_i;$$

therefore, there is at most one such t, namely, that given in (25). Conversely, given the n elements c_i in A, this formula (25) does define a function $t:R^n \to A$ with $t(\boldsymbol{\varepsilon}_i) = c_i$; it remains only to show that this function is linear. To do this, take any two elements $\boldsymbol{\xi}$ and $\boldsymbol{\eta}$ of R^n and any two scalars κ and λ and calculate

$$\begin{aligned} t\left[\kappa\left(\sum \xi_i\boldsymbol{\varepsilon}_i\right) + \lambda\left(\sum \eta_i\boldsymbol{\varepsilon}_i\right)\right] &= t\left[\sum (\kappa\xi_i + \lambda\eta_i)\boldsymbol{\varepsilon}_i\right] \\ &= \sum (\kappa\xi_i + \lambda\eta_i)c_i = \kappa\sum \xi_i c_i + \lambda\sum \eta_i c_i \\ &= \kappa t\left(\sum \xi_i\boldsymbol{\varepsilon}_i\right) + \lambda t\left(\sum \eta_i\boldsymbol{\varepsilon}_i\right). \end{aligned}$$

This proves that $t[\kappa\boldsymbol{\xi} + \lambda\boldsymbol{\eta}] = \kappa t(\boldsymbol{\xi}) + \lambda t(\boldsymbol{\eta})$, as required.

This result described the universal property of the n unit vectors as a property of the *list* $\boldsymbol{\varepsilon}_1, \cdots, \boldsymbol{\varepsilon}_n$ of n unit vectors, taken in order. We can also formulate this property in terms of the *set* $\{\boldsymbol{\varepsilon}_1, \cdots, \boldsymbol{\varepsilon}_n\}$ of n vectors, disregarding order. Then a function $f:\{\boldsymbol{\varepsilon}_1, \cdots, \boldsymbol{\varepsilon}_n\} \to A$ on this set to any R-module determines n elements $f\boldsymbol{\varepsilon}_1 = c_1, \cdots, f\boldsymbol{\varepsilon}_n = c_n$ of that R-module. Hence, Theorem 8 states in effect that any such function $f:\{\boldsymbol{\varepsilon}_1, \cdots, \boldsymbol{\varepsilon}_n\} \to A$ extends to a unique linear map $t:R^n \to A$; namely, the map t with $t(\sum \xi_i\boldsymbol{\varepsilon}_i) = \sum \xi_i(f\boldsymbol{\varepsilon}_i)$. Because of this property, we say that R^n is a "free" module on the set $X = \{\varepsilon_1, \cdots, \varepsilon_n\}$ of its unit vectors. Here is the general definition:

DEFINITION. Let X be a subset of an R-module F and $j:X \to F$ the insertion of X in F. Then F is called a free module *on X (and X a set of* free generators *for the module F) when to every function $f:X \to A$ to an R-module A there is exactly one linear map $t:F \to A$ with $t \circ j = f$, as in the commutative diagram below:*

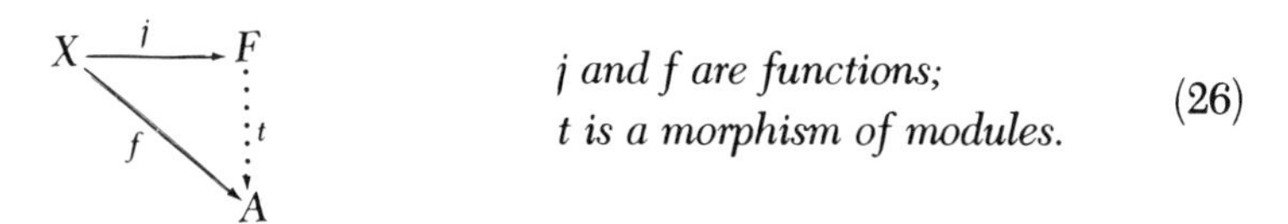

In this language, Theorem 8 becomes

THEOREM 9. *The module R^n is free on the set $\{\varepsilon_1, \cdots, \varepsilon_n\}$ of its n unit elements.*

Any set X of free generators is also called a "basis" for the free module. Properties of these bases will be important in the study of vector spaces in §VI.1. For this reason, and for general background, we will now examine bases for free modules.

The free module R^n is spanned by its free generators $\varepsilon_1, \cdots, \varepsilon_n$. This holds in general: *If F is free on the subset X, then X spans F.* For let D be the submodule of F spanned by X, $p:F \to F/D$ the projection to the quotient module, and $0:F \to F/D$ the zero morphism, sending all elements of F to $0 = D + 0$ in F/D. Both p and 0 have $pj = 0j = 0:X \to F/D$. Hence, by the uniqueness of t in (26), $p = 0$, so $F/D = 0$. In other words, $D = F$, so that X does span F.

We note that one and the same module F may be free on several different bases. For example, R^3 is freely generated by the three elements $(0, 0, 1)$, $(0, 1, -1)$, $(1, -1, 0)$ because any element ξ in R^3 can be written in exactly one way as a linear combination

$$(\xi_1, \xi_2, \xi_3) = \xi_1(1, -1, 0) + (\xi_2 + \xi_1)(0, 1, -1) + (\xi_3 + \xi_2 + \xi_1)(0, 0, 1)$$

of these three elements, so any linear map $t:R^3 \to A$ is determined by its values on them. Moreover, a free module on a finite set of given size may be constructed in many different ways. For example, in R^3 the subset of all elements $\xi = (\xi_1, \xi_2, \xi_3)$ with $\xi_2 = \xi_3$ (this is the "plane" through the origin with equation $y = z$) is a free module on a basis $(1, 0, 0)$ and $(0, 1, 1)$ of two elements.

Since the free module F on a given basis X is a universal, the uniqueness theorem for universals (Theorem IV.1) states that a free module F is uniquely determined, up to an isomorphism, by its basis X. Specifically, if two insertions $j:X \to F$ and $j':X \to F'$ make both F and F' free modules with generators X, then there is $t:F \to F'$ with $t \circ j = j'$ and also $t':F' \to F$ with $t' \circ j' = j$. By the uniqueness part of the definition, both composites $t' \circ t$ and $t \circ t'$ are identities, so $t:F \to F'$ is an isomorphism with inverse t'. (This argument is really just a particular case of the proof of the uniqueness theorem for any universal.)

The free modules R^n are special cases of "function modules". If A is an R-module, so is the set A^n of all lists $\mathbf{a}:\mathbf{n} \to A$ of n elements of A. More generally, if X is any set, the *function module* A^X is the set of all functions $f:X \to A$ with the usual "pointwise" module operations $(f, g) \mapsto f + g$ and $(\kappa, f) \mapsto \kappa f$ defined by

$$(f + g)(x) = f(x) + g(x), \qquad (\kappa f)(x) = \kappa(fx), \qquad x \in X. \quad (27)$$

The module axioms for these operations follow at once.

For $A = R$, note in particular the function module R^X. This module contains for each $x \in X$ an element ε_x defined for all $y \in X$ by the equations

$$\begin{aligned} \varepsilon_x(y) &= 1, \qquad && \text{if } y = x, \\ &= 0, && \text{if } y \neq x. \end{aligned}$$

If $X = \mathbf{n}$, these ε_x are exactly the previous unit elements of R^n. We let $R^{(X)}$ denote the submodule of R^X spanned by these elements ε_x, for all $x \in X$. In other words, $R^{(X)}$ is the set of all R-linear combinations of finitely many elements ε_x. This module $R^{(X)}$ turns out (Exercise 5) to be the free module on the set $\{\varepsilon_x | x \in X\}$. Since $\varepsilon_x \mapsto x$ is a bijection $\{\varepsilon_x | x \in X\} \cong X$, this process constructs a free module on *any* set X—an important result.

For example, take X to be the infinite set $\mathbf{N}$ of natural numbers. An element $f \in R^{\mathbf{N}}$ is a function $f:\mathbf{N} \to R$, so can be exhibited as the "infinite list" of its values

$$(f_0, f_1, f_2, \cdots).$$

The unit element ε_n is the particular infinite list with 1 at position n as its only non-zero entry, and the submodule $R^{(\mathbf{N})}$ spanned by all these elements ε_n consists of all these infinite lists which have only finitely many non-zero entries.

If F is any free module, say on the set X, the diagram (26) in its definition shows that $t \mapsto t \circ j$ is a bijection

$$\hom(F, A) \cong A^X \quad (28)$$

from the "hom-set" of all morphisms $t:F \to A$ to the function set A^X. This bijection is just a special case of the representation Theorem IV.2 for universals. In the present case, this set hom (F, A) is an abelian group $\mathrm{Hom}_R(F, A)$ under pointwise addition (Theorem 1), and so is A^X. Since the bijection $t \mapsto t \circ j$ carries pointwise sums to pointwise sums, it is an isomorphism

$$\mathrm{Hom}_R(F, A) \cong A^X$$

of abelian groups. In case R is a commutative ring K, more is true: It is an isomorphism

$$\mathrm{Hom}_K(F, A) \cong A^X$$

of K-modules. When X is finite, say $X = \mathbf{n}$, this result may be formulated as follows:

Proposition 10. *If F is a free K-module on n free generators $b_1, \cdots, b_n$, while A is any K-module, there is an isomorphism*

$$\mathrm{Hom}_K(F, A) \cong A^n \tag{29}$$

of K-modules which assigns to each morphism $t: F \to A$ of K-modules the list $t(b_1), \cdots, t(b_n)$ in A^n.

The following property of free modules is especially useful.

Theorem 11. *If $p: A \to A'$ is an epimorphism of R-modules, each morphism $t: F \to A'$ with domain a free R-module F can be written as a composite $t = p \circ s$ for some morphism $s: F \to A$ of R-modules.*

Proof: The diagram is as shown below:

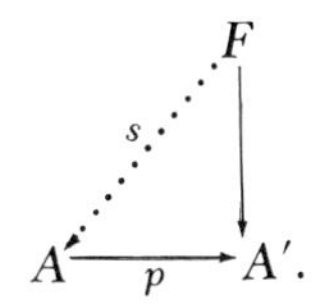

F is free on some set X and the horizontal map p is an epimorphism; we wish to find a linear map s which makes the diagram commute, as in $p \circ s = t$. (We also say that t "factors through" p or that t "lifts" to s.)

For each $x \in X$, $t(x)$ is an element of A'. Since the epimorphism p is surjective as a function, we can choose to each $x \in X$ some element $a_x \in A$ with $p(a_x) = t(x)$. Since F is free on X, there is a linear map $s: F \to A$ with $s(x) = a_x$ for every $x \in F$. Then $(p \circ s)(x) = t(x)$ for each free generator x so the composite $p \circ s$ must be t, as desired.

EXERCISES

1. Show that R^2 is free on the set with two elements (0, 1) and (1, 1).

2. Show that R^3 is free on each of the following sets of three elements:

(a) (0, 0, 1), (0, 1, 1), (1, 1, 1).

(b) (1, 0, 1), (0, 1, 1), (1, 1, 0).

3. Show that every subgroup of **Z** is a free **Z**-module.

4. If the ring R is commutative and $a \in R$ is not a zero divisor, show that the principal ideal of all multiples of a is a free R-module on one generator.

5. The "support" of a function $f:X \to R$ is the subset $\{x|x \in X$ and $f(x) \neq 0\}$ of X.

(a) Show that the submodule $R^{(X)}$ as defined in the text is exactly the set of all $f:X \to R$ with finite support.

(b) Prove that $R^{(X)}$ is free on the set of all ε_x for $x \in X$.

6. Prove that every R-module A is an epimorphic image of a free R-module; explicitly, if X is the set of all non-zero elements of A, show that A is an epimorphic image of $R^{(X)}$.

7. Let a set X be given. To each morphism $t:A \to A'$ of R-modules show that the exponential function t^X of (IV.19) is a morphism $t^X:A^X \to A'^X$ of modules. Show also that $(1_A)^X$ is the identity morphism, and that $(s \circ t)^X = (s^X) \circ (t^X)$ whenever the composite $s \circ t$ is defined. (These properties state that $A \mapsto A^X$ and $t \mapsto t^X$ define a functor on R-modules to R-modules.)

8. Let an R-module A be given. To each function $h:X \to X'$ show that the exponential function A^h of (IV.20) is a morphism $A^h:A^{X'} \to A^X$ of modules, that A^1 is the identity morphism, and that $A^{(h' \circ h)} = A^h \circ A^{h'}$ whenever the composite $h' \circ h$ is defined. (These properties state that $X \mapsto A^X$ and $h \mapsto A^h$ define a contravariant functor on sets to R-modules.)

9. Show by an example that the "lifted" map s of Theorem 11 is not unique. (*Hint:* s depends on the choice of the elements a_x used in the proof.)

6. Biproducts

The product of two sets X_1 and X_2 is a set $X_1 \times X_2$, their cartesian product, equipped with two functions, the "projections"

$$X_1 \xleftarrow{p_1} X_1 \times X_2 \xrightarrow{p_2} X_2,$$

which are universal among pairs of functions from a set Y to X_1 and X_2. The coproduct of two sets X_1 and X_2 is a set $X_1 \mathbin{\dot\cup} X_2$, their disjoint union (§IV.7), equipped with two functions, the "injections"

$$X_1 \to X_1 \mathbin{\dot\cup} X_2 \leftarrow X_2$$

which are universal among pairs of functions from X_1 and X_2 to a set Y.

These universality properties have been used to define "product" and "coproduct" in any concrete category (§IV.7). Also, we have constructed the product of two groups and of two rings, and we have indicated (§IV.7) that the product of two abelian groups ($\mathbf{Z}$-modules) is also a coproduct. This is the case for R-modules too: Given two R-modules A_1 and A_2, there is a single R-module $A_1 \oplus A_2$ which is a product of A_1 and A_2 relative to suitable projections and at the same time is a coproduct of A_1 and A_2 relative to suitable injections. This module will therefore be called the "biproduct" of A_1 and A_2. In other books it goes by many other names: "direct sum", "direct product", or "cartesian product".

Let A_1 and A_2 be two R-modules. On the set $A_1 \oplus A_2$ of all ordered pairs (a_1, a_2) of elements $a_i \in A_i$ define termwise module operations by setting

$$(a_1, a_2) + (a_1', a_2') = (a_1 + a_1', a_2 + a_2'), \qquad \kappa(a_1, a_2) = (\kappa a_1, \kappa a_2) \quad (30)$$

for all elements $a_i, a_i' \in A_i$ and all scalars $\kappa \in R$. Under this addition, $A_1 \oplus A_2$ is just the product of the abelian groups A_1 and A_2, hence is itself an abelian group; under these scalar multiples by $\kappa \in R$, the usual termwise verification shows that the remaining module axioms (1)–(4) hold. Hence, $A_1 \oplus A_2$ is an R-module, the *biproduct* of the R-modules A_1 and A_2. For example, the biproduct $R \oplus R$ is just the free module R^2, while the biproduct $R^m \oplus R^n$ of the two free modules R^m and R^n is evidently isomorphic to the free module R^{m+n}.

Next construct the *projections* p_i and the *injections* e_i of the biproduct $A_1 \oplus A_2$.

Lemma 1. *For the biproduct $A_1 \oplus A_2$ of two R-modules A_1 and A_2 there are four morphisms*

$$A_1 \underset{e_1}{\overset{p_1}{\leftrightarrows}} A_1 \oplus A_2 \underset{e_2}{\overset{p_2}{\rightleftarrows}} A_2 \quad (31)$$

of R-modules whose composites satisfy the conditions

$$\begin{aligned} p_1e_1 &= 1, \quad p_1e_2 = 0, \qquad 1 \text{ the identity on } A_1; \\ p_2e_1 &= 0, \quad p_2e_2 = 1, \qquad 1 \text{ the identity on } A_2; \end{aligned} \quad (32)$$

$$e_1p_1 + e_2p_2 = 1, \qquad 1 \text{ the identity on } A_1 \oplus A_2. \quad (33)$$

Proof: The functions p_i and e_i defined for $i = 1, 2$ by

$$p_1(a_1, a_2) = a_1, \quad p_2(a_1, a_2) = a_2, \quad e_1(a_1) = (a_1, 0), \quad e_2(a_2) = (0, a_2)$$

are clearly morphisms which satisfy (32). The composites e_1p_1 and e_2p_2 are both endomorphisms of $A_1 \oplus A_2$, so they have a pointwise sum, defined as in §2; for this sum,

$$(e_1 p_1 + e_2 p_2)(a_1, a_2) = e_1(a_1) + e_2(a_2) = (a_1, 0) + (0, a_2) = (a_1, a_2);$$

this sum is therefore the identity morphism on $A_1 \oplus A_2$, as required by (33).

Next we show that the projections p_1 and p_2 of the biproduct do give a product in the category of all R-modules.

Theorem 12. *If B is any R-module and $t_i : B \to A_i$ are two morphisms of modules $(i = 1, 2)$, there is a unique morphism $t : B \to A_1 \oplus A_2$ of modules such that $p_1 \circ t = t_1$ and $p_2 \circ t = t_2$.*

Proof: Given p_i and t_i, we must show that the diagram

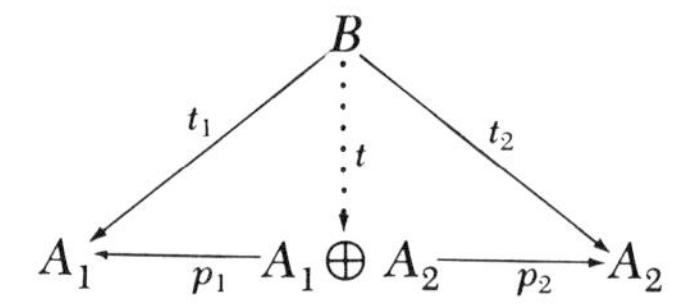

can be filled in with a unique vertical morphism t so as to be commutative. Now this commutativity $p_1 t = t_1$ and $p_2 t = t_2$ implies

$$e_1 t_1 + e_2 t_2 = e_1 p_1 t + e_2 p_2 t = (e_1 p_1 + e_2 p_2) t = t,$$

by (33) and the distributive law (13) for sum and composites of morphisms. Hence, t, if it exists, must be $t = e_1 t_1 + e_2 t_2$. Conversely, this sum $e_1 t_1 + e_2 t_2$ is a morphism $B \to A_1 \oplus A_2$. By (32) and the distributive law again, it satisfies

$$p_1(e_1 t_1 + e_2 t_2) = p_1 e_1 t_1 + p_1 e_2 t_2 = t_1 + 0 t_2 = t_1,$$
$$p_2(e_1 t_1 + e_2 t_2) = p_2 e_1 t_1 + p_2 e_2 t_2 = 0 t_1 + t_2 = t_2;$$

hence, $t = e_1 t_1 + e_2 t_2$ has the required properties.

This theorem tells us that $A_1 \oplus A_2$ is a product object because it asserts that the pair (p_1, p_2) of projections from $A_1 \oplus A_2$ is a universal element for the functor $\mathcal{C} = \hom(-, A_1) \times \hom(-, A_2)$. This functor (cf. §IV.7) is a contravariant functor on R-modules to sets; for fixed R-modules A_1 and A_2, it assigns to each R-module B the set

$$\mathcal{C}(B) = \hom(B, A_1) \times \hom(B, A_2)$$

of all pairs (t_1, t_2) of morphisms $t_i : B \to A_i$. Also, the assignment $t \mapsto (t_1, t_2)$

suggested by the theorem is a bijection

$$\hom(B, A_1 \oplus A_2) \cong \hom(B, A_1) \times \hom(B, A_2); \tag{34}$$

indeed, this bijection is just the representation theorem (Theorem IV.2 and Equation (IV.11)) for the universal element (p_1, p_2) of the functor $\mathcal{C}$. In the present case more is true: Each set hom $(B, -)$ of morphisms may be made into an abelian group $\mathrm{Hom}_R(B, -)$, by defining addition as in §2. For these abelian groups the following holds.

Corollary. *For R-modules B, A_1, and A_2 there is an isomorphism*

$$\mathrm{Hom}_R(B, A_1 \oplus A_2) \cong \mathrm{Hom}_R(B, A_1) \oplus \mathrm{Hom}_R(B, A_2) \tag{35}$$

of abelian groups, given in terms of the projections p_1 and p_2 of the biproduct $A_1 \oplus A_2$ by assigning to each $t: B \to A_1 \oplus A_2$ the pair $(p_1 \circ t, p_2 \circ t)$. If $R = K$ is a commutative ring, this assignment is an isomorphism of K-modules.

Note that the symbol $\oplus$ on the left of (35) stands for the biproduct of R-modules, while the same symbol on the right denotes the biproduct of abelian groups (**Z**-modules)—or of K-modules, in the commutative case.

Proof: By the distributive law for morphisms, $p_i(t + t') = p_i t + p_i t'$, hence $t \mapsto (p_1 t, p_2 t)$ is a morphism of addition. Since it is a bijection, it is an isomorphism of abelian groups. In case $R = K$ is commutative, each abelian group Hom_K may be regarded as a K-module (Theorem 3) and $t \mapsto (p_1 t, p_2 t)$ is also a morphism of K-modules, just as required.

Next we show that $A_1 \oplus A_2$, equipped with the injections e_1 and e_2, is a coproduct of A_1 and A_2 in the category of R-modules.

Theorem 13. *If C is any R-module and $s_i: A_i \to C$ are two morphisms of modules $(i = 1, 2)$, there is a unique morphism $s: A_1 \oplus A_2 \to C$ of modules such that $s \circ e_1 = s_1$ and $s \circ e_2 = s_2$.*

Proof: Given e_i and s_i, we must show that the diagram

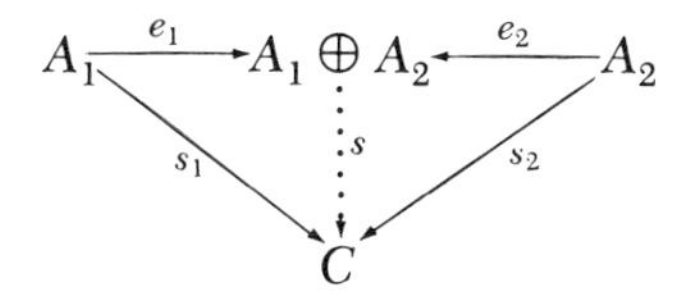

can be filled in with a unique morphism s (at the dotted arrow) so as to be commutative. Now this commutativity $se_1 = s_1$ and $se_2 = s_2$ implies

$$s_1 p_1 + s_2 p_2 = se_1 p_1 + se_2 p_2 = s(e_1 p_1 + e_2 p_2) = s,$$

by the identity (33) and the distributive law. Hence, s, if it exists, must be $s = s_1 p_1 + s_2 p_2$. Conversely, this sum $s_1 p_1 + s_2 p_2$ is a morphism $A_1 \oplus A_2 \to C$. By (13) and the identities (32),

$$(s_1 p_1 + s_2 p_2)e_1 = s_1 p_1 e_1 + s_2 p_2 e_1 = s_1,$$

$$(s_1 p_1 + s_2 p_2)e_2 = s_1 p_1 e_2 + s_2 p_2 e_2 = s_2.$$

Hence, it is the morphism s required for the theorem.

This theorem states that $A_1 \oplus A_2$ is a coproduct object because it asserts that the pair (e_1, e_2) of injections to $A_1 \oplus A_2$ is a universal element for the functor $\mathfrak{F} = \hom(A_1, -) \times \hom(A_2, -)$. This functor is a covariant functor on R-modules to sets; for fixed modules A_1 and A_2, it assigns to each R-module C the set

$$\mathfrak{F}(C) = \hom(A_1, C) \times \hom(A_2, C)$$

of all pairs (s_1, s_2) of morphisms $s_i : A_i \to C$, $i = 1, 2$. Moreover, the assignment $s \mapsto (s_1, s_2)$ is the bijection

$$\hom(A_1 \oplus A_2, C) \cong \hom(A_1, C) \times \hom(A_2, C)$$

of the representation theorem. Just as in the previous case, this gives the

Corollary. *For R-modules A_1, A_2, and C there is an isomorphism*

$$\mathrm{Hom}_R(A_1 \oplus A_2, C) \cong \mathrm{Hom}_R(A_1, C) \oplus \mathrm{Hom}_R(A_2, C) \tag{36}$$

of abelian groups, given in terms of the injections e_1 and e_2 of the biproduct $A_1 \oplus A_2$ by the assignment $s \mapsto (s \circ e_1, s \circ e_2)$. If $R = K$ is commutative, this is an isomorphism of K-modules.

Now we have proved that the set $A_1 \oplus A_2$ of all pairs (a_1, a_2) has the properties both of a product module (universal pair of projections) and of a coproduct module (universal pair of injections). Either property will determine $A_1 \oplus A_2$ uniquely up to an isomorphism, by the usual uniqueness theorem for universals (Theorem IV.1). For that matter, the proof of the two universality theorems above did not make use of the elements (a_1, a_2) of the module $A_1 \oplus A_2$, but only of the identities (32) and (33) on the projections and injections together. Therefore these identities characterize the biproduct $A_1 \oplus A_2$ up to isomorphism (Exercise 4).

We will also use iterated biproducts. For example, given three R-modules A_1, A_2, and A_3, there is an evident module isomorphism $A_1 \oplus (A_2 \oplus A_3) \cong$

$(A_1 \oplus A_2) \oplus A_3$ between the two iterated biproducts, because both iterated biproducts have a universal triple of projections (Exercise 3). In §VII.2 we will show how the universality properties of projections and injections of a biproduct lead to the basic idea of representing a linear map $t:R^n \to R^m$ by a matrix.

Here is a simple case of quotients of biproducts.

Proposition 14. *If D_1 and D_2 are submodules of A_1 and A_2, respectively, then $(A_1 \oplus A_2)/(D_1 \oplus D_2) \cong (A_1/D_1) \oplus (A_2/D_2)$.*

Proof: Let $p_i:A_i \to A_i/D_i$ be the projection on the quotient module, for $i = 1, 2$. Then $q(a_1, a_2) = (p_1a_1, p_2a_2)$ defines an epimorphism $q:A_1 \oplus A_2 \to (A_1/D_1) \oplus (A_2/D_2)$. Its kernel is $D_1 \oplus D_2$; hence, its codomain is isomorphic to $(A_1 \oplus A_2)/(D_1 \oplus D_2)$, just as required.

Much as for the direct product of groups (Proposition II.13), the biproduct of modules can be described not by morphisms but by submodules. Note first that the biproduct $A_1 \oplus A_2$ contains two submodules

$$A_1' = \{\text{all } (a_1, 0) \text{ for } a_1 \in A_1\} \cong A_1,\ A_2' = \{\text{all } (0, a_2) \text{ for } a_2 \in A_2\} \cong A_2.$$

Their intersection $A_1' \cap A_2'$ is 0; their sum $A_1' + A_2'$ is $A_1 \oplus A_2$. These facts characterize the biproduct, in the following sense.

Proposition 15. *If a module B has submodules A_1, A_2 with*

$$A_1 \cap A_2 = 0, \qquad A_1 + A_2 = B, \tag{37}$$

there is a module isomorphism $k:A_1 \oplus A_2 \cong B$ which makes each composite $ke_i:A_i \to B$ the insertion of the submodule $A_i \subset B$.

Proof: The insertions $A_1 \to B \leftarrow A_2$ of the given submodules give a diagram to which the universal property of Theorem 13 applies, so there is a morphism $k:A_1 \oplus A_2 \to B$ with each ke_i the insertion; explicitly, $k(a_1, a_2) = k(e_1a_1 + e_2a_2) = a_1 + a_2$. If the resulting element is zero, $a_1 = -a_2$ is in $A_1 \cap A_2$, hence is zero by hypothesis. Therefore, k is a monomorphism. On the other hand, the second hypothesis $A_1 + A_2 = B$ ensures that k is an epimorphism, hence an isomorphism, as required.

Under the conditions (37) we call B the *direct sum* of its submodules A_1 and A_2. A submodule $A_1 \subset B$ is called a *direct summand of B* if there exists another submodule A_2 which together with A_1 satisfies (37).

Lemma 2. *If F is a free module and $t{:}B \to F$ an epimorphism, then B is the direct sum of* Ker t *and a submodule F' isomorphic to F.*

Proof: Since F is free and t an epimorphism, there is, as in Theorem 11, a morphism $t'{:}F \to B$ with $t \circ t' = 1_F$, as displayed in the commutative diagram

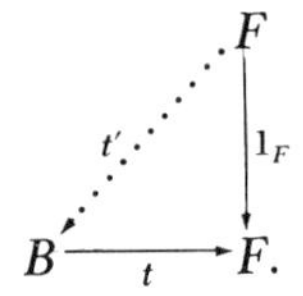

Let F' be the image of t'. As $tt' = 1$, $\operatorname{Ker} t \cap F' = 0$. Any $b \in B$ is $b = (b - t'tb) + t'tb$, where $t'tb \in F'$ and $(b - t'tb) \in \operatorname{Ker} t$; this means that $\operatorname{Ker} t + F' = B$. Therefore, B is the direct sum of its submodules Ker t and F', as asserted, while t' restricts to an isomorphism $F \cong F'$.

It follows that F' is a free module on the t'-images of the free generators of F, and that $B \cong (\operatorname{Ker} t) \oplus F$.

EXERCISES

1. Establish a module isomorphism $A_1 \oplus A_2 \cong A_2 \oplus A_1$.
2. Show that the biproduct of any two free modules is free.
3. (a) For three given R-modules A_1, A_2, and A_3 define a contravariant functor $\mathcal{C}$ on R-modules to sets which will have the form $\mathcal{C} = \hom(-, A_1) \times \hom(-, A_2) \times \hom(-, A_3)$ and construct a universal element$(p_1, p_2, p_3) \in \mathcal{C}(A_1 \oplus (A_2 \oplus A_3))$.

   (b) Find a universal element in $\mathcal{C}((A_1 \oplus A_2) \oplus A_3)$ for the same functor and deduce a module isomorphism $A_1 \oplus (A_2 \oplus A_3) \cong (A_1 \oplus A_2) \oplus A_3$.

   (c) State and prove the universal properties of $A_1 \oplus \cdots \oplus A_n$.
4. Prove: If the diagram of R-modules and R-linear maps

$$A_1 \underset{e'_1}{\overset{p'_1}{\leftrightarrows}} D \underset{e'_2}{\overset{p'_2}{\rightleftarrows}} A_2$$

satisfies the identities (32) and (33), then D is isomorphic to $A_1 \oplus A_2$; more explicitly, there is an isomorphism $k{:}A_1 \oplus A_2 \cong D$ with $p'_i k = p_i$ and $ke_i = e'_i$ for $i = 1, 2$. (*Hint:* $k = e'_1 p_1 + e'_2 p_2$, $k^{-1} = e_1 p'_1 + e_2 p'_2$.)

5. Show that (33), $p_1 e_1 = 1$, and $p_2 e_2 = 1$ imply the remaining identities of (32).
6. If B has three submodules A_1, A_2, and A_3, with $A_1 + A_2 + A_3 = B$ and both $A_1 \cap (A_2 + A_3)$ and $A_2 \cap A_3$ zero, establish an appropriate isomorphism $k{:}A_1 \oplus A_2 \oplus A_3 \cong B$.
7. Extend the result of Exercise 6 from 3 factors to n.
8. Show that every $\mathbf{Z}$-submodule of $\mathbf{Z} \oplus \mathbf{Z}$ is a free $\mathbf{Z}$-module.

7. Dual Modules

In this section we will show how to associate to every module another module, called its "dual", and to every morphism of modules an appropriate "dual" morphism. The description of these dual modules will be the key to the precise understanding of many important distinctions in algebra (coefficients versus unknowns, points versus hyperplanes, rows versus columns, covariance versus contravariance).

Hitherto all our modules have been left R-modules, with each scalar multiplier written on the left. It will turn out that the dual of a left R-module is necessarily a right R-module, and vice versa. Here a *right R-module* is an abelian group A together with a function $A \times R \to R$, written $(a, \kappa) \mapsto a\kappa$, which satisfies the right-hand version of the module axioms (1)–(4); they are

$$(a + b)\kappa = a\kappa + b\kappa, \quad a(\kappa + \lambda) = a\kappa + a\lambda, \quad a(\kappa\lambda) = (a\kappa)\lambda, \quad a1 = a.$$

For example, the ring R itself is both a left R-module and a right R-module. All the properties developed hitherto for left modules clearly apply, *mutatis mutandis*, to right modules.

There is a systematic way of turning a right module into a left module over the opposite ring (§III.2):

Proposition 16. *If R^{op} is the ring opposite to R, every right R-module A becomes a left R^{op}-module A^{op}, where A^{op} is the same abelian group as A, but with left scalar multiples $\triangledown: R \times A \to A$ defined by $\kappa \triangledown a = a\kappa$ for all $a \in A$ and all $\kappa \in R$.*

Proof: The opposite ring R^{op} has the same elements as R but the opposite multiplication $\square: R^{\text{op}} \times R^{\text{op}} \to R^{\text{op}}$, defined for elements κ, λ as $\kappa \square \lambda = \lambda\kappa$ (cf. §II.2, Rule 6). With this product $\square$ and the scalar multiples $\triangledown$, each of the right module axioms above turns into the corresponding axiom for a left module; only the associative law (3) requires special note. For this law, any $a \in A$, and any scalars $\kappa, \lambda \in R$, the definitions of $\square$ and $\triangledown$ give

$$\begin{aligned}(\kappa \square \lambda) \triangledown a &= a(\kappa \square \lambda) = a(\lambda\kappa) \\ &= (a\lambda)\kappa = \kappa \triangledown (a\lambda) = \kappa \triangledown (\lambda \triangledown a);\end{aligned}$$

this is the associative law for the product $\square$ and the scalar multiple $\triangledown$.

For a commutative ring K, $K^{\text{op}} = K$. This gives the

Corollary. *Over a commutative ring K, each right K-module is also a left K-module with left scalar multiples $\kappa a = a\kappa$.*

Hence, for modules over a commutative ring we can and will write the scalar factors κ on the left or on the right, as may be convenient.

Now we construct dual modules. Let A be a right R-module over any ring R, and regard the ring R as a right R-module. For each scalar κ and each morphism $f:A \to R$ of *right* R-modules the *left* scalar multiple $\kappa f:A \to R$, defined for each $a \in A$ as $(\kappa f)a = \kappa(fa)$, is also a morphism of right R-modules; indeed, the function κf is clearly a morphism of addition, while, for any scalar λ,

$$(\kappa f)(a\lambda) = \kappa[\, f(a\lambda)] = \kappa[(fa)\lambda] = [\kappa(fa)]\lambda = [(\kappa f)a]\lambda.$$

This states that κf is also a morphism of right scalar multiples. Thus the set

$$A^* = \mathrm{Hom}_R\,(A, R) = \{f \mid f : A \to R \text{ is a morphism of right } R\text{-modules}\} \tag{38}$$

is closed under the sum and the left scalar multiples defined by

$$(f + g)(a) = fa + ga, \qquad (\kappa f)(a) = \kappa(fa) \tag{39}$$

for all $a \in A$ and all $\kappa \in R$. The usual pointwise verification shows that A^* is a left R-module under these operations. It is called the *dual* (sometimes the "conjugate") of the right module A.

In general for two right R-modules A and B the set $\mathrm{Hom}_R\,(A, B)$ is, as already noted (Theorem 1), just an abelian group. However, when $B = R$ the set $A^* = \mathrm{Hom}_R\,(A, R)$ is more than an abelian group, since we have just shown that it is a left R-module. This is because the scalar multiplication $r \mapsto \kappa r$ of an element $r \in R$ by a scalar κ on the left naturally carries over to a multiplication $f \mapsto \kappa f$ of f by κ on the left. This is because R is a left R-module. However, f is a morphism to R as a right module with right scalar multiples $r \mapsto r\lambda$. In other words, the ring R is both a left R-module (under $r \mapsto \kappa r$) and a right R-module (under $r \mapsto r\lambda$) and these operations commute, because $\kappa(r\lambda) = (\kappa r)\lambda$. For these reasons, R is sometimes called an "R-bimodule". Similarly, each free module R^n has an R-bimodule structure.

For a right module A, we have indicated why the dual is a left module A^*. Similarly, the dual C^* of a left R-module C is a right R-module.

The dual A^* of a module is usually quite different from A. As an exceptional case, however, we may calculate the dual of a free right module of the form R^n. Each list $\boldsymbol{\alpha}:\mathbf{n} \to R$ of n scalars yields a function $\theta\boldsymbol{\alpha}$ defined for each $\boldsymbol{\xi} \in R^n$ by

$$(\theta\boldsymbol{\alpha})(\xi_1, \cdots, \xi_n) = \alpha_1\xi_1 + \cdots + \alpha_n\xi_n. \tag{40}$$

Now $\theta\boldsymbol{\alpha}$ is in $(R^n)^*$ because $\theta\boldsymbol{\alpha}:R^n \to R$ is a morphism of right R-modules. Under this morphism the ith unit element $\boldsymbol{\varepsilon}_i = (0, \cdots, 1, \cdots, 0)$ of R^n has image $(\theta\boldsymbol{\alpha})\boldsymbol{\varepsilon}_i = \alpha_i$. But R^n is a free right R-module on the n unit elements $\boldsymbol{\varepsilon}_1, \cdots, \boldsymbol{\varepsilon}_n$ as free generators, and by Theorem 8 a morphism $f:R^n \to R$ on a free module is completely determined by the images $f\boldsymbol{\varepsilon}_i$ of the generators. Therefore, every element f of the dual $(R^n)^*$ has the form $\theta\alpha$ for some unique $\boldsymbol{\alpha}$, so that θ is a bijection $R^n \cong (R^n)^*$. For two lists $\boldsymbol{\alpha}$ and $\boldsymbol{\alpha}'$ the pointwise sum is $\theta\boldsymbol{\alpha} + \theta\boldsymbol{\alpha}' = \theta(\boldsymbol{\alpha} + \boldsymbol{\alpha}')$, while for a left scalar multiple, $\theta(\lambda\boldsymbol{\alpha}) = \lambda(\theta\boldsymbol{\alpha})$. Hence the bijection θ is an isomorphism of left modules:

$$\theta : ({}_RR)^n \to ((R_R)^n)^*.$$

The notation ${}_RR$ on the left here designates the ring R regarded as a *left* R-module; that is, with $\alpha \mapsto \lambda\alpha$ the multiple of $\alpha \in R$ by a scalar λ. On the right, R_R denotes R regarded as a right R-module, with $\xi \mapsto \xi\kappa$ the multiplication by a scalar κ. Thus $(R_R)^n$ is also a right module, and its dual $(R_R^n)^*$, as above, is a left module, as usual. We have proved

Proposition 17. *The dual of the right R-module $(R_R)^n$ is the left R-module $({}_RR)^n$. More exactly, there is an isomorphism θ of left R-modules, as above, which assigns to each list $\boldsymbol{\alpha}:\mathbf{n} \to R$ in the left module $({}_RR)^n$ the morphism $\theta\boldsymbol{\alpha}:(R_R)^n \to R$ of right R-modules defined by (40).*

For $n = 1$, this result states that ${}_RR \cong (R_R)^*$ under the map θ which sends each $\alpha \in R$ into the right linear map $\xi \mapsto \alpha\xi$ of R_R to itself. When R is a commutative ring K, right modules are identified with left modules (Corollary 16), so the Proposition states that $K^n \cong (K^n)^*$. In particular, if K is a field F, one has $F^n \cong (F^n)^*$. If we recall that F-modules are just vector spaces, this states that a finite-dimensional vector space with a given basis $\varepsilon_1, \cdots, \varepsilon_n$ is isomorphic to its own dual, the isomorphism depending on the basis.

Here is another interpretation of this duality. In the basic formula (40), think of the list $\boldsymbol{\alpha} = (\alpha_1, \cdots, \alpha_n)$ as a "row", and of the list $\boldsymbol{\xi}$ as a "column". The application (40) of the linear function $\theta\boldsymbol{\alpha}$ to the column $\boldsymbol{\xi}$ then can be rewritten as

$$(\alpha_1, \cdots, \alpha_n)\begin{bmatrix} \xi_1 \\ \cdot \\ \cdot \\ \cdot \\ \xi_n \end{bmatrix} = \alpha_1\xi_1 + \cdots + \alpha_n\xi_n \in R;$$

or "row times column gives a scalar". This is a first instance of the row-by-column multiplication of matrices (to be defined formally in §VII.1.) The right-hand side of this formula is a linear homogeneous expression in the "variables" ξ_i with "constant" coefficients α_i. In other words, if we consider a homogeneous linear equation over a field

$$\alpha_1\xi_1 + \cdots + \alpha_n\xi_n = 0,$$

the "solutions" $\boldsymbol{\xi}$ lie in one space, F^n, and the coefficients $\boldsymbol{\alpha}$ in its dual,

$(F^n)^*$. Thus the relation between a vector space and its dual is like that between unknowns and coefficients in such linear homogeneous equations, or that between rows and columns.

The dual of a non-zero module may turn out to be zero. For example, take $R = \mathbf{Z}$ to be the ring of integers, while the $\mathbf{Z}$-module (= abelian group) A is $\mathbf{Z}_n$, the cyclic group of order some natural number $n > 0$. The elements of the dual $(\mathbf{Z}_n)^*$ are morphisms $\mathbf{Z}_n \to \mathbf{Z}$ of abelian groups. But since $\mathbf{Z}$ has no non-zero elements of finite order, the only such morphism is the morphism zero. Hence, $(\mathbf{Z}_n)^* = 0$.

Not only modules, but also morphisms have duals. Let $t:B \to A$ be a morphism of right R-modules. Its *dual* is the function $t^*:A^* \to B^*$ defined for each element $f:A \to R$ of the dual module A^* as the composite

$$t^*(f) = f \circ t:B \to R. \tag{41}$$

Note that it goes backward; thus $t:B \to A$ yields $t^*:A^* \to B^*$, as displayed in the diagram:

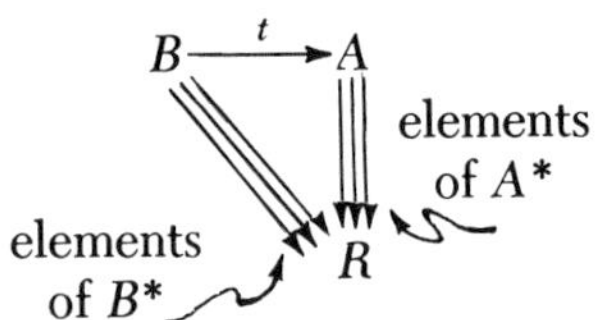

PROPOSITION 18. *The dual of a morphism of right R-modules is a morphism of left R-modules.*

Proof: We must show $t^*:A^* \to B^*$ left-linear for the pointwise operations already defined on the dual modules A^* and B^*. By the distributive laws (13),

$$t^*(f + g) = (f + g) \circ t = f \circ t + g \circ t = t^*(f) + t^*(g),$$

so that t^* is additive. For all elements $b \in B$ and all scalars κ,

$$[t^*(\kappa f)](b) = [(\kappa f) \circ t](b) = \kappa[(f \circ t)(b)] = [\kappa(t^* f)](b),$$

so that $t^*(\kappa f) = \kappa(t^* f)$ for all scalars κ. Thus t^* is left-linear, as required.

Consider next the duals of sums and composites of linear maps. For the sums one has

PROPOSITION 19. *For right R-modules B and A the assignment $t \mapsto t^*$ is a morphism of abelian groups*

$$\mathrm{Hom}_R\,(B, A) \to \mathrm{Hom}_R\,(A^*, B^*). \tag{42}$$

If $R = K$ is commutative, it is a morphism of K-modules.

Proof: For morphisms $t_1, t_2: B \to A$ of right R-modules the definitions and (13) show that $(t_1 + t_2)^* = t_1^* + t_2^*$; hence, $t \mapsto t^*$ is a morphism of addition, as in the first assertion of the proposition. For $R = K$ commutative, all K-modules (in particular, both A and A^*) may be written as left K-modules, while $\text{Hom}_K(B, A)$ is also a K-module under the pointwise scalar multiples λt defined for each scalar λ, as in Theorem 3, by $(\lambda t)(b) = \lambda(tb)$. To prove $t \mapsto t^*$ a morphism of K-modules we must prove the equality $(\lambda t)^* = \lambda t^*$ for these scalar multiples. Now, to prove that an equality $(\lambda t)^* = \lambda t^*: A^* \to B^*$ of functions it suffices to prove that the values are equal for all $f \in A^*$, as in $(\lambda t)^* f = (\lambda t^*) f$. Here both sides are functions on B to K, so it again suffices to prove that they have the same value at each $b \in B$. This is routine: We verify in succession that

$$\begin{aligned}[(\lambda t)^* f](b) &= [f \circ (\lambda t)](b) = f[(\lambda t)b] = f[\lambda(tb)] \\ &= \lambda[f(tb)] = \lambda[(f \circ t)(b)] = \lambda[(t^* f)(b)] \\ &= [\lambda(t^* f)](b) = [(\lambda t^*) f](b).\end{aligned}$$

At each step in the argument we apply the (only) relevant definition or property.

Next we study the duals of composite maps.

Proposition 20. *If 1_A is the identity morphism of a right module A, and if the composite $t \circ s$ of two morphisms of right modules is defined, then*

$$(1_A)^* = 1_{(A^*)}, \qquad (t \circ s)^* = s^* \circ t^*. \tag{43}$$

Proof: The first is immediate. As for the second, $t \circ s$ defined means that t, s, their duals, and their composites may be displayed in the following commutative diagrams:

For each $f \in A^*$, $(s^* \circ t^*)f = s^*(f \circ t) = f \circ t \circ s = (t \circ s)^* f$, so that (43) holds.

The construction of duals thus assigns to each right module A a left module A^* and to each morphism t of right modules a morphism t^* of left modules, all so as to satisfy (43). This means that $A \mapsto A^*$, $t \mapsto t^*$ is a contravariant functor, in the sense defined in (§IV.7), from right R-modules to left R-modules. Symmetrically, the dual of a left module is a right module;

this process gives a contravariant functor from left R-modules to right R-modules.

Next we study duals of biproducts.

Proposition 21. *If $A_1 \oplus A_2$ with injections e_i and projections p_i is a biproduct of right R-modules, there is an isomorphism*

$$\theta : (A_1 \oplus A_2)^* \cong A_1^* \oplus A_2^* \tag{44}$$

of left R-modules which assigns to each morphism $f: A_1 \oplus A_2 \to R$ the pair $\theta f = (f \circ e_1, f \circ e_2)$ and whose inverse assigns to each pair of morphisms $f_i: A_i \to R$ the element $\theta^{-1}(f_1, f_2) = f_1 p_1 + f_2 p_2 : A_1 \oplus A_2 \to R$.

Proof: Since $A_1 \oplus A_2$ is a coproduct with injections e_1 and e_2, their universality (Theorem 13) shows that each f is determined by the pair of composites $f \circ e_i$; thus θ as described is a bijection, with the indicated inverse. This bijection θ preserves sums and left scalar multiples, hence is a module isomorphism, just as asserted.

This isomorphism may be expressed diagrammatically. Take the whole diagram (31) for the biproduct $A_1 \oplus A_2$, apply * to all its modules and all its maps (thus turning the maps around) and compare with the diagram for the biproduct $A_1^* \oplus A_2^*$, with projections p_i' and injections e_i':

$$\begin{array}{ccccc}
A_1^* & \underset{e_1^*}{\overset{p_1^*}{\rightleftarrows}} & (A_1 \oplus A_2)^* & \underset{e_2^*}{\overset{p_2^*}{\leftrightarrows}} & A_2^* \\
\downarrow 1 & & \downarrow \theta & & \downarrow 1 \\
A_1^* & \underset{e_1'}{\overset{p_1'}{\leftrightarrows}} & A_1^* \oplus A_2^* & \underset{e_2'}{\overset{p_2'}{\rightleftarrows}} & A_2^* .
\end{array}$$

The isomorphism θ of (44) then becomes an isomorphism of the first biproduct diagram to the second one, in the sense that

$$\theta p_i^* = e_i', \qquad p_i'\theta = e_i^*, \qquad i = 1, 2. \tag{45}$$

These equations can be checked from the definition; thus $p_i'\theta(f) = p_i'(fe_1, fe_2) = fe_i = e_i^* f$. To summarize: The dual takes the biproduct to the biproduct but turns the projections p_i into injections and vice versa.

The connection between a right R-module and its dual left R-module A^* can be expressed in another way. To elements $f \in A^*$ and $a \in A$ assign the value $f(a)$ of f at a; this determines a function $A^* \times A \to R$ which we write in symmetric form as $(f, a) \mapsto \langle f, a \rangle$; thus $\langle f, a \rangle$ is defined by

$$\langle f, a \rangle = f(a) \qquad \text{for } f : A \to R \text{ and } a \in A.$$

In this notation, the right-linearity of f is the condition

$$\langle f, a_1\kappa_1 + a_2\kappa_2 \rangle = \langle f, a_1 \rangle \kappa_1 + \langle f, a_2 \rangle \kappa_2 \tag{46}$$

for all $a_i \in A$ and all scalars $\kappa_i \in R$, while the pointwise definitions (39) of the left module operations on the dual module A^* become

$$\langle \kappa_1 f_1 + \kappa_2 f_2, a \rangle = \kappa_1 \langle f_1, a \rangle + \kappa_2 \langle f_2, a \rangle \tag{47}$$

for all $f_i : A \to A$ and all scalars κ_i. If we hold a fixed, the assignment $f \mapsto \langle f, a \rangle$ gives a function $\omega(a) : A^* \to R$ whose values are

$$[\omega(a)](f) = \langle f, a \rangle, \qquad f \in A^*, \quad a \in A.$$

Now we interpret the two conditions (46) and (47). The condition (47) states for each a that the function $\omega(a) : A^* \to R$ is a morphism of left R-modules, hence an element $\omega(a) \in (A^*)^*$ of the dual of the dual (the double dual). The condition (46) states that the assignment $a \mapsto \omega(a)$ takes right linear combinations in A to (pointwise) right linear combinations in the double dual, so is a morphism $\omega : A \to (A^*)^*$ of right R-modules. If we note that this morphism $\omega = \omega_A$ depends on the given R-module A, we have proved

Proposition 22. *For each right R-module A there is a morphism*

$$\omega_A : A \to (A^*)^* \tag{48}$$

of right R-modules, with $[\omega_A(a)](f) = f(a)$ for each $a \in A$ and each $f \in A^*$.

Now any morphism $t : B \to A$ of right R-modules yields a dual morphism $t^* : A^* \to B^*$ of left R-modules and hence, in turn, a double dual morphism $(t^*)^* : (B^*)^* \to (A^*)^*$, in the *same* direction as the original t.

Proposition 23. *For any morphism $t : B \to A$ of right modules, the morphisms ω_B and ω_A of (48) yield a commutative diagram*

$$\begin{array}{ccc} B & \xrightarrow{\omega_B} & (B^*)^* \\ {\scriptstyle t}\downarrow & & \downarrow{\scriptstyle (t^*)^*} \\ A & \xrightarrow{\omega_A} & (A^*)^*. \end{array} \tag{49}$$

Proof: A direct check from the definitions shows that $\omega_A \circ t = (t^*)^* \circ \omega_B$, so this diagram is commutative, as required.

Commutativity conditions like (49) will be studied systematically in Chapter XV. They play an important role in describing why a morphism ω_A, defined for all modules A, is "natural", that is, independent of artificial choices.

The contravariant functor $A \mapsto A^* = \mathrm{Hom}_R(A, R)$, $t \mapsto t^*$ suggests that any fixed right R-module D should also yield a contravariant functor with $A \mapsto \mathrm{Hom}_R(A, D)$. Recall that the abelian group $\mathrm{Hom}_R(A, D)$ has as its elements the morphisms $f: A \to D$ of right R-modules. Then each morphism $t: B \to A$ of right R-modules determines a function

$$t^* : \mathrm{Hom}_R(A, D) \to \mathrm{Hom}_R(B, D), \qquad t^*(f) = f \circ t \qquad \text{for } f : A \to D,$$

which is a morphism of abelian groups. One may then check (Exercise 5) that the assignments $A \mapsto \mathrm{Hom}_R(A, D)$, $t \mapsto t^*$ yields a contravariant functor on right R-modules to abelian groups (or to K-modules, in case $R = K$ is commutative). This functor is usually called the functor $\mathrm{Hom}_R(-, D)$, with D fixed, while t^* is written as $\mathrm{Hom}_R(t, D)$. Symmetrically, each fixed module D yields a *covariant* functor $\mathrm{Hom}_R(D, -)$ on R-modules to abelian groups; it assigns to each module A the abelian group $\mathrm{Hom}_R(D, A)$ and to each morphism $t: A \to A'$ of modules the morphism $t_*: \mathrm{Hom}_R(D, A) \to \mathrm{Hom}_R(D, A')$ of abelian groups defined by $t_*(f) = t \circ f$ for each $f: D \to A$.

EXERCISES

1. For $\mathbf{Z}_m$ regarded as a $\mathbf{Z}$-module, show $(\mathbf{Z}_m)^* = 0$, while for $\mathbf{Z}_m$ regarded as a $\mathbf{Z}_n$-module for n any multiple of m show that $(\mathbf{Z}_m)^* \cong \mathbf{Z}_m$.

2. If X is any set, show that the dual of the free right module $R^{(X)}$ is the left function module R^X. (*Hint:* Use the freeness of $R^{(X)}$ to describe all $f: R^{(X)} \to R$.)

3. Show ω_A an isomorphism for $A = R$ and also for $A = R^n$.

4. If D is a submodule of A show that the projection $p: A \to A/D$ yields a monomorphism $p^*: (A/D)^* \to A^*$. Conclude that $(A/D)^*$ is isomorphic to a submodule of A^* and determine this submodule.

5. (a) For D a fixed right R-module, prove in detail that $\mathrm{Hom}_R(-, D)$ is a contravariant functor on right R-modules to abelian groups.

(b) Prove similarly that $\mathrm{Hom}_R(D, -)$ is a (covariant) functor.

6. For D a fixed module and $t: A \to A'$ a monomorphism of modules prove that $t_*: \mathrm{Hom}_R(D, A) \to \mathrm{Hom}_R(D, A')$ is a monomorphism of abelian groups.

7. (a) If A is an (additive) abelian group and $h: R \to \mathrm{End}_{\mathbf{Z}}(A)$ a morphism of rings, show that $\kappa a = [h(\kappa)]a$ for $\kappa \in R$ and $a \in A$ makes A a left R-module.

(b) Prove that every left R-module can be obtained in this fashion from exactly one abelian group A and one morphism h of rings.

8. If $u: R' \to R$ is a morphism of rings and A is a left R-module, show that the definition $\kappa' a = (u\kappa')a$ for $\kappa' \in R'$ and $a \in A$ makes A a left R'-module (called the module obtained from A by "pull-back" along u).

9. Show in detail that (49) commutes.

CHAPTER VI

Vector Spaces

THIS CHAPTER considers finite-dimensional vector spaces and linear transformations between them. A *vector space* V is a module over a field F; a *linear transformation* is a map preserving linear combinations (that is, is a morphism of F-modules); a *vector* is an element of some vector space. A vector space is *finite-dimensional* when it can be spanned by a finite list of its elements. Such spaces have properties not true for modules over arbitrary rings: Each finite-dimensional vector space V is a free module and is isomorphic to its dual space V^*. Any list of free generators of V is called a "basis" of V; the number n of elements in a basis is the "dimension" of V. In general, the same space V has many different bases, all with the same number of elements; at the end of the chapter we present computational methods for passing from one basis to another, and show how these methods are related to the solution of systems of linear equations. However, all the fundamental concepts can be defined without using bases, and many of the theorems are proved without reference to a basis. This approach emphasizes the geometric meaning of these theorems as statements about the vectors in a space and not about their coordinates. Each n-tuple $(\xi, \cdots, \xi_n)$ of scalars is a vector in the vector space F^n, but a vector v is just an element of *some* vector space. It need *not* be an n-tuple of *any* sort, row or column.

In this chapter the field F—the "field of scalars" $\kappa, \lambda, \cdots,$ —is fixed (except in some exercises) and "vector space" always means "vector space over F". In particular, we denote by V, V', W, right vector spaces (right F-modules); their duals V^*, V'^*, and W^* will then be written as *left* vector spaces (left F-modules), just as in §V.7. This arrangement has the advantage that our formulas will also apply in the noncommutative case because all the theorems of this chapter (except two which are noted) are in fact true, with the same proofs, if the field F is replaced by any division ring.

When the scalars form a field F, so that the multiplication of scalars is commutative, any right scalar multiple $v\kappa$ can be rewritten as a left scalar multiple $\kappa v = v\kappa$, as in Corollary V.16. However, we normally use right scalars because then the fact that a linear transformation $t: V \to W$ preserves scalar multiples is expressed by the simple formula $t(v\kappa) = (tv)\kappa$, in which the three symbols t, v, and κ *stay in the same order* on both sides.

1. Bases and Coordinates

A "list" of vectors of V is a basis of V when it "spans" V *and* is "linearly independent" in V. To define these terms, we note that V is to be a right vector space over the field F, n is a natural number, and $\mathbf{n} = \{1, \cdots, n\}$ the set of the first n positive integers. A function $\mathbf{v}:\mathbf{n} \to V$, sending each $i \in \mathbf{n}$ to some $v_i \in V$, is called a *list* $(v_1, \cdots, v_n)$ of n vectors in V, or often an "n-tuple" of vectors. Such lists will be denoted as usual by boldface letters. For example, the field F is itself a vector space, and a function $\boldsymbol{\xi}:\mathbf{n} \to F$ is a list of n scalars $\xi_1, \cdots, \xi_n$. The set F^n of all such lists $\boldsymbol{\xi}$ is a vector space under termwise addition and (right) scalar multiple.

Vector space operations on the n vectors of the given list $\mathbf{v}$ yield various vectors v in V as linear combinations

$$v = v_1\xi_1 + \cdots + v_n\xi_n. \tag{1}$$

with coefficients ξ_i from F. These expressions underlie several definitions of central importance:

The list $\mathbf{v}$ *spans* the space V when every vector $v \in V$ can be expressed in at least one way as a linear combination (1) of the vectors of $\mathbf{v}$.

The list $\mathbf{v}$ is a *basis* for the space V when every vector $v \in V$ has exactly one such expression as a linear combination of vectors of $\mathbf{v}$.

The list $\mathbf{v}$ is *linearly independent* in V when every vector $v \in V$ has at most one such expression (1).

Thus the list $\mathbf{v}$ is linearly independent when, for all lists ξ and η of n scalars each,

$$\sum v_i\xi_i = \sum v_i\eta_i \quad \Rightarrow \quad \xi_1 = \eta_1, \xi_2 = \eta_2, \cdots, \xi_n = \eta_n. \tag{2}$$

Since $\Sigma v_i\xi_i = \Sigma v_i\eta_i$ implies $\Sigma v_i(\xi_i - \eta_i) = 0$, it is enough to require, for each list ξ of n scalars, that

$$\sum v_i\xi_i = 0 \quad \Rightarrow \quad \xi_1 = \xi_2 = \cdots = \xi_n = 0. \tag{3}$$

A list $\mathbf{v}$ of vectors of V is called *linearly dependent* in V when it is not linearly independent; that is, when $\Sigma v_i\xi_i = 0$ for some list ξ of scalars ξ_i not all zero. For example, if any one vector in a list $\mathbf{v}$ is a scalar multiple of another, as in $v_2 = v_1\kappa$, the list is linearly dependent, for $v_1(-\kappa) + v_2 1 = 0$ has coefficients not all zero.

Note that a list $\mathbf{v}$ is a basis for V precisely when $\mathbf{v}$ spans V and is linearly independent in V.

The operation "linear combination", as used in these definitions, can also be formulated as a morphism defined on the free vector space F^n.

Proposition 1. *Each list* v *of n vectors in V determines a linear transformation* $L_{\mathrm{v}}:F^n \to V$ *by*

$$L_{\mathrm{v}}(\xi) = v_1\xi_1 + \cdots + v_n\xi_n. \tag{4}$$

Relative to this transformation,

$$\begin{aligned} \mathrm{v}\ \textit{spans}\ V &\Leftrightarrow L_{\mathrm{v}} \text{ is an epimorphism,} \\ \mathrm{v}\ \textit{is linearly independent}\text{ in } V &\Leftrightarrow L_{\mathrm{v}} \text{ is a monomorphism,} \\ \mathrm{v} \text{ is a } \textit{basis} \text{ of } V &\Leftrightarrow L_{\mathrm{v}} \text{ is an isomorphism.} \end{aligned}$$

Proof: To show L_{v} linear, we must prove that it preserves sums and scalar multiples. If $\boldsymbol{\xi}$ and $\boldsymbol{\eta}$ are two lists of n scalars, the sum $\boldsymbol{\xi} + \boldsymbol{\eta}$ is the list of the scalars $\xi_i + \eta_i$, while

$$L_{\mathrm{v}}(\boldsymbol{\xi} + \boldsymbol{\eta}) = \sum_i v_i(\xi_i + \eta_i) = \sum_i v_i\xi_i + \sum_i v_i\eta_i = L_{\mathrm{v}}(\boldsymbol{\xi}) + L_{\mathrm{v}}(\boldsymbol{\eta}).$$

If κ is any scalar, then $\boldsymbol{\xi}\kappa$ is the list of the scalars $\xi_i\kappa$, and

$$L_{\mathrm{v}}(\boldsymbol{\xi}\kappa) = \sum_i v_i(\xi_i\kappa) = \left(\sum_i v_i\xi_i\right)\kappa = L_{\mathrm{v}}(\boldsymbol{\xi})\kappa.$$

Given that L_{v} is linear, the statement that L_{v} is an epimorphism means exactly that the list v spans, and similarly for bases. Also, L_{v} a monomorphism means exactly that $L_{\mathrm{v}}\boldsymbol{\xi} = L_{\mathrm{v}}\boldsymbol{\eta}$ implies $\boldsymbol{\xi} = \boldsymbol{\eta}$, as in (2). This is equivalent to saying that L_{v} has kernel 0, as stated in (3).

For any list v, the image of the transformation L_{v} is called the subspace of V *spanned* by the list v, more briefly the *span* of v. The space is said to be of *finite type* (as for modules, in §V.3) when it is spanned by some finite list.

In real 3-space $\mathbf{R}^3$ each vector $v = (x, y, z)$ can be uniquely written as a linear combination of the three unit vectors (1, 0, 0), (0, 1, 0), (0, 0, 1), as

$$v = (x, y, z) = (1, 0, 0)x + (0, 1, 0)y + (0, 0, 1)z.$$

Hence, this list of unit vectors is a basis for $\mathbf{R}^3$, while x, y, and z are the "coordinates" relative to this basis. For a different basis, the same vector would have different coordinates.

DEFINITION. The coordinates *of a vector* $v \in V$ *relative to a basis* $\mathbf{b}$ *of* V *are the n scalars* ξ_i *of the unique list* $\boldsymbol{\xi}$ *used in the expression*

$$v = b_1\xi_1 + \cdots + b_n\xi_n \tag{5}$$

of v *as a linear combination of the* b_i. *Each scalar* ξ_i *is called the* ith coordinate *of* v, *relative to* $\mathbf{b}$ (*in brief*, rel $\mathbf{b}$).

This definition, with Proposition 1, gives

Theorem 2. *If* $\mathbf{b}:\mathbf{n} \to V$ *is a basis of the vector space* V, *then the function sending each vector* $v \in V$ *to the list* $\boldsymbol{\xi}$ *of its coordinates*, rel $\mathbf{b}$, *is an isomorphism* $V \cong F^n$ *of vector spaces. The inverse of this isomorphism is the linear transformation*

$$L_{\mathbf{b}} : F^n \cong V, \tag{6}$$

assigning to each list $\boldsymbol{\xi}$ *the vector* $\Sigma b_i \xi_i$, *as in* (4).

Thus a vector space V with a finite basis is isomorphic to a vector space F^n of n-tuples of scalars—and isomorphic in many ways, one isomorphism for each choice of a basis. Now in §V.5 we noted that F^n is a free F-module, with free generators the unit vectors

$$\boldsymbol{\varepsilon}_1 = (1, 0, \cdots, 0), \ \boldsymbol{\varepsilon}_2 = (0, 1, \cdots, 0), \cdots, \ \boldsymbol{\varepsilon}_n = (0, 0, \cdots, 1). \tag{7}$$

The isomorphism $L_{\mathbf{b}}$ of (6) takes these unit vectors to the basis vectors $b_1, \cdots, b_n$. Hence

Corollary. *A vector space* V *with a basis* $\mathbf{b}$ *is a free* F-*module with the vectors of* $\mathbf{b}$ *as free generators.*

By the definition of a free module, this means: Given a list of n vectors $w_1, \cdots, w_n$ in any vector space W, there is a unique linear transformation $t: V \to W$ with $t(b_i) = w_i$ for $i = 1, \cdots, n$. We recall the proof of this fact: By linearity, any such t must be given by the formula

$$t\left(\sum_{i=1}^{n} b_i \xi_i\right) = \sum_{i=1}^{n} w_i \xi_i.$$

Conversely, this formula defines a function $t: V \to W$ which is linear and has $t(b_i) = w_i$.

If a list $\mathbf{v} = (v_1, \cdots, v_n)$ is a basis, so is any list of the same vectors in a different order. The list $\mathbf{v}$ in a different order can be written as $(v_{\sigma 1}, \cdots, v_{\sigma n})$, where $\sigma:\mathbf{n} \to \mathbf{n}$ is some permutation on the set $\mathbf{n}$, so that the permuted list is simply the composite function $\mathbf{v} \circ \sigma:\mathbf{n} \to V$. Since addition in V is associative and commutative, any linear combination $\Sigma v_i \xi_i$ of the vectors of the original list may be written as $\Sigma v_i \xi_i = \Sigma v_{\sigma i} \xi_{\sigma i}$, so is a linear combination of the vectors of the permuted list $\mathbf{v} \circ \sigma$. By this argument, if $\mathbf{v}:\mathbf{n} \to V$ is any list of n vectors and $\sigma:\mathbf{n} \to \mathbf{n}$ any permutation, then $\mathbf{v}$ spans

V (or, is a basis of V; or, is linearly independent in V) if and only if the permuted list $v \circ \sigma$ spans V (or, is a basis of V; or, is linearly independent in V, as the case may be).

A basis has been defined as a list $(b_1, \cdots, b_n)$ of vectors *in order*. Because of this result about permutations, one can also say that the set $\{b_1, \cdots, b_n\}$—the same vectors, but in no particular order—is a basis.

Up to this point, we could have treated right R-modules over any ring R and not just modules over a field. The definitions of *span*, *linear independence*, and *basis* apply exactly as they are stated to lists $\mathbf{a}:\mathbf{n} \to A$ of elements in such an R-module A. The analog of Theorem 2 holds in this case; just as in its Corollary, an R-module with a basis is a free module. Conversely, *any* R-module A free on a finite subset $X \subset A$ has X as basis (cf. Exercise 11). When R is not a field, however, an R-module need not have *any* basis at all; for instance, the $\mathbf{Z}$-modules $\mathbf{Z}_n$ do not. When R is a field, the existence of a basis is a consequence of the following proposition, whose proof involves the field axioms (the existence of inverses) at a crucial point.

PROPOSITION 3. *A list* $\mathbf{v}$ *of vectors is linearly dependent in* V *if and only if some one vector* v_k *of the list is zero or a linear combination of the previous vectors of the list. When this is the case, removal of the vector* v_k *gives a new list with the same span as* $\mathbf{v}$.

Proof: If v_k is a linear combination $v_1\eta_1 + \cdots + v_{k-1}\eta_{k-1}$ of previous vectors, then

$$v_1\eta_1 + v_2\eta_2 + \cdots + v_{k-1}\eta_{k-1} + \eta_k(-1) = 0,$$

so there is a linear combination with at least one coefficient (to wit, -1) not zero, so the list $\mathbf{v}$ is indeed linearly dependent. Moreover, let w be any vector in the subspace spanned by all the $v_1, \cdots, v_n$, so that $w = \Sigma v_i\xi_i$. Replacing v_k here by its expression $v_1\eta_1 + \cdots + v_{k-1}\eta_{k-1}$ represents w as a linear combination of the shorter list $(v_1, \cdots, v_{k-1}, v_{k+1}, \cdots, v_n)$ with v_k removed, as asserted.

Conversely, suppose that the list $\mathbf{v}$ is linearly dependent; then $\Sigma_i v_i\xi_i = 0$ for scalars ξ_i not all zero. Let k be the last index with a coefficient $\xi_k \neq 0$, so that $v_1\xi_1 + \cdots + v_k\xi_k = 0$. Since F is a field, ξ_k^{-1} exists in F; multiply by ξ_k^{-1} and solve for v_k as

$$v_k = v_1(-\xi_1\xi_k^{-1}) + \cdots + v_{k-1}(-\xi_{k-1}\xi_k^{-1});$$

the vector v_k is indeed a linear combination of previous vectors unless $k = 1$, in which case this equation shows that v_k must be zero.

COROLLARY. *Any vector space of finite type has a basis.*

Proof: Since V is of finite type, it is spanned by some finite list $\mathbf{v}$ of n vectors. If this list $\mathbf{v}$ happens to be also linearly independent, it is a basis of V. If not, $\mathbf{v}$ is linearly dependent; by the proposition we remove vectors one by one from this list till we get a shorter list, still spanning V, which is independent.

Bases for vector spaces are notably useful in the study of linear differential equations with constant coefficients. Consider, for instance, one equation $ax'' + bx' + cx = 0$, where $a, b, c \in \mathbf{R}$ are constants and x' is the first derivative of the function $t \mapsto x(t)$ on $\mathbf{R}$ to $\mathbf{R}$ (assumed to have derivatives of all orders). The sum of two solutions x and y and a multiple of a solution by a real scalar are solutions, so the set of all solutions is a vector space over $\mathbf{R}$. For an equation of second order, such as this, this vector space has a basis consisting of two solutions; for example, the differential equation $x'' + k^2x = 0$ has the solutions $x = \sin kt$ and $x = \cos kt$ as basis for the space of solutions.

EXERCISES

1. In any vector space, prove: A list of just one vector is linearly independent if and only if the vector is non-zero; a list of two vectors is linearly dependent if and only if each is a scalar multiple of the other.

2. Show that two vectors (ξ_1, ξ_2) and (η_1, η_2) are linearly independent in F^2 if and only if $\xi_1\eta_2 - \xi_2\eta_1 \neq 0$.

3. For any choice of scalars κ, λ, and μ show that the vectors $(1, \kappa, \lambda)$, $(0, 1, \mu)$, and $(0, 0, 1)$ form a basis of F^3.

4. Do $(1, 2, 3)$, $(2, 3, 4)$, and $(3, 4, 5)$ form a basis of $\mathbf{Q}^3$?

5. (a) Show that the list $(1, 1, 0)$, $(1, 0, 1)$, $(0, 1, 1)$ is a basis of $\mathbf{Q}^3$.
   (b) Find the coordinates of the unit vectors of $\mathbf{Q}^3$ relative to this basis.

6. Find all bases of $(\mathbf{Z}_2)^3$; of $(\mathbf{Z}_3)^2$.

7. If $u\kappa + v\lambda + w\mu = 0$, where $\kappa\mu \neq 0$, show that the vectors u and v span the same subspace as do v and w.

8. If $\mathbf{w}$ is a linearly independent list of n vectors of V, show that a vector v is a linear combination of the list $\mathbf{w}$ if and only if the list $v, w_1, \cdots, w_n$ is linearly dependent.

9. If u, v, and w are three linearly independent vectors in a vector space over $\mathbf{Q}$, prove that $v + w$, $w + u$, and $u + v$ are also linearly independent. Is this true over every field?

10. If $t: V \to W$ is linear and $\mathbf{v}$ is a list of n vectors in V, prove the following properties of the composite list $t \circ \mathbf{v} = (t(v_1), \cdots, t(v_n))$:
    (a) If $\mathbf{v}$ spans V, then $t \circ \mathbf{v}$ spans $\operatorname{Im} t$.
    (b) If $t \circ \mathbf{v}$ is linearly independent in W, then $\mathbf{v}$ is linearly independent in V.

(c) If $\mathbf{v}$ is independent in V and t is a monomorphism, then $t \circ \mathbf{v}$ is independent in W.

11. If F is a free R-module on a set X, as in (V.26), prove that every finite subset of X is linearly independent.

2. Dimension

Simple algebraic manipulation shows that *any* basis of F^2 must consist of exactly two vectors, and any basis of F^3 of exactly three vectors. In this section we prove this and more: If a vector space V has some one basis of n vectors, then any basis of V has the same number n of vectors. This number is called the "dimension" of the vector space V. Here is the basic result:

THEOREM 4. *If a vector space V is spanned by some list of n vectors and also has a list $\mathbf{v}$ of m linearly independent vectors, then $m \leqslant n$. Moreover, V can be spanned by a list of n vectors containing the given list $\mathbf{v}$.*

Proof: Given n, we shall prove both conclusions together by induction on m. For $m = 0$, the result is immediate. Suppose then that the conclusion holds for all lists of m independent vectors, and let $\mathbf{v}$ be a list of $m + 1$ linearly independent vectors. The first m vectors of this list $\mathbf{v}$ are still linearly independent, so, by the induction assumption, $n - m \geqslant 0$ and V is spanned by some list

$$(v_1, v_2, \cdots, v_m, w_1, \cdots, w_{n-m}) \tag{8}$$

of n vectors including $v_1, \cdots, v_m$. In particular, v_{m+1}, as a vector in V, must be a linear combination of these spanning vectors, say as

$$v_{m+1} = v_1\xi_1 + \cdots + v_m\xi_m + w_1\eta_1 + \cdots + w_{n-m}\eta_{n-m}. \tag{9}$$

Now we can prove $m + 1 \leqslant n$. If not, $m = n$, there are no vectors w_i in (9), and this formula expresses v_{m+1} in terms of $v_1, \cdots, v_m$, a contradiction to the hypothesis that the list $(v_1, \cdots, v_m, v_{m+1})$ is linearly independent.

Now adjoin the vector v_{m+1} to the list (8) of n vectors, getting a new list

$$(v_1, v_2, \cdots, v_m, v_{m+1}, w_1, \cdots, w_{n-m})$$

of $n + 1$ vectors which still spans V (because the shorter list (8) did). Moreover, by the relation (9) above, these $n + 1$ vectors are linearly dependent. By Proposition 3, we can remove some one vector from this list; namely, the first one which is a linear combination of previous vectors. The vector thereby removed is surely not one of the $v_1, \cdots, v_{m+1}$, for these vectors are known to be linearly independent. Therefore, the vector removed must be one in the list **w**, say the vector w_j. We now have a new list

$$(v_1, v_2, \cdots, v_{m+1}, w_1, \cdots, w_{j-1}, w_{j+1}, \cdots, w_{n-m})$$

of n vectors still spanning V and containing the whole list **v**. This completes the induction, hence the proof.

The algorithm used in this proof is called the "exchange process".

COROLLARY 1. (*Invariance of Dimension*). *Any two bases for a vector space V of finite type have the same number of elements.*

Proof: Let $(b_1, \cdots, b_m)$ and $(c_1, \cdots, c_n)$ be two bases of m and n vectors, respectively. Since **b** is linearly independent and **c** spans V, the theorem implies $m \leqslant n$. Since **c** is independent and **b** spans, $n \leqslant m$. Together, these conclusions give $n = m$, as desired.

The number n of vectors in any one basis of V is called the *dimension* of V, written $\dim V = n$, while $\dim V$ infinite means that V has no (finite) basis. Thus $\dim F^n = n$. A vector space of finite type is also called a *finite-dimensional* space. The theorem also implies

COROLLARY 2. *In a vector space V of dimension n:*

(*i*) *Any list of $n + 1$ vectors of V is linearly dependent.*
(*ii*) *No list of $n - 1$ vectors of V can span V.*

THEOREM 5. *In a finite-dimensional vector space V:*

(*i*) *Any linearly independent list of vectors is part of a basis.*
(*ii*) *Any list of vectors spanning V has a part which is a basis.*

Proof: To prove (ii), start with a list spanning V and remove dependent vectors by Proposition 3 till the resulting list is independent, and hence a basis. As for (i), let the list **v** be independent, while some other list spans V, because V is finite-dimensional. Theorem 4 then yields a list containing **v** which spans V. Remove dependent vectors, as before, until this list becomes a basis containing **v**.

COROLLARY. *In a vector space V of dimension n:*

(*i*) *Any list of n linearly independent vectors is a basis.*
(*ii*) *Any list of n vectors spanning V is a basis.*

Proof: For (i), let **v** be a list of n independent vectors. By the theorem, **v** is part of a basis; by the invariance of dimension, this basis has exactly n vectors, hence must be just the original list **v**. The proof of (ii) is similar.

By the Corollary of Theorem 2, any finite-dimensional vector space is a free F-module. This fact has many useful and important consequences, for example:

Proposition 6. *Any monomorphism $s: V \to V'$ with a finite-dimensional codomain V' has a linear left inverse. Any epimorphism $V \to V'$ with a finite-dimensional domain V has a linear right inverse.*

First, observe that if $\mathbf{v}: \mathbf{m} \to V$ is a list of vectors of V and $t: V \to V'$ a linear map, the composite function $t\mathbf{v}: \mathbf{m} \to V'$ is a list of vectors of V'; we say that t *carries* (or "takes") the list $\mathbf{v} = (v_1, \cdots, v_m)$ to the *composite list* $t\mathbf{v} = (tv_1, \cdots, tv_m)$.

Proof: Let V' have dimension n. Since s is a monomorphism, it takes lists independent in V to lists independent in V', so V is also of finite dimension m, with $m \leqslant n$. Take a basis $\mathbf{b}$ of m vectors in V. The composite list $s\mathbf{b}$ is still independent in V'; hence it extends to a basis of V', say $sb_1, \cdots, sb_m, b'_{m+1}, \cdots, b'_n$. Now this is a list of free generators for the free F-module V', so we can define a linear map $t: V' \to V$ by specifying that $t(sb_i) = b_i$ for $i \in \mathbf{m}$ and $tb'_j = 0$ for each $j = m + 1, \cdots, n$. Then $(t \circ s)b_i = b_i$, so $t \circ s$ must be the identity $V \to V$, and thus t is the required left inverse.

The proof of the second conclusion is left to the reader.

Appendix. There is a parallel treatment for (possibly) infinite-dimensional vector spaces V. In this treatment, a basis B is a possibly infinite set of vectors of V such that every vector $v \in V$ has exactly one expression as a linear combination of a finite number of vectors of B. Using the axiom of choice (§I.2), one can prove that *every* vector space V has a basis B, and that for any two bases B and B' of the same space there is a bijection $B \cong B'$.

Also, a subset $X \subset V$ is said to be linearly independent in V when every finite subset of X is linearly independent, in the sense of our previous definition.

EXERCISES

1. Show that any part of a linearly independent list is linearly independent.

2. If $\mathbf{w}$ is a list of vectors in V and if some part of $\mathbf{w}$ spans V, show that $\mathbf{w}$ spans V.

3. If the vector v is not in the subspace S, but is in the subspace spanned by S and the vector w, show that w is in the subspace spanned by S and v.

4. Prove the second half of Proposition 6.

5. If $s: V \to V'$ is a monomorphism with $\dim V' > \dim V$ (and both dimensions finite), construct two different left inverses for s.

6. Prove that a subset $B \subset V$ is a basis of V, as defined in the Appendix, if and only if B spans V and is linearly independent, as defined in the appendix.

7. (*Existence of a basis for every vector space* V.) Call a subset $M \subset V$ *maximal independent* in V when M is linearly independent but no larger subset containing M is independent.

*(a) From the axiom of choice (or Zorn's lemma), prove that there is a maximal independent subset M in every vector space V.

(b) Prove that such an M is always a basis of V.

3. Constructions for Bases

Each of the module constructions of Chapter V, when applied to given finite-dimensional vector spaces, will produce a new such space. In certain cases we now show how to compute the dimension of this new space and how to obtain a basis for this new space from bases for the given spaces. For example: Two finite-dimensional vector spaces V and V' are isomorphic if and only if they have the same dimension; if $t: V \cong V'$ is an isomorphism, it carries each basis $\mathbf{b}$ of V to a basis $t \circ \mathbf{b}$ of V'.

PROPOSITION 7. *If S is a subspace of a finite-dimensional vector space V, each basis of S is part of a basis of V. Hence,* $\dim S \leqslant \dim V$; *moreover,* $\dim S < \dim V$ *whenever S is a proper subspace of V.*

Proof: Any list of vectors of S linearly independent in S is *a fortiori* linearly independent in V. By Theorem 5, each such list is thus part of a basis of V. Moreover, a basis for S can be a basis for the whole space V only if every vector of V is already in S.

In the next few theorems, we have occasion to "combine" two lists $\mathbf{v} = (v_1, \cdots, v_m)$ and $\mathbf{w} = (w_1, \cdots, w_r)$ of vectors into a single list which will be written as

$$\mathbf{v} \vee \mathbf{w} = (v_1, \cdots, v_m, w_1, \cdots, w_r).$$

First, we shall give a result about the dimensions of the kernel and the image of a linear map t.

THEOREM 8. *If $t: V \to V'$ is a linear transformation with a finite-dimensional domain, then*

$$\dim V = \dim (\mathrm{Ker}\, t) + \dim (\mathrm{Im}\, t). \tag{10}$$

In more detail, if a basis $\mathbf{v}$ *for* Ker t *is part of a basis* $\mathbf{v} \vee \mathbf{w}$ *for V, then* $t \circ \mathbf{w}$ *is a basis for* Im t.

If $\mathbf{v}$ has $m = \dim (\mathrm{Ker}\, t)$ elements and $\mathbf{w}$ has r elements, the more detailed statement implies $\dim (\mathrm{Im}\, t) = r$ and $\dim V = m + r$, which is (10). The *rank* and the *nullity* of a linear map t are defined by

$$\text{rank } t = \dim (\text{Im } t), \qquad \text{nullity } t = \dim (\text{Ker } t). \tag{11}$$

In this language, (10) states that "rank plus nullity equals dimension of domain". (Sometimes Ker t is called the "null space" and Im t the "range" of t.)

Proof: Since any vector $v \in V$ is a linear combination

$$v = \sum v_i \xi_i + \sum w_j \eta_j$$

for suitable lists ξ and η of scalars, and since $tv_i = 0$, each vector tv is $\Sigma(tw_j)\eta_j$. Hence, the list $t \circ \mathbf{w}$ spans Im t. On the other hand, $\Sigma(tw_j)\eta_j = 0$ means $t(\Sigma w_j \eta_j) = 0$; hence, $\Sigma w_j \eta_j \in \text{Ker } t$. But $\mathbf{v}$ is a basis for Ker t, so $\Sigma w_j \eta_j = \Sigma v_i \xi_i$ for some list ξ of scalars. But $\mathbf{v} \vee \mathbf{w}$ linearly independent makes all the η_j and all the ξ_i zero. Hence, the list $t \circ \mathbf{w}$ is linearly independent in Im t. Since it is independent and also spans Im t, it is a basis, as asserted.

Corollary 1. *If two vector spaces V and V' have the respective finite dimensions n and n', then the rank r of any linear map $t: V \to V'$ is at most the smaller of n and n'. For each such map t there is a basis* $\mathbf{b}$ *of V and a basis* $\mathbf{b}'$ *of V' such that*

$$tb_1 = b_1', \cdots, tb_r = b_r', tb_{r+1} = 0, \cdots, tb_n = 0.$$

In other words, basis vectors can be so chosen in both spaces that the map t carries basis vectors into basis vectors or the vector 0.

Proof: The inequalities $r \leqslant n$ and $r \leqslant n'$ follow from (10) and (11). To get the indicated bases, we use the notation of the theorem, set $\mathbf{b} = \mathbf{w} \vee \mathbf{v}$, and make $t\mathbf{w}$ part of a basis $\mathbf{b}'$ of V'.

Corollary 2. *Let V and V' be vector spaces of the same finite dimension. Then any epimorphism $t: V \to V'$, and also any monomorphism $t: V \to V'$, is necessarily an isomorphism.*

This corollary does not hold for all modules of finite type over an arbitrary ring R. For example, it fails for finite abelian groups, with $R = \mathbf{Z}$.

Proof: Set $\dim V = n = \dim V'$ and use (10). If $\dim (\text{Im } t) = n$, (10) implies that $\dim (\text{Ker } t) = 0$, so t is an isomorphism. If Ker $t = 0$, (10) implies that $\dim (\text{Im } t) = n$, so Im t must be all of V', and t is again an isomorphism.

Corollary 3. *If S is any subspace of a finite-dimensional vector space V, the corresponding quotient space V/S has the dimension*

$$\dim (V/S) = \dim V - \dim S. \tag{12}$$

Proof: Apply the theorem to the projection $p: V \to V/S$.

COROLLARY 4. *Any subspace S of a finite-dimensional vector space V is a direct summand of V; that is, there is a subspace T of V with $S + T = V$ and $S \cap T = 0$; moreover, $T \cong V/S$.*

Proof: Take a basis of S and adjoin vectors $w_1, \cdots, w_m$ to get a basis of V. The subspace T of V spanned by the adjoined vectors $w_1, \cdots, w_m$ has $S + T = V$ and $S \cap T = 0$, as required. Moreover, the theorem applied to the projection $p: V \to V/S$ states that the list $S + w_1, \cdots, S + w_m$ of cosets is a basis of V/S. Hence, the assignment $\Sigma w_j \eta_j \mapsto \Sigma(S + w_j)\eta_j$ is an isomorphism $T \cong V/S$.

Note that this subspace T depends upon the choice of the vectors w_i adjoined to the original basis of S.

PROPOSITION 9. *If a vector space W is the direct sum of two finite-dimensional subspaces V_1 and V_2, then any basis $\mathbf{b}'$ of V_1 and any basis $\mathbf{b}''$ of V_2 combine to form a basis $\mathbf{b}' \vee \mathbf{b}''$ of W. Hence, the direct sum W is finite-dimensional, and* $\dim W = \dim V_1 + \dim V_2$.

Proof: As in (V.37), W the direct sum of V_1 and V_2 means $V_1 \cap V_2 = 0$ and $V_1 + V_2 = W$, so each vector $w \in W$ can be written uniquely as a sum $w = v_1 + v_2$ of vectors $v_i \in V_i$. By the description of a basis, v_1 is uniquely a linear combination of the vectors of the list $\mathbf{b}'$, and v_2 a similar combination of $\mathbf{b}''$. Hence, each w is uniquely a linear combination of the list $\mathbf{b}' \vee \mathbf{b}''$, and the proposition follows.

Since each biproduct $V_1 \oplus V_2$ is a direct sum, the proposition also determines the dimension of a biproduct:

COROLLARY 1. *If $\mathbf{b}'$ and $\mathbf{b}''$ are bases for the finite-dimensional vector spaces V_1 and V_2, then in the biproduct $V_1 \oplus V_2$ with $e_i: V_i \to V_1 \oplus V_2$ as injections the lists $e_1\mathbf{b}'$ and $e_2\mathbf{b}''$ combine to give a basis $e_1\mathbf{b}' \vee e_2\mathbf{b}''$ for $V_1 \oplus V_2$. In particular, $V_1 \oplus V_2$ is finite-dimensional, and*

$$\dim (V_1 \oplus V_2) = \dim V_1 + \dim V_2. \tag{13}$$

COROLLARY 2. *If S and T are finite-dimensional subspaces of a vector space, then the subspace $S + T$ is also finite-dimensional, and*

$$\dim S + \dim T = \dim (S \cap T) + \dim (S + T). \tag{14}$$

Proof: The quotient space $(S + T)/S \cap T$ is by (V.37) the direct sum of its subspaces $S/S \cap T$ and $T/S \cap T$. Hence, the previous results on the dimensions of quotients and of direct sums yield

$$\begin{aligned}\dim (S + T) &= \dim ((S + T)/S \cap T) + \dim (S \cap T) \\ &= \dim (S/S \cap T) + \dim (T/S \cap T) + \dim (S \cap T) \\ &= \dim (S) + \dim (T) - \dim (S \cap T), \qquad \text{Q.E.D.}\end{aligned}$$

Given V and any set X, the functions $f: X \to V$ are the vectors of a new space, the *function space* V^X, defined just as in (V.27).

Proposition 10. *If V is a finite-dimensional vector space, so is the function space V^m for each positive integer m; indeed,*

$$\dim (V^m) = m \dim V. \tag{15}$$

The proof is by induction on m. For $m = 1$, (15) is immediate. Now assume (15) for some m. An element of the function space V^{m+1} is a function $f: \{1, \cdots, m + 1\} \to V$. To each such function assign the pair $(f', f(m + 1))$, where f' is the function f with domain restricted to the subset $\mathbf{m} = \{1, \cdots, m\}$. This assignment is a bijection

$$V^{m+1} \cong V^m \oplus V.$$

Since the operations "sum" and "scalar multiple" on the function space are defined termwise, this bijection is an isomorphism. This isomorphism and (13) yield $\dim (V^{m+1}) = \dim V^m + \dim V = (m + 1) \dim V$, as required to complete the induction.

Given a basis of V, one may also construct a basis of each V^m (see Exercise 12).

Corollary. *If V and V' are finite-dimensional vector spaces, so is* Hom (V, V'), *and*

$$\dim [\text{Hom } (V, V')] = (\dim V)(\dim V'). \tag{16}$$

(In case the field F is replaced by a division ring D, as explained in the introduction of this chapter, this corollary does not apply, because $\text{Hom}_D (V, V')$ is then just an abelian group and not a D-module.)

Proof: Let V have dimension n. Then V is a free F-module with free generators the vectors $b_1, \cdots, b_n$ of any basis $\mathbf{b}$. By Proposition V.10, Hom (V, V') is then isomorphic to $(V')^n$, which by the theorem has dimension $n(\dim V') = (\dim V)(\dim V')$.

EXERCISES

1. Let S and T be the subspaces of $\mathbf{Q}^4$ spanned, respectively, by the vectors

$$S:\quad (1, -1, 2, -3), (1, 1, 2, 0), (3, -1, 6, -6);$$

$$T:\quad (0, -2, 0, -3), (1, 0, 1, 0).$$

Find the dimensions of S, of T, of $S \cap T$, and of $S + T$.

2. If S and T are distinct two-dimensional subspaces of a three-dimensional space, show that their intersection $S \cap T$ has dimension 1. What does this mean geometrically?

3. If W is the direct sum of its subspaces V_1 and V_2 while $\mathbf{v}'$ and $\mathbf{v}''$ are lists of vectors in V_1 and V_2, respectively, prove:

(a) If $\mathbf{v}'$ spans V_1 and $\mathbf{v}''$ spans V_2, then the combined list $\mathbf{v}' \vee \mathbf{v}''$ spans W.

(b) If $\mathbf{v}'$ and $\mathbf{v}''$ are each linearly independent, then so is $\mathbf{v}' \vee \mathbf{v}''$.

4. Given linear transformations $s: U \to V$ and $t: V \to W$ with U and V finite-dimensional vector spaces, prove that

$$\text{rank}\,(t \circ s) \leqslant \text{rank}\,(t),\ \text{rank}\,(t \circ s) \leqslant \text{rank}\,(s),\ \text{nullity}\,(t \circ s) \geqslant \text{nullity}\,(s).$$

5. In Exercise 4 give an example to show that nullity $(t \circ s)$ need not be as large as nullity (t).

6. (a) If three subspaces S, T, and T' of a finite-dimensional V satisfy $S \cap T = S \cap T'$, $S + T = S + T'$, and $T \subset T'$, prove that $T = T'$.

(b) Prove the same result, without assuming that dim V is finite.

7. Given a linear transformation $t: V \to V'$, a list $\mathbf{v}$ of vectors in Ker t, and a list $\mathbf{w}$ of vectors in V, show that any two of the following properties imply the third: (i) $\mathbf{v}$ is a basis for Ker t; (ii) $t \circ \mathbf{w}$ is a basis for Im t; (iii) $\mathbf{v} \vee \mathbf{w}$ is a basis for V.

8. Show that neither statement of Corollary 2 of Theorem 8 is true for infinite-dimensional spaces.

9. Give a direct proof of (14), constructing a basis for $S + T$ from suitable bases for $S \cap T$, S, and T.

***10.** The "codimension" of a subspace S of a vector space V is defined to be dim (V/S), provided the latter is finite. Prove: If S and T are subspaces of finite codimension in V, then $S \cap T$ also has finite codimension and

$$\text{codim}\, S + \text{codim}\, T = \text{codim}\,(S \cap T) + \text{codim}\,(S + T).$$

11. If dim V is finite, deduce the result of Exercise 10 from (14).

12. Prove Proposition 10 as follows. Take a basis $b_1, \cdots, b_n$ for V, define for each $i = 1, \cdots, m$ and $j = 1, \cdots, n$ a function $h_{ij}: \mathbf{m} \to V$ by $h_{ij}(i) = b_j$ and $h_{ij}(k) = 0$ when $i \neq k$, and prove that a basis of V^m is given by the list

$$(h_{11}, \cdots, h_{1n}, h_{21}, \cdots, h_{2n}, \cdots, h_{m1}, \cdots, h_{mn}).$$

***13.** Show that the vector space Hom (V, V') is not of finite dimension as soon as one of V or V' is not of finite dimension and the other is non-zero.

14. If V is any non-zero vector space, while Y is an infinite set, prove that neither of the function spaces V^Y or $V^{(Y)}$ defined as in (V.27) can be finite-dimensional.

4. Dually Paired Vector Spaces

When a vector space V is considered as a right F-module, its dual is the vector space $V^* = \mathrm{Hom}_F(V, F)$, regarded as a left F-module. The elements of V^* are by definition the F-linear transformations $f: V \to F$—also called *linear forms* on V. Because V is a free module, these linear forms can be explicitly constructed as follows, when $\dim V$ is finite.

THEOREM 11. *Let* **b** *be a basis of n vectors for the vector space V. Then to each list* $\boldsymbol{\mu}$ *of n scalars there is exactly one linear form* $f: V \to F$ *with* $f \circ \mathbf{b} = \boldsymbol{\mu}$; *namely, the function f defined for each list* $\boldsymbol{\xi}$ *of n scalars as*

$$f\left(\sum_{i=1}^{n} b_i \xi_i\right) = \sum_{i=1}^{n} \mu_i \xi_i. \tag{17}$$

Moreover, the assignment $f \mapsto f \circ \mathbf{b}$ *is an isomorphism* $V^* \cong F^n$ *of left vector spaces.*

Proof: Since V is free on $b_1, \cdots, b_n$, each possible linear form $f: V \to F$ is uniquely determined by the n scalars $fb_1, \cdots, fb_n$. In (17), these are the scalars $\mu_1, \cdots, \mu_n$. See also Theorem V.8.

COROLLARY 1. *Any finite-dimensional vector space V has the same dimension as its dual space* V^*.

Proof: The isomorphism $V^* \cong F^n$ shows that V^* has dimension $n = \dim V$.

Here $V^* \cong F^n$ is an isomorphism of left modules, while the basis **b** gives as in (6) an isomorphism $L_{\mathbf{b}}: F^n \cong V$ of right modules—where F^n, like F, may be considered either as a right F-module or as a left F-module. Because multiplication of scalars is commutative, left modules can be rewritten as right modules (§V.7), so this yields an isomorphism $V^* \cong F^n \cong V$: *Every finite-dimensional vector space is isomorphic to its dual.* (This does not make sense for vector spaces over noncommutative division rings.) Note that the isomorphism $V^* \cong V$ depends on the choice of basis.

This theorem will also construct an explicit basis for V^*, as follows:

COROLLARY 2. *If V has a basis* $b_1, \cdots, b_n$, *its dual* V^* *has a basis*

$x_1, \cdots, x_n$, *consisting of the linear forms* $x_i: V \to F$ *defined for* $i \in \mathbf{n}$ *by*

$$x_i(b_i) = 1, \tag{18}$$

$$x_i(b_j) = 0, \qquad i \neq j, j = 1, \cdots, n. \tag{19}$$

Proof: The isomorphism $f \mapsto f \circ \mathbf{b}$ of the theorem carries the forms x_i to the unit vectors $\boldsymbol{\varepsilon}_i = (0, \cdots, 1, \cdots, 0)$ of (7), which together constitute a basis of F^n. Hence, $x_1, \cdots, x_n$ is a basis of V^*.

By (18) and (19), $x_i(\Sigma b_j \xi_j) = \xi_i$. Hence, the linear form $x_i: V \to F$ is just the function sending each vector v of V to its ith coordinate ξ_i, relative to $\mathbf{b}$. In brief, given a basis, the "coordinate functions" relative to that basis are elements of the dual space and form a basis there.

This basis $x_1, \cdots, x_n$ of V^* is called the *dual basis* to the given basis $\mathbf{b}$. With the following *Kronecker delta notation* for $i, j \in \mathbf{n}$,

$$\begin{aligned} \delta_{ij} &= 1 \qquad \text{if } i = j, \\ &= 0 \qquad \text{if } i \neq j, \end{aligned}$$

the conditions (18) and (19) that a basis $\mathbf{x}$ of V^* be dual to a basis $\mathbf{b}$ of V read

$$x_i b_j = \delta_{ij}, \qquad i, j = 1, \cdots, n. \tag{20}$$

Here are two basic facts about linear forms:

LEMMA 1. *To each non-zero linear form f on V there is at least one vector v with $f(v) = 1$.*

This is a triviality. Since $f: V \to F$ is a function, $f \neq 0$ means that $f(u) \neq 0$ for some vector u; a suitable scalar multiple v of u then has $f(v) = 1$.

The "symmetrical" property holds, provided that dim V is finite:

LEMMA 2. *To any non-zero vector u of a finite-dimensional vector space V, there exists at least one linear form f with $f(u) = 1$.*

Proof: Since $u \neq 0$, it is linearly independent, hence part of a basis of V. Take f to be the corresponding linear form in the dual basis.

An element $f \in V^*$ is by definition a function $v \mapsto f(v)$ of a "variable" $v \in V$. If we hold v fixed and let f vary, then $f \mapsto f(v)$ becomes a function $V^* \to F$. This function was labeled $\omega v: V^* \to F$ in (V.48). Thus ω maps V into its double dual V^{**}.

COROLLARY 3. *Any finite-dimensional vector space V is isomorphic to its double dual V^{**} under the morphism $\omega: V \to V^{**}$ which sends each vector v*

of V into the linear form $\omega v: V^* \to F$, where ωv is defined for each $f \in V^*$ by $(\omega v)(f) = f(v)$.

Proof: That $v \mapsto \omega v$ as defined is a linear map $V \to V^{**}$ is immediate (and already known by Proposition V.22). Lemma 2 implies that Ker $(\omega) = 0$. Since dim $V =$ dim V^{**} also, it follows that ω is an isomorphism, as required.

Note that the map $\omega: V \to V^{**}$ is defined without reference to any basis, while a basis was used to construct the isomorphism $V \cong V^*$ after Corollary 1. Also, ω is defined conceptually, in the "same" way for all V. For these reasons, it is called a "natural" map. See §XV.5, which gives a formal definition of naturality in terms of the equation (V.49).

To bring out the parallel between Lemma 1 and Lemma 2, we will write $\langle f, v\rangle$ for the value $f(v)$ of the form $f \in V^*$ on the vector $v \in V$. This gives a function

$$(f, v) \mapsto \langle f, v\rangle, \qquad V^* \times V \to F.$$

To emphasize the symmetry of V and its dual V^* in this situation, we write W for the dual space V^*, so that $\langle w, v\rangle$ is a function $W \times V \to F$. This function of two arguments w and v is linear in each argument, in the sense that the equations

$$\langle \kappa_1 w_1 + \kappa_2 w_2, v\rangle = \kappa_1\langle w_1, v\rangle + \kappa_2\langle w_2, v\rangle, \tag{21}$$

$$\langle w, v_1\kappa_1 + v_2\kappa_2\rangle = \langle w, v_1\rangle\kappa_1 + \langle w, v_2\rangle\kappa_2, \tag{22}$$

hold for all scalars $\kappa_i \in F$ and all vectors $w, w_i \in W, v, v_i \in V$. The first equation is (V.47), and merely records the definition of the linear combination $\kappa_1 w_1 + \kappa_2 w_2$ of two linear forms $w_i \in V^*$. The second equation (22) is (V.46), and records the fact that each $\langle w, -\rangle$ is a linear form on V. In this language, Lemma 1 and Lemma 2 becomes symmetrical statements about fixed vectors $w_0 \in W$ or $v_0 \in V$:

(i) If $\langle w_0, v\rangle = 0$ for all $v \in V$, then $w_0 = 0$.
(ii) If $\langle w, v_0\rangle = 0$ for all $w \in W$, then $v_0 = 0$.

More generally, if S is any subset of V we can define its *annihilator* to be the subset of the dual space $W = V^*$,

$$\text{Annih } S = \{w \mid w \in W \quad \text{and} \quad \langle w, s\rangle = 0 \quad \text{for all} \quad s \in S\},$$

consisting of those linear forms w which carry every s to 0. Similarly, if T is a subset of W, its annihilator is the subset

$$\text{Annih } T = \{v \mid v \in V \quad \text{and} \quad \langle t, v\rangle = 0 \quad \text{for all} \quad t \in T\}$$

of V. Because of the linearity (21) and (22), it follows that Annih T is a subspace of V and Annih S a subspace of W. Also conditions (i) and (ii) can now be written as

$$\text{Annih } V = 0 \qquad \text{and} \qquad \text{Annih } W = 0. \tag{23}$$

Theorem 12. *If a function* $\langle\ \rangle : W \times V \to F$ *on finite-dimensional vector spaces* W *and* V *over* F *has the properties* (21), (22), *and* (23), *then it determines isomorphisms* $W \cong V^*$ *and* $V \cong W^*$.

Functions with these three properties are sometimes called "dual pairings".

Proof: For each vector $v \in V$ the assignment $w \mapsto \langle w, v\rangle$ is a function, written $\langle -, v\rangle : W \to F$, which is left-linear on W by (21). Thus the assignment $v \mapsto \langle -, v\rangle$ is a function on V to the dual W^*, and by (22) it is a right linear transformation $V \to W^*$. The assumption Annih $W = 0$ of (23) states that this function has kernel 0 (is a monomorphism); hence by Proposition 7 and Corollary 11.1,

$$\dim V \leqslant \dim W^* = \dim W.$$

Symmetrically, for each $w \in W$ the assignment $w \mapsto \langle w, -\rangle$ is a right linear monomorphism $W \to V^*$, so that $\dim W \leqslant \dim V^* = \dim V$. Combined with the previous inequality this gives $\dim W = \dim V$ and implies by Corollary 8.2 that both monomorphisms are isomorphisms $V \cong W^*$ and $W \cong V^*$. This also yields $V \cong V^{**}$, as in the isomorphism $\omega : V \to V^{**}$ of Corollary 3 of Theorem 11.

Theorem 13. *For subspaces* S *of a finite-dimensional vector space* V *and* T *of its dual* $W = V^*$ *there are isomorphisms*

$$S^* \cong V^*/\text{Annih } S, \qquad (V/S)^* \cong \text{Annih } S, \tag{24}$$

$$T^* \cong W^*/\text{Annih } T, \qquad (W/T)^* \cong \text{Annih } T. \tag{25}$$

Proof: By the symmetry of $\langle\ ,\ \rangle : W \times V \to F$ it suffices to consider the case of the subspace S. For each $w \in W$, the function $\langle w, -\rangle : V \to F$, if restricted to elements $s \in S$ in the subspace $S \subset V$, is a linear transformation denoted by

$$\langle w, -\rangle | S : S \to F; \qquad s \mapsto \langle w, s\rangle \in F.$$

This assignment $w \mapsto \langle w, -\rangle | S$, called restriction to S, is a linear map

$$\text{res}_S : W = V^* \to S^*.$$

Now any basis of the subspace $S \subset V$ is by Theorem 5 part of a basis of V, so any linear map $g : S \to F$ can be extended to a map $g' : V \to F$ linear on all of V (say by making g' zero on the additional part of the chosen basis of V). This observation means that every g in S^* is the restriction to S of some $g' = \langle w, -\rangle$, so the map res_S above is an epimorphism. Its kernel is Annih S, by the very definition of this annihilator. Therefore, by (V.20)—the "main theorem" for quotient modules—there is an isomorphism $S^* \cong V^*/\text{Annih } S$, as in the first of (24).

To treat the second isomorphism of (24), consider any $h: V/S \to F$ in the dual space $(V/S)^*$. Composing h with the projection $p: V \to V/S$ yields a linear map $h \circ p: V \to F$ which annihilates S. On the other hand, any linear map $f: V \to F$ which annihilates S must, by the main theorem on quotient modules, factor through the projection $p: V \to V/S$ as $f = h \circ p$ for a unique h. Therefore, $h \mapsto h \circ p$ is the desired isomorphism $(V/S)^* \cong$ Annih S.

Proposition 14. *For any subspace S of a finite-dimensional vector space V,*

$$\dim S + \dim (\text{Annih } S) = \dim V, \tag{26}$$

$$\text{Annih } (\text{Annih } S) = S. \tag{27}$$

Proof: If $\dim V = n$ and $\dim S = k$, then $\dim (V/S) = n - k$, and by the second equation of (24), $\dim (\text{Annih } S) = \dim (V/S)^* = n - k$. This is (26), as desired. Repeating the argument gives $\dim (\text{Annih } (\text{Annih } S)) = k$. But if $s \in S$, then $\langle w, s\rangle = 0$ for all $w \in \text{Annih } S$, so $s \in \text{Annih } (\text{Annih } S)$ and therefore $S \subset \text{Annih } (\text{Annih } S)$. Since these spaces have the same dimension k, they must by Proposition 7 be equal, as asserted in (27).

Note that this result depends essentially upon the fact that V is of finite dimension.

For a subspace T of $W = V^*$ the equation corresponding to (26) is

$$\dim (\text{Annih } T) = \dim V - \dim T. \tag{28}$$

This conceptual result is really a statement of the fundamental fact about systems of homogeneous linear equations. For if $f \in V^*$ and $v \in V$ the equation $fv = 0$ is a homogeneous linear equation f in n unknowns (the coordinates of v) as in §V.7. Thus if $f_1, \cdots, f_k$ is a list of k elements $f_i \in V^*$ and T the subspace of V^* which they span, then Annih T is just the set of all $v \in V$ with

$$f_1 v = 0, \cdots, f_k v = 0;$$

in other words, Annih T is the set (actually, the *subspace*) of all solutions of these k simultaneous homogeneous linear equations. Thus (28) states that the number of linearly independent solutions is $\dim V$, the number of unknowns, minus $\dim T$, the maximum number of linearly independent equations (cf. §6 below).

EXERCISES

1. If V is a finite-dimensional vector space, show that the vector spaces $\text{Hom}(V, F)$ and $\text{Hom}(F, V)$ are dually paired by $(f, h) \mapsto (f \circ h)(1)$, where $f: V \to F$ and $h: F \to V$.

2. Let S_1, S_2 be subspaces of a finite-dimensional vector space V, while $\langle\ ,\ \rangle: W \times V \to F$ is a dual pairing. Prove:

   (a) $S_1 \subset S_2 \quad \Rightarrow \quad \text{Annih}\ S_1 \supset \text{Annih}\ S_2$.

   (b) $\text{Annih}\ (S_1 + S_2) = (\text{Annih}\ S_1) \cap (\text{Annih}\ S_2)$.

   (c) $\text{Annih}\ (S_1 \cap S_2) = (\text{Annih}\ S_1) + (\text{Annih}\ S_2)$.

3. In Theorem 12, show that the hypothesis (23) can be replaced by either one of the following two conditions:

   (a) $\text{Annih}\ W = 0$ and $\dim V \geqslant \dim W$.

   (b) The map $v \mapsto \langle -, v \rangle$ is an isomorphism $V \cong W^*$.

5. Elementary Operations

Next we develop means of testing whether a given list of vectors is linearly independent, spans, or is a basis.

An *elementary operation* on a list of n vectors in V is any one of the following three processes:

(I) Interchange any two vectors of the list.
(II) Multiply one vector of the list by a non-zero scalar.
(III) Add a scalar multiple of one vector of the list to another one.

The effect on a list of a typical operation of each type may be exhibited as

$$\begin{aligned}
&\text{I} \quad (v_1, v_2, v_3, \cdots, v_n) \Rightarrow (v_2, v_1, v_3, \cdots, v_n).\\
&\text{II}_\kappa \quad (v_1, v_2, \cdots, v_n) \Rightarrow (v_1\kappa, v_2, \cdots, v_n), \qquad \kappa \neq 0.\\
&\text{III}_\lambda \quad (v_1, v_2, \cdots, v_n) \Rightarrow (v_1 + v_2\lambda, v_2, \cdots, v_n).
\end{aligned}$$

Here $\kappa, \lambda \in F$.

Since a list $\mathbf{v}$ of n vectors of V is an element of the function space V^n, each elementary operation may be regarded as a function $\phi: V^n \to V^n$; in fact, as a linear transformation $V^n \to V^n$. The inverse of an elementary operation is another such operation; for instance, in the typical cases listed above,

$$\text{I} \circ \text{I} = 1, \qquad \text{II}_\kappa \circ \text{II}_{(\kappa^{-1})} = 1, \qquad \text{III}_\lambda \circ \text{III}_{(-\lambda)} = 1.$$

Therefore, each elementary operation is an automorphism of V^n.

Two lists $\mathbf{v}$ and $\mathbf{v}'$ of n vectors each from V are called "equivalent" when there is a composite ϕ of elementary operations for which $\phi(\mathbf{v}) = \mathbf{v}'$. Since the inverse of each elementary operation is elementary, the relation of equivalence so defined is reflexive, symmetric, and transitive.

THEOREM 15. *If ϕ is any composite of elementary operations on V^n, while* $\mathbf{v}$ *is a list of n vectors from V, then the span of* $\mathbf{v}$ *is the span of* $\phi\mathbf{v}$, *and* $\phi\mathbf{v}$ *is linearly independent in V if and only if* $\mathbf{v}$ *is. Consequently,*

$$\mathbf{v} \text{ is a basis for } V \quad \Leftrightarrow \quad \phi\mathbf{v} \text{ is a basis for } V.$$

Proof: It suffices to consider the case when ϕ is elementary. In each of cases I, II, and III, each linear combination of the original list $\mathbf{v}$ may be rewritten as a linear combination of the new list $\phi\mathbf{v}$; thus

$$\begin{aligned}
&(\mathrm{I}) \quad v_1\xi_1 + v_2\xi_2 + \cdots = v_2\xi_2 + v_1\xi_1 + \cdots .\\
&(\mathrm{II}_\kappa) \quad v_1\xi_1 + v_2\xi_2 + \cdots = (v_1\kappa)(\kappa^{-1}\xi_1) + v_2\xi_2 + \cdots .\\
&(\mathrm{III}_\lambda) \quad v_1\xi_1 + v_2\xi_2 + \cdots = (v_1 + v_2\lambda)\xi_1 + v_2(\xi_2 - \lambda\xi_1) + \cdots .
\end{aligned}$$

Hence, any vector which is a linear combination of the list $\mathbf{v}$ is also a linear combination of $\phi\mathbf{v}$, and if there is a non-zero list ξ of scalars with $\sum v_i\xi_i = 0$, there is such a list for $\phi\mathbf{v}$, Q.E.D.

How much may a list of vectors be simplified by successive elementary operations? To see this, we choose a fixed basis $\mathbf{b}$ of m vectors for the space V and express each vector v_j of the given list in terms of the basis b_i as $v_j = \sum_i b_i A_{ij}$, for suitable scalars $A_{ij} \in F$. Then the list $v_1, \cdots, v_n$ of vectors can be replaced by the matrix A over F. Here A is a function $A:\mathbf{m} \times \mathbf{n} \to F$ which can be exhibited as the $m \times n$ rectangular array

$$A = \begin{bmatrix} A_{11} & \cdots & A_{1n} \\ \vdots & & \vdots \\ A_{m1} & \cdots & A_{mn} \end{bmatrix}$$

of these scalars. Each column is the list of coordinates (rel $\mathbf{b}$) of the corresponding vector. This amounts to replacing V by F^m and each vector v_j by a column (a list of m scalars).

Note: Matrices A are always written with a capital A, and usually as $A = \|a_{ij}\|$, with (scalar) entries a_{ij}. Since we regard each matrix as a *function* on $\mathbf{m} \times \mathbf{n}$, we write the entries of A as the values of that function, and so as capitals: $A = \|A_{ij}\|$.

The three elementary operations on the list of vectors now become operations on the columns of the matrix A:

(I) Interchange any two columns.
(II) Multiply one column by a non-zero scalar.
(III) Add a scalar multiple of one column to another column.

Two $m \times n$ matrices A and B over F are called *column equivalent* when the first can be carried to the second by a sequence of such elementary operations; thus column equivalence is a reflexive, symmetric, and transitive relation (between matrices of the same size over the same field).

How much can the first row of the matrix be simplified by such operations? It may be that the first row consists exclusively of 0's; in this case the operations leave it all zero. Otherwise, the first row has at least one entry, say A_{1j}, which is not zero. Then by an operation (I) we can make the jth column the first column, while by operation (II), multiplying this column by A_{1j}^{-1}, we can make its first entry 1, so now the first row is $(1, \cdots)$. Next, by subtracting suitable multiples of that new first column, we can make the rest of the entries in the first row all 0. As a result of all this, the first row has one of the two forms

$$(0, 0, \cdots, 0), \qquad (1, 0, \cdots, 0).$$

In the first of these cases, we can operate on the columns to simplify the *second* row until it has one of these two forms. In the second case, we leave the first column unchanged and operate on the *remaining* $n - 1$ columns so as to simplify the last $n - 1$ entries in the second row. As a result, the first two rows are now in one of the forms

$$\begin{pmatrix} 0, 0, \cdots, 0 \\ 0, 0, \cdots, 0 \end{pmatrix}, \quad \begin{pmatrix} 0, 0, \cdots, 0 \\ 1, 0, \cdots, 0 \end{pmatrix}, \quad \begin{pmatrix} 1, 0, \cdots, 0 \\ *, 0, \cdots, 0 \end{pmatrix}, \quad \begin{pmatrix} 1, 0, \cdots, 0 \\ *, 1, \cdots, 0 \end{pmatrix},$$

where the *'s designate scalar entries which may or may not be zero.

In each of *these* cases, we next operate similarly on the remaining columns not yet headed by a 1 so as to simplify the third row, with possible results (0's to the right omitted)

$$\begin{bmatrix} 0 \\ 0 \\ 0 \end{bmatrix}, \begin{bmatrix} 0 \\ 0 \\ 1 \end{bmatrix}, \begin{bmatrix} 0 \\ 1 \\ * \end{bmatrix}, \begin{bmatrix} 0 & 0 \\ 1 & 0 \\ * & 1 \end{bmatrix}, \begin{bmatrix} 1 \\ * \\ * \end{bmatrix}, \begin{bmatrix} 1 & 0 \\ * & 0 \\ * & 1 \end{bmatrix}, \begin{bmatrix} 1 & 0 \\ * & 1 \\ * & * \end{bmatrix}, \begin{bmatrix} 1 & 0 & 0 \\ * & 1 & 0 \\ * & * & 1 \end{bmatrix}.$$

This continues till the rows or the columns are exhausted; the matrix is then in *echelon form*, described as follows:

(i) For some r, only the first r columns are non-zero,

(ii) In each non-zero column, the top non-zero entry is 1.

(iii) For $j + 1 \leqslant r$, the top 1 in column $j + 1$ is on a lower row than was the top 1 in column j.

The last condition can be reformulated as "Each non-zero column has more zeros above the first 1 than did the preceding column". It means that the leading 1's step down "in echelon". We have proved

Theorem 16. *By successive elementary operations on its columns, any rectangular matrix can be put into echelon form.*

Here is the real advantage of putting a matrix in echelon form:

Theorem 17. *If a matrix is in echelon form, its non-zero columns are linearly independent.*

Proof: Label the non-zero columns $w_1, \cdots, w_r$. If there is a linear relation $\sum_j w_j \kappa_j = 0$, look at the first non-zero row. Only the first column w_1 has an entry there, and that entry is 1. The corresponding entry (or coordinate) of the column vector $\sum w_j \kappa_j$ is therefore just κ_1. Hence, $\kappa_1 = 0$. Now look at the next row, containing a leading entry 1. Since κ_1 is already 0, the corresponding entry in the column $\sum w_j \kappa_j$ is κ_2. Therefore, $\kappa_2 = 0$. Continuing, all the κ's are zero, so the columns are linearly independent, as asserted.

This proof amounted to finding all the ways of expressing the column zero as a linear combination of columns in echelon. The same idea can be used to express other columns as such linear combinations. Let E_{ij} be an $m \times n$ matrix in echelon form, while $u_1, \cdots, u_n$ is another column of scalars. To express the column $\mathbf{u}$ as a linear combination of the columns E_{-j} of E amounts to finding solutions $x_1, \cdots, x_n$ of the m equations

$$\begin{aligned} E_{11}x_1 + \cdots + E_{1n}x_n &= u_1, \\ &\vdots \\ E_{m1}x_1 + \cdots + E_{mn}x_n &= u_m. \end{aligned} \tag{29}$$

Now E in echelon form means that the first non-zero row of E (say, the kth row) is the row $(1, 0, \cdots, 0)$. Hence, the first k equations of the system read $0 = u_1, \cdots, 0 = u_{k-1}$ and $x_1 = u_k$. Thus the equations have no solutions unless $u_1 = \cdots = u_{k-1} = 0$, and then x_1 must be u_k. Continuing, the first equation (say, the lth) which actually involves x_2 will have the form $E_{l1}x_1 + x_2 = u_l$. This equation determines the only possible value of x_2, and so on. Thus, if there are r non-zero columns in E, the values of $x_1, \cdots, x_r$ are determined uniquely (while the values of $x_{r+1}, \cdots, x_n$ can be arbitrary). Depending on the u's, these values x_i may or may not satisfy the remaining (not yet used) equations of (29). This process determines whether or not the column $\mathbf{u}$ is a linear combination of the columns of E, and finds such a combination if there is one.

By Theorem 15, elementary operations do not change the linear independence of a list of vectors nor the span of that list. Hence, the echelon form can be used to answer a variety of questions about lists of vectors. In the following cases, we assume that each list $\mathbf{v}$ is written as a matrix—that is, as a list of the columns of coordinates of $v_1, \cdots, v_n$ relative to some basis.

To test a list $\mathbf{v}$ of n vectors for linear independence: Reduce the list to echelon form $\mathbf{w}$. Since any list with at least one vector zero is linearly dependent, the original list $\mathbf{v}$ is linearly independent if and only if none of the vectors in $\mathbf{w}$ is zero.

To test whether a list $\mathbf{v}$ of n vectors is a basis for a space of dimension n, it suffices to test for independence.

To determine the dimension of the span of a list **v** of n vectors: Reduce to echelon form. By Theorem 17, the required dimension is the number of non-zero echelon vectors.

To determine a basis for the span of a list **v**: Reduce **v** to the form $(w_1, \cdots, w_r, 0, \cdots, 0)$ with r non-zero vectors in echelon. These vectors then form a basis, by Theorem 17.

To determine whether a vector u is in the span of a list **v**: As above, find the dimension of the span of the list $(v_1, \cdots, v_r, u)$. This dimension is that of $(v_1, \cdots, v_n)$ if and only if u is in the span of $(v_1, \cdots, v_n)$.

To express a vector u as a linear combination of the vectors of a list **v**, when that is possible: First reduce **v** to echelon form **w**. As observed in Theorem 15, this process will give an expression for each vector w_i of the new list as a linear combination of the vectors of **v**. From the echelon form of **w** one now calculates, as described at (29) above, whether or not u is a linear combination of **w**, and, if so, what the coefficients x_i in this combination may be. From this expression of u in terms of **w** and the formulas above for **w** in terms of **v**, one thus has an expression of u in terms of **v**, if this is possible.

To determine the rank of a linear map $t: V \to V'$: Take a basis **b** of V and a basis **b**$'$ for $(V')^*$. Since Im (t) is spanned by the list t**b**, test as above for the dimension of the span of t**b**; this dimension is the desired rank. Moreover, the echelon form for t**b** will also yield a basis for Im (t).

To determine the nullity of a linear map $t: V \to V'$: Test as above for the rank of t and apply "rank plus nullity equals dimension of domain".

This nullity is, by definition, the dimension of Ker t. To get not just the dimension, but a basis of Ker t, another observation is needed. Each elementary operation ϕ on lists of n vectors applies not just to lists of n vectors from one space V; it applies in the same way to lists of n vectors from any vector space over the same field F. In other words, an elementary operation ϕ really assigns to each vector space V a function $\phi_V: V^n \to V^n$. This function commutes with every linear map, in the following sense.

Proposition 18. *If ϕ is a composite of elementary operations on lists of n vectors, while $t: V \to V'$ is any linear map, then for each list* **v** *of n vectors of V,*

$$t(\phi_v \mathbf{v}) = \phi_{V'}(t\mathbf{v}). \tag{30}$$

It suffices to prove this when ϕ is a single elementary operation. It is evident when ϕ interchanges two vectors (type I). Operations of type II and type III take linear combinations of vectors. Since a linear map t preserves linear combinations, the result (30) is immediate for ϕ of these types.

With this result, one can determine a basis for the kernel of t. Take a basis $b_1, \cdots, b_n$ of V and find a composite ϕ of elementary operations reducing the list $t \circ \mathbf{b}$ to the form $(c_1, \cdots, c_r, 0, \cdots, 0)$, with r non-zero vectors in echelon. Apply the same composite operation ϕ to $\mathbf{b}$ to get $\phi\mathbf{b} = \mathbf{b}'$, a new basis of V (in practice, ϕ and $\mathbf{b}'$ are constructed together, step by step). By the proposition above, $t\mathbf{b}' = t(\phi\mathbf{b}) = \phi(t\mathbf{b})$, so, by the form of $t \circ \mathbf{b}$ given above, $t: V \to V'$ now has the form

$$tb_1' = c_1, \cdots, tb_r' = c_r, \qquad tb_{r+1}' = 0, \cdots, tb_n' = 0$$

(just as in Corollary 1 of Theorem 8). Since $\mathbf{c}$ is in echelon, the first r vectors are linearly independent and so r is the rank of t. Hence, the last $n - r$ vectors $b_{r+1}', \cdots, b_n'$ of the list $\mathbf{b}'$ form a basis for Ker t.

Finally, we observe by the following result that there are "enough" elementary operations.

Theorem 19. *For any two bases* $\mathbf{b}$ *and* $\mathbf{c}$ *of a finite-dimensional vector space there is a sequence of elementary operations taking the list* $\mathbf{c}$ *to the list* $\mathbf{b}$.

Proof: Express the list $\mathbf{c}$ by the matrix of its coordinates relative to the basis $\mathbf{b}$, and then use elementary operations to bring $\mathbf{c}$ into echelon form $\mathbf{c}'$. By Theorem 15, $\mathbf{c}'$ is still a basis, so the matrix E in echelon form (with columns the vectors of $\mathbf{c}'$) must have the triangular form

$$\begin{bmatrix} 1 & 0 & \cdots & 0 \\ * & 1 & \cdots & 0 \\ \cdot & & & \\ \cdot & & & \\ \cdot & & & \\ * & * & \cdots & 1 \end{bmatrix};$$

1's along the main diagonal (upper left to lower right), zeros above this diagonal, and (possibly) non-zero entries below it. Additional elementary operations on the list $\mathbf{c}'$ will now make all entries below the main diagonal zero; thus add a multiple of the second column to the first column so that this first column gets a zero in the second row, and so on, making all the other starred entries zero, starting at the top. After all these operations, the basis $\mathbf{c}''$ represented by the new columns is identical to $\mathbf{b}$. This procedure has found a sequence of elementary operations carrying the basis $\mathbf{c}$ to the basis $\mathbf{b}$, as desired.

Now we can also construct an inverse for an endomorphism $t: V \to V$ of rank the dimension n of V. Take a basis $\mathbf{b}$ of V. As t has rank n, its image is V, so $t\mathbf{b}$ is still a basis of V. By Theorem 19 there is a sequence ϕ of elementary operations with $\phi(t\mathbf{b}) = \mathbf{b}$. By (30), $t(\phi\mathbf{b}) = \mathbf{b}$. But V is a free module with free generators the vectors of $\mathbf{b}$, so we can construct a linear map $s: V \to V$ with $s \circ \mathbf{b}$ the list $\phi\mathbf{b}$. This means that $(t \circ s)\mathbf{b} = t(\phi\mathbf{b}) = \mathbf{b}$, so $t \circ s$ is the identity on each basis vector, and therefore $t \circ s = 1$. This implies that s is the desired inverse of t.

Note: An echelon matrix can be further reduced (cf. the proof of Theorem 19). Let B be a matrix already in echelon. In any non-zero column, consider the first non-zero entry 1; if all the entries in the row to the left of each such 1 are zero, the matrix B is said to be in *reduced echelon* form or "Hermite normal form". For applications, see Exercises 11 and 12.

EXERCISES

1. Exhibit the forms of all the 5×4 matrices in echelon.

2. In the definition of equivalence of lists, show that it suffices to use only the operations I, II, and III_λ for $\lambda = 1$.

3. Reduce each of the following matrices to column equivalent echelon form:

$$\text{(a)} \begin{bmatrix} 1 & 2 & 0 \\ -1 & -4 & 3 \\ 3 & 1 & 2 \end{bmatrix}. \quad \text{(b)} \begin{bmatrix} 1 & 4 & 7 & -6 \\ 6 & 0 & 2 & 3 \\ -2 & 4 & 0 & -3 \\ 5 & -2 & 2 & 3 \end{bmatrix}. \quad \text{(c)} \begin{bmatrix} 2 & 0 & 4 & 2 \\ -1 & 2 & -2 & -3 \\ 3 & 1 & 3 & 4 \\ 2 & 4 & 9 & 5 \end{bmatrix}.$$

4. Test the following lists of vectors for linear independence:

(a) $(1, 0, 1), (0, 2, 2), (3, 7, 1)$ in $\mathbf{Q}^3$ and $\mathbf{R}^3$.

(b) $(1, 0, 0), (0, 1, 1), (1, 1, 1)$ in $\mathbf{R}^3$ and $(\mathbf{Z}_5)^3$.

(c) $(1, 2, 1, 2, 1), (2, 1, 2, 1, 2), (1, 0, 1, 1, 0)$, and $(0, 1, 0, 0, 1)$ in $\mathbf{Q}^5$.

In each case of linear dependence, find a linearly independent list which spans the same subspace.

5. In $\mathbf{Q}^6$, test each of the following lists of vectors for independence, and in each case find a basis for the subspace which they span:

(a) $(2, 4, 3, -1, -2, 1), (1, 1, 2, 1, 3, 1), (0, -1, 0, 3, 6, 2)$.

(b) $(2, 1, 3, -1, 4, -1), (-1, 1, -2, 2, -3, 3), (1, 5, 0, 4, -1, 7)$.

6. In Exercise 5, find a basis for the subspace spanned by the combined list ((a) and (b) together).

7. In each of the following examples, a linear transformation $t: \mathbf{Q}^4 \to \mathbf{Q}^4$ is determined by requiring that it carry the unit vectors in $\mathbf{Q}^4$ (in order) into the columns of the indicated matrix. Find the rank and nullity of t:

$$\text{(a)} \begin{bmatrix} 1 & 2 & 1 & 2 \\ 3 & 2 & 3 & 2 \\ -1 & -3 & 0 & 4 \\ 0 & 4 & -1 & -3 \end{bmatrix}. \quad \text{(b)} \begin{bmatrix} 0 & 1 & 2 & 3 \\ -1 & 2 & 1 & 0 \\ 3 & 0 & -1 & -2 \\ 5 & -3 & -1 & 1 \end{bmatrix}.$$

8. In each part of Exercise 7, determine a basis for the kernel of t.

9. (a) If $\mathbf{v}: \mathbf{m} \to V$ and $\mathbf{w}: \mathbf{m} \to V$ are two linearly independent lists of m vectors each with the same span, show that there is a composite ϕ of elementary operations with $\phi(\mathbf{v}) = \mathbf{w}$.

(b) Prove that any two lists $\mathbf{v}$ and $\mathbf{w}$, each of m vectors in V, have the same span if and only if there is a composite ϕ of elementary operations with $\phi(\mathbf{v}) = \mathbf{w}$.

10. (a) Exhibit the forms of all 5×4 matrices in reduced echelon form.

(b) Prove that any matrix is column equivalent to a matrix in reduced echelon form.

11. Prove that no two different $n \times m$ reduced echelon matrices can be column equivalent.

12. Show that each subspace of F^n has exactly one basis which is in reduced echelon form.

13. Show that any elementary operation of type I can be accomplished by a composite of at most four elementary operations of types II and III. (*Hint:* Try 2×2 matrices.)

6. Systems of Linear Equations

Let $A:\mathbf{m} \times \mathbf{n} \to F$ be an $m \times n$ matrix and $\boldsymbol{\kappa}$ a list of m scalars of the field F. The *system of linear equations* with coefficients A and constants $\boldsymbol{\kappa}$ in n "unknowns" x_i is

$$\sum_{j=1}^{n} A_{1j}x_j = \kappa_1, \quad \cdots, \quad \sum_{j=1}^{n} A_{mj}x_j = \kappa_m. \tag{31}$$

A *solution* of this system is a list $\boldsymbol{\xi}:\mathbf{n} \to F$ of n scalars with $\sum_j A_{ij}\xi_j = \kappa_i$ for all $i \in \mathbf{m}$. The system is *homogeneous* if all κ_i are 0.

If $\boldsymbol{\varepsilon}_i$ is the usual basis of unit vectors of F^n, the unknowns x_i may be interpreted as the elements of the dual basis of $(F^n)^*$, so the left-hand side of the ith equation is a linear form $f_i = \sum_j A_{ij}x_j:F^n \to F$. A solution of the system is then a vector $\boldsymbol{\xi} \in F^n$ with $f_i(\boldsymbol{\xi}) = \kappa_i$ for all $i \in \mathbf{m}$ because $x_j(\boldsymbol{\xi}) = \xi_j$.

The forms $f_1, \cdots, f_m$ span a subspace T of the dual space $(F^n)^*$, of dimension at most m. A solution of the *homogeneous system* (31) (a system with all $\kappa_i = 0$) is exactly a vector in the subspace Annih $(T) \subset F^n$. The set of all solutions is the subspace Annih (T), so the maximum number of linearly independent solutions is dim (Annih T). As for any vector space, this dimension is independent of the choice of a basis used to represent it and is determined by Theorem 13 on annihilators as follows.

THEOREM 20. *A system of m linearly independent homogeneous linear equations in n unknowns with coefficients in a field F has $n - m$ linearly independent solutions over that field. Given $n - m$ such solutions, any other solution is a linear combination of these.*

Here the m equations are said to be *linearly independent* when the m corresponding forms $f_1, \cdots, f_m$ (that is, the m rows of the matrix A) are linearly independent.

In case the m equations of the system are not linearly independent, they may be replaced by fewer equations which are independent (Exercise 10).

Two useful cases of this theorem are

Corollary 1. *Over any field, n linearly independent homogeneous linear equations in n unknowns have only the zero solution.*

Corollary 2. *Any m homogeneous linear equations in $n > m$ unknowns always have a non-zero solution.*

Now consider a nonhomogeneous system (31). A solution of this system is a vector $\boldsymbol{\eta} \in F^n$ for which

$$f_1(\boldsymbol{\eta}) = \kappa_1, f_2(\boldsymbol{\eta}) = \kappa_2, \cdots, f_m(\boldsymbol{\eta}) = \kappa_m.$$

This may be written $t(\boldsymbol{\eta}) = \boldsymbol{\kappa}$, where $t: F^n \to F^m$ is the linear transformation given by $\boldsymbol{\eta} \mapsto (f_1(\boldsymbol{\eta}), \cdots, f_m(\boldsymbol{\eta}))$. Hence, the set of all solutions is just the inverse image $t^*(\boldsymbol{\kappa})$; when $\boldsymbol{\kappa} \neq 0$, the set of solutions is not a subspace of F^n. There may be no solution; indeed, a solution exists if and only if $\boldsymbol{\kappa}$ is in Im (t). If there is a solution $\boldsymbol{\eta}$, the difference $\boldsymbol{\xi} = \boldsymbol{\eta} - \boldsymbol{\eta}'$ of any two solutions $\boldsymbol{\eta}$ and $\boldsymbol{\eta}'$ has $t(\boldsymbol{\xi}) = 0$. This proves

Theorem 21. *If $\boldsymbol{\eta}'$ is any one solution of a system (31) of linear equations, then every solution $\boldsymbol{\eta}$ of the system can be expressed uniquely as $\boldsymbol{\eta} = \boldsymbol{\eta}' + \boldsymbol{\xi}$, where $\boldsymbol{\xi}$ is a solution of the corresponding homogeneous system (with all $\kappa_i = 0$).*

Corollary. *If a system of n linear equations in the same number n of unknowns with coefficients in a field has n linearly independent left-hand sides, it has exactly one solution.*

Indeed, the linear independence of $f_1, \cdots, f_n$ means by Corollary 20.1 that Ker $t = 0$ and hence by Corollary 8.2 that t is an isomorphism, so there is exactly one solution $\boldsymbol{\eta} = t^{-1}(\boldsymbol{\kappa})$.

Elementary operations may be used to calculate the solutions of a system. Each linear equation $f = \boldsymbol{\kappa}$ may be regarded as a vector $(f, \boldsymbol{\kappa})$ in the space $(F^n)^* \oplus F$ of dimension $n + 1$, and a system of linear equations as a list of such vectors.

Lemma. *An elementary operation on the list $(f_1, \kappa_1), \cdots, (f_m, \kappa_m)$ does not alter the set of solutions of the corresponding equations*

$$f_1 = \kappa_1, \cdots, f_m = \kappa_m.$$

The proof is immediate. For example, an operation of type III_λ replaces the first two equations by $(f_1 + \lambda f_2, \kappa_1 + \lambda\kappa_2)$ and (f_2, κ_2); if $f_1(\boldsymbol{\eta}) = \kappa_1$ and $f_2(\boldsymbol{\eta}) = \kappa_2$, then $(f_1 + \lambda f_2)(\boldsymbol{\eta}) = \kappa_1 + \lambda\kappa_2$ and $f_2(\boldsymbol{\eta}) = \kappa_2$, and conversely.

Note that the elementary operations apply to *both* components f and $\boldsymbol{\kappa}$ of $(f, \boldsymbol{\kappa})$; that is, to both sides of the equations. These operations may be visualized as operations on the rows of the $m \times (n+1)$ matrix

$$\begin{bmatrix} A_{11} & \cdots & A_{1n} & \kappa_1 \\ \cdot & & & \\ \cdot & & & \\ \cdot & & & \\ A_{m1} & \cdots & A_{mn} & \kappa_m \end{bmatrix}. \tag{32}$$

As before, successive elementary operations will put the matrix A in echelon form *for rows*. This process, called "Gauss elimination", gives a new system with the same solutions. The fact that this system is in echelon may be described thus: Let just the first r new equations have non-zero left-hand sides, and let the leading unknown in the ith equation be x_{l_i} (the ith row of the matrix has its first non-zero entry a 1 in column l_i). Then $l_1 < l_2 < \cdots < l_r$, and the system is

$$\begin{aligned} x_{l_1} + \cdots \qquad\qquad\qquad\qquad &= \kappa_1' \\ x_{l_2} + \cdots \qquad\qquad &= \kappa_2' \\ &\;\vdots \\ x_{l_r} + \cdots &= \kappa_r' \\ 0 &= \kappa_{r+1}', \cdots, \\ 0 &= \kappa_m', \end{aligned} \tag{33}$$

with new constants κ_i' on the right. (All operations are applied to both sides of the equations; the original matrix A of coefficients has been reduced by elementary operations on its *rows* to a form which is in echelon for the rows.)

If the m rows of the original matrix A are not linearly independent, there will remain some $m - r$ equations with no unknowns, of the form $0 = \kappa_i'$, as exhibited. If any of the scalars κ_i', $i = r+1, \cdots, m$ are non-zero, the original equations have no solution. If all these scalars are zero, one may find all solutions as follows. Pick values at will for the $n - r$ unknowns x_j with $j \neq l_1, \cdots, l_r$; then in succession solve the rth equation for x_{l_r}, the $(r-1)$st for $x_{l_{r-1}}, \cdots,$ and finally solve the first equation for x_{l_1}. In the homogeneous case, for example, set one of these $x_j = 1$ and the others $= 0$, and construct the solution as described above; this gives $n - r$ solutions which form a basis for the space of all solutions.

EXERCISES

1. Find a basis for the subspace of all those vectors of $\mathbf{Q}^4$ which satisfy:
   (a) $x_1 - x_2 = 0 = x_3 - x_4$.
   (b) $3x_1 - 2x_2 + 4x_3 + x_4 = 0 = x_1 + x_2 - 3x_3 - 2x_4$.
   (c) $x_1 + x_2 + x_3 = x_2 + x_3 + x_4 = x_3 + x_4 + x_1 = x_4 + x_1 + x_2 = 0$.
2. Find a basis for the set of all rational solutions of each of the

following systems of equations:

$$\text{(a)}\quad \begin{aligned} x + 2y - 4z &= 0,\\ 3x + y - 2z &= 0. \end{aligned} \qquad \text{(b)}\quad \begin{aligned} x + y + z + t &= 0,\\ 2x + 3y - z + t &= 0,\\ 3x + 4y + 2t &= 0. \end{aligned}$$

3. Do Exercise 2 if the equations are taken to be congruences modulo 5.

4. Solve the following simultaneous congruences:

$$x + 2y - z \equiv 4 \pmod 5, \qquad x + y + z \equiv 1 \pmod 5.$$

5. In any field F, prove that two simultaneous equations $ax + by = e$, $cx + dy = f$, with $ad - bc \neq 0$, have the unique solution

$$x = (ed - fb)/(ad - bc), \qquad y = (af - ce)/(ad - bc).$$

6. If a system of simultaneous linear equations with coefficients in a field F has no solutions in F, prove that it also has no solutions in any larger field $F' \supset F$.

7. If a system of homogeneous linear equations has coefficients in a field F, show that any basis for the solutions in F is also a basis for the solutions in any larger field $F' \supset F$.

8. Prove that two systems (31) of m linear equations in n unknowns with coefficients in the same field have the same solutions if and only if the two $m \times (n + 1)$ matrices (32) attached to these systems are row equivalent.

9. Use Gauss elimination to prove Theorem 20.

10. (*Generalization of Theorem* 20.) Prove that a system of m homogeneous linear equations in n unknowns and with coefficients in F has precisely $n - r$ linearly independent solutions over F, where r is the maximum number of linearly independent equations in the given system.

11. Prove that the system (31) of nonhomogeneous linear equations has a solution if and only if the rank of its matrix A of coefficients equals the rank of the matrix (A, κ) of (32). (Here the *rank* of A is the dimension of the space spanned by the columns of A.)

CHAPTER VII

Matrices

Explicit calculations with a linear transformation $t: V \to V'$ usually depend on a representation of that transformation by a matrix of scalars. Relative to given bases of V and V', each $t: V \to V'$ has such a matrix representation. This chapter develops ways of replacing operations on linear transformations by operations on the corresponding matrices, as well as methods of describing the change in the matrix representing a transformation caused by a change in the choice of bases.

The connection between linear transformations and matrices can be described in two ways: By free R-modules for any ring R or by bases and biproducts for vector spaces over a field.

The first way, to be given in §1, uses only the free R-modules R^n, one for each natural number n. Each of these modules has a standard basis of unit vectors, so each linear map $t: R^n \to R^m$ can be described completely by listing the images under t of the unit vectors of R^n. Each image is an element of R^m, thus is a column of m scalars, and the whole list of n columns is then an $m \times n$ matrix A. In this way, each linear map t "is" a matrix A. Moreover, the composition of linear maps gives the multiplication of matrices.

The second way, to be given in §3 below, considers linear transformations $t: V \to V'$ between any two finite-dimensional vector spaces V and V' over the same field F. Choosing a basis $\mathbf{b}$ of n vectors for V and a basis $\mathbf{b}'$ of m vectors for V' then amounts to giving isomorphisms $V \cong F^n$ and $V' \cong F^m$. This replaces t by a linear map $F^n \to F^m$, that is, by an $m \times n$ matrix A. This is the matrix A of t, *relative* to the given bases $\mathbf{b}$ and $\mathbf{b}'$. Changing to different bases would give a different matrix for t.

This verbal description of "the matrix of t relative to bases $\mathbf{b}$ and $\mathbf{b}'$ " can be supplemented by formulas. Since choosing a basis $\mathbf{b}$ in V amounts to representing V as a biproduct (direct sum) of n copies of the field F, the desired formulas rest on properties of the biproducts, to be described in §2. These properties, although often not stated so explicitly, allow a perspicuous description of the relation of t to its various matrix representations.

1. Matrices and Free Modules

The $m \times n$ matrix A of elements of the ring R,

$$A = \begin{bmatrix} A_{11} & & A_{1j} & & A_{1n} \\ A_{21} & \cdots & A_{2j} & \cdots & A_{2n} \\ \cdot & & \cdot & & \cdot \\ \cdot & & \cdot & & \cdot \\ \cdot & & \cdot & & \cdot \\ A_{m1} & & A_{mj} & & A_{mn} \end{bmatrix}, \qquad \begin{array}{l} A_{ij} \in R, \\ i = 1, \cdots, m, \\ j = 1, \cdots, n, \end{array}$$

is, formally, a function $A : \mathbf{m} \times \mathbf{n} \to R$, via $(i, j) \mapsto A_{ij}$. Each column of A is a list of m scalars of R, hence is an element of the free module R^m. With the m unit vectors $\varepsilon_1, \cdots, \varepsilon_m$ of R^m as basis, the jth column of A is the linear combination (with scalars on the *right*)

$$\varepsilon_1 A_{1j} + \varepsilon_2 A_{2j} + \cdots + \varepsilon_m A_{mj}.$$

Thus the matrix A is a list of n columns (n elements of R^m). Now R^n is the free R-module on its n unit vectors $\varepsilon_1, \cdots, \varepsilon_n$, so the matrix A determines a linear transformation $t_A : R^n \to R^m$, taking the ε_j in order into the columns of A. Thus t_A is determined by

$$t_A(\varepsilon_j) = \varepsilon_1 A_{1j} + \cdots + \varepsilon_m A_{mj} = \sum_{i=1}^{m} \varepsilon_i A_{ij}, \qquad j = 1, \cdots, n. \quad (1)$$

Moreover, every linear $t : R^n \to R^m$ has this form for exactly one matrix A. (*Note:* In (1), ε_j with subscript j on the left stands for a unit element, better written $\varepsilon_j^{(n)}$, of the module R^n, while ε_i on the right, with subscript i, refers to a unit element in the module R^m. We depend on the subscripts $i \neq j$ to make this distinction.) The discussion has proved

THEOREM 1. *Each $m \times n$ matrix A of elements from the ring R determines by* (1) *a morphism $t_A : R^n \to R^m$ of right R-modules. The assignment $A \mapsto t_A$ is a bijection*

$$R^{m \times n} \cong \hom_R (R^n, R^m)$$

from the set $R^{m \times n}$ of all $m \times n$ matrices over R to the set of all morphisms $t : R^n \to R^m$ of right R-modules. Given any such t, the corresponding matrix A is the matrix whose columns are the images $t\varepsilon_1, \cdots, t\varepsilon_n$ of the unit vectors ε_j of R^n.

More briefly: The *columns of the matrix of a transformation are the transforms of the unit vectors.*

Now the set $R^{m \times n}$ of all $m \times n$ matrices is a right R-module under termwise operations: To add two matrices, add corresponding entries; to multiply a matrix on the right by a scalar, multiply each entry on the right by that scalar. (This is just a special case of the module operations defined in §V.5 for the function modules R^X.) These module operations on matrices correspond exactly to operations on the corresponding morphisms:

Theorem 2. *The bijection $A \mapsto t_A$ of the previous theorem is an isomorphism*

$$R^{m \times n} \cong \mathrm{Hom}_R\,(R^n, R^m) \tag{2}$$

of abelian groups. If $R = K$ is commutative, it is an isomorphism of K-modules.

Proof: For two $m \times n$ matrices A and B, $(A + B)_{ij} = A_{ij} + B_{ij}$. The corresponding morphism t_{A+B} of (1) thus is

$$t_{A+B}(\varepsilon_j) = \sum_i \varepsilon_i A_{ij} + \sum_i \varepsilon_i B_{ij} = t_A(\varepsilon_j) + t_B(\varepsilon_j).$$

The sum $t_A + t_B$ of morphisms is also a pointwise sum

$$(t_A + t_B)(\varepsilon_j) = t_A(\varepsilon_j) + t_B(\varepsilon_j).$$

Therefore, $t_{A+B} = t_A + t_B$ on the free generators ε_j and hence on all of R^n, so the bijection $A \mapsto t_A$ is a morphism of addition, as required.

For each scalar κ, $(A\kappa)_{ij} = A_{ij}\kappa$ for all $i \in \mathbf{m}$ and $j \in \mathbf{n}$. The corresponding morphism $t_{(A\kappa)}$ of (1) thus is

$$t_{(A\kappa)}(\varepsilon_j) = \sum_i \varepsilon_i A_{ij}\kappa = \left[\, t_A(\varepsilon_j)\right]\kappa.$$

But for K commutative each $\mathrm{Hom}_K\,(-, -)$ is a K-module when the multiple $t\kappa$ of a morphism t by a scalar κ is defined pointwise (this for right modules, just as for left modules in Theorem V.3). This means that $(t_A\kappa)(\varepsilon_j) = [t_A(\varepsilon_j)]\kappa$ for every ε; and hence that $t_{(A\kappa)} = (t_A)\kappa$. Thus $A \mapsto t_A$ is a morphism of K-modules, as required.

The multiplication of matrices is defined as follows. If A is an $m \times n$ matrix and B an $n \times q$ matrix (both over the same ring R), their *product* $C = AB$ is the $m \times q$ matrix with entries

$$\begin{aligned} C_{ik} &= A_{i1}B_{1k} + A_{i2}B_{2k} + \cdots + A_{in}B_{nk} \\ &= \sum_{j=1}^{n} A_{ij}B_{jk}, \qquad i \in \mathbf{m}, \quad k \in \mathbf{q}. \end{aligned} \tag{3}$$

This matrix product AB is defined *only* when the number n of columns in A equals the number of rows in B. The entry in row i and column k of AB is thus formed from row i of A and column k of B, both of length n, by multiplying corresponding terms and adding as indicated in the following schematic diagram:

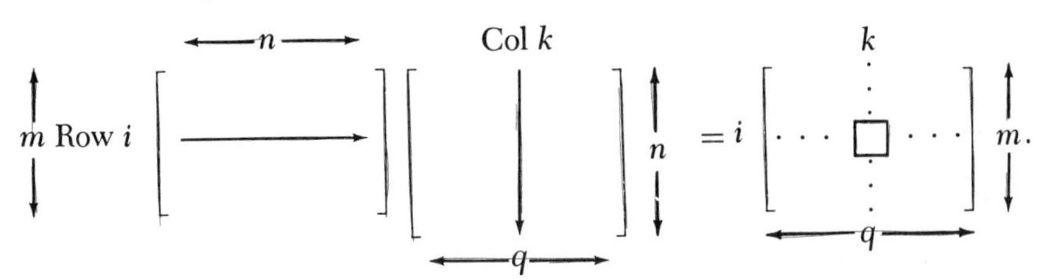

The whole point of this definition of matrix product is that it matches the composition of linear maps.

Theorem 3. *For rectangular matrices A and B over R, the composite morphism $t_A \circ t_B$ is defined precisely when the matrix product AB is defined, and $t_{AB} = t_A \circ t_B$.*

In this "whenever defined" sense, the bijection $A \mapsto t_A$ of Theorem 1 is thus a morphism of multiplication.

Proof: For any $m \times n$ matrix A and any $n \times q$ matrix B, the corresponding morphisms $t_B : R^q \to R^n$ and $t_A : R^n \to R^m$ do have a composite $t_A \circ t_B$, with values on the unit elements $\boldsymbol{\varepsilon}_k$ of R^q given by

$$(t_A \circ t_B)(\boldsymbol{\varepsilon}_k) = t_A(t_B\boldsymbol{\varepsilon}_k) = t_A\left(\sum_{j=1}^{n} \boldsymbol{\varepsilon}_j B_{jk}\right) = \sum_j (t_A\boldsymbol{\varepsilon}_j)B_{jk}$$

$$= \sum_j \left(\sum_{i=1}^{m} \boldsymbol{\varepsilon}_i A_{ij}\right) B_{jk} = \sum_{i=1}^{m} \boldsymbol{\varepsilon}_i \left(\sum_{j=1}^{n} A_{ij}B_{jk}\right).$$

This is exactly $t_{AB}(\boldsymbol{\varepsilon}_k)$, hence the theorem.

By making this calculation first, we could have "discovered" by this formula how the matrix product should be defined.

We now know that the function $A \mapsto t_A$ from matrices to morphisms of free modules is an isomorphism for addition, scalar multiple, and multiplication. We can therefore "transfer" to matrices all the identities valid for morphisms of modules. In particular, this proves

Corollary 1. *Multiplication of matrices is associative and distributive (on either side) over addition.*

Specifically, the triple product of matrices is associative whenever it is defined: If A is $m \times n$, B is $n \times q$, and C is $q \times r$, then $A(BC) = (AB)C$.

The detailed statement of distributivity is similar; for example, if A is an $m \times n$ matrix, while B and B' are both $n \times q$ matrices, then $A(B + B') = AB + AB'$.

All these laws could also be proved directly from the explicit definition (3) of the matrix product. By that means one can also prove

$$\kappa(AB) = (\kappa A)B, \qquad (A\kappa)B = A(\kappa B), \qquad A(B\kappa) = (AB)\kappa,$$

for any scalar κ, any $m \times n$ matrix A, and any $n \times q$ matrix B.

For each n, the $n \times n$ *identity matrix*

$$I = I_n = \begin{bmatrix} 1 & 0 & \cdots & 0 \\ 0 & 1 & \cdots & 0 \\ \vdots & \vdots & & \vdots \\ 0 & 0 & \cdots & 1 \end{bmatrix} \tag{4}$$

is the matrix δ with entries the Kronecker deltas δ_{ij}. The corresponding morphism $t_I : R^n \to R^n$ is the identity morphism. Moreover, the matrices I_n are identities for matrix multiplication: If A is any $m \times n$ matrix, then $I_m A = A = A I_n$.

Now restrict attention to $n \times n$ square matrices. The product of two matrices is then always defined, and the results above specialize to yield

Corollary 2. *The set $R^{n \times n}$ of all $n \times n$ matrices over R is a ring under (termwise) addition and matrix multiplication, with the $n \times n$ identity matrix as unit. The bijection $A \mapsto t_A$ is an isomorphism of this ring to the ring* End (R^n) *of all endomorphisms of the free R-module R^n.*

The complete algebra of the rectangular matrices of all sizes is not a ring, because the matrix product AB is defined only when the sizes of A and B match as above. This algebra is a category. Specifically, it is the concrete category (see §IV.5) whose objects are the free right R-modules R^n for all natural numbers n, with morphisms from R^n to R^m the set (actually the abelian group) $\mathrm{Hom}_R\,(R^n, R^m)$ of all linear transformations $t : R^n \to R^m$. It is convenient here to replace each such $t = t_A$ by the corresponding $m \times n$ matrix A, so that the morphisms of the category *are* the matrices $A : R^n \to R^m$. In particular, the morphisms $t : R \to R^m$, where R has the unit element 1, are the one-column matrices $t(1) \in R^m$; in other words, $\mathrm{Hom}_R\,(R, R^m)$ is R^m. In particular, $\mathrm{Hom}_R\,(R, R)$ is identified with R. Moreover, given any two matrices

$$B : R^k \to R^l, \qquad A : R^n \to R^m$$

the product AB is defined only when $l = n$; that is, only when the codomain of B is the domain of A, just as in the definition of the composite in a category.

In this category each hom-set is an abelian group under addition, composition is distributive on both sides, and any two objects R^n and R^m have a biproduct R^{n+m}. Such categories, which arise in other connections, are called "additive categories", to reflect this additive structure in each $\mathrm{Hom}_R(R^n, R^m)$. Also, all the interest resides in the morphisms, not the objects. One might have taken the objects to be just the natural numbers n, with underlying sets R^n—and then "forget" the underlying sets, to get an "abstract" category which is no longer "concrete" (see Chapter XV).

The situation with the algebra of matrices may be summarized without explicitly using categorical concepts, because the composition of morphisms in the category is just matrix product. Thus:

A morphism $n \to m$ is an $m \times n$ matrix A.

A morphism $1 \to m$ is an $m \times 1$ matrix or *column*; it is just an element of the right module R^m.

A morphism $1 \to 1$ is a 1×1 matrix or scalar (element of R).

A morphism $n \to 1$ is an $1 \times n$ matrix or *row*; as a list $\mathbf{n} \to R$ it is an element of the left module R^n; as a linear map $R^n \to R$ of right modules it is an element of the dual left module $(R^n)^*$.

With these observations, the product of any two such matrices has the evident meaning, as follows: The matrix product of a column ξ and a scalar κ is

$$\begin{bmatrix} \xi_1 \\ \cdot \\ \cdot \\ \cdot \\ \xi_n \end{bmatrix} [\kappa] = \begin{bmatrix} \xi_1\kappa \\ \cdot \\ \cdot \\ \cdot \\ \xi_n\kappa \end{bmatrix};$$

this is just the multiple of ξ by κ in the right module R^n. Similarly, the matrix product of a scalar κ by a row $\boldsymbol{\mu}$ is the left multiple by κ, for

$$[\kappa][\mu_1, \cdots, \mu_n] = [\kappa\mu_1, \cdots, \kappa\mu_n].$$

The product of a row $\boldsymbol{\mu}$ by a column ξ is a scalar

$$[\mu_1, \cdots, \mu_n] \begin{bmatrix} \xi_1 \\ \cdot \\ \cdot \\ \cdot \\ \xi_n \end{bmatrix} = [\mu_1\xi_1 + \cdots + \mu_n\xi_n].$$

If we regard $\boldsymbol{\mu}$ as an element of the dual module $(R^n)^*$, as in Proposition V.17, this scalar $\boldsymbol{\mu}\xi$ is just the value of the linear form $\boldsymbol{\mu}: R^n \to R$ at the argument ξ. Finally, the product of a matrix A by a column ξ is the result of applying the morphism t_A to ξ:

THEOREM 4. *When each element ξ of the free module R^n is written as a one-column matrix, then for each $m \times n$ matrix A over R the corresponding linear transformation $t_A : R^n \to R^m$ is given by matrix multiplication as $\xi \mapsto A\xi$.*

Proof: If we write ξ as a linear combination $\xi = \sum_j \boldsymbol{\varepsilon}_j \xi_j$ of the unit elements and apply the definition (1) of the linear transformation t_A, we obtain

$$t_A(\xi) = \sum_{j=1}^{n} t_A(\boldsymbol{\varepsilon}_j)\xi_j = \sum_{i=1}^{m} \boldsymbol{\varepsilon}_i \left(\sum_{j=1}^{n} A_{ij}\xi_j \right).$$

This equation states that the ith coordinate of $t_A(\xi)$ is the product of the ith row of A by the column ξ, and hence that $t_A(\xi)$, regarded as a column, is the matrix product $t_A(\xi) = A\xi$. The display is

$$\begin{bmatrix} \xi_1 \\ \cdot \\ \cdot \\ \cdot \\ \xi_n \end{bmatrix} \mapsto \begin{bmatrix} A_{11} & \cdots & A_{1n} \\ \cdot & & \\ \cdot & & \\ \cdot & & \\ A_{m1} & \cdots & A_{mn} \end{bmatrix} \begin{bmatrix} \xi_1 \\ \cdot \\ \cdot \\ \cdot \\ \xi_n \end{bmatrix} = \begin{bmatrix} A_{11}\xi_1 + \cdots + A_{1n}\xi_n \\ \cdot \\ \cdot \\ \cdot \\ A_{m1}\xi_1 + \cdots + A_{mn}\xi_n \end{bmatrix}.$$

This explicit display of a linear transformation is often effective. If one writes $\boldsymbol{\eta}$ for the column of coordinates of $t_A(\xi)$, the display becomes a set of m linear equations

$$\eta_i = A_{i1}\xi_1 + \cdots + A_{in}\xi_n, \qquad i = 1, \cdots, m,$$

which give the coordinates η_i of the transformed element in terms of the coordinates ξ_j of the original element. Calculations with such equations in terms of coordinates (usually with x_j for ξ_j and y_i for η_i) are often used instead of matrices.

This display again shows that the images of the unit elements $\boldsymbol{\varepsilon}_1, \cdots, \boldsymbol{\varepsilon}_n$ under t_A are just the columns of A. Here is a geometric example of this observation. In the real coordinate plane $\mathbf{R}^2$, let r_θ be the transformation $\mathbf{R}^2 \to \mathbf{R}^2$ which rotates the plane counterclockwise about the origin through the angle θ.

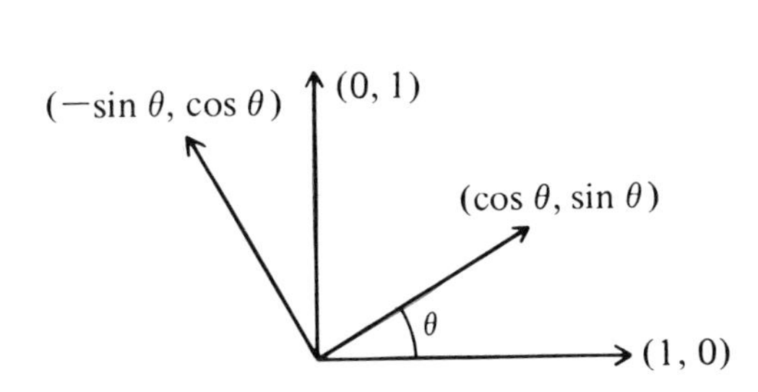

This transformation is linear. By the very definition of the trigonometric functions, its effect upon the unit (column) vectors is

$$r_\theta\begin{bmatrix} 1 \\ 0 \end{bmatrix} = \begin{bmatrix} \cos\theta \\ \sin\theta \end{bmatrix}, \qquad r_\theta\begin{bmatrix} 0 \\ 1 \end{bmatrix} = \begin{bmatrix} -\sin\theta \\ \cos\theta \end{bmatrix}.$$

Hence, the matrix of r_θ is

$$\begin{bmatrix} \cos\theta & -\sin\theta \\ \sin\theta & \cos\theta \end{bmatrix}$$

and its action upon any point ξ is, according to Theorem 4,

$$r_\theta\begin{bmatrix} \xi_1 \\ \xi_2 \end{bmatrix} = \begin{bmatrix} \cos\theta & -\sin\theta \\ \sin\theta & \cos\theta \end{bmatrix}\begin{bmatrix} \xi_1 \\ \xi_2 \end{bmatrix} = \begin{bmatrix} \xi_1\cos\theta - \xi_2\sin\theta \\ \xi_1\sin\theta + \xi_2\cos\theta \end{bmatrix}.$$

These are the customary formulas for rotation of a point with coordinates (ξ_1, ξ_2). Now follow the rotation r_θ by another rotation r_ϕ. The result should be rotation through the angle $\theta + \phi$; the formula (3) for the product of the matrices gives (in part)

$$\begin{bmatrix} \cos\phi & -\sin\phi \\ \sin\phi & \cos\phi \end{bmatrix}\begin{bmatrix} \cos\theta & -\sin\theta \\ \sin\theta & \cos\theta \end{bmatrix} = \begin{bmatrix} \cos\phi\cos\theta - \sin\phi\sin\theta & \cdots \\ \sin\phi\cos\theta + \cos\phi\sin\theta & \cdots \end{bmatrix}.$$

Comparing the result with the known matrix of $r_{\theta+\phi}$ gives

$$\cos(\phi + \theta) = \cos\phi\cos\theta - \sin\phi\sin\theta,$$
$$\sin(\phi + \theta) = \sin\phi\cos\theta + \cos\phi\sin\theta.$$

This shows that the familiar addition formulas for the trigonometric functions cos and sin express the effect of composition in the group of rotations about the origin.

Here is another operation on matrices. If A is an $m \times n$ matrix, its *transpose* A^T is the $n \times m$ matrix obtained from A by interchanging rows and columns. In other words.

$$(A^T)_{ji} = A_{ij}, \qquad i \in \mathbf{m}, \quad j \in \mathbf{n}. \tag{5}$$

To find the transpose C^T of the product matrix $C = AB$ over a commutative ring, observe that the definitions (3) and (5) yield

$$(C^T)_{ki} = C_{ik} = \sum_j A_{ij}B_{jk} = \sum_j (B^T)_{kj}(A^T)_{ji} = (B^TA^T)_{ki}.$$

This proves that $(AB)^T = B^TA^T$; in words, transposition carries the product of matrices into the opposite of the product, so is an "antiautomorphism" of matrix multiplication.

Matrix Notation. Since A designates an $m \times n$ matrix, which we regard as a function on $\mathbf{m} \times \mathbf{n}$, the entries of A are the values of the function at $i \in \mathbf{m}$, $j \in \mathbf{n}$, and so are written here as A_{ij} with a *capital* letter A, although most authors write the entries of A with the corresponding lowercase letter as a_{ij}.

Right Scalars. In the formulas here and below, the scalar multiples κ appear on the *right* and not (as often used) on the left. In other words, we have written the formulas to apply to right R-modules, and in particular to vector spaces as right modules over a field F. This is done so that the formulas will also apply to modules over a division ring D, in which the multiplication is not commutative—as in the division ring of quaternions to be discussed in Chapter VIII.

This use of right modules fits with the convention of writing a function f systematically on the left of its argument, as in $f(x)$ or in $t(\ )$ for a linear transformation t. Consequently, a matrix A, regarded as a function t_A, is written to the left of its argument ξ. As a result, the equation stating that t_A preserves the right scalar multiples by κ reads

$$A(\xi\kappa) = (A\xi)\kappa$$

with no change in the order of the three letters, left and right. Had we tried to use scalars on the left, $A(\kappa\xi) = \kappa(A\xi)$ for 2×2 A would give

$$\begin{pmatrix} a & b \\ c & d \end{pmatrix}\begin{pmatrix} \kappa x \\ \kappa y \end{pmatrix} = \begin{pmatrix} a\kappa x + b\kappa y \\ c\kappa x + d\kappa y \end{pmatrix} = \begin{pmatrix} \kappa a x + \kappa b y \\ \kappa c x + \kappa d y \end{pmatrix} = \kappa\begin{pmatrix} a & b \\ c & d \end{pmatrix}\begin{pmatrix} x \\ y \end{pmatrix},$$

an identity which breaks down when $a\kappa \neq \kappa a$. In other words, to preserve associativity, we must use either linear transformations on the left and scalars on the right, or transformations right and scalars left. We have chosen the former.

EXERCISES

1. Prove directly from the definition (3) that the multiplication of matrices is associative.

2. Prove directly from the definitions that the multiplication of matrices is distributive (on either side) over addition.

3. Describe geometrically the linear transformation $t_A : \mathbf{R}^2 \to \mathbf{R}^2$ determined by each of the following 2×2 matrices A:

$$\begin{bmatrix} \cos\theta & \sin\theta \\ -\sin\theta & \cos\theta \end{bmatrix}, \quad \begin{bmatrix} 1 & 0 \\ 2 & 1 \end{bmatrix}, \quad \begin{bmatrix} 3 & 3 \\ 0 & 3 \end{bmatrix}, \quad \begin{bmatrix} -1 & 0 \\ 0 & -1 \end{bmatrix}, \quad \begin{bmatrix} 2 & 0 \\ 0 & -1 \end{bmatrix}.$$

4. In the 3-space $\mathbf{R}^3$, in the coordinates (x, y, z) of analytic geometry, find the matrix of each of the following linear transformations $\mathbf{R}^3 \to \mathbf{R}^3$:

   (a) Reflection in the (x, y)-plane.

(b) Reflection in the plane with equation $x = y$.

(c) Vertical projection on the (x, y)-plane.

5. (a) Compute the products BE_1, BE_2, BE_3, E_2E_3, where

$$B = \begin{bmatrix} 1 & 2 & 1 \\ 1 & 3 & 2 \\ 1 & 4 & 6 \end{bmatrix}, \quad E_1 = \begin{bmatrix} 0 & 0 & 1 \\ 0 & 1 & 0 \\ 1 & 0 & 0 \end{bmatrix}, \quad E_2 = \begin{bmatrix} 1 & 0 & \kappa \\ 0 & 1 & 0 \\ 0 & 0 & 1 \end{bmatrix}, \quad E_3 = \begin{bmatrix} \lambda & 0 & 0 \\ 0 & \mu & 0 \\ 0 & 0 & \nu \end{bmatrix};$$

(b) If A is any 3×3 matrix, how is AE_3 related to A?

(c) Describe the effect caused by multiplying any 3×3 matrix on the right by E_1; by E_2.

6. (a) Find all 3×3 matrices, with entries in an integral domain, which commute with the matrix E_3 of Exercise 5 (λ, μ, ν distinct).

(b) The same when $\lambda = \mu$, but $\lambda \neq \nu$.

7. Prove that $R^{m \times n}$ is a free R-module with free generators all those $m \times n$ matrices which have some one entry 1 and all other entries zero.

8. For each $n \times n$ matrix A over a field, prove that there is some number $m \leqslant n^2$ and scalars κ_i for $i = 0, \cdots, m-1$ such that

$$A^m + A^{m-1}\kappa_{m-1} + \cdots + A\kappa_1 + I\kappa_0 = 0.$$

9. (a) If A is an $n \times n$ matrix of real numbers and d a positive real number with $|A_{ij}| \leqslant d$ for every i, j, prove that every entry of the kth power matrix A^k is bounded by $n^{k-1}d^k$.

(b) Deduce that the series $\exp(A) = I + A + A^2/2! + \cdots + A^n/n! + \cdots$ is convergent for any $n \times n$ real matrix A.

(c) Prove that $\exp(A(\kappa + \mu)) = \exp(A\kappa)\exp(A\mu)$.

10. (a) Let $A: \mathbf{N} \times \mathbf{N} \to R$ be an infinite matrix with only a finite number of non-zero entries in each column. For the free module $R^{(N)}$ of §V.5 construct an endomorphism $t_A: R^{(N)} \to R^{(N)}$ and show that $A \mapsto t_A$ is a bijection from the set of these infinite matrices to the set $\hom(R^{(N)}, R^{(N)})$.

(b) If the infinite matrices $A, B: \mathbf{N} \times \mathbf{N} \to R$ both satisfy the condition of (a), show that their product AB is defined, satisfies the same condition, and that $t_{AB} = t_A \circ t_B: R^{(N)} \to R^{(N)}$.

2. Matrices and Biproducts

Matrices arise naturally from linear maps between biproducts. Consider, for example, a morphism $t: C_1 \oplus C_2 \to C_1' \oplus C_2'$ between two biproducts of right R-modules. The projections p_i and injections e_i of the biproducts yield, as in (V.31), two biproduct diagrams

$$\begin{array}{ccccc} C_1 & \underset{e_1}{\overset{p_1}{\leftrightarrows}} & C_1 \oplus C_2 & \underset{e_2}{\overset{p_2}{\rightleftarrows}} & C_2 \\ & & \vdots\, t & & \\ & & \downarrow & & \\ C_1' & \underset{e_1'}{\overset{p_1'}{\leftrightarrows}} & C_1' \oplus C_2' & \underset{e_2'}{\overset{p_2'}{\rightleftarrows}} & C_2' \end{array}$$

and satisfy (in both cases) the identities (cf. (V.31))

$$p_i e_j = \delta_{ij}, \qquad e_1 p_1 + e_2 p_2 = 1_{A_1 \oplus A_2}. \tag{6}$$

The universality of the injections e_1 and e_2 of $C_1 \oplus C_2$ yields (Theorem V.13) a bijection $t \mapsto (te_1, te_2)$. Here each composite $te_j : C_j \to C_1' \oplus C_2'$ is a morphism to $C_1' \oplus C_2'$, so the universality of the projections p_1' and p_2' of $C_1' \oplus C_2'$ yields (Theorem V.12) a bijection $te_j \mapsto (p_1' te_j, p_2' te_j)$. Thus each t determines four morphisms $t_{ij} = p_i' te_j$, which we may arrange as a 2×2 matrix T

$$t \mapsto T = \begin{bmatrix} t_{11} & t_{12} \\ t_{21} & t_{22} \end{bmatrix}, \qquad t_{ij} = p_i' te_j : C_j \to C_i',$$

where T has entries not from *one* ring but from four different sets hom (C_j, C_i'). Now this assignment $t \mapsto T$ is a bijection from the set $\hom_R(C_1 \oplus C_2, C_1' \oplus C_2')$ of all such morphisms t to the set of all 2×2 matrices with entries $t_{ij} \in \hom_R(C_j', A_i')$ for $i, j = 1, 2$. The same universality theorems give the inverse of this bijection by the formula

$$\begin{bmatrix} t_{11} & t_{12} \\ t_{21} & t_{22} \end{bmatrix} \mapsto e_1' t_{11} p_1 + e_1' t_{12} p_2 + e_2' t_{21} p_1 + e_2' t_{22} p_2;$$

the right-hand side is the sum of four morphisms $e_i' t_{ij} p_j : C_1 \oplus C_2 \to C_1' \oplus C_2'$.

Now consider this 2×2 matrix for a composite morphism

$$C_1'' \oplus C_2'' \xrightarrow{s} C_1 \oplus C_2 \xrightarrow{t} C_1' \oplus C_2'.$$

The entries of this matrix are the composites $(t \circ s)_{ik} = p_i' tse_k''$, where the e_1'' and e_2'' are the injections of the biproduct $C_1'' \oplus C_2''$. But by the biproduct identities (6), the identity morphism of the middle module $C_1 \oplus C_2$ is $1 = e_1 p_1 + e_2 p_2$, so we may write

$$(ts)_{ik} = p_i' t(e_1 p_1 + e_2 p_2) se_k'' = (p_i' te_1)(p_1 se_k'') + (p_i' te_2)(p_2 se_k'');$$

by the definition of the matrices for t and for s, this sum is

$$(ts)_{ik} = t_{i1} s_{1k} + t_{i2} s_{2k}.$$

This shows that the 2×2 matrix for the composite $t \circ s$ is the matrix product

$$\begin{bmatrix} t_{11} & t_{12} \\ t_{21} & t_{22} \end{bmatrix} \begin{bmatrix} s_{11} & s_{12} \\ s_{21} & s_{22} \end{bmatrix} = \begin{bmatrix} t_{11}s_{11} + t_{12}s_{21} & t_{11}s_{12} + t_{12}s_{22} \\ t_{21}s_{11} + t_{22}s_{21} & t_{21}s_{12} + t_{22}s_{22} \end{bmatrix}. \tag{7}$$

Therefore, the bijection from morphisms t to 2×2 matrices T is itself a morphism of multiplication.

Here is a simple example. Given two morphisms $t_i : C_i \to C_i'$, the "diagonal" matrix

$$\begin{bmatrix} t_1 & 0 \\ 0 & t_2 \end{bmatrix} \tag{8}$$

determines by our bijection a morphism $t:C_1 \oplus C_2 \to C_1' \oplus C_2'$; namely, the morphism t with $p_i'te_i = t_i$ for $i = 1, 2$ and $p_i'te_j = 0$ for $i \neq j$. This particular morphism t is usually written as $t = t_1 \oplus t_2$. From rule (7) for matrix multiplication it follows that

$$(t_1 \oplus t_2)(s_1 \oplus s_2) = t_1s_1 \oplus t_2s_2, \tag{9}$$

whenever the composites $t_i \circ s_i$ are defined. Also, if t_1 and t_2 are identity morphisms, so is $t_1 \oplus t_2$.

Much the same formalism will apply to n-fold biproducts $C_1 \oplus \cdots \oplus C_n$ of n right R-modules C_i. For such a biproduct there are injection and projection morphisms, as in

$$C_i \xrightarrow{e_i} C_1 \oplus \cdots \oplus C_n \xrightarrow{p_j} C_j, \qquad i, j = 1, \cdots, n, \tag{10}$$

which satisfy the identities

$$p_je_i = \delta_{ij}, \qquad e_1p_1 + e_2p_2 + \cdots + e_np_n = 1. \tag{11}$$

Each morphism $t:C_1 \oplus \cdots \oplus C_n \to C_1' \oplus \cdots \oplus C_m'$ from an n-fold biproduct to an m-fold biproduct determines an $m \times n$ matrix of morphisms $t_{ij}:C_j \to C_i'$ by $t_{ij} = p_i'te_j$ for $i \in \mathbf{m}, j \in \mathbf{n}$, and each such matrix of morphisms $t_{ij}:C_j \to C_i'$ determines t by $t = \Sigma_{ij}e_i't_{ij}p_j$. In particular, the free module R^n is just a biproduct $R \oplus \cdots \oplus R$ with n factors, so each $t:R^n \to R^m$ is described by an $m \times n$ matrix of morphisms $t_{ij}:R \to R$. This can be viewed as a matrix of scalars t_{ij}, for each morphism $s:R \to R$ of right R-modules is determined by the scalar $s(1)$. When $t = t_A$, as in §1, this matrix of scalars t_{ij} is, as the reader may readily show, just the matrix A.

A free R-module of finite type is a biproduct of copies of R. Hence, given two such free R-modules C and C' this biproduct description will yield for each linear map $t:C \to C'$ a corresponding matrix, relative to given free generators (bases) of C and C'. We will carry this out only in the case of vector spaces (modules over a field) where every module is automatically free. In §1 we used the category with objects the free R-modules R^n and morphisms $A:R^n \to R^m$ the $m \times n$ matrices. Here we will use an equivalent category of vector spaces.

Given a field F, we consider the concrete category (§IV.5) of all finite-dimensional right vector spaces V over F. The objects of this category are these vector spaces V, each with the set of its elements as underlying set. For two such spaces V and W the set hom (V, W) of morphisms $V \to W$ is the vector space Hom (V, W) of all linear transformations $t: V \to W$. In particular, when $V = F$, Hom (F, W) is isomorphic to W under the map sending $t: F \to W$ to the vector $t(1) \in W$. We will replace each $t: F \to W$ by this vector. This means in particular that the linear maps $F \to F$ are just the scalars $\lambda \in F$; in other words, λ *is* "scalar multiplication by λ." With this, the morphisms of the category are all the linear transformation $t: V \to W$, including:

$$\begin{gathered}\text{elements } w \in W \text{ as transformations } F \to W \text{ by } \kappa \mapsto w\kappa,\\ \text{scalars } \lambda \in F \text{ as transformations } F \to F \text{ by } \kappa \mapsto \lambda\kappa,\\ \text{elements } f \in V^* \text{ as transformations } V \to F.\end{gathered}$$

The composites of such elements then have the expected meanings; for example, the composite $w \circ \lambda: F \to W$ is the right multiple of the vector w by the scalar λ, while $\lambda \circ f$ is the left multiple of $f \in V^*$ by λ.

This approach will be used to visualize dual bases. If **b** is a basis for the n-dimensional vector space V while **x** is the dual basis for V^*, the elements of the lists **b** and **x** can be displayed as morphisms in a diagram,

$$F \overset{\mathbf{b}}{\underset{\cdots}{\rightrightarrows}} V \overset{\mathbf{x}}{\underset{\cdots}{\rightrightarrows}} F, \tag{12}$$

with n arrows on each side. Each composite $x_i \circ b_j$ is a scalar, while each composite $b_i \circ x_j$ is an endomorphism of V.

Theorem 5. *The basis* **b** *of V and its dual* **x** *of V^* satisfy*

$$x_i b_j = \delta_{ij}, \qquad b_1 x_1 + b_2 x_2 + \cdots + b_n x_n = 1, \tag{13}$$

where $i, j \in$ **n**, *δ_{ij} is the Kronecker delta, and* 1 *at the right is the identity $V \to V$.*

We call (**x**, **b**) a *dual basis pair* for V, to emphasize the symmetry.

Proof: The first equation of (13) is just the definition (VI.20) of the dual basis vectors x_i. The second equation, applied to a vector $v \in V$ with coordinates $\xi_1, \cdots, \xi_n$ relative to the basis **b**, is simply the statement that $v = b_1\xi_1 + \cdots + b_n\xi_n$; this is the definition of coordinates.

Conversely, it can be shown that these conditions (13) suffice to make **b** a basis and **x** its dual basis (Exercise 1).

Just as in (11) above, and as in Exercise V.6.4, these conditions (13) also state that the diagram (12) represents V as the biproduct $F \oplus \cdots \oplus F$ of n copies of the field F. This is just Theorem VI.2, that each basis **b** gives an isomorphism $F^n \cong V$.

EXERCISES

1. If lists **b** and **x** of n vectors b_j and n forms x_i of a vector space V satisfy the equations (13), prove that **b** is a basis of V and **x** its dual basis.

2. If **b** is a list of three vectors and **x** a list of three linear forms of V, show that the conditions

$$x_1 b_1 = 1,\ x_2 b_2 = 1,\ x_3 b_3 = 1,\ x_1 b_2 = 0, \quad \text{and} \quad b_1 x_1 + b_2 x_2 + b_3 x_3 = 1_V$$

suffice to make (**x**, **b**) a dual basis pair for V.

3. In Exercise 2, show that the hypothesis $x_1 b_2 = 0$ can be omitted if the coefficient field satisfies $1 + 1 \neq 0$.

3. The Matrix of a Map

Let $t: V \to V'$ be a linear transformation from an n-dimensional vector space V to an m-dimensional vector space V' (over the same field F). Given bases **b** for V and **b**′ for V', the transformation t will be described by a matrix. Since V is free on $b_1, \cdots, b_n$, t is determined by the list $tb_1, \cdots, tb_n$ of n vectors of V'; each of these vectors tb_j is in its turn determined by m coordinates $x_i'(tb_j)$. All told, t is determined by mn scalars $x_i' tb_j$ which are the entries of an $m \times n$ matrix, to be denoted in this section by Mt or by $M(t): \mathbf{m} \times \mathbf{n} \to F$, and to be called the matrix of t (rel **b**, **b**′).

This may be restated in terms of biproducts, as follows. Pick a dual basis pair (**x**, **b**) for V; as remarked in §2, this pair presents V as a biproduct of n copies of the field F. Similarly, a dual basis pair (**x**′, **b**′) for V' presents V' as a biproduct of m copies of F. Thus t, as a linear transformation from a biproduct to a biproduct, will be determined by mn linear maps $F \to F$ (that is, by mn scalars) as displayed in the schematic figure below; this figure is just a diagram of morphisms from the category of finite-dimensional vector spaces, as described in §2:

$$\begin{array}{ccc} V & \xrightarrow{\ t\ } & V' \\ \mathbf{b} \Uparrow & & \Downarrow \mathbf{x}' \\ F & \Rrightarrow & F. \end{array} \tag{14}$$

Explicitly, the *matrix* of $t: V \to V'$ (rel **b**, **b**′) is the function $Mt: \mathbf{m} \times \mathbf{n} \to F$ defined for arguments i and j by

$$(Mt)_{ij} = x_i' tb_j, \qquad i = 1, \cdots, m;\ j = 1, \cdots, n. \tag{15}$$

Each entry of this matrix is thus a composite of linear maps

$$F \xrightarrow{b_j} V \xrightarrow{t} V' \xrightarrow{x_i'} F, \tag{16}$$

hence a transformation $F \to F$, to be identified with the corresponding scalar via the isomorphism $\mathrm{Hom}_F(F, F) \cong F$.

Theorem 6. *Given bases* $\mathbf{b}$ *and* $\mathbf{b}'$ *of n and m elements, respectively, for the vector spaces V and V′ over F, the function assigning to each linear map* $t: V \to V'$ *its matrix Mt* (rel $\mathbf{b}, \mathbf{b}'$) *is a bijection*

$$M = M_{\mathbf{b}'}^{\mathbf{b}} : \hom(V, V') \cong F^{m \times n} \tag{17}$$

from the set of all linear maps $t: V \to V'$ *to the set* $F^{m\times n}$ *of all* $m \times n$ *matrices over F. Moreover, the bijection M is itself a linear transformation (between the vector spaces* Hom (V, V') *and* $F^{m \times n}$).

Proof: That M is a bijection and linear is a consequence of the properties already established for maps between biproducts. A direct proof of this fact may be given by constructing the following inverse to the function M of (17). Given an $m \times n$ matrix A, we regard each (scalar) entry A_{hk} as a transformation $A_{hk}: F \to F$, form the composite transformations

$$V \xrightarrow{x_k} F \xrightarrow{A_{hk}} F \xrightarrow{b'_h} V', \qquad h \in \mathbf{m}, \quad k \in \mathbf{n},$$

and add them all to get a single linear transformation

$$M^{-1}A = \sum_{h=1}^{m} \sum_{k=1}^{n} b'_h A_{hk} x_k : V \to V', \tag{18}$$

called the "map of the matrix". Now the matrix of this transformation has the entries

$$[M(M^{-1}A)]_{ij} = x'_i\Big(\sum_h \sum_k b'_h A_{hk} x_k\Big) b_j = A_{ij},$$

where the last equation holds by the biproduct identities $x'_i b'_h = \delta_{ih}$, $x_k b_j = \delta_{kj}$. Hence, $M(M^{-1}A) = A$. On the other hand, by the definitions and the distributive laws,

$$\begin{aligned} M^{-1}(Mt) &= \sum_h \sum_k b'_h (x'_h t b_k) x_k \\ &= \Big(\sum_{h=1}^{m} b'_h x'_h\Big) t \Big(\sum_{k=1}^{n} b_k x_k\Big) = 1t1 = t, \end{aligned}$$

this by the biproduct identities $\sum b_k x_k = 1$. Hence $M^{-1}(Mt) = t$. Therefore, the function M^{-1} is the desired two-sided inverse for M, and M is a bijection. Finally, the formula (15) shows that M takes linear combinations to linear combinations, so is an isomorphism of vector spaces.

This construction of the matrix of a map t will be used so much that it is helpful to reformulate it in several ways. First, the matrix Mt of a map may be displayed as

$$Mt = \begin{bmatrix} x'_1tb_1 & \cdots & x'_1tb_j & \cdots & x'_1tb_n \\ \vdots & & \vdots & & \vdots \\ x'_itb_1 & \cdots & x'_itb_j & \cdots & x'_itb_n \\ \vdots & & \vdots & & \vdots \\ x'_mtb_1 & \cdots & x'_mtb_j & \cdots & x'_mtb_n \end{bmatrix}.$$

In words, the matrix Mt, rel $(\mathbf{b}, \mathbf{b}')$, of a map $t: V \to V'$ has

(i, j)-entry the ith $\mathbf{b}'$-coordinate of t(jth basis vector),
jth column the column of $\mathbf{b}'$-coordinates of tb_j,
ith row the row of $\mathbf{x}$-coordinates of x'_it.

The first two statements simply rephrase the definition (15) of Mt. So does the third, for each $x'_it: V \to F$ is a linear form on V, hence an element of V^*; relative to the basis $\mathbf{x}$ of V^*, the coordinates of any element $f: V \to F$ of V^* are the scalars $fb_1, \cdots, fb_n$.

As a special case, one may speak of the matrix of a vector v (rel $\mathbf{b}$), because a vector is just a linear map $v: F \to V$. Hence, choosing the multiplicative unit $1 \in F$ as the basis for the vector space F, the matrix $M(v)$ has entries x_iv; in other words, *the matrix of a vector* v (rel $\mathbf{b}$) *is the column of* $\mathbf{b}$-*coordinates of* v.

Here are two other ways to describe the map $M^{-1}A$ of a matrix A.

Corollary 1. *The map* (rel $\mathbf{b}, \mathbf{b}'$) *of an* $m \times n$ *matrix* A *is that linear transformation* $t: V \to V'$ *determined by giving its effect on the basis vectors* $b_1, \cdots, b_n$ *of* V *as*

$$t(b_j) = \sum_{i=1}^{m} b'_iA_{ij}, \qquad j = 1, \cdots, n. \tag{19}$$

This map t *is also determined by giving its composite with each coordinate form* $x'_i: V' \to F$ *of the dual basis of* V' *as*

$$x'_i \circ t = \sum_{j=1}^{n} A_{ij}x_j, \qquad i = 1, \cdots, m. \tag{20}$$

In classical books, linear transformations t were usually written with "variables" x'_i and x_j in the form (20), but not explicitly mentioning t.

Proof: The definition (18) of $t = M^{-1}A$ and the identities $x_kb_j = \delta_{kj}$ yield (19), while the definition and the identities $x'_ib'_h = \delta_{ih}$ yield (20).

Equations (19) between vectors of V' describe t in terms of bases; Equations (20) between linear forms on V describe t in terms of coordinates. Either system gives a rule for finding the matrix of a given transformation t. For example: *Write the coordinates* $(x'_1, \cdots, x'_m)$ *of each transformed vector* tv *as linear expressions in the coordinates* $x_1, \cdots, x_n$ *of* v. *In these expressions* $x'_i = \sum A_{ij}x_j$, *the coefficients* A_{ij} *are the entries of the matrix of* t.

If we replace the vector v by the column of its coordinates $x_j v$, and similarly for tv, the coordinate Equations (20) become

$$\begin{bmatrix} x'_1(tv) \\ \vdots \\ x'_m(tv) \end{bmatrix} = \begin{bmatrix} A_{11} & \cdots & A_{1n} \\ \vdots & & \vdots \\ A_{m1} & \cdots & A_{mn} \end{bmatrix} \begin{bmatrix} x_1 v \\ \vdots \\ x_n v \end{bmatrix} \tag{21}$$

just as in the matrix description of §1. This parallel can be stated more explicitly if we recall that each basis $\mathbf{b}$ of V gives as in (VI.6) an isomorphism $L_{\mathbf{b}}:F^n \cong V$, where $(L_{\mathbf{b}})^{-1}v$ is defined to be the column of $\mathbf{b}$-coordinates of v.

Corollary 2. *If V and V' are vector spaces of dimensions n and m with bases* $\mathbf{b}$ *and* $\mathbf{b}'$, *respectively, then the map* $M^{-1}A$ (rel $\mathbf{b}$, $\mathbf{b}'$) *corresponding to an* $m \times n$ *matrix* A *may be expressed in terms of the standard map* $t_A:F^n \to F^m$ *and the isomorphisms* $L_{\mathbf{b}}$ *as*

$$M^{-1}A = L_{\mathbf{b}'}t_A(L_{\mathbf{b}})^{-1} : V \to V'. \tag{22}$$

Proof: The conclusion states the commutativity of the diagram

$$\begin{array}{ccc} V & \xrightarrow{t = M^{-1}A} & V' \\ {\scriptstyle (L_{\mathbf{b}})^{-1}}\downarrow & & \downarrow{\scriptstyle (L_{\mathbf{b}'})^{-1}} \\ F^n & \xrightarrow[t_A]{} & F^m. \end{array}$$

This commutativity is just the Equation (21), for $(L_{\mathbf{b}'})^{-1}tv$ is the column of $\mathbf{b}'$-coordinates of tv, as stated on the left of Equation (21), while $t_A(L_{\mathbf{b}})^{-1}v$ is the matrix product of A by the column of $\mathbf{b}$-coordinates of v, as on the right of (21). Given bases $\mathbf{b}$ and $\mathbf{b}'$, this equation (22) describes completely the bijection between $m \times n$ matrices A and linear transformations $t: V \to V'$. It is often used (implicitly) as a starting point for this subject.

The transpose of a matrix is useful for describing dual morphisms. Recall by (V.41) that each linear transformation $t: V \to V'$ yields a dual transformation $t^*: V'^* \to V^*$, which assigns to each linear form $f: V' \to F$ on V' the linear form $t^*(f) = f \circ t: V \to F$ on V. Now the description (20) of t in terms of its matrix A (rel $\mathbf{b}$, $\mathbf{b}'$) becomes

$$t^*(x'_i) = \sum_{j=1}^{n} A_{ij}x_j, \qquad i = 1, \cdots, m. \tag{23}$$

These equations describe the dual transformation $t^*: V'^* \to V^*$ in terms of its effect upon bases $\mathbf{x}$ and $\mathbf{x}'$ of these dual spaces—and with the same matrix A. However, in this equation the scalars A_{ij} appear to the left of the vectors $x_j \in V^*$—as they should, since V^* is a left vector space. Because the field F is commutative, this left vector space may be changed as in §V.7 to a right vector space by defining each right scalar multiple $f\kappa$ to be κf, for $f \in V^*$. This process changes Equation (23) to

$$t^*(x_i') = \sum_{j=1}^{n} x_j A_{ij}, \qquad i = 1, \cdots, m.$$

This equation specifies the effect of $t^*: V'^* \to V^*$ upon each basis element x_i' of the right vector space V'^*. Here we can read off the matrix of t^*, rel $\mathbf{x}$, $\mathbf{x}'$, by rule (19) above. Since $x_j A_{ij} = x_j (A^T)_{ji}$, it is the transposed matrix A^T; in other words,

$$M(t^*) = [M(t)]^T. \tag{24}$$

This result (unlike the others in this chapter) does not apply to modules over a division ring.

EXERCISES

1. Given Equations (19) for $t: V \to V'$ in terms of bases, derive directly from them Equations (20) for t in terms of coordinates.
2. Given Equations (20) for $t: V \to V'$ in terms of coordinates, derive directly from them Equations (19) for t in terms of bases.
3. Relative to the bases of unit vectors, show for t_A as in (1) that the matrix of $t_A: F^n \to F^m$ is A.
4. If W and W' are left vector spaces of dimensions n and m and with bases $\mathbf{c}$ and $\mathbf{c}'$, respectively, show that each $n \times m$ matrix B defines a linear transformation $s: W \to W'$ by $s(c_i) = \Sigma_j B_{ij} c_j'$, for $i = 1, \cdots, n$.
5. Describe the matrix of a linear form $f: V \to F$ relative to a basis $\mathbf{b}$ of V.

4. The Matrix of a Composite

Next we establish the basic fact that taking the matrix of a map is a morphism of multiplication, just as in Theorem 3.

THEOREM 7. *The matrix of a composite map $t \circ s$ is the product of the matrices of the factors:*

$$M(ts) = (Mt)(Ms).$$

In more detail, given three finite-dimensional vector spaces with

$$\begin{array}{llll} \textit{Dimensions:} & q & n & m, \\ \textit{Vector spaces:} & V'' \xrightarrow{s} & V \xrightarrow{t} & V', \\ \textit{Bases:} & \mathbf{b}'' & \mathbf{b} & \mathbf{b}', \end{array}$$

the matrices of t and s relative to the bases displayed satisfy

$$M_{\mathbf{b}'}^{\mathbf{b}''}(ts) = M_{\mathbf{b}'}^{\mathbf{b}}(t)M_{\mathbf{b}}^{\mathbf{b}''}(s). \tag{25}$$

The proof is immediate. Write A for $M(t)$ and B for $M(s)$. By the definition of the matrix product AB and the definition of A and B via dual basis pairs,

$$(AB)_{ik} = \sum_{j=1}^{n} A_{ij}B_{jk} = \sum_{j=1}^{n} (x_i'tb_j)(x_jsb_k'') = x_i't\left(\sum_{j=1}^{n} b_jx_j\right)sb_k'' = x_i'(ts)b_k'';$$

the result is the (ik)-entry in the matrix of $t \circ s$.

An especially important case is that of endomorphisms $t: V \to V$ of a single vector space V. The set End $(V) =$ Hom (V, V) of all such endomorphisms is both a vector space under addition and scalar multiples and a ring, with unit the identity $V \to V$, under the binary operations of addition and composition. Now pick one basis, a basis $\mathbf{b}$ of V; for an endomorphism $V \to V$ we *always use the same basis in domain and codomain*. If V is n-dimensional, each t yields an $n \times n$ matrix Mt, rel $\mathbf{b}$. Now the set $F^{n \times n}$ of all $n \times n$ matrices over F is, like End V, both a vector space and a ring. The results above and Theorem 6 prove:

THEOREM 8. *Relative to any basis* $\mathbf{b}$ *of an n-dimensional vector space V, the assignment* $t \mapsto Mt$ *on endomorphisms to matrices is a bijection* $M{:}\mathrm{End}\,(V) \cong F^{n \times n}$ *which is an isomorphism both of vector spaces and of rings.*

In particular, the identity endomorphism is represented for any basis by the $n \times n$ identity matrix I_n, relative to any basis.

Next consider transformations $t: V \to V'$ when both spaces $V = V_1 \oplus V_2$ and $V' = V_1' \oplus V_2'$ are biproducts, with V_i of dimension n_i and V_i' of dimension m_i for $i = 1, 2$. With the usual diagrams for the injections e_i and the projections p_i of each biproduct, the results of §2 show that each linear transformation $t: V_1 \oplus V_2 \to V_1' \oplus V_2'$ is determined by a 2×2 matrix

$$t \mapsto \begin{bmatrix} t_{11} & t_{12} \\ t_{21} & t_{22} \end{bmatrix} = \begin{bmatrix} p_1'te_1 & p_1'te_2 \\ p_2'te_1 & p_2'te_2 \end{bmatrix}$$

of linear transformations $t_{ij}: V_j \to V_i'$. Now pick bases $\mathbf{b}^{(1)}$ and $\mathbf{b}^{(2)}$ in V_1 and V_2, $\mathbf{b}'^{(1)}$ and $\mathbf{b}'^{(2)}$ in V_1' and V_2', so that each t_{ij} is represented (rel $\mathbf{b}^{(j)}$, $\mathbf{b}'^{(i)}$) by the $m_i \times n_j$ matrix $M(t_{ij})$ of scalars. The two lists $e_1\mathbf{b}^{(1)}$ and $e_2\mathbf{b}^{(2)}$ combine as in Corollary VI.9.1 to give a single list, written $e_1\mathbf{b}^{(1)} \vee e_2\mathbf{b}^{(2)}$, which is a basis for $V = V_1 \oplus V_2$; there is an analogous basis $e_1'\mathbf{b}'^{(1)} \vee e_2'\mathbf{b}'^{(2)}$ for V'. The matrix of t relative to these combined bases is just the matrix formed by piecing together the four "blocks" Mt_{ij} as in

$$M(t) = \begin{bmatrix} Mt_{11} & Mt_{12} \\ Mt_{21} & Mt_{22} \end{bmatrix} = \begin{array}{c} \leftarrow n_1 \rightarrow\leftarrow n_2 \rightarrow \\ \begin{array}{|c|c|} \hline \quad & \quad \\ \hline \quad & \quad \\ \hline \end{array} \end{array} \begin{array}{c} \uparrow m_1 \downarrow \\ \uparrow m_2 \downarrow \end{array}$$

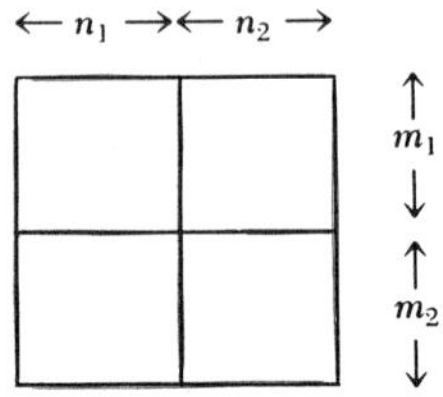

We say: The square matrix $M(t)$ has been split into (four) *blocks*.

For a composite of two such transformations

$$V_1'' \oplus V_2'' \xrightarrow{s} V_1 \oplus V_2 \xrightarrow{t} V_1' \oplus V_2',$$

(7) showed that the rule for multiplying matrices of transformations s_{ij} and t_{ij} is identical with the rule for multiplying matrices of scalars:

$$\begin{bmatrix} t_{11} & t_{12} \\ t_{21} & t_{22} \end{bmatrix} \begin{bmatrix} s_{11} & s_{12} \\ s_{21} & s_{22} \end{bmatrix} = \begin{bmatrix} t_{11}s_{11} + t_{12}s_{21} & t_{11}s_{12} + t_{12}s_{22} \\ t_{21}s_{11} + t_{22}s_{21} & t_{21}s_{12} + t_{22}s_{22} \end{bmatrix}.$$

Since M is a morphism of multiplication and addition, the same formula holds when every t_{ij} is replaced by Mt_{ij} and every s_{ij} by Ms_{ij}. Since every matrix of scalars is the matrix of a map, the formula still holds when every t_{ij} or s_{ij} is replaced by *any* block matrix of the appropriate size. The result can therefore be formulated directly as a rule for multiplying matrices in blocks, as follows. If an $m \times n$ matrix A and an $n \times q$ matrix B are each split into four blocks $A^{(ij)}$ and $B^{(ij)}$, with the *columns* of A split by $n = n_1 + n_2$ and the *rows* of B by the same $n = n_1 + n_2$, then the matrix product AB is

$$\begin{bmatrix} A^{(11)} & A^{(12)} \\ A^{(21)} & A^{(22)} \end{bmatrix} \begin{bmatrix} B^{(11)} & B^{(12)} \\ B^{(21)} & B^{(22)} \end{bmatrix} = \begin{bmatrix} A^{(11)}B^{(11)} + A^{(12)}B^{(21)} & A^{(11)}B^{(12)} + A^{(12)}B^{(22)} \\ A^{(21)}B^{(11)} + A^{(22)}B^{(21)} & A^{(21)}B^{(12)} + A^{(22)}B^{(22)} \end{bmatrix}, \tag{26}$$

formed just as if the blocks were scalars. Moreover, the dimensions fit:

$$\left[\begin{array}{c|c} \xrightarrow{n_1} & \\ \hline & \end{array}\right]\left[\begin{array}{c|c} \downarrow n_1 & \\ \hline & \end{array}\right].$$

The same *block multiplication rule* applies to a subdivision of the rows (or the columns) into more than two parts. An $m \times n$ matrix A may be cut into blocks by grouping the rows into k consecutive sets, corresponding to a representation of the codomain of $t = M^{-1}A$ as a biproduct of k vector spaces, or else by grouping the columns, to correspond to a biproduct representation of the domain of t, or by grouping both rows and columns. The matrix product AB of two matrices may then be computed using row by column multiplication of blocks, provided only that the n columns of A are grouped in the same pattern as are the n rows of B. (This condition reflects the use of the same biproduct decompositions in that vector space which is both the codomain of $M^{-1}B$ and the domain of $M^{-1}A$.)

The *direct sum* of two matrices A and B is the matrix of blocks (a "block diagonal matrix")

$$A \oplus B = \begin{bmatrix} A & 0 \\ 0 & B \end{bmatrix}, \tag{27}$$

where the zeros denote matrices of suitable size, with entries all zeros. The corresponding linear transformation $t_{A \oplus B}$ may be obtained from t_A and t_B as the direct sum $t_A \oplus t_B$ in the sense of (9).

Here is some convenient terminology: A square matrix A is called:

Diagonal:	when $A_{ij} = 0$	for all $i \neq j$,
Scalar:	when $A_{ij} = \kappa\delta_{ij}$	for all i, j and some κ,
Triangular:	when $A_{ij} = 0$	for all $j < i$,
Strictly triangular:	when $A_{ij} = 0$	for all $j \leqslant i$;

thus a matrix is strictly triangular if all the entries on or below the principal diagonal are zero. In the 3×3 case these types of matrices may be displayed as follows (with *'s for possibly non-zero entries):

Diagonal	*Scalar*	*Triangular*	*Strictly Triangular*
$\begin{bmatrix} * & 0 & 0 \\ 0 & * & 0 \\ 0 & 0 & * \end{bmatrix}$,	$\begin{bmatrix} \kappa & 0 & 0 \\ 0 & \kappa & 0 \\ 0 & 0 & \kappa \end{bmatrix}$,	$\begin{bmatrix} * & * & * \\ 0 & * & * \\ 0 & 0 & * \end{bmatrix}$,	$\begin{bmatrix} 0 & * & * \\ 0 & 0 & * \\ 0 & 0 & 0 \end{bmatrix}$.

EXERCISES

1. Use block multiplication to find AB, BA, AC, AD, and BD if

$$A = \begin{bmatrix} 2 & -1 & 0 & 0 \\ -4 & 2 & 0 & 0 \\ 0 & 0 & 3 & 1 \\ 0 & 0 & 0 & 2 \end{bmatrix}, \quad B = \begin{bmatrix} 1 & 0 & 0 & 0 \\ 1 & 1 & 0 & 0 \\ 1 & 3 & 1 & 0 \\ 2 & 4 & 0 & 1 \end{bmatrix}, \quad C = \begin{bmatrix} 2 \\ 1 \\ 3 \\ 4 \end{bmatrix}, \quad D = \begin{bmatrix} 0 & 1 \\ 1 & 0 \\ 3 & 0 \\ 0 & 3 \end{bmatrix}.$$

2. If D is a diagonal matrix and all terms on the diagonal are distinct, what square matrices commute with D?

3. A square matrix A is called *nilpotent* if some power of A is 0. Prove that any strictly triangular matrix is nilpotent.

4. If $t: V \to V$ is an endomorphism, and $\mathbf{b}$ is any basis of V, show that $M(t)$ (rel $\mathbf{b}$) is a scalar matrix if and only if t is the morphism of "scalar multiple by some κ".

5. Show that the $n \times n$ scalar matrices over F form a ring isomorphic to F, and that the $n \times n$ diagonal matrices form a ring isomorphic to the n-fold product (§III.2) of the ring F with itself.

6. Let $\mathbf{b}$ be a basis of V, S_i the subspace spanned by b_i for $i \in \mathbf{n}$. For $t: V \to V$ linear, prove that $M(t)$ is diagonal if and only if $t_*(S_i) \subset S_i$ for all $i \in \mathbf{n}$.

7. Find a description like that of Exercise 6 for triangular matrices.

8. (a) Show that all $n \times n$ triangular matrices over F form a ring (under matrix addition and multiplication, with unit the identity matrix) and that all $n \times n$ strictly triangular matrices constitute a (two-sided) ideal in this ring.

(b) Identify the quotient ring (triangular matrices modulo strictly triangular matrices).

5. Ranks of Matrices

The rank of a linear transformation t has been defined in §VI.3 as the dimension of its image, while the nullity of t is the dimension of Ker t. Now define the *rank* and the *nullity* of an $m \times n$ matrix A over a field F to be the rank and the nullity of the linear transformation $t_A : F^n \to F^m$ of §1. This rank and this nullity are those of *any* linear transformation $t: V \to V'$ with matrix A relative to some pair of bases $\mathbf{b}$ and $\mathbf{b}'$, for, by Corollary 6.2, $t = L_{\mathbf{b}'} t_A L_{\mathbf{b}}^{-1}$, and in this equation both $L_{\mathbf{b}'}$ and $L_{\mathbf{b}}$ are isomorphisms, and so preserve dimensions of spaces. In brief, *the rank of t is the same as the rank of every one of the matrices representing t.*

THEOREM 9. *A linear transformation $t: V \to V'$ between finite-dimensional vector spaces V and V' has the same rank as its dual $t^*: V'^* \to V^*$.*

Proof: By definition of t^*, each $t^*f = ft$ for $f: V \to F$. Thus $f \in \text{Ker } t^*$ if and only if $ft = 0$; that is, if and only if f annihilates the image of t. Thus

$$\text{nullity } t^* = \dim(\text{Ker } t^*) = \dim(\text{Annih Im } t).$$

Now rank t by definition is the dimension of Im t, so by the formula (VI.26) for the dimension of an annihilator

$$\text{nullity } t^* = \dim V' - \dim(\text{Im } t) = \dim V' - \text{rank } t.$$

However, by (VI.10), rank plus nullity (of t^*) is the dimension of the domain.

This is

$$\text{rank } t^* = \dim V' - \text{nullity } t^* = \text{rank } t.$$

By its definition, the rank of a matrix A is the maximum number of linearly independent columns of A (that is, the "column rank"), and the nullity of A is the maximum number of linearly independent columns ξ which satisfy the m simultaneous linear equations $\sum_j A_{ij}\xi_j = 0$. As for maps in (VI.10),

$$\text{rank } (A) + \text{nullity } (A) = n \qquad (A \text{ is } m \times n). \tag{28}$$

The rank of the transposed matrix A^T is the maximum number of linearly independent rows of A, so is sometimes called the "row rank" of A.

Theorem 10. *For any matrix* A, rank (A) = rank (A^T).

Proof: This result claims that the maximum number of linearly independent columns of A is the same as the maximum number of linearly independent rows of A. This can be proved by long direct manipulations (Exercise 10 of §7), but the reason will be clearer if we translate to a "geometric" form. The rank of A is the rank of t_A, regarded as a map between *right* vector spaces. We have just seen in (24) that the transposed matrix A^T represents the dual map $(t_A)^*$ when viewed as a map between right vector spaces. The rank of A^T is thus the rank of $(t_A)^*$. Our theorem on matrices is thus a corollary of the previous theorem on transformations.

EXERCISES

1. Prove that rank $(A + B) \leqslant$ rank A + rank B, for A and B both $m \times n$.

2. Show that the rank of a product of (rectangular) matrices never exceeds the rank of either factor.

3. Prove that rank $(A \oplus B)$ = rank A + rank B for any two matrices A and B.

4. Establish a corresponding rule for the nullity of $A \oplus B$.

5. If A and B are $n \times n$ matrices of ranks r and s, prove that the rank of AB is never less than $(r + s) - n$. (*Hint:* Use transformations.)

***6.** (a) Prove *Sylvester's law of nullity*: The nullity of a product AB never exceeds the sum of the nullities of the factors and is never less than the nullity of either factor if B is square.

(b) Give examples to show that both these limits can be attained by the nullity of AB.

7. When is nullity A = nullity A^T?

8. Prove Theorem 9 by choosing bases suited to t as in Corollary 1 of Theorem VI.8.

6. Invertible Matrices

We now ask: When does a square matrix A over a field F have a matrix over F as inverse?

Theorem 11. *The following properties of a matrix A in the ring $F^{n\times n}$ of all $n \times n$ matrices over F are equivalent:*

(*i*) *A has a left inverse in $F^{n\times n}$.*
(*ii*) *A has a right inverse in $F^{n\times n}$.*
(*iii*) *Nullity A = 0.*
(*iv*) *Rank A = n.*
(*v*) *A has a two-sided inverse A^{-1} in $F^{n\times n}$.*

When these conditions hold, any two matrix inverses of A(left, right, or two-sided) are equal.

For the proof we simply translate these properties into properties of the corresponding linear transformations, say the transformation $t = t_A : F^n \to F^n$ of §1. Property (i) becomes "t has a linear left inverse"; this implies that t is a monomorphism, which is (iii). But by Corollary VI.8.2, a monomorphism between spaces of the same finite dimensions is an isomorphism, hence has a two-sided linear inverse, and therefore *a fortiori* a right inverse. This has proved (i) ⇒ (iii) ⇒ (v) ⇒ (ii). Similarly, a linear right inverse for t_A makes t_A an epimorphism, hence an isomorphism, so (ii) ⇒ (iv) ⇒ (v) ⇒ (i). Since A^{-1} in (v) is the matrix of the two-sided inverse of t, it must be unique, Q.E.D.

This theorem, and its proof, holds also for square matrices over a division ring. Over an arbitrary ring R the theorem does not hold; for example, a monomorphism $\mathbf{Z}^n \to \mathbf{Z}^n$ of $\mathbf{Z}$-modules need not be an isomorphism. In other words condition (iii) on a matrix of integers does not imply condition (v) when F is replaced by $\mathbf{Z}$.

A square matrix A over a field F with the properties listed in this theorem is said to be *invertible* over F (or, "nonsingular"); its inverse A^{-1} satisfies

$$A^{-1}A = I = AA^{-1}. \tag{29}$$

This implies that A^{-1} is also invertible, with $(A^{-1})^{-1} = A$. The transpose A^T of an invertible matrix A is invertible, with inverse

$$(A^T)^{-1} = (A^{-1})^T; \tag{30}$$

for a proof, just apply to (29) the rule $(AB)^T = B^TA^T$ for the transpose of a product. Also, the product of two $n \times n$ invertible matrices A and B is invertible, with inverse given by

$$(AB)^{-1} = B^{-1}A^{-1}. \tag{31}$$

COROLLARY. *Any square matrix with linearly independent columns (or, with linearly independent rows) is invertible.*

Proof: To say that the columns of the $n \times n$ matrix A are independent is to say that rank $A = n$, and hence that A is invertible, by part (iv) of the theorem. To say that the rows of A are independent is to say that rank $(A^T) = n$, and hence by Theorem 10 that rank $A = n$ and that A is invertible.

For each natural number $n > 0$ the set of all invertible $n \times n$ matrices over F is a group, called the *general linear group* $GL(n, F)$. This group is isomorphic (via M) to the group of automorphisms of any n-dimensional vector space V over F. It contains many interesting subgroups. The symmetric group S_n is an example. For each bijection $\sigma: \mathbf{n} \to \mathbf{n}$ we construct the linear map $t_\sigma: F^n \to F^n$ which makes the corresponding permutation of the unit vectors, as $t_\sigma(\varepsilon_i) = \varepsilon_{\sigma i}$ for each $i \in \mathbf{n}$. For two permutations $\rho, \sigma \in S_n$, $t_\rho(t_\sigma \varepsilon_i) = t_\rho(\varepsilon_{\sigma i}) = \varepsilon_{\rho\sigma i} = t_{\rho\sigma}(\varepsilon_i)$. Therefore, $t_\rho \circ t_\sigma = t_{\rho\sigma}$, so $\sigma \mapsto t_\sigma$ is a morphism $S_n \to \text{Aut}(F^n)$ of groups. The matrix P_σ of t_σ (rel ε) is called a *permutation matrix*. Since its ith column consists of the ε-coordinates of $t_\sigma(\varepsilon_i)$, each column and each row of $P = P_\sigma$ has exactly one nonzero entry 1. Any matrix with this last property is a permutation matrix for some permutation σ, and the $n \times n$ permutation matrices form a multiplicative group isomorphic to the symmetric group S_n.

To find the inverse of a matrix A explicitly, one may use the elementary operations of §VI.5 on the columns of A. First, define an *elementary matrix* E to be any (square) matrix obtained by applying one of the elementary operations to the columns of the identity matrix. For example, the three typical operations "interchange the first two columns", "multiply the first column by $\kappa \neq 0$", and "add λ times the second column to the first" when applied to the identity matrix I_2 yield the three 2×2 elementary matrices

$$\begin{bmatrix} 0 & 1 \\ 1 & 0 \end{bmatrix}, \quad \begin{bmatrix} \kappa & 0 \\ 0 & 1 \end{bmatrix}, \quad \begin{bmatrix} 1 & 0 \\ \lambda & 1 \end{bmatrix}. \tag{32}$$

Now observe the effect of multiplying any 2×2 matrix A on the right by one of these elementary matrices. In the first case,

$$\begin{bmatrix} a_1 & b_1 \\ a_2 & b_2 \end{bmatrix}\begin{bmatrix} 0 & 1 \\ 1 & 0 \end{bmatrix} = \begin{bmatrix} a_1 0 + b_1 1 & a_1 1 + b_1 0 \\ a_2 0 + b_2 1 & a_2 1 + b_2 0 \end{bmatrix} = \begin{bmatrix} b_1 & a_1 \\ b_2 & a_2 \end{bmatrix};$$

the columns are interchanged. In the second case,

$$\begin{bmatrix} a_1 & b_1 \\ a_2 & b_2 \end{bmatrix}\begin{bmatrix} \kappa & 0 \\ 0 & 1 \end{bmatrix} = \begin{bmatrix} a_1\kappa + b_1 0 & a_1 0 + b_1 1 \\ a_2\kappa + b_2 0 & a_2 0 + b_2 1 \end{bmatrix} = \begin{bmatrix} a_1\kappa & b_1 \\ a_2\kappa & b_2 \end{bmatrix};$$

the first column is multiplied by κ. In the third case,

$$\begin{bmatrix} a_1 & b_1 \\ a_2 & b_2 \end{bmatrix}\begin{bmatrix} 1 & 0 \\ \lambda & 1 \end{bmatrix} = \begin{bmatrix} a_1 + b_1\lambda & b_1 \\ a_2 + b_2\lambda & b_2 \end{bmatrix};$$

λ times the second column is added to the first. In each case the elementary operation on the matrix A amounts to multiplying A on the right by the corresponding elementary matrix. This observation suggests

Proposition 12. *If an elementary operation (on lists of n vectors and hence on matrices with n columns) carries the $n \times n$ identity matrix I to the elementary matrix E, then it carries any $m \times n$ matrix A to the matrix product AE.*

Proof: The row by column definition of the matrix product AB states that the jth column of the product matrix AB is the matrix A times the jth column of the matrix B. Hence, to operate on the columns of a product AB it suffices to carry out the same operation on the columns of the second matrix B, then multiply by A. In particular, $A = AI$, so it suffices to carry out the operations on the columns of the identity matrix I, and this is the statement of the Proposition.

Proposition 13. *A (square) matrix over a field is invertible if and only if it can be written as a product of elementary matrices.*

Proof: Since an elementary operation does not change the rank of a matrix and since each $n \times n$ elementary matrix E is obtained by such an operation upon the identity matrix of rank n, each such E has rank n, and hence is invertible. Any product of elementary matrices is therefore invertible.

Conversely, let an $n \times n$ matrix P be invertible. Then P has rank n, so its columns are independent and thus a basis for F^n. By Theorem VI.19, there is a sequence of elementary operations taking this basis to any other basis, in particular to the basis given by the unit vectors (the columns of the identity matrix I). In matrix language this states that there is a sequence $\phi_1, \phi_2, \cdots, \phi_k$ of elementary column operations reducing the matrix P to the identity. If $E_1, E_2, \cdots, E_k$ are the corresponding elementary matrices, the previous proposition states that $PE_1E_2 \cdots E_k = I$, hence that

$$E_1 E_2 \cdots E_k = P^{-1}. \tag{33}$$

As any invertible matrix Q can be written as the inverse P^{-1} of another one, this shows that any invertible matrix is a product of elementary matrices.

This Equation (33) expresses the inverse P^{-1} as a product of elementary matrices and hence as the result of a sequence $IE_1 \cdots E_k$ of elementary column operations upon I. This may be stated as follows:

Corollary 1. *If a square matrix P is reducible to the identity matrix by a sequence of elementary column operations, then P is invertible and the same sequence of column operations applied to the identity matrix I will yield P^{-1}.*

Corollary 2. *Two $m \times n$ matrices A and B are column equivalent if and only if there is an invertible $n \times n$ matrix P with $AP = B$.*

Proof: If $AP = B$, write P as a product $P = E_1 E_2 \cdots E_k$ of elementary matrices. By the previous proposition, the corresponding operations then carry A to B, so that A and B are indeed equivalent under column operations.

Conversely, to say that A is column equivalent to B means that there is a sequence of elementary (column) operations carrying A to B and hence a matrix $P = E_1 E_2 \cdots E_k$, surely invertible, with $AP = B$.

Proposition 13 also contains the following special geometric result:

Corollary 3. *Any invertible linear transformation $t:\mathbf{R}^2 \to \mathbf{R}^2$ of the two-dimensional real vector space $\mathbf{R}^2$ can be written as a product of the following types of transformations: Shears parallel to an axis, reflections, and compressions (or elongations) parallel to an axis (cf. Example V.2.2).*

Proof: Represent t by an invertible 2×2 matrix P and then P as a product of elementary matrices. There are five types of 2×2 elementary matrices ($\kappa \neq 0$):

$$\begin{bmatrix} 0 & 1 \\ 1 & 0 \end{bmatrix}, \quad \begin{bmatrix} \kappa & 0 \\ 0 & 1 \end{bmatrix}, \quad \begin{bmatrix} 1 & 0 \\ \lambda & 1 \end{bmatrix}, \quad \begin{bmatrix} 1 & 0 \\ 0 & \kappa \end{bmatrix}, \quad \begin{bmatrix} 1 & \lambda \\ 0 & 1 \end{bmatrix}.$$

In each case, the corresponding endomorphism $t:\mathbf{R}^2 \to \mathbf{R}^2$ takes the unit vectors $\boldsymbol{\varepsilon}_1$ and $\boldsymbol{\varepsilon}_2$ into the columns of the matrix, as displayed. Thus the first transformation just interchanges $\boldsymbol{\varepsilon}_1$ and $\boldsymbol{\varepsilon}_2$, hence is the reflection of the plane in a 45° line through the origin. The second multplies $\boldsymbol{\varepsilon}_1$ by $\kappa \neq 0$ and leaves $\boldsymbol{\varepsilon}_2$ fixed, hence is a compression (or elongation) parallel to the x_1-axis. The third is $\boldsymbol{\varepsilon}_1 \mapsto \boldsymbol{\varepsilon}_1 + \lambda\boldsymbol{\varepsilon}_2$, $\boldsymbol{\varepsilon}_2 \mapsto \boldsymbol{\varepsilon}_2$, and is a shear parallel to the x_2-axis. The last two are like the second and third cases, with the roles of the two axes interchanged.

EXERCISES

1. (a) Exhibit all 3×3 permutation matrices.

(b) Exhibit the isomorphism between these matrices and the permutations of the symmetric group S_3.

2. Exhibit the inverse of each 2×2 elementary matrix, as displayed in the proof of Corollary 13.3.

3. (a) Prove that $A = \begin{bmatrix} a & b \\ c & d \end{bmatrix}$ is invertible if and only if $ad - bc \neq 0$.

(b) Show that, if A is invertible, its inverse is $\Delta^{-1}\begin{bmatrix} d & -b \\ -c & a \end{bmatrix}$, where $\Delta = ad - bc$.

4. Exhibit the inverse of each of the 3×3 matrices E_1, E_2, and E_3 of Exercise 5 of §1.

5. Exhibit inverses (if any) for the following matrices (over $\mathbf{Q}$):

$$\begin{bmatrix} 1 & 2 & -2 \\ -1 & 3 & 0 \\ 0 & -2 & 1 \end{bmatrix}, \quad \begin{bmatrix} 1 & 3 & -1 \\ 0 & 2 & 1 \\ -1 & 0 & -1 \end{bmatrix}, \quad \begin{bmatrix} 1 & 0 & 3 \\ 2 & 4 & 1 \\ 1 & 3 & 0 \end{bmatrix}.$$

6. If A satisfies $A^2 - A + I = 0$, prove that A^{-1} exists and is $I - A$.

7. (a) Compute a formula for the inverse of a 2×2 triangular matrix.

(b) The same for 3×3 matrices. (*Hint:* Try a triangular inverse.)

(c) Prove that every triangular matrix over a field with no zero terms on the diagonal has a triangular inverse.

8. Given A, B, A^{-1}, B^{-1}, and C, find the multiplicative inverses of

$$\text{(a)} \begin{bmatrix} A & 0 \\ 0 & B \end{bmatrix}; \quad \text{(b)} \begin{bmatrix} A & C \\ 0 & B \end{bmatrix}; \quad \text{(c)} \begin{bmatrix} A & 0 \\ C & B \end{bmatrix}.$$

9. Prove that an $m \times n$ matrix A over a field F has an $n \times m$ left inverse over F if and only if rank $A = n$ and an $n \times m$ right inverse if and only if rank $A = m$.

10. Show that every $n \times n$ elementary matrix may be obtained by applying an elementary operation to the *rows* of the identity matrix I.

11. Using the results of Exercise 10, show that each elementary operation on the rows of a matrix A amounts to replacing A by EA, for some elementary matrix E.

12. Write each of the following matrices as a product of elementary matrices:

$$\text{(a)} \begin{bmatrix} 3 & 6 \\ 2 & 1 \end{bmatrix}; \quad \text{(b)} \begin{bmatrix} 4 & -2 \\ 3 & -5 \end{bmatrix}; \quad \text{(c) the first matrix of Exercise 5.}$$

13. Represent the transformation $x' = 2x - 5y$, $y' = -3x + y$ of $\mathbf{R}^2$ to R^2 as a product of shears, compressions, and reflections.

14. For a three-dimensional space, state and prove an analog of Corollary 13.3.

15. Prove that any invertible 2×2 matrix can be represented as a product of matrices of the form

$$\begin{bmatrix} 0 & 1 \\ 1 & 0 \end{bmatrix}, \quad \begin{bmatrix} 1 & 1 \\ 0 & 1 \end{bmatrix}, \quad \begin{bmatrix} \kappa & 0 \\ 0 & 1 \end{bmatrix},$$

where $\kappa \neq 0$ is any scalar. What does this result mean geometrically?

16. (a) Show that $GL(2, \mathbf{Z}_p)$ with p prime has order $(p^2 - 1)(p^2 - p)$.
(b) Show that $GL(3, \mathbf{Z}_p)$ has order $p^3(p^3 - 1)(p^2 - 1)(p - 1)$.

17. Prove that $GL(n, F)$ for $n > 1$ is never abelian.

7. Change of Bases

We now examine exactly how the matrix of a linear map depends on the choice of bases. Let $\mathbf{b}$ and $\mathbf{c}$ be two bases for an n-dimensional vector space V; in more detail, let $(\mathbf{x}, \mathbf{b})$ and $(\mathbf{y}, \mathbf{c})$ be two dual basis pairs for V, displayed as linear maps below:

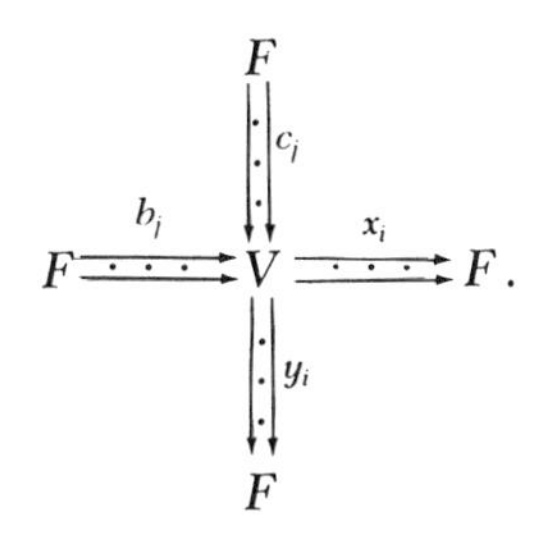

Thus for each vector v, $x_i(v)$ or $y_i(v)$ is the ith coordinate of v relative to the basis $\mathbf{b}$ or the basis $\mathbf{c}$. Now define the *change matrix* P (from $\mathbf{b}$ to $\mathbf{c}$) to be the $n \times n$ matrix of scalars

$$P_{ij} = y_i b_j, \qquad i, j = 1, \cdots, n. \tag{34}$$

Thus each entry P_{ij} is the ith "new" coordinate of the jth "old" basis vector. By the identities on dual basis pairs,

$$y_i = y_i\left(\sum_{j=1}^{n} b_j x_j\right) = \sum_j P_{ij} x_j, \qquad b_j = \left(\sum_{i=1}^{n} c_i y_i\right) b_j = \sum_i c_i P_{ij}.$$

This argument proves

Theorem 14. *If P is the change matrix from an "old" dual basis pair $(\mathbf{x}, \mathbf{b})$ to a "new" dual basis pair $(\mathbf{y}, \mathbf{c})$ for the same space, then the new coordinates are expressed in terms of P and the old coordinates as*

$$y_i = \sum_{j=1}^{n} P_{ij} x_j, \qquad i \in \mathbf{n}, \tag{35}$$

while the old basis vectors are expressed in terms of the new ones and P as

$$b_j = \sum_{i=1}^{n} c_i P_{ij}, \qquad j \in \mathbf{n}. \tag{36}$$

Put differently: Write each old basis vector b_j as a linear combination of the new basis vectors $c_1, \cdots, c_n$. The coefficients in these expressions are the entries P_{ij} of the change matrix from **b** to **c**.

Now consider successive changes of bases.

THEOREM 15. *If* **b**, **c**, *and* **d** *are three bases for the same finite-dimensional vector space V, while the change matrix from* **b** *to* **c** *is P, that from* **c** *to* **d** *is Q, and that from* **b** *(all the way) to* **d** *is R, then R is the matrix product $R = QP$.*

Proof: The definition (34) states that the change matrix P from **b** to **c** is the matrix of the identity map $1_V: V \to V$ relative to these two bases. Thus $P = M_{\mathbf{c}}^{\mathbf{b}}(1_V)$ and $Q = M_{\mathbf{d}}^{\mathbf{c}}(1_V)$; by (25) this gives $QP = M_{\mathbf{d}}^{\mathbf{b}}(1_V) = R$.

COROLLARY 1. *The change matrix P from a basis* **b** *to a basis* **c** *is invertible; its inverse P^{-1} is the change matrix from* **c** *to* **b**.

Proof: If Q is the change matrix from **c** to **b**, then, by the theorem, QP is the change matrix from **b** to **b**, so must be the identity matrix. But $QP = I$ makes P invertible and $Q = P^{-1}$, as asserted.

We now prove that every invertible matrix P is a change matrix:

COROLLARY 2. *If* **c** *is a basis of the n-dimensional vector space V and P is any invertible $n \times n$ matrix of scalars, there is exactly one basis* **b** *of V with P as its change matrix (to* **c**).

In other words, given **c**, the assignment $\mathbf{b} \mapsto P$ is a bijection from the set of all bases of V to the set of all $n \times n$ invertible matrices.

Proof: Given P, we define a list **b** of n vectors in V by $b_j = \Sigma_i c_i P_{ij}$. If we multiply the jth equation by $(P^{-1})_{jk}$ and add over all $j \in \mathbf{n}$, then

$$\sum_j b_j (P^{-1})_{jk} = \sum_i c_i \left[\sum_j P_{ij} (P^{-1})_{jk} \right] = \sum_i c_i \delta_{ik} = c_k.$$

This states that each c_k, and hence all of V, is in the span of **b**. As a spanning list of n vectors in an n-dimensional space, **b** is a basis. As $b_j = \Sigma c_i P_{ij}$ by construction, its change matrix to **c** is P, as desired.

Next, we compute the effect of a change of basis on the matrix of a map.

THEOREM 16. *Let $t: V \to V'$ be a linear transformation between finite-dimensional vector spaces V and V', and let* **b** *and* **c** *be bases of V,* **b**$'$ *and* **c**$'$

bases of V'. If P is the change matrix from **b** *to* **c** *and P' that from* **b**$'$ *to* **c**$'$, *then the matrices of t, namely*

$$M(t) = A \qquad (\text{rel } \mathbf{b}, \mathbf{b}'), \qquad M(t) = B \qquad (\text{rel } \mathbf{c}, \mathbf{c}')$$

are related by $B = P'AP^{-1}$.

Proof: As in the proof of Theorem 15, the change matrices are matrices of the identity map, $P^{-1} = M_{\mathbf{b}}^{\mathbf{c}}(1_V)$ and $P' = M_{\mathbf{c}'}^{\mathbf{b}'}(1_{V'})$. Hence, by (25),

$$P'AP^{-1} = M_{\mathbf{c}'}^{\mathbf{b}'}(1_{V'})M_{\mathbf{b}'}^{\mathbf{b}}(t)M_{\mathbf{b}}^{\mathbf{c}}(1_V) = M_{\mathbf{c}'}^{\mathbf{c}}(1_{V'}t1_V) = B.$$

In the important special case of an endomorphism $t: V \to V$ there is but *one* basis to choose (or to change), so two matrices A and B for t relative to different bases are related by $B = PAP^{-1}$. This important relation between matrices is named as follows.

DEFINITION. Two $n \times n$ matrices A and B (over the same field F) are similar *(over F) if and only if there is an invertible $n \times n$ matrix P with entries in F for which*

$$B = PAP^{-1}. \tag{37}$$

In this language, Theorem 16 and Corollary 15.2 yield

COROLLARY 1. *An endomorphism $t: V \to V$ of a finite-dimensional vector space V is represented by matrices A and B, relative to (possibly) different bases of V, if and only if B is similar to A.*

The relation of similarity between matrices is clearly reflexive, symmetric, and transitive. It will be studied extensively in Chapters IX and XI.

Now return to the case of a transformation between different spaces.

DEFINITION. Two $m \times n$ matrices A and B are equivalent *over F if and only if there are invertible square matrices P and Q over F with*

$$B = QAP^{-1}. \tag{38}$$

Clearly Q is $m \times m$, P is $n \times n$, and the relation of equivalence is reflexive, symmetric, and transitive (hence, in the more general terminology of §I.3, an equivalence relation).

COROLLARY 2. *A linear map $t: V \to V'$ between finite-dimensional vector spaces V and V' is represented by matrices A and B, relative to (possibly) different choices of bases in V and V', if and only if A and B are equivalent.*

But we have known since Corollary VI.8.1 that for any $t: V \to V'$ of rank r one can choose bases $\mathbf{c}$ and $\mathbf{c}'$ in V and V', respectively, so that

$$tc_1 = c'_1, \cdots, tc_r = c'_r, \qquad tc_{r+1} = 0, \cdots, tc_n = 0.$$

Relative to these two bases the matrix of t has the simple block form

$$D^{(r)} = \begin{bmatrix} I_r & 0_{r,\, n-r} \\ 0_{m-r,\, r} & 0_{m-r,\, n-r} \end{bmatrix}, \qquad I_r \text{ the } r \times r \text{ identity matrix,} \quad (39)$$

where $0_{i,\,j}$ denotes the $i \times j$ matrix with all entries 0. This proves

THEOREM 17. *Any $m \times n$ matrix of rank r over a field is equivalent over that field to the matrix D with all entries zero except for the first r entries 1 along the main diagonal; the entries of $D = D^{(r)}$ are thus*

$$D_{ii} = 1, \qquad i = 1, \cdots, r; \qquad D_{ij} = 0, \qquad \textit{for } i \neq j \textit{ or } i = j > r.$$

COROLLARY 1. *Two $m \times n$ matrices over a field are equivalent there if and only if they have the same rank.*

Proof: If the matrices A and B are equivalent, they are matrices for the same linear map $V \to V'$, hence have the same rank. Conversely, if A and B both have rank r, they are both equivalent to the special $m \times n$ matrix $D^{(r)}$ displayed above.

We have also proved

COROLLARY 2. *Two $m \times n$ matrices over a field are equivalent there if and only if they are equivalent there to the same matrix $D^{(r)}$ of (39).*

These corollaries describe the relation of equivalence between $m \times n$ matrices in terms of an "invariant" r and a "canonical form" $D^{(r)}$. Two matrices are equivalent if and only if they have the same rank (the same value of the invariant); every matrix is equivalent to exactly one of the canonical forms.

Similar terminology applies to any equivalence relation E on a set X. If T is another set, a function $f: X \to T$ is called an *invariant* of E when $xEy \Rightarrow f(x) = f(y)$; it is called a *complete invariant* for E when $xEy \Leftrightarrow f(x) = f(y)$. A list $f_1, \cdots, f_k$ of functions $f_i: X \to T_i$ is called a *complete system of invariants* for E when each f_i is an invariant for E and the assignment $x \mapsto (f_1x, \cdots, f_kx)$ is a complete invariant $X \to T_1 \times \cdots \times T_k$. We usually search for explicit and computable invariants, with values, say, in sets T of numbers or polynomials, but we note that there always exists a complete invariant for any E; namely, the projection $p:X \to X/E$ to the quotient set by E. The universality of this projection states that any invariant f of E can be written in the form of a composite $f = f' \circ p$.

A set of *canonical forms* for an equivalence relation E on X is a subset C of X such that there is to each $x \in X$ exactly one $c \in C$ with xEc. This amounts to requiring that the projection $p:X \to X/E$, restricted to $C \subset X$, is a bijection $C \cong X/E$.

In the present case, for equivalence between matrices of given size $m \times n$ over a given field F, the rank is a complete numerical invariant and the set of all $m \times n$ matrices of the form $D^{(r)}$ is a set of canonical forms. Invariants for the relation of similarity between square matrices—both numerical invariants and polynomial ones—will be found in Chapter IX, and Chapter XI will present a complete system of invariants as well as two different canonical forms for matrices under similarity.

The relation of equivalence between matrices can be expressed in another way.

Theorem 18. *Two $m \times n$ matrices A and B are equivalent if and only if there is a sequence of elementary operations on rows and columns carrying A to B.*

Proof: By Corollary 13.2, there is a sequence of column operations carrying A to B if and only if $B = AP$ for some invertible P. Now matrix transposition carries columns to rows, column operations to row operations, and AP to P^TA^T, with $P^T = Q$ invertible. Hence there is a sequence of elementary row operations carrying A to B if and only if $B = QA$ for some invertible Q. Combining these conclusions, row and column operations together carry A to QAP. Since every invertible is the inverse of an invertible, the result may also be written as QAP^{-1}, to agree with the definition (38) of equivalence between matrices.

An $n \times n$ invertible matrix P may be interpreted in two ways. If V is an n-dimensional vector space with a given basis $\mathbf{b}$, then either

(i) P is the change matrix from $\mathbf{b}$ to a new basis $\mathbf{c}$, or
(ii) P is the matrix, relative to $\mathbf{b}$, of an automorphism $s:V \to V$.

Theorems about invertible matrices thus have two corresponding interpretations. As an example, take Corollary 1 of Theorem 17, in the following form:

Corollary 3. *Two $m \times n$ matrices A and B over a field F have the same rank if and only if there are $m \times m$ and $n \times n$ invertible matrices P' and P over F such that $P'AP^{-1} = B$.*

For vector spaces V and V' of dimensions n and m with the respective bases $\mathbf{b}$ and $\mathbf{b}'$, the two interpretations of this Corollary are as follows:

(i) Let $t: V \to V'$ be the linear transformation with matrix A, rel $\mathbf{b}, \mathbf{b}'$. Interpret P as the change matrix from $\mathbf{b}$ to $\mathbf{c}$ in V and P' as the change matrix from $\mathbf{b}'$ to $\mathbf{c}'$ in V'. Then $P'AP^{-1}$ is the matrix of the *same* transformation t relative to the *new* bases $\mathbf{c}$ and $\mathbf{c}'$. Therefore: Two $m \times n$ matrices A and B represent the same linear transformation relative to different choices of bases (in both spaces) if and only if they have the same rank.

(ii) Let $t_A, t_B: V \to V'$ be the two linear transformations with matrices A and B, respectively, both relative to the *same* pair of bases $\mathbf{b}$ and $\mathbf{b}'$. Interpret P as the matrix relative to $\mathbf{b}$ of an automorphism $s: V \to V$ and P' as the matrix, rel $\mathbf{b}'$, of an automorphism $s': V' \to V'$. The matrix equation $B = P'AP^{-1}$ now states that $t_B = s't_As^{-1}$, so the Corollary now becomes: Two linear transformations $t_1, t_2: V \to V'$ have the same rank if and only if there are automorphisms s of V and s' of V' such that $t_1 = s't_2s^{-1}$, as in the commutative diagram below:

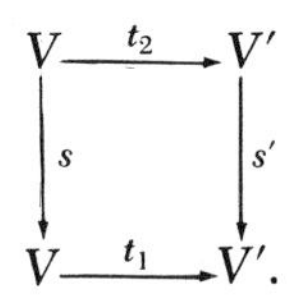

In the first interpretation, each vector is described by new coordinates, hence is given a different "name"; we call this an *alias* interpretation. In the second case, each vector is mapped (by s or s') to a different vector, hence is given an "alibi"; we call this an *alibi* interpretation. In Chapter X we will encounter other examples of such "alias" and "alibi" interpretations.

EXERCISES

1. In $V = \mathbf{Q}^2$ or $\mathbf{Q}^3$ let the "old basis" $\mathbf{b}$ be the basis of unit vectors, while the new basis is given by the vectors listed below. In each case find the change matrix P, the formula for the new coordinates in terms of the old coordinates, and the formula for the old coordinates in terms of the new. In cases (a) and (b), draw a figure:

(a) $(1, 1), (1, -1)$. (c) $(1, 1, 0), (1, 0, 1), (0, 1, 1)$.
(b) $(2, 3), (-2, -4)$. (d) $(1, 2, 3), (1, -1, 1), (0, 1, 0)$.

2. Prove Theorem 15 on successive changes of bases by using formulas like (36).

3. Prove Theorem 16 on the matrices of a map by using formulas $tb_j = \Sigma b_i' A_{ij}$, $tc_j = \Sigma c_i' B_{ij}$ and (36).

4. Give a direct matric proof that row and column operations will reduce any matrix to one of the canonical forms $D^{(r)}$.

5. Reduce the following matrices over $\mathbf{Q}$ to canonical form under equivalence:

$$\begin{bmatrix} 1 & 3 & 5 \\ 2 & 1 & 2 \\ 4 & 2 & 1 \end{bmatrix}, \quad \begin{bmatrix} 1 & -1 & 2 \\ 3 & 2 & -2 \\ 4 & 1 & 4 \\ -1 & 5 & 1 \end{bmatrix}, \quad \begin{bmatrix} 1 & 2 & 3 & 4 \\ 2 & 3 & 4 & 1 \\ 3 & 4 & 1 & 2 \\ 4 & 1 & 2 & 3 \end{bmatrix}.$$

6. If $F \subset F'$ are fields, while A and B are $m \times n$ matrices with entries in F, show that A is equivalent to B over F' if and only if this is so over F.

7. Prove that any matrix of rank r is the sum of r matrices of rank 1.

8. Prove (geometrically) that any linear map $t: V \to V'$ of rank r is the sum of r linear maps of rank 1.

9. Prove: An $m \times n$ matrix A has rank at most 1 if and only if it can be written as a product $A = BC$, where B is $m \times 1$ and C is $1 \times n$.

10. (a) Show that a row operation $A \mapsto PA$ does not alter (i) the row rank of A; (ii) the nullity of A; (iii) the column rank of A.

(b) Deduce that row and column operations leave both row and column ranks fixed.

(c) Deduce further without using Theorem 10 that row rank (A) = column rank (A). (*Hint:* Use the echelon form or the result of Exercise 4.)

11. Show that the set $\{0, 1, \cdots, n-1\}$ is a set of canonical forms for the equivalence relation of congruence modulo n on the set $\mathbf{Z}$ of integers.

12. Find a canonical form for the set of permutations $\sigma \in S_n$ under the relation "σ is conjugate to τ in S_n".

***13.** Find a canonical form for $m \times n$ matrices under the relation of "column equivalence".

14. Show that each of the following assignments defines an action of the indicated group on the set $F^{n \times n}$ of all $n \times n$ matrices A. In each case, describe the meaning of the relation of equivalence under this action, as defined in §II.7:

(a) $(P, A) \mapsto AP^{-1}$, for $P \in GL(n, F)$.

(b) $(P, A) \mapsto PAP^{-1}$, for $P \in GL(n, F)$.

(c) $((P, Q), A) \mapsto QAP^{-1}$, for $(P, Q) \in GL(n, F) \times GL(n, F)$.

8. Eigenvectors and Eigenvalues

Further analysis of a linear transformation t of a space V *into itself* makes special use of those vectors v transformed into scalar multiples of themselves, as in

$$t(v) = v\lambda, \qquad v \neq 0. \tag{40}$$

DEFINITION. Let $t: V \to V$ be any linear transformation of a finite-dimensional vector space V into itself. An eigenvector *of t is a vector $v \neq 0$ such that $t(v)$ is a scalar multiple of v, while a scalar λ is an* eigenvalue *of t when there is a vector $v \neq 0$ with $t(v) = v\lambda$.*

Thus each eigenvector v determines an eigenvalue λ, as in (40), while each eigenvalue λ must arise in this way from at least one eigenvector. If λ is any eigenvalue of t, the set of all vectors v of V with $t(v) = v\lambda$ is a nonzero subspace of V which is called the *eigenspace* of λ; it consists of the zero vector and all the eigenvectors belonging to λ.

An eigenvalue ("Eigenwert" in German) is also called a *characteristic value* or a *proper value*, while eigenvectors are correspondingly named *characteristic vectors* or *proper vectors*.

If A is an $n \times n$ matrix, define an eigenvector or an eigenvalue of the matrix A to be one for the corresponding linear transformation $t_A: F^n \to F^n$. If A is the matrix of a map t relative to a basis $\mathbf{b}$, the eigenvalues of A are exactly those of t. Indeed, t is given in terms of the matrix A as $t(\Sigma_i b_i\xi_i) = \Sigma_{i\,j} b_i A_{i\,j}\xi_j$, so the vector $\Sigma b_i\xi_i \neq 0$ is an eigenvector of t (with the eigenvalue λ), if and only if $\lambda\xi_i = \Sigma_j A_{i\,j}\xi_j$ for each $i = 1, \cdots, n$; that is, if and only if the list $\boldsymbol{\xi}$ is an eigenvector of the matrix A for the same eigenvalue λ.

Consider a diagonal matrix such as the one below.

$$\begin{bmatrix} 3 & 0 & 0 \\ 0 & 3 & 0 \\ 0 & 0 & -1 \end{bmatrix}.$$

Then $(1, 0, 0) \mapsto (3, 0, 0)$, $(0, 1, 0) \mapsto (0, 3, 0)$ and $(0, 0, 1) \mapsto (0, 0, -1)$ in the corresponding linear transformation on F^3. Hence, the unit vectors are eigenvectors with the eigenvalues 3, 3, and -1, so the diagonal entries of the matrix are the eigenvalues. For that matter, any list $(\xi_1, \xi_2, 0) \neq 0$ is an eigenvector with eigenvalue 3; these vectors, with 0, form the eigenspace of 3.

This can be stated more generally.

THEOREM 19. *A linear transformation $t: V \to V$ has a diagonal matrix D relative to some basis* $\mathbf{b}$ *of the space V if and only if the eigenvectors of t span V. When this is the case, the eigenvalues of t are exactly the diagonal entries of D, and each occurs on the diagonal a number of times equal to the dimension of its eigenspace.*

Proof: First suppose that V is spanned by eigenvectors, and hence spanned by linearly independent eigenvectors $b_1, \cdots, b_n$ with the corresponding eigenvalues $\lambda_1, \cdots, \lambda_n$. Then $t(b_i) = b_i\lambda_i$ for $i = 1, \cdots, n$; these equations state that the matrix of t (rel $\mathbf{b}$) is the diagonal matrix with diagonal entries $\lambda_1, \cdots, \lambda_n$. Conversely, if t has a diagonal matrix with diagonal entries $\lambda_1, \cdots, \lambda_n$, all relative to some basis $\mathbf{b}$, the equations of t are $tb_i = b_i\lambda_i$; they state that the basis vectors are eigenvectors.

Now let the matrix of t be diagonal relative to some basis $\mathbf{b}$, so that $tb_i = b_i\lambda_i$, and consider any eigenvector $v = \Sigma b_i\xi_i \neq 0$ of t with eigenvalue

λ. Then $t(v) = v\lambda$ means that $\Sigma b_i\lambda_i\xi_i = \Sigma b_i\xi_i\lambda$, and hence that $\lambda_i\xi_i = \xi_i\lambda$ for every i. If $\xi_i \neq 0$, this gives $\lambda_i = \lambda$. Therefore, any eigenvalue λ of t must be one of the given "diagonal" entries $\lambda_1, \cdots, \lambda_n$. Moreover, an eigenvector $\Sigma b_i\xi_i$ can actually involve two or more different b_i only if they belong to equal diagonal entries λ_i. For, suppose that only the first m diagonal entries $\lambda_1 = \cdots = \lambda_m$ are equal. Then the eigenvectors belonging to $\lambda_1 = \lambda_m$ are exactly the linear combinations $b_1\xi_1 + \cdots + b_m\xi_m$. They constitute the eigenspace of λ_1, and its dimension m is exactly the number of times this eigenvalue appears on the diagonal, just as asserted in the theorem.

COROLLARY. *An $n \times n$ matrix A is similar to a diagonal matrix if and only if the eigenvectors of A span F^n.*

This is an immediate consequence of the theorem, for two $n \times n$ matrices are similar if and only if they represent the same transformation $F^n \to F^n$ relative to (possibly) different bases.

But do not conclude from this that every matrix is similar to a diagonal one. Take, for example, the 2×2 matrix displayed below, and with $a \neq 0$ in any field F:

$$\begin{bmatrix} 1 & a \\ 0 & 1 \end{bmatrix}.$$

In the two-dimensional space F^2, and relative to the basis of unit vectors, this matrix represents the shear $(x, y) \mapsto (x + ay, y)$. What vectors (x, y) can be eigenvectors? If $x + ay = \lambda x$ and $y = \lambda y$, the second of these equations gives $\lambda = 1$ or $y = 0$, and $y = 0$ in the first equation gives $\lambda = 1$. Hence, the only eigenvalue is $\lambda = 1$, and the only eigenvectors are multiples of $(1, 0)$. They cannot span the space, so the matrix cannot be similar to a diagonal matrix. A different argument shows that the matrix $\begin{bmatrix} 0 & 1 \\ -1 & 0 \end{bmatrix}$ has *no* real eigenvalues.

EXERCISES

1. (a) Find a diagonal matrix similar to $\begin{bmatrix} -3 & 4 \\ 2 & -1 \end{bmatrix}$ over the field $\mathbf{Q}$.

(b) Find a diagonal matrix similar to $\begin{bmatrix} 1 & 1 \\ 1 & -1 \end{bmatrix}$ over the real field $\mathbf{R}$.

2. Compute the eigenvalues and the eigenvectors of the following matrices over the field $\mathbf{C}$ of complex numbers:

(a) $\begin{bmatrix} 2 & 4 \\ 5 & 3 \end{bmatrix}$. (b) $\begin{bmatrix} 3 & 2 \\ -2 & 3 \end{bmatrix}$. (c) $\begin{bmatrix} 1 & 2 \\ 2 & -2 \end{bmatrix}$. (d) $\begin{bmatrix} -1 & 2i \\ -2i & 2 \end{bmatrix}$.

3. For each matrix A given in Exercise 2 find, when possible, an invertible matrix P over $\mathbf{C}$ for which PAP^{-1} is diagonal.

4. The matrix $\begin{bmatrix} 0 & 1 \\ -1 & 0 \end{bmatrix}$ represents a clockwise rotation of the plane through 90°. Show that this matrix has no real eigenvalues but two complex eigenvalues, and find its complex eigenvectors.

5. (a) Find the complex eigenvalues of the real matrix representing a rotation of the plane through an angle θ.

(b) Prove that the matrix representing a rotation of the plane through an angle θ $(0 < \theta < \pi)$ is not similar to any real diagonal matrix.

6. Show that a square matrix A has 0 as eigenvalue if and only if A is not invertible.

7. Prove that the eigenvalues of a triangular matrix are the diagonal entries of that matrix.

8. Show that the scalar λ is the only eigenvalue of each of the following matrices, but that no two of these matrices are similar:

$$\begin{bmatrix} \lambda & 0 & 0 \\ 1 & \lambda & 0 \\ 0 & 1 & \lambda \end{bmatrix}, \qquad \begin{bmatrix} \lambda & 0 & 0 \\ 1 & \lambda & 0 \\ 0 & 0 & \lambda \end{bmatrix}, \qquad \begin{bmatrix} \lambda & 0 & 0 \\ 0 & \lambda & 0 \\ 0 & 0 & \lambda \end{bmatrix}.$$

9. In an $n \times n$ matrix A the sum of the entries of each row is 1. Prove that 1 is an eigenvalue of A.

10. (a) If the $n \times n$ matrix A over F has n linearly independent lists $\xi^{(i)}$, $i = 1, \cdots, n$, as eigenvectors in F^n, prove that the matrix P with the n columns $\xi^{(1)}, \cdots, \xi^{(n)}$ is invertible and that $P^{-1}AP$ is diagonal.

(b) Using this result, find P so that PAP^{-1} is diagonal for each of the matrices of Exercise 1.

11. (a) If A is a 2×2 matrix over a field F, show that $\lambda \in F$ is an eigenvalue of A if and only if

$$\lambda^2 - (A_{11} + A_{22})\lambda + (A_{11}A_{22} - A_{12}A_{21}) = 0.$$

(b) Deduce from this the results of Exercises 4 and 5a on eigenvalues.

12. If an $n \times n$ matrix over F has n *different* eigenvalues (in F), show that the corresponding eigenvectors are linearly independent (and hence span F^n).

CHAPTER VIII

Special Fields

THE BULK of this chapter will be devoted to three special fields which play a central role in mathematics: The real field **R**, its subfield **Q** of rational numbers, and its extension **C** (the complex field). In these fields, order and related limit concepts are fundamental. Hence, we shall begin by establishing the basic properties of order and limits in **Q**, **R**, and **C**. In each case, we shall derive some of the most important algebraic facts stemming from these properties.

The rest of the chapter will be concerned with the division ring of real quaternions, and with the fields of extended formal power series and p-adic numbers. These also have important and unique special properties.

1. Ordered Domains

We will begin our study of special fields by a review of some general properties of order relations in rings.

DEFINITION. A ring R with $0 \neq 1$ is ordered *when there is given a nonvoid subset $P \subset R$, called the set of* positive *elements of R, such that*

(*i*) $a \in P$ *and* $b \in P \quad \Rightarrow \quad a + b \in P$ *and* $ab \in P$;
(*ii*) *for each $a \in R$ exactly one of the following alternatives holds*:

$$a \in P, \quad a = 0 \quad \text{or} \quad (-a) \in P \qquad (\textit{trichotomy}).$$

An ordered ring which is also an integral domain is called an ordered domain; *one which is also a field is called an* ordered field.

For example, the ring **Z** of integers is an ordered domain, if one takes for P the set of all positive integers. Likewise, as the reader will recognize (although we cannot prove it formally yet), the real field **R** is an ordered field if P is the set of all positive real numbers. Actually, there is no other way to make **Z** or **R** into ordered rings, as we shall see below after Theorem 3.

PROPOSITION 1. *In any ordered ring R, all squares of non-zero elements are positive.*

Proof: Suppose that $a \neq 0$ in R. By trichotomy, either $a \in P$ or $(-a) \in P$. Since P is closed under multiplication, $a^2 = (-a)^2 \in P$ in either case, as asserted.

It is a corollary that the unit $1 = 1^2$ of R is always positive, while -1 is never positive.

Proposition 2. *Any ordered commutative ring R is an integral domain of characteristic ∞.*

Proof: First, suppose that $a \neq 0$, $b \neq 0$ were zero divisors, with $ab = 0$. Then $(\pm a)(\pm b) = 0$, but, by the trichotomy assumption, one of $\pm a$ and one of $\pm b$ is in P; hence, some one of the four products $(\pm a)(\pm b)$ is in P, a contradiction to $0 \notin P$. Hence, R is an (ordered) integral domain.

Second, since $1 \in P$, it follows by repeated application of part (i) of the definition that $1 + \cdots + 1$ to any number of terms must be in P, and hence cannot be 0. Therefore, by its definition (§III.1), the characteristic of R is ∞.

The result above can be sharpened considerably, using the following notion of an order-morphism of rings.

DEFINITION. If R and R' are ordered rings, with positive subsets P and P' respectively, an order-morphism of rings *is a morphism $h: R \to R'$ of rings such that $x \in P$ implies $h(x) \in P'$ (that is, $h_* P \subset P'$).*

One easily shows that any order-morphism h of rings is a monomorphism. For, if Ker $h \neq 0$, then it contains both a and $-a$ for some $a \neq 0$; however, either $a \in P$ or $-a \in P$. In either case, $h(x) = 0 \notin P'$ for some $x = \pm a$ in P, a contradiction. Moreover, $x \in P$ if and only if $h(x) \in P'$.

Theorem 3. *For any ordered ring R', the unital morphism $\mu: \mathbf{Z} \to R'$ is an order monomorphism of rings.*

Proof: In §III.1 we constructed the unital morphism $\mu: \mathbf{Z} \to R'$. By Proposition 2, this is a monomorphism of rings. Again, $\mu(1 + \cdots + 1) = 1' + \cdots + 1'$ for any number of terms, where $1'$ is the unit of R'. As $1' \in P'$ in R' (by Proposition 1), $1' + \cdots + 1' \in P'$ by the closure of P' under addition. That is, μ is an order-morphism; hence, it is an order monomorphism of rings.

This shows that there is only one way to make $\mathbf{Z}$ into an ordered ring. The same is true for the real field $\mathbf{R}$, for by Proposition 1 any non-zero square must be positive in any ordering of $\mathbf{R}$, while (see §6 below) the usual positive real numbers all have real square roots, hence must be positive in every ordering of **R**. However, it is not true for the ring $\mathbf{Z}[\sqrt{2}\,]$ of all numbers of the form $m + n\sqrt{2}$ $(m, n \in \mathbf{Z})$ which can be made into an ordered domain in at least two ways.

Namely, one can say that $m + n\sqrt{2}$ is in P when $m + n\sqrt{2}$ is positive as a real number. One can verify that this is the case when m and n satisfy one of the following conditions:

$$(\alpha)\ m > 0 \text{ and } m^2 > 2n^2 \qquad \text{or} \qquad (\beta)\ n > 0 \text{ and } 2n^2 > m^2.$$

One can also verify, assuming only that **Z** is an ordered ring, that the set P so defined does make $\mathbf{Z}[\sqrt{2}\,]$ into an ordered ring.

On the other hand, since $m + n\sqrt{2} \mapsto m - n\sqrt{2}$ is an automorphism of this ring, one can say that $m + n\sqrt{2}$ is "positive" when $m + (-n)\sqrt{2}$ satisfies (α) or (β). This gives a different set P_1, consisting of all pairs (m, n) which satisfy either (α) above or (γ) $n < 0$ and $2n^2 > m^2$. This P_1 also makes $\mathbf{Z}[\sqrt{2}\,]$ into an ordered ring, but with respect to a different order.

From the set P of positive elements in any ordered ring, one may define a binary relation *less than*, by the condition

$$a < b \quad \Leftrightarrow \quad (b - a) \in P. \tag{1}$$

We also write $b > a$ for $a < b$ (thus $>$ is the converse of the relation $<$), and $a \leqslant b$ (or $b \geqslant a$) for "$a < b$ or $a = b$".

Proposition 4. *In any ordered ring R, the relation $<$ defined by* (1) *is:*

Trichotomous: *Given $a, b \in R$, exactly one of the alternatives $a < b$, $a = b$, $b < a$ holds,* (2)

Transitive: $a < b$ *and* $b < c \Rightarrow a < c$, (3)

Isotonic for sums: $b < c \Rightarrow (a + b) < (a + c)$, (4)

Isotonic for positive factors: $a > 0$ *and* $b < c \Rightarrow ab < ac$. (5)

Proof: Substituting in the definition of $<$, condition (2) asserts that either $(b - a) \in P$, or $(b - a) = 0$, or $-(b - a) \in P$, which is condition (ii) on P. Condition (3) follows since $c - a = (c - b) + (b - a)$, which is in P if $(c - b) \in P$ and $(b - a) \in P$, by the closure of P under addition. Condition (4) follows, inasmuch as $(a + c) - (a + b) = c - b$; and (5) follows since $ac - bc = (a - b)c \in P$ if $a \in P$ and $(c - b) \in P$, by the closure of P under multiplication.

It can be proved that, conversely, if "$<$" is any binary "order" relation on a ring R satisfying conditions (2)–(5), and if P is the set of all $a \in R$ with $a > 0$ in this order, then P satisfies conditions (i) and (ii) stated at the beginning of the section, provided $0 \neq 1$ in R. In other words, conditions (2)–(5) can also be used as the axioms for ordered rings.

The absolute value concept can be introduced in any ordered ring, as follows:

DEFINITION. In an ordered ring, the absolute value $|a|$ *of* $a \in R$ *is zero if* $a = 0$, *and otherwise is the positive one of the pair* a *and* $-a$.

Obviously, $|0| = 0$ and $|a| > 0$ if $a \neq 0$. Moreover, by appropriate consideration of the cases $a \in P$, $a = 0$, and $-a \in P$ one can prove the familiar rules

$$|ab| = |a| \cdot |b|, \qquad |a + b| \leqslant |a| + |b|. \tag{6}$$

The second rule (the "triangle inequality") can also be proved from (4), by adding together the inequalities $-|a| \leqslant a \leqslant |a|$ and $-|b| \leqslant b \leqslant |b|$. (Note that (4) holds for the relation $\leqslant$ as well as for the relation $<$.)

Characterization of **Z**. We now recall from §I.6 the fact that the set **P** of all positive integers is *well-ordered*, in the sense that every nonvoid subset $S \subset \mathbf{P}$ has a *least* member a; that is, an element $a \in S$ such that $a \leqslant x$ for all $x \in S$. We shall now show that this property characterizes **Z** up to an order-isomorphism of rings.

Theorem 5. *Let R be any ordered ring whose positive elements are well-ordered in R. Then R is order-isomorphic to* **Z**, *the ring of integers.*

Proof: Consider the unital order-monomorphism of rings, $\mu: \mathbf{Z} \to R$, with image $U = \mu_* \mathbf{Z} \subset R$. Suppose that R contained an element x not in this image U. Then $(-x) \in R$ would not be in U either; yet one of x and $(-x)$ must be in the subset P of positive elements of R. Hence, the set of positive elements not in U would be non-empty and so, being well-ordered, would contain a *least* element m. This $m \neq 1$, since $1 \in U$. If $m > 1$, then $(m - 1) < m$ would also be in P, although not in U, contrary to the choice of m as least such. Therefore, $m < 1$, and so $0 = m \cdot 0 < m \cdot m < m \cdot 1 = m < 1$ by (5). Hence, m^2 is positive but not in U, since all positive elements of $U = \mu_* \mathbf{Z}$ are $\geqslant 1$. Yet $m^2 < m$, again contradicting the choice of m and so completing the proof.

EXERCISES

1. Let D be any ordered domain, with positive subset P, and let S be any subdomain of D. Show that S is an ordered domain with the "positive subset" $S \cap P$.

2. Show the following rules valid in any ordered ring:
   (a) $a + x < a + y \iff x < y$.
   (b) $a - x < a - y \iff x > y$.
   (c) If $a < 0$, then $ax < ay \iff x > y$.

(d) $a > 0$ and $ax > ay \Rightarrow x > y$.
(e) $a < b \Leftrightarrow a^3 < b^3$.

3. (a) List a complete set of properties for the binary relation $x \leqslant y$ in an ordered commutative ring.

(b) Show that an integral domain with a binary relation having the properties of your list is necessarily an ordered domain.

4. Prove in detail the rules (6).

5. Show that $a^{2n+1} = b^{2n+1}$ (n a positive integer) implies $a = b$ in any ordered ring.

6. In any ordered domain D, show that $x^2 - xy + y^2 \geqslant 0$ for all $x, y \in D$.

***7.** Make the ring $\mathbf{Z}[\sqrt{3}\,] \cong \mathbf{Z}[x]/(x^2 - 3)$ into an ordered ring by defining P suitably and without assuming anything about the real field.

8. Prove that $\mathbf{Z}[\sqrt{2}\,]$ becomes an ordered ring when the positive elements are either of the sets P or P_1 described in the text.

9. Prove that an ordered ring R is necessarily nontrivial (that is, has $0 \neq 1$).

2. The Ordered Field Q

In §III.5 we showed how to embed any integral domain D in its field of quotients $Q(D)$. Applying this construction to the domain $\mathbf{Z}$, we get the rational field $\mathbf{Q} = Q(\mathbf{Z})$. We now make $\mathbf{Q}$ into an ordered field.

THEOREM 6. *Let D be any ordered domain with positive subset P, and let $Q(D)$ be its field of quotients. Then there is one and only one way to define an order (that is, a positive subset P_1 in $Q(D)$), so that the insertion $D \to Q(D)$ is a morphism of order. This order is defined by the condition*

$$(a/b) \in P_1 \qquad \text{if and only if} \qquad ab \in P \text{ in } D. \tag{7}$$

Proof: The field $Q(D)$ as constructed in §III.5 contains D as a subdomain, while every element of $Q(D)$ is a quotient a/b in $Q(D)$ of elements $a, b \in D$ with $b \neq 0$.

Suppose first that $Q(D)$ is ordered by some set P_1 of positive elements. Then $a/b = ab(1/b)^2$. Since $(1/b)^2$ is necessarily in P_1, $a/b \in P_1$ if and only if $ab \in P_1$. But if the insertion $D \to Q(D)$ is a morphism of order, $ab \in P_1$ in $Q(D)$ if and only if $ab \in P$ in D. This shows that P_1 must satisfy (7).

An equality $a/b = a'/b'$ in $Q(D)$ implies $ab' = a'b$ in D. Multiplied by bb', this yields $abb'^2 = a'b'b^2$, hence $ab \in P$ if and only if $a'b' \in P$. Thus, given the set P of positive elements in D, we can define P_1 in $Q(D)$ by $a/b \in P_1$ if and only if $ab \in P$, as proposed in (7), consistent with equality in $Q(D)$.

Conversely, define P_1 by (7), and take a/b and c/d in P_1. Their sum $a/b + c/d = (ad + bc)/bd$ is again in P_1, since $ab \in P$ and $cd \in P$ imply $d^2ab \in P$ and $b^2cd \in P$, whence

$$bd(ad + bc) = (d^2ab + b^2cd) \in P,$$

which gives $[(a/b) + (c/d)] \in P_1$ by (7). The same hypotheses imply $adbc = (ab)(cd) \in P$, whence $(a/b)(c/d) \in P_1$ by (7). Finally, P_1 satisfies the trichotomy condition (ii) of §1; indeed, of the conditions $ab \in P$, $ab = 0$, and $(-a)b \in P$ (that is, $-(a/b) = (-a)/b \in P_1$), one and only one holds by trichotomy for D.

In the special case $D = \mathbf{Z}$, condition (7) evidently gives the rational field $\mathbf{Q}$ with its usual notion of positivity, and hence [by (1)] with its usual order relation.

COROLLARY. *In any ordered field F, the prime subfield is order-isomorphic to $\mathbf{Q}$.*

Proof: By Proposition 2, F has characteristic ∞, so its prime subfield (§III.11) is isomorphic to $\mathbf{Q}$ as a field. This must be an order-isomorphism, since by the previous theorem the only order on $\mathbf{Q}$ is the usual one.

There are many interesting ordered rings besides the real field and its subfields; for example, for polynomials in an indeterminate x we have

PROPOSITION 7. *Let D be any ordered domain with positive subset P. In $D[x]$, define the set $P^* \subset D$ by*

$$(a_0 + a_1x + \cdots + a_nx^n) \in P^*, \qquad a_n \neq 0, \tag{8}$$

if and only if $a_n \in P$ in D. Then $D[x]$, with positive subset P^*, is an ordered domain.

We leave the proof to the reader. The reader can also easily verify the following additional rules for inequalities, valid in any ordered field:

$$0 < 1/a \quad \Leftrightarrow \quad a > 0, \tag{9}$$

$$a/b < c/d \quad \Leftrightarrow \quad abd^2 < b^2cd, \tag{10}$$

$$0 < a < b \quad \Rightarrow \quad 0 < 1/b < 1/a, \tag{11}$$

$$a_1^2 + a_2^2 + \cdots + a_n^2 > 0 \text{ unless } a_1 = a_2 = \cdots = a_n = 0. \tag{12}$$

EXERCISES

1. Prove formulas (9)–(12) of the text.

2. Prove that $a < b < 0$ implies $0 > 1/a > 1/b$ in any ordered field.

3. For any $a < b$ in an ordered field, prove that there are infinitely many x satisfying $a < x < b$.

4. Prove that in no ordered field do the positive elements form a well-ordered set.

5. Prove that $|(|x| - |y|)| \leqslant |x - y|$ in any ordered domain.

6. Prove that $|xx' + yy'| \leqslant \sqrt{(x^2 + y^2)(x'^2 + y'^2)}$ in any ordered field in which all positive elements have square roots.

7. Prove Proposition 7.

8. (a) If D is an ordered domain, F an ordered field, and $h:D \to F$ an order-morphism of rings, show that there is a unique order-morphism $h':Q(D) \to F$ of fields with $h = h' \circ j$, where $j:D \to Q(D)$ is the insertion.

(b) Construct a suitable functor for which j, by part (a), is a universal element.

3. Polynomial Equations

Over any field, it is important to know how to factor a general polynomial into its monic irreducible factors. It was shown in Corollary III.19.2 that the linear factors $(x - c_i)$ of any polynomial $f(x)$ are given by the constants c_i such that $f(c_i) = 0$. Since quadratic and cubic polynomials are irreducible unless they have a linear factor, this also enables one to determine which quadratic and cubic polynomials are *irreducible* in $\mathbf{Q}[x]$.

PROPOSITION 8. *If the polynomial*

$$f = f_0 + f_1 x + \cdots + f_{k-1}x^{k-1} + f_k x^k, \qquad \text{all } f_i \in \mathbf{Z}, \quad f_k \neq 0, \quad (13)$$

has a rational root, then it has one of the form $x = r/s$, where r and s are integral divisors of f_0 and f_k, respectively.

Proof: Suppose that (13) is satisfied for some $b = m/n \in \mathbf{Q}$. Dividing out the g.c.d. of m and n, we obtain b in "lowest terms" as a quotient $b = r/s$ of relatively prime integers r and s. Substituting into $s^k f(r/s) = 0$, we get

$$0 = f_0 s^k + f_1 s^{k-1} r + \cdots + f_{k-1} s r^{k-1} + f_k r^k, \qquad f_k \neq 0.$$

Here s divides every term except the last, so $s \mid f_k r^k$. Since $(s, r) = 1$, this implies $s \mid f_k$. Likewise, r divides every term except the first, and therefore $r \mid (-\sum_{i=1}^{k} f_i s^{k-i} r^i) = f_0 s^k$. Since $(r, s) = 1$, this implies $r \mid f_0$, Q.E.D.

COROLLARY. *Any rational root of a monic polynomial with coefficients in* **Z** *must belong to* **Z**.

As a trivial example, the equation $x^2 = p$ has no solution in **Q**, for any prime p. More generally, the equation $x^2 = n$ ($n \in \mathbf{Z}$) cannot have a solution in **Q** unless, in the prime factorization of n, the exponent of every prime factor happens to be even.

Proposition 8 can be applied to any polynomial with rational coefficients. To apply it, it suffices to multiply through by the least common multiple of the denominators of the coefficients (the "least common denominator"), getting a polynomial with integral coefficients. One can apply Proposition 8 to the polynomial so obtained.

For the present, we note only that any reducible quadratic or cubic polynomial must have a linear factor. Hence, the method described above also gives an effective test for the *irreducibility* of quadratic and cubic polynomials over the rational field **Q**. Another sufficient condition for irreducibility is the following.

PROPOSITION 9. *Over an ordered domain D, any quadratic polynomial $ax^2 + bx + c$ with negative discriminant $b^2 - 4ac < 0$ is irreducible.*

Proof: For any reducible quadratic polynomial

$$ax^2 + bx + c = (\alpha x + \beta)(\gamma x + \delta) = \alpha\gamma x^2 + (\alpha\delta + \beta\gamma)x + \beta\delta,$$

direct substitution gives $b^2 - 4ac = (\beta\gamma - \alpha\delta)^2 \geqslant 0$.

EXERCISES

1. Test the following equations for rational roots:

(a) $3x^3 - 7x = 5$. (b) $5x^3 + 8x^2 + 6x = 4$.
(c) $8x^5 + 3x^2 = 17$. (d) $6x^3 - 3x = 18$.

2. Prove that there is no integer $n > 1$ such that $30x^n = 91$ has a rational root.
3. Determine all rational numbers x such that $3x^2 - 7x$ is an integer.
4. Determine the reducibility or irreducibility over $\mathbf{Z}_3$ and $\mathbf{Z}_5$ of:
(a) $x^2 - 2$. (b) $x^3 + x + 2$.
5. List all monic irreducible quadratic polynomials over $\mathbf{Z}_5$.
6. Determine all monic irreducible cubic polynomials over $\mathbf{Z}_3$.
7. Prove that if $f_0 + f_1x + f_2x^2 + \cdots + f_nx^n$ is irreducible over a field F, then so is $f_n + f_{n-1}x + f_{n-2}x^2 + \cdots + f_0x^n$.
8. Show that if the polynomial $f(x)$ is irreducible over a field F, then so is $f(x + a)$ for any $a \in F$.
9. Decompose the polynomial $x^4 - 5x^2 + 6$ into irreducible factors over **Q**, over the field $\mathbf{Q}[\sqrt{2}\,]$, and also over the real field **R**.

***10.** (a) If f is a monic polynomial with coefficients in $\mathbf{Z}$, show that its irreducibility mod p implies its irreducibility over $\mathbf{Q}$ (p a prime).

(b) Show that if f is monic and reducible over $\mathbf{Z}$, every factor must reduce mod p to a factor of the same degree over $\mathbf{Z}_p$.

(c) Use the results above to test the irreducibility over $\mathbf{Q}$ of

$$x^3 + 6x^2 + 5x + 25, \qquad x^3 + 6x^2 + 11x + 8, \qquad x^4 + 8x^3 + x^2 + 2x + 5.$$

(*Hint:* Try small primes $p = 2, 3, 5$, etc.)

4. Convergence in Ordered Fields

We now recall the definition of limit in terms of order. In any ordered field F (or abelian group), the infinite sequence $a_1, a_2, a_3, \cdots$ is said to *converge* to the limit $a \in F$ (in symbols, $\lim a_n = a$) when, for any $\varepsilon > 0$ in F, there exists a positive integer $n = n(\varepsilon)$ such that

$$|a_k - a| < \varepsilon \qquad \text{for all } k \geqslant n. \tag{14}$$

Convergence thus defined has some familiar and easily proved properties, including the following.

Lemma 1. *If* $\lim a_n = a$ *and* $\lim a_n = b$, *then* $a = b$.

Proof: Let $\varepsilon = |a - b|/2$. If $a \neq b$, then $\varepsilon > 0$ yet $|a_k - a| < \varepsilon$ and $|a_k - b| < \varepsilon$ cannot both hold. Hence, (14) cannot hold for a at the same time that it holds for b.

Lemma 2. *If* $\lim a_n = a$ *and* $\lim b_n = b$, *then*

$$\lim (a_n + b_n) = a + b. \tag{15}$$

Proof: By hypothesis,for any $\varepsilon/2 > 0$ there exist m_1 and m_2 such that $|a_k - a| < \varepsilon/2$ for all $k \geqslant m_1$, and $|b_k - b| < \varepsilon/2$ for all $k \geqslant m_2$. Hence, if we define $m = m_1 + m_2$:

$$|(a_k + b_k) - (a + b)| < \varepsilon \qquad \text{for all } k \geqslant m. \tag{16}$$

Theorem 10. *In any ordered field* F, $\lim a_n = a$ *implies* $\lim f(a_n) = f(a)$, *for any polynomial* $f = f_0 + f_1x + \cdots + f_kx^k$.

In the language of the calculus, Theorem 10 asserts that *any polynomial function is continuous*. To establish it, we first prove:

Lemma 3. *Given a polynomial* $g \in F[x_1, \cdots, x_r]$ *over an ordered field* F, *let* ϕ *be the polynomial obtained from* g *by replacing each coefficient by* *its absolute value. Then*

$$|x_j| \leqslant d_j \text{ for } j = 1, \cdots, r \textit{ implies } |g(x_1, \cdots, x_r)| \leqslant \phi(d_1, \cdots, d_r). \tag{17}$$

Proof: Since $|a| \leqslant a'$ and $|b| \leqslant b'$ imply $|ab| \leqslant a'b'$, a simple induction argument shows that each term of g in (17) is bounded in magnitude by the corresponding term of ϕ. Adding up the terms of g and ϕ, the result now follows from repeated use of the triangle inequality (6).

LEMMA 4. *For given $c > 0$ and a and f as in Theorem* 10, *there exists a constant $M \in F$ such that $|f(a + h) - f(a)| \leqslant M|h|$ for all $h \in F$ with $|h| \leqslant c$.*

Proof: In $f(x + h) - f(x)$, the terms in x alone cancel out, so there exists a polynomial $g(x, h)$ over F such that $f(x + h) - f(x) = hg(x, h)$. We now let M be the element $\phi(|a|, c)$ of F obtained from $g(x, h)$ by substituting for each coefficient of g its absolute value, substituting $|a|$ for x, and substituting c for h. From Lemma 3, it follows that $|g(a, h)| \leqslant M$ for all h with $|h| \leqslant c$. Therefore,

$$|f(a + h) - f(a)| = |h| \cdot |g(a, h)| \leqslant |h| \cdot M, \qquad \text{Q.E.D.}$$

Proof of theorem: Let $\varepsilon > 0$ be given; choose M as in Lemma 4, and then $\eta = \varepsilon/(M + 1)$. Since $\lim a_n = a$, we can find $m \in \mathbf{N}$ so large that $k > m$ implies $|a_k - a| < \eta$. It will then follow from Lemma 4 that

$$|f(a_k) - f(a)| < \varepsilon = (M + 1)\eta \qquad \text{for all } k > m.$$

This completes the proof.

We next define a *Cauchy sequence* in an ordered field F to be an infinite sequence denoted $\{x_n\} = x_1, x_2, x_3, \cdots$ such that, for any $\varepsilon > 0$ in F, there exists an integer m such that

$$j > m, k > m \quad \Rightarrow \quad |x_j - x_k| < \varepsilon. \tag{18}$$

LEMMA 5. *If a sequence $\{x_n\}$ in an ordered field F has a limit $a = \lim x_n$, then it is a Cauchy sequence.*

Proof: Given $\varepsilon > 0$, by (14) there exists an integer m such that $|x_j - a| < \varepsilon/2$ and $|x_k - a| < \varepsilon/2$ for all $j, k > m$. For this m, (18) clearly holds.

Caution: Although the preceding properties of limits are true in all ordered fields, many other familiar properties of limits of real numbers are not. Thus we will construct, in §9, an ordered field in which it is *not* true that the sequence $1, \frac{1}{2}, \frac{1}{3}, \cdots$ converges to zero.

Using Cauchy sequences, the field $\mathbf{R}$ of real numbers can be constructed from the field $\mathbf{Q}$ by a method due to G. Cantor (Exercises 2–6 below).

EXERCISES

1. Show that if $\{x_n\}$ and $\{y_n\}$ are Cauchy sequences of rational numbers, then so are:

(a) $\{x_n + y_n\}$. (b) $\{x_n y_n\}$.

(c) $\{-x_n\}$. (d) if no $y_n = 0$ and $\lim y_n \neq 0$, $\{1/y_n\}$.

2. Define $\{x_n\} \equiv \{x'_n\}$ to mean that $\lim (x_n - x'_n) = 0$. Show that this is an equivalence relation on sequences in any ordered field.

3. For Cauchy sequences, and using the fact (§5) that every Cauchy sequence is bounded, show that the equivalence relation of Exercise 2 has the property that $\{x_n\} \equiv \{x'_n\}$ implies

$$\{x_n + y_n\} \equiv \{x'_n + y_n\} \quad \text{and} \quad \{x_n y_n\} \equiv \{x'_n y_n\}.$$

4. Let S be the subset of Cauchy sequences in $\mathbf{Q}$, and let $\equiv$ be the equivalence relation of Exercise 2.

(a) Show that $S/\equiv$ is an integral domain under the operations of addition and multiplication defined as in Exercise 1(a) and 1(b).

(b) Show that $p: S \to S/\equiv$ is a morphism of rings.

5. Show that the ring $S/\equiv$ of Exercise 4 is a field.

6. Defining $\{a_n\} > 0$ in $S/\equiv$ to mean that there is a $b > 0$ with $a_n > b$ for all sufficiently large n, show that the integral domain $S/\equiv$ of Exercise 4 is an ordered domain.

7. Prove in detail that the sequence $1, \frac{1}{2}, \frac{1}{3}, \cdots$ converges to zero in $\mathbf{Q}$.

8. Let $x_1 = 1$, and let $x_{n+1} = (x_n{}^2 + 2)/2x_n$ in $\mathbf{Q}$. Show that $\{x_n\}$ is a Cauchy sequence in $\mathbf{Q}$, but that $\lim x_n = a$ for no $a \in \mathbf{Q}$. (*Hint:* $\lim x_n = \sqrt{2}$.)

5. The Real Field R

The real field $\mathbf{R}$ is distinguished from other ordered fields by being *complete*. Completeness can be defined in various ways, for example by the condition (converse of Lemma 5) that every Cauchy sequence has a limit. We prefer to adopt another, equivalent definition which involves the concept of greatest lower bound.

DEFINITION. A lower bound *to a set* S *of elements of an ordered domain* D (*or any other poset*) *is an element* b *of* D (*which need not be in* S) *such that* $b \leqslant s$ *for all* $s \in S$. *A* greatest lower bound (*g.l.b.*) *of* S *is a lower bound* b *of* S *such that* $c \leqslant b$ *for every other lower bound* c *of* S. *The concepts of* upper bound *and* least upper bound (*l.u.b.*) *of* S *in* D *are defined dually, by interchanging* $<$ *with* $>$ *throughout.*

In case $S = \{a, b\}$ is a set consisting of the two elements a and b only, then g.l.b. $\{a, b\} = a \wedge b$ is the *meet* of a and b as already defined in §IV.6, while l.u.b. $\{a, b\} = a \vee b$ is their *join*.

PROPOSITION 11. *Let $x_1 \geqslant x_2 \geqslant x_3 \geqslant \cdots$ be a nonincreasing sequence of elements of an ordered field F, with g.l.b. $\{x_n\} = a$. Then $\lim x_n = a$.*

Proof: Let $\varepsilon > 0$ be given. Since $a + \varepsilon > a$, $a + \varepsilon$ cannot be a lower bound to $\{x_n\}$. Hence, $x_n < a + \varepsilon$ for some $n = n(\varepsilon)$; but as the sequence is nonincreasing and a is a lower bound, this implies for all $k \geqslant n(\varepsilon)$ that $a \leqslant x_k < a + \varepsilon$, and so $|x_k - a| < \varepsilon$, proving (14).

The converse also holds; its proof is left as an exercise.

We now define an ordered domain D to be *complete* when every nonempty set S of positive elements of D has a g.l.b. in D. Our basic axiom on the real field, whose existence we assume (cf. Exercise 12 below), is the following:

AXIOM. *The real numbers form a complete ordered field* **R**.

Some properties of **R** follow almost immediately from this axiom. For example, we have:

PROPOSITION 12. *In* **R**, *every nonempty subset S which has a lower bound has a g.l.b., and dually, every nonempty subset T which has an upper bound has a l.u.b.*

Proof: Suppose that S has a lower bound b. If $1 - b$ is added to each number s of S, there results a set S* of positive numbers $s - b + 1$. By our postulate, this set S* has a g.l.b. c^*. The number $c = c^* + b - 1$ is then a g.l.b. for the original set S, as may be readily verified.

Dually, if the set T has an upper bound a, the set of all negatives $-t$ of elements of T has a lower bound $-a$, hence, by the previous proof, has a greatest lower bound b^*. The number $a^* = -b^*$ is then a least upper bound of the given set T, Q.E.D.

PROPOSITION 13 (*Archimedean Law*). *Given $a > 0$ and $b > 0$ in* **R**, *there exists a natural number n such that $na > b$.*

Proof: Suppose the conclusion false for two particular real numbers a and b, so that, for every n, $b \geqslant na$. The set S of all the multiples na then has the upper bound b, so that it has also a least upper bound b^*. Therefore, $b^* \geqslant na$ for every n, so that also $b^* \geqslant (m + 1)a$ for every m. This implies

$b^* - a \geqslant ma$, so that $b^* - a$ is an upper bound for the set S of all the multiples na of a, although it is smaller than the given least upper bound b. This contradiction proves the proposition.

COROLLARY. *Given real numbers a and b, with $a > 0$, there exists a (unique) integer q with $b = qa + r$, $0 \leqslant r < a$.*

We omit the proof. Note that Proposition 13 does not hold in the (incomplete) ordered field $\mathbf{R}[[t]]$ of real formal power series, since $t > 0$ and $1 > 0$, yet $nt < 1$ for all integers n (see §9 below).

By the Corollary of Theorem 6, the prime subfield of **R** is isomorphic to **Q**. If we identify it with **Q** by this isomorphism, we can display **Q** as a *dense* subset of **R**, in the following sense.

THEOREM 14. *Between any two real numbers $c > d$, there exists a rational number m/n with $c > m/n > d$.*

Proof: By hypothesis, $c - d > 0$, so the Archimedean law yields a positive integer n such that $n(c - d) > 1$, or $1/n < c - d$. Now let m be the smallest integer such that $m > nd$; then $(m - 1)/n \leqslant d$, so that

$$m/n = (m - 1)/n + 1/n < d + (c - d) = c. \tag{19}$$

Since $m/n > d$, this completes the proof.

COROLLARY. *Every real number is the l.u.b. of a set of rational numbers.*

Proof: For a given real number c, let S denote the set of all rationals $m/n \leqslant c$. Then c is an upper bound of S; by the theorem, no smaller real number d could be an upper bound of S, hence c is the *least* upper bound of S.

Completeness and Cauchy Sequences. We conclude this section by showing that our definition of completeness by g.l.b. implies completeness in terms of Cauchy sequences.

PROPOSITION 15. *In* **R**, *every Cauchy sequence $\{x_n\}$ converges to some limit.*

Proof: We first observe that every Cauchy sequence $\{x_n\}$ is bounded, and that this is true in any ordered field F. For, setting $\varepsilon = 1$ in the definition of a Cauchy sequence, we have, for the corresponding $n = n(1)$,

$$\min\{x_1, x_2, \cdots, x_{n(1)} - 1\} \leqslant x_k \leqslant \max\{x_1, x_2, \cdots, x_{n(1)} + 1\}.$$

for all x_k in the sequence. Hence, in **R**, the elements $a_n = \text{l.u.b. } \{x_k | k \geqslant n\}$ exist by Proposition 12. But the sequence $\{a_n\}$ is nonincreasing by definition. Hence, by Proposition 11, $\lim a_n = a$ for suitable $a \in \mathbf{R}$. Since $\{x_n\}$ is a Cauchy sequence, it follows that for any $\varepsilon > 0$,

$$|x_k - a| \leqslant |x_k - a_k| + |a_k - a| < \varepsilon \qquad \text{for all } k \geqslant n,$$

some sufficiently large natural number n (we omit the details). This shows that $\lim x_k = a$ in **R**, as claimed.

EXERCISES

1. Show that there is no ordered domain D in which *every* nonempty set has a l.u.b. (*Hint:* Show that D itself can have no upper bound.)
2. Show that the ordered domain **Z** is complete.
3. Let S and T be sets of real numbers with the respective least upper bounds b and c. What is the least upper bound (i) of the set $S + T$ of all sums $s + t$ (for $s \in S$, $t \in T$), and (ii) of the set-union $S \cup T$?
4. Show that an element a in an ordered field F is a least upper bound for a set S if and only if (i) $s \leqslant a$ for all $s \in S$, and (ii) for each positive $e \in F$ there is an $s \in S$ with $|s - a| < e$.
5. Show that between any two real numbers $c < d$, there exists a rational cube $(m/n)^3$ such that $c < (m/n)^3 < d$.
6. Show that if $n > 1$ is a fixed natural number, then between any two real numbers $c < d$ there lies a rational number of the form m/n^k, where $m \in \mathbf{Z}$ and $k \in \mathbf{N}$.
7. Given a, b, c, d positive real numbers, show that $a/b = c/d$ if and only if $na > mb \Rightarrow nc > md$ and $na < mb \Rightarrow nc < md$, for all positive integers m and n.
8. Prove the converse of Proposition 11.
9. Prove the Corollary of Proposition 13.
10. Prove that every subfield of **R** satisfies Proposition 13.

*11. Prove that if an ordered field F satisfies Proposition 13, there is an order-monomorphism of rings from F to **R**. (*Hint:* Apply the condition of Theorem 14 to the elements of F, to represent each element of F (and of **R**) as the l.u.b. of a suitable subset of **Q**.)

*12. Show that the ordered domain of Exercise 6 of §4 is complete.

*13. Prove that any two complete ordered fields are order-isomorphic.

*6. Polynomials over R

Trivially, linear polynomials are irreducible over any field. We have shown (Proposition 9) that quadratic polynomials with negative discriminant are irreducible over any ordered field. Thus, over the real field **R**, both of the preceding sets of polynomials are irreducible. In §§6 and 7, we shall show that these are the *only* irreducible polynomials over **R** (the only primes of **R**$[x]$). This is one of the most important results of classical algebra.

In the present section, we shall establish some weaker related results. Many of these are corollaries of

THEOREM 16. *Let f be a real polynomial; let $a < b$; and let d satisfy $f(a) < d < f(b)$. Then the equation $f(x) = d$ has a solution (root) $x = c$ with $a < c < b$.*

Proof: Let S be the set of real numbers x between a and b which satisfy $f(x) \leqslant d$. Since $f(a) < d$, S is nonvoid, and it has b as upper bound; hence, it has a real *least* upper bound c. We shall show that $f(c) = d$.

For this purpose, it is clearly sufficient to exclude the possibilities $f(c) < d$ and $f(c) > d$. Moreover, by Lemma 4 of §4 (in changed notation), there is an $M > 0$ such that the inequality $|f(x) - f(c)| \leqslant M|x - c|$ holds for all x with $a \leqslant x \leqslant b$.

Now suppose that $f(c) < d$, and let $h = [d - f(c)]/M$ or $(b - c)/2$, whichever is less. Then $f(c + h) \leqslant d$ by the above inequality, contradicting our definition of c as an upper bound to S. On the other hand, $f(c) > d$ would imply $f(c - h) > d$ for all positive h with $c - h \geqslant a$ and $h \leqslant [f(c) - d]/2M$. This would contradict our definition of c as the *least* upper bound of S, for it shows that $c - h$, for any such h, would be a smaller upper bound. There remains only the possibility that $f(c) = d$, Q.E.D.

From the theorem one readily proves

COROLLARY 1. *If f is a polynomial with positive coefficients and no constant term, and if $d > 0$, then $f(x) = d$ has a positive real root.*

COROLLARY 2. *If f is of odd degree, then $f(x) = c$ has a real root for every real number c.*

As a special case of Corollary 1, we obtain the converse of Proposition 1, §1: In **R**, a number is nonnegative if *and only if* it has a square root.

In turn, this result has a further consequence, when considered in the light of the obvious fact that having a square root is preserved under any automorphism of rings. Since $a \geqslant b$ in **R** if and only if $a - b$ has a square root, we conclude that any *ring*-automorphism of **R** is also an *order*-automorphism.

On the other hand, the restriction of any ring-automorphism α of **R** to its prime field **Q** must be the identity $1_{\mathbf{Q}}$. Hence, for any $a \in \mathbf{R}$, α must map the set L of all rational numbers x with $x \leqslant a$ onto itself. Since (as just shown) α is necessarily an order-automorphism of **R**, it must preserve least upper bounds. Consequently, $\alpha(a) = \alpha(\text{l.u.b. } L) = \text{l.u.b. } \alpha(L) = a$. We have proved the following result.

Proposition 17. *The only order-automorphism of the field* $\mathbf{R}$ *is the identity: In symbols,* $\text{Aut } \mathbf{R} = \{1_{\mathbf{R}}\}$.

Theorem 16, though formulated as a theorem about polynomials, is actually true for continuous functions $f:\mathbf{R} \to \mathbf{R}$ in general. The same is true of the following result, formulated as a theorem about polynomials in n variables.

Minimum Property. Given a real polynomial function $f:\mathbf{R}^n \to \mathbf{R}$, and a closed bounded set $S \subset \mathbf{R}^n$, there is a point $\mathbf{c} = (c_1, \cdots, c_n) \in S$ such that $f(\mathbf{c}) \leqslant f(\mathbf{x})$ for all $\mathbf{x} \in S$, that is, which *minimizes* $f(\mathbf{x})$ on S.

Explanation. A set $S \subset \mathbf{R}^n$ is called *bounded* when, for some fixed $M > 0$, $|x_i| < M$ for all $\mathbf{x} = (x_1, \cdots, x_n) \in S$ and $i = 1, \cdots, n$. It is called *closed* when $\lim x_i^{(m)} = a_i$ for $i = 1, \cdots, n$, all $x_i^{(m)} \in S$, imply $(a_1, \cdots, a_n) \in S$.

Though we shall use this Minimum Property below, we shall omit its proof as belonging to analysis.

EXERCISES

1. Show that, for any positive real number a and any positive integer n, the equation $x^n = a$ has one and only one positive real root $\sqrt[n]{a}$.

2. Show that $x^4 - x = c$ has two real roots for every real $c > -\frac{3}{8}$.

3. Show that a monic polynomial of even degree assumes a least real value k, and every real value $c > k$ at least twice.

4. Prove Corollaries 1 and 2 of Theorem 16.

5. (a) If a and b are positive reals, show that $ax^{n+1} > bx^n$ for all sufficiently large positive x.

   (b) Given a polynomial f with positive leading coefficient, show how to obtain a real number M such that $f(x) > 0$ for all $x > M$.

*6. (a) Show that, in the ring $\mathbf{Q}[[t]]$ as defined in §9 below, $x^2 = 1 + t$ for

$$(*)\ x = 1 + \tfrac{1}{2}t + \tfrac{1}{2}\left(-\tfrac{1}{2}\right)t^2/2! + \tfrac{1}{2}\left(-\tfrac{1}{2}\right)\left(-\tfrac{3}{2}\right)t^3/3! + \cdots .$$

   (b) Show that, in $\mathbf{R}[[t]]$, the evaluation of (*) for any fixed $t \in \mathbf{R}$ gives a Cauchy sequence.

*7. (a) Using (*), compute $\sqrt{5}$ to four decimal places.

   (b) Compute $\sqrt{2}$ to six decimal places. (*Hint:* $(5\sqrt{2}/7)^2 = 1 + 1/49$.)

7. The Complex Plane

The complex field $\mathbf{C}$ can be defined in various ways, all of which yield isomorphic fields. We shall define it here as a quotient ring in the sense of

§III.13. Specifically, in the polynomial ring $\mathbf{R}[t]$ in an indeterminate t take the principal ideal $(t^2 + 1)$ and define $\mathbf{C}$ to be the quotient ring $\mathbf{R}[t]/(t^2 + 1)$. Since the polynomial $t^2 + 1$ is irreducible over $\mathbf{R}$, this ring $\mathbf{C}$, as defined, is a field. Every element of $\mathbf{C}$ is uniquely represented by a linear polynomial (a remainder on division by $t^2 + 1$). Thus, if we write i for the coset of t, every element of $\mathbf{C}$ can be written uniquely in the form $x + yi$, where x and y are real numbers (*not* indeterminates). The ring operations on these elements are

$$(x + yi) + (x' + y'i) = (x + x') + (y + y')i, \tag{20}$$

$$(x + yi)(x' + y'i) = (xx' - yy') + (xy' + yx')i. \tag{21}$$

Moreover, as a corollary of Theorem III.8, we have

Theorem 18. *The complex field* $\mathbf{C}$ *contains a subfield isomorphic to* $\mathbf{R}$, *and a square root of* -1. *In any field* F *which contains* $\mathbf{R}$ *and a square root* j *of* -1, *the subfield generated by* $\mathbf{R}$ *and* j *must be isomorphic to* $\mathbf{C}$.

Note that there is no positive subset which makes $\mathbf{C}$ into an ordered field, because -1 is a negative number which cannot have a square root in any ordered field (Proposition 1). Note also that the multiplicative inverse of $x + yi \neq 0$ in $\mathbf{C}$ can be computed explicitly by the formula

$$(x + yi)^{-1} = [x/(x^2 + y^2)] + [-y/(x^2 + y^2)]i. \tag{22}$$

To visualize the complex field, picture the complex number $z = x + yi$ as the point $P = (x, y)$ of the Cartesian plane. This defines the *complex plane*. If one pictures z as standing for the vector $\mathbf{OP}$ from the origin $O = (0, 0)$ to P, then complex addition as defined by (20) becomes vector addition by the parallelogram law.

Again, consider the function which assigns to each complex number $z = x + yi$ its *complex conjugate* $z^* = x - yi$. Geometrically, this amounts to a reflection of the complex plane in the *real axis* (the x-axis $y = 0$); it is a continuous transformation. What is much more interesting, it is an automorphism of $\mathbf{C}$.

Proposition 19. *The assignment* $z \mapsto z^*$ *is a ring-automorphism of* $\mathbf{C}$ *which leaves all real numbers fixed:*

$$(z_1 + z_2)^* = z_1^* + z_2^*, \qquad (z_1 z_2)^* = z_1^* z_2^*, \qquad (z^*)^* = z \tag{23}$$

for all z_1, z_2, z *in* $\mathbf{C}$.

The verification of the first two formulas of (23) is straightforward, using (20) and (21); the third formula of (23) is trivial. Although $\mathbf{C}$ has many other

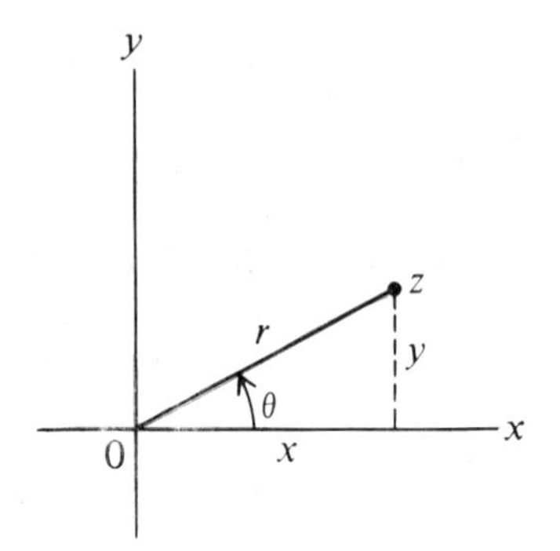

FIGURE VIII-1

automorphisms (interchanging $\pm\sqrt{2}$, for example), it has no other *continuous* automorphisms.

Most remarkable are the connections between the multiplication in **C** and trigonometric formulas, which correspond to the geometrical representation above. Namely, let $r = \sqrt{x^2 + y^2}$ be the (nonnegative) length of the segment joining the origin to the point z, while θ is the angle, with $0 \leqslant \theta < 2\pi$ from the x-axis to this segment (see Figure VIII-1), so that

$$|z| = r = (x^2 + y^2)^{1/2}, \qquad \arg z = \theta = \tan^{-1}(y/x). \tag{24}$$

One calls r the *absolute value* of the complex number z and θ the *argument* of z. The polar coordinates $r \geqslant 0$ and θ determine x and y by

$$x = r\cos\theta, \qquad y = r\sin\theta, \qquad z = r(\cos\theta + i\sin\theta), \tag{25}$$

the usual laws for the transformation from polar to rectangular coordinates. One also writes (25) in the form $z = re^{i\theta}$, since the usual Taylor series expansion gives

$$e^{i\theta} = 1 + i\theta + \frac{(-1)\theta^2}{2!} + \frac{(-i)\theta^3}{3!} + \cdots = \cos\theta + i\sin\theta.$$

THEOREM 20. *The absolute values of a product of complex numbers is the product of the absolute values of the factors; its argument is the sum of the arguments of the factors. In symbols:*

$$|zz'| = |z|\cdot|z'|, \qquad \arg zz' = \arg z + \arg z' \pmod{2\pi}. \tag{26}$$

Proof: As in (25), $z = r(\cos\theta + i\sin\theta)$, $z' = r'(\cos\theta' + i\sin\theta')$. Substituting in (21), we get

$$zz' = rr'[(\cos\theta\cos\theta' - \sin\theta\sin\theta') + i(\cos\theta\sin\theta' + \sin\theta\cos\theta')];$$

by well-known trigonometric formulas this is equivalent to

$$zz' = rr'\left[\cos(\theta + \theta') + i \sin(\theta + \theta')\right].$$

This gives the result (26).

Not only the multiplicative, but the additive properties of (inequalities on) absolute values are valid for complex as well as real numbers. That is,

$$|z| > 0 \qquad \text{unless } z = 0, \quad |0| = 0; \tag{27}$$

$$|z + z'| \leqslant |z| + |z'|. \tag{28}$$

To prove these, we note that formula (20) means that the sum $z + z'$ may be found by drawing (Figure VIII-2) the parallelogram with three vertices at z, 0, and z'; the fourth vertex will be $z + z'$. Formulas (27) and (28) now follow from the identity between absolute values and geometrical lengths.

The complex solutions of the equation $z^n = 1$ are called the nth *roots of unity*. By formula (26) they can be located in the complex plane; they are the n complex numbers z_k with $|z_k| = 1$ and $\arg z_k = 2\phi k/n$, $k = 0, 1, \cdots, n - 1$, so they are also the n vertices of a regular polygon of n sides inscribed in the unit circle $|z| = 1$ and with one vertex at the point $z = 1$. By (26), $z_h z_k = z_{h+k}$, all subscripts being taken mod n; moreover, $z_n = 1$. In other words, there are exactly n complex nth roots of unity; they form a group under multiplication and the assignment $k \mapsto z_k$ is an isomorphism of the additive cyclic group $\mathbf{Z}_n$ to the multiplicative group of nth roots of unity.

Since $z_1{}^k = z_k$, the complex number $z_1 = e^{2\pi i/n}$ generates this cyclic group; the other generators are the z_m with $(m, n) = 1$. These generators are called the "primitive" nth roots of unity; they are the numbers whose order is exactly n in the multiplicative group of all non-zero complex numbers.

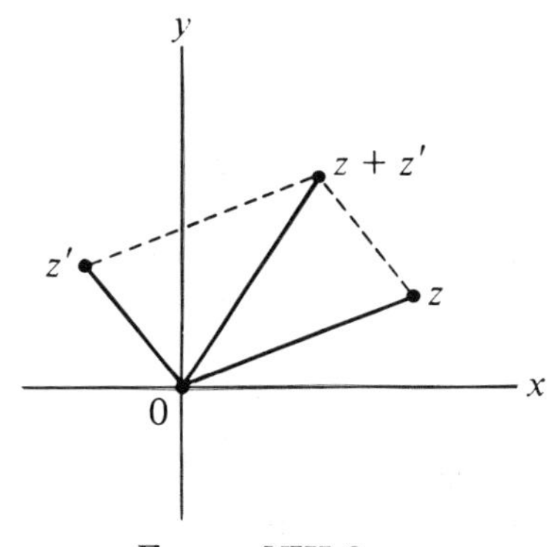

FIGURE VIII-2

Fundamental Theorem of Algebra. The importance of the complex field stems from the following basic result, often called the "Fundamental Theorem of Algebra".

THEOREM 21. *Every nonconstant polynomial f with coefficients in* **C** *has a root in* **C**.

There are many proofs of Theorem 21; all of them use non-algebraic ideas. We will confine ourselves here to explaining some of its algebraic consequences.

By Corollary III.19.2, Theorem 21 implies that every complex polynomial has a complex linear factor. This proves

COROLLARY 1. *A polynomial of* $\mathbf{C}[z]$ *is irreducible if and only if it is linear.*

Now applying the unique factorization theorem for polynomials, in the indeterminate z, over an arbitrary field, we obtain

COROLLARY 2. *Each polynomial f of degree m in* $\mathbf{C}[z]$ *can be written in one and (apart from permutations of the factors) only one way in the form*

$$f = c(z - z_1)(z - z_2) \cdots (z - z_m), \tag{29}$$

where $c \neq 0$ and $z_1, \cdots, z_m$ are complex numbers.

Clearly, the z_j in (29) are the roots of $f(z)$; if any z_j occurs exactly r times, it is called a root of *multiplicity* r.

Next, we consider the effect of the automorphism $z_i \mapsto z_i^*$ on the representation (29). If $f \in \mathbf{R}[z]$ (that is, if its coefficients are *real*), then this automorphism must permute the $(z - z_j)$ in (29) among themselves, and $c = c^*$ must be real, since $c^*(z - z_1^*)(z - z_2^*) \cdots (z - z_m^*) = f^*(z) = f(z)$. Hence, we have

PROPOSITION 22. *The nonreal complex roots of any real polynomial occur in conjugate pairs, paired roots having equal multiplicity.*

Further, a pair of complex conjugate linear factors $z - z_j$ and $z - z_j^*$ in (29), with $z_j = a + bi$, $b \neq 0$, can be multiplied to give

$$(z - (a + bi))(z - (a - bi)) = z^2 - 2az + (a^2 + b^2),$$

which is a real quadratic polynomial which has a negative discriminant $4a^2 - 4(a^2 + b^2) = -4b^2 < 0$. This proves

Theorem 23. *Any real polynomial can be uniquely factored into (real) linear factors and real quadratic factors with negative discriminant.*

Comparing with Proposition 9, we obtain the following:

Corollary. *A polynomial of* $\mathbf{R}[x]$ *is irreducible if and only if it is linear, or quadratic with negative discriminant.*

EXERCISES

1. Is the quotient-ring $\mathbf{C}[z]/(z^2 + 1)$ an integral domain? Justify your answer.

2. Factor in $\mathbf{C}[z]$:
(a) $z^2 + z + 1 + i$. (b) $z^4 + z^3 - z - 1$.

3. (a) Show that if $f = f_0 + f_1 z + f_2 z^2 + \cdots + f_{2n} z^{2n}$ satisfies $f_k = f_{2n-k}$ for all k, then $f(z) = z^n g(z + z^{-1})$, where g is a polynomial of degree n.

(b) Find the irreducible factors of $z^4 + 3z^3 + 4z^2 + 3z + 1$ in $\mathbf{C}[z]$ and in $\mathbf{R}[z]$.

*(c) Generalize the result of (a) to polynomials of degree $2n - 1$.

4. Show that the equation $z^n = c$ has n distinct complex roots for any $c \neq 0$ in $\mathbf{C}$, and that these constitute the vertices of a regular polygon.

5. Let $b^2 < 4ac$, where $a, b, c \in \mathbf{R}$. Construct an isomorphism $\mathbf{C} \cong \mathbf{R}[x]/(ax^2 + bx + c)$.

***6.** Let $f = z^3 + f_1 z + f_0$ have roots z_1, z_2, z_3. Show that

$$[(z_1 - z_2)(z_2 - z_3)(z_1 - z_3)]^2 = -4f_1^3 - 27f_0^2.$$

(The right side of this equation is called the *discriminant* of f.)

***7.** Use partial fractions to show that any rational function over the field $\mathbf{C}$ can be written as the sum of a polynomial and rational functions, in which each numerator is a constant and each denominator is a power of a linear function.

***8.** Using the relation $\cos\theta = (e^{i\theta} + e^{-i\theta})/2$, show that there is a polynomial T_n such that $\cos n\theta = T_n(\cos\theta)$. (This is called the *Chebyshev polynomial* of degree n.)

*8. The Quaternions

We stated in §VI.3 that most of the properties of vector spaces remained true for scalars from a division ring (noncommutative field). As the most interesting example of such a division ring, we now construct the ring of quaternions. The construction uses the field $\mathbf{C}$ of complex numbers and the automorphism $z = x + iy \mapsto z^* = x - iy$ of $\mathbf{C}$, which carries each $z \in \mathbf{C}$ to its complex conjugate z^*.

Let Q denote a two-dimensional (left) vector space over $\mathbf{C}$ with a selected basis of two elements, written $\mathbf{1}$ and $\mathbf{j}$. Elements of Q thus have the form $q = z_1\mathbf{1} + z_2\mathbf{j}$, $r = w_1\mathbf{1} + w_2\mathbf{j}$, and so on. We will define the *product* of two such "quaternion" numbers as

$$qr = (z_1w_1 - z_2w_2^*)\mathbf{1} + (z_1w_2 + z_2w_1^*)\mathbf{j}. \tag{30}$$

In particular, the basis vectors $\mathbf{1}$ and $\mathbf{j}$ of Q multiply by the rules

$$q\mathbf{1} = \mathbf{1}q = q, \qquad \text{all } q \in Q; \qquad \mathbf{j}^2 = -1, \qquad \mathbf{j}w = w^*\mathbf{j}. \tag{31}$$

The last rule shows that multiplication is not commutative.

We wrote w for $w\mathbf{1} + 0\mathbf{j}$ in (31) because the basis vector $\mathbf{1}$ is a unit for the product (30); more generally, we will write $z_1 + z_2\mathbf{j}$ for $z_1\mathbf{1} + z_2\mathbf{j}$. Quaternions of the form $z + 0\mathbf{j}$ clearly form an isomorphic copy of $\mathbf{C}$ under (30), as in the second statement of

Theorem 24. *The additive group of the vector space Q with the product (30) is a division ring containing a copy of $\mathbf{C}$ as a subring.*

Proof: Direct calculation shows that the product defined above is associative, has $\mathbf{1}$ as unit, and is distributive on the left or the right over the vector space sum. Therefore, Q is a ring. It remains to prove that every non-zero quaternion q has a two-sided multiplicative inverse. To prove this, first define the "conjugate" q^* of q to be $(z_1 + z_2\mathbf{j})^* = z_1^* - z_2\mathbf{j}$. From the definition of the product, one then sees that $(qr)^* = r^*q^*$, so $q \mapsto q^*$ is an isomorphism of Q to the opposite ring Q^{op}. Next, define the "norm" of q to be $N(q) = qq^*$, so

$$N(z_1 + z_2\mathbf{j}) = (z_1 + z_2\mathbf{j})(z_1^* - z_2\mathbf{j}) = z_1z_1^* + z_2z_2^*.$$

Then $N(q)$ also equals the product q^*q. More important, for any $z \in \mathbf{C}$, zz^* is a nonnegative real number, positive when $z \neq 0$, and hence $N(q)$ is real and positive when $q \neq 0$. Since $N(r) = rr^*$ is real, it commutes with every quaternion (see (30) below), so

$$N(qr) = qr(qr)^* = qrr^*q^* = qq^*rr^* = N(q)N(r).$$

Therefore, $N: Q \to \mathbf{R}$ is a morphism of multiplication. If $q \neq 0$,

$$q[q^*(Nq)^{-1}] = (Nq)(Nq)^{-1} = [q^*(Nq)^{-1}]q,$$

so $q \neq 0$ has a two-sided inverse $q^{-1} = q^*(Nq)^{-1}$, and Q is a division ring.

The division ring Q is called the *ring of quaternions*.

Now observe that for r *real* (that is, for $r = w_1 + w_2\mathbf{j}$ with $w_1 = u$ real and $w_2 = 0$), (30) gives

$$qr = z_1u + z_2u\mathbf{j} = uz_1 + uz_2\mathbf{j} = rq. \tag{32}$$

Conversely (Exercise 6), $qr = rq$ for all $q \in Q$ implies $r \in \mathbf{R}$; that is, the subfield $\mathbf{R}$ of $\mathbf{C}$ in Theorem 24 is the *center* of Q.

Corollary. *The quaternions are a four-dimensional vector space over* $\mathbf{R}$ *with basis* $\mathbf{1}$, $\mathbf{i}$, $\mathbf{j}$, *and* $\mathbf{k} = \mathbf{ij}$.

Proof: Q is given as a vector space over $\mathbf{C}$, with scalar multiples zq. Using only the multiples by real z makes Q a vector space over $\mathbf{R}$. For $z_1 = x_0 + \mathbf{i}x_1$ and $z_2 = x_2 - \mathbf{i}x_3$ each quaternion q has the form

$$q = z_1 + z_2\mathbf{j} = x_0 + x_1\mathbf{i} + x_2\mathbf{j} + x_3\mathbf{k} = x_0 + \mathbf{i}x_1 + \mathbf{j}x_2 + \mathbf{k}x_3$$

for all x_i real, so Q has the basis $\mathbf{1}$, $\mathbf{i}$, $\mathbf{j}$, and $\mathbf{k}$, as asserted.

From the start, Q might have been described as a four-dimensional vector space over $\mathbf{R}$ with this basis and with product determined by the ring axioms and the rules

$$\mathbf{i}^2 = \mathbf{j}^2 = \mathbf{k}^2 = -1, \qquad \mathbf{ij} = \mathbf{k} = -\mathbf{ji}, \qquad \mathbf{jk} = \mathbf{i} = -\mathbf{kj}, \qquad \mathbf{ki} = \mathbf{j} = -\mathbf{ik}.$$

These rules specify the product of any two of these basis elements.

EXERCISES

1. Prove that the multiplication of quaternions is associative.
2. Prove that there are infinitely many quaternions q with $q^2 = -1$.
3. Let $q = x_0 + \mathbf{i}x_1 + \mathbf{j}x_2 + \mathbf{k}x_3$ be a quaternion.
   (a) Prove that $q^* = x_0 - \mathbf{i}x_1 - \mathbf{j}x_2 - \mathbf{k}x_3$ and $N(q) = x_0^2 + x_1^2 + x_2^2 + x_3^2$.
   (b) Show that q and q^* are both roots of the quadratic polynomial $t^2 - 2x_0t + N(q)$ in the indeterminate t.
4. (a) Show that the eight quaternions $\pm\mathbf{1}$, $\pm\mathbf{i}$, $\pm\mathbf{j}$, and $\pm\mathbf{k}$ form a multiplicative group (called the "quaternion group").
   (b) Show that this group has the defining relations $\mathbf{i}^4 = 1$, $\mathbf{i}^2 = \mathbf{j}^2$, and $\mathbf{j}^{-1}\mathbf{ij} = \mathbf{i}^{-1}$, as in Exercise II.5.8.
5. If the integers m and n are both sums of four squares of integers, prove that the product mn is also a sum of four squares. (*Hint:* Use the quaternion norm.)
6. The "center" of any ring R is the set of all $x \in R$ such that $xa = ax$ for all $a \in R$.
   (a) Show that the center of any ring R is a subring of R.
   (b) Show that if R is a division ring, its center is a field.
   (c) Prove in detail that the center of Q is precisely $\mathbf{R}$.
   (d) Infer that the embedding of $\mathbf{R}$ in Q, unlike that of $\mathbf{C}$, is preserved by all automorphisms of Q.

9. Extended Formal Power Series

In the calculus, one deals with real and complex *power series* of the form

$$f(x) = \sum_{k=0}^{\infty} f_k x^k = f_0 + f_1 x + f_2 x^2 + \cdots + f_n x^n + \cdots, \tag{33}$$

and ascribes to each such series a *radius of convergence*

$$R = 1/\lim inf \sqrt[n]{|f_n|} .$$

More generally, a power series like (33) can be considered purely formally, as just a list of coefficients; that is, as a function $n \mapsto f_n$ on the set $\mathbf{N}$ of nonnegative integers to the real numbers, the complex numbers, or any commutative ring K. Two such functions $f, g: \mathbf{N} \to K$ can be added and multiplied by formulas (27) and (28) of Chapter III. This defines from any such K a new commutative ring $K[[x]]$, called the *formal power series ring* in x over K. In this ring, the polynomials form the subring $K[x]$ of those formal power series which have only finitely many non-zero coefficients f_k.

If $K = F$ is a field, we can rewrite any non-zero $f \in F[[x]]$ in the form

$$f(x) = f_m x^m \left[1 + \sum_{k=1}^{\infty} (f_{m+k}/f_m) x^k \right], \qquad f_m \neq 0. \tag{34}$$

The m in (34) is called the *order* $o(f)$ of f; evidently $o(fg) = o(f) + o(g)$. If $o(f) > 0$ in (34), the equation $fg = 1$ cannot be solved for g in $F[[x]]$, so f is not invertible. Conversely, if $o(f) = 0$ (i.e., if $f_0 \neq 0$), then by (III.28) the formulas $g_0 = f_0^{-1}$ and

$$g_{n+1} = -(f_1 g_n + \cdots + f_{n+1} g_0)/f_0, \qquad n \in \mathbf{N},$$

gives an inverse $g(x) = [f(x)]^{-1} \in F[[x]]$ to $f(x)$. This proves the following result.

Proposition 25. *Over any field F, the formal power series* (33) *form an integral domain $F[[x]]$ whose invertibles are the f with $f_0 \neq 0$ (that is with $o(f) = 0$).*

We next prove

Proposition 26. *Over any commutative ordered ring K, the formal power series* (33) *constitute an ordered ring $K[[x]]$ if $f > 0$ is defined to mean that the first non-zero f_k is positive.*

Proof: Let the first non-zero coefficients of $f > 0$ and $g > 0$ in $K[[x]]$ be f_m and g_n, respectively. Then $f_m > 0$ and $g_n > 0$ in K, by definition; hence, the first non-zero coefficient $f_m g_n$ of fg is positive, and so $fg > 0$. The same is true of the first non-zero coefficient of $f + g$; this is $f_m + g_n$ if $m = n$, f_m if $m < n$, and g_n if $m > n$.

When $K = F$ is an ordered *field*, $F[[x]]$ is an ordered domain since $f_m g_n > 0$ as above. Hence, its field of quotients, $Q(F[[x]])$, must be an ordered field by Theorem 6. But if $g = g_n x^n[1 + \sum_{k=1}^{\infty}(g_{n+k}/g_n)x^k] \neq 0$, Proposition 25 gives an explicit solution of the equation $gh = f$ in the form

$$h = h_{m-n}x^{m-n}\left[1 + \sum_{k=1}^{\infty} h_{m-n+k}x^k\right], \qquad h_{m-n} = f_m/g_n, \qquad (35)$$

where $m - n$ may now be negative. There follows

Theorem 27. *The field $Q(F[[x]])$ of quotients of formal power series over the field F consists of 0 and the extended formal power series h of (35). It is an ordered field if F is, with $h > 0$ if and only if $h_{m-n} > 0$ in (35).*

The field $Q(F[[x]])$ is usually denoted by $F((x))$.

Note that the preceding ordered field is always *nonarchimedean*, since $x > 0$ yet $nx < 1$ for all $n \in \mathbf{N}$ (cf. Proposition 13). It follows that the sequence $1, \frac{1}{2}, \frac{1}{3}, \cdots, \frac{1}{n}$ does not converge to zero as $n \to \infty$. Note also that when $F = \mathbf{C}$, those series (34) having a positive radius of convergence represent actual functions $\tilde{f}$, whose order $m = o(f)$ is the order (that is, the multiplicity) of the zero of $\tilde{f}$ at $z = 0$. Likewise, if $m < n$, (35) for $m < n$ represents a function having a pole of order $n - m$ at $z = 0$ in the sense of classical analysis.

EXERCISES

1. Compute the formal power series inverses of:

(a) $1 + x^3$. (b) $1 - x^3$. (c) $1 + x + x^2$.
(d) $1 - x + x^2$. (e) $1 + 2x + x^2$.

2. Show that, over an integral domain D, $f(x)$ in (33) is invertible in $D[[x]]$ if and only if f_0 has an inverse in D.

3. (a) Let $f(x)$ in (33) have a positive radius of convergence. Show that $f \in \mathbf{R}[[x]]$ is positive if and only if the function $f(x)$ has positive values for all sufficiently small positive x.

(b) Prove a similar result for $\mathbf{R}((x))$.

Let F be any field. Define the "valuation" $v(f)$ of a non-zero $f \in F((x))$ as $2^{-o(f)}$, and let $v(0) = 0$.

4. Show that $v(f)$ has the following properties, analogous to those of $|f|$:

(i) $v(0) = 0$, while $v(f) > 0$ if $f \neq 0$.
(ii) $v(f + g) \leqslant \max[v(f), v(g)] \leqslant v(f) + v(g)$.
(iii) $v(fg) = v(f)v(g)$.

5. If $f_n \to f$ in $F((x))$ is defined to mean that $v(f_n - f) \to 0$ in $\mathbf{R}$, show that $F[[x]]$ is the set of "limits" of sequences of polynomials.

In an ordered ring R, define $f \ll g$ to mean that $nf \leqslant g$ for every integer n, and $|f|$ to be $\pm f$, whichever is positive.

6. (a) Show that $f_1 \ll g$ and $f_2 \ll g$ imply $f_1 + f_2 \ll g$.

(b) With the same hypotheses, show that, in the notation of Theorem 3, $k_1 f_1 + k_2 f_2 \ll g$ for all $k_1, k_2 \in \mu(\mathbf{Z})$.

(c) Show that $|f_i| \ll |g_i|$, $i = 1, \cdots, n$, implies $\Pi|f_i| \ll \Pi|g_i|$.

7. Show that, in an ordered field, $fh \ll gh$ and $h > 0$ imply $f \ll g$.

Define $x \sim y$ in an ordered ring (in words, x and y have the same "order of magnitude") when for some positive integer n, $|f| \leqslant n|g|$ and $|g| \leqslant n|f|$.

8. Show that in an archimedean ring, $x \sim y$ for all non-zero x, y.

9. Show that if $f, g \neq 0$ in $\mathbf{R}((x))$, then $f \sim g$ if and only if $o(f) = o(g)$.

***10.** Show that there is an monomorphism of ordered rings from a given ordered field F to $\mathbf{R}$ if and only if F is archimedean.

11. Show that an ordered ring is archimedean if $y > 0$ and $x \ll y$ imply $x = 0$.

10. Valuations and p-adic Numbers

There are other ways of "completing" the rational field $\mathbf{Q}$ than that described in §5. Thus one can construct for each prime p a field of "p-adic numbers" using similar ideas. To explain this construction, we first define a *valuation* on an integral domain D as a function $v: D \to \mathbf{R}$ having the following properties of ordinary absolute values:

$$\begin{aligned} &\text{(i) } v(0) = 0, \qquad \text{while } v(a) > 0 \text{ if } a \neq 0. \\ &\left.\begin{aligned} &\text{(ii) } v(ab) = v(a)v(b) \\ &\text{(iii) } v(a+b) \leqslant v(a) + v(b) \end{aligned}\right\} \quad \text{for all } a, b \in D. \end{aligned}$$

For any such valuation, the "distance" $d(x, y) = v(x - y)$ will then share with ordinary distance the following properties:

M1. $d(x, x) = 0$, while $d(x, y) > 0$ if $x \neq y$.
M2. $d(x, y) = d(y, x)$ (symmetry).
M3. $d(x, y) + d(y, z) \geqslant d(x, z)$ (triangle inequality).

Indeed, M1 follows directly from (i) and M3 from (iii). Finally, (ii) implies $v(1) = 1$ and so $v(-1)^2 = 1$; in turn, this implies $v(x - y) = v(y - x)$, that is, M2.

A set with a distance function satisfying M1–M3 for all x, y, z is called a *metric space*, and one can complete *any* metric space in two stages, as follows.† First define a *Cauchy sequence* as an infinite sequence $\{x_n\}$ such

†For a general explanation of the completion of metric spaces, see E. T. Copson, *Metric Spaces* (Cambridge Tracts no. 57) Cambridge Univ. Press, 1968.

that $\lim_{m, n \to \infty} d(x_m, x_n) = 0$. Then call two Cauchy sequences *concurrent* (in symbols, $\{x_n\} \equiv \{y_n\}$) when $\lim_{n \to \infty} d(x_n, y_n) = 0$. Modulo the equivalence relation of concurrence, the Cauchy sequences of D form a *complete* metric space D^* in which every Cauchy sequence has a limit, and D is embedded in D^* as the set of *constant* sequences $\{x, x, x, \cdots\}$.

Now for any prime number p, we can define a new valuation on the field $\mathbf{Q}$, as follows. For each non-zero $x \in \mathbf{Q}$, there are integers μ, r, and s with

$$x = p^\mu r/s, \qquad (p, r) = (p, s) = (r, s) = 1; \tag{36}$$

there is just one way to do this so that $s > 0$. Finally, define the *p-adic order* of x as the integer $\mu = o(x)$ in (36).

LEMMA 1. *For the p-adic order defined by (36), the function* $|x|_p = p^{-\mu}$ *for* $x \neq 0$, *with* $|0|_p = 0$, *is a valuation* $v_p(x)$ *on* $\mathbf{Q}$.

Proof: Property (i) holds trivially, while (ii) follows from the prime factorization theorem in $\mathbf{Z}$. Next,

(iii′) $v_p(a + b) \leqslant \max [v_p(a), v_p(b)]$,

since if $a = p^\mu a'$ and $b = p^\mu b'$ where a' and b' have denominators relatively prime to p, then $a + b = p^\mu(a' + b')$, where the denominator of $a' + b'$ is relatively prime to p. Clearly, (iii′) implies (iii).

Now fix p, and note that every *positive* integer $n \in \mathbf{Z}^+$ can be uniquely written as a *p-adic pseudo-polynomial*

$$n = a_0 + a_1 p + a_2 p^2 + \cdots + a_r p^r, \qquad 0 \leqslant a_k < p, \quad a_r \neq 0. \tag{37}$$

Thus, if $p = 3$, we have $7 = 1 + 2p$ and $13 = 1 + p + p^2$.

p-adic Series. The central idea of Hensel, the creator of p-adic numbers, was to extend the semiring of p-adic pseudo-polynomials (37) to include formal power series such as those discussed in §9.

DEFINITION. A p-adic integer α is an infinite "p-adic series" with integral coefficients a_k of the form

$$\alpha = a_0 + a_1 p + a_2 p^2 + \cdots + a_r p^r + \cdots, \qquad 0 \leqslant a_k < p. \tag{38}$$

Also, $\bar{Z}_p$, will denote the set of all p-adic integers.

A non-zero *p-adic number* is an "extended p-adic series"

$$\gamma = p^\mu \left(\sum_{k=0}^{\infty} b_k p^k \right) = \sum_{k=\mu}^{\infty} a_k p^k, \qquad a_k = b_{k-\mu}, \quad a_0 \neq 0, \tag{39}$$

again with coefficients b_k and a_k integers, $0 \leqslant b_k < p$.

Note that, trivially, every non-zero p-adic integer is also an extended p-adic series with $v_p(x) = p^{-\mu}$, where a_μ is the first non-zero a_k.

Proposition 28. *Under the valuation $v_p(\gamma) = p^{-\mu}$, the partial sums of any extended p-adic series form a Cauchy sequence. Two such Cauchy sequences are concurrent if and only if they are identical.*

Proof: It is obvious that identical series (39) are concurrent. Conversely, if two series (39) first differ in some kth term, corresponding partial sums up to p^n differ by p^{-k} for every $n \geqslant k$.

To illustrate Hensel's ideas, we next show how to represent -1 and $\frac{1}{13}$ as p-adic series. Though -1 cannot be written in the form (37), it is the sum of the p-adic series

$$-1 = (p-1)/(1-p) = (p-1)(1 + p + p^2 + \cdots + p^k + \cdots).$$

Indeed, $|(-1) - \Sigma_{k=0}^{n}(p-1)p^k|_p = p^{-n-1} \to 0$ as $n \to \infty$. Therefore, in the p-adic valuation of Lemma 1 with $p = 5$,

$$4, 24, 124, 624, 3124, 15624, \cdots \to -1.$$

Likewise, for $p = 3$, since $3^3 = 2 \cdot 13 + 1$, we have

$$\begin{aligned}\tfrac{1}{13} &= 1 - \tfrac{24}{26} = 1 + (2p + 2p^2)/(1 - p^3)\\ &= 1 + 2p + 2p^2 + 2p^4 + 2p^5 + 2p^7 + 2p^8 + \cdots .\end{aligned}$$

Note that the coefficients a_k are *recurrent* with period 3, in the sense that $a_k = a_{k+3}$ for all $k > 1$. Computing the first few partial sums, we obtain

$$1, 7, 25, 187, 673, 5047, \cdots \to \tfrac{1}{13}.$$

In particular, as might have been foreseen, the sum of the terms through $2p^7$ is congruent to $\frac{1}{13}$ mod 3^8: $13 \times 5047 = 10 \cdot 3^8 + 1$.

Each p-adic integer α of (38), if truncated at p^n, determines an element $g_n\alpha = a_0 + a_1p + \cdots + a_{n-1}p^{n-1}$ of the quotient ring $\mathbf{Z}_{p^n} = \mathbf{Z}/(p^n)$, and every element of that quotient ring can be written as the coset, mod p^n, of a unique element $a_0 + a_1p + \cdots + a_{n-1}p^{n-1}$ with $0 \leqslant a_i < p$. These functions g_n compare the p-adic integers to the sequence of rings

$$\mathbf{Z}_p \xleftarrow{f_2} \mathbf{Z}_{p^2} \leftarrow \cdots \leftarrow \mathbf{Z}_{p^{n-1}} \xleftarrow{f_n} \mathbf{Z}_{p^n} \leftarrow \cdots \tag{40}$$

and ring homomorphisms f_n, where each f_n sends the coset $a + (p^n)$ to $a + (p^{n-1})$. Moreover, the functions g_n on the set $\bar{Z}_p$ of all p-adic integers are consistent with the f_n in the sense that $g_{n-1} = f_n g_n$ for all $n > 1$, so that all the upper triangular diagrams below are commutative:

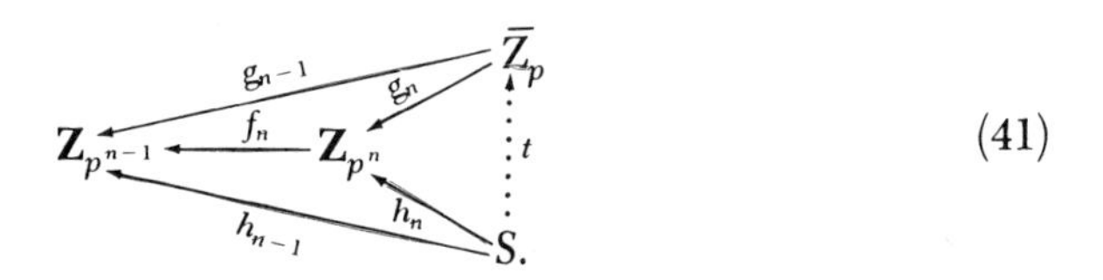

(41)

Lemma 2. *If S is any set and $h_n : S \to \mathbf{Z}_{p^n}$ a list of functions, consistent in the sense that $h_{n-1} = f_n h_n$ for all $n > 1$, then there is a unique function $t : S \to \bar{Z}_p$ such that $h_n = f_n t$ for all n.*

Proof: For each $s \in S$, $h_n s$ is some expansion $a_0 + \cdots + a_{n-1}p^{n-1}$ (mod p^n). By consistency, it agrees (mod p^{n-1}) with $h_{n-1}s$. Hence, all these expansions combine to determine a single element $ts = a_0 + \cdots + a_{n-1} p^{n-1} + \cdots$ in $\bar{Z}_p$. Clearly $f_n t = h_n$, as required.

This lemma states that the functions g_n of (41) form a universal consistent list. As with other universals (Theorem IV.1) this universal property determines the set $\bar{Z}_p$ of p-adic integers uniquely, up to isomorphism. (One then says that the set $\bar{Z}_p$ is the (inverse) "limit" of the diagram (40) of sets.) This property will also determine the sums and products of p-adic integers, so as to make $\bar{Z}_p$ into a ring which is the inverse limit of the diagram (40) of rings.

Proposition 29. *There are unique binary operations + and × on the set $\bar{Z}_p$ of p-adic integers such that, for every n and all $\alpha, \beta \in \bar{Z}_p$,*

$$g_n(\alpha + \beta) = g_n\alpha + g_n\beta, \qquad g_n(\alpha\beta) = (g_n\alpha)(g_n\beta). \tag{42}$$

Proof: In (40), each f_n is a morphism of rings, so the functions $(\alpha, \beta) \mapsto g_n(\alpha) + g_n(\beta)$ are consistent. Therefore, by Lemma 2, there is a unique function + satisfying (42), as in the commutative square

$$\begin{array}{ccc} \bar{Z}_p \times \bar{Z}_p & \xrightarrow{g_n \times g_n} & \mathbf{Z}_{p^n} \times Z_{p^n} \\ \vdots{+} & & \downarrow{+} \\ \bar{Z}_p & \xrightarrow[g_n]{} & Z_{p^n}. \end{array}$$

The case of multiplication is similar.

This definition of addition may be phrased directly: To add two p-adic integers, truncate them at each p^n, add the results, modulo p^n; the resulting sums for all p^n fit together as a p-adic integer.

Theorem 30. *Under the definition (42) of sum and product, the p-adic integers form an integral domain $\bar{Z}_p$ whose invertibles are those p-adic integers α of (38) with $a_0 > 0$ (that is, with $a_0 \not\equiv 0 \pmod{p}$).*

Proof: Since every $\mathbf{Z}_{p^n}$ is a commutative ring, the usual ring identities hold for $g_n(\alpha + \beta)$ for all n, and therefore hold for $\alpha + \beta$. Again, if $a_k p^k$ is the first non-zero term of α and $b_m p^m$ that of β, the term $a_k b_m p^{k+m}$ is non-zero in $\alpha\beta$; hence, the ring $\bar{Z}_p$ is an integral domain.

Finally, to show that any α with $a_0 > 0$ is invertible, we *compute* $\gamma = 1/\alpha$ by successively solving for $c_0, c_1, c_2, \cdots \quad (0 \leqslant c_k < p)$:

$$a_0 c_0 \equiv 1 \pmod{p}; \qquad \text{that is,} \qquad a_0 c_0 - 1 = p b_1$$

$$a_0 c_1 + a_1 c_0 + b_1 \equiv 0 \pmod{p}; \qquad \text{that is,}$$

$$a_0 c_0 + (a_0 c_1 + a_1 c_0) p - 1 = p^2 b_2$$

$$\vdots$$

$$a_0 c_k + a_1 c_{k-1} + \cdots + a_k c_0 + b_k \equiv 0 \pmod{p}; \qquad \text{that is,}$$

$$a_0 c_0 + (a_0 c_1 + a_1 c_0) p + \cdots + (a_0 c_k + \cdots + a_k c_0) p^k - 1 = p^{k+1} b_{k+1}.$$

This is possible since a_0 is invertible in $\mathbf{Z}/(p^r)$ if it is invertible in $\mathbf{Z}/(p)$.

Corollary. *The p-adic numbers (39) form a field, the field of quotients of $\bar{\mathbf{Z}}_p$.*

Proof: The field of quotients of $\bar{\mathbf{Z}}_p$ consists of 0 and the α/β (α, $\beta \neq 0$ in $\bar{\mathbf{Z}}_p$). But by the theorem, we can write each non-zero α and β as the product of a power of p times an invertible α', β'. Hence we can write α/β in the form $p^\mu \alpha'(1/\beta')$, where $\alpha'(1/\beta')$ is itself an invertible p-adic integer; that is, comes in the form (39).

p-adic Arithmetic. We next give a brief introduction to p-adic arithmetic. Think of p-adic numbers as analogous to real numbers written to the base p, but with three differences: (i) they are *written from right to left*; (ii) the "integral" part comes *after* the "decimal point" and can have infinitely many digits; and (iii) the "decimal" part $\Sigma_{k=1}^{\mu} a_k p^{-k}$ *precedes* the decimal point and has only a finite number of digits.

Rule (i) makes it easy to add and multiply p-adic integers: Just "carry" to the right instead of to the left. Thus, if $p = 7$:

$$(1 + 3p + 6p^2 + 4p^3 + 2p^4) + (4 + 5p + 2p^2 + p^3 + 3p^4) = 5 + p + 2p^2 + 6p^3 + 5p^4$$

and

$$(3 + 4p + 5p^2)(5 + 4p + 3p^2) = 1 + 6p + 5p^2 + 4p^3 + 6p^4 + 2p^5.$$

In another notation, listing the coefficients as *n-tuples* gives

$$(1, 3, 6, 4, 2) + (4, 5, 2, 1, 3) = (5, 1, 2, 6, 5)$$

and

$$(3, 4, 5) \times (5, 4, 3) = (1, 6, 5, 4, 6, 2).$$

To compute the p-adic series for a p-adic integer $-\alpha$ from that for α, first form $-1 = (p - 1, p - 1, p - 1, \cdots)$, whence $-1 - \alpha = (p - 1 - a_0, p - 1 - a_1, p - 1 - a_2, \cdots)$ is the 'complement' of α, and then add 1 to $-1 - \alpha$. Finally , knowing how to add, multiply, and form additive inverses of arbitrary p-adic *integers* α and β, it is easy to do the same for general p-adic *numbers* $p^{-\mu}\alpha$ and $p^{-\nu}\beta$. For $(p^{-\mu}\alpha)(p^{-\nu}\beta) = p^{-\mu-\nu}\alpha\beta$ and (suppose that $\nu \geqslant \mu$)

$$p^{-\mu}\alpha \pm p^{-\nu}\beta = p^{-\nu}(p^{\nu-\mu}\alpha \pm \beta).$$

To form reciprocals is more interesting.

Theorem 31. *A p-adic number is rational (that is, in* $\mathbf{Q}$*) if and only if its p-adic series is recurrent, in the sense that there are a k and a period m such that $a_{n+m} = a_n$ for all $n > k$.*

Proof: If the p-adic number x is recurrent with the period m, then $x - p^m x = (1 - p^m)x$ is a polynomial, hence an ordinary integer $r \in \mathbf{Z}$, and so $x = r/(1 - p^m)$ is rational.

Conversely, let $x \in \mathbf{Q}$ be given; we can write $x = r/s$, where $(s, p) = 1$ and $p^m = ks + 1$. Then†

$$x = (-kr)/(-ks) = (-kr)/(1 - p^m) = a + b/(1 - p^m),$$

†This proof was suggested to us by Neal Koblitz; the case $b = 0$ is trivial.

where b is the least positive residue of $-kr \bmod (p^m - 1)$ and $a \in \mathbf{Z}$. Adding an integer a to $b + bp^m + bp^{2m} + \cdots$ clearly changes only a finite number of digits, hence also gives a recurrent series.

EXERCISES

1. In the metric space defined by any valuation satisfying (iii′):

(a) Show that every triangle is isosceles.

(b) Show that in the "open disc" of all x satisfying $d(x, c) < r$ for a fixed "center" c, $r \in \mathbf{R}$, every point is a center of the disc.

2. Show that for any odd prime p:

$$-\tfrac{1}{2} = [(p-1)/2](1 + p + p^2 + p^3 + \cdots) = c + cp + cp^2 + cp^3 + \cdots,$$

where $c = (p-1)/2$.

3. Show that, in the field of 17-adic numbers,

$$\tfrac{1}{3} = 6 + 11p + 5p^2 + 11p^3 + 5p^4 + 11p^5 + \cdots .$$

4. (a) Compute the 5-adic expansion of $1/4$, of $1/6$, of $1/31$.

(b) Compute the 7-adic expansion of $1/12$ and $1/171$.

5. Let $(s', p) = (s, p) = 1$. Show that the p-adic series of r/s and r'/s' agree through terms in p^{n-1} if and only if $r's \equiv rs' \pmod{p^n}$.

6. (a) Show that $\sqrt{-1}$ has the 5-adic expansion $2 + p + 2p^2 + p^3 + \cdots$.

(b) Show that the 3-adic expansion of $\sqrt{7}$ is $1 + p + p^2 + 2p^4 + \cdots$.

***7.** Show that $\sqrt{-1}$ exists in $\bar{Z}_p$ if and only if $p \equiv 1 \pmod 4$.

CHAPTER IX

Determinants and Tensor Products

THE DETERMINANT of a square matrix A over a commutative ring K will be defined as a certain function which is linear in each column of A, and for this reason is said to be a "multilinear function" of these columns. This determinant turns out to be the same multilinear function of the rows of A. Moreover, determinants provide an effective means of finding eigenvalues of the matrix A. Such multilinear functions of several arguments are the main subject of this chapter, which culminates in the construction of a universal multilinear function called the "tensor product".

The tensor product of two modules (or two vector spaces) will prove to have many uses: To describe the bilinear operation of multiplication in a linear algebra (§12); to explain the change of rings (§11), whereby a matrix of real numbers or a real vector space is regarded as a matrix of complex numbers or a complex vector space; to formulate "multilinear algebra" and "tensor algebra" (Chapter XVI). A related and useful notion is that of an exact sequence of modules (§9).

1. Multilinear and Alternating Functions

A "bilinear" function is a function h of two variables, linear in each variable separately. An example is the evaluation function which assigns to each f in the dual V^* of a vector space V and to each v in V the value $\langle f, v\rangle = f(v)$ of f taken at v. For fixed f, this function is linear in v as in (V.46),

$$\langle f, v_1\kappa_1 + v_2\kappa_2\rangle = \langle f, v_1\rangle\kappa_1 + \langle f, v_2\rangle\kappa_2,$$

because f is linear. For fixed v, this function is linear in f as in (V.47),

$$\langle \kappa_1 f_1 + \kappa_2 f_2, v\rangle = \kappa_1\langle f_1, v\rangle + \kappa_2\langle f_2, v\rangle,$$

by the definition of the linear combination $\kappa_1 f_1 + \kappa_2 f_2$ of the f's. All told, $\langle f, v\rangle$ is bilinear. In general: Let C and D be two K-modules and $C \times D$ their cartesian product as sets.

DEFINITION. A K-bilinear function h on $C \times D$ to a right K-module E is a function $h: C \times D \to E$ such that always

$$h(c_1\kappa_1 + c_2\kappa_2, d) = h(c_1, d)\kappa_1 + h(c_2, d)\kappa_2, \tag{1}$$

$$h(c, d_1\kappa_1 + d_2\kappa_2) = h(c, d_1)\kappa_1 + h(c, d_2)\kappa_2. \tag{2}$$

Equation (1) states for each fixed $d \in D$ that the function $c \mapsto h(c, d)$ is a K-linear map $C \to E$; Equation (2) states for each $c \in C$ that the function $d \mapsto h(c, d)$ is a K-linear map $D \to E$.

For example, consider a ring R as an abelian group (a **Z**-module). The product in the ring is a function $R \times R \to R$. The distributive laws assert that the product $(\lambda, \mu) \mapsto \lambda\mu$ is additive in $\lambda \in R$ for fixed μ and additive in $\mu \in R$ for fixed λ. Now any additive function is automatically **Z**-linear; hence, the product in a ring R is a **Z**-bilinear map on $R \times R$ to R. A **Z**-bilinear map is often called "biadditive".

Again for example, the "inner product" of two vectors (x_1, x_2) and (y_1, y_2) in the real coordinate plane $\mathbf{R}^2$ is the real number defined by the formula

$$(x_1, x_2)(y_1, y_2) = x_1 y_1 + x_2 y_2.$$

For fixed $\mathbf{y}$, this expression is **R**-linear in the vector (x_1, x_2); for fixed $\mathbf{x}$, it is **R**-linear in (y_1, y_2). Hence, the inner product is an **R**-bilinear function $\mathbf{R}^2 \times \mathbf{R}^2 \to \mathbf{R}$.

Determinants provide another example of bilinearity. For instance, the determinant $|A|$ of a 2×2 matrix A with entries in K is the scalar

$$\begin{vmatrix} A_{11} & A_{12} \\ A_{21} & A_{22} \end{vmatrix} = A_{11}A_{22} - A_{21}A_{12}. \tag{3}$$

Now consider $A \mapsto |A|$ as a function of the columns of the matrix A; since each column is an element of the free module K^2, this makes the determinant a function $K^2 \times K^2 \to K$. Moreover, each term in formula (3) for $|A|$ involves exactly one factor A_{i1} from the first column and exactly one factor A_{j2} from the second column. Hence, the determinant of a 2×2 matrix is a bilinear function $K^2 \times K^2 \to K$ of the columns. Moreover, the determinant has two other simple properties. The determinant of the identity matrix is 1, and the determinant of A vanishes when the two columns of A are equal.

Conversely, any bilinear function $A \mapsto d(A)$ of the columns of A which has these two properties must be this determinant $d(A) = |A|$. To prove this, take the two unit elements $\boldsymbol{\varepsilon}_1 = (1, 0)$ and $\boldsymbol{\varepsilon}_2 = (0, 1)$ as the free generators of K^2, so that the jth column of A is $\boldsymbol{\varepsilon}_1 A_{1j} + \boldsymbol{\varepsilon}_2 A_{2j}$. Writing d as a function of these two columns and using bilinearity gives

$$\begin{aligned} d(A) &= d(\boldsymbol{\varepsilon}_1 A_{11} + \boldsymbol{\varepsilon}_2 A_{21}, \boldsymbol{\varepsilon}_1 A_{12} + \boldsymbol{\varepsilon}_2 A_{22}) \\ &= d(\boldsymbol{\varepsilon}_1, \boldsymbol{\varepsilon}_1)A_{11}A_{12} + d(\boldsymbol{\varepsilon}_1, \boldsymbol{\varepsilon}_2)A_{11}A_{22} + d(\boldsymbol{\varepsilon}_2, \boldsymbol{\varepsilon}_1)A_{21}A_{12} + d(\boldsymbol{\varepsilon}_2, \boldsymbol{\varepsilon}_2)A_{21}A_{22}. \end{aligned}$$

But d vanishes for two columns alike, so $d(\boldsymbol{\varepsilon}_1, \boldsymbol{\varepsilon}_1) = 0 = d(\boldsymbol{\varepsilon}_2, \boldsymbol{\varepsilon}_2)$, while

$$0 = d(\boldsymbol{\varepsilon}_1 + \boldsymbol{\varepsilon}_2, \boldsymbol{\varepsilon}_1 + \boldsymbol{\varepsilon}_2) = d(\boldsymbol{\varepsilon}_1, \boldsymbol{\varepsilon}_2) + d(\boldsymbol{\varepsilon}_2, \boldsymbol{\varepsilon}_1).$$

Therefore, $d(\boldsymbol{\varepsilon}_2, \boldsymbol{\varepsilon}_1) = -d(\boldsymbol{\varepsilon}_1, \boldsymbol{\varepsilon}_2)$, while $d(\boldsymbol{\varepsilon}_1, \boldsymbol{\varepsilon}_2) = 1$ because $\boldsymbol{\varepsilon}_1$ and $\boldsymbol{\varepsilon}_2$ are the columns of the identity matrix I. Thus the formula above simplifies to

$$d(A) = A_{11}A_{22} - A_{21}A_{12},$$

exactly as in (3).

To establish the corresponding result for $n \times n$ matrices, some additional terminology will be convenient.

Let $C_1 \times \cdots \times C_n$ be the cartesian product of n K-modules C_i, for $n \geqslant 1$, and E another K-module. Define then a *K-multilinear function* $h: C_1 \times \cdots \times C_n \to E$ to be a function $(c_1, \cdots, c_n) \mapsto h(c_1, \cdots, c_n)$ which is linear as a function of each argument $c_i \in C_i$ when all the remaining arguments $c_1, \cdots, c_{i-1}, c_{i+1}, \cdots, c_n$ are held fixed. Such a function h is called simply "multilinear" or "n-linear", when the ring K of scalars is clear from the context; thus 2-linear means *bilinear* and 3-linear "trilinear". A multilinear function with the ring K of scalars as codomain is called a *multilinear form* or an *n-linear form*.

Let $C^n = C \times \cdots \times C$ be the cartesian product of a K-module C with itself n times. A multilinear function $h: C^n \to E$ is called *alternating* provided $h(c_1, \cdots, c_n)$ vanishes whenever two of its arguments c_i and c_j are equal; that is, provided that

$$c_i = c_j \text{ for some } i \neq j \quad \Rightarrow \quad h(c_1, \cdots, c_n) = 0.$$

Proposition 1. *Each value $h(c_1, \cdots, c_n)$ of an alternating multilinear function is multiplied by (-1) when any two arguments c_i and c_j $(i \neq j)$ are interchanged.*

Proof: Suppose, for example, that c_1 and c_2 are interchanged, and ignore the remaining arguments. Bilinearity of h implies that

$$h(c_1 + c_2, c_1 + c_2) = h(c_1, c_1) + h(c_1, c_2) + h(c_2, c_1) + h(c_2, c_2).$$

By hypothesis, $h(c, c)$ vanishes when any argument c is repeated. Hence, this equation becomes $0 = h(c_1, c_2) + h(c_2, c_1)$ or $h(c_1, c_2) = -h(c_2, c_1)$, just as asserted.

Corollary. *For any alternating multilinear function $h: C^n \to E$ and any permutation $i \mapsto \sigma_i$ on $\mathbf{n}$ of parity $(-1)^{\operatorname{sgn} \sigma}$,*

$$h(c_{\sigma_1}, \cdots, c_{\sigma_n}) = (-1)^{\operatorname{sgn} \sigma} h(c_1, \cdots, c_n). \tag{4}$$

Proof: Any permutation σ is a product of transpositions. By the proposition, any one transposition of the arguments of h changes the sign, while σ is odd or even according as the total number of transpositions is odd or even (Corollary II.17.1). In particular, the sign in (4) is $+1$ if and only if σ is in the alternating group of all even permutations (Corollary II.17.3).

A function h of n arguments with the property (4) for all permutations σ is often called "skew-symmetric". In this terminology the Corollary reads: Any alternating multilinear function is skew-symmetric. Note that the converse statement need not be true (Exercise 4).

EXERCISES

1. (a) If a, b, c, and d are four given real numbers, show that

$$((x_1, x_2), (y_1, y_2)) \mapsto ax_1 y_1 + bx_1 y_2 + cx_2 y_1 + dx_2 y_2$$

defines an $\mathbf{R}$-bilinear function $\mathbf{R}^2 \times \mathbf{R}^2 \to \mathbf{R}$.

(b) Prove that every $\mathbf{R}$-bilinear function $\mathbf{R}^2 \times \mathbf{R}^2 \to \mathbf{R}$ has this form.

2. Let $h: C \times D \to E$ be a K-bilinear function.

(a) If $t: E \to E'$ is K-linear, prove that $t \circ h: C \times D \to E'$ is K-bilinear.

(b) If $u: C' \to C$ and $v: D' \to D$ are both K-linear maps, show that the composite $h \circ (u \times v): C' \times D' \to E$ is K-bilinear.

3. Extend the results of Exercise 2 to multilinear and to alternating multilinear functions.

4. (a) Show that the product $\mathbf{Z}_2 \times \mathbf{Z}_2 \to \mathbf{Z}_2$, $(x, y) \mapsto xy$, for the ring $\mathbf{Z}_2$ is $\mathbf{Z}_2$-bilinear and skew-symmetric but not alternating.

(b) If V and W are vector spaces over a field F of characteristic not 2, prove that any skew-symmetric bilinear function $h: V \times V \to W$ is necessarily alternating.

5. If $h: C^n \to E$ is a multilinear function with $h(c_1, \cdots, c_n) = 0$ whenever $c_i = c_{i+1} (i = 1, \cdots, n-1)$, prove that h is alternating.

2. Determinants of Matrices

We now study multilinear functions on the free module K^n to K. Regard an element c of K^n as a column of n scalars; then an element $(c_1, \cdots, c_n)$ of $K^n \times \cdots \times K^n = (K^n)^n$ is an $n \times n$ matrix A; indeed, the matrix A with ith column c_i. Thus the values of a function $d: (K^n)^n \to K$ may be written as $d(A)$ or as $d(c_1, \cdots, c_n)$, with c_i the ith column of A.

THEOREM 2. *There is exactly one alternating n-linear form $d: (K^n)^n \to K$ with $d(I) = 1$ for I the $n \times n$ identity matrix. More generally, for each scalar κ there is exactly one alternating n-linear form $d_\kappa: (K^n)^n \to K$ with $d_\kappa(I) = \kappa$.*

When $d(I) = 1$, the value $d(A)$ of d for an $n \times n$ matrix A will be called the *determinant* of A and written as $|A|$; we will show that $|A|$ is given by the usual formula for a determinant.

Proof: First we suppose that some such function d is given. We write $\boldsymbol{\varepsilon}_1, \cdots, \boldsymbol{\varepsilon}_n$ for the usual unit elements in the free module K^n, so that the identity matrix I is the matrix with columns $\boldsymbol{\varepsilon}_1, \cdots, \boldsymbol{\varepsilon}_n$, and the hypothesis $d(I) = 1$ reads $d(\boldsymbol{\varepsilon}_1, \cdots, \boldsymbol{\varepsilon}_n) = 1$. By (4),

$$\sigma \in S_n \quad \Rightarrow \quad d(\boldsymbol{\varepsilon}_{\sigma_1}, \cdots, \boldsymbol{\varepsilon}_{\sigma_n}) = (-1)^{\operatorname{sgn} \sigma} d(\boldsymbol{\varepsilon}_1, \cdots, \boldsymbol{\varepsilon}_n) = (-1)^{\operatorname{sgn} \sigma}.$$

In terms of $\boldsymbol{\varepsilon}$, the jth column of an $n \times n$ matrix A is the element $\boldsymbol{\varepsilon}_1 A_{1j} + \cdots + \boldsymbol{\varepsilon}_n A_{nj} = \sum_i \boldsymbol{\varepsilon}_i A_{ij}$ of K^n, so

$$d(A) = d\Big(\sum_i \boldsymbol{\varepsilon}_i A_{i1}, \cdots, \sum_i \boldsymbol{\varepsilon}_i A_{ij}, \cdots, \sum_i \boldsymbol{\varepsilon}_i A_{in} \Big).$$

We can expand this expression by multilinearity. Expanding at the jth place above gives n terms, one for each summand of $\sum \boldsymbol{\varepsilon}_i A_{ij}$. Expanding simultaneously at the 1st, 2nd, $\cdots$, nth places thus gives n^n terms. Each of these terms is determined by choosing one summand $\boldsymbol{\varepsilon}_i A_{ij}$ for each index j; that is, by a function $f: \mathbf{n} \to \mathbf{n}$ specifying this choice as $i = f_j$; that is, as $\boldsymbol{\varepsilon}_{f_j} A_{f_j j}$. The whole expansion expresses $d(A)$ as a sum of n^n terms,

$$d(A) = \sum_f d(\boldsymbol{\varepsilon}_{f_1}, \cdots, \boldsymbol{\varepsilon}_{f_n}) A_{f_1 1} \cdots A_{f_n n}.$$

Now if any two arguments in $d(\boldsymbol{\varepsilon}_{f_1}, \cdots, \boldsymbol{\varepsilon}_{f_n})$ are equal, the result is zero because d is alternating. There remain only the $n!$ terms for those functions f with all $f_1, \cdots, f_n$ different; that is, with $f: \mathbf{n} \to \mathbf{n}$ a permutation, say $f = \sigma$. Putting in the values of $d = d(\boldsymbol{\varepsilon}_{\sigma_1}, \cdots)$ already determined in these cases gives $d(A)$ as a sum of $n!$ terms, one for each permutation $i \mapsto \sigma i = \sigma_i$:

$$d(A) = \sum_{\sigma \in S_n} (-1)^{\operatorname{sgn} \sigma} A_{\sigma_1 1} \cdots A_{\sigma_n n}. \tag{5}$$

This shows that there is at most one function d satisfying the hypotheses with $d(I) = 1$. Similarly, d_κ with $d_\kappa(I) = \kappa$ is given by (5) with an extra factor κ on the right.

Conversely, for each $n \times n$ matrix A over the commutative ring K, define the *determinant* $d(A) = |A|$ by this formula (5). This formula does define a function $(K^n)^n \to K$. We check its properties, as follows.

First, $|A|$ as defined is a linear function of the jth column when all the other columns are fixed. Indeed, each term in the formula (5) for $|A|$ has exactly one factor $A_{\sigma_j j}$ from the jth column, so is a linear function of that column. Hence $|A|$, as a sum of such linear functions, is linear.

Next suppose that A has two columns alike; for simplicity, suppose the first two columns alike, so that $A_{i1} = A_{i2}$ for every i. Every permutation

σ in the symmetric group S_n is even or odd; the transposition (1 2) is odd, and every odd permutation τ can be written uniquely as $\tau = \rho(1\ 2)$ for some even ρ. Then $(-1)^{\operatorname{sgn} \tau} = -1$, while $\tau 1 = \rho 2$, $\tau 2 = \rho 1$, and $\tau i = \rho i$ for $i > 2$. Thus, the τ-term in the formula (5) for $|A|$ is

$$-A_{\tau_1 1} \cdots A_{\tau_n n} = -A_{\rho_2 1} A_{\rho_1 2} A_{\rho_3 3} \cdots A_{\rho_n n}.$$

But $A_{\rho_2 1} = A_{\rho_2 2}$ and $A_{\rho_1 2} = A_{\rho_1 1}$, so that, except for sign, this is exactly the term in (5) for the even permutation ρ. Therefore, the terms in (5) cancel in pairs, so $|A| = 0$. This proves that $A \mapsto |A|$ is alternating.

Finally, if A is the identity matrix I, every term in the expansion (5) of $|I|$ vanishes except for the one corresponding to the identity permutation. This one term gives 1, so $|I| = 1$. This completes the proof that the explicit formula (5) for the determinant $|A|$ does give the required alternating multilinear function.

COROLLARY. *Any alternating multilinear function $h:(K^n)^n \to K$ is a scalar multiple of the determinant.*

Proof: The function $A \mapsto h(I)|A|$ is alternating and multilinear and agrees with h at I. By the second part of the theorem it is the function h.

Several elementary properties of determinants follow directly from the explicit formula. For example, the formula shows that the determinant of a diagonal matrix is the product of the diagonal entries. It also proves

THEOREM 3. *If A^T is the transpose of A, then $|A^T| = |A|$.*

Proof: Consider a typical term $A_{\sigma_1 1} \cdots A_{\sigma_n n}$ in the formula (5) for $|A|$. Since K is commutative, we may permute these n factors $A_{\sigma_j j} = A_{ij}$, so that the first subscripts $\sigma_1, \cdots, \sigma_n$ (the row indices) come in order; this one factor is thus $A_{\sigma_j j} = A_{i(\sigma^{-1})i} = A_{i\tau_i}$ for $\tau = \sigma^{-1}$, and the whole term is $A_{1\tau_1} \cdots A_{n\tau_n}$. Since the permutations σ and $\tau = \sigma^{-1}$ have the same parity,

$$|A| = \sum_{\tau \in S_n} (-1)^{\operatorname{sgn} \tau} A_{1\tau_1} \cdots A_{n\tau_n}. \tag{6}$$

Since the transposed matrix A^T has entries $(A^T)_{ij} = A_{ji}$, this formula may be written as

$$|A| = \sum_{\tau \in S_n} (-1)^{\operatorname{sgn} \tau} (A^T)_{\tau_1 1} \cdots (A^T)_{\tau_n n}.$$

But, by the original formula (5) applied to $|A^T|$, this is just $|A^T|$, Q.E.D.

To evaluate determinants explicitly, there are three handy rules.

RULE I. Interchanging two columns of A multiplies $|A|$ by -1.

This is just the fact that $|A|$ is alternating, hence skew-symmetric.

RULE II. Multiplying one column of A by a scalar κ multiples $|A|$ by that scalar.

This rule is an immediate consequence of the linearity of $|A|$ as a function of that column.

RULE III. Adding a multiple of one column of A to another column leaves $|A|$ unchanged.

Write the determinant as a function $d(c_1, \cdots, c_n)$ of n columns c_i. Then adding λ times column 2 to column 1 (and leaving the other columns unchanged and unmentioned) gives

$$d(c_1 + \lambda c_2, c_2) = d(c_1, c_2) + d(c_2, c_2)\lambda = d(c_1, c_2) + 0.$$

These three rules correspond to the three types of elementary operations (§VI.5) on the columns of the matrix A. For each such operation ϕ, they show that $|\phi A|$ is a non-zero constant (-1, κ, or 1, as the case may be) times $|A|$. In particular, if the matrix A has entries from a field, these rules provide a systematic way to evaluate $|A|$: Use successive elementary row and column operations to reduce A to echelon or diagonal form, keeping track of the effect on $|A|$ of each operation.

PROPOSITION 4. *A square matrix A with entries in a field is invertible if and only if $|A| \neq 0$.*

Proof: If $|A| \neq 0$, the rules above show that $|A|$ remains non-zero after elementary row and column operations. If A is invertible, Proposition VII.13 shows that such operations will reduce A to the diagonal matrix I, with $|I|$ and hence $|A|$ non-zero. If A is not invertible, such operations reduce A to a matrix with some column zero, and hence with determinant zero.

This result may be restated thus: A matrix A is invertible as an element of the ring of all $n \times n$ matrices over a *field* F if and only if its determinant $|A|$ is invertible in F. Put in this form, the result also holds for matrices with entries in any commutative ring (Theorem 6 below).

A final important rule gives the determinant of a matrix product:

THEOREM 5. *For $n \times n$ matrices B and A over K, $|BA| = |B|\,|A|$.*

Proof: For a fixed index $k \in \mathbf{n}$, let A_{-k} be the kth column of A; it is an $n \times 1$ matrix. The definition $(BA)_{ik} = \sum_j B_{ij}A_{jk}$ of the matrix product may then be written $(BA)_{-k} = B(A_{-k})$, so the kth column of BA is a linear function of the kth column of A. Now hold B fixed and consider $|BA|$ as a function of the columns of A. Since the determinant of any matrix is multilinear and alternating in the columns, $|BA|$ is a multilinear and alternating function of the columns of A. By the Corollary of Theorem 2, $|BA|$ for fixed B must be a scalar multiple of $|A|$, say $|BA| = \kappa|A|$. To determine the multiple κ, set $A = I$; this gives $\kappa = |B|$, so $|BA| = |B|\,|A|$, as required.

In §4 we shall give another, more conceptual, proof of this theorem.

EXERCISES

1. Compute the determinants of the matrices of Exercise 5, §VII.6.

2. For κ a scalar, L a column of scalars, 0 a row of zeros, and B a square matrix, show that

$$\begin{vmatrix} \kappa & 0 \\ L & B \end{vmatrix} = \kappa|B|.$$

3. Using Exercise 2, prove that the determinant of a triangular matrix is the product of its diagonal entries.

4. Show that the determinant of any permutation matrix is ± 1.

5. If E is an elementary matrix, prove without using Theorem 5 that $|EA| = |E|\,|A|$.

6. (a) Obtain the following expansion of the "Vandermonde" determinant:

$$\begin{vmatrix} 1 & x_1 & x_1^2 \\ 1 & x_2 & x_2^2 \\ 1 & x_3 & x_3^2 \end{vmatrix} = (x_2 - x_1)(x_3 - x_1)(x_3 - x_2).$$

(*Hint:* Both sides are polynomials of degree 3 in $\mathbf{Z}[x_1, x_2, x_3]$ with the same coefficients for $x_2x_3^2$, hence—why?—they must be equal.)

(b) Generalize this result to the 4×4 case.

(c) Generalize to the $n \times n$ case, by proving that if $A_{ij} = x_i^{j-1}$, then $|A| = \Pi_{i>j}(x_i - x_j)$.

7. Deduce the following expansions:

$$\begin{vmatrix} 1 & x_1 & x_2 \\ 1 & y_1 & x_2 \\ 1 & y_1 & y_2 \end{vmatrix} = (y_1 - x_1)(y_2 - x_2),$$

$$\begin{vmatrix} 1 & x_1 & x_2 & x_3 \\ 1 & y_1 & x_2 & x_3 \\ 1 & y_1 & y_2 & x_3 \\ 1 & y_1 & y_2 & y_3 \end{vmatrix} = (y_1 - x_1)(y_2 - x_2)(y_3 - x_3).$$

8. Theorem 5 states that determinants can be multiplied "row by column". Prove that determinants also can be multiplied "row by row".

9. If A and C are square matrices, prove for block matrices that

$$\begin{vmatrix} A & B \\ 0 & C \end{vmatrix} = |A|\,|C|.$$

10. Give another proof that $|BA| = |B|\,|A|$ by applying the result of Exercise 9 as follows: Construct from B and A the $2n \times 2n$ matrices

$$\begin{bmatrix} B & 0 \\ -I & A \end{bmatrix}, \quad \begin{bmatrix} B & BA \\ -I & 0 \end{bmatrix}, \quad \begin{bmatrix} BA & B \\ 0 & I \end{bmatrix},$$

and find elementary operations not changing the determinant which carry the first matrix through the second to the third one.

11. If Ω is the $n \times n$ matrix with the entries $\Omega_{ij} = \omega^{ij}$, where ω is a primitive complex nth root of unity (that is, where ω is of order n in the multiplicative group of **C**), show that $|\Omega|^2 = \pm\, n^n$, with the sign $+1$ when $n \equiv 1$ or $2 \pmod 4$.

12. If A and B are $n \times n$ matrices over a field with A invertible, show that $A + \lambda B$ is invertible for all but a finite list of scalars λ.

3. Cofactors and Cramer's Rule

The formula (5) for the determinant of a 3×3 matrix is

$$|A| = A_{11}A_{22}A_{33} + A_{12}A_{23}A_{31} + A_{13}A_{21}A_{32} - A_{11}A_{23}A_{32} - A_{12}A_{21}A_{33} - A_{13}A_{22}A_{31}. \tag{7}$$

Consider how $|A|$ depends on one of the columns, say the second. On rearrangement:

$$|A| = (A_{23}A_{31} - A_{21}A_{33})A_{12} + (A_{11}A_{33} - A_{13}A_{31})A_{22} + (A_{13}A_{21} - A_{11}A_{23})A_{32},$$

$$|A| = -\begin{vmatrix} A_{21} & A_{23} \\ A_{31} & A_{33} \end{vmatrix} A_{12} + \begin{vmatrix} A_{11} & A_{13} \\ A_{31} & A_{33} \end{vmatrix} A_{22} - \begin{vmatrix} A_{11} & A_{13} \\ A_{21} & A_{23} \end{vmatrix} A_{32}. \tag{8}$$

Each term A_{i2} in the second column has as coefficient a 2×2 determinant with a suitable sign.

More generally, consider the determinant $|A|$ of an $n \times n$ matrix A as a function of the jth column A_{-j}. Each term in the definition (5) of $|A|$ contains exactly one of the $A_{1j}, \cdots, A_{nj}$ as a factor; collecting all the terms with factor A_{1j} into one term $(\cdots)A_{1j} = \alpha_{1j}A_{1j}$ for some scalar α_{1j}, and similarly collecting terms in $A_{2j}, \cdots, A_{nj}$ yields an expression

$$|A| = \alpha_{1j}A_{1j} + \cdots + \alpha_{nj}A_{nj}, \qquad j = 1, \cdots, n, \tag{9}$$

for $|A|$; the scalar coefficients α_{ij} are called the *cofactors* of A. This expression exhibits $|A|$ as a linear function of the jth column.

The cofactor α_{ij} can be given explicitly. To begin with, α_{ij} can involve only those entries of the matrix A not in the ith row and not in the jth column. If $A^{(ij)}$ denotes the $(n-1) \times (n-1)$ matrix obtained from A by striking out row i and column j, we have

Lemma 1. *Each cofactor α_{ij} of an $n \times n$ matrix A is the scalar*

$$\alpha_{ij} = (-1)^{i+j}|A^{(ij)}|, \qquad i, j \in \mathbf{n}. \tag{10}$$

Often, $|A^{(ij)}|$ is called the *minor* of A at the position (i, j). In general, a "minor" of a rectangular matrix B is the determinant $|M|$ of any square matrix M obtained by crossing out rows and columns in B.

Proof: For $i = j = 1$, the cofactor α_{11} is the total coefficient of A_{11} in the expansion of $|A|$. Now a term with permutation σ in this expansion (5) involves A_{11} precisely when $\sigma(1) = 1$; together these terms are

$$A_{11} \sum (-1)^{\operatorname{sgn} \sigma} A_{\sigma_2 2} \cdots A_{\sigma_n n}.$$

The sum is taken over all permutations $\sigma \in S_n$ with $\sigma(1) = 1$; this amounts to summing over *all* the permutations on $\{2, \cdots, n\}$. Therefore, the coefficient of A_{11} in the expression above is exactly the determinant $|A^{(11)}|$, as claimed in (10) for $i = j = 1$.

Now consider the cofactor α_{ij} for any i and j. Interchange row i with each previous row in succession till row i is first, and then similarly move column j to first position. These operations do not alter $|A^{(ij)}|$, because the relative position of rows and columns in $A^{(ij)}$ is unaffected, but they do change the sign of $|A|$ (and hence the sign of α_{ij}) a total of $(i - 1) + (j - 1) = i + j - 2$ times. Since $\alpha_{11} = |A^{(11)}|$, this gives the result (10) complete with sign $(-1)^{i+j-2} = (-1)^{i+j}$.

This sign $(-1)^{i+j}$ may be described as the sign to be found in the (i, j) position on a $\pm$ checkerboard, which starts with $+$ in the upper left-hand corner.

This *expansion* of $|A|$ *by cofactors* of the jth column, as given by the formulas (9) and (10), is a convenient way of computing $|A|$, especially when the ring of scalars is a field: By elementary operations, arrange that any convenient column of A has one entry 1 and all other entries 0. Then cross out the row and column of this entry 1; except for the sign, the computation of $|A|$ is reduced to that of the $(n - 1) \times (n - 1)$ matrix remaining.

Lemma 2. *For each pair of indices $j, k = 1, \cdots, n$,*

$$\sum_{i=1}^{n} \alpha_{ij} A_{ik} = \delta_{jk}|A|. \tag{11}$$

Proof: By (9),

$$\alpha_{1j}A_{1j} + \cdots + \alpha_{nj}A_{nj} = |A|;$$

this is (11) for $k = j$. For $k \neq j$, apply the same expansion to a matrix B which is just like A except that its jth column (as well as it kth column) is the kth column of A. Then striking out column j in A or in B gives the same result, hence $B^{(ij)} = A^{(ij)}$ for this j and every i, and the expansion for $|B|$ by cofactors of the jth column reads

$$\alpha_{1j}A_{1k} + \cdots + \alpha_{nj}A_{nk} = |B|.$$

But B has its jth and kth columns alike, so $|B| = 0$; this proves (11) for $k \neq j$.

Now regard $\boldsymbol{\alpha}$ as an $n \times n$ matrix with entries the cofactors α_{ij} (and call its transpose $\boldsymbol{\alpha}^T$ the "classical adjoint" of A). The left-hand side of (11) then gives the entries of the matrix product $\boldsymbol{\alpha}^T A$; on the right-hand side, each δ_{jk} is the (j, k) entry in the $n \times n$ identity matrix I.

Theorem 6. *For a square matrix A with entries in any commutative ring K, let $A^{(ij)}$ be the matrix obtained from A by striking out row i and column j, while $\boldsymbol{\alpha}$ is the square matrix with entries $\boldsymbol{\alpha}_{ij} = (-1)^{i+j}|A^{(ij)}|$. Then*

$$\boldsymbol{\alpha}^T A = |A|I. \tag{12}$$

In particular, if $|A|$ is invertible in K, then A has a two-sided inverse matrix A^{-1} over K, given by

$$A^{-1} = |A|^{-1}\boldsymbol{\alpha}^T. \tag{13}$$

In words: The inverse of A may be formed by transposing the matrix of cofactors after dividing each by the scalar $|A|$.

Proof: Since I has entries δ_{jk}, formula (11) implies (12). Hence, if $|A|$ is invertible in K, the matrix A does have the left inverse given by (13). Columns may be replaced by rows throughout this discussion. Fix attention on the ith row of A. Each term of $|A|$ has exactly one factor from this row, and the total coefficient α_{ij} of each A_{ij} is exactly the same cofactor as before. Hence,

$$|A| = A_{i1}\alpha_{i1} + \cdots + A_{in}\alpha_{in}, \qquad i = 1, \cdots, n. \tag{14}$$

This is the expansion of $|A|$ by cofactors of the ith row. As before, this gives $A\boldsymbol{\alpha}^T = |A|I$. In case $|A|$ is invertible in K, this shows that $|A|^{-1}\boldsymbol{\alpha}^T$ is also a right inverse of A, as claimed.

Corollary 1. *The following properties of an $n \times n$ matrix A in the ring $K^{n\times n}$ are equivalent: (i) A has a left inverse in $K^{n\times n}$; (ii) A has a right inverse in $K^{n\times n}$; (iii) A has a two-sided inverse in $K^{n\times n}$; (iv) $|A|$ is invertible in K. When these conditions hold, any two inverses of A (left, right, or two-sided) are equal, and equal to the matrix A^{-1} of (13).*

Proof: By the theorem, condition (iv) implies each of the other conditions. Conversely, suppose that A has some sort of matrix inverse, say a left-inverse B. Then $BA = I$ gives $|B|\,|A| = |I| = 1$, hence $|A|$ invertible in K. Also, $B = B(AA^{-1}) = (BA)A^{-1} = A^{-1}$, for A^{-1} as in (13), Q.E.D.

For K a field F, part of this result (the equivalence of conditions (i), (ii), and (iii) had already been established in Theorem VII.11. The rest of the corollary, restated, reads: *A is invertible in $K^{n\times n}$* (on either or both sides), *if and only if $|A|$ is invertible in K.*

COROLLARY 2 (*Cramer's Rule*). *If n linear equations*

$$\sum_{j=1}^{n} A_{ij}x_j = \kappa_i, \qquad i = 1, \cdots, n,$$

in n unknowns with coefficients A_{ij} and κ_i in a commutative ring K have an invertible matrix A of coefficients, they have a unique solution

$$x_j = (\alpha_{1j}\kappa_1 + \cdots + \alpha_{nj}\kappa_n)/|A|, \qquad j = 1, \cdots, n, \tag{15}$$

in K, where each α_{ij} is the cofactor of A_{ij} in A.

Proof: Regard the list $(x_1, \cdots, x_n)$ of "unknowns" as a column X, and the list $(\kappa_1, \cdots, \kappa_n)$ of scalars as a column $\boldsymbol{\kappa}$. The n equations may then be written as a single matrix equation $AX = \boldsymbol{\kappa}$. Since A is invertible, the solution is the matrix product $X = A^{-1}\boldsymbol{\kappa}$. Putting in the formula (13) for the inverse gives the conclusion.

The numerator of this formula for x_j may itself be written as a determinant: It is the expansion by cofactors of the jth column of the determinant of the matrix obtained from A by replacing the jth column of A by the column $\boldsymbol{\kappa}$.

The general linear group $GL(n, K)$ can now be defined for the commutative ring K just as in the case of §VII.6 when K is a field; it is the group of all those $n \times n$ matrices P over K which are invertible in the ring $K^{n\times n}$. By Theorem 5, $|AB| = |A|\,|B|$, so the determinant is a morphism $P \mapsto |P|$ on the group $GL(n, K)$ to the multiplicative group of invertible scalars of K. The kernel of this morphism is thus a normal subgroup $SL(n, K)$, called the *special linear group*. It consists of all those $n \times n$ matrices U over K for which $|U| = 1$. Such a matrix is called a *unimodular matrix*. In case $K = \mathbf{R}$, these matrices have a geometric interpretation in terms of the standard "inner product" in $\mathbf{R}^n$ (see §X.5 below). For example, a 2×2 real matrix A is unimodular if and only if the corresponding endomorphism $t_A : \mathbf{R}^2 \to \mathbf{R}^2$ preserves oriented "areas". (For details, see Birkhoff–Mac Lane, *A Survey of Modern Algebra*, 4th ed. (New York: Macmillan, 1977), pp. 327–330.)

EXERCISES

1. (a) If $A = \begin{bmatrix} 1 & -1 & 0 \\ -1 & 0 & 1 \\ 2 & 1 & 1 \end{bmatrix}$ is a matrix of rational numbers, compute $|A|$ both by the minors of the first row and by the minors of the first column, and compare the results.

(b) Compute $|A|$ on the assumption that the entries of A are in $\mathbf{Z}_2$.

2. Use the classical adjoint to compute the inverse of the matrix A of Exercise 1(a).

3. If a square matrix A is invertible, prove that $|A^{-1}| = |A|^{-1}$.

4. Show that $|\boldsymbol{\alpha}| = |A|^{n-1}$ for A an $n \times n$ matrix, $n \geqslant 2$, where α is the classical adjoint of A.

5. Prove that the (classical) adjoint of the adjoint of an $n \times n$ matrix A is $|A|^{n-2}A$, when $n \geqslant 2$.

6. Using Cramer's rule, solve the system of congruences

$$\begin{aligned} x - 2y + z &\equiv 5 \pmod{13}, \\ 2x + 2y &\equiv 7 \pmod{13}, \\ 5x - 3y + 4z &\equiv 1 \pmod{13}. \end{aligned}$$

7. (a) Show that the pair of homogeneous linear equations

$$a_1x + b_1 y + c_1z = 0, \qquad a_2x + b_2 y + c_2z = 0,$$

with coefficients a_i, b_i, c_i in a field has a solution

$$x = \begin{vmatrix} b_1 & c_1 \\ b_2 & c_2 \end{vmatrix}, \qquad y = \begin{vmatrix} c_1 & a_1 \\ c_2 & a_2 \end{vmatrix}, \qquad z = \begin{vmatrix} a_1 & b_1 \\ a_2 & b_2 \end{vmatrix}.$$

(b) When is this solution a basis for the whole space of solutions?

(c) Derive similar formulas for three equations in 4 unknowns.

8. The *determinant rank* d of a rectangular matrix B with entries in a field is defined to be the largest natural number for which B has a $d \times d$ minor $|M| \neq 0$; when $B = 0$, define d to be 0.

(a) Prove that the determinant rank is not altered by elementary row and column operations.

(b) Prove that the rank of B equals its determinant rank.

***9.** Consider the rank of the classical adjoint $\boldsymbol{\alpha}$ of an $n \times n$ matrix A with entries in a field. Prove that rank $(A) = n$ implies rank $(\boldsymbol{\alpha}) = n$, rank $(A) = n - 1$ implies rank $(\boldsymbol{\alpha}) = 1$, and rank $(A) < n - 1$ implies rank $(\boldsymbol{\alpha}) = 0$.

4. Determinants of Maps

Alternating multilinear functions can be used to establish some basic properties of free modules of finite type over a commutative ring. As a first example, we show that the number n of generators of a free K-module C of finite type is an invariant; that is, depends *only* on the module and not on the choice of the generators. This number n will be called the *rank* of the module. Since each list $b_1, \cdots, b_n$ of n free generators (that is, each basis) of C gives an isomorphism $K^n \cong C$ by the assignment $(\xi_1, \cdots, \xi_n) \mapsto b_1\xi_1 + \cdots + b_n\xi_n$, the invariance of the rank may be stated as follows:

THEOREM 7. *An isomorphism $K^n \cong K^m$ of K-modules implies $n = m$.*

The proof rests on the following

LEMMA 1. *If $m > n$, every alternating multilinear form $(K^n)^m \to K$ is zero.*

Proof: Suppose that $h: C^m \to K$ is an alternating m-linear form, where C is the module $C = K^n$. Write each element $c \in C$ as a linear combination of the usual n unit elements ε_i of C. Since h is multilinear, each value $h(c_1, \cdots, c_m)$ with arguments $c_i \in C$ can be expressed as a linear combination of values $h(e_1, \cdots, e_m)$, where each argument e_i is one of the ε_j's. But there are m arguments e_i and only $n < m$ unit elements ε_j, so some two of the arguments e_i must be the same ε_j. But h is alternating, so all these values of h and thus h itself must be zero, Q.E.D.

Proof of Theorem: Suppose, instead, that $n \neq m$, say that $m > n$. Write C for K^m. Then $C = K^m$ has a basis of m elements, so by Theorem 2 the determinant is a non-zero alternating m-linear map $C^m \to K$. But, by $C \cong K^n$ and Lemma 1, every such form must be zero. This contradiction shows that we must have had $m = n$, Q.E.D.

Since the rank of a vector space is just its dimension, this argument gives a new proof for the invariance of the dimension of any vector space of finite type. It is instructive to compare this proof with the previous one (Theorem VI.4). The former proof used the exchange process of Theorem VI.4, and so is "constructive"; it describes a process for turning any one basis of a vector space into any other basis. The present proof is "intrinsic", in the sense that the dimension n of the space V is described directly, as the largest natural number m for which there can exist a non-zero alternating multilinear form on V^m.

Each $n \times n$ square matrix A determines a corresponding endomorphism $t_A: K^n \to K^n$, so this endomorphism t_A should have a "determinant" equal to $|A|$. Actually, the determinant of an endomorphism t of the free module K^n can be defined intrinsically; that is, without using matrices.

LEMMA 2. *To every endomorphism $t: C \to C$ of a free K-module C of rank n there is a unique scalar $|t|$ such that the equation*

$$d(tc_1, \cdots, tc_n) = d(c_1, \cdots, c_n)|t| \tag{16}$$

holds for all alternating n-linear forms $d: C^n \to K$ and all elements $c_i \in C$.

This unique scalar $|t|$ is called the *determinant of the endomorphism t.*

Proof: Choose any one non-zero alternating n-linear form $d_0:C^n \to K$. By the Corollary of Theorem 2, any other such form d is a scalar multiple of d_0; hence, it suffices to find $|t|$ which satisfies (16) just for the form d_0. Now the assignment $(c_1, \cdots, c_n) \mapsto d_0(tc_1, \cdots, tc_n)$ is an alternating n-linear form on C, hence is a scalar multiple of d_0. If we call this multiple $|t|$, we get (16) for $d = d_0$ and hence for all d.

THEOREM 8. *For any two endomorphisms $s, t:C \to C$, $|s \circ t| = |s|\,|t|$.*

Proof: By the formula (16) defining $|s|$ and $|t|$,

$$d(stc_1, \cdots, stc_n) = d(tc_1, \cdots, tc_n)|s| = d(c_1, \cdots, c_n)|t|\,|s|.$$

Hence, $|st| = |t|\,|s| = |s|\,|t|$, since K is commutative.

Next, we show that the function $A \mapsto t_A$ from square matrices to endomorphisms preserves determinants.

THEOREM 9. *For each square matrix A over K, $|A| = |t_A|$.*

Proof: As in §VII.1, $t_A:K^n \to K^n$ is the endomorphism with $t_A\varepsilon_j = \sum_i \varepsilon_i A_{ij}$. In the definition (16) of $|t_A|$, pick for d the unique alternating n-linear form d with $d(\varepsilon_1, \cdots, \varepsilon_n) = 1$. Then, by the definition of $|A|$,

$$\begin{aligned} d(t_A\varepsilon_1, \cdots, t_A\varepsilon_n) &= d\Big(\sum_i \varepsilon_i A_{i1}, \cdots, \sum_i \varepsilon_i A_{in}\Big) \\ &= |A| = d(\varepsilon_1, \cdots, \varepsilon_n)|A|. \end{aligned}$$

Therefore, by (16), $|t_A| = |A|$.

COROLLARY 1. *If $t:V \to V$ is an endomorphism of a finite-dimensional vector space over a field F and Mt is its matrix relative to some basis of V, then $|t| = |Mt|$.*

Proof: Set $A = Mt$. The given basis yields an isomorphism $L:F^n \cong V$, while, by (VII.22), $t = Lt_A L^{-1}$, as displayed below:

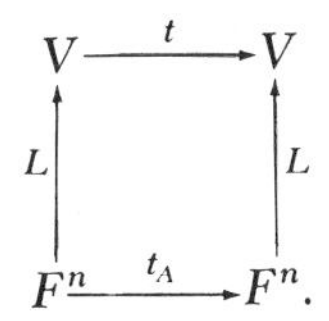

Since L is an isomorphism and each multilinear $d:V \to F$ yields a multilinear function $d \circ L:F^n \to F$, the definition (16) shows $|t| = |t_A|$, while by the theorem, $|t_A| = |A|$, Q.E.D.

Corollary 2. *Two similar matrices have the same determinant.*

Proof: Two matrices are similar, by Corollary 1 of Theorem VII.16, if and only if they represent the same endomorphism relative to (possibly) different bases, Q.E.D.

There is a more computational proof of the same corollary. If P is any invertible matrix, then $1 = |I| = |PP^{-1}| = |P|\,|P^{-1}|$, so $|P^{-1}| = |P|^{-1}$. Any matrix similar to A has the form PAP^{-1} for P invertible, so its determinant is $|PAP^{-1}| = |P|\,|A|\,|P^{-1}| = |A|\,|P|\,|P|^{-1} = |A|$.

Here is another description of the determinant of an endomorphism. First, consider a free K-module D on *one* free generator. If $d \neq 0$ is any one generator of D, every other element can be written uniquely as a scalar multiple $d\kappa$. Hence any endomorphism $D \to D$ is just multiplication by a scalar.

For any module C, let $\mathrm{Alt}_n\,(C)$ be the set of all alternating multilinear forms $h : C^n \to K$. This set is itself a module under pointwise operations. Each morphism $t : C' \to C$ of modules gives a corresponding function $t^n : C'^n \to C^n$, and the composite $h \circ t^n : C'^n \to K$ is an alternating multilinear form on C'. Moreover, $h \mapsto h \circ t^n$ is a morphism

$$\mathrm{Alt}_n(t) : \mathrm{Alt}_n(C) \to \mathrm{Alt}_n(C')$$

of modules. Now suppose that C is free of rank n and t an endomorphism of C. By Theorem 2, $\mathrm{Alt}_n(C)$ is then a free K-module of rank 1 and $\mathrm{Alt}_n(t)$, as an endomorphism $\mathrm{Alt}_n(C) \to \mathrm{Alt}_n(C)$ of a rank 1 module, amounts to multiplication in this module by some unique scalar. This scalar is $|t|$.

The sometimes-puzzling question of the orientation of a real vector space may be clarified by using the signs of determinants. Let V be an n-dimensional vector space over $\mathbf{R}$. If $\mathbf{b}$ and $\mathbf{b}'$ are two bases for V, we write $\mathbf{b} \sim \mathbf{b}'$ when the change matrix P from $\mathbf{b}$ to $\mathbf{b}'$ has positive determinant $|P| > 0$. Since the product of two matrices with positive determinant has positive determinant, this relation $\sim$ is reflexive, symmetric, and transitive. Since a non-zero real determinant $|P|$ is either positive or negative, this relation $\mathbf{b} \sim \mathbf{b}'$ divides all the bases of V into just two equivalence classes. Now define an *oriented* vector space V over $\mathbf{R}$ to be a finite-dimensional vector space V together with a selection of one of these two equivalence classes of bases. Call the bases in this selected equivalence class "positively" oriented and those in the other class "negatively" oriented. Hence, a given real vector space V has just two orientations; moreover, to give an orientation for V it is enough to specify that some one basis is a positively oriented one. This agrees with the familiar practice of orienting real three-dimensional space by specifying some one ("right-handed") set of axes as positively oriented.

An automorphism t of an oriented vector space V "preserves orientation" when it carries one (and hence every) positively oriented basis of V into a positively oriented basis. It is equivalent to require that the determinant of t be positive.

EXERCISES

1. Show explicitly that Alt_n is a contravariant functor on modules to modules.

2. (a) Prove that $\mathrm{Alt}_2 (K^3)$ is a free K-module of rank 3.

(b) If a list of two elements of K^3 is regarded as a 3×2 matrix A, show that the determinants of the three 2×2 submatrices of A (the 2×2 minors) give a basis of three elements for $\mathrm{Alt}_2 (K^3)$.

3. If $m < n$, prove that there exists a non-zero alternating multilinear function $(K^n)^m \to K$.

4. (a) Show that an automorphism of an oriented real vector space V preserves orientation if and only if its matrix A (relative to *any* basis of V) has a positive determinant.

(b) Show that the $n \times n$ real matrices A with $|A| > 0$ form a subgroup of index 2 in $GL(n, \mathbf{R})$ and determine the corresponding quotient group.

5. For D a principal ideal domain, prove that $D^n \cong D^m$ implies $n = m$ by using dimensions of vector spaces over the field $D/(p)$, for p any prime in D.

6. Prove that $K^n \cong K^m$ implies $n = m$ as follows. If $n > m$, an isomorphism must have the form t_A with left inverse t_B, where A is $m \times n$ and B $n \times m$. In the block product.

$$I_n = BA = \begin{bmatrix} B_1 \\ B_2 \end{bmatrix} [A_1 \quad A_2],$$

where B_1 and A_1 are $m \times m$, show $B_1{}^{-1}$ exists, and hence $B_2A_2 = 0 = I$ [cf. W. G. Leavitt, *Am. Math. Monthly*, **71**, 1112–1113 (1964)].

7. Show that Theorem VI.11 and its corollaries for finite-dimensional vector spaces also apply to free modules of finite rank over a commutative ring K.

5. The Characteristic Polynomial

In this section and the next we consider more carefully the basic relation of similarity between square matrices over a field F as defined in §VII.7. Similar matrices have the same eigenvalues; we now show how determinants enter the study of eigenvalues.

Theorem 10. *A scalar λ is an eigenvalue of the matrix A if and only if*

$$|A - \lambda I| = 0. \tag{17}$$

The proof consists in observing that each of the following statements is logically equivalent to the next one:

λ is an eigenvalue of A.
There is some column $\xi = (\xi_1, \cdots, \xi_n) \neq 0$ with $A\xi = \xi\lambda$.
There is some column $\xi \neq 0$ with $(A - \lambda I)\xi = 0$.
The matrix $A - \lambda I$ has positive nullity.
The matrix $A - \lambda I$ is not invertible.
The determinant $|A - \lambda I| = 0$. Q.E.D.

Now consider this determinant $|A - \lambda I|$. In the 2×2 case,

$$|A - \lambda I| = \begin{vmatrix} A_{11} - \lambda & A_{12} \\ A_{21} & A_{22} - \lambda \end{vmatrix} = \lambda^2 + c_1\lambda + c_0$$

is a monic quadratic polynomial in λ with scalar coefficients $c_1 = -A_{11} - A_{22}$ and $c_0 = |A|$. In the 3×3 case,

$$|A - \lambda I| = \begin{vmatrix} A_{11} - \lambda & A_{12} & A_{13} \\ A_{21} & A_{22} - \lambda & A_{23} \\ A_{31} & A_{32} & A_{33} - \lambda \end{vmatrix} = -\lambda^3 + c_2\lambda^2 + c_1\lambda + c_0$$

is a polynomial in λ of degree 3 with scalar coefficients c_i. The constant term c_0 is just the value of $|A - \lambda I|$ for $\lambda = 0$, while the term involving λ^2 can only come from the product $(A_{11} - \lambda)(A_{22} - \lambda)(A_{33} - \lambda)$ of the diagonal entries of $A - \lambda I$. Hence,

$$c_0 = |A|, \qquad c_2 = A_{11} + A_{22} + A_{33}. \tag{18}$$

Finally, a term of degree 1 in λ can arise from $|A - \lambda I|$ only as a diagonal entry times its cofactor in A; hence,

$$c_1 = -\begin{vmatrix} A_{22} & A_{23} \\ A_{32} & A_{33} \end{vmatrix} - \begin{vmatrix} A_{11} & A_{13} \\ A_{31} & A_{33} \end{vmatrix} - \begin{vmatrix} A_{11} & A_{12} \\ A_{21} & A_{22} \end{vmatrix}. \tag{19}$$

In the $n \times n$ case, each entry of $A - \lambda I$ is linear in λ or a constant. Since the determinant is a multilinear function of the columns, $|A - \lambda I|$ must be a polynomial of degree n in λ of the form

$$|A - \lambda I| = (-1)^n\lambda^n + c_{n-1}\lambda^{n-1} + \cdots + c_1\lambda + c_0. \tag{20}$$

This expression, when regarded as a polynomial over the field F in λ regarded as an indeterminate, is called the *characteristic polynomial* of the matrix A, written $c = \Sigma c_i\lambda^i$. The previous theorem may now be restated as a property of the characteristic polynomial.

Corollary. *The eigenvalues of A are the zeros of the characteristic polynomial $|A - \lambda I|$ of A.*

It follows that an $n \times n$ matrix A over a field has at most n different eigenvalues. Also, by the fundamental theorem of algebra, a matrix of complex numbers has at least one complex number as eigenvalue.

In the expansion (20) of the characteristic polynomial, the constant term c_0 is $|A|$, while the coefficient c_{n-1} of λ^{n-1} comes from the diagonal of $A - \lambda I$ and is given by

$$(-1)^{n+1}c_{n-1} = A_{11} + A_{22} + \cdots + A_{nn}.$$

This sum of the diagonal entries of A is known as the *trace* of the matrix A. The trace, like the other coefficients in the characteristic polynomial, is an invariant under the relation of similarity. This is a consequence of

THEOREM 11. *Similar matrices have the same characteristic polynomial.*

Proof: Let the two similar matrices be A and PAP^{-1}, with P invertible. Since $|P|$ and $|P^{-1}| = |P|^{-1}$ are scalars, they commute. Hence, the rule for multiplying determinants gives

$$\begin{aligned} |PAP^{-1} - \lambda I| &= |PAP^{-1} - \lambda PIP^{-1}| = |P(A - \lambda I)P^{-1}| \\ &= |P|\,|A - \lambda I|\,|P|^{-1} = |A - \lambda I|, \end{aligned}$$ Q.E.D.

EXERCISES

1. Find two real 2×2 matrices which have the same characteristic polynomial yet are not similar.

2. Compute the eigenvalues and the eigenvectors of the following matrices of complex numbers:

$$\text{(a)} \begin{bmatrix} -1 & 2 & 2 \\ 2 & 2 & 2 \\ -3 & -6 & -6 \end{bmatrix}; \quad \text{(b)} \begin{bmatrix} 3 & 2 & 2 \\ 1 & 4 & 1 \\ -2 & -4 & -1 \end{bmatrix}; \quad \text{(c)} \begin{bmatrix} 4 & 9 & 0 \\ 0 & -2 & 8 \\ 0 & 0 & 7 \end{bmatrix}.$$

3. Find all 2×2 matrices over $\mathbf{Q}$ with eigenvalues 1 and -1.

4. Find a necessary and sufficient condition that the (complex) eigenvalues of a 2×2 matrix of (real or complex) numbers are equal.

5. Show that any real 2×2 matrix with negative determinant is similar to a diagonal matrix. Interpret geometrically.

6. Prove that a matrix A and its transpose A^T have the same characteristic polynomial.

7. In the characteristic polynomial $|A - \lambda I|$, show that the coefficient of λ^{n-2} is $(-1)^n\Sigma(A_{ii}A_{jj} - A_{ij}A_{ji})$, where the sum is taken over all $i, j \in \mathbf{n}$ with $i < j$.

8. Prove that the characteristic polynomial of a triangular matrix T is $(T_{11} - \lambda)(T_{22} - \lambda) \cdots (T_{nn} - \lambda)$.

9. For square matrices A and B, show that the characteristic polynomial of the matrix $\begin{bmatrix} A & 0 \\ C & B \end{bmatrix}$ is the product of those for A and B.

10. Show that the matrix products AB and BA have the same trace.

***11.** If the characteristic polynomial of an $n \times n$ matrix A over F is a product of linear factors in $F[\lambda]$, prove that A is similar to a triangular matrix. (*Hint:* Choose a basis of F^n with one vector characteristic and use induction on n.)

6. The Minimal Polynomial

Similar matrices have the same determinant and the same trace. Hence, the determinant and the trace of an $n \times n$ matrix are functions $F^{n \times n} \to F$ which are invariants (as defined in §VII.7) for the relation of similarity between $n \times n$ matrices over F. These invariants take values in the field of scalars (and so may be called "numerical" invariants). A different sort of invariant for similarity is the function $F^{n \times n} \to F[\lambda]$ which assigns to each matrix A its characteristic polynomial. We now construct another such "polynomial" invariant, the "minimal polynomial" of A.

If $t: V \to V$ is an endomorphism of the finite-dimensional vector space V over F, so is each polynomial expression $\Sigma f_i t^i$ in t with coefficients f_i in F. More explicitly, the subring of End (V) generated by t and the scalar multiples of 1_A is a commutative ring containing F and one new element t, while the polynomial ring $F[x]$ is a "universal" commutative ring containing F and one new element, the indeterminate x. Therefore (by Theorem III.15), there is a unique morphism $E_t: F[x] \to \text{End}(V)$ of rings with $E_t \kappa = \kappa$ for all $\kappa \in F$ and $E_t x = t$. This morphism just replaces x by t, so takes each polynomial $f = \Sigma f_i x^i$ of degree k into the endomorphism

$$f(t) = E_t f = f_0 + f_1 t + \cdots + f_k t^k : V \to V. \tag{21}$$

If V has dimension n, then End $(V) \cong F^{n \times n}$ has dimension n^2, and hence the $n^2 + 1$ powers $1, t, t^2, \cdots, t^{n^2}$ are linearly dependent. This means that there is a non-zero polynomial f of degree at most n^2 with $f(t) = 0$, and hence, an element $f \neq 0$ in the kernel of the morphism E_t. The kernel of $E_t: F[x] \to \text{End}(V)$ is thus a non-zero ideal in the polynomial ring. Like every ideal in that ring, it is principal, so consists of all the multiples of some single polynomial m. We can choose m monic (that is, with leading coefficient 1); m is then unique. It may be described directly as that monic polynomial m of least degree in $F[x]$ with $m(t) = 0$, and so is called the *minimal polynomial* of t. We have proved

THEOREM 12. *Let V be a finite-dimensional vector space over F. For each linear transformation $t: V \to V$, the polynomials $f \in F[x]$ with $f(t) = 0$ are the multiples of a unique monic polynomial m.*

A similar process applies to an $n \times n$ matrix A. Given A in the matrix ring $F^{n\times n}$, there is a unique morphism $E_A : F[x] \to F^{n\times n}$ of rings with $E_A x = A$ and $E_A \kappa = \kappa I$ for each scalar κ, where I is the identity matrix. This morphism assigns to each polynomial $f = \sum f_i x^i$ the $n \times n$ matrix $E_A f = f(A)$,

$$f(A) = f_0 I + f_1 A + \cdots + f_k A^k. \tag{22}$$

Theorem 12, with t replaced by A, describes the polynomials f with $f(A) = 0$ as the multiples of a single monic polynomial m, the *minimal polynomial of the matrix A*.

Now pass from endomorphisms to square matrices. For each basis **b** of V, the function which assigns to each linear transformation $t: V \to V$ its matrix $Mt = A$, rel **b**, is an isomorphism of rings. This proves

COROLLARY 1. *The minimal polynomial of a linear transformation $t: V \to V$ is identical to the minimal polynomial of any matrix representing t.*

Since similar matrices represent the same endomorphism relative to different bases, the following consequence is immediate:

COROLLARY 2. *Similar matrices have the same minimal polynomial.*

This may also be proved by matrix calculations. For P invertible, $PA^iP^{-1} = (PAP^{-1})^i$, and hence

$$f(PAP^{-1}) = \sum_i f_i(PAP^{-1})^i = P\Big(\sum_i f_i A^i\Big)P^{-1} = Pf(A)P^{-1}.$$

Therefore, $f(A) = 0$ if and only if $f(PAP^{-1}) = 0$.

An interesting example is given by nilpotent transformations. A linear transformation $t: V \to V$ is called *nilpotent* when $t^k = 0$ for some positive integer k. The minimal polynomial of t is thus some divisor of the polynomial x^k, so has the form x^n for $n \leqslant k$. Suppose, in particular, that $\dim V = n$ and $t: V \to V$ is nilpotent with minimal polynomial x^n for the same n. Then $t^{n-1} \neq 0$, so there is some vector $v \in V$ with $t^{n-1}(v) \neq 0$. Under these conditions, we claim that the n vectors $v, t(v), \cdots, t^{n-1}(v)$ are linearly independent. For suppose instead that $v\kappa_0 + (tv)\kappa_1 + \cdots + (t^{n-1}v)\kappa_{n-1} = 0$ with scalar coefficients κ_i not all zero. If κ_j is the first non-zero coefficient, t^{n-j-1} applied to this equation gives $(t^{n-1}v)\kappa_j = 0$, in contradiction to $t^{n-1}v \neq 0$. If the independent vectors $v, t(v), \cdots, t^{n-1}(v)$ are taken as a basis of V, the map t is represented by a matrix in which the ith column is, as usual, the column of coordinates of $t(t^{i-1}v)$; this matrix is therefore the

$n \times n$ matrix

$$\begin{bmatrix} 0 & 0 & \cdots & 0 & 0 \\ 1 & 0 & \cdots & 0 & 0 \\ 0 & 1 & \cdots & 0 & 0 \\ \vdots & \vdots & & \vdots & \vdots \\ 0 & 0 & \cdots & 1 & 0 \end{bmatrix}. \tag{23}$$

Its only non-zero entries are 1's along the diagonal just below the principal diagonal. This $n \times n$ matrix has minimal polynomial x^n; its characteristic polynomial is also x^n.

The result leads to a way of constructing an $n \times n$ matrix whose minimal polynomial is an arbitrary given monic polynomial

$$g = g_0 + g_1 x + \cdots + g_{n-1} x^{n-1} + x^n$$

of degree n. For example, if $n = 4$, this matrix, called the *companion matrix* of g, is

$$M_g = \begin{bmatrix} 0 & 0 & 0 & -g_0 \\ 1 & 0 & 0 & -g_1 \\ 0 & 1 & 0 & -g_2 \\ 0 & 0 & 1 & -g_3 \end{bmatrix}. \tag{24}$$

All its entries are zero except for entries 1 along the diagonal just below the principal diagonal and entries in the last column, each the negative of the corresponding coefficient of g. This last statement describes the companion matrix M_g for a monic polynomial g of any degree n.

PROPOSITION 13. *The companion matrix M_g of a monic polynomial g of degree n is an $n \times n$ matrix with minimal polynomial g and characteristic polynomial $(-1)^n g$.*

Proof: Let M_g act as a linear transformation $t(\xi) = M_g \xi$ on the space F^n of lists ξ of n scalars. The ith column of M_g gives the image under t of the ith unit vector ε_i, so that

$$t\varepsilon_1 = \varepsilon_2, \quad t\varepsilon_2 = \varepsilon_3, \quad \cdots, \quad t\varepsilon_{n-1} = \varepsilon_n,$$
$$t\varepsilon_n = -\varepsilon_1 g_0 - \varepsilon_2 g_1 - \cdots - \varepsilon_n g_{n-1}.$$

By the first $n - 1$ equations, $\varepsilon_{i+1} = t^i \varepsilon_1$ for each $i < n$. Putting these values in the last equation gives

$$t^n \varepsilon_1 = -\left(g_0 + g_1 t + \cdots + g_{n-1} t^{n-1}\right)\varepsilon_1.$$

Hence, $g(t)\varepsilon_1 = 0$ and so $g(t)\varepsilon_{i+1} = g(t)t^i\varepsilon_1 = 0$ for every i. Therefore, $g(t) = 0$. If f is a polynomial of lower degree, $(\sum f_i t^i)\varepsilon_1$ is the linear combination $\sum f_i \varepsilon_{i+1}$, so is zero only when every coefficient $f_i = 0$. Therefore, g is indeed the minimal polynomial for t and hence for M_g.

As for the characteristic polynomial of the 4×4 companion matrix M_g, the determinant

$$|M_g - \lambda I| = \begin{vmatrix} -\lambda & 0 & 0 & -g_0 \\ 1 & -\lambda & 0 & -g_1 \\ 0 & 1 & -\lambda & -g_2 \\ 0 & 0 & 1 & -\lambda - g_3 \end{vmatrix}$$

may be expanded, say by minors of the last column, to give exactly $g(\lambda)$. For general n, it is $(-1)^n g(\lambda)$.

In Chapter XI we shall prove that every matrix is similar to a direct sum of such companion matrices for various polynomials g.

Consider next the relation between the minimal polynomial m and the characteristic polynomial c of a square matrix A. For $A = M_g$ a companion matrix, $\pm c = g = m$; for A a direct sum $A = M_g \oplus M_g$, $c = g^2$ and $m = g$, so m divides c. For A diagonal, m divides c (Exercise 6). We now prove that m always divides c; since $m(A) = 0$, we can state this result as $c(A) = 0$. ("Every matrix satisfies its characteristic equation".)

Theorem 14 (*Cayley–Hamilton*). *Every square matrix A over a field F has $c(A) = 0$ for c the characteristic polynomial of A.*

Proof: Let A be a matrix with entries in the field F, while $F[\lambda]$ is the ring of all polynomials in an indeterminate λ with coefficients in the field F. A matrix with entries in the ring $F[\lambda]$ will be termed (in this proof) a λ-matrix. For example, $A - \lambda I$ is such a matrix, with diagonal entries the linear polynomials $A_{ii} - \lambda$ and off-diagonal entries the scalars A_{ij} (polynomials of degree 0). The multiples $\lambda A, \lambda^2 A, \cdots$ are also λ-matrices. In any non-zero λ-matrix D we let k denote the highest degree of any non-zero polynomial which is an entry of D. Then D can be written in terms of matrices $D^{(i)}$ of scalars as a sum

$$D = D^{(0)} + D^{(1)}\lambda + \cdots + D^{(k)}\lambda^k. \tag{25}$$

Next observe that the familiar factorization

$$a^i - \lambda^i = (a^{i-1} + a^{i-2}\lambda + \cdots + \lambda^{i-1})(a - \lambda), \qquad i = 2, 3, 4 \cdots,$$

for polynomials has an analog

$$A^i - \lambda^i I = (A^{i-1} + \lambda A^{i-2} + \cdots + \lambda^{i-1} I)(A - \lambda I),$$

which is a valid equation for matrices. To verify it, simply "multiply out" the right-hand side and cancel. Therefore, letting $L^{(i)}$ designate the λ-matrix $A^{i-1} + \lambda A^{i-2} + \cdots + \lambda^{i-1} I$, with $L^{(1)} = I$ and $L^{(0)} = 0$, we thus have

$$A^i - \lambda^i I = L^{(i)}(A - \lambda I), \qquad i = 0, 1, 2, \cdots . \tag{26}$$

Now consider the characteristic polynomial

$$|A - \lambda I| = c(\lambda) = c_0 + c_1\lambda + \cdots + c_n\lambda^n$$

of A (with $c_n = \pm 1$). For the matrix $A - \lambda I$ the transposed matrix of cofactors is a λ-matrix B; by (12) of Theorem 6,

$$B(A - \lambda I) = c(\lambda)I, \qquad c(\lambda) = |A - \lambda I|. \tag{27}$$

In $c(A)$, replace each power A^i by (26) to get

$$c(A) = \sum_{i=0}^{n} c_i A^i = \sum_{i=0}^{n} c_i \lambda^i I + \sum_{i=0}^{n} c_i L^{(i)}(A - \lambda I).$$

The first term on the right is $c(\lambda)I$; putting in B from (27) gives

$$c(A) = B(A - \lambda I) + \sum_{i=0}^{n} c_i L^{(i)}(A - \lambda I) = D(A - \lambda I), \tag{28}$$

where the coefficient $D = B + \Sigma c_i L^{(i)}$ is another λ-matrix. If $D \neq 0$, it has some degree k in λ, so we expand it as in (25) to get

$$c(A) = (D^{(0)} + D^{(1)}\lambda + \cdots + D^{(k)}\lambda^k)(A - \lambda I)$$

with $D^{(k)} \neq 0$ a non-zero matrix of scalars. Multiplying out gives a non-zero term $-D^{(k)}\lambda^{k+1}$ of positive degree $k + 1$ in λ, a contradiction, since the left-hand side $c(A)$ is a matrix of constants (degree 0 in λ). Hence, $D = 0$; by (28) this implies $c(A) = 0$, completing the proof.

Now by Theorem 12 every polynomial f with $f(A) = 0$ is a multiple of the minimal polynomial of A. Hence

Corollary. *The minimal polynomial of any $n \times n$ matrix divides the characteristic polynomial, hence has degree at most n.*

Previously (Exercise VII.1.8) the best that we knew was that the degree of the minimal polynomial was at most n^2.

EXERCISES

1. If V is an n-dimensional vector space and $t: V \to V$ a nilpotent transformation, prove that $t^n = 0$ (that is, that the minimal polynomial of t is a divisor of x^n).

2. Show that every non-zero 2×2 nilpotent matrix is similar to

$$\begin{bmatrix} 0 & 0 \\ 1 & 0 \end{bmatrix}.$$

3. Show that every 3×3 nilpotent matrix is similar to one of

$$\begin{bmatrix} 0 & 0 & 0 \\ 1 & 0 & 0 \\ 0 & 1 & 0 \end{bmatrix}, \quad \begin{bmatrix} 0 & 0 & 0 \\ 1 & 0 & 0 \\ 0 & 0 & 0 \end{bmatrix}, \quad \begin{bmatrix} 0 & 0 & 0 \\ 0 & 0 & 0 \\ 0 & 0 & 0 \end{bmatrix}.$$

4. Prove that every 4×4 nilpotent matrix of rank 3 is similar to the companion matrix of the polynomial x^4.

5. Show that every 2×2 matrix A with $A^2 = I$, over a field with $2 \neq 0$, is similar to a diagonal matrix.

6. Show by direct calculation (without using Theorem 14) that the characteristic polynomial of any triangular matrix is a multiple of its minimal polynomial.

7. If λ is an eigenvalue of A and f any polynomial, show that $f(\lambda)$ is an eigenvalue of $f(A)$.

8. Show that the companion matrix of the polynomial $x^4 - 1$ over the field $\mathbf{C}$ of complex numbers is similar (over $\mathbf{C}$) to a diagonal matrix.

9. For the 4×4 matrix displayed below show that the eigenvalues λ_1, λ_2, λ_3, and λ_4 are given by

$$\lambda_k = a + bi^k + ci^{2k} + di^{3k}$$

for $k = 1, 2, 3, 4$, and $i = \sqrt{-1}$:

$$\begin{bmatrix} a & d & c & b \\ b & a & d & c \\ c & b & a & d \\ d & c & b & a \end{bmatrix}.$$

(*Hint:* Use Exercise 7 and the companion matrix of $x^4 - 1$.)

10. Generalize Exercise 9 to $n \times n$ matrices of a similar form.

11. If A is invertible with characteristic polynomial $c(\lambda)$ of degree n, show that the transposed matrix of cofactors of A is

$$\alpha^T = -\left[c_1 I + c_2 A + \cdots + c_{n-1} A^{n-2} + (-1)^n A^{n-1}\right].$$

12. In the notation of Exercise 11, show that the characteristic polynomial of A^{-1} is

$$(-1)^n \left[\lambda^n + \frac{c_1}{|A|}\lambda^{n-1} + \cdots + \frac{(-1)^n}{|A|}\right].$$

13. Prove that any 3×3 matrix with one eigenvalue λ_0 and characteristic polynomial $(\lambda - \lambda_0)^3$ is similar to one of the three matrices of §VII.8, Exercise 8, and show that the three matrices there have different minimal polynomials.

7. Universal Bilinear Functions

Now we return to the study of bilinear functions $h:A \times B \to C$, where A, B, and C are right modules over a commutative ring K. Given A and B, we wish to construct a "universal" such bilinear function; that is, a bilinear function $h_0:A \times B \to D$ with codomain a suitable module D such that every bilinear $h:A \times B \to C$ can be written as $h = th_0$ for a unique linear map $t:D \to C$. This will be done in general in §8 by constructing the module D as a "tensor product" of A and B; in this section we describe some specific universal bilinear functions.

First a simple example. For each right K-module A the scalar multiples $(a, \kappa) \mapsto a\kappa$ define a function $h_0:A \times K \to A$. For fixed a it is linear in κ; for fixed κ, it is linear in a, since the commutative law for the product in K gives

$$(a_1\lambda_1 + a_2\lambda_2)\kappa = a_1(\lambda_1\kappa) + a_2(\lambda_2\kappa) = (a_1\kappa)\lambda_1 + (a_2\kappa)\lambda_2.$$

Therefore, h_0 is bilinear. If $h:A \times K \to C$ is *any* bilinear function with the same domain $A \times K$, then $a \mapsto h(a, 1)$ is a linear transformation $t:A \to C$. Moreover,

$$h(a, \kappa) = h(a, 1)\kappa = (ta)\kappa = t(a\kappa) = (t \circ h_0)(a, \kappa).$$

In other words, the arbitrary bilinear function h is expressed as a composite $h = t \circ h_0$ with the fixed bilinear function h_0, as in the commutative diagram

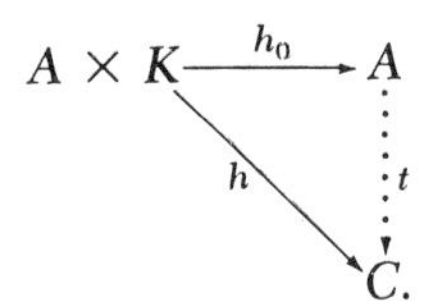

Moreover, this condition uniquely determines t. For suppose that $t':A \to C$ were some other linear transformation with $t' \circ h_0 = h$. Then $t(a) = h(a, 1) = t'h_0(a, 1) = t'(a)$ for all $a \in A$, so $t' = t$. Therefore, $h_0:A \times K \to A$ is a universal bilinear function on A and K.

More generally, for A and the free module K^n there is a universal bilinear function $h_0:A \times K^n \to A^n$. Indeed, let $\varepsilon_1, \cdots, \varepsilon_n$ be the usual unit elements of K^n, so that an element of the free module K^n can be written as a sum $\varepsilon_1\xi_1 + \cdots + \varepsilon_n\xi_n$ with coefficients $\xi_i \in K$. Then

$$h_0\left(a, \sum_i \varepsilon_i\xi_i\right) = (a\xi_1, \cdots, a\xi_n), \qquad a \in A \tag{29}$$

defines a bilinear function $h_0 : A \times K^n \to A^n$ with codomain the module A^n of all lists of n elements of A. If $h : A \times K^n \to C$ is any bilinear function with the same domain, a linear transformation $t : A^n \to C$ is defined by

$$t(a_1, \cdots, a_n) = h(a_1, \varepsilon_1) + \cdots + h(a_n, \varepsilon_n).$$

Then $th_0(a, \Sigma\varepsilon_i\xi_i) = \Sigma h(a\xi_i, \varepsilon_i) = h(a, \Sigma\varepsilon_i\xi_i)$, so $t \circ h_0 = h$. As in the previous case, this condition $t \circ h_0 = h$ uniquely determines the linear map t. Therefore, h_0 is universal for bilinear functions on $A \times K^n$.

EXERCISES

1. Construct a universal bilinear function $K^2 \times K^2 \to K^4$.
2. Construct a universal trilinear function $K \times B \times K \to B$.
3. Show that the image of the universal bilinear function $h_0 : K^2 \times B \to B^2$ need not be a submodule of B^2 (say in the case $B = K^2$ and $K = \mathbf{Z}$).
4. For the ring $K = \mathbf{Z}$ of integers:
   (a) Construct a universal $\mathbf{Z}$-bilinear function $\mathbf{Z}_3 \times \mathbf{Z}_3 \to \mathbf{Z}_3$.
   (b) Construct a universal $\mathbf{Z}$-bilinear function $\mathbf{Z}_6 \times \mathbf{Z}_{10} \to \mathbf{Z}_2$.
   (c) If $(m, n) = 1$, show that any $\mathbf{Z}$-bilinear function on $\mathbf{Z}_m \times \mathbf{Z}_n$ is zero.

8. Tensor Products

For any two right modules A and B over a commutative ring K we now construct a universal K-bilinear function on $A \times B$; more precisely, we simultaneously construct a new K-module $A \otimes B$ and a K-bilinear map $A \times B \to A \otimes B$ which is universal among bilinear functions from $A \times B$ to a K-module.

First, we construct a free module F with the set $A \times B$ as free generators. Take F to be the set of all those functions $f : A \times B \to K$ which have only a *finite* number of non-zero values. Under termwise addition and termwise scalar multiples, these functions f do form a K-module. For elements $a \in A$, $b \in B$ we may write $[a, b]$ for the special function $f \in F$ which is 1 on $(a, b) \in A \times B$ and 0 elsewhere. Then every function f is a finite linear combination

$$f = \sum_{(a,b)} [a, b] f(a, b).$$

This amounts to showing that the $[a, b]$ form a (possibly infinite) basis of F. Define a function $u:A \times B \to F$ by $u(a, b) = [a, b]$, for all elements a, b.

If C is any module and $h:A \times B \to C$ any function, the formula

$$sf = \sum_{(a, b)} h(a, b)f(a, b)$$

defines a linear transformation $s:F \to C$; in particular, $s[a, b] = h(a, b)$. Comparison with the previous formula shows that s is the only linear transformation with $s[a, b] = h(a, b)$ for all $a \in A$ and $b \in B$. In other words, it is the unique linear map which makes

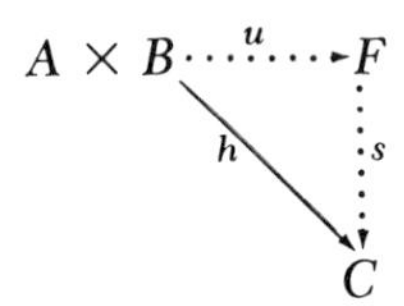

commute. This states that F is the free module on $A \times B$.

However, u is by no means bilinear; for example, the element $[a_1, b] + [a_2, b]$ in F is never the element $[a_1 + a_2, b]$. What we now do is to take a "biggest possible" quotient module F/S of F so that the composite $A \times B \to F \to F/S$ will be bilinear. To do this, let S be the submodule of F spanned by all the elements

$$[a_1\kappa_1 + a_2\kappa_2, b] - [a_1, b]\kappa_1 - [a_2, b]\kappa_2, \tag{30}$$

$$[a, b_1\kappa_1 + b_2\kappa_2] - [a, b_1]\kappa_1 - [a, b_2]\kappa_2 \tag{31}$$

for every choice of scalars κ_i and elements in A and in B. Write $A \otimes B$ for the quotient module F/S and $(a, b) \mapsto a \otimes b$ for the composite of the insertion $u:A \times B \to F$ and the projection $p:F \to F/S$. Thus

$$a \otimes b = p[a, b] \in A \otimes B$$

stands for the coset $S + [a, b]$ in F/S. Now the projection p takes all elements of S to zero; in particular, the elements (30) and (31). Therefore,

$$(a_1\kappa_1 + a_2\kappa_2) \otimes b = (a_1 \otimes b)\kappa_1 + (a_2 \otimes b)\kappa_2,$$
$$a \otimes (b_1\kappa_1 + b_2\kappa_2) = (a \otimes b_1)\kappa_1 + (a \otimes b_2)\kappa_2.$$

These equations state that the function

$$\otimes : A \times B \to A \otimes B$$

defined by $(a, b) \mapsto a \otimes b$ is bilinear; in other words, S has been chosen just big enough to make the composite $A \times B \to F \to F/S$ bilinear.

We have constructed a K-module $A \otimes B$ and a bilinear function $A \times B \to A \otimes B$, as proposed; because the elements $[a, b]$ span F, this module $A \otimes B$ is spanned by the elements $a \otimes b$ in the image of this bilinear function. By linearity they satisfy $a \otimes 0 = 0$ and $a \otimes (-b) = -(a \otimes b)$ for all $a \in A$ and $b \in B$. The "main theorem" on tensor products now asserts that this function $A \times B \to A \otimes B$ is universal:

Theorem 15. *To each K-bilinear function $h: A \times B \to C$ there is exactly one K-linear transformation $t: A \otimes B \to C$ with $t(a \otimes b) = h(a, b)$.*

In other words, every bilinear function h can be had by following the special bilinear function $\otimes$ by a suitable (uniquely determined) K-linear map t.

Proof: We are given the solid arrows in the diagram

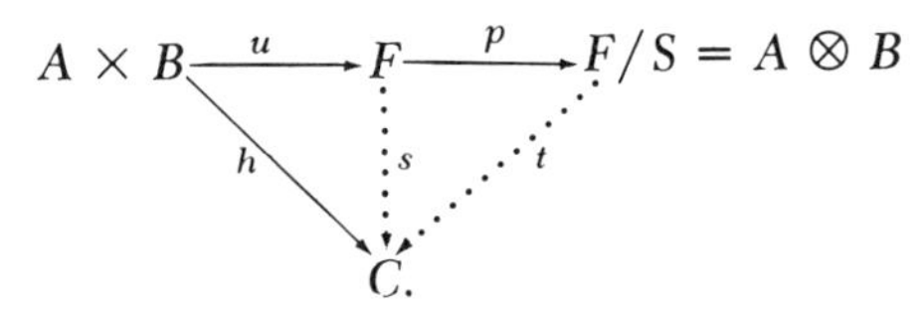

Since F is free, there is a unique K-linear map s with $s \circ u = h$ as shown. Now h bilinear implies that s carries every one of the generators (30) and (31) of the submodule S into zero. Since the kernel of s contains the whole submodule S, the universal property of the projection $p: F \to F/S$ now produces a K-linear function $t: F/S \to C$, again unique, with $t \circ p = s$. All told, $tpu = h$, so $t(a \otimes b) = h(a, b)$ for all a and b, and the map t is that required in the theorem.

Any element of $A \otimes B$ is a finite sum $\Sigma a_i \otimes b_i$ for some $a_i \in A$, $b_i \in B$. As the elements $a \otimes b$ span $A \otimes B$, a linear map t on $A \otimes B$ is fully determined by giving the values $t(a \otimes b)$; indeed, *to define a linear function t on $A \otimes B$ it suffices* by this theorem *to require that $t(a \otimes b)$ be any expression $h(a, b)$ which is bilinear* in a and b.

Suppose now that $h_0: A \times B \to D$ is some other universal K-bilinear function on the same K-modules A and B. The usual uniqueness property for universals shows that there is an isomorphism $\theta: A \otimes B \cong D$ of K-modules with $\theta(a \otimes b) = h_0(a, b)$ for all $a \in A, b \in B$. Thus any two universal bilinear functions on $A \times B$ have isomorphic codomains $A \otimes B \cong D$. For example, in the last section the commutativity of K was used to show that $(a, \kappa) \mapsto a\kappa$ is a universal bilinear function $A \times K \to A$ with codomain A. Hence, in this case there is for each K-module A an isomorphism

$$\phi : A \otimes K \cong A, \qquad \phi(a \otimes \kappa) = a\kappa. \tag{32}$$

Similarly, by the universality of h_0 in (29) there is an isomorphism

$$\theta : A \otimes K^n \cong A^n, \qquad \theta(a \otimes (\xi_1, \cdots, \xi_n)) = (a\xi_1, \cdots, a\xi_n) \qquad (33)$$

for all $\xi_i \in K$ and $a \in A$. This isomorphism θ shows that the K-module $A \otimes K^n$ "is" the function module A^n (up to isomorphism, which is all that matters). In finding this isomorphism and elsewhere in dealing with the tensor product modules $A \otimes B$, we *never* need to use the free module F and the submodule S from which $A \otimes B$ was constructed, but only the universality of $\otimes : A \times B \to A \otimes B$.

One useful consequence is: *Every tensor product of free modules is free.* This we prove for free modules of finite type as follows (for other types, see Exercise 14).

Proposition 16. *If A and C are free K-modules on the finite sets $\{b_1, \cdots, b_m\}$ and $\{c_1, \cdots, c_n\}$, respectively, then $A \otimes C$ is a free K-module on the set $\{b_i \otimes c_j | i \in \mathbf{m}, j \in \mathbf{n}\}$ of mn elements.*

In particular, for finite-dimensional vector spaces V and W,

$$\dim (V \otimes W) = (\dim V)(\dim W). \qquad (34)$$

The proof amounts to the following construction of all bilinear functions on $A \times C$:

Lemma. *For mn elements d_{ij} in any K-module D, there is a unique bilinear $h : A \times C \to D$ with $h(b_i, c_j) = d_{ij}$ for all $i \in \mathbf{m}$, $j \in \mathbf{n}$.*

Proof: If such an h exists, bilinearity shows that

$$h\left(\sum b_i\kappa_i, \sum c_j\lambda_j\right) = \sum_i \sum_j d_{ij}\kappa_i\lambda_j$$

for all lists κ_i and λ_j of scalars. Thus h is unique if it exists. Conversely, the h given by this formula is bilinear, and has the required values $h(b_i, c_j) = d_{ij}$ for all i and j, Q.E.D.

Now we can prove the Proposition. First take F to be some free K-module with a basis of mn elements e_{ij}. Then to each bilinear $h : A \times C \to D$ there is a unique linear map $t : F \to D$ with $t(e_{ij}) = h(b_i, c_j)$ for all i and j. By the lemma, there is also a bilinear map $w : A \times C \to F$ with $w(b_i, c_j) = e_{ij}$ for all i and j. Thus $t : F \to D$ is the unique linear t with $(t \circ w)(b_i, c_j) = h(b_i, c_j)$. By the uniqueness assertion of the lemma, this means that t is the unique linear map with $t \circ w = h$. This states exactly that w is a universal bilinear function on $A \times C$, hence that $w = \otimes$ and its codomain F *is* the tensor product $A \otimes C$. By construction, it is a free module on the basis $w(b_i, c_j) = b_i \otimes c_j$, as asserted in the proposition.

Now consider a tensor product of A and a biproduct module $B_1 \oplus B_2$, with its injections $e_i : B_i \to B_1 \oplus B_2$ given by $e_1(b_1) = (b_1, 0)$ and $e_2(b_2) = (0, b_2)$ as in (V.31).

Proposition 17. *For K-modules A, B_1, and B_2, there is an isomorphism*

$$\theta : A \otimes (B_1 \oplus B_2) \cong (A \otimes B_1) \oplus (A \otimes B_2) \tag{35}$$

of K-modules, uniquely determined by the requirement that

$$\theta(a \otimes (b_1, b_2)) = (a \otimes b_1, a \otimes b_2), \qquad a \in A, \quad b_1 \in B_1, b_2 \in B_2. \tag{36}$$

Proof: The expression $(a \otimes b_1, a \otimes b_2)$ on the right of (36) is bilinear in its arguments a and (b_1, b_2). Hence, the universality of $\otimes$ provides a unique linear function θ from $A \otimes (B_1 \oplus B_2)$ to $(A \otimes B_1) \oplus (A \otimes B_2)$ satisfying (36). To show θ an isomorphism, we construct its inverse, using universality arguments on the diagram

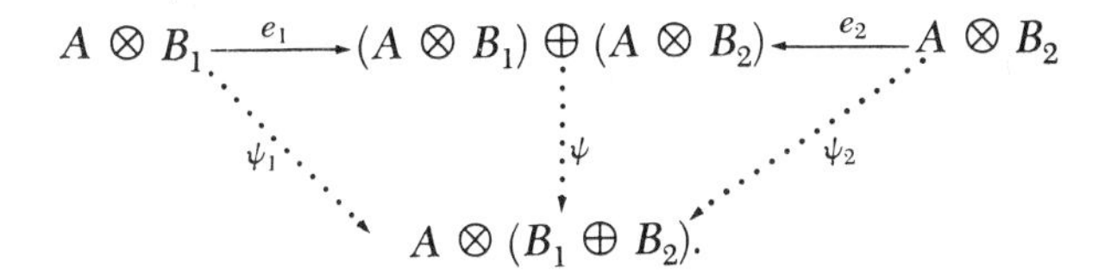

Since $a \otimes (b_1, 0)$ or $a \otimes (0, b_2)$ is bilinear in its arguments a and b_i, and since $\otimes : A \times B_i \to A \otimes B_i$ is universal, there are linear functions ψ_i, as displayed above, with

$$\psi_1(a \otimes b_1) = a \otimes (b_1, 0), \qquad \psi_2(a \otimes b_2) = a \otimes (0, b_2).$$

Since the injections e_1 and e_2 of the biproduct, as displayed, are universal, these two functions ψ_1 and ψ_2 combine to give a single linear function ψ on $(A \otimes B_1) \oplus (A \otimes B_2)$ with $\psi e_1 = \psi_1$ and $\psi e_2 = \psi_2$, hence

$$\psi(a_1 \otimes b_1, a_2 \otimes b_2) = a_1 \otimes (b_1, 0) + a_2 \otimes (0, b_2) \in A \otimes (B_1 \oplus B_2).$$

From these formulas and the definition of $+$ in $B_1 \oplus B_2$ we calculate that

$$\theta\psi(a_1 \otimes b_1, a_2 \otimes b_2) = (a_1 \otimes b_1, 0) + (0, a_2 \otimes b_2) = (a_1 \otimes b_1, a_2 \otimes b_2),$$
$$\psi\theta(a \otimes (b_1, b_2)) = a \otimes (b_1, 0) + a \otimes (0, b_2) = a \otimes (b_1, b_2);$$

both composites $\theta\psi$ and $\psi\theta$ thus are the identity, so θ has a two-sided inverse ψ and therefore is an isomorphism.

A parallel argument will show that

$$(A_1 \oplus A_2) \otimes B \cong (A_1 \otimes B) \oplus (A_2 \otimes B). \tag{37}$$

Alternatively, this may be deduced from (35) and the evident isomorphism

$$A \otimes B \cong B \otimes A. \tag{38}$$

Given two K-linear transformations $s:A \to A'$ and $t:B \to B'$ we now construct

$$s \otimes t : A \otimes B \to A' \otimes B', \tag{39}$$

a K-linear transformation called the *tensor product* of s and t. First, since $\otimes: A' \times B' \to A' \otimes B'$ is bilinear, the assignment $(a, b) \mapsto sa \otimes tb$ is a bilinear function $A \times B \to A' \otimes B'$. Since $\otimes: A \times B \to A \otimes B$ is a universal bilinear function on $A \times B$, there is a unique linear transformation $s \otimes t: A \otimes B \to A' \otimes B'$ with $(s \otimes t)(a \otimes b) = sa \otimes tb$ for all $a \in A$ and all $b \in B$. From this uniqueness of $s \otimes t$ it follows that $1_A \otimes 1_B$ is an identity map when 1_A and 1_B are identities and, if composites $s' \circ s$ and $t' \circ t$ are defined, that

$$(s' \circ s) \otimes (t' \circ t) = (s' \otimes t') \circ (s \otimes t). \tag{40}$$

If A is fixed, these properties show that $B \mapsto A \otimes B$ and $t \mapsto 1_A \otimes t$ is a functor, "tensor product on the left by A", on K-modules to K-modules. Similarly, for B fixed, $A \mapsto A \otimes B$ and $s \mapsto s \otimes 1_B$ is a functor, "tensor product on the right by B".

The tensor product of two K-modules A and B is often written as $A \otimes_K B$ to distinguish it from the tensor product $A \otimes_{\mathbf{Z}} B$ of the abelian groups A and B, regarded just as $\mathbf{Z}$-modules. In other words, $\otimes_K: A \times B \to A \otimes_K B$ is universal for *K-bilinear* functions, and $\otimes_{\mathbf{Z}}: A \times B \to A \otimes_{\mathbf{Z}} B$ for *biadditive* functions $A \times B \to C$.

For S a submodule of A and module elements $s \in S$ and $b \in B$ the symbol $s \otimes b$ has two meanings: Either $\otimes$ denotes the universal bilinear function on $S \times B$, so that $s \otimes b$ is an element of $S \otimes_K B$, or $\otimes$ denotes the universal bilinear function on $A \times B$, so that $s \otimes b$ is an element of a different module $A \otimes_K B$. Although S is a submodule of A, there is no reason why the first module $S \otimes_K B$ should be a submodule of the second one, $A \otimes_K B$ (see Exercise 1b for an example). We can, however, raise a related question, by replacing the inclusion *relation* $S \subset A$ by the insertion *map* $i: S \to A$. Now i is a monomorphism; we ask whether the tensor product map $i \otimes 1_B: S \otimes_K B \to A \otimes_K B$ is a monomorphism. An affirmative answer would mean that $i \otimes 1_B$ carries $S \otimes_K B$ isomorphically onto a submodule of $A \otimes_K B$ and each element $s \otimes b$ (in the first sense) into the element $s \otimes b$ (in the second sense). If A and B are finite-dimensional vector spaces, the answer is affirmative, as follows:

Proposition 18. *If S, V, and W are finite-dimensional vector spaces over a field F, then if a linear transformation $t: S \to V$ is a monomorphism, so is $t \otimes 1_W: S \otimes_F W \to V \otimes_F W$.*

Proof: A monomorphism $t:S \to V$ of vector spaces must by Proposition VI.6 have a linear left inverse $u:V \to S$, so that $u \circ t = 1$. Then by (40) $(u \otimes 1_W)(t \otimes 1_W) = 1_{S \otimes W}$, so $t \otimes 1_W$ also has a left inverse and therefore is also a monomorphism.

All this is straightforward; the surprise is that this Proposition does *not* hold for modules (even of finite type) over a commutative ring K. One can find two abelian groups A and B and a subgroup $S \subset A$ so that the insertion $i:S \to A$ does not give a monomorphism $S \otimes_{\mathbf{Z}} B \to A \otimes_{\mathbf{Z}} B$ (see Exercise 1 for an example). Put the matter differently: The insertion $i:S \to A$ and the identity map $1_S:S \to S$ have the same graph (the set of all pairs (s, s) for $s \in S$), but the corresponding tensor product maps $i \otimes 1_B$ and $1_S \otimes 1_B$ can differ (for example, $i \otimes 1_B = 0$, $1_S \otimes 1_B$ an identity map). This gives another reason why we have insisted from the start that these two count as different functions $i \neq 1_S$: Their tensor products with other functions can be very different.

This curious result (which is the origin of the "torsion products" of modules used in homological algebra) raises other questions. For example, if $t:A \to C$ is an epimorphism, is $t \otimes 1_B$ an epimorphism? The answer (Exercise 9.4) is yes.

EXERCISES

1. (a) For cyclic $\mathbf{Z}$-modules $\mathbf{Z}_m$ and $\mathbf{Z}_n$ with generators a and b, show that $\mathbf{Z}_m \otimes_{\mathbf{Z}} \mathbf{Z}_n \cong \mathbf{Z}_{(m,n)}$ is cyclic with generator $a \otimes b$ of order the g.c.d. (m, n).

(b) Show that the monomorphism $i:\mathbf{Z}_2 \to \mathbf{Z}_4$ has $i \otimes 1:\mathbf{Z}_2 \otimes \mathbf{Z}_2 \to \mathbf{Z}_4 \otimes \mathbf{Z}_2$ the zero map.

2. If A is a finite abelian group and $\mathbf{Q}$ the additive group of rational numbers, prove that $A \otimes_{\mathbf{Z}} \mathbf{Q} = 0$.

3. (a) Show that $(a, b, c) \mapsto a \otimes (b \otimes c)$ is a universal trilinear function $A \times B \times C \to A \otimes (B \otimes C)$ for given K-modules A, B, and C.

(b) Prove from this that $A \otimes (B \otimes C) \cong (A \otimes B) \otimes C$.

4. Use universality to show that $a \otimes b \mapsto b \otimes a$ gives the isomorphism (38).

5. Establish the isomorphism (37) in detail.

6. Give an example to show $A \otimes_K B \neq A \otimes_{\mathbf{Z}} B$. Which is "bigger"?

7. Show that there can be an element in a tensor product $A \otimes B$ which cannot be written as a single term $a \otimes b$ for any elements $a \in A$, $b \in B$ (take $A = K^2 = B$).

8. Show that the universality of $\otimes$ implies that $A \otimes B$ is spanned by the elements $a \otimes b$.

9. Prove (40).

10. Let A be a right module and B a left module over a not necessarily commutative ring R. For G any additive abelian group, consider biadditive functions $h:A \times B \to G$ with the additional property that $h(ar, b) = h(a, rb)$ for all $a \in A$, $r \in R$, and $b \in B$. Construct a universal such function h_0. (The codomain G of h_0 is an abelian group, usually written $A \otimes_R B$ and called the "tensor product" of A and B over R.)

11. Take bases **b** and **b**$'$ for finite-dimensional vector spaces V and V' over a field F. If endomorphisms $t: V \to V$ and $t': V' \to V'$ have the square matrices A and A' relative to the bases **b** and **b**$'$ of n and m elements, respectively, describe (as a block matrix) the matrix of $t \otimes t'$ relative to the basis of $V \otimes_F V'$ consisting of all tensor products $b_i \otimes b_j'$, taken in the order

$$(b_1 \otimes b_1', \cdots, b_1 \otimes b_m', b_2 \otimes b_1', \cdots, b_2 \otimes b_m', \cdots, b_n \otimes b_1', \cdots, b_n \otimes b_m').$$

12. For $t: B \to C$ and $s: A \to B$ morphisms of K-modules show that $(t, s) \mapsto t \circ s$ gives a linear transformation $\operatorname{Hom}(B, C) \otimes \operatorname{Hom}(A, B) \to \operatorname{Hom}(A, C)$.

13. Show that the free module F used in the proof of Theorem 15 is the module $K^{(A \times B)}$ defined in §V.5 as a submodule of the function module $K^{A \times B}$.

14. Let $K^{(X)}$ and $K^{(Y)}$ be free K-modules on sets X, Y of generators, while similarly $K^{(X \times Y)}$ has free generators the pairs (x, y) for $x \in X$, $y \in Y$, all as in Exercise V.5.5.

(a) Show that there is a bilinear map $h_0: K^{(X)} \times K^{(Y)} \to K^{(X \times Y)}$ with $h_0(x, y) = (x, y)$ for all $x \in X$ and $y \in Y$.

(b) Prove that h_0 is universal for bilinear maps on $K^{(X)} \times K^{(Y)}$.

(c) Conclude that the tensor product of any two free K-modules is free (and specify the free generators).

9. Exact Sequences

The behavior just noted for tensor products of monomorphisms is often described in a different terminology, that of "exact sequences". This terminology has proved to be very effective in handling more complicated phenomena of the same general type in topology and elsewhere. There are such exact sequences of morphisms of vector spaces, or of abelian groups or, more generally, of modules over any ring R, commutative or not.

A sequence (s, t) of two morphisms

$$A \xrightarrow{s} B \xrightarrow{t} C \tag{41}$$

of R-modules is said to be *exact* or *exact at B* when $\operatorname{Im} s = \operatorname{Ker} t$. Since the composite $t \circ s$ is defined, this amounts to two requirements: The composite $t \circ s$ is the zero map ($\operatorname{Im} s \subset \operatorname{Ker} t$); every $b \in B$ with $t(b) = 0$ has the form $b = s(a)$ for some $a \in A$ ($\operatorname{Im} s \supset \operatorname{Ker} t$). For example, to say that a sequence $0 \to A \to B$ is exact (at A) is simply to say that $A \to B$ is a monomorphism, while $B \to C \to 0$ is exact (at C) if and only if $B \to C$ is an epimorphism.

A longer sequence

$$A_0 \xrightarrow{t_1} A_1 \xrightarrow{t_2} A_2 \xrightarrow{t_3} \cdots \to A_{n-1} \xrightarrow{t_n} A_n$$

of morphisms of R-modules is called an *exact sequence* when each sequence (t_i, t_{i+1}) is exact at A_i, for $i = 1, \cdots, n-1$. In particular, an exact sequence of the form

$$0 \to A \xrightarrow{s} B \xrightarrow{t} C \to 0, \tag{42}$$

with zero modules and hence zero morphisms at the ends, is called a *short exact sequence*. In this case, exactness means that s is a monomorphism, that $\operatorname{Im} s = \operatorname{Ker} t$, and that t is an epimorphism. For example, if S is any submodule of B, the insertion $S \to B$ and the projection $B \to B/S$ together yield a short exact sequence

$$0 \to S \to B \to B/S \to 0.$$

Up to isomorphism, any short exact sequence has this simple form. For, given (42), take in B the submodule $S = \operatorname{Im} s$. Then exactness of the short sequence (42) means, as in this figure,

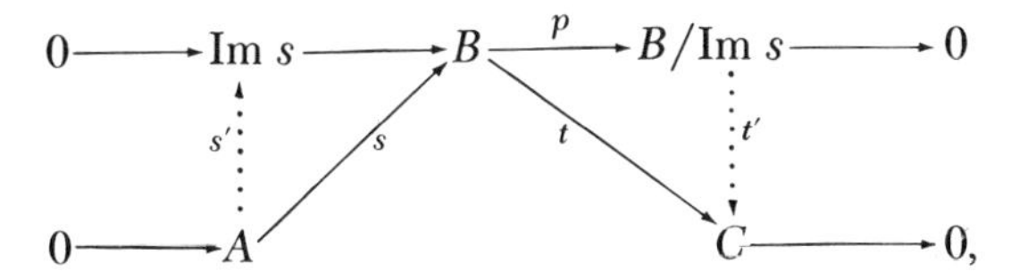

that s is the composite of an isomorphism $s' : A \cong \operatorname{Im} s$ and the insertion $\operatorname{Im} s \to B$, that this submodule $\operatorname{Im} s$ is the kernel of the epimorphism t, and hence that t is the projection p followed by an isomorphism t'. Both vertical maps s' and t' in the figure are thus isomorphisms.

Any biproduct $A_1 \oplus A_2$ yields a short exact sequence

$$0 \to A_1 \xrightarrow{e_1} A_1 \oplus A_2 \xrightarrow{p_2} A_2 \to 0$$

with e_1 an injection and p_2 a projection of the biproduct. In this case p_2 has a linear right inverse, for the insertion $e_2 : A_2 \to A_1 \oplus A_2$ has $p_2 e_2 = 1$; similarly, e_1 has a linear left inverse p_1.

LEMMA (*The Short Five Lemma*). *If the following diagram*

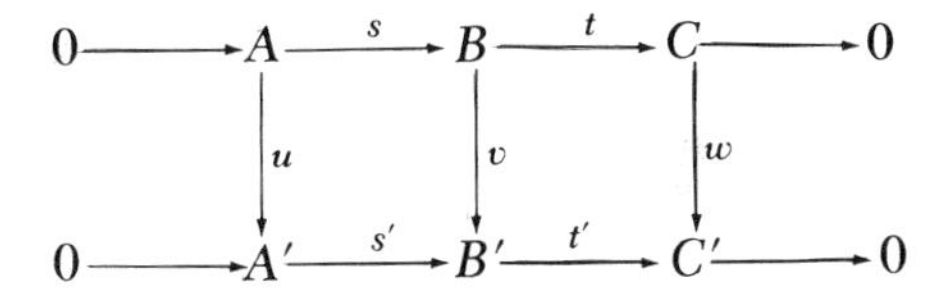

of morphisms of R-modules is commutative and has both rows short exact sequences, then:

(*i*) *u and w monomorphisms $\Rightarrow$ v is a monomorphism.*
(*ii*) *u and w epimorphisms $\Rightarrow$ v is an epimorphism.*

Proof: To show (i), consider any $b \in \operatorname{Ker} v$. By commutativity, $wtb = t'vb = t'0 = 0$, whence w a monomorphism gives $tb = 0$. By exactness, b is then sa for some $a \in A$. Then $s'ua = vsa = vb = 0$, by commutativity, so exactness of the lower row gives $ua = 0$. Since u is a monomorphism, a and hence $b = sa$ is zero, as required.

The proof of (ii) is left to the reader (Make a similar "diagram chase".)

The result in Proposition 18 about tensor products of monomorphisms of vector spaces may be expressed in exact sequence language as follows: If W is a finite-dimensional vector space and $0 \to S \to V$ an exact sequence of such spaces, then $0 \to S \otimes W \to V \otimes W$ is also exact. Here is another proposition of the same general nature expressed most directly in the exact sequence language.

Proposition 19. *If W is a finite-dimensional vector space and $V'' \xrightarrow{s} V \xrightarrow{t} V'$ an exact sequence of such spaces, then the sequence*

$$V'' \otimes W \xrightarrow{s \otimes 1} V \otimes W \xrightarrow{t \otimes 1} V' \otimes W$$

is exact.

The proof is left to the reader.

EXERCISES

1. (a) In a short exact sequence (42), prove that t has a linear right inverse t' with $tt' = 1_C$ if and only if s has a linear left inverse $s' : B \to A$ with $s's = 1_A$. (When either and hence both of these conditions obtain, the short exact sequence is said to be "split"—by t' or by s', as the case may be.)

(b) If (42) is a split short exact sequence, prove that $B \cong A \oplus C$.

2. Prove the second half of the Short Five Lemma.

3. Prove Proposition 19.

***4.** If $A' \xrightarrow{s} A \xrightarrow{t} A'' \to 0$ is a sequence of K-modules exact at A and A'', prove for each B that the induced sequence

$$A' \otimes B \xrightarrow{s \otimes 1} A \otimes B \xrightarrow{t \otimes 1} A'' \otimes B \to 0$$

is exact at $A \otimes B$ and $A'' \otimes B$.

***5.** Given an R-module D and a short exact sequence (42) of R-modules, prove each of the following sequences of abelian groups exact with $t_* =$

Hom (D, t) and $t^* = \text{Hom}\,(t, D)$ defined as in (IV.4):

(a) $0 \to \text{Hom}_R\,(D, A) \xrightarrow{s_*} \text{Hom}_R\,(D, B) \xrightarrow{t_*} \text{Hom}_R\,(D, C)$;

(b) $0 \to \text{Hom}_R\,(C, D) \xrightarrow{t^*} \text{Hom}_R\,(B, D) \xrightarrow{s^*} \text{Hom}_R\,(A, D)$.

6. In Exercise 5, show by examples that neither t_* nor s^* need be epimorphisms.

7. In Exercise 5a, show that D a free module makes t_* an epimorphism.

8. Exhibit the induced sequences (a) and (b) of Exercise 5, when $R = \mathbf{Z}$, the given short exact sequence is $0 \to m\mathbf{Z} \to \mathbf{Z} \to \mathbf{Z}/m\mathbf{Z} \to 0$, and D is one of $\mathbf{Z}$, $\mathbf{Z}_m$, or $\mathbf{Z}_n$ for some n with $(m, n) = 1$.

10. Identities on Tensor Products

Consider now the set Bilin $(A, B; C)$ of all K-bilinear maps $A \times B \to C$. The pointwise sum $h + k$ of two such bilinear maps is bilinear. Moreover, since K is commutative, the pointwise scalar multiple $h\kappa$ of a bilinear map by a scalar κ is also bilinear. It follows readily that the set Bilin $(A, B; C)$ is a K-module under these pointwise operations. This module turns out to be isomorphic to another one; we will prove, much as in (IV.24), that

$$\text{Bilin}\,(A, B; C) \cong \text{Hom}\,(A, \text{Hom}\,(B, C)). \tag{43}$$

First, the values of any function $h{:}A \times B \to C$ can be written as $h(a, b) = H_a(b)$, so that each $H_a = h(a, -)$ is a function $B \to C$. Now suppose that h is bilinear. The condition that h be linear in b for fixed $a \in A$ reads

$$h(a, b_1\kappa_1 + b_2\kappa_2) = h(a, b_1)\kappa_1 + h(a, b_2)\kappa_2$$

for all $b_i \in B$ and $\kappa_i \in K$. This is equivalent to

$$H_a(b_1\kappa_1 + b_2\kappa_2) = H_a(b_1)\kappa_1 + H_a(b_2)\kappa_2,$$

hence states that the function $H_a{:}B \to C$ is linear, so an element of Hom (B, C). Thus $a \mapsto H_a$ is a function $H{:}A \to \text{Hom}\,(B, C)$. The condition that h be linear in a for fixed $b \in B$ reads

$$h(a_1\kappa_1 + a_2\kappa_2, b) = h(a_1, b)\kappa_1 + h(a_2, b)\kappa_2$$

for all $a_i \in A$ and $\kappa_i \in K$. This is equivalent to

$$H_{a_1\kappa_1 + a_2\kappa_2}(b) = H_{a_1}(b)\kappa_1 + H_{a_2}(b)\kappa_2,$$

for all $b \in B$, and so states (in the addition for Hom (B, C)) that

$$H_{a_1\kappa_1 + a_2\kappa_2} = H_{a_1}\kappa_1 + H_{a_2}\kappa_2.$$

Therefore $H{:}A \to \text{Hom}\,(B, C)$ is a linear map, so is an element $H \in \text{Hom}\,(A, \text{Hom}\,(B, C))$.

THEOREM 20. *For K-modules A, B, and C there is an isomorphism*

$$\text{Bilin}\,(A, B; C) \cong \text{Hom}\,(A, \text{Hom}\,(B, C))$$

of K-modules given by assigning to each bilinear $h : A \times B \to C$ the linear function $H : A \to \text{Hom}\,(B, C)$ with $H_a(b) = h(a, b)$. Similarly $h \mapsto H^*$, with $H^*_b(a) = h(a, b)$, gives a K-module isomorphism

$$\text{Bilin}\,(A, B; C) \cong \text{Hom}\,(B, \text{Hom}\,(A, C)). \tag{44}$$

Proof: First, we show that the assignment $h \mapsto H$ is a bijection by constructing an inverse. Given any $H : A \to \text{Hom}\,(B, C)$, define h by $h(a, b) = H_a(b)$; then the linearity of H itself shows $h(a, b)$ linear in a, while the linearity of each $H_a : B \to C$ shows $h(a, b)$ linear in b. Therefore, h is bilinear, and $H \mapsto h$ is the desired inverse.

Second, $h \mapsto H$ is a morphism of modules (and hence an isomorphism) because the module operations on both h and H are defined pointwise.

The derivation of (44) is analogous.

COROLLARY. *For three K-modules A, B, and C there is an isomorphism*

$$\eta : \text{Hom}\,(A \otimes B, C) \cong \text{Hom}\,(A, \text{Hom}\,(B, C)) \tag{45}$$

of K-modules which assigns to each linear function $t : A \otimes B \to C$ the function $\eta t : A \to \text{Hom}\,(B, C)$ with $[(\eta t)a](b) = t(a \otimes b)$, for $a \in A$, $b \in B$.

Proof: The universality of the tensor product states that every bilinear $h : A \times B \to C$ has the form $h(a, b) = t(a \otimes b)$ for a unique linear function $t : A \otimes B \to C$. In other words, $t \mapsto h$ is a bijection

$$\text{Hom}\,(A \otimes B, C) \cong \text{Bilin}\,(A, B; C).$$

(This is just the "representation theorem" for a universal; see Theorem IV.2.) This bijection is also an isomorphism of K-modules, for in both Bilin $(-, -)$ and in Hom $(-, -)$ the module operations are defined "pointwise", while the bijection $t \mapsto h$ carries pointwise operations on t to the corresponding operations on h. This isomorphism, followed by that of the Theorem, is the desired isomorphism η of the Corollary. This important isomorphism η can be used to give a different definition of the tensor product $A \otimes B$ in terms of Hom (Exercise 5).

EXERCISES

1. Combination of the two bijections of Theorem 20 gives an isomorphism

$$\text{Hom}\,(A, \text{Hom}\,(B, C)) \cong \text{Hom}\,(B, \text{Hom}\,(A, C))$$

of modules. Describe this isomorphism directly.

2. For K-modules B and C a function $\zeta:B^* \otimes C \to \operatorname{Hom}(B, C)$ is defined as follows: Given $f \in B^*$ and $c \in C$, $\zeta(f \otimes c)$ is $b \mapsto cf(b)$:

(a) Prove that ζ is a morphism of modules.

(b) Prove ζ an isomorphism when $B = K^n$ (try $n = 1$ first).

3. (Conceptual description of the trace of a matrix.)

(a) Show that $(f, c) \mapsto f(c)$ defines a linear map $E:C^* \otimes C \to K$. (This map is called "evaluation" or "contraction".)

(b) For C a free K-module of finite type, use contraction and the isomorphism ζ of Exercise 2, as in

$$\operatorname{Hom}(C, C) \xrightarrow{\zeta^{-1}} C^* \otimes C \xrightarrow{E} K,$$

to define for each $t:C \to C$ a scalar $\operatorname{tr}(t) = E\zeta^{-1}(t)$. If t is represented, relative to any basis of C, by a matrix A, show that $\operatorname{tr}(t)$ is the trace of the matrix A, as defined in §5.

4. Show that the Corollary of Theorem 20 can be formulated in terms of universality, by defining a suitable morphism $e:\operatorname{Hom}(B, C) \otimes B \to C$ and showing that e is universal among arrows $- \otimes B \to C$, as in (IV.23).

5. Use the η of (45) to define a morphism $A \to \operatorname{Hom}(B, A \otimes B)$ which is universal among morphisms $A \to \operatorname{Hom}(B, -)$. Show that this property determines $A \otimes B$ up to isomorphism.

11. Change of Rings

Any square matrix A of real numbers is also a matrix of complex numbers, because of the inclusion $\mathbf{R} \subset \mathbf{C}$ of the real field in the complex field. This reinterpretation of the matrix A is useful, because A may have complex eigenvalues which are not real. This also means that the corresponding linear transformation $t_A:\mathbf{R}^n \to \mathbf{R}^n$ may acquire complex eigenvectors when regarded as a linear transformation $t_A:\mathbf{C}^n \to \mathbf{C}^n$ on the extended vector space $\mathbf{C}^n$.

More generally, given a subring K of a commutative ring K', each free K-module $C \cong K^n$ can be extended to a free K'-module $C' \cong K'^n$; if the elements of C relative to a basis $\mathbf{b}$ are the linear combinations $\sum b_i\kappa_i$ with coefficients $\kappa_i \in K$, then C' has as elements the linear combinations $\sum b_i\kappa_i'$ of the *same* basis elements b_i but now with coefficients κ_i' in the larger ring K'; this process is called a "change of rings" from K to K'. This description of the extended module C' appears to depend on the choice of a basis $\mathbf{b}$ in C. Nevertheless, the extended module can actually be described in an invariant way, not using any basis, in terms of tensor products, written $\otimes_K$, of K-modules.

First, note that $K \subset K'$ means that any right K'-module B' is automatically a right K-module: Just use only the multiples $(b, \kappa) \mapsto b\kappa$ of elements $b \in B'$ by scalars $\kappa \in K \subset K'$. In particular, the larger ring K' is a right K-module. Given any K-module A we can therefore take the tensor product

$\otimes_K$ over K to form the K-module $A \otimes_K K'$, as well as the K-linear map

$$u : A \to A \otimes_K K', \qquad u(a) = a \otimes 1 \qquad \text{for } a \in A. \tag{46}$$

We call u the *K'-extension* of the K-module A.

LEMMA. *The K-module $A \otimes_K K'$ becomes a right K'-module when the scalar multiple of each $a \otimes \kappa'$ by any $\lambda' \in K'$ is defined by*

$$(a \otimes \kappa')\lambda' = a \otimes \kappa'\lambda', \qquad \kappa', \lambda' \in K', \quad a \in A. \tag{47}$$

In particular, for $\lambda' = \lambda \in K$, this is $u(a)\lambda = u(a\lambda)$ for all a.

Proof: Fix λ'. Then $(a, \kappa') \mapsto a \otimes \kappa'\lambda'$ is a K-bilinear function $A \times K' \to A \otimes_K K'$, so gives a unique K-linear function $t_{\lambda'}:A \otimes_K K' \to A \otimes_K K'$ with $t_{\lambda'}(a \otimes \kappa') = a \otimes \kappa'\lambda'$. Now $t_{\lambda'}$ is the intended operation of scalar multiplication by λ' as in Equation (47); it remains only to check the various module axioms for this operation. Consider, for instance, the distributive law $(-)(\lambda' + \mu') = (-)\lambda' + (-)\mu'$. In the notation above, this law is $t_{\lambda'+\mu'} = t_{\lambda'} + t_{\mu'}$. But both functions $t_{\lambda'+\mu'}$ and $t_{\lambda'} + t_{\mu'}$ send the element $a \otimes \kappa'$ into $a \otimes (\kappa'\lambda' + \kappa'\mu')$. Hence, by the universality of $(a, \kappa') \mapsto a \otimes \kappa'$, these functions are equal. The other module axioms are proved similarly.

The K'-extension u of (46) is thus a map of the K-module A into the K'-module $A \otimes_K K'$.

THEOREM 21. *If A is a K-module, while K is a subring of the commutative ring K', then the K'-extension u of (46) is universal among K-linear maps from A to K'-modules.*

In other words, if $r:A \to B'$ is a K-linear map of A into any K'-module B', there is exactly one K'-linear map $r':A \otimes_K K' \to B'$ such that $r = r' \circ u$, as in the diagram

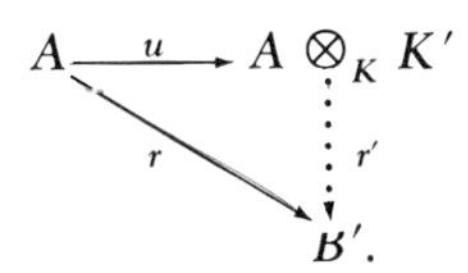

Proof: For all $\kappa' \in K'$ and $a \in A$, the definition of u and the commutativity of the diagram would give

$$r'(a \otimes \kappa') = r'(a \otimes 1)\kappa' = (r'ua)\kappa' = (ra)\kappa'.$$

Hence r', if it exists, is uniquely determined by r. Conversely, $(a, \kappa') \mapsto (ra)\kappa'$ is K-bilinear, hence there is a (unique) K-linear map r', as shown, with $r'(a \otimes \kappa') = (ra)\kappa'$ for all κ' and a. Now the definition of scalar multiples by

λ' in the lemma shows at once that this map r' is also K'-linear as well, Q.E.D.

In case $A = K$, this extension u is just the insertion $K \to K \otimes_K K' \cong K'$. If $A = K^n$ is the free module on n free generators $\varepsilon_1, \cdots, \varepsilon_n$, then, by (33), $K^n \otimes_K K' \cong (K')^n$, so the extended module is again free, and indeed a free K'-module on the corresponding generators $\varepsilon_1 \otimes 1, \cdots, \varepsilon_n \otimes 1$. Thus, in case A is a free module of finite type, the extension u is a monomorphism, and we may use u to identify the given K-module A with its image u_*A, a K-submodule of $A \otimes_K K'$. In particular, this applies to vector spaces. If F is a subfield of F', each finite-dimensional vector space V over F determines a vector space $V \otimes_F F'$ of the same dimension over F', the extension $u: V \to V \otimes_F F'$ is a monomorphism, and so may be used to identify V with an F-subspace of the F'-vector space $V \otimes_F F'$.

If a K-module A is not free, the extension u need not be a monomorphism. For example, $\mathbf{Z}$ is a subring of $\mathbf{Q}$, and $\mathbf{Z}_n$ is a $\mathbf{Z}$-module (but not a free one); its $\mathbf{Q}$-extension $\mathbf{Z}_n \otimes_{\mathbf{Z}} \mathbf{Q}$ is zero (see Exercise 8.2).

Next, K-linear maps may be extended as follows:

Theorem 22. *To any K-linear map $t: A \to B$ between K-modules with K'-extensions u and v there is a unique K'-linear map $t': A \otimes_K K' \to B \otimes_K K'$ with $t' \circ u = v \circ t$.*

Proof: We are required to construct t' so that the diagram below will

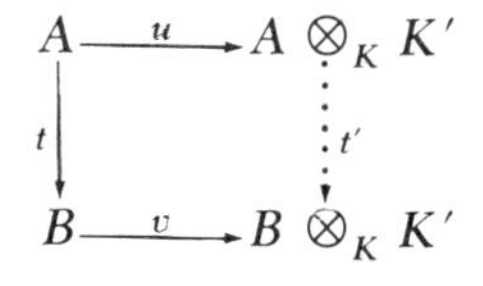

commute. But the composite $v \circ t$ in this diagram is a K-linear function on the K-module A to the K'-module $B \otimes_K K'$. Since u is a universal such function, t' does exist (and uniquely). Indeed, t' is just the tensor product $t \otimes 1_{K'}$, of the functions t and 1.

In particular, if M is an $n \times n$ matrix of real numbers, the real linear map $t_M: \mathbf{R}^n \to \mathbf{R}^n$ with matrix M extends to the complex linear map $\mathbf{C}^n \to \mathbf{C}^n$ with the same matrix. As observed above, it may thereby acquire more (complex) eigenvectors.

EXERCISES

1. In the lemma, give a proof that all the module axioms for scalar multiples hold in $A \otimes_K K'$.

2. If A is a $\mathbf{Z}$-module (an abelian group) show that all elements of finite order in A lie in the kernel of the extensions $u: A \to A \otimes_{\mathbf{Z}} \mathbf{Q}$.

3. Show that the extension of a biproduct $A_1 \oplus A_2$ of K-modules has codomain a biproduct of K'-modules.

4. If A is a K-module, show that the elements of the $K[x]$-module $A \otimes_K K[x]$ may be written as polynomials in x with coefficients in A.

5. Let p be a prime number, $\mathbf{Z}_{(p)}$ the ring of all rational numbers with denominator prime to p (Proposition III.32), and A a $\mathbf{Z}$-module.

(a) In any $\mathbf{Z}_{(p)}$-module A', show that every element of finite additive order has order some power of p.

(b) If $a \in A$ has finite order prime to p, show that a is in the kernel of the extension $A \to A \otimes_{\mathbf{Z}} \mathbf{Z}_{(p)}$.

6. Show that $A \mapsto A \otimes_K K'$ and $t \mapsto t' = t \otimes 1_{K'}$ define a functor on K-modules to K'-modules.

7. If $f:K \to K'$ is any morphism of commutative rings, any right K'-module C' becomes a K-module with scalar multiples defined by $(c', \kappa) \mapsto c'(f\kappa)$. Construct a universal K-linear map of a given K-module A into a suitable K'-module.

12. Algebras

The set of endomorphisms of a given module, the set of all $n \times n$ matrices over a given ring, and the set of all quaternions are all examples of algebraic systems which are simultaneously rings and modules. Suitable systems of this sort will be called "algebras"; more exactly, linear algebras over our commutative ring K of scalars.

DEFINITION. A K-algebra A *is a right* K*-module which is also a ring in which addition in the ring is addition in the module and*

$$(a_1a_2)\kappa = (a_1\kappa)a_2 = a_1(a_2\kappa) \tag{48}$$

for all a_1 *and* $a_2 \in A$ *and all* $\kappa \in K$. A morphism $A \to A'$ *of* K*-algebras* A, A' *is a function on the set* A *to the set* A' *which is a morphism both of* K*-modules and of rings.*

Thus, a K-algebra is a ring and a K-module, with condition (48) relating these two structures. As usual, the identity function of each K-algebra A is a morphism $A \to A$ of K-algebras, and the composite of two morphisms of K-algebras, when defined, is also such a morphism. With these morphisms, the K-algebras form a concrete category.

Here are some examples of algebras.

Matrices. The set $K^{n \times n}$ of all $n \times n$ matrices over K is a K-algebra under the usual operations (cf. §VII.1) of matrix addition, multiplication of a matrix by a scalar, and matrix multiplication.

Endomorphisms. If C is a given K-module, the set $\operatorname{End}_K(C)$ of all K-module endomorphisms $t:C \to C$ is a ring under addition and composition of morphisms (Corollary V.2) and a K-module; with these structures (48) holds, so $\operatorname{End}(C)$ is a K-algebra. Moreover, the bijection $A \mapsto t_A$ of §VII.1

from matrices to endomorphisms of the free module K^n is an isomorphism $K^{n \times n} \cong \text{End}(K^n)$ of algebras.

Commutative Algebras. For any K, the polynomial rings $K[x]$ and $K[x, y]$ and the ring $K[[x]]$ of formal power series in x are K-algebras.

Rings with Center Containing K. The "center" C of a ring R is the set of all $c \in R$ with $cr = rc$ for every r in the ring. The center so defined is a (commutative) subring of R. If K is any subring of the center C, then the whole ring R is a K-algebra: The multiples $(r, \kappa) \mapsto r\kappa$ make R a (right) K-module; it is already a ring, and the hypothesis that κ is in the center gives the axiom (48) needed for an algebra.

Rings. Every ring R is an algebra over the ring $\mathbf{Z}$ of integers, because the additive group of R is a $\mathbf{Z}$-module under the integral multiples $r \mapsto rn$ for $n \in \mathbf{Z}$ and these multiples satisfy (48). In particular, the finite rings $\mathbf{Z}_n$ are $\mathbf{Z}$-algebras. This example shows that a K-algebra A need not contain K as a subring!

R-*Algebras.* The complex numbers $\mathbf{C}$ and the quaternions Q are both algebras (of dimensions 2 and 4, respectively) over the real field $\mathbf{R}$. They are *division algebras*; that is, to each element $a \neq 0$ there is an inverse a^{-1} with $a^{-1}a = 1$. A celebrated theorem of Frobenius (1878) and B. Peirce (1875) states that the reals, the complex numbers, and the quaternions are the only (finite-dimensional) division algebras over $\mathbf{R}$.

A very interesting classical problem concerns the determination of all algebras over a field F which are finite-dimensional as vector spaces over F. This determination is complete for the "semi-simple" algebras. In particular, a *simple* algebra $A \neq 0$ over a field F is defined to be a finite-dimensional F-algebra such that any epimorphism $A \to A'$ of algebras is either an isomorphism $A \cong A'$ or the map $A \to 0$. The algebra of all $n \times n$ matrices with entries in a division algebra D is always a simple algebra. A famous theorem of Wedderburn (1906) asserts that every simple algebra over a field is such an algebra of matrices over a division algebra. It is a corollary that the only finite-dimensional *real* simple algebras are the matrix algebras $\mathbf{R}^{n \times n}$, $\mathbf{C}^{n \times n}$, and $Q^{n \times n}$.

The definition of a K-algebra can be reformulated in terms of tensor products in a suggestive way which will also be useful in Chapter XVI on multilinear algebra. In a K-algebra the distributive law for multiplication together with (48) states that the product $(a_1, a_2) \mapsto a_1a_2$ is a K-bilinear map $A \times A \to A$. By the main theorem on tensor products, there is a unique linear map $\pi: A \otimes A \to A$ such that the product is simply $a_1a_2 = \pi(a_1 \otimes a_2)$. On the other hand, since A is K-module, the assignment $\kappa \mapsto 1\kappa$ (with 1 the unit of A) is a K-linear map $u: K \to A$.

A triple product $a(bc)$ can be expressed in terms of the map π as

$$a(bc) = \pi(a \otimes bc) = \pi(a \otimes \pi(b \otimes c)) = \pi(1 \otimes \pi)(a \otimes b \otimes c),$$

where 1 is the identity map $1: A \to A$, while $1 \otimes \pi: A \otimes (A \otimes A) \to A \otimes A$

is a tensor product of maps, as in (39). The triple product in the other association may be written similarly as

$$(ab)c = \pi(ab \otimes c) = \pi(\pi(a \otimes b) \otimes c) = \pi(\pi \otimes 1)(a \otimes b \otimes c).$$

Hence, if we identify $A \otimes (A \otimes A)$ with $(A \otimes A) \otimes A$, the associative law for multiplication states that the following diagram commutes:

$$\begin{array}{ccc} A \otimes A \otimes A & \xrightarrow{1 \otimes \pi} & A \otimes A \\ \downarrow{\scriptstyle \pi \otimes 1} & & \downarrow{\scriptstyle \pi} \\ A \otimes A & \xrightarrow{\pi} & A. \end{array} \tag{49}$$

On the other hand, the fact that 1 is a right unit for the product in A means that the map $u:K \to A$ satisfies

$$a\kappa = (a1)\kappa = a(1\kappa) = \pi(a \otimes 1\kappa) = \pi(a \otimes u(\kappa)) = \pi(1 \otimes u)(a \otimes \kappa).$$

But $a\kappa$ can be written as $\phi(a \otimes \kappa)$, where $\phi:A \otimes K \cong A$ is the standard isomorphism noted in (32). The result is the commutativity of the right-hand square in the following diagram:

$$\begin{array}{ccccc} K \otimes A & \xrightarrow{\phi'} & A & \xleftarrow{\phi} & A \otimes K \\ \downarrow{\scriptstyle u \otimes 1} & & \downarrow{\scriptstyle 1_A} & & \downarrow{\scriptstyle 1 \otimes u} \\ A \otimes A & \xrightarrow{\pi} & A & \xleftarrow{\pi} & A \otimes A. \end{array} \tag{50}$$

The commutativity of the left-hand square similarly states that 1 is a left unit for the multiplication in A. With these two steps we have "translated" some of the axioms for an algebra into diagrammatic form (where the elements are replaced by arrows!). Here is the result:

Theorem 23. *Let A be a K-module equipped with two K-linear maps*

$$\pi : A \otimes A \to A, \qquad u : K \to A$$

such that the diagrams (49) and (50) commute. Then A is a K-algebra with the product $a_1a_2 = \pi(a_1 \otimes a_2)$ and multiplicative unit $u(1)$, and every K-algebra arises in this way.

Given a free K-module A of rank n with basis $\mathbf{b}$, each linear map $\pi:A \otimes A \to A$ is determined by the n^2 elements $\pi(b_i \otimes b_j)$ of A and hence by the n^3 scalars $\gamma_{i\,j}^{\,k}$ with $\pi(b_i \otimes b_j) = \Sigma_k b_k \gamma_{i\,j}^{\,k}$. These n^3 scalars, called the "multiplication constants" of A, must satisfy additional conditions if the multiplication π is to be associative and to have a unit. For example, the algebra of all $n \times n$ matrices over K is a free K-module with basis the n^2 matrices $E^{(ij)}$, where $E^{(ij)}$ has entry 1 at (ij) and zero elsewhere. The multiplication constants are given by $E^{(ij)}E^{(kl)} = 0$ or $E^{(il)}$ according as $j \neq k$ or $j = k$.

Algebras will be considered again in Chapter XVI.

EXERCISES

1. Give the proof of Theorem 23.

2. For each additive monoid S, show that the monoid ring $K^{(S)}$ of Exercise III.6.6 is a K-algebra.

3. Show that all $n \times n$ triangular matrices over K form an algebra.

4. If S is a submodule of C, show that the set of all those endomorphisms t of C with $t_*(S) \subset S$ form a subalgebra of End (C).

5. (a) If $f:A \to A'$ is a morphism of K-algebras, show that the kernel of f is both an ideal in A and a K-submodule of A (that is, an *algebra*-ideal of A).

(b) In a K-algebra A let J be a subset of A which is both an ideal and a K-submodule. Prove that J is the kernel of a morphism of algebras and construct a morphism f with $f_*J = 0$ and universal with this property.

(c) Show that a finite-dimensional algebra A over a field F is simple if and only if it has no (algebra)-ideals beside A and the set $\{0\}$.

6. Prove that the algebra $F^{n \times n}$ of $n \times n$ matrices over F is simple. (*Hint:* Suppose that the epimorphism $A \to A'$ of algebras has a kernel containing a matrix M with a non-zero entry in row i and column j. Multiply M by pre- and post-factors to obtain the matrix $E^{(ij)}$ in the kernel, non-zero only in row i and column j.)

7. Prove that the center of $K^{n \times n}$ is the set of all scalar multiples of the identity matrix.

CHAPTER X

Bilinear and Quadratic Forms

A FORM is a special sort of function. Thus a linear form on a vector space V over a field F has been defined as a function $f: V \to F$ which is linear; in terms of coordinates ξ_i for V, this amounts to requiring that f can be expressed as a polynomial function $\xi \mapsto \sum \kappa_i \xi_i$ in which every non-zero term $\kappa_i \xi_i$ has degree 1. Quadratic forms on V will be treated in this chapter; they will be described as certain functions $q: V \to F$, and the description will amount to requiring that q can be expressed in terms of coordinates as a polynomial function $\xi \mapsto \sum A_{i\,j} \xi_i \xi_j$ in which every non-zero term $A_{i\,j} \xi_i \xi_j$ has degree 2. In general, a form will be a function $V \to F$ which can be expressed in coordinates as a polynomial function which is "homogeneous" (all non-zero terms of the same degree).

One important quadratic form $q: \mathbf{R}^2 \to \mathbf{R}$ is defined by $q(x, y) = x^2 + y^2$; its square root $\sqrt{x^2 + y^2}$ is the length of the vector from the origin to (x, y) in the real vector space $\mathbf{R}^2$. Many real vector spaces come equipped with a length function defined in this way from a quadratic form, and there are analogous lengths defined for vectors in a complex vector space. This chapter is devoted to the study of such vector spaces, called "inner product spaces", and to the related problems of reducing *any* quadratic form on such a space to a canonical form for equivalence (in the sense of §VII.7) under the automorphisms of the space. The chapter begins with a preliminary, the study of bilinear forms.

1. Bilinear Forms

A *bilinear form* on the finite-dimensional vector spaces V and V' (over the same field F) is a bilinear function $h: V \times V' \to F$ in the sense of §IX.1. In this section we wish to find the simplest expression in terms of coordinates for a given bilinear form h; it will turn out that this problem can be answered by using the results already obtained for the equivalence of rectangular matrices.

Let V and V' have dimensions n and n', respectively, and choose bases $\mathbf{b}$ and $\mathbf{b}'$ there. Define the *matrix of the form* $h : V \times V' \to F$ relative to $\mathbf{b}$ and $\mathbf{b}'$ to be the $n \times n'$ matrix A over F with entries

$$A_{ik} = h(b_i, b'_k), \qquad i \in \mathbf{n}, \quad k \in \mathbf{n}'. \tag{1}$$

This matrix determines the form. To see this, we write vectors $v \in V$ and $v' \in V'$ in terms of the respective bases as $v = \sum b_i \xi_i$ and $v' = \sum b'_k \xi'_k$, with lists $\boldsymbol{\xi}$ and $\boldsymbol{\xi}'$ of scalars. Then the bilinearity of h shows that h is determined by the matrix A as

$$h(v, v') = \sum_{i=1}^{n} \sum_{k=1}^{n'} \xi_i h(b_i, b'_k)\xi'_k = \sum_i \sum_k \xi_i A_{ik} \xi'_k. \tag{2}$$

In other words, the matrix of a form h is the matrix of coefficients in the formula for $h(v, v')$ in terms of the coordinates of the vectors v and v'. The function $h \mapsto A$ sending each form to its matrix (rel $\mathbf{b}, \mathbf{b}'$) is clearly an isomorphism of the vector space of all bilinear forms $V \times V' \to F$ to the vector space $F^{n \times n'}$ of matrices.

This isomorphism may also be described by tensor products. By the main theorem on tensor products, each bilinear form $h: V \times V' \to F$ may be written as a composite $h(v, v') = \bar{h}(v \otimes v')$ with the tensor product for a unique linear form $\bar{h}: V \otimes V' \to F$ as in the commutative diagram

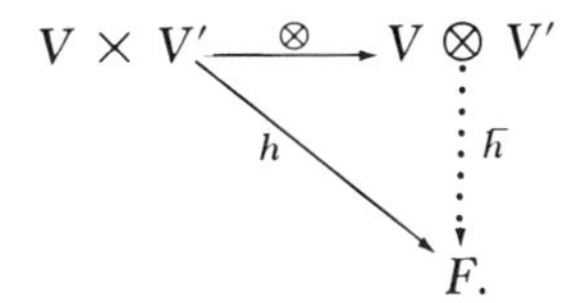

The $n \times n'$ elements $b_i \otimes b'_k$ are, as in Proposition IX.16, a basis for $V \otimes V'$, and the matrix A of the form h is just the array of the values of $\bar{h}$ on this basis.

Bilinear forms may also be generated by products of linear forms. If $f: V \to F$ and $f': V' \to F$ are linear forms on V and V', their product (taken "pointwise" in F) is the function $ff': V \times V' \to F$ given by $(v, v') \mapsto (fv)(f'v')$; it is a bilinear form. If κ is any scalar, the scalar multiple $(ff')\kappa$ is also a bilinear form $(v, v') \mapsto (fv)(f'v')\kappa$. It is written more symmetrically as $f\kappa f'$. In this chapter the scalar factor in a *scalar multiple will be written either on the right or the left as may be convenient* (and as is possible, since F is commutative, by the Corollary of Proposition V.16).

Now take the basis $\mathbf{x}$ of V^* dual to $\mathbf{b}$, and $\mathbf{x}'$ dual to $\mathbf{b}'$. Thus $x_i: V \to F$ is the linear form assigning to each vector $\sum b_i \xi_i$ its ith coordinate ξ_i, while any product $x_i \kappa x'_k$ is a bilinear form; in particular, the bilinear form $x_i A_{ik} x'_k$ is $(v, v') \mapsto \xi_i A_{ik} \xi'_k$. Thus the formula above for the function h in terms of its matrix (rel $\mathbf{b}, \mathbf{b}'$) may be written in terms of the functions x_i as

$$h = \sum_{i=1}^{n} \sum_{k=1}^{n'} x_i A_{ik} x'_k. \tag{3}$$

Again, the matrix A of h (rel $\mathbf{b}, \mathbf{b}'$) is the matrix of coefficients in the formula for h. This formula may be written as a matrix product $h = \mathbf{x}^T A \mathbf{x}'$, with h a

1×1 matrix and $\mathbf{x}$ and $\mathbf{x}'$ columns (column matrices) of linear forms. Hence, the transpose $\mathbf{x}^T$ is a row, so that the matrix product $\mathbf{x}^T A \mathbf{x}'$ is defined. Displayed, this is

$$[h] = [x_1, \cdots, x_n] \begin{bmatrix} A_{11} & \cdots & A_{1n'} \\ \vdots & & \vdots \\ A_{n1} & \cdots & A_{nn'} \end{bmatrix} \begin{bmatrix} x'_1 \\ \vdots \\ x'_{n'} \end{bmatrix}.$$

We next consider the effect of a change of basis on the matrix of a form.

Theorem 1. *Let the bilinear form $h: V \times V' \to F$ have the matrix A relative to bases $\mathbf{b}$, $\mathbf{b}'$ and the matrix B relative to (possibly different) bases $\mathbf{c}$ and $\mathbf{c}'$. If P is the change matrix from $\mathbf{b}$ to $\mathbf{c}$ and P' that from $\mathbf{b}'$ to $\mathbf{c}'$, while P^T is the transpose of P, then*

$$A = P^T B P'. \tag{4}$$

Proof: Write the new coordinates $\mathbf{y}$ and $\mathbf{y}'$ as columns, so that h rel $\mathbf{c}$, $\mathbf{c}'$ is the matrix product $h = \mathbf{y}^T B \mathbf{y}'$. The new coordinates are expressed in terms of P and the old coordinates as $y_i = \sum_j P_{ij} x_j$; this can be written as a matrix product $\mathbf{y} = P\mathbf{x}$. The transpose is then the matrix product in the opposite order, $\mathbf{y}^T = \mathbf{x}^T P^T$. Since $\mathbf{y}' = P'\mathbf{x}'$ also, we may calculate h (using matrix products) to be

$$h = \mathbf{y}^T B \mathbf{y}' = (\mathbf{x}^T P^T) B (P' \mathbf{x}') = \mathbf{x}^T (P^T B P') \mathbf{x}'.$$

This expresses h in terms of the old coordinates; the matrix $P^T B P'$ of the coefficients in this expression must be identical to the matrix A of (1), Q.E.D.

Corollary 1. *Two $n \times n'$ matrices represent the same bilinear form relative to (possibly) different bases (in V and in V') if and only if they are equivalent as matrices.*

Proof: For matrices, A equivalent to B means that $A = QBP'$ for Q and P' invertible. Then $P = Q^T$ is also invertible, hence is a change matrix, and $A = P^T B P'$ is given by the formula (4) found above, so represents the same bilinear form as does B relative to a suitable basis. Conversely, if A and B are two matrices for the same bilinear form, the theorem gives $A = P^T B P'$ with P and P' invertible, hence A equivalent to B.

By Corollary 1 of Theorem VII.17, two matrices are equivalent if and only if they have the same rank. Hence, we may define the *rank* of a bilinear form h to be the rank of any one (and hence of every one) of the matrices of h. Also, by Theorem VII.17, any matrix A of rank r is equivalent to one with r entries 1 on the diagonal, and all other entries zero. This proves

Corollary 2. *For any bilinear form $h: V \times V' \to F$ of rank r there is a choice of coordinates* y *in* V *and* y′ *in* V' *so that*

$$h = y_1 y_1' + \cdots + y_r y_r'. \tag{5}$$

In a numerical case, the rank may be found by taking any matrix A for h and reducing it to diagonal form by successive elementary operations on rows and columns.

EXERCISES

1. Prove that any bilinear form of rank r is a sum of r bilinear forms of rank 1.

2. Prove: Two bilinear forms $g, h: V \times V' \to F$ have the same rank if and only if there are automorphisms t of V and t' of V' with $h(v, v') = g(tv, t'v')$ for all vectors $v \in V$ and $v' \in V'$. (*Hint:* This is an "alibi", while Corollary 1 above is the corresponding "alias", in the sense of §VII.7.)

3. Show that the rank of a bilinear form h may be described directly as follows. Define the "left radical" L_h of h to be the set of all those vectors $v \in V$ for which $h(v, v') = 0$ for *every* vector $v' \in V'$.

(a) Show that L_h is a subspace of V.

(b) If h has the expression (5) of Corollary 2, prove that

$$L_h = (\text{Ker } y_1) \cap \cdots \cap (\text{Ker } y_r).$$

(c) Conclude that rank $h = \dim V - \dim L_h$.

(d) Define the "right radical" R_h of h as a subspace of V' and find a bilinear form k so that h factors as the composite

$$V \times V' \xrightarrow{p \times p'} (V/L_h) \times (V'/R_h) \xrightarrow{k} F,$$

where p and p' are projections on the quotient spaces.

2. Symmetric Matrices

Consider bilinear forms $h: V \times V \to F$ for a single finite-dimensional vector space V and, for each basis **b**, the corresponding square matrix A, as in (1). Call h

Symmetric:	when $h(v, v') = h(v', v)$,
Skew-symmetric:	when $h(v, v') = -h(v', v)$,
Alternating:	when $h(v, v) = 0$;

each for all vectors $v, v' \in V$. In the treatment of determinants (Proposition IX.1) we have already shown that any alternating form is automatically skew-symmetric, while a skew-symmetric form is necessarily alternating unless the field has characteristic 2.

There are corresponding definitions for matrices: A square matrix A is *symmetric* when $A^T = A$ and *skew-symmetric* when $A^T = -A$; that is, when $A_{ik} = -A_{ki}$ for every pair of indices i and k.

Lemma. *A bilinear form $h: V \times V \to F$ is symmetric or skew-symmetric if and only if its matrix A (relative to any one basis* $\mathbf{b}$ *of V) is symmetric or skew-symmetric, respectively.*

Proof: If the form h is symmetric, then, for all i and k,

$$A_{ik} = h(b_i, b_k) = h(b_k, b_i) = A_{ki} = (A^T)_{ik}.$$

Conversely, let A be symmetric. Since the transpose of a 1×1 matrix is itself, we can express the values $h(v, v')$ as

$$\begin{aligned} h(v, v') = h(v, v')^T &= (\mathbf{x}^T A \mathbf{x}')^T = \mathbf{x}'^T A^T \mathbf{x}^{TT} \\ &= \mathbf{x}'^T A \mathbf{x} = h(v', v), \end{aligned}$$

where $\mathbf{x}$ is short for $\mathbf{x}(v)$. By this equation, the form h is a symmetric one. The skew-symmetric case is similar.

If C is any square matrix, $C + C^T$ is symmetric and $C - C^T$ is skew-symmetric. Hence, provided $2 \neq 0$ so that $\frac{1}{2}$ exists in the coefficient field, we can write any square matrix C as

$$C = \tfrac{1}{2}(C + C^T) + \tfrac{1}{2}(C - C^T).$$

Over a field of characteristic not 2, any square matrix is the sum of a symmetric matrix and a skew-symmetric matrix. Over a field of characteristic 2—say, over $\mathbf{Z}_2$—a symmetric matrix is also skew-symmetric, and the preceding result is false.

Each bilinear form $h: V \times V \to F$ is represented, relative to a (single) basis $\mathbf{b}$ for the space V, by a square matrix A, and, as in Theorem 1, changing $\mathbf{b}$ to $\mathbf{c}$ by a change matrix P will replace A by P^TAP; observe that A symmetric implies P^TAP symmetric, and similarly for skew-symmetry. We call two symmetric matrices A and B *congruent* when there is an invertible matrix P (of the same size) with $A = P^TBP$. Thus two symmetric matrices represent the same symmetric bilinear form (relative to different bases), if and only if they are congruent.

In order to study the relation of congruence, it will be convenient to use not just the bilinear form $(v, v') \mapsto h(v, v')$, but also the corresponding function $v \mapsto h(v, v)$. This is a "quadratic" function of the vector v. The next section will discuss such functions.

EXERCISES

1. Prove that the matrix products AA^T and A^TA are always symmetric.
2. If the matrix A is skew-symmetric, prove A^2 symmetric.
3. Prove: If A and B are symmetric, then AB is symmetric if and only if $AB = BA$.
4. Determine the symmetry (or skew-symmetry) of the matrix $AB - BA$ in the following cases:
   (a) A and B both symmetric.
   (b) A and B both skew-symmetric.
   (c) A symmetric and B skew-symmetric.
5. Prove that every skew-symmetric matrix over $\mathbf{Z}_2$ is symmetric.
6. Exhibit a matrix over $\mathbf{Z}_2$ which is not the sum of a symmetric and a skew-symmetric matrix.
7. Show that the relation of congruence between matrices is reflexive, symmetric, and transitive.
8. If n is odd and the characteristic of the field is not 2, show that any $n \times n$ skew-symmetric matrix A has determinant 0.
9. For a 4×4 skew-symmetric matrix A show that

$$|A| = (A_{12}A_{34} - A_{13}A_{24} + A_{14}A_{23})^2.$$

3. Quadratic Forms

For any field F, the assignment $\kappa \mapsto \kappa^2$ for $\kappa \in F$ defines a "homogeneous" quadratic form $q:F \to F$. It has two characteristic properties: first, $q(-\kappa) = q(\kappa)$; second, the expression

$$q(\kappa + \lambda) - q(\kappa) - q(\lambda) = (\kappa + \lambda)^2 - \kappa^2 - \lambda^2 = 2\kappa\lambda$$

is a bilinear function of κ and λ. For any finite-dimensional vector space V over F, the square $q = x_i^2$ of any coordinate form x_i of V is a function $q:V \to F$ with the same two characteristic properties. These two properties will be used to define quadratic forms in general.

DEFINITION. If V is a finite-dimensional vector space over a field F of characteristic not 2, a quadratic form q *on V is a function $q:V \to F$ such that $q(-v) = q(v)$ for all v, and such that the function h defined by*

$$2h(u, v) = q(u + v) - q(u) - q(v) \tag{6}$$

gives a bilinear form $h:V \times V \to F$. The rank *of q is defined to be the rank of h.*

The bilinear form h of (6) is evidently symmetric; we say that it is obtained by "polarizing" the quadratic form q.

Here are some elementary consequences of this definition. The additivity $h(u, v + w) = h(u, v) + h(u, w)$ states that

$$q(u + v + w) - q(u) - q(v + w)$$
$$= q(u + v) - q(u) - q(v) + q(u + w) - q(u) - q(w).$$

This can be written more symmetrically as the condition

$$q(u + v + w) - q(u + v) - q(u + w) - q(v + w)$$
$$+ q(u) + q(v) + q(w) = 0 \quad (7)$$

for all vectors u, v, and $w \in V$. In particular, if we set $u = v = w = 0$ in (7) we get $q(0) = 0$, while if we set $v = u$ and $w = -u$ we get

$$q(u) - q(2u) - q(0) - q(0) + q(u) + q(u) + q(-u) = 0.$$

Since $q(-u) = q(u)$, we have proved that any quadratic form q has

$$q(2u) = 4q(u), \qquad q(0) = 0. \quad (8)$$

Now we show that every symmetric bilinear form may be obtained by polarizing a unique quadratic form.

Theorem 2. *If V is a finite-dimensional vector space over a field F of characteristic not 2, each symmetric bilinear form $h: V \times V \to F$ defines a quadratic form $q: V \to F$ by the formula*

$$q(v) = h(v, v), \qquad v \in V. \quad (9)$$

This is the only quadratic form q with $2h(u, v) = q(u + v) - q(u) - q(v)$.

Proof: The function q defined by (9) satisfies

$$q(-v) = h(-v, -v) = (-1)^2 h(v, v) = q(v);$$

moreover, since h is bilinear and symmetric,

$$q(u + v) - q(u) - q(v) = h(u + v, u + v) - h(u, u) - h(v, v)$$
$$= h(u, v) + h(v, u) = 2h(u, v).$$

Hence, q is quadratic in the sense of our definition.

Conversely, (6) implies $2h(v, v) = q(2v) - q(v) - q(v)$; by (8) this is $2q(v)$. Hence, since $2 \neq 0$, $h(v, v) = q(v)$. Thus when h is defined from q by (6), it also determines q as in (9).

Now take a fixed basis $\mathbf{b}$ of V. Relative to $\mathbf{b}$, we get a bijection from quadratic forms via symmetric bilinear forms to symmetric matrices. Explicitly, the *matrix A of a quadratic form q*, rel $\mathbf{b}$, is defined to be the matrix of the associated bilinear form h. Thus A has the entries

$$A_{ij} = \tfrac{1}{2}[q(b_i + b_j) - q(b_i) - q(b_j)], \qquad i, j = 1, \cdots, n;$$

in particular, $A_{ii} = q(b_i)$. Conversely, given the symmetric matrix A, the corresponding quadratic form (rel $\mathbf{b}$) is the form q with $q(v) = h(v, v)$, expressed as in (3) by

$$q = \sum_{i=1}^{n} \sum_{j=1}^{n} x_i A_{ij} x_j. \tag{10}$$

Here the coordinates x_i and x_j are (linear) functions of the vector $v \in V$, so that each term $x_i A_{ij} x_j$ and hence the whole sum is, like q, a function of v. For $n = 2$, this is

$$q = x_1 A_{11} x_1 + x_1 A_{12} x_2 + x_2 A_{21} x_1 + x_2 A_{22} x_2.$$

This is a sum of terms of degree 2; the cross-product term $x_1 x_2 = x_2 x_1$ occurs twice with the same coefficient $A_{12} = A_{21}$, so that

$$q = A_{11} x_1^2 + 2A_{12} x_1 x_2 + A_{22} x_2^2.$$

Thus each quadratic form q on V, expressed relative to coordinates x, appears as a polynomial in these coordinate forms $x_1, \cdots, x_n$ in which every term has degree 2 (a "homogeneous" quadratic polynomial). The matrix A for q has as diagonal entries A_{ii} the coefficients of the squares x_i^2, while each off-diagonal entry A_{ij}, for $i \neq j$, is half the coefficient of the crossed product $x_i x_j$. (Thus each crossed product term is written symmetrically as $A_{ij} x_i x_j + A_{ji} x_j x_i$.)

This discussion gives a direct way of finding the matrix A of q from the quadratic expression for the form q. In former treatments a quadratic form was defined to be an expression $\Sigma x_i A_{ij} x_j$; our definition emphasizes the fact that a quadratic form is really a function, and that this function has different expressions as a quadratic polynomial, one expression for each coordinate system.

Theorem 3. *Let V be a vector space of dimension n over a field of characteristic not 2. For any quadratic form q of rank r on V one can find coordinates $\mathbf{y}$ in V and non-zero scalars $\alpha_1, \cdots, \alpha_r$ such that*

$$q = \alpha_1 y_1^2 + \cdots + \alpha_r y_r^2. \tag{11}$$

This result is usually stated as "Every quadratic form q over a field not of characteristic 2 can be diagonalized". This can be done by successively "completing the square". Formally, the proof is by induction on the dimension n, so suppose that every quadratic form on a vector space of dimension less than n can be diagonalized. Let V have a basis $\mathbf{b}$ of n elements. If $q = 0$, the result is immediate, so we may assume that $q \neq 0$.

First, we will find a new basis $\mathbf{c}$ with $q(c_1) \neq 0$. In case $q(b_i) \neq 0$ for some i, we can get this new basis from $\mathbf{b}$ by simply interchanging b_1 and b_i. Otherwise, $q(b_i) = 0$ for all i. Since q and hence its matrix is assumed non-zero, there is some pair of indices $i \neq j$ with $h(b_i, b_j) \neq 0$, where h is the polarized bilinear form of q. Permuting again, we can assume $h(b_1, b_2) \neq 0$, but $q(b_1) = q(b_2) = 0$. Then, by the definition of the polarized bilinear form h,

$$q(b_1 + b_2) = q(b_1) + q(b_2) + 2h(b_1, b_2) = 2h(b_1, b_2) \neq 0,$$

since the characteristic of F is not 2. The new basis $c_1 = b_1 + b_2$, $b_2, \cdots, b_n$ now has $q(c_1) \neq 0$.

In the coordinates $\mathbf{y}$ for the new basis $\mathbf{c}$, q is expressed as

$$q = A_{11}\, y_1{}^2 + 2 \sum_{i=2}^{n} A_{1i} y_1\, y_i + \text{terms not involving } y_1.$$

Here $q(c_1) = A_{11} \neq 0$, so we can complete the square on the terms displayed, getting

$$q = A_{11}\left(y_1 + A_{11}{}^{-1} \sum_{i=2}^{n} A_{1i} y_i \right)^2 + \text{(other) terms not involving } y_1.$$

This suggests changing to the new coordinates $z_1 = y_1 + A_{11}{}^{-1}\Sigma_{i=2}^{n} A_{1i} y_i$, $z_2 = y_2, \cdots, z_n = y_n$; the corresponding change matrix is triangular with diagonal entries 1, hence invertible, so the z's are coordinates. In these coordinates, $q = A_{11} z_1{}^2 + q'$, where q' involves only $z_2, \cdots, z_n$, so q' is in effect a quadratic form on the $(n - 1)$-dimensional space Ker z_1. By the induction assumption, q' can be diagonalized, giving the final result (11).

This theorem for quadratic forms can also be rewritten as one for the corresponding symmetric bilinear forms (polarize q) and for the corresponding symmetric matrices as in §2 (take the matrix of q):

Theorem 4. *For a finite-dimensional vector space V over F of characteristic not 2, there is to each symmetric bilinear form $h: V \times V \to F$ of rank r a choice of coordinates $\mathbf{y}$ in V and scalars $\alpha_1, \cdots, \alpha_r$, none zero, such that*

$$h(v, v') = \alpha_1\, y_1(v) y_1(v') + \cdots + \alpha_r y_r(v) y_r(v')$$

for all vectors v and v' in V.

Corollary. *Over a field of characteristic not 2, every symmetric matrix is congruent to a diagonal matrix.*

Here B congruent to A means that $B = P^TAP$ for some nonsingular P, as in §2. This relation of congruence is reflexive, symmetric, and transitive, and the rank of a matrix is an invariant, in the sense of §VII.7, for this relation. However, we do not yet have a canonical form, in the sense of §VII.7, for symmetric matrices under the relation of congruence.

When $r = n$ (rank = dimension), a quadratic form q is called *nondegenerate*.

EXERCISES

1. If q is a quadratic form, κ a scalar and v a vector, prove that $q(v\kappa) = [q(v)]\kappa^2$.

2. Find the symmetric matrix associated with each of the following quadratic forms over the field **Q** of rational numbers:

(a) $2x^2 + 3xy + 6y^2$.
(b) $8xy + 4y^2$.
(c) $x^2 + 2xy + 4xz + 3y^2 + yz + 7z^2$.
(d) $4xy$.
(e) $x^2 + 4xy + 4y^2 + 2xz + z^2 + 2yz$.

3. Diagonalize each of the quadratic forms of Exercise 2. (*Hint:* Complete the square!)

4. Diagonalize $2x^2 + xy + 3y^2$ over the field $\mathbf{Z}_5$.

5. Over the field $\mathbf{Z}_5$, show that every quadratic form can be expressed relative to suitable coordinates as $\Sigma\alpha_i y_i^2$ with $\alpha_i = 0$, 1, or 2.

6. Over **Q**, show that the quadratic form $x_1^2 + x_2^2$ can be transformed by change of coordinates into either $9y_1^2 + 4y_2^2$ or $2z_1^2 + 8z_2^2$.

7. For V a two-dimensional vector space over a field F of characteristic 2, show that the quadratic form xy cannot be diagonalized.

8. Prove that a function $q: V \to F$ is a quadratic form if and only if $q(2v) = 4q(v)$ for all $v \in V$ and $(u, v) \mapsto q(u + v) - q(u) - q(v)$ is bilinear (characteristic $F \neq 2$).

9. Let $h: V \times V \to F$ be a skew-symmetric bilinear form, with V finite-dimensional and characteristic $F \neq 2$. Show that there is an integer k and coordinates y in V such that, when $y_i = y_i(v)$ and $y_i' = y_i(v')$,

$$h(v, v') = (y_1 y_2' - y_2 y_1') + (y_3 y_4' - y_4 y_3') + \cdots + (y_{2k-1} y_{2k}' - y_{2k} y_{2k-1}').$$

10. Recast the result of Exercise 9 as a theorem about skew-symmetric matrices.

4. Real Quadratic Forms

When a quadratic form over a general field is diagonalized, the diagonal coefficients are not unique. But over the field **R** of real numbers, we can do better because we can make all the diagonal entries ± 1; these entries are then unique:

Theorem 5. *Let q be a quadratic form on an n-dimensional vector space V over $\mathbf{R}$. Then there are integers s and r with $0 \leqslant s \leqslant r \leqslant n$ depending only on q, and there are coordinate forms $z_1, \cdots, z_n$ for V so that*

$$q = z_1^2 + \cdots + z_s^2 - z_{s+1}^2 - \cdots - z_r^2. \tag{12}$$

Proof: Diagonalize q, then permute coordinates till $q = \Sigma \alpha_i y_i^2$ has all positive coefficients first. If the real number α_i is positive, it has a square root, say $\beta_i \neq 0$. Make the change of coordinates $z_i = \beta_i y_i$; the ith term becomes $\alpha_i y_i^2 = \beta_i^2 y_i^2 = z_i^2$. On the other hand, if $\alpha_j < 0$, then $\alpha_j = -\beta_j^2$ and $\alpha_j y_j^2 = -\beta_j^2 y_j^2 = -z_j^2$. This gives the desired expression (12).

The crucial point remains: To show that the numbers s and r depend only on the given quadratic form q. Now r is the total number of non-zero diagonal terms, hence is the rank of q; we already know that this is an invariant. It remains to treat s, the number of positive diagonal terms.

Suppose then that the same form q, relative to the coordinates x for some different basis, has a diagonal expression

$$q = x_1^2 + \cdots + x_t^2 - x_{t+1}^2 - \cdots - x_r^2 \tag{13}$$

with t positive terms. If $t \neq s$, we can suppose $t < s$. We construct the subspaces

$$T = \{v | x_1(v) = \cdots = x_t(v) = 0\}, \qquad \dim T = n - t,$$
$$S = \{v | z_{s+1}(v) = \cdots = z_n(v) = 0\}, \quad \dim S = s,$$

with the indicated dimensions. Their sum $S + T$ has at most the dimension n of the whole space, so the intersection $S \cap T$ has dimension [cf. (VI.14)]

$$\dim (S \cap T) = \dim S + \dim T - \dim (S + T)$$
$$\geqslant s + n - t - n = s - t > 0.$$

Hence, there is a non-zero vector v in $S \cap T$. The second formula (13) for q and the definition of T show that $q(v) \leqslant 0$. The first formula (12) for q and the definition of S show that $q(v) > 0$, a contradiction. Hence $s = t$ is an invariant, as asserted in the theorem.

The number s (or sometimes the numbers of + and − signs in the diagonal form) is called the *signature* of the form q (or of its matrix). The invariance of this signature, as just proved, is Sylvester's *law of inertia*. Since two symmetric matrices represent the same quadratic form if and only if the matrices A and B are congruent ($A = P^T BP$, with P invertible), this result can also be reformulated as one about matrices: Any real symmetric matrix is congruent to exactly one diagonal matrix with diagonal entries $+1$, -1, and 0, and with the diagonal entries arranged so that all $+1$'s precede all -1's, which in turn precede all zeros. The set of these diagonal matrices is a set of canonical forms for real symmetric matrices under the relation of congruence. Alternatively, the rank r (total number of non-zero diagonal terms) and the signature s (number of $+1$ terms) together form a complete system of invariants for real symmetric matrices under congruence.

This result deals with a single quadratic form and a change of basis to simplify the expression of that form in terms of other coordinates; in this sense it is an "alias" result (§VII.7). But the change matrix P in question can be read as the matrix of an automorphism; this yields a corresponding "alibi" result:

Corollary. *If p and q are two real quadratic forms on the same finite-dimensional real vector space V, there is an automorphism $s: V \to V$ with $p = q \circ s$ if and only if p and q have the same rank and the same signature.*

Note that $p = q \circ s$ for some automorphism s means that p and q are equivalent under the full linear group.

Proof: Observe first that any quadratic form $q: V \to \mathbf{R}$ and any linear transformation $t: V \to V$ have a composite $q \circ t: V \to \mathbf{R}$ which is a quadratic form. Now if q has the matrix A relative to some basis $\mathbf{b}$, while an automorphism $s: V \to V$ has the invertible matrix P relative to the same basis, then (with $\mathbf{x}$ for the column of coordinates $\mathbf{x}(v)$)

$$q(sv) = [\mathbf{x}(sv)]^T A [\mathbf{x}(sv)] = [P\mathbf{x}]^T A [P\mathbf{x}] = \mathbf{x}^T (P^T A P)\mathbf{x}.$$

This equation states that the matrix of $q \circ s$, relative to $\mathbf{b}$, is the matrix $P^T A P$, hence is congruent to A. The corollary is now immediate from the matrix version of the theorem.

A real quadratic form q is called *positive definite* when $v \neq 0 \Rightarrow q(v) > 0$, *positive semidefinite* when $q(v) \geqslant 0$ for all $v \in V$, and *negative definite* when $v \neq 0 \Rightarrow q(v) < 0$. Thus in n dimensions the corresponding canonical forms are

Positive definite:	$x_1^2 + \cdots + x_n^2$,	
Positive semidefinite:	$x_1^2 + \cdots + x_r^2$,	$r < n$,
Negative definite:	$-x_1^2 - \cdots - x_n^2$.	

A real symmetric matrix A is called *positive definite*, if it is the matrix of a positive definite quadratic form q. Since this form q in suitable coordinates is $q = x_1^2 + \cdots + x_n^2$, with matrix the identity matrix I, it follows that each positive definite A must be congruent to I; that is, $A = P^TIP$. Restated, this is

Proposition 6. *A real symmetric matrix A is positive definite if and only if there exists an invertible real matrix P with $A = P^TP$.*

Quadratic forms for $n = 2$ are related to conic sections. In a two-dimensional vector space V, each quadratic form q determines the set (or "locus") C_q consisting of all vectors v with $q(v) = 1$. Call C_q a *central conic*. Choose coordinates x and y to make q canonical. In case q has rank 2, we get one of the three loci

$$x^2 + y^2 = 1, \qquad x^2 - y^2 = 1, \qquad -x^2 - y^2 = 1.$$

They are, respectively, a circle, a rectangular hyperbola, and no locus. Put differently: By choosing a suitable basis, any ellipse looks like a circle and any hyperbola looks rectangular, but no ellipse ever becomes a hyperbola; the signatures are different.

Similarly, in three dimensions for q of rank 3, the signature s distinguishes between ellipsoids ($s = 3$), hyperboloids of one sheet ($s = 2$), hyperboloids of two sheets ($s = 1$), and no locus.

The signature is useful in studying maxima and minima of functions of several real variables. A *critical point* for a differentiable function $z = f(x, y)$ is defined to be a point (x_0, y_0) where both first partial derivatives $f'_x(x_0, y_0)$ and $f'_y(x_0, y_0)$ vanish; this means that there are no first-degree terms in the expansion of $f(x, y)$ by Taylor series in powers of $x - x_0$ and $y - y_0$. Taking $x_0 = y_0 = 0$ for simplicity, this expansion (assumed to exist and be convergent) is

$$f(x, y) = f(0, 0) + \tfrac{1}{2}\left[Ax^2 + 2Bxy + Cy^2\right] + \cdots,$$

with the remaining terms of higher degree, while the coefficients are the second partial derivatives

$$A = f''_{xx}(0, 0), \qquad B = f''_{xy}(0, 0), \qquad C = f''_{yy}(0, 0)$$

at the critical point $(0, 0)$. For nearby values of x and y, the higher-degree terms in the expansion may be neglected, so the function f is approximated by a constant $f(0, 0)$ plus a quadratic form q in x and y. We suppose this form to be of rank 2 (the critical point is then said to be *nondegenerate*). If q is positive definite, values of $f(x, y)$ near the critical point $(0, 0)$ must exceed $f(0, 0)$, so f has a minimum there. If q is negative definite, f has a maximum at $(0, 0)$. If q is neither of these, it has signature 1, and its values may be either positive or negative. The critical point $(0, 0)$ is then neither a maximum nor a minimum, but a *saddle point* on the surface $z = f(x, y)$ (like the height of a pass between two mountain peaks, where motion in one direction increases the altitude z, in another direction decreases z). Thus maxima, minima, and saddle points are distinguished by the signature of a quadratic form.

EXERCISES

1. Find the rank and the signature of each of the following real quadratic forms:

(a) $9x_1^2 + 12x_1x_2 + 79x_2^2$. (b) $2x_1^2 - 12x_1x_2 + 18x_2^2$.

(c) $-2x_1^2 - 4x_1x_2 + 22x_2^2 + 12x_2x_3 + 6x_3x_1 - x_3^2$.

2. Let q be a quadratic form of rank r on an n-dimensional vector space V over the field $\mathbf{C}$ of complex numbers. Prove that there are coordinates $\mathbf{z}$ in V for which $q = z_1^2 + \cdots + z_r^2$.

3. Let p and q be two quadratic forms on the same n-dimensional vector space V over $\mathbf{C}$. Prove that p and q are equivalent under the full linear group $GL(n, \mathbf{C})$ if and only if they have the same rank. (*Hint:* This is an alibi translation of the alias result of Exercise 2.)

4. Show that $ax^2 + 2bxy + cy^2$ is positive definite if and only if $a > 0$ and $ac - b^2 > 0$.

5. List all the types of real quadratic forms of rank 4 in 4 dimensions and describe qualitatively the geometric locus corresponding to each.

6. Describe the possible types of critical points for functions of three real variables.

5. Inner Products

Many geometrical questions about real vector spaces involve both vector algebra and two additional concepts: The "length" of a vector and the "angle" between two vectors. For example, in the two-dimensional space $\mathbf{R}^2$, the length $|\boldsymbol{\xi}|$ of a vector $\boldsymbol{\xi} = (\xi_1, \xi_2)$ is determined by the usual pythagorean formula

$$|\boldsymbol{\xi}| = \left(\xi_1^2 + \xi_2^2\right)^{1/2}. \tag{14}$$

This formula uses the positive definite quadratic form $\xi_1^2 + \xi_2^2$. Its polarized symmetric bilinear form assigns to each pair of vectors $\boldsymbol{\xi} = (\xi_1, \xi_2)$ and $\boldsymbol{\eta} = (\eta_1, \eta_2)$ in $\mathbf{R}^2$ their *inner product*, written as

$$\langle \boldsymbol{\xi}, \boldsymbol{\eta} \rangle = \xi_1\eta_1 + \xi_2\eta_2; \tag{15}$$

since it is a scalar, it is often called the "scalar product" of the vectors $\boldsymbol{\xi}$ and $\boldsymbol{\eta}$ and is written as $\boldsymbol{\xi} \cdot \boldsymbol{\eta}$ in physics texts.

If θ is the angle between the vectors ξ and η, as shown below

$$\eta - \xi,$$

the law of cosines applied to the triangle with sides ξ, η, and $\eta - \xi$ is

$$|\eta - \xi|^2 = |\eta|^2 + |\xi|^2 - 2|\xi|\,|\eta| \cos \theta.$$

On the other hand, $|\eta - \xi|^2 = \langle \eta - \xi, \eta - \xi \rangle$, expanded by bilinearity, becomes

$$\langle \eta, \eta \rangle - 2\langle \eta, \xi \rangle + \langle \xi, \xi \rangle = |\eta|^2 + |\xi|^2 - 2\langle \eta, \xi \rangle.$$

Comparing these two expressions for $|\eta - \xi|^2$ gives

$$\langle \eta, \xi \rangle = |\eta|\,|\xi| \cos \theta,$$

so the inner product of the two vectors is the product of their lengths by the cosine of the angle θ between them. The corresponding formula holds in three dimensions.

In the n-dimensional space $\mathbf{R}^n$ we may define the inner product of two lists ξ and η of n real numbers each by the "standard" formula

$$\langle \xi, \eta \rangle = \xi_1\eta_1 + \cdots + \xi_n\eta_n. \tag{16}$$

This makes $\mathbf{R}^n$ an inner product space in the sense of the following general definition:

DEFINITION. An inner product space U (also called a "euclidean vector space") is a vector space over the field $\mathbf{R}$ *of real numbers equipped with a symmetric bilinear form h whose associated quadratic form is positive definite.*

The value $h(u, v)$ of this symmetric form h on vectors $u, v \in U$ will be written as $\langle u, v \rangle$ and will be called the *inner product* of u and v. In this notation, the definition above requires that the inner product have the following three properties:

Symmetry: $\langle u, v \rangle = \langle v, u \rangle$ for all $u, v \in U$;

Left linearity: $\langle u_1\kappa_1 + u_2\kappa_2, v \rangle = \langle u_1, v \rangle \kappa_1 + \langle u_2, v \rangle \kappa_2$ for all real scalars κ_i;

Positive definiteness: $u \neq 0 \Rightarrow \langle u, u \rangle > 0$.

The linearity of $\langle u, v \rangle$ in the second argument v follows by symmetry from the linearity in the first argument.

In any inner product space, length, perpendicularity, and angle can be defined.

The *length* (or, the "norm") $|u|$ of any vector u is defined to be the nonnegative square root,

$$|u| = \langle u, u \rangle^{1/2}; \tag{17}$$

this is possible because the nonnegative real number $\langle u, u \rangle$ has a real square root. For any scalar κ, it follows that $|u\kappa| = |u|\,|\kappa|$, while $u \neq 0$ implies $|u| > 0$. As in the polarization process (6), the symmetric bilinear form $\langle u, v \rangle$ may be expressed in terms of the quadratic form $\langle u, u \rangle$ as

$$2\langle u, v \rangle = |u + v|^2 - |u|^2 - |v|^2. \tag{18}$$

Two vectors u and v are called *orthogonal* or "perpendicular" (in symbols, $u \perp v$) when their inner product is $\langle u, v \rangle = 0$. By symmetry of the inner product, $u \perp v$ implies $v \perp u$, so the relation of orthogonality is symmetric. By linearity, the set of all vectors u orthogonal to a given vector v is a subspace, called the *orthogonal complement* of v. A vector u is said to be *orthogonal* to a subspace S of U if it is orthogonal to every vector in S; by linearity again, it suffices that u be orthogonal to every vector in a basis of S. The set of all vectors u orthogonal to a given subspace S is another subspace $S^{\perp}$, the *orthogonal complement* of S.

Any subspace S of an inner product space U is itself an inner product space. For, the inner product $\langle u, v \rangle$ of vectors in U, when restricted to vectors of S, defines a bilinear form $S \times S \to \mathbf{R}$ which is evidently symmetric and positive definite. $\mathbf{R}$ itself is a (one-dimensional) inner product space for the quadratic form $x \mapsto x^2$, so the length in this space $\mathbf{R}$ is just the usual absolute value of a real number.

Before defining angles, we prove an important inequality:

Theorem 7. *If u and v are vectors in any inner product space, then*

$$|\langle u, v \rangle| \leqslant |u| \cdot |v| \qquad \textit{(the Schwarz inequality)}. \tag{19}$$

Proof: For every scalar κ, positive definiteness gives

$$\langle \kappa u + v, \kappa u + v \rangle = \kappa^2 |u|^2 + 2\kappa \langle u, v \rangle + |v|^2 \geqslant 0.$$

Now hold u and v fixed, and regard this expression $a\kappa^2 + b\kappa + c$ as a quadratic in κ with coefficients $a = |u|^2$, $b = 2\langle u, v \rangle$, and $c = |v|^2$; since this quadratic is never negative, its discriminant $b^2 - 4ac$ cannot be positive, so

$$4\langle u, v \rangle^2 \leqslant 4|u|^2|v|^2.$$

Taking the positive square root of both sides yields the inequality (19).

This inequality implies that $-1 \leqslant \langle u, v\rangle/(|u|\cdot|v|) \leqslant +1$, so the quotient $\langle u, v\rangle/(|u|\cdot|v|)$ is the cosine of exactly one angle between 0° and 180°. We define this to be the *angle* between the vectors u and v. In particular, this angle is 90° when $\langle u, v\rangle = 0$; that is, when $u \perp v$; by the law of cosines, this angle is the usual one in the spaces $\mathbf{R}^2$ and $\mathbf{R}^3$ with the standard inner product.

Corollary 1. *For vectors u and v in an inner product space,*

$$|u + v| \leqslant |u| + |v|. \tag{20}$$

Proof: By bilinearity and symmetry,

$$|u + v|^2 = \langle u + v, u + v\rangle = \langle u, u\rangle + 2\langle u, v\rangle + \langle v, v\rangle.$$

We apply the Schwarz inequality to the middle term on the right to get

$$|u + v|^2 \leqslant |u|^2 + 2|u|\,|v| + |v|^2 = (|u| + |v|)^2, \qquad \text{Q.E.D.}$$

This law (20) may be written as $|u - v| \leqslant |u| + |v|$. The zero vector, u, and v together form a triangle whose third side has length $|u - v|$, so this law restates the familiar geometric fact that the length of one side of a triangle is never greater than the sum of the lengths of the other two sides.

As in §VIII.10, a *metric space* X is defined to be a set X together with a function $\rho: X \times X \to \mathbf{R}$ which assigns to each pair of points $x, y \in X$ a nonnegative real number $\rho(x, y)$, called the *distance* from x to y, in such a way that $\rho(x, y) = \rho(y, x)$ for all x and y, $\rho(x, y) = 0$ if and only if $x = y$, and

$$\rho(x, y) + \rho(y, z) \geqslant \rho(x, z) \qquad (\textit{the triangle axiom}).$$

Now in an inner product space, $|u - v|$ may be regarded as the distance from u to v (think of the distance between the endpoints of the vectors u and v). From the corollary above one then has

Corollary 2. *An inner product space is a metric space when the distance between two vectors u and v is taken to be the length $|u - v|$.*

Corollary 3. *For real numbers x_1, x_2, y_1, and y_2,*

$$|x_1 y_1 + x_2 y_2| \leqslant (x_1^2 + x_2^2)^{1/2}(y_1^2 + y_2^2)^{1/2}.$$

This is just the Schwarz inequality for the space $\mathbf{R}^2$ with the standard inner product (15). There is an analogous inequality for $2n$ real numbers $x_1, \cdots, x_n, y_1, \cdots, y_n$.

EXERCISES

1. In any inner product space, prove that

$$|u - v|^2 + |u + v|^2 = 2(|u|^2 + |v|^2).$$

2. Show that $\mathbf{R}^2$ becomes an inner product vector space when the inner product is $\langle \xi, \eta \rangle = \xi_1\eta_1 + 2\xi_1\eta_2 + 2\xi_2\eta_1 + 5\xi_2\eta_2$.

3. If u and $v \neq 0$ are vectors in an inner product space, find a formula for the shortest vector of the form $u + v\kappa$ and prove this vector orthogonal to v. (Draw a figure!)

4. If $|u| = |v|$, prove that $(u - v) \perp (u + v)$. Draw a figure.

5. In any inner product space, prove that $||u| - |v|| \leqslant |u - v|$.

6. In $\mathbf{R}^3$, define the *outer product* of two vectors ξ and η to be the vector

$$\xi \times \eta = (\xi_2\eta_3 - \xi_3\eta_2, \xi_3\eta_1 - \xi_1\eta_3, \xi_1\eta_2 - \xi_2\eta_1).$$

(a) For four vectors $u, v, u', v' \in \mathbf{R}^3$ prove that

$$\langle u \times v, u' \times v' \rangle = \langle u, u' \rangle\langle v, v' \rangle - \langle u, v' \rangle\langle v, u' \rangle.$$

(b) Derive the Schwarz inequality in $\mathbf{R}^3$ from (a) (set $u = u'$, $v = v'$).

(c) Prove that $u \times (v \times w) = \langle u, w \rangle v - \langle u, v \rangle w$.

6. Orthonormal Bases

In any finite-dimensional inner product space U the quadratic form $\langle u, u \rangle = |u|^2$ is positive definite. Hence, for a suitable choice of a basis $\mathbf{b}$ in U the matrix of $\langle u, u \rangle$ is the identity matrix I; this means exactly, for the corresponding coordinates $\xi_i = x_i(u)$, that

$$\langle u, u \rangle = \xi_1^{\,2} + \xi_2^{\,2} + \cdots + \xi_n^2. \tag{21}$$

If v is a second vector with coordinates $\eta_1, \cdots, \eta_n$, their inner product is

$$\langle u, v \rangle = \xi_1\eta_1 + \xi_2\eta_2 + \cdots + \xi_n\eta_n. \tag{22}$$

Now a basis $\mathbf{b}$ in U gives this standard inner product precisely when the matrix of $\langle u, u \rangle$ rel $\mathbf{b}$ is I; that is, when

$$\langle b_i, b_j \rangle = \delta_{ij}, \qquad i, j = 1, \cdots, n. \tag{23}$$

These equations state that each basis vector b_i has length 1 while any two different basis vectors b_i and b_j are orthogonal.

Any list $\mathbf{c}$ of k vectors $c_1, \cdots, c_k$ in U is called *orthonormal* when each vector of the list has length 1, while any two different vectors are orthogonal. This amounts to the requirement that $\langle c_i, c_j \rangle = \delta_{ij}$ for $i, j = 1, \cdots, k$. Each orthonormal list is linearly independent, for $\sum c_i\kappa_i = 0$ with scalars κ_i implies for each j that

$$0 = \langle 0, c_j \rangle = \left\langle \sum_i c_i\kappa_i, c_j \right\rangle = \sum_i \langle c_i, c_j \rangle \kappa_i = \kappa_j,$$

so all the scalars κ_j are zero. In particular, an orthonormal list of $n = \dim U$ vectors is linearly independent, hence is a basis of U. In this case we speak of an *orthonormal basis*.

We showed in (23) that every finite-dimensional inner product space has an orthonormal basis. The following Gram–Schmidt orthogonalization process will explicitly construct such bases.

Lemma. *If $e_1, \cdots, e_k$ is a list of k linearly independent vectors in a finite-dimensional inner product space U, there is an orthonormal list* $\mathbf{c}$ *which spans the same subspace, and such that each partial list $c_1, \cdots, c_i$ spans the same subspace as does $e_1, \cdots, e_i$, for each $i = 1, \cdots, k$.*

The proof is by induction on k. If $k = 1$, $e_1 \neq 0$ can fail to be orthonormal only if its length $|e_1|$ is not 1, so we simply "normalize" e_1, taking $c_1 = e_1\kappa_1$ with $\kappa_1 = |e_1|^{-1}$. Now suppose that the lemma is true for lists of k vectors, and take some list $\mathbf{e}$ of $k + 1$ vectors. By the induction assumption, there is an orthonormal list $c_1, \cdots, c_k$ with the same span S as $e_1, \cdots, e_k$. The last vector $e = e_{k+1}$ of the given list cannot be in this subspace S (the list $\mathbf{e}$ is independent!), so we try to "split" e into a part $\Sigma c_i\mu_i$ in S and a part $u = e - \Sigma c_i\mu_i$ orthogonal to S. Indeed, u orthogonal to S means for each $j = 1, \cdots, k$ that

$$0 = \langle u, c_j\rangle = \langle e, c_j\rangle - \left\langle \sum_{i=1}^{k} c_i\mu_i, c_j \right\rangle = \langle e, c_j\rangle - \sum_{i=1}^{k} \langle c_i, c_j\rangle\mu_i = \langle e, c_j\rangle - \mu_j.$$

Hence, choosing each scalar μ_j as $\langle e, c_j\rangle$ gives a vector $v = \Sigma c_j\mu_j$ in S and a vector $u = e - v$ orthogonal to S. Since $e \notin$ S, $u \neq 0$, so we can multiply u by a scalar to get a vector c_{k+1} also orthogonal to S and with $|c_{k+1}| = 1$. With $c_1, \cdots, c_k$, it gives the required orthonormal list.

Theorem 8. *In a finite-dimensional inner product space U, any orthonormal list is part of an orthonormal basis. In particular, U has an orthonormal basis.*

Proof: The given orthonormal list $c_1, \cdots, c_k$ is part of a basis $c_1, \cdots, c_k, e_{k+1}, \cdots, e_n$; to this basis apply the Gram–Schmidt process; it will not change the first k vectors, and so yields an orthonormal basis containing them, Q.E.D.

If $\mathbf{b}$ is an orthonormal basis of U and if each vector $u \in U$ is written in terms of the corresponding coordinates $\xi_i = x_i(u)$ as $u = \Sigma b_i\xi_i$, then for each b_j,

$$\langle b_j, u\rangle = \left\langle b_j, \sum b_i\xi_i \right\rangle = \sum_{i=1}^{n} \langle b_j, b_i\rangle\xi_i = \xi_j; \tag{24}$$

that is, the jth coordinate form $x_j : U \to \mathbf{R}$ is just the inner product $\langle b_j, - \rangle$. In other words, the *coordinates of a vector are its inner products with the* (orthonormal) *basis vectors*. Also, $u \mapsto (\xi_1, \cdots, \xi_n)$ preserves the inner product, by (22), so is an isomorphism of inner product spaces. Hence the

Corollary. *Each orthonormal basis* $\mathbf{b}$ *of an* n*-dimensional inner product space* U *yields an isomorphism of inner product spaces,* $U \cong \mathbf{R}^n$*, where* $\mathbf{R}^n$ *has the standard inner product* (16).

Now consider the dual vector space of an inner product space U. If v is a fixed vector of U, the function "inner product with v" is a linear form $\langle v, - \rangle : U \to F$, hence an element of the dual vector space U^*—namely, the element $u \mapsto \langle v, u \rangle$.

Theorem 9. *For each finite-dimensional inner product space* U*, the function sending each vector* $v \in U$ *to the linear form* $\langle v, - \rangle$ *in the dual space* U^* *is an isomorphism of vector spaces* $U \cong U^*$.

Proof: Since $\langle v, u \rangle$ is linear in u, $\langle v, - \rangle$ is an element of U^*; since $\langle v, u \rangle$ is linear in v, the assignment $v \mapsto \langle v, - \rangle$ is a linear map $U \to U^*$. Its kernel is zero, since $\langle v_0, u \rangle = 0$ for one v_0 and all u gives in particular $\langle v_0, v_0 \rangle = |v_0|^2 = 0$, hence $v_0 = 0$. Since U and U^* have the same finite dimension, the monomorphism $U \to U^*$ is an isomorphism, Q.E.D.

This proof amounts to showing that the inner product $\langle u, v \rangle$ is a dual pairing $U \times U \to \mathbf{R}$ in the sense of §VI.4.

It is convenient to identify the inner product space U with its dual U^* according to this isomorphism. This means that each vector v is identified with the linear form $\langle v, - \rangle$. Hence each vector $v \in U$ is also a linear form $v : U \to \mathbf{R}$. For example, if $\mathbf{b}$ is an orthonormal basis of U, then by (24) each vector b_i in this basis is identified with the corresponding vector x_i in the dual basis, so *an orthonormal basis is its own dual basis*.

This identification of U with its dual leads to the following important operation on linear transformations:

Theorem 10. *If* $t : U \to U'$ *is a linear transformation between two finite-dimensional inner product spaces* U *and* U'*, there is a unique linear transformation* $t^* : U' \to U$ *(backward!) such that, for all vectors* $v' \in U'$ *and* $u \in U$,

$$\langle t^* v', u \rangle = \langle v', tu \rangle. \tag{25}$$

Proof: In discussing dual modules we constructed in (V.41) to each t a "dual" linear transformation $t^* : U'^* \to U^*$. When the dual spaces U'^* and U^* are identified with U' and U, as proposed above, this dual transformation becomes a linear map $t^*: U' \to U$. But the map so constructed does satisfy the condition (25) of the theorem, for

$$\begin{aligned}
\langle t^*v', u\rangle &= (t^*v')(u), && \text{by the identification } U = U^*, \\
&= (v' \circ t)(u), && \text{by definition of } t^*, \\
&= v'(tu), && \text{by the definition of composite}, \\
&= \langle v', tu\rangle, && \text{by the identification } U' = U'^*.
\end{aligned}$$

This shows that a t^* with property (25) exists. To prove t^* unique, note merely that (25) specifies the inner product of the vector $v_1 = t^*v'$ with *every* vector $u \in U$, while $\langle v_1, u\rangle = \langle v_2, u\rangle$ for every u implies that $\langle v_1 - v_2, u\rangle = 0$ and hence, by positive definiteness for $u = v_1 - v_2$, that $v_1 - v_2 = 0$.

The map t^* defined by (25) is called the *adjoint* of t. By the symmetry of the inner product, $t^{**} = t$.

Proposition 11. *Let* **b** *and* **b**$'$ *be finite orthonormal bases in U and U', respectively. The matrix* (rel **b**, **b**$'$) *of each linear transformation $t: U \to U'$ is the transpose of the matrix* (rel **b**$'$, **b**) *of its adjoint $t^*: U' \to U$.*

Proof: The matrix A of t is defined by $A_{ij} = x_i'(tb_j)$. But by the identification of x_i' with $\langle b_i', -\rangle$ this may be written in the notation of (VII.17) as

$$A_{ij} = \langle b_i', tb_j\rangle, \qquad A = M_{\mathbf{b}'}^{\mathbf{b}}(t). \tag{26}$$

For the same reason, the matrix B of t^* is $B_{ji} = \langle b_j, t^*b_i'\rangle$. Now the symmetry of the inner product and the definition (25) of the adjoint give

$$B_{ji} = \langle b_j, t^*b_i'\rangle = \langle t^*b_i', b_j\rangle = \langle b_i', tb_j\rangle = A_{ij}.$$

Hence, $A = B^T$, as asserted.

This result agrees with that already noted for dual maps in (VII.24).

For a single finite-dimensional inner product space U, a linear transformation $t: U \to U$ is called *self-adjoint* (or, "symmetric") when $t^* = t$; that is, when $\langle tv, u\rangle = \langle v, tu\rangle$ for all pairs of vectors $u, v \in U$. The proposition just proved implies that t is self-adjoint if and only if its matrix A (relative to some orthonormal basis) is symmetric. Note that its matrix $B = PAP^{-1}$ relative to any other orthonormal basis will then also be symmetric.

Since each finite-dimensional inner product space U is its own dual U^*, that dual space is itself an inner product space; in particular, each linear form $f \in U^*$ has a length $|f|$ in this dual inner product space. This length $|f|$ is usually called the *norm* of f; it may be described directly in terms of the function $f: U \to \mathbf{R}$ as follows:

Proposition 12. *On a finite-dimensional inner product space U, the norm of a linear form $f: U \to \mathbf{R}$ is the maximum of all the real numbers $|f(u)|/|u|$ for all vectors $u \neq 0$ in U, and is also the maximum of all the real numbers $|f(u)|/|u|$ for all those $u \in U$ with $|u| = 1$.*

Proof: By the isomorphism of Theorem 9, each linear form f is $f = \langle v, - \rangle$ for a unique vector $v \in U$, and by this identification the norm $|f|$ is the length $|v|$. Hence, $f(u) = \langle v, u \rangle$ for every $u \in U$, and the Schwarz inequality yields

$$|f(u)| = |\langle v, u \rangle| \leqslant |v|\,|u| = |f|\,|u|$$

for all u, with equality when $u = v$. Therefore, $|f|$ is the maximum of the real numbers $|f(u)|/|u|$ for $|u| \neq 0$. Since multiplying u by a positive scalar multiplies both $|f(u)|$ and $|u|$ by that scalar, it is enough to take the maximum over the vectors u with $|u| = 1$. (These are just the vectors on the "unit sphere" of the space U.)

This description of the norm of a linear form can be extended so as to define norms for linear transformations of U (Exercise 7). Such norms are often used in higher analysis, especially for infinite-dimensional spaces with an inner product or with a definition of the "length" of each vector.

EXERCISES

1. In the space $\mathbf{R}^4$ with the standard inner product (16), find orthonormal bases for the subspace spanned by each of the following lists:

(a) $(1, 1, 0, 0)$, $(0, 2, 3, 0)$, and $(0, 0, 4, 1)$.

(b) $(3, 0, 0, 0)$, $(2, 1, 1, 0)$, and $(0, -2, 1, 2)$.

(*Hint: First* find orthogonal bases, *then* normalize to length 1.)

2. Find a basis for the orthogonal complement of each of the subspaces of Exercise 1.

3. In $\mathbf{R}^3$ with the standard inner product, find a basis for the orthogonal complement of the vector $(1, -2, 1)$.

4. If $S^{\perp}$ is the orthogonal complement of a subspace S of a finite-dimensional inner product space U, prove that

$$\dim S^{\perp} + \dim S = \dim U; \qquad (S^{\perp})^{\perp} = S.$$

5. (a) If S is a subspace of a finite-dimensional inner product space U, show that every vector u has a unique representation $u = v + w$ with $v \in S$ and $w \in S^{\perp}$. (Call v the *orthogonal projection* of u on S.)

(b) Find the orthogonal projection of $(1, 2, 3)$ on the subspace of $\mathbf{R}^3$ spanned by $(1, 0, 1)$.

(c) Find the orthogonal projection of $(0, 0, 0, 5)$ on each of the subspaces of Exercise 1.

6. If the Gram–Schmidt process takes a basis **e** of a finite-dimensional inner product space U to an orthonormal basis **c**, show that the change matrix from **e** to **c** is triangular.

7. (a) If $t: U \to U'$ is a linear transformation between two finite-dimensional inner product spaces, show that there is a real number M such that $|t(u)| \leqslant M|u|$ for all $u \in U$ with $|u| = 1$, with equality $|t(u_0)| = M|u_0|$ for at least one vector u_0. Call M the "norm" $M = |t|$ of t. (*Hint:* Under $U \cong \mathbf{R}^n$, the set $\{u \mid |u| = 1\}$ is mapped to a set closed and bounded in the sense of §VIII.6; use the minimum property of §VIII.6 to show that the continuous function $u \mapsto |t(u)|/|u|$ takes on its least upper bound on $|u| = 1$.)

(b) For *all* $u \in U$, prove that $|t(u)| \leqslant |t|\,|u|$.

7. Orthogonal Matrices

In the plane $\mathbf{R}^2$, a rotation about the origin through a given angle or a reflection in a given line through the origin is a linear transformation $\mathbf{R}^2 \to \mathbf{R}^2$ which preserves all inner products, lengths, and angles. A transformation with these properties is said to be "orthogonal".

DEFINITION. A linear transformation $t: U \to U'$, for U and U' inner product spaces, is orthogonal *when*

$$\langle tu, tv \rangle = \langle u, v \rangle \tag{27}$$

for all vectors $u, v \in U$.

For t orthogonal, $tu = 0$ implies $u = 0$, so every orthogonal transformation t is a monomorphism of vector spaces. Since lengths and angles are defined in terms of the inner product, any orthogonal transformation preserves lengths ($|t(u)| = |u|$) and angles (the angle between $t(u)$ and $t(v)$ is that between u and v). In particular, $u \perp v$ implies $t(u) \perp t(v)$. Therefore an orthogonal transformation carries an orthonormal list of vectors in U into an orthonormal list in U'. Since the inner product can be defined in terms of lengths [by (18)], $t: U \to U'$ is orthogonal if and only if $|t(u)| = |u|$ for every $u \in U$. The composite of two orthogonal maps, when defined, is clearly orthogonal.

For a single finite-dimensional inner product space U, any orthogonal map $t: U \to U$ is an isomorphism, so its inverse is orthogonal; moreover, the set of all orthogonal maps $t: U \to U$ (all automorphisms of the inner product space) is a group, the *orthogonal group* of U.

THEOREM 13. *Let U and U' be inner product spaces and* **b** *an orthonormal basis of n vectors in U. Then a linear map $t: U \to U'$ is orthogonal if and only if the list $tb_1, \cdots, tb_n$ is orthonormal in U'.*

Proof: The condition is clearly necessary for the orthogonality of t. Hence assume that the list $t\mathbf{b}$ is orthonormal and consider any vector $u = \sum b_i \xi_i$ in U. Since $\mathbf{b}$ is orthonormal, the length of u satisfies $|u|^2 = \sum \xi_i^2$. Now

$$|tu|^2 = \langle tu, tu \rangle = \left\langle \sum tb_i\xi_i, \sum tb_j\xi_j \right\rangle = \sum_{i,j} \xi_i \langle tb_i, tb_j \rangle \xi_j.$$

But $t\mathbf{b}$ is orthonormal, so this gives $|tu|^2 = \sum \xi_i^2$ and thus $|u| = |tu|$; the linear map t is indeed orthogonal.

COROLLARY. *For finite-dimensional inner product spaces U and U', there is an orthogonal isomorphism $U \cong U'$ if and only if* $\dim U = \dim U'$.

For, if U and U' have the same dimension, choose a orthonormal basis in each, and let $t: U \to U'$ be the linear map taking the first basis to the second.

To illustrate, we determine all orthogonal maps $t: \mathbf{R}^2 \to \mathbf{R}^2$, where the plane $\mathbf{R}^2$ has the standard inner product. A vector (ξ_1, ξ_2) of length 1 in $\mathbf{R}^2$ has $\xi_1^2 + \xi_2^2 = 1$, hence has the form $(\cos\theta, \sin\theta)$ for a unique angle θ in the range $0 \leqslant \theta < 2\pi$, in radians. There are only two vectors of length 1 orthogonal to this vector $(\cos\theta, \sin\theta)$; namely, the vectors $(-\sin\theta, \cos\theta)$ and $(\sin\theta, -\cos\theta)$. Now the unit vectors $(1, 0)$ and $(0, 1)$ are an orthonormal basis in $\mathbf{R}^2$, so any orthogonal $t: \mathbf{R}^2 \to \mathbf{R}^2$ is determined by the images of the unit vectors in one of the following two ways t or t':

$$\begin{aligned} t(1,0) &= (\cos\theta, \sin\theta), & t'(1,0) &= (\cos\theta, \sin\theta),\\ t(0,1) &= (-\sin\theta, \cos\theta); & t'(0,1) &= (\sin\theta, -\cos\theta). \end{aligned}$$

The first map t is counterclockwise rotation about the origin through the angle θ, while the second map t' may be described as reflection in the $(1, 0)$ axis followed by the previous rotation. (It may also be described as reflection in the line through the origin at an angle of $\theta/2$ with the $(1, 0)$ axis.) Hence, any orthogonal transformation of the plane is either a rotation or a reflection. The matrix of the map t, relative to the basis of unit vectors, has columns the transforms of the unit vectors, so is one of the following matrices:

$$\begin{bmatrix} \cos\theta & -\sin\theta \\ \sin\theta & \cos\theta \end{bmatrix}, \qquad \begin{bmatrix} \cos\theta & \sin\theta \\ \sin\theta & -\cos\theta \end{bmatrix}. \tag{28}$$

Such matrices are called *orthogonal*: the columns are normal and orthogonal for the standard inner product in $\mathbf{R}^2$.

DEFINITION. An $n \times n$ matrix A of real numbers is orthogonal *when its columns are orthonormal in the standard inner product of* $\mathbf{R}^n$.

This definition is so chosen that a transformation is orthogonal when its matrix is. In detail, let $\mathbf{b}$ be any orthonormal basis for a finite-dimensional inner product space U. Then, by the theorem above, a linear transformation $t:U \to U$ is orthogonal if and only if its matrix (rel $\mathbf{b}$) is an orthogonal matrix.

The conditions that an $n \times n$ matrix A be orthogonal read

$$\sum_{i=1}^{n} A_{ij}A_{ij} = 1, \qquad \sum_{i=1}^{n} A_{ij}A_{ik} = 0, \qquad j \neq k, \quad j, k \in \mathbf{n} \tag{29}$$

(column j has length 1 and is orthogonal to every other column k). They may be recast in many equivalent forms:

Theorem 14. *The following conditions on a square matrix A are equivalent:*

(*i*) *A is orthogonal,*
(*ii*) $A^TA = I, \quad AA^T = I,$
(*iii*) *Each row of A has length 1, and any two different rows are orthogonal.*

Proof: The conditions (29) read $\Sigma_i (A^T)_{ji}A_{ik} = \delta_{jk}$; they state that $A^TA = I$; in words, *the transpose of an orthogonal matrix is its* (left) *inverse*. Since any left inverse of a square matrix is a right inverse, and conversely, this condition becomes $AA^T = I$. The latter matrix product, written out, is

$$\sum_{k=1}^{n} A_{ik}A_{ik} = 1, \qquad \sum_{k=1}^{n} A_{ik}A_{jk} = 0, \qquad i \neq j. \tag{30}$$

These equations state that A has orthonormal rows, as in (iii) of the theorem. Moreover, *every orthogonal matrix is invertible.*

Since $(A^T)^T = A$, this theorem shows in particular that the inverse A^T of any orthogonal matrix is orthogonal. The set O_n of all $n \times n$ orthogonal matrices is therefore a group, called the *orthogonal group* in n dimensions. Since orthogonal transformations correspond to orthogonal matrices, the orthogonal group O_n is isomorphic to the orthogonal group of any inner product space U of n dimensions.

The determinant of an orthogonal matrix is ± 1, for $|A^T| = |A|$ and $A^TA = I$ imply $|A|^2 = 1$. The orthogonal matrices of determinant $+1$ form a subgroup SO_n of O_n called the *special orthogonal group*. It is a subgroup of index 2 in O_n, hence (§II.9) a normal subgroup of O_n.

Proposition 15. *Let $\mathbf{b}$ be a finite orthonormal basis for the inner product space U. Then a basis $\mathbf{c}$ of U is orthonormal if and only if the change matrix P from $\mathbf{b}$ to $\mathbf{c}$ is orthogonal.*

Proof: The change matrix Q from $\mathbf{c}$ to $\mathbf{b}$ is the square matrix having entries $Q_{ij} = x_i c_j$, with jth column the coordinates (rel $\mathbf{b}$) of the jth vector of the list **c**. Therefore, Q is orthogonal if and only if these vectors **c** are orthonormal. But Q^{-1} is the change matrix from **b** to **c**, hence the conclusion.

Let U be an inner product space. A "similarity transformation" $t: U \to U$ is a linear transformation which multiplies the lengths of all vectors by some real scalar $\kappa \neq 0$, so that $|tu| = \kappa|u|$ for every vector $u \in U$. Under composition, the set of all similarity transformations on U is a group, called the "similarity group" of U; it contains the orthogonal group of U as a subgroup; in fact, as a normal subgroup, for the orthogonal group is the kernel of that morphism of groups which maps each similarity transformation to the corresponding scalar κ (in the multiplicative group of non-zero real numbers).

Equivalence under the action of the orthogonal group on an inner product space U has the usual meaning (§II.7). Thus two vectors u and u' of U are *orthogonally equivalent* when there is an orthogonal transformation $t: U \to U$ with $t(u) = u'$, and two linear forms $f, f': U \to \mathbf{R}$ are *orthogonally equivalent* when there is an orthogonal transformation $t: U \to U$ with $f = f' \circ t$. Two vectors are orthogonally equivalent if and only if they have the same length, as one may prove by choosing suitable orthonormal bases. The corresponding result holds for forms:

Proposition 16. *Two linear forms on a finite-dimensional inner product space are orthogonally equivalent if and only if they have the same norm.*

Proof: Since $|f|$ is the maximum of $|f(u)|/|u|$, by Proposition 12, two orthogonally equivalent linear forms $f, f': U \to \mathbf{R}$ necessarily have the same norm. Conversely, suppose that f and f' are given with the same norm. By the isomorphism of Theorem 9, we may write f as $\langle v, - \rangle$ and f' as $\langle v', - \rangle$ where $|f| = |v|$ and hence where $|v| = |v'|$. Therefore, there is an orthogonal transformation $t: U \to U$ with $tv' = v$, and so

$$f = \langle v, - \rangle = \langle tv, t - \rangle = \langle v', t - \rangle = f' \circ t.$$

Thus, f and f' are orthogonally equivalent.

This result actually provides a canonical form for a linear form f under the action of the orthogonal group of U: Given a linear form $f: U \to \mathbf{R}$, there is an orthonormal basis of U such that $f = x_n \kappa$, where $\kappa = |f|$ and x_n is the last linear form in the dual basis **x**.

EXERCISES

1. For each of the 2×2 matrices of (28), show explicitly that the transpose is the inverse.
2. Find orthogonal 3×3 matrices with the following first rows: $(1/3, 2/3, 2/3)$, $(0.6, 0.8, 0)$, $(5/13, 0, 12/13)$.

3. Prove that the direct sum of two orthogonal matrices is orthogonal.

4. Show that SO_2 consists of the rotations in O_2.

5. Show that $A \mapsto (A^{-1})^T$ is an automorphism of the general linear group. For which A is $A = (A^{-1})^T$?

6. Find all 3×3 "orthogonal" matrices over the field $\mathbf{Z}_2$.

7. Show that two pairs (u, v) and (u', v') of vectors in the inner product space U are equivalent under the orthogonal group if and only if $|u| = |u'|$, $|v| = |v'|$, and $\langle u, v\rangle = \langle u', v'\rangle$.

8. Show that an $n \times n$ matrix A is orthogonal if and only if t_A leaves the form $x_1^2 + \cdots + x_n^2$ invariant.

9. A Lorentz transformation is defined to be a linear transformation $t:\mathbf{R}^4 \to \mathbf{R}^4$ which leaves the quadratic form $x_1^2 + x_2^2 + x_3^2 - x_4^2$ invariant. Show that a 4×4 real matrix A defines a Lorentz transformation t_A if and only if $|A| \neq 0$ and $A^{-1} = DA^TD^{-1}$, where D is the special diagonal matrix with the diagonal entries 1, 1, 1, and -1.

10. Give necessary and sufficient conditions on AA^T that a matrix A represent a similarity transformation relative to a given orthonormal basis.

11. Show that a similarity transformation multiplies all inner products by the same factor.

***12.** If U is a finite-dimensional inner product space, show that a non-zero linear transformation $t: U \to U$ is a similarity transformation if and only if it preserves orthogonality ($u \perp v$ implies $tu \perp tv$).

8. The Principal Axis Theorem

A self-adjoint linear transformation $t: U \to U$ on an inner product space U is represented, relative to an orthonormal basis of U, by a symmetric matrix A. The problem of this section is to represent t by a simpler matrix by choosing a suitable new orthonormal basis of U. This will be done by using eigenvalues of A.

LEMMA. *The characteristic polynomial of a real symmetric matrix A is a product of real linear factors.*

Proof: The $n \times n$ matrix A of real numbers may be regarded as a matrix of complex numbers (which happen all to be real); this amounts to making a change of rings, in the sense of §IX.11, from $\mathbf{R}$ to $\mathbf{C}$. By the fundamental theorem of algebra, the characteristic polynomial $|A - \lambda I|$ of A has complex roots λ_i, so is a product of linear factors $\lambda - \lambda_i$. Moreover, each λ_i is an eigenvalue of the complex-linear transformation $t_A: \mathbf{C}^n \to \mathbf{C}^n$. If λ denotes any one of these eigenvalues and $\xi \in \mathbf{C}^n$ a corresponding eigenvector, regarded as a column matrix, the equation $A\xi = \xi\lambda$ states that ξ is an eigenvector. Now λ and each component ξ_i of ξ is a complex number; write λ^* and ξ_i^* for their complex conjugates, and ξ^* for the column of the ξ_i^*. In the matrix product $A\xi = \xi\lambda$ take the complex conjugate of every term. Since A is real and since $\kappa \mapsto \kappa^*$ is an automorphism of fields, this gives the matrix equation $A\xi^* = \lambda^*\xi^*$; transposed, this is $\xi^{*T}A = \xi^{*T}\lambda^*$, where ξ^* is in the row $(\xi_1^* \cdots \xi_n^*)$. Now we can calculate the 1×1 matrix $\xi^{*T}A\xi$ (a scalar) in two ways:

$$\xi^{*T}A\xi = \xi^{*T}(A\xi) = \xi^{*T}\xi\lambda = \lambda(\xi^{*T}\xi),$$
$$\xi^{*T}A\xi = (\xi^{*T}A)\xi = (\xi^{*T}\lambda^*)\xi = \lambda^*(\xi^{*T}\xi).$$

But $\xi \neq 0$ implies $\xi^{*T}\xi = \Sigma\xi_i^*\xi_i > 0$, so the equality of these two results implies $\lambda = \lambda^*$. This means that λ is real, as required.

This lemma can be stated more briefly as: *Every eigenvalue of a real symmetric matrix is real.* If $\lambda_1, \cdots, \lambda_k$ are the distinct eigenvalues (all real) of A, the characteristic polynomial of A is

$$|A - \lambda I| = (-1)^n(\lambda - \lambda_1)^{m_1}(\lambda - \lambda_2)^{m_2} \cdots (\lambda - \lambda_k)^{m_k}. \tag{31}$$

Here the exponent m_i is the *multiplicity* of the eigenvalue λ_i, and

$$n = m_1 + m_2 + \cdots + m_k \qquad (A \text{ is } n \times n).$$

If A is similar to any diagonal matrix D, that matrix must have the same characteristic polynomial, so the diagonal entries of D must be $\lambda_1, \cdots, \lambda_k$, each *appearing with its multiplicity* (that is, λ_i appears m_i times). As in Theorem VII.19, the multiplicity m_i may then also be described as the dimension of the eigenspace of λ_i. These eigenvalues are the eigenvalues of the self-adjoint t represented by A.

THEOREM 17. (*Principal Axis Theorem*). *In a finite-dimensional inner product space U, any self-adjoint $t: U \to U$ has a diagonal matrix relative to some orthonormal basis of U. The diagonal entries are the eigenvalues of t; they are real, and each appears as many times as its multiplicity.*

The proof is by induction on the dimension n of U, the case $n = 1$ being trivial. For U of dimension n, the Lemma shows that t has a real eigenvalue λ_1. Pick a corresponding eigenvector $c_1 \neq 0$, so that $tc_1 = c_1\lambda_1$. Normalize c_1, so that $|c_1| = 1$. Let $S \subset U$ be the orthogonal complement of c_1. Since t is self-adjoint, $\langle tu, c_1\rangle = \langle u, tc_1\rangle = \langle u, c_1\lambda_1\rangle = \lambda_1\langle u, c_1\rangle$ for any vector $u \in U$. In particular, $u \perp c_1$ implies $t(u) \perp c_1$, so t restricted to S is a linear transformation $t_S: S \to S$. Also, t_S is self-adjoint (because $t_S(v) = t(v)$ for every $v \in S$). The induction assumption applied to S now produces an orthonormal basis $c_2, \cdots, c_n$ for S for which t_S is diagonal. Adjoining c_1 gives an orthonormal basis for all of U, with t diagonal, as required.

This theorem can be translated into a result about matrices. Suppose that $t:U \to U$ is represented, relative to some orthonormal basis **b** of U, by a matrix A: since t is self-adjoint, A is symmetric. If **c** is any other orthonormal basis of U, the change matrix P from **b** to **c** must be orthogonal, and the matrix of t (rel **c**) is PAP^{-1}. By the theorem we can choose P so that PAP^{-1} is diagonal. Hence, the theorem now reads as follows:

Theorem 18. *For any real symmetric matrix A there is an orthogonal matrix P with PAP^{-1} diagonal. The (real) diagonal entries are the eigenvalues of A, each appearing as many times as its multiplicity.*

The "geometric" proof given for the previous theorem can be translated into a direct proof, matrix style, for this theorem. This theorem often is stated: "Any real symmetric matrix is orthogonally equivalent to a diagonal matrix". Here A *orthogonally equivalent* to B means that $B = PAP^{-1}$ for some orthogonal P.

Now P orthogonal means that $P^{-1} = P^T$; so $B = PAP^T$. Therefore, orthogonally equivalent symmetric matrices B and A are congruent, in the sense defined in §2 and relevant to quadratic forms; indeed, if the quadratic form $q:U \to \mathbf{R}$ has matrix A, relative to the orthonormal basis **b**, then the change by P from **b** to the basis **c** replaces A by P^TAP. In this guise the theorem reads:

Theorem 19. *For any quadratic form $q:U \to \mathbf{R}$ on an n-dimensional inner product space U there exists a choice of orthonormal coordinates x_i for which q is diagonal:*

$$q = \lambda_1 x_1{}^2 + \cdots + \lambda_n x_n{}^2. \tag{32}$$

The coefficients $\lambda_1, \cdots, \lambda_n$ are the eigenvalues of (the matrix of) q, each appearing as many times as to its multiplicity.

All three of these theorems are immediately equivalent to each other. The last form explains the phrase "principal axes". For example, if $n = 2$, the quadratic one q determines as in §4 a locus or "central conic" consisting of all those vectors u with $q(u) = 1$. The diagonal form is $\lambda x^2 + \mu y^2 = 1$. If λ and μ are both positive, this is the standard form $(x^2/a^2 + y^2/b^2 = 1)$ of an ellipse, and the eiegenvalues λ and μ determine the lengths a and b of the semiaxes of the ellipse. Indeed, the orthonormal basis needed to diagonalize q consists of unit vectors along the "principal axes" of the ellipse. If λ is positive and μ negative, one similarly has the principal axes for a hyperbola.

There is also a direct proof for this version of the theorem in terms of properties of the quadratic form q. First examine the expression (32). By permuting the coordinates we can assume that the eigenvalues are arranged in order of magnitude, as $\lambda_1 \geqslant \lambda_2 \geqslant \cdots \geqslant \lambda_n$. Now restrict the function $q: U \to \mathbf{R}$ to the unit sphere S^{n-1}, consisting of all vectors u with $|u| = 1$, and with equation $x_1^2 + \cdots + x_n^2 = 1$. From (32) it is clear that λ_1 is the maximum value of q on S^{n-1}, and that q takes on this maximum at the first basis vector b_1 (with $x_1 = 1$, $x_2 = \cdots = x_n = 0$).

This analysis suggests the proof. Any real quadratic form q on U is a continuous function on S^{n-1}; the set S^{n-1} is closed and bounded in the sense of §VIII.6. Therefore, the least upper bound of q on S^{n-1} is its value at some vector b_1 on S^{n-1}. Choosing b_1 as the first basis vector in an orthonormal basis of U, an application of the usual rules for the maximum of the function q will show that q has in these coordinates the expression

$$q(u) = \lambda_1 x_1^2 + q'(u), \qquad \text{where } q' \text{ depends only on } x_2, \cdots, x_n.$$

An induction on n then yields the desired canonical form (32).

Consider in more detail a basis $\mathbf{c}$ for which a self-adjoint transformation $t: U \to U$ becomes diagonal. If the eigenvalue λ_1 appears m_1 times on the diagonal, the corresponding m_1 basis vectors, say $c_1, \cdots, c_{m_1}$, are all eigenvectors for the eigenvalue λ_1; as in Theorem VII.19, they span the eigenspace U_1 of λ_1 (the subspace of all u with $tu = u\lambda_1$). If λ_i is a different eigenvalue, its eigenspace U_i is spanned by different basis vectors, so U_1 is orthogonal to U_i. All told, this gives the following version of the theorem (called the "spectral theorem"; the *spectrum* of any linear transformation t on a finite-dimensional space is defined to be the set of its eigenvalues).

Theorem 20 (*Spectral Theorem*). *If the self-adjoint linear transformation $t: U \to U$ for U finite-dimensional has the distinct eigenvalues λ_i with corresponding eigenspaces U_i, $i = 1, \cdots, k$, then U is the direct sum of its subspaces U_i, U_i and U_j are orthogonal for $i \neq j$, and the dimension of U_i is the multiplicity of the eigenvalue λ_i.*

Put differently: $U \cong U_1 \oplus \cdots \oplus U_k$, and t restricted to U_i is the linear transformation "multiplication by the scalar λ_i."

We now have four versions of the principal axis theorem; there are still others, adapted (like Theorem 20) to infinite-dimensional generalizations.

Proposition 21. *Any invertible real matrix C can be written as a matrix product $C = SQ$ with S symmetric and positive definite and Q orthogonal.*

Proof: Since C is invertible, the real symmetric matrix CC^T is positive definite by Proposition 6. By the principal axis theorem, $CC^T = PEP^T$, where E is diagonal and P orthogonal. As the diagonal entries of E are positive, we can extract the positive square root of each to get $E = D^2$ with

D diagonal. Thus $S = PDP^T$ is symmetric and positive definite, while

$$CC^T = S^2.$$

If we show $S(C^T)^{-1}$ orthogonal, we are done. But this is immediate:

$$S(C^T)^{-1}\left[(C^T)^{-1}\right]^T S^T = S(CC^T)^{-1}S^T = SS^{-2}S = I.$$

EXERCISES

1. Find an orthogonal matrix P reducing each of the following symmetric matrices to a diagonal matrix.

$$\text{(a)} \begin{bmatrix} 5 & -3 \\ -3 & 5 \end{bmatrix}; \qquad \text{(b)} \begin{bmatrix} 2 & 2\sqrt{3} \\ 2\sqrt{3} & -2 \end{bmatrix}.$$

2. If $t: U \to U$ is self-adjoint and u_1, u_2 are eigenvectors to different eigenvalues λ_i and λ_2 of t, prove directly that $u_1 \perp u_2$.

3. Give a direct proof of the matrix form (Theorem 18) of the principal axis theorem, showing that for any eigenvalue λ_1 there is an orthogonal matrix Q with

$$QAQ^{-1} = \begin{bmatrix} \lambda_1 & * \\ O & D \end{bmatrix},$$

and proceeding by induction.

4. Prove that every real skew-symmetric matrix A has the form $A = PBP^{-1}$ where P is orthogonal and B^2 is diagonal.

5. To the quadratic form $9x_1^2 - 9x_2^2 + 18x_3^2$ apply the orthogonal transformation

$$3x_1 = 2y_1 - y_2 + 2y_3, \quad 3x_2 = -y_1 + 2y_2 + 2y_3, \quad 3x_3 = 2y_1 + 2y_2 - y_3.$$

For the resulting expression, show directly that the vector with y-coordinates $(\frac{2}{3}, \frac{2}{3}, -\frac{1}{3})$ yields the maximum value when $y_1^2 + y_2^2 + y_3^2 = 1$.

***6.** (a) If $C = SQ$ with S symmetric and Q orthogonal, prove that $S^2 = CC^T$.

(b) For C real and invertible, show that there is only one positive definite symmetric matrix S which satisfies $S^2 = CC^T$ and hence $C = SQ$.

***7.** (a) Show that any real 3×3 matrix A has a real eigenvector.

(b) If A is an orthogonal 3×3 matrix, show that there is an orthogonal matrix P such that

$$PAP^T = \begin{bmatrix} \pm 1 & 0 \\ 0 & B \end{bmatrix},$$

with the 2×2 matrix B orthogonal.

(c) Using known results about 2×2 orthogonal matrices, show that any 3×3 orthogonal matrix A with $|A| > 0$ has an eigenvalue $+1$ and is a rigid rotation. (This is Euler's theorem.)

9. Unitary Spaces

In the next two sections we study inner products and related concepts for vector spaces over the field $\mathbf{C}$ of all complex numbers. The automorphism $\mathbf{C} \to \mathbf{C}$ which maps each complex number $\kappa = \alpha + i\beta$ (α, β real) to its complex conjugate $\kappa^* = \alpha - i\beta$ will play a special role. For instance, if V is a vector space over $\mathbf{C}$, a function $g: V \to \mathbf{C}$ will be called a *conjugate linear form* on V when, for all $\kappa_i \in \mathbf{C}$ and all $v_i \in V$,

$$g(v_1\kappa_1 + v_2\kappa_2) = g(v_1)\kappa_1^* + g(v_2)\kappa_2^*. \tag{33}$$

These conjugate linear forms will be used in addition to the (usual) linear forms $f: V \to \mathbf{C}$, which satisfy $f(v_1\kappa_1 + v_2\kappa_2) = f(v_1)\kappa_1 + f(v_2)\kappa_2$. The linear forms f are the elements of the dual space V^*, while the conjugate linear forms g can be regarded as the elements of an "adjoint" space to V (Exercise 13).

In $\mathbf{R}^n$, the basic fact that the standard inner product gives a positive definite quadratic form rests on the observation that the square of a nonzero real number is positive. For a complex number $\kappa \neq 0$ the corresponding positive quantity is the product $\kappa\kappa^*$ of κ and its conjugate. Thus, in the complex vector space $\mathbf{C}^n$ of all lists $\xi = (\xi_1, \cdots, \xi_n)$ of n complex numbers the *standard inner product* of two lists ξ and η is the complex number

$$\langle \xi, \eta \rangle = \xi_1\eta_1^* + \xi_2\eta_2^* + \cdots + \xi_n\eta_n^*. \tag{34}$$

This is positive definite, in the sense that $\xi \neq 0$ implies $\langle \xi, \xi \rangle$ real and positive. With this inner product, $\mathbf{C}^n$ is a unitary space according to the following definition.

DEFINITION. A unitary space U is a vector space over the field $\mathbf{C}$ *of complex numbers equipped with a function $U \times U \to \mathbf{C}$ which assigns to vectors u and v a complex number $\langle u, v \rangle$ as their hermitian* inner product *so that the following properties hold (for all vectors u, u_1, u_2, and v and all scalars κ_1 and κ_2):*

Left linearity: $\langle u_1\kappa_1 + u_2\kappa_2, v \rangle = \langle u_1, v \rangle\kappa_1 + \langle u_2, v \rangle\kappa_2$;
Conjugate symmetry: $\langle v, u \rangle = \langle u, v \rangle^*$;
Definiteness: $u \neq 0$ *implies that* $\langle u, u \rangle$ *is real and positive.*

Thus left linearity states that, for each v, $u \mapsto \langle u, v \rangle$ is a linear form $U \to \mathbf{C}$; by conjugate symmetry, this implies for each $u \in U$ that $v \mapsto \langle u, v \rangle$ is a conjugate linear form; in other words, that

$$\langle u, v_1\kappa_1 + v_2\kappa_2 \rangle = \langle u, v_1 \rangle\kappa_1^* + \langle u, v_2 \rangle\kappa_2^*. \tag{35}$$

A form $\langle u, v\rangle$ with these two linearity properties is sometimes called "sesquilinear".

The complex numbers themselves form a one-dimensional unitary space, with inner product $\langle \kappa, \lambda\rangle = \kappa\lambda^*$; this is just the space $\mathbf{C}^1$ with the standard inner product (34).

In any unitary space U, the *length* of a vector $|u|$ is again defined to be the nonnegative (real) square root $[\langle u, u\rangle]^{1/2}$. The inner product may again be expressed in terms of lengths (this is the process of "polarization"), for expanding both $|u + v|^2$ and $|u + iv|^2$ by linearity gives

$$2\langle u, v\rangle = |u + v|^2 + i|u + iv|^2 - (1 + i)(|u|^2 + |v|^2).$$

Two vectors u and v are called *orthogonal* (in symbols $u \perp v$) when $\langle u, v\rangle = 0$; by conjugate symmetry, this is equivalent to $v \perp u$. Any vector subspace S of U is a unitary space, with inner product the restriction to S $\times$ S of the inner product in U. The *orthogonal complement* of a subspace S is again the set of all $u \in U$ with $u \perp s$ for every $s \in$ S, and is itself a (unitary) subspace of U.

Theorem 22. *For vectors u and v in a unitary space,*

$$|\langle u, v\rangle| \leqslant |u|\cdot|v| \qquad (\textit{the Schwarz inequality}). \tag{36}$$

Proof: As in (VIII.25), we may write the complex number $\langle u, v\rangle$ as $r(\cos\theta + i\sin\theta) = re^{i\theta}$, with θ and $r = |\langle u, v\rangle|$ real. Now for any real t, positive definiteness of the inner product gives

$$0 \leqslant \langle ut + ve^{i\theta}, ut + ve^{i\theta}\rangle = |u|^2t^2 + \langle u, v\rangle e^{-i\theta}t + \langle u, v\rangle^* e^{i\theta}t + |v|^2.$$

But $\langle u, v\rangle e^{-i\theta} = r = \langle u, v\rangle^* e^{i\theta}$, so this becomes

$$0 \leqslant |u|^2t^2 + 2|\langle u, v\rangle|t + |v|^2.$$

The condition that this real quadratic polynomial in t be non-negative is exactly (36), Q.E.D.

A list $\mathbf{c}$ of k vectors $c_1, \cdots, c_k$ in U is called *orthonormal* when each has length 1 and every two are orthogonal; as in the real case this means that $\langle c_i, c_j\rangle = \delta_{ij}$ for all i and j, and implies linear independence. If the number k of vectors in the list $\mathbf{c}$ is the dimension of U the list is a basis of U, called an *orthonormal basis*. The Gram–Schmidt process applies: If $\mathbf{c}$ is an orthonormal list, and u is any vector, then $u - \sum_i c_i\langle u, c_i\rangle$ is orthogonal to each of $c_1, \cdots, c_k$, and, if not 0, may be "normalized" to have length 1. This process proves

Theorem 23. *Any orthonormal list in a finite-dimensional unitary space is part of an orthonormal basis. In particular, every such unitary space has an orthonormal basis.*

Let $\mathbf{b}$ be an orthonormal basis of the n-dimensional unitary space U, so that $\langle b_i, b_j\rangle = \delta_{ij}$. Each vector $u \in U$ then has the form $u = \sum b_i\xi_i$ for a (unique) list ξ of complex scalars. By left linearity,

$$\langle u, b_j\rangle = \left\langle \sum b_i\xi_i, b_j \right\rangle = \xi_j.$$

Hence, the linear form $\langle -, b_j\rangle$ (inner product with b_j) is exactly the linear form $x_j: U \to \mathbf{C}$ which assigns to each vector $\sum b_i\xi_i$ its jth coordinate ξ_j, taken rel $\mathbf{b}$. If $v = \sum b_j\eta_j$ is a second vector of U, the inner product $\langle u, v\rangle$ is, by linearity and conjugate linearity,

$$\left\langle \sum_i b_i\xi_i, \sum_j b_j\eta_j \right\rangle = \xi_1\eta_1^* + \cdots + \xi_n\eta_n^*.$$

This is just the standard inner product formula (34) for $\mathbf{C}^n$; in other words, each orthonormal basis $\mathbf{b}$ of U yields an isomorphism (of vector spaces and inner products) $\mathbf{C}^n \cong U$, via $\xi \mapsto \sum b_i\xi_i$. A unitary space U cannot be identified with its dual vector space (as in Theorem 9 for inner product spaces); instead of linear forms (elements of the dual), we make the following use of conjugate linear forms.

Lemma. *If $g: U \to \mathbf{C}$ is a conjugate linear form on a finite-dimensional unitary space U, then there is a unique vector $u \in U$ with $g(v) = \langle u, v\rangle$ for all $v \in U$.*

Proof: First, we show that there can be at most one such u. If $\langle u, v\rangle = \langle u', v\rangle$ for all v, then $\langle u - u', v\rangle = 0$ for all v. By positive definiteness (set $v = u - u'$ in this formula) this implies $u = u'$, as asserted.

To show that u exists for a given g, take an orthonormal basis $\mathbf{b}$ and set $u = \sum_i b_i g(b_i)$. Then for any vector $v = \sum b_j\eta_j$,

$$\langle u, v\rangle = \sum_i g(b_i)\left\langle b_i, \sum_j b_j\eta_j \right\rangle = \sum_i g(b_i)\eta_i^*.$$

But g is conjugate linear, so the result is $\langle u, v\rangle = g(\sum b_i\eta_i) = g(v)$, as required.

Using this lemma, we can construct an "adjoint" for a linear map as follows:

Theorem 24. *If $t: U \to U'$ is a linear transformation between finite-dimensional unitary spaces U and U', there is a unique linear transformation $t^*: U' \to U$ such that*

$$\langle t^*u', v\rangle = \langle u', tv\rangle \qquad \text{for all } u' \in U', \quad v \in U. \tag{37}$$

Proof: For each u', the assignment $v \mapsto \langle u', tv\rangle$ is a conjugate linear form on U, so by the lemma there is a unique vector u in U with

$\langle u, v\rangle = \langle u', tv\rangle$. The assignment $u' \mapsto u$ defines the function $t^*: U' \to U$ satisfying (37); it remains to prove this unique function linear. For vectors u_1' and u_2' and scalars κ_1 and κ_2, the definition of t^* and left linearity of the inner product imply that

$$\langle t^*(u_1'\kappa_1 + u_2'\kappa_2), v\rangle = \langle u_1'\kappa_1 + u_2'\kappa_2, tv\rangle = \langle u_1', tv\rangle\kappa_1 + \langle u_2', tv\rangle\kappa_2;$$

the same facts also give

$$\begin{aligned}\langle (t^*u_1')\kappa_1 + (t^*u_2')\kappa_2, v\rangle &= \langle t^*u_1', v\rangle\kappa_1 + \langle t^*u_2', v\rangle\kappa_2\\ &= \langle u_1', tv\rangle\kappa_1 + \langle u_2', tv\rangle\kappa_2.\end{aligned}$$

The results are equal, so, by the uniqueness assertion of the lemma, we have $t^*(u_1'\kappa_1 + u_2'\kappa_2) = (t^*u_1')\kappa_1 + (t^*u_2')\kappa_2$, and hence t^* is linear, as required.

This map t^*, as defined by (37), is called the *adjoint* of t. If $t: U \to U$ is an endomorphism with $t^* = t$, it is called *self-adjoint*.

The asterisk has been used here both to denote the complex conjugate of a complex number and the adjoint of a transformation. To justify this, let us for the moment write $\bar{\kappa}$ for the ordinary complex conjugate of the complex number κ. The conjugate linearity of the inner product of vectors v and u in a unitary space then gives $\langle v, u\kappa\rangle = \langle v, u\rangle\bar{\kappa} = \langle v\bar{\kappa}, u\rangle$. Hence, the linear map "scalar multiple by κ" has as adjoint the linear map "scalar multiple by $\bar{\kappa}$". This shows (after the fact!) that our use of * for the complex conjugate is consistent with the use of * for the adjoint.

Given orthonormal bases $\mathbf{b}$ and $\mathbf{b}'$ of n and m vectors, respectively, for unitary spaces U and U', the $m \times n$ matrix A (rel $\mathbf{b}$, $\mathbf{b}'$) of a linear map $t: U \to U'$ is defined, as always, from the corresponding coordinates $\mathbf{x}'$ as $A_{ij} = x_i'(tb_j)$; since $x_i' = \langle -, b_i'\rangle$, this is $A_{ij} = \langle tb_j, b_i'\rangle$. Correspondingly, the matrix B (rel $\mathbf{b}'$, $\mathbf{b}$) of the adjoint t^* is the $n \times m$ matrix B with the entries

$$B_{ji} = \langle t^*b_i', b_j\rangle = \langle b_i', tb_j\rangle = \langle tb_j, b_i'\rangle^* = (A_{ij})^*.$$

So define the *hermitian conjugate* A^{*T} of the $m \times n$ matrix A of complex numbers by

$$(A^{*T})_{ji} = (A_{ij})^*; \tag{38}$$

thus A^{*T} is the matrix obtained from A by transposing *and* taking the conjugate of each entry (so is the "conjugate transpose" of A). The calculation just made has proved

Proposition 25. *Relative to the same pair of orthonormal bases, the matrix of the adjoint of a map is the hermitian conjugate of the matrix of the map.*

If U and U' are unitary spaces a *unitary transformation* $s: U \to U'$ (also called a *morphism of unitary spaces*) is a linear transformation of U to U' such that $\langle su, sv \rangle = \langle u, v \rangle$ for all vectors $u, v \in U$. This condition in the finite-dimensional case is equivalent to $\langle s^*su, v \rangle = \langle u, v \rangle$, so s^*s is the adjoint of the identity map, hence is itself the identity. Thus s is unitary if and only if $s^*s = 1$. Hence, any unitary transformation is a monomorphism; moreover, by definition any unitary transformation preserves lengths.

Now consider just a single finite-dimensional unitary space U. Each endomorphism $t: U \to U$ then has as adjoint an endomorphism $t^*: U \to U$; by the definition (37) and conjugate symmetry, it satisfies both the equations

$$\langle t^*u, v \rangle = \langle u, tv \rangle, \qquad \langle u, t^*v \rangle = \langle tu, v \rangle \tag{39}$$

for all vectors $u, v \in U$. In case t is unitary, it is a monomorphism, as just noted. By finite dimensionality, it is then an isomorphism, with $t^*t = 1$, and its adjoint t^* is therefore its inverse—and this adjoint is therefore also unitary.

There is a corresponding terminology for square matrices of complex numbers. A square matrix H is called *hermitian* when $H^{*T} = H$; this means that H is the matrix—as always here, relative to some orthonormal basis—of a self-adjoint map. A square matrix L is called *unitary* when $L^{*T}L = I$; since products of matrices correspond to composites of maps, this means that L is the matrix of a unitary endomorphism. As in the real case, $L^{*T}L = I$ is equivalent to the requirement that $LL^{*T} = I$, or to the requirement that the rows (the columns) of L are orthonormal in the standard inner product. The set U_n of all $n \times n$ unitary matrices under matrix multiplication is a group, the *unitary group*. For unitary change matrices, the analog of Proposition 15 holds.

EXERCISES

1. For vectors u and v in a unitary space, prove that

$$|u + v|^2 + |u - v|^2 = 2|u|^2 + 2|v|^2.$$

2. Prove the pythagorean law in a unitary space: If $u \perp v$, then

$$|u + v|^2 = |u|^2 + |v|^2.$$

3. For complex numbers x_1, x_2, y_1, and y_2, prove that

$$(x_1 y_1^* + x_2 y_2^*)(x_1^* y_1 + x_2^* y_2) \leqslant (x_1 x_1^* + x_2 x_2^*)(y_1 y_1^* + y_2 y_2^*).$$

4. In $\mathbf{C}^3$, find an orthonormal basis for the subspace spanned by the vector $(1, i, 1 + i)$ and for its orthogonal complement.

5. Which of the following matrices are unitary or hermitian?

$$\begin{bmatrix} (1+i)/2 & (1-i)/2 \\ (1-i)/2 & (1+i)/2 \end{bmatrix}, \qquad \begin{bmatrix} 2 & 1-3i \\ 1+3i & 5 \end{bmatrix}, \qquad \begin{bmatrix} 1 & i \\ i & 1 \end{bmatrix}.$$

6. For ω a primitive nth root of unity, show that the $n \times n$ matrix D with $D_{ij} = \omega^{ij} n^{-1/2}$ is unitary.

7. State and prove the analog of Theorem 14 for unitary matrices.

8. From the definition of an adjoint, prove that $(s \circ t)^* = t^* \circ s^*$.

9. Give another method of polarization, by proving that

$$4\langle u, v\rangle = |u + v|^2 - |u - v|^2 + i|u + iv|^2 - i|u - iv|^2.$$

10. Show that a unitary space is a metric space if the distance from u to v is defined to be $|u - v|$.

11. If the list $c_1, \cdots, c_k$ is orthonormal, while $\kappa_i = \langle u, c_i\rangle$, prove that $|\kappa_1|^2 + \cdots + |\kappa_k|^2 \leqslant |u|^2$. (This is Bessel's inequality; the κ_i are called the "Fourier coefficients" of the vector u relative to the list **c**.)

12. (a) If V is a vector space of dimension n over the field **C** of complex numbers, show under the operations of addition and multiplication by real scalars that V is a vector space of dimension $2n$ over the field **R** of real numbers, and that scalar multiplication by i defines an **R**-linear endomorphism $t: V \to V$ with $t^2 = -1_V$.

(b) Conversely, if W is a finite-dimensional vector space over **R**, while $t: W \to W$ is **R**-linear with minimal polynomial $x^2 + 1$ over **R**, show that the dimension of W over **R** is even and that W becomes a vector space over **C** when iw is defined to be tw.

13. (a) If V is a vector space over **C**, show that the set $V^{\#}$ of all conjugate linear forms $g: V \to \mathbf{C}$ is also a vector space over **C** under the linear combinations defined (pointwise) by $(\lambda_1 g_1 + \lambda_2 g_2)(v) = \lambda_1 g_1(v) + \lambda_2 g_2(v)$.

(b) For $t: V \to V'$ linear, construct $t^{\#}: V'^{\#} \to V^{\#}$ so that the assignments $V \mapsto V^{\#}$ and $t \mapsto t^{\#}$ define a functor (on the category of vector spaces over **C** to itself).

(c) If U is unitary and finite-dimensional, show that $(\varphi u)(v) = \langle u, v\rangle$ defines an isomorphism $\varphi: U \cong U^{\#}$ of vector spaces.

(d) Use these constructions to give a proof of Theorem 24 like that of Theorem 10 for inner product spaces.

10. Normal Matrices

If U is a unitary space, a linear transformation $t: U \to U$ is said to be *normal* when it commutes with its adjoint: $tt^* = t^*t$. Similarly, a square matrix A of complex numbers is *normal* when $AA^{*T} = A^{*T}A$. Hence, t is normal if and only if its matrix, relative to some one (and thus to any) orthonormal basis of U, is normal. We will now show that these normal endomorphisms t are exactly those which can be diagonalized.

THEOREM 26. *If U is a finite-dimensional unitary space, then a linear transformation $t: U \to U$ is normal if and only if it has a diagonal matrix relative to some orthonormal basis of U.*

Proof: If D is a diagonal matrix, then clearly $DD^{*T} = D^{*T}D$; hence, if D is the matrix of t for an orthonormal basis, $tt^* = t^*t$, and t is normal.

Conversely, we will prove that a normal endomorphism t can be diagonalized, by induction on the dimension n of U. So assume the theorem to be true for all unitary spaces of dimension less than n, and consider $t: U \to U$ normal for a unitary space U of dimension n. By the fundamental theorem of algebra, the characteristic polynomial of t has a complex root λ, and λ is an eigenvalue of t. Pick a corresponding eigenvector v, taking v of length 1; then $\langle v, v\rangle = 1$ and $t(v) = v\lambda$.

We claim that $t^*(v) = v\lambda^*$; in other words, v is also an eigenvector for the adjoint transformation t^*. For consider the vector $u = t^*(v) - v\lambda^*$. The square of its length may be expanded, using the definition (37) of the adjoint t^* and (39), as

$$\begin{aligned}\langle u, u\rangle &= \langle t^*v - v\lambda^*, t^*v - v\lambda^*\rangle \\ &= \langle t^*v, t^*v\rangle - \langle v\lambda^*, t^*v\rangle - \langle t^*v, v\lambda^*\rangle + \langle v\lambda^*, v\lambda^*\rangle \\ &= \langle v, tt^*v\rangle - \langle t(v\lambda^*), v\rangle - \langle v, t(v\lambda^*)\rangle + \langle v, v\rangle\lambda\lambda^*.\end{aligned}$$

In the first term on the right, we use the hypothesis $tt^* = t^*t$, the rule $tv = v\lambda$, and the definition of the adjoint again to obtain

$$\begin{aligned}\langle u, u\rangle &= \langle tv, tv\rangle - \langle v\lambda\lambda^*, v\rangle - \langle v, v\lambda\lambda^*\rangle + \lambda\lambda^* \\ &= \lambda\lambda^* - \lambda\lambda^* - \lambda\lambda^* + \lambda\lambda^* = 0.\end{aligned}$$

By definiteness, u then is zero, and so $t^*(v) = v\lambda^*$.

The *orthogonal complement* of the eigenvector v is the subspace W of all those vectors $w \in U$ with $\langle w, v\rangle = 0$. Then $t(W) \subset W$ follows from the calculation that

$$\langle tw, v\rangle = \langle w, t^*v\rangle = \langle w, v\lambda^*\rangle = \langle w, v\rangle\lambda = 0.$$

Therefore, t restricted to W is a normal endomorphism of W; by the induction assumption, this restriction can be diagonalized for some orthonormal basis $b_2, \cdots, b_n$ of W. Since $t(v) = v\lambda$ and $v \perp b_i$ also, the list $v, b_2, \cdots, b_n$ is an orthonormal basis of U for which the matrix of t is diagonal, as required.

As in the case of any diagonal matrix for an endomorphism of a vector space (Theorem VII.19), the diagonal matrix D for t has as diagonal entries the eigenvalues of t, each appearing a number of times equal to its multiplicity.

Since a self-adjoint transformation ($t^* = t$) is automatically normal ($tt^* = t^*t$), Theorem 26 includes a complex analog of the principal axis theorem:

COROLLARY 1. *Any self-adjoint linear transformation $t: U \to U$ on a finite-dimensional unitary space U has a real diagonal matrix relative to some orthonormal basis of U.*

Corollary 2. *All eigenvalues of a self-adjoint t are real.*

Proof: The adjoint of a diagonal matrix D is the diagonal matrix D^{*T} with diagonal entries the complex conjugates of those of D. Thus $t = t^*$ gives $D = D^*$, and all the diagonal entries (the eigenvalues of t) must be real.

These results may be restated as theorems about matrices, as follows. Two $n \times n$ matrices A and B of complex numbers are said to be *unitary-equivalent* when there is an $n \times n$ unitary matrix L with $LAL^{*T} = B$. Since the adjoint L^{*T} of a unitary matrix L is its inverse, unitary-equivalent matrices are necessarily similar matrices: $LAL^{-1} = B$. Since the change matrix from one orthonormal basis to another such basis is a unitary matrix, two $n \times n$ matrices A and B represent the same linear transformation $t: U \to U$ relative to (possibly) different orthonormal bases of U if and only if they are unitary-equivalent. Then Theorem 26 takes the following form:

Corollary 3. *A square matrix A of complex numbers is unitary-equivalent to a diagonal matrix if and only if it is normal* ($AA^{*T} = A^{*T}A$).

In particular, the "principal axis" case (Corollary 1) is:

Corollary 4. *For every hermitian matrix H there is a unitary matrix L with LHL^{*T} diagonal and real.*

Corollary 5. *All eigenvalues of a hermitian matrix are real.*

Finally, the principal axis theorem has a "quadratic form" version, provided that we employ the appropriate notion of "form". For each unitary space U, let a *hermitian bilinear form* h be a function $h: U \times U \to \mathbf{C}$ with the two properties (for all u's, v's and κ's):

Left linearity: $\quad h(u_1\kappa_1 + u_2\kappa_2, v) = h(u_1, v)\kappa_1 + h(u_2, v)\kappa_2;$

Conjugate symmetry: $\quad h(v, u) = [h(u, v)]^*.$

Thus the inner product is itself a hermitian bilinear form; as in that case these two properties imply conjugate linearity in the second argument:

$$h(u, v_1\kappa_1 + v_2\kappa_2) = h(u, v_1)\kappa_1^* + h(u, v_2)\kappa_2^*.$$

Let $\mathbf{b}$ be an orthonormal basis of n vectors for the unitary space U. The matrix of the hermitian bilinear form h, rel $\mathbf{b}$, is the $n \times n$ matrix H with entries $H_{ij} = h(b_i, b_j)$. The conjugate symmetry of the form h implies that this matrix H is hermitian, and the matrix H determines the form h, by linearity and conjugate linearity, as

$$h\left(\sum b_i\xi_i, \sum b_j\eta_j\right) = \sum_{i,j=1}^{n} \xi_i H_{ij} \eta_j^*. \tag{40}$$

In other words, a hermitian bilinear form "looks" like an ordinary bilinear form, but with the second set of coordinates replaced by their conjugates. Conversely, each $n \times n$ hermitian matrix H defines a hermitian bilinear form in U, relative to the orthonormal basis $\mathbf{b}$, via this formula (40). If we write $\boldsymbol{\xi}^T$ for the row of coordinates ξ_i, and $\boldsymbol{\eta}$ for the row η_i, the form may be written as a matrix product $\boldsymbol{\xi}^T H \boldsymbol{\eta}^*$. Since a (unitary) change of coordinates gives new coordinates $\boldsymbol{\eta}'$ with $\boldsymbol{\eta} = \boldsymbol{\eta}' L$ and L unitary, such a change replaces the matrix H by the unitary-equivalent matrix LHL^{*T}. In this language, the principal axis theorem reads as follows:

Corollary 6. *If h is a hermitian form on an n-dimensional unitary space there is a choice of orthonormal coordinates* $\mathbf{x}$ *for which h is diagonal; in the coordinates $\boldsymbol{\xi} = \mathbf{x}(u)$ and $\boldsymbol{\eta} = \mathbf{x}(v)$,*

$$h(u, v) = \lambda_1\xi_1\eta_1^* + \lambda_2\xi_2\eta_2^* + \cdots + \lambda_n\xi_n\eta_n^*.$$

The coefficients $\lambda_1, \cdots, \lambda_n$ are the eigenvalues of (the matrix of) h. Each appears a number of times equal to its multiplicity.

EXERCISES

1. Give a direct matrix proof of the fact that every eigenvalue of a hermitian matrix is real.

2. Prove that every eigenvalue of a unitary matrix is a complex number of absolute value 1.

3. For a normal linear transformation, show that eigenvectors u, v with distinct eigenvalues are orthogonal.

4. If h is a hermitian bilinear form on the unitary space U, define a function $q: U \to \mathbf{C}$ by $q(u) = h(u, u)$. Show how q is expressed in terms of coordinates and prove ("polarization") that the function q determines the original function h. Prove also that $q(u)$ is always a real number.

5. Prove that any invertible matrix of complex numbers can be written uniquely as a product HL with H hermitian and L unitary.

***6.** If H and K are $n \times n$ hermitian matrices, prove that there is a unitary matrix L for which LHL^{*T} and LKL^{*T} are both diagonal, if and only if $HK = KH$.

CHAPTER XI

Similar Matrices and Finite Abelian Groups

THIS CHAPTER is devoted primarily to the proof of two major theorems: Every finite abelian group is a biproduct of cyclic groups; every square matrix over a field F is similar to a direct sum of companion matrices. The first theorem completely describes the "structure" of any finite abelian group; the second theorem gives a complete answer to the problem of finding a canonical form for square matrices under the relation of similarity. Both theorems turn out to be special cases of a single theorem about modules over a principal ideal domain D. On the one hand, each finite abelian group is a module over the principal ideal domain $\mathbf{Z}$. On the other hand, an $n \times n$ matrix A over the field F determines a linear map $t_A : F^n \to F^n$; from the linear map t_A one can construct linear maps which are polynomials in t_A, and so make the vector space F^n into a module over the principal ideal domain $F[x]$ of all polynomials in x. Thus both problems lead to the study of a module of finite type over a principal ideal domain D. Our main theorem will then assert that any such module is a biproduct of cyclic modules, in an essentially unique way. A second proof (§8) of this theorem will give explicit means of finding these cyclic modules.

These results, interpreted for matrices, show that every square matrix A is similar to a direct sum of companion matrices for monic polynomials $g_1, \cdots, g_k$. These polynomials can be chosen so that each one divides the next. They are called the "invariant factors" of A because two matrices are similar if and only if they lead to the same such list of monic polynomials. Alternatively, these invariant factors may be decomposed into irreducibles to give a list of "elementary divisors" which again determine the matrix A up to similarity.

1. Noetherian Modules

For finite $\mathbf{Z}$-modules, mathematical induction on the order of the module is a powerful tool. For R-modules of finite type, a comparable tool, the ascending chain condition, has already been of use in the proof that prime factorization exists in a principal ideal domain (§III.10).

DEFINITION. An R-module A satisfies the ascending chain condition (*ACC*) on submodules *when for each ascending sequence*

$$S_1 \subset S_2 \subset S_3 \subset \cdots \subset A \tag{1}$$

of submodules of A there is an index m with $S_m = S_{m+1} = S_{m+2} \cdots$.

Thus *ACC* means: Every ascending sequence of submodules is ultimately constant.

Recall that an R-module is of *finite type* (§V.3) when it can be spanned by some finite list of its elements.

THEOREM 1. *A right R-module A satisfies the ascending chain condition for submodules if and only if every submodule of A is of finite type.*

Proof: First, assume *ACC*, and suppose S a submodule of A not of finite type. Then no finite list of its elements can span S. Choose an element $s_1 \in S$; there must be an element $s_2 \in S$ not in the submodule S_1 spanned by s_1. Similarly, given elements $s_1, \cdots, s_n \in S$, let S_n be the proper submodule of S which they span. There is an $s_{n+1} \in S$ not in S_n; with S_n it spans a submodule S_{n+1} properly containing S_n. This process continued (by the axiom of choice) produces a sequence $S_1 \subset S_2 \subset \cdots$ of submodules which is not ultimately constant, a contradiction to *ACC*.

Conversely, suppose that every submodule of A is of finite type. Given any ascending sequence of submodules S_n of A, the union of all the sets S_n is clearly a submodule of A; as such, it is spanned by a finite list $a_1, \cdots, a_k$. Each of these elements a_i is in some module of the given sequence, say in S_{n_i}. If m is the largest of these indices $n_1, \cdots, n_k$, then all of $a_1, \cdots, a_k$ lie in S_m, so this module S_m is already the whole union, and the given sequence is thus constant from m on, Q.E.D.

A right R-module A is called *noetherian* when it satisfies the *ACC* for submodules (hence, when all its submodules are of finite type). Now each ring R is a right module over itself; call the ring R *right noetherian* when R is a noetherian right R-module. But a submodule of the right R-module R is the same thing as a right ideal in R; thus the ring R is right noetherian if and only if it satisfies the *ACC* for right ideals. In the commutative case (which is our chief concern), R *noetherian* means that the *ACC* holds for (two-sided) ideals in R. For example, Proposition III.22 states in this terminology that any principal ideal domain is noetherian.

Now consider sequences of R-modules

$$0 \to A \xrightarrow{t} B \xrightarrow{u} C \to 0,$$

where t and u are morphisms of R-modules. Such a sequence is exact, as defined in §IX.9, when

$$\operatorname{Ker} t = 0, \qquad \operatorname{Im} t = \operatorname{Ker} u, \qquad \operatorname{Im} u = C;$$

in other words, when A is isomorphic (via t) to a submodule S of B and C is isomorphic (via u) to the quotient module $C \cong B/S$.

Lemma. *In a short exact sequence* $0 \to A \xrightarrow{t} B \xrightarrow{u} C \to 0$ *of* R*-modules,* A *and* C *noetherian imply* B *noetherian, and conversely.*

Proof: To an ascending sequence $S_1 \subset S_2 \subset \cdots$ of submodules of B, consider the image sequence $u_* S_1 \subset u_* S_2 \subset \cdots$ in C and the inverse image sequence $t^* S_1 \subset t^* S_2 \subset \cdots$ in A. Since C and A are noetherian, there is an index m beyond which both of these sequences are constant; this means that

$$k \geqslant m \quad \Rightarrow \quad u_* S_m = u_* S_k \quad \text{and} \quad t^* S_m = t^* S_k.$$

But these conditions suffice to make $S_m = S_k$ for all $k \geqslant m$. For we are given $S_m \subset S_k$; if $b \in S_k$, then $ub \in u_* S_k = u_* S_m$, so there is $b' \in S_m$ with $ub = ub'$. This means that $b - b'$ is in the kernel of u, hence by exactness that $b - b' = ta$ for some $a \in A$. Then $a \in t^* S_k = t^* S_m$ implies that ta is in $t_* t^* S_m = S_m$. Thus $b = b' + ta'$, as the sum of two elements in S_m, is in S_m. Hence, $S_m = S_k$ for each $k \geqslant m$, and the sequence is indeed constant beyond m.

Conversely, suppose that the middle module B in the short exact sequence is noetherian. Then A is also noetherian, because every submodule of A is isomorphic by t to a submodule of B. On the other side, if C has an ascending sequence of submodules, their inverse images under u form an ascending sequence of submodules of B. Such a sequence is ultimately constant in B, and by exactness so is the original sequence in C. This proves the converse part of the lemma.

Any noetherian module (with all its submodules) is of finite type. The converse holds when the ring in question is noetherian:

Theorem 2. *If* R *is a right noetherian ring, any right* R*-module of finite type is noetherian.*

Proof: Projecting a free module R^n upon one of its factors gives an exact sequence $0 \to R^{n-1} \to R^n \to R \to 0$ of right R-modules. Now R a noetherian ring means exactly that the module R is noetherian, hence by the lemma, applied to this sequence for $n = 2$, that R^2 is noetherian. An induction on n and the lemma then show R^n noetherian for all n.

Now take any R-module C of finite type, so that C is spanned, say by the n elements $c_1, \cdots, c_n$. Since R^n is a free module on the unit elements ε_i, there is a morphism $u: R^n \to C$ with $\varepsilon_i \mapsto c_i$ for each i. This morphism is an epimorphism u, so by the converse part of the lemma, R^n noetherian implies C noetherian, Q.E.D.

EXERCISES

1. Prove the first half of the lemma by constructing directly for each submodule S of B a finite list spanning S.

2. Prove that $\mathbf{Q}$ is not a noetherian $\mathbf{Z}$-module.

3. If F is a field, show that an F-module V is noetherian if and only if it is finite-dimensional (as a vector space).

4. Show that the biproduct of two noetherian modules is noetherian.

5. Prove that the R-module A satisfies the ACC for submodules if and only if there is in every non-empty set U of submodules of A a maximal element M. (Here M *maximal* in U means that M is properly contained in no other element of U.)

2. Cyclic Modules

This section and the next consider modules over a principal ideal domain D. For example, D might be $\mathbf{Z}$, in which case the D-modules are just abelian groups. This is a suggestive case to bear in mind throughout.

A *cyclic* D-module C is a module spanned (that is, "generated") by a single element c_0, so that the assignment $\kappa \mapsto c_0\kappa$ is an epimorphism $D \to C$ of D-modules. Therefore C is isomorphic to a quotient of the domain D by some submodule of D. Since a submodule of D is an ideal in D and every ideal in D is principal, a cyclic D-module is thus a module of the form $C \cong D/(\mu)$, where (μ) is the principal ideal consisting of all multiples of the scalar $\mu \in D$. If $\mu = 0$, $C \cong D$ is a free module on one generator c_0. If $\mu \neq 0$, the elements of C are all multiples $c_0\kappa$ with $\kappa \in D$, while $c_0\kappa = 0$ if and only if κ is a multiple of μ. We then say that C is cyclic with generator c_0 of *order* μ, just as in the special case of an abelian group ($D = \mathbf{Z}$). Any other element generating C has the same order (up to an invertible factor), for the principal ideal (μ) is the ideal of all those scalars κ with $c\kappa = 0$ for every c. Hence we may call μ the *order* of the module C.

For integers, $mx \equiv 0 \pmod{mn}$ implies that $x = ny$ for some integer y. In other words, every element x of order m in the cyclic group $\mathbf{Z}_{mn}$ of order mn is the n-fold ny of some other element y. The same property holds for cyclic modules:

LEMMA 1. *If the cyclic D-module C of order $\mu = \lambda\nu$ has an element c with $c\nu = 0$, then $c = c'\lambda$ for some element $c' \in C$.*

Proof: Let c_0 be a generator of C, so $c = c_0\kappa$ for some κ. Then $c\nu = 0$ gives $c_0\kappa\nu = 0$ and hence $\kappa\nu$ a multiple of the order $\mu = \lambda\nu$ of c_0. Canceling, κ is a multiple of λ, say $\kappa = \kappa'\lambda$, so $c = c_0\kappa = (c_0\kappa')\lambda = c'\lambda$ for some element $c' = c_0\kappa' \in C$, as required.

Proposition 3. *If a cyclic D-module C of order μ is a submodule of a noetherian D-module A with $a\mu = 0$ for all $a \in A$, then C is a direct summand of A.*

Proof: Since A is noetherian, it is of finite type, so can be spanned by C and a finite number k of elements $a_1, \cdots, a_k$. The proposition will be proved by induction on the number k. If $k = 0$, then $A = C$, and the result is immediate. So suppose that the proposition is true for every A spanned by C and k elements, and consider a module A spanned by C and $k + 1$ elements $a_1, \cdots, a_k, a_{k+1} = a$. Let A_0 be the submodule spanned by C and $a_1, \cdots, a_k$. By the induction assumption, C is a direct summand of A_0. The complementary summand is a submodule B_0 of A_0, as in Proposition V.15, with

$$A_0 = C + B_0, \qquad C \cap B_0 = \{0\}. \tag{2}$$

The whole module A has just one more generator a, so the quotient module A/A_0 is spanned by one coset $A_0 + a$ and thus must be cyclic of some order κ dividing μ, so that $\mu = \kappa\nu$. Then by the decomposition (2)

$$a\kappa = c + b_0, \qquad c \in C, \quad b_0 \in B_0.$$

But every element a of A is annihilated by μ, so

$$0 = a\mu = a\kappa\nu = c\nu + b_0\nu.$$

Since $C \cap B_0 = \{0\}$, this implies that $c\nu = 0$; by Lemma 1, this last equation, with $\mu = \kappa\nu$, means that $c = c'\kappa$ for some $c' \in C$. Then $a\kappa = c'\kappa + b_0$, so A can be generated by A_0 and the element $a' = a - c'$ with $a'\kappa = b_0$.

Now take B to be the submodule of A spanned by B_0 and this element a'. We claim that

$$A = C + B, \qquad C \cap B = \{0\},$$

thus displaying C as a direct summand of A. The first of these equations just restates the fact that A is spanned by A_0 and a'. As for the second, suppose that C and B had a common element, which could then be written as

$$c = a'\alpha + b, \qquad c \in C, \quad \alpha \in D, \quad b \in B_0.$$

Then $a'\alpha = c - b$ is in A_0, while a' (like a) has order κ modulo A_0. Therefore, $\alpha = \kappa\alpha_1$ for some scalar α_1, and $c = a'\kappa\alpha_1 + b$ is in B_0. But, by (2), $C \cap B_0 = \{0\}$, so $c = 0$ and thus $C \cap B = \{0\}$, Q.E.D.

EXERCISES

(In these exercises, D is a principal ideal domain.)

1. Let C be a cyclic D-module of order μ.

(a) Prove that every submodule of C is cyclic with order a divisor of μ.

(b) For each principal ideal (λ) of D with $(\lambda) \supset (\mu)$, show that C has exactly one submodule which is cyclic of order λ.

2. Prove that any quotient module of a cyclic D-module is cyclic—and describe the possible orders of such a quotient.

3. If $D = F[x]$, with F a field, show that a cyclic D-module of order $f(x)$, $f \in F[x]$, is also a vector space over F of dimension the degree of the polynomial f.

3. Torsion Modules

The main theorem of the chapter treats noetherian D-modules in which every element has finite order (is a "torsion element").

An element a of a D-module A is a *torsion element* when $a\kappa = 0$ for some $\kappa \neq 0$ in D. The set of all $\kappa \in D$ with $a\kappa = 0$ is an ideal in D, hence a principal ideal (μ). This element μ, called the *order* of a, has $a\mu = 0$ and $a\kappa = 0$ only for multiples κ of μ; it is unique up to invertible factors from D, and is the order of the cyclic submodule spanned by a, as defined in §2.

A *torsion module* A is a D-module in which every element is a torsion element. Let A be a torsion module of finite type. If the elements $a_1, \cdots, a_k$ spanning A have the respective orders $\mu_1, \cdots, \mu_k$, then the product $\mu = \mu_1 \cdots \mu_k$ is non-zero and has $a\mu = 0$ for every $a \in A$. The set of all μ in D with $a\mu = 0$ for every a in A is again an ideal in D, hence a principal ideal (ν). The generator ν of this ideal, which is unique up to invertible factors from D, will be called the *minimal annihilator* of A.

For example, if D is the domain $\mathbf{Z}$ of integers, a torsion $\mathbf{Z}$-module is an abelian group in which every element has finite order. A torsion $\mathbf{Z}$-module of finite type is therefore just a finite abelian group A, and the minimal annihilator of A is simply the least common multiple of the orders of the elements of A.

Proposition 4. *If torsion elements a_1 and a_2 of a D-module A have relatively prime orders μ_1 and μ_2, then $a_1 + a_2$ has order $\mu_1\mu_2$.*

Proof: To say that μ_1 and μ_2 are *relatively prime* in D is to say that their greatest common divisor is 1, in other words, that the ideal (μ_1, μ_2) is D. Hence there are scalars κ_1 and κ_2 with $1 = \mu_1\kappa_1 + \mu_2\kappa_2$. Then, as $a_2\mu_2 = 0$,

$$(a_1 + a_2)\mu_2\kappa_2 = a_1\mu_2\kappa_2 + a_2\mu_2\kappa_2 = a_1\mu_2\kappa_2 = a_1(1 - \mu_1\kappa_1) = a_1,$$

and similarly a_2 is a multiple of $a_1 + a_2$. If ν is the order of $a_1 + a_2$, this shows that $a_1\nu$ and $a_2\nu$ are both zero, so ν must be a multiple of both μ_1 and μ_2, hence a multiple of $\mu_1\mu_2$. On the other hand, $(a_1 + a_2)\mu_1\mu_2 = 0$, so ν must be an invertible times $\mu_1\mu_2$, Q.E.D.

Now we can prove the promised decomposition theorem.

THEOREM 5. *For any torsion module A of finite type over a principal ideal domain D there is a list $\mu_1, \cdots, \mu_k$ of scalars of D, each a multiple of the next, and an isomorphism*

$$A \cong C_1 \oplus \cdots \oplus C_k$$

of A to a biproduct of cyclic modules C_i of orders μ_i, $i = 1, \cdots, k$.

The essential device in the proof will be to extract a cyclic submodule C_1 of A as one of biggest possible order (that is, with order a multiple of the order of any element of A). We first show that there *is* such a submodule.

LEMMA 1. *A torsion D-module A of finite type with minimal annihilator ν has an element of order ν.*

Proof: The principal ideal domain D is a unique factorization domain. Hence, one can write ν as $\nu = u_0 p_1^{e_1} \cdots p_t^{e_t}$ with u_0 invertible in D, the exponents e_i positive integers, and $p_1, \cdots, p_t$ prime elements of D, no two associates of each other. The scalars $\nu_i = \nu p_i^{-1}$ are not multiples of ν; since ν is the minimal annihilator of A, there must be for each $i = 1, \cdots, t$ an element $a_i \in A$ with $a_i\nu_i \neq 0$. Set $b_i = a_i(\nu p_i^{-e_i})$, so that $b_i p_i^{e_i} = 0$, but $b_i p_i^{e_i - 1} \neq 0$. Therefore b_i is an element of order $p_i^{e_i}$. Also, $p_i^{e_i}$ and $p_1^{e_1} \cdots p_{i-1}^{e_{i-1}}$ are relatively prime, so repeated use of Proposition 4 show that $b_1 + \cdots + b_t$ has order ν, hence is the desired element of A.

Proof of Theorem: Given the torsion module A with minimal annihilator ν, Lemma 1 provides a cyclic submodule $C_1 \subset A$ of order $\mu_1 = \nu$. By Proposition 3, this submodule C_1 is a direct summand of A, so we may write A as the sum $A = C_1 + A_2$ for some other submodule A_2 with $C_1 \cap A_2 = 0$ (and hence with $A \cong C_1 \oplus A_2$); moreover, the minimal annihilator ν_2 of A_2 is a divisor of the minimal annihilator ν of A. Now A_2, as a submodule of a noetherian module A over a noetherian ring D, is also noetherian, hence of finite type. Therefore, Lemma 1 applies to A_2 and gives a cyclic submodule C_2 of A_2 with order $\mu_2 = \nu_2$ a divisor of μ_1. Again C_2 is, by Proposition 3, a direct summand of A_2, so $A_2 = C_2 + A_3$ for some submodule A_3 with $A_3 \cap C_2 = 0$. This process gives $A = C_1 + A_2$, $A_2 = C_2 + A_3$ and so on; if the process ever stops, it gives the desired biproduct of cyclic modules. But the process must stop, for otherwise $C_1, C_1 + C_2, C_1 + C_2 + C_3, \cdots$

would be a properly ascending infinite sequence of submodules of the noetherian module A, a contradiction.

This proof is the point at which we made use of the results of §1, stating that a module of finite type over a noetherian ring (here, the principal ideal domain D) satisfies the ascending chain condition. In special cases, this condition can be replaced by ordinary induction arguments. For example, if $D = \mathbf{Z}$, a torsion module A of finite type is a finite abelian group, and one may prove the theorem by induction on the order of A.

Now we examine the extent to which the decomposition into cyclic modules is unique. The cyclic submodules C_i themselves are not unique; indeed, the proof shows that one can use for C_1 *any* cyclic submodule of A with order the minimal annihilator ν; there is a similar freedom of choice for the subsequent C's. However, the orders μ_i of the successive cyclic factors *are* uniquely determined, up to invertible factors:

THEOREM 6. *If the torsion D-module A of finite type has two decompositions*

$$C_1 \oplus \cdots \oplus C_k \cong A \cong C'_1 \oplus \cdots \oplus C'_l \tag{3}$$

into biproducts of non-zero cyclic modules C_i, C'_j *of the respective orders* μ_i *and* μ'_j *such that* $\mu_{i+1} \mid \mu_i$ *and* $\mu'_{j+1} \mid \mu'_j$ *for all* i *and* j, *then* $l = k$ *and* $(\mu_1) = (\mu'_1), \cdots, (\mu_k) = (\mu'_k)$.

In other words, the module A uniquely determines the list $(\mu_1), \cdots, (\mu_k)$ of principal ideals of D. This list of ideals (or of the scalars μ_i) is called the list of *invariant factors* of the torsion module A.

Let us first illustrate the argument to be given in case D is the ring $\mathbf{Z}$ of integers. Any (left) $\mathbf{Z}$-module (abelian group) of order, say, 32 is a biproduct of cyclic groups; $A = \mathbf{Z}_8 \oplus \mathbf{Z}_4$ and $A' = \mathbf{Z}_8 \oplus \mathbf{Z}_2 \oplus \mathbf{Z}_2$ are two such biproducts. Now $8A = 0 = 8A'$, so both groups A and A' have the same minimal annihilator 8. On the other hand, $2A \cong \mathbf{Z}_4 \oplus \mathbf{Z}_2$ is not cyclic, while $2A' \cong \mathbf{Z}_4$ is, so there can be no isomorphism $A \cong A'$. This could also be deduced by noting that the elements of order 2 in A form a subgroup isomorphic to $\mathbf{Z}_2 \oplus \mathbf{Z}_2$, while those of order 2 in A' form a subgroup $\mathbf{Z}_2 \oplus \mathbf{Z}_2 \oplus \mathbf{Z}_2$ of different order; again, this shows that $A \not\cong A'$.

This particular argument for the two abelian groups A and A' depends on using as "tests" either the image $2A$ or the kernel (all elements of order 2) of the endomorphism $A \to A$ given by multiplication by 2. For a right module A over any principal ideal domain D, each scalar κ gives a similar endomorphism $A \to A$, scalar multiplication by κ. Both the image $A\kappa$ and the kernel, call it ${}^{(\kappa)}A$, of this endomorphism are submodules of A; together they fit in a short exact sequence

$$0 \to {}^{(\kappa)}A \to A \to A\kappa \to 0. \tag{4}$$

These submodules ${}^{(\kappa)}A$ and $A\kappa$ are defined for every D-module A, and do not depend on the representation of such an A (of finite type) as a biproduct of cyclic modules. However, these submodules may be readily exhibited for biproducts by the evident formulas

$$(B \oplus B')\kappa \cong B\kappa \oplus B'\kappa, \qquad {}^{(\kappa)}(B \oplus B') \cong {}^{(\kappa)}B \oplus {}^{(\kappa)}B'. \tag{5}$$

Similar formulas hold for biproducts with more than two factors. These submodules may be calculated for cyclic modules as follows.

Lemma 2. *Let $C = D/(\mu)$ be a cyclic module of order μ. If κ is relatively prime to μ, $C\kappa = C$ and ${}^{(\kappa)}C = 0$. If κ is a divisor of μ, say $\mu = \kappa\kappa'$, then $C\kappa$ and ${}^{(\kappa)}C$ are both cyclic, indeed $C\kappa \cong D/(\kappa')$ and ${}^{(\kappa)}C \cong D/(\kappa)$.*

Proof: If κ and μ are relative prime, there exist scalars λ and ν with $\lambda\kappa + \nu\mu = 1$. Then, for all c in C, $c = c1 = (c\lambda)\kappa + (c\mu)\nu = (c\lambda)\kappa$, so $c \in C\kappa$, while $c\kappa = 0$ implies $c = 0$. In the other case, with $\mu = \kappa\kappa'$, $C \cong D/(\kappa\kappa')$, so $C\kappa \cong (\kappa)/(\kappa\kappa')$; under the map $c\kappa \mapsto c$, this quotient module is isomorphic to $(1)/(\kappa') = D/(\kappa')$. Similarly, ${}^{(\kappa)}C = (\kappa')/(\kappa\kappa') \cong D/(\kappa)$.

With this preparation we turn to the proof of the theorem, first proving for the two decompositions of A that $k = l$. If not, take $k > l$. Since each cyclic factor C_i in the first decomposition is non-zero the order μ_k of C_k is not an invertible in D and so is divisible by at least one prime p of the domain. Choose such a prime p; it divides every one of $\mu_1, \cdots, \mu_k$, so by the formulas (5) and Lemma 2

$${}^{(p)}A \cong {}^{(p)}C_1 \oplus \cdots \oplus {}^{(p)}C_k \cong D/(p) \oplus \cdots \oplus D/(p) \quad (k \text{ factors}). \tag{6}$$

In the second given decomposition, this same prime p will divide all the orders $\mu_1', \cdots, \mu_l'$ up to some point, say up to μ_h' with $h \leqslant l < k$. Hence, from the second decomposition of A and Lemma 2,

$${}^{(p)}A \cong {}^{(p)}C_1' \oplus \cdots \oplus {}^{(p)}C_l' \cong D/(p) \oplus \cdots \oplus D/(p) \quad (h \text{ factors}). \tag{7}$$

In (6) and (7) the same module ${}^{(p)}A$ is represented in two ways as a biproduct of modules $D/(p)$ with $k \neq h$ factors. For abelian groups, with $D = \mathbf{Z}$, this is clearly impossible. To show in general that it is impossible, we first note that any D-module A annihilated by p can be regarded as a $D/(p)$-module. For a multiple of $a \in A$ by a coset $(p) + \kappa$ in $D/(p)$ can be defined as $a[(p) + \kappa] = a\kappa$; since $ap = 0$, the definition is independent of the choice of the element κ in the coset $(p) + \kappa$. In this way, (6) and (7) become isomorphisms of $D/(p)$-modules. But p a prime, by Proposition III.30, makes $D/(p)$ a field. Thus the two formulas (6) and (7) present the vector space ${}^{(p)}A$ over the field $D/(p)$ as a space of dimension k and as a space of dimension $h < k$, in contradiction to the invariance of dimension.

Both decompositions (3) of A now have the same number $k = l$ of factors. If the two lists of orders of these cyclic factors are not the same, pick the first index $t + 1$ where they differ, so that $(\mu_1) = (\mu_1'), \cdots, (\mu_t) = (\mu_t')$ but $(\mu_{t+1}) \neq (\mu_{t+1}')$. By choice of notation, we can then assume that $\kappa = \mu_{t+1}$ is not a multiple of μ_{t+1}'. Now form the submodule $A\kappa$. In the first decomposition the choice of $\kappa = \mu_{t+1}$ gives $C_{t+1}\kappa = \cdots = C_k\kappa = 0$, so by (5)

$$A\kappa \cong C_1\kappa \oplus \cdots \oplus C_t\kappa$$

has at most t cyclic factors. In the second decomposition, κ is not a multiple of μ_{t+1}', so $C_{t+1}'\kappa$ is not the zero module, and by (5)

$$A\kappa \cong C_1'\kappa \oplus \cdots \oplus C_{t+1}'\kappa \oplus \cdots$$

has at least $t + 1$ non-zero cyclic factors. But these two formulas are in contradiction, by the part of the theorem already proved, applied to the D-module $A\kappa$. The proof is complete.

For $D = \mathbf{Z}$, this theorem gives the following definitive description of finite abelian groups.

Corollary. *For each finite abelian group $A \neq 0$ there is exactly one list $m_1, \cdots, m_k$ of integers $m_i > 1$, each a multiple of the next, for which there is an isomorphism*

$$A \cong \mathbf{Z}_{m_1} \oplus \cdots \oplus \mathbf{Z}_{m_k}.$$

In this description, the first integer m_1 is the least positive integer $m = m_1$ with $mA = 0$, while the product $m_1 m_2 \cdots m_k$ is the order of A.

As in the general case, the last statement holds because the order of a biproduct $B \oplus C$ is the product of the orders of B and C.

The list $m_1, \cdots, m_k$ is called the list of *invariant factors* of the abelian group A. The corollary implies that two finite abelian groups are isomorphic if and only if they have the same invariant factors. This list is therefore a complete set of invariants for finite abelian groups under the relation of isomorphism.

With this result, one can list all nonisomorphic abelian groups of a specified order n; indeed, one need only find all possible lists $m_1, \cdots, m_k$ with product n and $m_{i+1} | m_i$ for $i = 1, \cdots, k - 1$. In particular, these conditions imply that every prime factor of n is a prime factor of m_1. For example, the possible abelian groups of order 36 are

$$\mathbf{Z}_{36}, \qquad \mathbf{Z}_{18} \oplus \mathbf{Z}_2, \qquad \mathbf{Z}_{12} \oplus \mathbf{Z}_3, \qquad \mathbf{Z}_6 \oplus \mathbf{Z}_6, \tag{8}$$

and no two of these groups are isomorphic.

EXERCISES

1. List all nonisomorphic abelian groups of order 64 and of order 96.
2. List all nonisomorphic abelian groups of order 32; for each pair of your list given an explicit argument to show the groups not isomorphic.
3. In the biproduct $A = \mathbf{Z}_4 \oplus \mathbf{Z}_2$ (a) Find all possible cyclic subgroups of order 4; (b) Find all presentations of A as a direct sum of cyclic subgroups.
4. Do Exercise 3 for the group $A = \mathbf{Z}_4 \oplus \mathbf{Z}_2 \oplus \mathbf{Z}_2$.
5. If a finite abelian group A has minimal annihilator $m \in \mathbf{Z}$, show by examples that a cyclic subgroup of A of order a proper divisor of m need not be a direct summand of A.
6. If C and C' are cyclic D-modules of orders λ and μ, respectively, show that the invariant factors of the D-module $C \oplus C'$ are the least common multiple of λ and μ and the greatest common divisor of λ and μ.
7. If A has invariant factors $\mu_1, \cdots, \mu_k$, while a submodule $S \subset A$ has invariant factors $\lambda_1, \cdots, \lambda_t$, prove that $t \leqslant k$ and $\lambda_i | \mu_i$ for $i = 1, \cdots, t$.
8. (a) For κ fixed in D and $f: A \to A'$ a linear map of D-modules, show that $f_*(A\kappa) \subset A'\kappa$ and $f_*({}^{(\kappa)}A) \subset {}^{(\kappa)}A'$.

   (b) Deduce that $A \mapsto A\kappa$ and $A \mapsto {}^{(\kappa)}A$ yield functors (on modules to modules).
9. For each scalar κ in D, show that a D-module A with $A\kappa = 0$ may be regarded as a $D/(\kappa)$-module, and conversely that every $D/(\kappa)$-module may be so obtained from a D-module A with $A\kappa = 0$.

4. The Rational Canonical Form for Matrices

The domain $F[x]$ of all polynomials in an indeterminate x with coefficients in a field F is a principal ideal domain; the decomposition theorem will thus apply to $F[x]$-modules. Now a module A over the polynomial ring $F[x]$ is an abelian group together with operations of scalar multiplication by all the polynomials in $F[x]$. The multiples by constants in F make A a vector space V over F, while the multiplication by the indeterminate x is an endomorphism $t: V \to V$ of that space. This observation will allow us to "translate" facts about $F[x]$-modules into facts about endomorphisms t of vector spaces, as follows.

Proposition 7. *Let F be a field. Each right $F[x]$-module A determines both a vector space $V = V_A$ over F (V_A is the additive group A with scalar* *multiples just by the elements of the field* $F \subset F[x]$*) and an endomorphism* $t_x: V \to V$ *of this space (with* t_x *the operation of scalar multiple by* x*). The assignment* $A \mapsto (V_A, t_x)$ *is a bijection from all modules over* $F[x]$ *to all pairs* (V, t)*, with* V *a vector space over* F *and* $t: V \to V$ *an endomorphism of that space.*

Proof: Given A, V_A as described is clearly a vector space and t_x an endomorphism. Vice versa, let $t: V \to V$ be given. In each polynomial of $F[x]$, substitute t for x, just as in the discussion (§IX.6) of the minimal polynomial of the endomorphism t. This substitution process is a morphism E_t from the polynomial ring $F[x]$ to the ring of all endomorphisms of V:

$$E_t : F[x] \to \text{End } V, \qquad E_t x = t, \qquad E_t \kappa = \kappa \qquad \text{for } \kappa \in F.$$

Since $F[x]$ is commutative, the definition $vf = (E_t f)v$ for each $f \in F[x]$ makes V a right $F[x]$-module A.

These constructions $A \mapsto (V_A, t_x)$ and $(V, t) \mapsto A$ are clearly inverses of each other, hence yield the asserted bijection.

Now we study the linear transformation t by using the associated $F[x]$-module A. First we make a "translation" of concepts from one language to the other.

Corollary 1. *If* V *is finite-dimensional, the corresponding* $F[x]$*-module* A *is a torsion module of finite type. The minimal polynomial for the transformation* t *is the minimal annihilator for the* $F[x]$*-module* A*.*

Both the minimal polynomial and the minimal annihilator will always be taken as monic polynomials. This agreement makes each unique.

Proof: To say that a monic polynomial $m \in F[x]$ is the minimal polynomial for t is to say that $m(t) = 0$ and that $f(t) = 0$ for any polynomial $f \in F[x]$ implies $m|f$. For the corresponding module A, this states that the set of all polynomials f with $Af = 0$ is precisely the principal ideal generated by the polynomial m; hence that m is the minimal annihilator of A.

In the bijection $A \mapsto (V, t)$, each $F[x]$-submodule S of A becomes a subspace S of V with $t_* S \subset S$ (that is, t carries S into itself). Such subspaces S of V, with $t_* S \subset S$, are called the *t-invariant* subspaces. Moreover, a *t-cyclic* subspace is a t-invariant subspace which is spanned (as a vector space) by a single vector together with the images of this vector under successive powers of t. These definitions yield

Corollary 2. *Submodules of* A *are identical with* t*-invariant subspaces of* V*, and cyclic submodules with* t*-cyclic subspaces.*

Corollary 3. *Let g be a monic polynomial of $F[x]$. Then an $F[x]$-module C is cyclic of order g if and only if the vector space $V = V_C$ has a basis for which the matrix of t_x is the companion matrix of g.*

Proof: Let g have degree d, while C is cyclic with generator b of order g. As an $F[x]$-module, C is generated by b; hence, as a vector space V, it is spanned by $b, tb, t^2b, \cdots$. But $g(t)b = 0$, so V is actually spanned by the first d of these elements; since g is minimal with $g(t)b = 0$, these d elements $b, tb, \cdots, t^{d-1}b$ are linearly independent over F, so form a basis of V. The endomorphism t carries each vector t^ib of this basis to the next and the last to a linear combination of them all. This means exactly that the matrix of t relative to this basis is the $d \times d$ companion matrix M_g of the polynomial g, as described in (IX.24). Conversely, an endomorphism t with matrix M_g clearly has the corresponding $F[x]$-module a cyclic one, of order g.

With these translations, the decomposition theorem for $F[x]$-modules takes the following form.

Theorem 8. *Let $t: V \to V$ be an endomorphism of a finite-dimensional vector space V over the field F. Then there exists exactly one list $g_1, \cdots, g_k$ of nonconstant monic polynomials of $F[x]$, each a multiple of the next, and with the property that V has at least one basis for which the matrix of t is*

$$M_{g_1} \oplus \cdots \oplus M_{g_k}, \tag{9}$$

the direct sum of the companion matrices of the polynomials g_i.

Note that we do not claim that the indicated basis of V is unique, but only that the monic *polynomials* g_i are unique. These polynomials for t are exactly the invariant factors of §3 for the corresponding module A: they are thus called the *invariant factors* of the endomorphism t.

These invariant factors have the following properties:

Corollary 1. *The first invariant factor of the endomorphism t is its minimal polynomial. The product $g_1 g_2 \cdots g_k$ of the invariant factors of t is (except perhaps for sign) the characteristic polynomial of t; in particular, the sum of the degrees of the invariant factors of $t: V \to V$ is the dimension of the vector space V.*

Proof: The first assertion is immediate. As for the second, the characteristic polynomial of a companion matrix M_g is $\pm g$, the sign depending on the parity of the degree of g (Proposition IX.13)—the sign comes in simply because

$$|M_g - \lambda I| = \pm|\lambda I - M_g|.$$

Hence, the characteristic polynomial of the direct sum $M_{g_1} \oplus \cdots \oplus M_{g_k}$ is, up to sign, the product $g_1 g_2 \cdots g_k$; in particular, the degree of this product is the dimension of the space.

This gives a new proof of the Cayley–Hamilton theorem (Theorem IX.14): *Every endomorphism $t: V \to V$ satisfies its characteristic equation.* For every endomorphism t satisfies its minimal polynomial g_1. But by Corollary 1 above, the characteristic polynomial c is $\pm g_1 g_2 \cdots g_k$, so $g_1(t) = 0$ implies $c(t) = 0$, Q.E.D.

Theorem 8 can also be put in matrix language, as follows.

COROLLARY 2. *Any square matrix A with entries in a field F is similar over F to exactly one direct sum $M_{g_1} \oplus \cdots \oplus M_{g_k}$ of companion matrices, where the $g_1, \cdots, g_k$ are nonconstant monic polynomials of $F[x]$, each a multiple in $F[x]$ of the next.*

Proof: If A is $n \times n$, apply the theorem to $t_A: F^n \to F^n$.

The polynomials $g_1, \cdots, g_k$ are called the *invariant factors* of the matrix A; each is an invariant for the relation of similarity between square matrices (of the same size).

COROLLARY 3. *Two $n \times n$ matrices over a field F are similar if and only if they have the same invariant factors.*

In Corollary 2, the matrix $M_{g_1} \oplus \cdots \oplus M_{g_k}$ is called the *rational canonical form* for the matrix A under the relation of similarity. It is "canonical", because each A is similar to exactly one such form; it is "rational" because the similarity takes place in the given field F, and not in some larger field. An important application of this fact is given in Exercise 7.

EXERCISES

1. Prove that any nilpotent matrix is similar to a matrix in which all the entries are zero except for certain entries 1 on the diagonal next below the main diagonal.

2. Find all possible rational canonical forms over the field $\mathbf{Q}$ for matrices described as follows:

(a) 4×4, minimal polynomial $(x+1)^2$.

(b) 6×6, minimal polynomial $(x+2)^2(x-1)$.

(c) 7×7, minimal polynomial $(x^2+1)(x-3)$.

(d) 6×6, characteristic polynomial $(x^4-1)(x^2-1)$.

3. Prove that the minimal polynomial of the direct sum of two matrices B and C is the least common multiple of the minimal polynomials of B and C.

4. Prove that a square matrix is similar to a companion matrix if and only if its minimal polynomial is (up to sign) equal to its characteristic polynomial.

5. (a) If a 6×6 diagonal matrix D has diagonal entries $(\lambda, \lambda, \lambda, \mu, \mu, \nu)$, with λ, μ, and ν three distinct scalars, show that the invariant factors of D are $(x - \lambda)(x - \mu)(x - \nu)$, $(x - \lambda)(x - \mu)$, and $(x - \lambda)$.

(b) Describe the invariant factors of any diagonal matrix.

6. If a matrix of rational numbers is regarded as a matrix of real numbers, prove that it has the same invariant factors (for $\mathbf{R}$ as for $\mathbf{Q}$).

7. If F is a subfield of F', each matrix over F may also be regarded as a matrix over F'. Prove that two $n \times n$ matrices with entries in F are similar over F' if and only if they are similar over F.

8. Prove that the characteristic polynomial of a matrix is a divisor of some power of the minimal polynomial of that matrix.

5. Primary Modules

Now we know that every finite abelian group is a biproduct of cyclic groups. But a decomposition such as $\mathbf{Z}_{12} \cong \mathbf{Z}_4 \oplus \mathbf{Z}_3$ suggests that a cyclic group itself can be further decomposed when its order (here 12) is the product of relatively prime factors (here 4 and 3). The same observation applies to cyclic modules over any principal ideal domain D; the essential point is the following "relatively prime" factorization.

Lemma 1. *If a module A over a principal ideal domain D has an annihilator $\nu \in D$ which is the product $\nu = \kappa\lambda$ of relatively prime factors κ and λ, then A is the direct sum of its submodules ${}^{(\kappa)}A$ and ${}^{(\lambda)}A$, and hence $A \cong {}^{(\kappa)}A \oplus {}^{(\lambda)}A$.*

Proof: As in (4), the submodule ${}^{(\kappa)}A$ consists of all the elements $a \in A$ with $a\kappa = 0$. Since D is a principal ideal domain while κ and λ are relatively prime in D, there are scalars κ' and λ' with $\kappa\kappa' + \lambda\lambda' = 1$, and so every element of A has the form $a = a\kappa\kappa' + a\lambda\lambda'$. Therefore, $a\kappa = 0 = a\lambda$ imply $a = 0$, so ${}^{(\kappa)}A \cap {}^{(\lambda)}A = 0$. Since $\nu = \kappa\lambda$ and $a\nu = 0$ for all a, $(a\kappa)\lambda$ and $(a\lambda)\kappa$ are zero, so $a\kappa \in {}^{(\lambda)}A$, $a\lambda \in {}^{(\kappa)}A$ and $a = (a\kappa)\kappa' + (a\lambda)\lambda'$ shows then that $A = {}^{(\lambda)}A + {}^{(\kappa)}A$; hence, A is the direct sum of ${}^{(\kappa)}A$ and ${}^{(\lambda)}A$.

For any D-module A with minimal annihilator ν, each decomposition of ν into several relatively prime factors will give, by iteration of this lemma, a direct sum decomposition of A. The finest such decomposition comes by using the prime factorization of ν, into powers of distinct (and nonassociate) primes.

Let p be a prime element of D. Define a *p-module* to be a D-module P in which every element has order some power of p, and call a D-module *primary* if it is a p-module for some prime p of D. In any module A, two elements a and b of the respective orders p^d and p^e have a sum $a + b$ of order some divisor of p^{d+e}. Hence, the set $T_p(A)$ of all elements in A of order some power of p is a submodule of A; indeed, it is the largest submodule of A which is a p-module.

Two prime elements p and p' of D are either associates or relatively prime. In a decomposition into prime factors, each factor p' may be replaced by an associate p. Hence, the minimal annihilator ν of a D-module may be written as a product $\nu = u_0 p_1^{e_1} \cdots p_h^{e_h}$ of an invertible u_0 by prime powers $p_i^{e_i}$, with no two of $p_1, \cdots, p_k$ associates; the invertible u_0 may then be dropped without changing the principal ideal (ν).

THEOREM 9 (*The Primary Decomposition Theorem*). *Any torsion module of finite type over a principal ideal domain D is a biproduct of primary modules. More explicitly, if such a module A has minimal annihilator $\nu = p_1^{e_1} \cdots p_k^{e_k}$, where $p_1, \cdots, p_k$ are primes of D, no two associates to each other, then*

$$A \cong T_{p_1}(A) \oplus \cdots \oplus T_{p_k}(A), \tag{10}$$

where each $T_{p_i}(A)$ is the largest p_i-submodule of A.

The proof, by induction on k, is a straightforward application of the lemma. For $k = 1$, A is already a p_1-module, and there is nothing to prove. If $k > 1$, ν has a factorization $\nu = \kappa\lambda$ with two factors $\kappa = p_1^{e_1} \cdots p_{k-1}^{e_{k-1}}$ and $\lambda = p_k^{e_k}$ relatively prime. Hence, the lemma shows A the direct sum of ${}^{(\kappa)}A$ and ${}^{(\lambda)}A$. But an element of A of order some power of p_k can have order at most $p_k^{e_k}$, for otherwise ν would not annihilate this element. Hence, the submodule ${}^{(\lambda)}A$ of elements of A annihilated by $\lambda = p_k^{e_k}$ is exactly the primary submodule $T_{p_k}(A)$. For similar reasons, $A' = {}^{(\kappa)}A$ contains each of $T_{p_1}(A), \cdots, T_{p_{k-1}}(A)$; indeed $T_{p_i}(A') = T_{p_i}(A)$ for $i \leqslant k-1$, and the minimal annihilator of A' is $\nu' = p_1^{e_1} \cdots p_{k-1}^{e_{k-1}}$. By the induction assumption,

$$A' \cong T_{p_1}(A') \oplus \cdots \oplus T_{p_{k-1}}(A') = T_{p_1}(A) \oplus \cdots \oplus T_{p_{k-1}}(A);$$

with $A \cong {}^{(\kappa)}A \oplus {}^{(\lambda)}A \cong A' \oplus T_{p_k}(A)$; this gives the conclusion (10).

The uniqueness of this decomposition (10) is immediate, for each factor in this biproduct is characterized directly for some prime p as the maximal p-submodule of A, Q.E.D.

Each of these primary factors is itself a torsion module of finite type, so the previous cyclic decomposition of Theorem 5 applies to each T_{p_i}, making it a biproduct

$$T_{p_i}(A) \cong C_{i,1} \oplus \cdots \oplus C_{i,h_i}, \qquad i = 1, \cdots, k, \tag{11}$$

of cyclic modules $C_{i,j}$. Each $C_{i,j}$ is a p_i-module, so must have order some power $p_i^{d_{ij}}$. Each order is a multiple of the next, so the exponents decrease; i.e., $d_{i1} \geqslant \cdots \geqslant d_{ih_i}$. But actually we can forget this; of any two powers of the same prime, one will divide the other, so any list of powers of one prime p_i can be arranged in essentially just one way so that each is a multiple of the next. By the uniqueness Theorem 6, we know that $T_{p_i}(A)$ uniquely determines this list, while A determines each of the $T_{p_i}(A)$. Substituting the k decompositions (11) into (10) now proves

Theorem 10 (*The Cyclic Primary Decomposition Theorem*). *Any torsion module A of finite type over a principal ideal domain D is isomorphic to a biproduct of primary cyclic modules. The list of orders of these factors is a list of prime powers; any other isomorphism of A to a biproduct of primary cyclic modules must have the same list of prime power orders, except perhaps for a permutation of the list or a replacement of each prime by an associate prime.*

The prime powers in this list are called the *elementary divisors* of A.

For abelian groups (that is, for $D = \mathbf{Z}$) this theorem states that any finite abelian group is a biproduct of cyclic groups of prime power orders, and that the list of these prime power orders is unique up to a permutation.

For example, there are three different abelian groups of order $24 = 2^3 \cdot 3$. They can be listed either as in the Corollary to Theorem 6 or in the primary form of Theorem 10, as follows:

Group		*Primary Form*	*Invariant Factors*	*Elementary Divisors*
$\mathbf{Z}_{24}$	$\cong$	$\mathbf{Z}_8 \oplus \mathbf{Z}_3$	24	8, 3
$\mathbf{Z}_{12} \oplus \mathbf{Z}_2$	$\cong$	$\mathbf{Z}_4 \oplus \mathbf{Z}_2 \oplus \mathbf{Z}_3$	12, 2	4, 2, 3
$\mathbf{Z}_6 \oplus \mathbf{Z}_2 \oplus \mathbf{Z}_2$	$\cong$	$\mathbf{Z}_2 \oplus \mathbf{Z}_2 \oplus \mathbf{Z}_2 \oplus \mathbf{Z}_3$	6, 2, 2	2, 2, 2, 3

Next take $D = F[x]$, and recall that a prime (= irreducible) element in this polynomial ring $F[x]$ is associate to a unique monic polynomial p, irreducible in $F[x]$. For this case, the cyclic primary decomposition theorem becomes the following theorem for a matrix (there is a corresponding theorem for an endomorphism $t: V \to V$ of a finite-dimensional vector space).

Corollary 1. *To each $n \times n$ matrix B over a field F there is a list L of powers p^e of monic irreducible polynomials p of $F[x]$ such that B is similar to the direct sum of the companion matrices of the powers p^e in this list. If B is similar over F to any direct sum of the companion matrices for some list L' of powers q^f of monic irreducibles q of $F[x]$, then the list L' is a permutation of the list L.*

Proof: Construct the endomorphism $t_B : F^n \to F^n$ and apply the theorem to the corresponding $F[x]$-module.

The corollary gives a canonical form, up to permutation of summands, for matrices under similarity. In this form, the list L is called the list of *elementary divisors* of the matrix B over F. Thus an elementary divisor of B over F is some power p^e of a monic irreducible polynomial, and the *same* power p^e may quite well occur more than once in the list of all elementary divisors. Two matrices over F are similar if and only if they have (after a permutation) the same list of elementary divisors over F; in this way the list of elementary divisors provides a complete invariant for the relation of similarity between square matrices over F.

The definition of elementary divisors has the following immediate consequences.

COROLLARY 2. *If p is a monic irreducible polynomial in $F[x]$, then the companion matrix of any power p^e has p^e as its only elementary divisor. The list of elementary divisors over F of a direct sum $B \oplus B'$ of two matrices is obtained by combining the lists of elementary divisors of B and B' over F. The characteristic polynomial of any square matrix B over F is, except perhaps for a factor -1, the product of the elementary divisors of B over F. The minimal polynomial of B is the least common multiple of its elementary divisors.*

For example, the companion matrices of the polynomials $x^2 + 2$, $(x - 3)^2$, and $x - 3$ are

$$C = \begin{bmatrix} 0 & -2 \\ 1 & 0 \end{bmatrix}, \qquad B = \begin{bmatrix} 0 & -9 \\ 1 & 6 \end{bmatrix}, \qquad A = [3],$$

respectively. Hence, any square matrix over $\mathbf{Q}$ with minimal polynomial $(x^2 + 2)(x - 3)^2$ is a direct sum of copies of C, B, and A, including at least one copy of C and one of B.

The elementary divisors of a matrix B, unlike the invariant factors, depend on the field F. For example, the matrix

$$B = \begin{bmatrix} 0 & -1 \\ 1 & 0 \end{bmatrix}$$

of rational numbers has the characteristic polynomial $1 + x^2$, which is irreducible over the field $\mathbf{Q}$ of rational numbers. Hence, B has just one elementary divisor over $\mathbf{Q}$, namely, $x^2 + 1$; indeed, B is the companion matrix of $x^2 + 1$. But now regard B as a matrix over the field $\mathbf{C}$ of complex numbers. Here $x^2 + 1 = (x + i)(x - i)$. Since the characteristic polynomial is the product of the elementary divisors, these elementary divisors over **C** must be $x + i$ and $x - i$. Moreover, B is similar (over the field **C**) to the direct sum of the two corresponding 1×1 companion matrices; that is, to the diagonal matrix with diagonal entries i and $-i$. Over the real field **R**, B is the matrix of a $90°$-rotation.

Consider the case when the irreducible polynomial p in each elementary divisor is a linear one, say $p = x - \lambda$ for some scalar λ. Let C be a cyclic p-module of order some power of this polynomial p, say $p^3 = (x - \lambda)^3$. Now translate, as in Proposition 7, to the vector space situation (V, t). As a vector space over F, $C \cong F[x]/(x - \lambda)^3$ has dimension 3. If c_0 generates C as a module, then the linear transformation t has $(t - \lambda)^3 c_0 = 0$, but $(t - \lambda)^2 c_0 \neq 0$. Thus the vectors $b_1 = c_0$, $b_2 = (t - \lambda)c_0$, and $b_3 = (t - \lambda)^2 c_0$ form a basis **b** for the vector space V. Moreover, $tb_1 = tc_0 = (t - \lambda)c_0 + \lambda c_0 = b_2 + \lambda b_1$. Similarly, $tb_2 = b_3 + \lambda b_2$ and $tb_3 = \lambda b_3$. These formulas show that the matrix of t (rel **b**) is

$$\begin{bmatrix} \lambda & 0 & 0 \\ 1 & \lambda & 0 \\ 0 & 1 & \lambda \end{bmatrix}. \tag{12}$$

(The ith column is the column of **b**-coordinates of tb_i.) Call such a matrix a 3×3 elementary Jordan matrix. In general, define an $e \times e$ *elementary Jordan matrix* J to be a matrix with entries all the same scalar λ on the main diagonal, entries 1 on the diagonal below the main one, and all the other entries zero. An argument like that above (for $e = 3$) will prove

Lemma 2. *If $p = x - \lambda$ is a monic linear polynomial in $F[x]$ and C is a cyclic p-module of order $(x - \lambda)^e$, with (V, t) the corresponding vector space with endomorphism t as in Proposition 7, then the matrix of t relative to a suitable basis of V is an $e \times e$ elementary Jordan matrix.*

The characteristic polynomial of this elementary Jordan matrix J is clearly $(-1)^e(x - \lambda)^e$, while its minimal polynomial is $(x - \lambda)^e$, as one may see by calculating successive powers of the matrix $J - \lambda I$. Thus the minimal polynomial of J is $(-1)^e$ times its characteristic polynomial, just as for companion matrices (Proposition IX.13).

Theorem 11 (*Jordan Canonical Form*). *Let the square matrix B over the field F have characteristic polynomial a product of linear factors. Then B is similar over F to a direct sum of elementary Jordan matrices, one for each elementary divisor of B. Explicitly, each elementary divisor of B is a power $(x - \lambda)^e$ of some monic linear polynomial $x - \lambda$; the corresponding matrix summand is the $e \times e$ elementary Jordan matrix with diagonal entries all equal to the scalar λ.*

The proof is immediate, from Theorem 10 and Lemma 2.

This theorem is especially important for the field **C** of complex numbers, for there (by the fundamental theorem of algebra) every non-zero polynomial is a product of linear factors. Hence, *every square matrix of complex numbers is similar to a direct sum of elementary Jordan matrices*. The decomposition is unique, except for permutation of the summands.

EXERCISES

1. List the invariant factors and the corresponding elementary divisors for all isomorphism types of abelian groups of order 144.

2. A 15×15 matrix of rational numbers has the invariant factors $x^6 - x^4 - 2x^2$, $x^5 - x^3 - 2x$, $x^4 - x^2 - 2$. What are its elementary divisors over **Q**, over **R**, and over **C**?

3. Given the list of invariant factors of a D-module A, describe how to construct from them the list of elementary divisors of A, and vice versa.

4. Prove that a square matrix with entries in a field F is similar to a diagonal matrix over F if and only if its elementary divisors over F are all linear.

5. A square matrix A is called "idempotent" when $A^2 = A$. Prove that two $n \times n$ idempotent matrices over a field F are similar if and only if they have the same rank.

6. Prove that two matrices of rational numbers are similar over **Q**, if and only if they have the same elementary divisors when regarded as matrices of complex numbers.

7. In Lemma 1, prove that ${}^{(\kappa)}A = A\lambda$.

8. If ν is the minimal annihilator of A in Lemma 1, show that ${}^{(\kappa)}A$ and ${}^{(\lambda)}A$ have the minimal annihilators κ and λ, respectively.

9. For p a prime element of a principal ideal domain D, prove that $T_p(A \oplus B) \cong T_p(A) \oplus T_p(B)$.

10. Show that the elementary divisors of a real symmetric matrix are all linear polynomials (cf. §X.8).

6. Free Modules

Any *torsion* module of finite type over the principal ideal domain D has been shown to be a biproduct of cyclic modules $D/(\mu)$. Next we will show that *any* D-module of finite type is a biproduct of cyclic modules, using as cyclic factors not only the torsion cyclic modules $D/(\mu)$, but also the cyclic module D itself. We already know that a free D-module F of finite type is isomorphic to the module D^n for some n and so to a biproduct of n copies of D. Moreover, we know by Theorem IX.7 that the number n depends only on F; it is the *rank* of F.

To represent given modules as biproducts the following result (Lemma V.6.2) will be useful.

Lemma. *If F is a free module and $t: B \to F$ is an epimorphism, then B is the direct sum of* Ker t *and a submodule F' isomorphic to F.*

In particular, if F is free on one generator g, one may take F' to be the free submodule of B on any element $g' \in B$ with $t(g') = g$. With this, we can show that submodules of free D-modules are free:

Theorem 12. *Over a principal ideal domain, any submodule S of a free module F of finite type is free with rank at most the rank of F.*

The proof is by induction on the rank n of F.

If $n = 1$, $F \cong D$, so the submodule S of F is just an ideal in D, hence a principal ideal $S = (\mu)$. If $\mu = 0$, $S = 0$ and S is free (on *no* free generators). If $\mu \neq 0$, then $\kappa \mapsto \mu\kappa$ is a D-module isomorphism $D \cong S = (\mu)$, so S is free on one generator; namely, the generator μ.

Now suppose that the theorem is true for all free modules of rank less than n, and take F free of rank n, say $F = D^n$. Then $(\xi_1, \cdots, \xi_n) \mapsto \xi_n$ is an epimorphism $t: D^n \to D$; its kernel is the submodule of all lists $(\xi_1, \cdots, \xi_{n-1}, 0)$, so is a free module on the $n - 1$ free generators $\varepsilon_1, \cdots, \varepsilon_{n-1}$. Let S be the given submodule of D^n and S_1 its image $t_*S \subset D$. Then t restricts to an epimorphism $t_1: S \to S_1$. Now Ker t_1 is a submodule of the free module Ker t, and hence a free module by the induction assumption. But $S_1 \subset D$ is free, so by the lemma $S \cong (\text{Ker } t_1) \oplus S_1$ is a biproduct of two free modules of ranks at most $n - 1$ and 1, respectively. Hence, S is free of rank at most n, just as required.

This proof amounts to an explicit construction of free generators for S. First, the submodule Ker $t_1 = S \cap \text{Ker } t$ consists of the elements of the form $(\xi_1, \cdots, \xi_{n-1}, 0)$ which lie in the submodule S; by the induction assumption it has $k \leqslant n - 1$ free generators $s_1, \cdots, s_k$. Now the image t_*S is the set of last coordinates ξ_n of all $\boldsymbol{\xi} \in S$. If all such $\xi_n = 0$, $S = \text{Ker } t_1$, and so S is free on $s_1, \cdots, s_k$. Otherwise, all the last coordinates ξ_n are multiples of some one scalar μ, so one can choose some $s_0 = (\eta_1, \cdots, \eta_{n-1}, \mu)$ in S with this scalar as its last coordinate. Then (as the proof above shows) S is free on the $k + 1 \leqslant n$ generators $s_1, \cdots, s_k, s_0$.

Theorem 13. *If D is a principal ideal domain, any D-module of finite type with no non-zero torsion elements is free.*

Proof: Take a list $\mathbf{c}$ spanning A. This list has the property

(*) To every $a \in A$ there is a scalar κ with $a\kappa$ in the span of $\mathbf{c}$.

Indeed, $\mathbf{c}$ spans A, so has the property (*) with each scalar $\kappa = 1$. We wish to construct a new list $\mathbf{c}$ which still has the property (*) and is linearly independent to boot. If a list **c** with property (*) is not linearly independent, there is a dependence relation such as $c_1\eta_1 + \cdots + c_n\eta_n = 0$ with some coefficient (say η_n for convenience) not zero. This means that $c_n\eta_n$ is a linear combination of the remaining elements $c_1, \cdots, c_{n-1}$. Hence for any $a \in A$, $a\kappa\eta_n$ is a linear combination of $c_1, \cdots, c_{n-1}$. Thus we get a shorter list $c_1, \cdots, c_{n-1}$ which still has property (*). We continue the process till the resulting list $b_1, \cdots, b_k$ has property (*) and is linearly independent.

By independence, the submodule $S \subset A$ spanned by the k elements $b_1, \cdots, b_k$ is clearly free on $b_1, \cdots, b_k$ as free generators. Moreover, by the property (*) of **b**, some multiple $c_i\kappa_i$ of each of the original spanning list c_i is in the submodule S. Therefore there is a scalar κ (say, $\kappa = \kappa_1 \cdots \kappa_n$) such that $a\kappa \in S$ for every $a \in A$. Consider the endomorphism $t:A \to A$ given by $a \mapsto a\kappa$. We have just seen that the image of t is contained in the free module S, so is itself free by the previous theorem. On the other hand, the kernel of t consists of all elements a of order dividing κ; by hypothesis there are no such elements except 0. Therefore, $t:A \to A$ is a monomorphism with image a free module, so A must be free.

Note: A module A with no non-zero torsion elements is often called a "torsion-free" module; in this language, the theorem just proved asserts that every torsion-free D-module of finite type is free.

We are now in a position to prove that any D-module of finite type is a biproduct of cyclic modules D or $D/(\mu)$ for various μ. In particular, (for $D = \mathbf{Z}$) any finitely generated abelian group A is a biproduct of cyclic groups.

THEOREM 14. *If D is a principal ideal domain, any D-module A of finite type has a decomposition*

$$A \cong F \oplus C_1 \oplus \cdots \oplus C_k, \tag{13}$$

where F is free, while $C_1, \cdots, C_k$ are cyclic modules of non-zero orders $\mu_1, \cdots, \mu_k$, each a multiple of the next. The rank n of F, the number k of cyclic summands, and the principal ideals $(\mu_1), \cdots, (\mu_k)$ are all invariants of A under isomorphism.

To say that n and the ideals (μ_i) are "invariants" of A means that any isomorphism $A \cong F' \oplus C_1' \oplus \cdots \oplus C_l'$ with F' free and C_i' cyclic of non-zero order μ_i', with each μ_i' a multiple of the next, implies rank $F' =$ rank F, $k = l$, and $(\mu_1') = (\mu_1), \cdots, (\mu_k') = (\mu_k)$. In topological applications (where the abelian group A is a homology group) the rank n of A is called the "Betti number" of A and the invariant factors $\mu_1, \cdots, \mu_k$ are called the "torsion coefficients" of A.

Proof: The set T of all torsion elements $a \in A$ is a submodule of A (it is often called the "torsion submodule" of A). It yields a short exact sequence

$$0 \to T \to A \xrightarrow{p} A/T \to 0. \qquad (14)$$

Now the quotient module A/T has no torsion elements except 0, for if a coset $T + a$ had order $\kappa \neq 0$, then $a\kappa \in T$ has order $\lambda \neq 0$, so $a(\kappa\lambda) = 0$ and a is a torsion element, hence in T. Moreover, A/T, as a quotient of a module A of finite type, is itself of finite type. The previous theorem thus ensures that A/T is a free module F (on a finite number of generators). Now that $A/T \cong F$ is free, the lemma proves $A \cong F' \oplus T$. The previous result that T is a biproduct $C_1 \oplus \cdots \oplus C_k$ of cyclic modules gives the desired decomposition (13).

In this decomposition, the number k and the ideals $(\mu_1), \cdots, (\mu_k)$ are invariants of A because they are invariants of T, which is determined by A in invariant fashion (as the submodule of all torsion elements). The number n of free generators of A/T is an invariant by Theorem IX.7.

EXERCISES

1. In the free abelian group $\mathbf{Z}^4$ find a set of free generators for the subgroup generated by each of the following lists of elements:
   (a) (6, 12, 0, 0), (4, 8, 0, 0).
   (b) (3, 2, 4, 3), (1, 2, 3, 2), (2, 2, 2, 3), (3, 0, 3, 1).
   (c) (2, 2, 2, 0), (3, 3, 0, 3), (4, 0, 4, 4), (0, 5, 5, 5).
2. Exhibit a submodule of a free $\mathbf{Z}[x]$-module which is not free.
3. Let (m, n) and (m', n') be any two elements of $\mathbf{Z}^2$, S the subgroup which they span, and $\Delta = mn' - m'n$ their determinant.
   (a) Show that (m, n) and (m', n') are linearly independent if and only if the subgroup S is isomorphic to $\mathbf{Z}^2$.
   (b) Show that this is also the case if and only if $\Delta \neq 0$.
   (c) Show that $\Delta \neq 0$ if and only if the quotient group $\mathbf{Z}^2/S$ is finite.
   (d) If $\Delta \neq 0$, show that the order of $\mathbf{Z}^2/S$ is $|\Delta|$.
4. For $D = \mathbf{Z}$, exhibit a $\mathbf{Z}$-module with no non-zero torsion elements which is not of finite type and not free. (This shows that the hypothesis "of finite type" is vital to Theorem 13.)

7. Equivalence of Matrices

Our next objective is that of finding explicit means of calculating the invariant factors of a given square matrix over a field. To do this we first develop a canonical form for the equivalence of rectangular matrices over the principal ideal domain $F[x]$.

For any integral domain D and any positive integer n, the general linear group $GL(n, D)$ is the group of all invertible elements P in the $n \times n$ matrix ring $D^{n \times n}$. Two $m \times n$ matrices M and N over D are *equivalent* when there are over D invertible matrices Q and P, $m \times m$ and $n \times n$, respectively, such that $N = QMP^{-1}$. In the special case when D is a field F, this is exactly the equivalence relation introduced in (VII.38); much as in that case, it can be interpreted in terms of free D-modules C and C' of rank n and m, respectively: Two matrices M and N are equivalent, if and only if they are matrices for the same linear transformation $s: C \to C'$. If D is a principal ideal domain, the image $s_* C$ is a submodule of C', hence is free. The rank of this image is the *rank* of the morphism and also the rank of the matrix M of s. Hence, just as in the case of matrices over a field, the rank of a matrix M over D is the maximum number of linearly independent columns, and equivalent matrices have the same rank r.

Now we consider the special domain $D = F[x]$ of all polynomials in an indeterminate x over a field F.

Theorem 15. *Each rectangular matrix M of rank r over the domain $F[x]$ is equivalent over $F[x]$ to a matrix of the form*

$$\begin{bmatrix} L & 0 \\ 0 & 0 \end{bmatrix}, \tag{15}$$

where L is an $r \times r$ diagonal matrix with non-zero diagonal entries the monic polynomials $d_1, \cdots, d_r \in F[x]$, each dividing the next, as in

$$d_1 | d_2, \cdots, d_{r-1} | d_r. \tag{16}$$

Before proving this theorem, we first study the equivalence of matrices by using the following operations on the columns of an $m \times n$ matrix over $F[x]$:

(I) Interchange any two columns.
(II) Multiply a column by an invertible element of $F[x]$.
(III) Add a polynomial multiple of one column to another column.

Call these the *elementary column operations*. They are just like the elementary column operations on matrices with entries in a field, except that in case (II) a column is multiplied only by an invertible of $F[x]$; that is, by a non-zero element of F (a constant polynomial). This provision ensures that the inverse of every elementary column operation is another such. If one of these elementary operations is applied to the $n \times n$ identity matrix I, the resulting matrix is said to be an *elementary matrix* E; it is an invertible matrix. To apply an elementary column operation to an $m \times n$ matrix M over D amounts to postmultiplying M by the corresponding elementary matrix E, as in $M \mapsto ME$, just as in the case when D is a field (Proposition

VII.12). Similarly, the elementary *row* operations amount to premultiplication $M \mapsto EM$ by elementary matrices E. Hence, any matrix obtained from M by elementary row and column operations is equivalent to M.

LEMMA. *To each non-zero $m \times n$ matrix M with entries polynomials of $F[x]$ there is a sequence of elementary row and column operations carrying M to the form*

$$\begin{bmatrix} d & 0 \\ 0 & N \end{bmatrix}, \tag{17}$$

where N is an $(m-1) \times (n-1)$ matrix of polynomials and the polynomial $d \neq 0$ divides every entry of N.

Proof: The aim of the proof is to use the division algorithm (Theorem III.19) repeatedly to take greatest common divisors of elements of M. For convenience in describing this proof, call the upper left-hand entry M_{11} of M the *corner* of M.

Since $M \neq 0$, we can assume, permuting rows and columns if need be, that M has a non-zero corner g. Now we try to make the rest of the first row all zeros. If the second column is headed by $f \neq 0$, the division algorithm gives polynomials q and r with $f - qg = r$ zero or of degree less than the degree of g. Then subtracting q times the first column from the second column gives a new first row $(g, r, \cdots)$. If $r \neq 0$, interchange the first and second columns to get a new corner r of smaller degree. If $r = 0$, work on the third column. Continuing with column operations eventually yields an equivalent matrix with a first row $(s, 0, \cdots, 0)$. Now use row operations to similarly reduce the first column to $(d, 0, \cdots, 0)$, possibly also changing the corner s to a polynomial d of smaller degree.

This produces an equivalent matrix of the form (17), except that the corner d may not divide all the entries of the residual block N. In case an entry in row i is not divisible by d, add that row to the first row, and then apply column operations as before to the result, to produce a new matrix like (17) but with a corner d of smaller degree. This process eventually stops, because the degree of the corner cannot decrease forever. When it stops, the corner d divides all entries of N, and we have the desired form (17).

This proof can be put more briefly (and less constructively) as follows: Examine all matrices equivalent to M. Pick one with a non-zero entry of lowest possible degree, and move that entry to the corner. That corner d must then divide every entry in the first row, because otherwise the division algorithm and a column operation produce an equivalent matrix with an entry r of smaller degree. So make the rest of the first row zero, and also the rest of the first column. In the resulting form (17), d must divide every entry of N, because otherwise there would be, as before, an equivalent matrix with an entry of smaller degree.

From this lemma, Theorem 15 follows by an immediate induction on r.

Proposition 16. *Any invertible square matrix with entries polynomials in $F[x]$ is a product of elementary matrices.*

Proof: We apply the last theorem to an $n \times n$ invertible matrix P. The sequence of row operations used may be expressed as premultiplications of P by elementary matrices $E_1, \cdots, E_k$, and similarly for the columns, so that

$$E_k \cdots E_1 P E_1' \cdots E_m' = L \tag{18}$$

with L diagonal and invertible over $F[x]$. But by Corollary IX.6.1, a matrix L is invertible over the commutative ring $F[x]$ if and only if its determinant $|L|$ is invertible in $F[x]$, i.e., is a non-zero element of F. But the determinant of the diagonal matrix L is just the product $d_1 \cdots d_n$ of its diagonal entries. Thus each d_i is a non-zero element of the field F. Multiplying the ith row by d_i^{-1} is another elementary operation; these operations make $L = I$. The modified equation $E \cdots P \cdots E' = I$ can now be solved for P as $P = E^{-1} \cdots E'^{-1}$. Since the inverse of each elementary matrix is again elementary, this gives P as the desired product.

Corollary. *Two $m \times n$ matrices M and N over the polynomial domain $F[x]$ are equivalent if and only if there is a sequence of elementary row and column operations carrying M to N.*

Proof: By definition M equivalent to N means that $N = QMP^{-1}$ for Q and P invertible square matrices of polynomials. By the proposition, both Q and P^{-1} can be written as products of elementary matrices, and these matrices represent the elementary row and column operations taking M to N. The converse is immediate.

Theorem 17. *Each rectangular matrix of integers, of rank r, is equivalent over $\mathbf{Z}$ to a diagonal matrix (15) with r positive non-zero diagonal entries $d_1, \cdots, d_r$, each dividing the next, as in (16).*

The proof is exactly like that for Theorem 15, using the division algorithm for integers rather than for polynomials. In the polynomial case, the process stopped because the degree of the corner polynomial could not decrease indefinitely. In the integer case, the positive integers in the corner evidently cannot decrease indefinitely.

The analog of Theorems 15 and 17 is true over any principal ideal domain D, but the proof is more complicated because there may be no division algorithm for D and Proposition 16 may not hold with $F[x]$ replaced by D.

EXERCISES

1. Reduce the following matrices of integers to canonical (diagonal) form

$$\begin{bmatrix} 2 & 8 & 6 \\ 4 & -2 & 10 \end{bmatrix}, \quad \begin{bmatrix} 3 & 4 & 5 \\ 5 & 6 & 7 \end{bmatrix}.$$

2. Over a principal ideal domain, prove that the greatest common divisor of all the entries of a matrix M is unchanged by any elementary operations on M.

3. Where does the greatest common divisor described in Exercise 2 appear in the diagonal form (15)?

4. Over a principal ideal domain, show that the greatest common divisor of all the 2×2 minors of a matrix M is unchanged by elementary operations on M.

5. Where does the greatest common divisor described in Exercise 4 appear in the diagonal form (15)?

*6. Prove that (15) is a canonical form: That any M is equivalent over $F[x]$ to exactly one matrix (15) with monic diagonal entries satisfying (16). (*Hint:* Use the results of Exercises 3, 5, $\cdots$.)

7. For Q the field of quotients of an integral domain D, regard the free module D^n as a subset of Q^n. Show that an endomorphism $t: Q^n \to Q^n$ carries D^n into D^n if and only if the matrix of t (relative to the basis of unit vectors in Q^n) has entries in D, and that t with this property has an inverse with this property if and only if the matrix of t is invertible in $D^{n \times n}$.

8. The Calculation of Invariant Factors

The diagonal form for matrices under equivalence will now be used to give a second proof that every D-module A of finite type is a biproduct of cyclic modules; the proof also provides an explicit way of calculating these cyclic modules, and hence of calculating the rank and the invariant factors of A. The proof uses the diagonal form (15) of Theorem 15 or Theorem 17, so is established here only for the domains $D = F[x]$ and $D = \mathbf{Z}$.

We can construct from each $m \times n$ matrix M over D a D-module A of finite type. First, as in §VII.1, M is the matrix of a linear map $s = s_M: D^n \to D^m$ between free D-modules with n and m free generators, respectively; define the module A to be the cokernel $A = D^m/(s_* D^n)$ of this map. This definition of A amounts to the requirement that the sequence

$$D^n \xrightarrow{s_M} D^m \to A \to 0 \qquad (19)$$

be exact (at D^m and at A).

An equivalence $M \mapsto QMP^{-1}$ of the matrix M corresponds to a change of free generators (by change matrices P and Q) in the modules D^n and D^m, so the associated module A is unchanged thereby. Every D-module of finite type arises in this way from at least one matrix M. Indeed, if A is spanned by m elements $a_1, \cdots, a_m$, the assignment $(\xi_1, \cdots, \xi_m) \mapsto \Sigma a_i \xi_i$ is an epimorphism $D^m \to A$. The kernel S of this epimorphism is a submodule of a free module D^m, hence by Theorem 12 is itself a free module, with $S \cong D^n$ for some n. This embeds A in a short exact sequence $0 \to D^n \to D^m \to A \to 0$. (The same module A may appear in other such sequences, with a different value of m, by a different choice of elements spanning A.)

Now take M in the diagonal form $L \oplus 0$ of (15) above. This matrix $L \oplus 0$ is the matrix of s relative to new bases in D^m and D^n. Let $b_1', \cdots, b_m'$ be this new basis in D^m and $b_1, \cdots, b_n$ in D^n. Then s is given by equations

$$sb_1 = b_1'd_1, \cdots, sb_r = b_r'd_r, \quad sb_{r+1} = 0, \cdots, sb_n = 0.$$

(the d_i are the "scalars"). The image of s in D^m is thus the submodule S spanned by the multiples $b_1'd_1, \cdots, b_r'd_r$ of the first r basis elements and the exact sequence (19) gives A as $A \cong D^m/S$. For example, if $r = 2 = m$, this is

$$A \cong (D \oplus D)/[(d_1) \oplus (d_2)] \cong D/(d_1) \oplus D/(d_2)$$

by Proposition V.14. For any r and m, A is similarly

$$A \cong D/(d_1) \oplus \cdots \oplus D/(d_r) \oplus D \cdots \oplus D; \tag{20}$$

it is a biproduct of r cyclic modules $D/(d_i)$ of orders d_i and $m - r$ copies of the cyclic module D.

This argument has established

Theorem 18. *For $D = F[x]$ or $\mathbf{Z}$, any D-module A of finite type is a biproduct of a finite number of cyclic D-modules.*

In particular, every abelian group of finite type is a biproduct of cyclic groups.

This is the previous Theorem 14, except that the factors in (20) are not in the order of the factors in (13). In (20) we can drop the zero modules $D/(d_i)$ with d_i invertible. This leaves, say, s noninvertible elements $d_1, \cdots, d_s$, so that, reversing the order,

$$A \cong D^{m-r} \oplus D/(d_s) \oplus \cdots \oplus D/(d_1),$$

with a free module of rank $m - r$ and s cyclic modules $D/(d_i)$, of non-zero orders $d_s, \cdots, d_1$ each a multiple of the next. The principal ideals $(d_s), \cdots, (d_1)$ are thus the invariant factors of A. Since equivalent matrices determine the same module A, these invariants are also invariants of the matrix. This proves

Theorem 19. *Every matrix of polynomials in $F[x]$ (or of integers) is equivalent to exactly one diagonal matrix of the form (15) and (16) with*

monic diagonal entries $d_1, \cdots, d_r$ *(respectively, with positive integers as diagonal entries).*

In other words, (15) and (16) give a canonical form for matrices of polynomials or of integers under equivalence.

In particular, an abelian group A which can be generated by m elements can be presented as a quotient group $\mathbf{Z}^m/S$ of a free abelian group $\mathbf{Z}^m$ on m generators. From this presentation, the explicit sequence of elementary operations described in §7 will thus yield the number $m - r$ for A (= the number of infinite cyclic factors in the canonical decomposition of A) and the successive invariant factors of A.

A corresponding result holds for the invariants of a module of finite type over the polynomial ring $F[x]$—that is, for the vector space case.

Theorem 20. *Let* $t: V \to V$ *be an endomorphism of a finite-dimensional vector space* V *over the field* F, *with matrix* C *relative to some basis* $\mathbf{e}$ *of* V. *If the matrix* $C - xI$ *of polynomials in* $F[x]$ *is reduced by equivalence in the polynomial domain* $F[x]$ *to the canonical form (15) with monic polynomials* $d_1, \cdots, d_r$, *as in (16), on the diagonal, and if those* d_i *which are* 1 *are discarded, the remaining* d_j, *with order inverted, are the invariant factors of the endomorphism* t.

Note that $C - xI$ is the matrix with off-diagonal entries C_{ij}, diagonal entries the linear polynomials $C_{ii} - x$, and determinant $|C - xI|$ the characteristic polynomial of C. In particular, $|C - xI| \neq 0$, so the same holds for any equivalent matrix, so that none of the diagonal entries d is zero.

Proof: The invariant factors of t are by definition those of the corresponding $F[x]$-module A. This module is constructed by the translation process of Proposition 7: Elements a of A are vectors of V, addition in A is addition in V, scalar multiples $a\kappa$ for $\kappa \in F$ are those of V, and the scalar multiple ax is the result of applying t as $ax = t(a)$. This module A is the quotient of a free $F[x]$-module, as follows: Take the given basis $e_1, \cdots, e_n$ of V, form the free $F[x]$-module $F[x]^n$ with n free generators $b_1, \cdots, b_n$, and define the $F[x]$-module epimorphism $p:(F[x])^n \to A$ by $pb_i = e_i$ for $i \in \mathbf{n}$. To express A as a cokernel as in (19) we want an exact sequence

$$(F[x])^n \xrightarrow{s} (F[x])^n \xrightarrow{p} A \to 0$$

of $F[x]$-modules. This amounts to writing the kernel of p as the image of another free module under some morphism s. The equations $te_j = \Sigma_i e_i C_{ij}$ express the fact that C is the matrix of t relative to the given vector space basis $\mathbf{e}$. Therefore,

$$p\left(b_j x - \sum_{i=1}^{n} b_i C_{ij}\right) = e_j x - \sum_i e_i C_{ij} = te_j - \sum e_i C_{ij} = 0.$$

The n elements $b_j x - \sum b_i C_{ij}$ thus lie in the kernel of p. Indeed, they span the kernel, as an $F[x]$-module, for modulo these n elements, any $b_j x^k \in (F[x])^n$ is congruent to a linear combination of $b_1, \cdots, b_n$ with coefficients in F and no such combination, save zero, is in Ker p. Therefore, Ker $(p) =$ Im (s), where $s:(F[x])^n \to (F[x])^n$ is the unique morphism of free $F[x]$-modules for which the images of the free generators b_j, $j = 1, \cdots, n$ are these n elements

$$sb_j = \sum_{i=1}^{n} b_i C_{ij} - b_j x = \sum_{i=1}^{n} b_i(C_{ij} - \delta_{ij} x).$$

These equations state that the matrix of s (rel $\mathbf{b}$) is the matrix $C - xI$ of the theorem. The elementary operations reducing this nonsingular matrix to diagonal form give monic polynomials $d_1, \cdots, d_n$ on the diagonal. Just as discussed above, this presents A as the biproduct

$$A \cong F[x]/(d_1) \oplus \cdots \oplus F[x]/(d_n)$$

of cyclic modules. If $d_i = 1$, the corresponding module is 0; drop it. The remaining polynomials d_i are the orders of the cyclic factors of A, which means that they are the invariant factors of the module A and hence of the original vector-space endomorphism t, Q.E.D.

To summarize: An endomorphism $t: V \to V$ of the F-vector space V makes V into an $F[x]$-module A, with $ax = t(a)$. Since V has finite dimension n as a vector space, the module A can be written as a quotient $F[x]^n/s_*(F[x])^n$ of a free $F[x]$-module on n generators by the image of a linear transformation s. Moreover, when the original endomorphism t is represented by a matrix C over F, the corresponding transformation s is represented by the matrix $C - xI$ of polynomials. Elementary operations on this latter matrix of polynomials produce a diagonal matrix of polynomials, in which each diagonal entry d_i corresponds to a cyclic factor of the $F[x]$-module A. In turn, these cyclic factors correspond to a new choice of basis in the vector space V, for which the matrix of t is a direct sum of the companion matrices for the polynomials d_i.

This process thus finds the rational canonical form for the given matrix C. It also yields the following useful test for similarity of square matrices.

Corollary. *Two $n \times n$ matrices C and C' over a field F are similar if and only if the corresponding matrices $C - xI$ and $C' - xI$ of polynomials in x are equivalent over the polynomial ring $F[x]$.*

EXERCISES

1. Find the invariant factors and the rational canonical form of each of the following matrices of rational numbers:

$$\begin{bmatrix} 0 & 4 & 2 \\ -1 & -4 & -1 \\ 0 & 0 & -2 \end{bmatrix}, \quad \begin{bmatrix} 1 & 2 & 0 & 0 \\ 0 & 1 & 2 & 0 \\ 0 & 0 & 1 & 2 \\ 0 & 0 & 0 & 1 \end{bmatrix}.$$

2. If C_g is the companion matrix of a monic polynomial g, show explicitly that $C_g - xI$ is equivalent over $F[x]$ to a diagonal matrix with diagonal entries $(g, 1, \cdots, 1)$.

CHAPTER XII

Structure of Groups

THE STRUCTURE of a finite abelian group is completely described by the elementary divisor theory of Chapter XI. For finite groups which are not abelian no such complete description is possible.

One may start by examining the normal subgroups of a finite group G and the corresponding quotient groups. A group G is said to be "simple" when it has no proper normal subgroups. The finite simple groups include the cyclic groups of prime order and the alternating groups A_n for $n \geqslant 5$ (to be proved simple in §9 below). A number of other finite simple groups are known, but as yet there is no complete tabulation of such groups.

If a finite group G is not simple, each of its proper normal subgroups N leads to a short exact sequence $1 \to N \to G \to G/N \to 1$ of morphisms of groups; we then call G an "extension" of N by G/N. This suggests the "extension problem", of describing the group G in terms of the smaller groups N and G/N and other data. One such extension of N by G/N is the product $N \times (G/N)$, also called the "direct" product; another is the "semi-direct" product to be discussed in §2. Other extensions are known, but the general case of the extension problem turns out to be intractable. Hence, we cannot follow a systematic program of first determining all finite simple groups, then all extensions of one such simple group by another, and so on.

Certain special classes of finite groups are more amenable because they are closer to abelian groups. These classes include the p-groups, in which every element has order some power of the prime p, more generally the nilpotent groups (§6), and still more generally the solvable groups (§7). Using these special classes of groups, one can assign certain invariants to any finite group. These include certain p-subgroups known as the Sylow subgroups (§5) and certain quotient groups known as composition factors (§8), obtained by considering series of successive normal subgroups within a group.

This chapter depends only on Chapter II and the results of §XI.3 on finite abelian groups.

1. Isomorphism Theorems

Each morphism $\phi: G \to H$ of groups has kernel N a normal subgroup of G (in symbols, $N \lhd G$) and image $\phi_* G$ a subgroup of H. By the main Theorem

II.26 on quotient groups, the projection $p:G \to G/N$ to the quotient group is universal with kernel N; hence, ϕ factors as $G \to G/N \to H$, and the image $\phi_* G$ is isomorphic to G/N. In case this image is normal in H, the quotient group $H/\phi_* G$ is called the *cokernel* of ϕ; the projection $q:H \to H/\phi_* G$ to the cokernel has the property that the composite $q\phi$ is trivial (maps all elements of G to 1) and q is universal with this property—just as the insertion $i:N \to G$ of the kernel of ϕ is universal among morphisms α to G with $\phi\alpha$ trivial.

Consider the effect of the morphism $\phi:G \to H$ on quotient groups. By Theorem II.22, the assignment $S \mapsto \phi_* S$ which sends each subgroup S of G to its image under ϕ in H is a bijection from the set of all those subgroups of G containing the kernel to the set of all those subgroups of H contained in the image. This bijection (sketched below)

$$\begin{array}{ccccccc}
 & & & & H & & \\
 & & & & \cup & & \\
 & & G & \to & \phi_* G & = & \operatorname{Im} \phi \\
 & & \cup & & \cup & & \\
 & & S & \to & \phi_* S & & \\
S/R & & \triangledown & & \triangledown & & (\phi_* S)/(\phi_* R) \\
 & & R & \to & \phi_* R & & \\
 & & \cup & & \cup & & \\
\operatorname{Ker} \phi & = & N & \to & 1 & & \\
 & & \cup & & & & \\
 & & 1 & & & &
\end{array}$$

is also an isomorphism of lattices (only a small indication of the whole lattice of subgroups between N and G is displayed). Under this bijection, quotients of corresponding "layers" are isomorphic, in the following sense.

THEOREM 1. *Let $\phi:G \to H$ be any morphism of groups with kernel $N \lhd G$, while R and S are subgroups of G with $N \subset R \subset S \subset G$. Then $R \lhd S$ if and only if $\phi_* R \lhd \phi_* S$; when this is the case, there is an isomorphism*

$$S/R \cong (\phi_* S)/(\phi_* R), \qquad Rs \mapsto (\phi_* R)(\phi s). \tag{1}$$

In particular, for $R = N$ this gives $N \lhd S$ and $S/N \cong \phi_* S$.

The proof is an application of the main theorem on quotient groups (Theorem II.26); Corollary 3 of that theorem states that any morphism of groups has image isomorphic to the quotient group, domain modulo kernel, hence $\phi_* G \cong G/N$ above. We also apply this theorem to the epimorphism $\phi_1:S \to \phi S$ given by restricting ϕ to S; then $\phi_* S \cong S/N$.

First suppose $\phi_* R \lhd \phi_* S$; construct the quotient group $\phi_* S/\phi_* R$, the corresponding projection $q:\phi_* S \to \phi_* S/\phi_* R$, and the composite $q \circ \phi_1:S \to$

$\phi_* S/\phi_* R$. It has kernel R; as the kernel of a morphism, R must be normal in S. By the main theorem again, one then has $S/R \cong \phi_* S/\phi_* R$. Moreover, this isomorphism sends each coset Rs of S/R to $q\phi_1(s)$; that is, to the coset $(\phi_* R)(\phi s)$ of the subgroup $\phi_* R \subset \phi_* S$. This proves (1).

Conversely, suppose that $R \lhd S$. Since the epimorphism $\phi_1 : S \to \phi_* S$ has kernel N, it is universal for this kernel. In other words (Theorem II.26), the projection $p : S \to S/R$ factors as $p = p' \circ \phi_1$, for a morphism $p' : \phi_* S \to S/R$, and the kernel of p' is $\phi_* R$. Hence, $\phi_* R \lhd \phi_* S$, just as asserted.

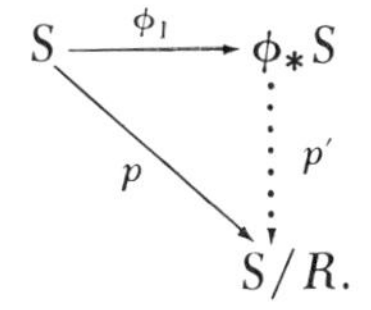

Corollary (*The Double-Quotient Isomorphism Theorem*). *If $N \lhd G$, then the subgroups of G/N all have the form R/N for some subgroup R with $N \subset R \subset G$. Also, R is normal in G if and only if R/N is normal in G/N; when this is so, the assignment $Rg \mapsto (R/N)(Ng)$ from cosets of R to cosets of R/N is an isomorphism*

$$G/R \cong (G/N)/(R/N).$$

More briefly: In a double quotient of groups, a common denominator N may be canceled.

Proof: Let $\phi : G \to G/N$ be the projection. Each subgroup of G/N has the form $\phi_* R$ for some R with $N \subset R \subset G$. But the image $\phi_* R$ consists of all cosets Nr of elements $r \in R$, hence is the quotient group R/N. The theorem, with $S = G$, now gives the corollary as stated.

Theorem 2 (*The Diamond Isomorphism Theorem*). *If S is any subgroup of G and N a normal subgroup of G, then the join $N \vee S$ is identical with the set NS of all products ns for $n \in N$, $s \in S$, the intersection $N \cap S$ is a normal subgroup of S, and the assignment $(N \cap S)s \mapsto Ns$ is an isomorphism*

$$\psi : S/(N \cap S) \cong (NS)/N.$$

It is convenient to visualize this theorem by the diagram shown below (a "diamond"):

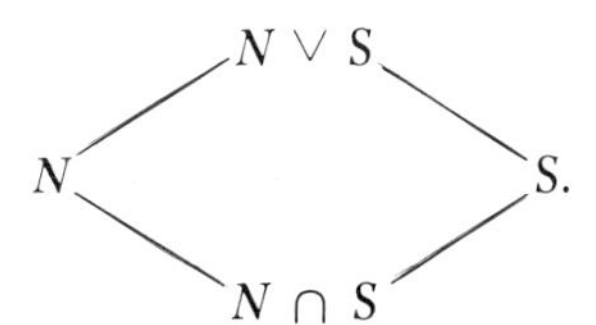

The theorem asserts that two opposite "sides" of this "diamond" are isomorphic. In case both N and S are normal in G, both pairs of opposite sides are isomorphic.

Proof: Proposition II.27 showed that if $N \lhd G$, then the join $N \vee S$ is the set NS of products. To construct the isomorphism, we start with the composite morphism α,

$$S \to G \to G/N$$

(insertion followed by projection). The kernel of α consists of those elements of S which project to 1, so is $S \cap N$. Hence, $(S \cap N) \lhd S$. The image of α is the group of all cosets of elements of S, hence is exactly NS/N. Thus α yields an epimorphism $S \to NS/N$ with kernel $N \cap S$; by the main theorem on quotient groups, the codomain is isomorphic to $S/(N \cap S)$, just as asserted.

These theorems apply in particular to abelian groups, with the simplification that all subgroups are automatically normal. Since the proofs use only formal properties of the kernel and the image and the universality of a projection $G \to G/N$, they also apply to modules over any ring R, as in the following statements.

Theorem 3. *Let $t: A \to C$ be a morphism of modules with kernel $N \subset A$. Then for any submodules L and S with $N \subset L \subset S \subset A$ there is an isomorphism*

$$S/L \cong (t_*S)/(t_*L), \qquad L + s \mapsto t_*(L) + t(s).$$

In particular, for $L = N$, $S/N \cong t_*S$. Also, $S/L \cong (S/N)/(L/N)$.

Theorem 4 (*The Diamond Isomorphism for Modules*). *For submodules S and T of a module A, the assignment $(S \cap T) + s \mapsto T + s$, for all $s \in S$, is an isomorphism*

$$S/(S \cap T) \cong (S + T)/T.$$

EXERCISES

1. In a product $G \times G$, the diagonal D is the subgroup of all pairs (g, g) for $g \in G$. Prove that G is abelian if and only if $D \lhd G \times G$.

2. Prove the double-quotient isomorphism theorem for rings: For $B \supset A$ two-sided ideals in a ring R there is an ideal B/A and an isomorphism

$$\theta : (R/A)/(B/A) \cong R/B$$

of rings which assigns to each coset $B/A + (A + x)$ the coset $B + x$.

2. Group Extensions

The "exact sequence" terminology, introduced for modules in §IX.9, is also useful in expressing interrelations between several groups and morphisms of groups. A sequence of two morphisms β and α of groups, as in

$$K \xrightarrow{\beta} G \xrightarrow{\alpha} H, \tag{2}$$

is said to be *exact* at G when $\operatorname{Im} \beta = \operatorname{Ker} \alpha$ (the image of the map coming into G is exactly the kernel of the map leaving G). Equivalently, exactness at G means both that $\alpha\beta(k) = 1$ for each $k \in K$ and that each $g \in G$ with $\alpha(g) = 1$ has the form $g = \beta(k)$ for some $k \in K$. For example, to say that a sequence $1 \to G \xrightarrow{\alpha} H$ is exact, with 1 the trivial group, is to say that α is a monomorphism, and to say that $K \xrightarrow{\beta} G \to 1$ is exact is to say that β is an epimorphism. A longer sequence $K \to G \to H \to L$ of groups and morphisms is similarly called *exact* when it is exact, in the sense described above, at each intermediate group—in this case, at the intermediate groups G and H.

A *short exact sequence* is a sequence of morphisms

$$1 \to K \xrightarrow{\beta} G \xrightarrow{\alpha} H \to 1 \tag{3}$$

of groups exact at K, at G, and at H. Exactness at K means that β is a monomorphism, at G that $\operatorname{Im}(\beta) = \operatorname{Ker}(\alpha)$, at H that α is an epimorphism; together, they imply $\beta_*K \lhd G$ and $G/\beta_*K \cong H$. Thus K is isomorphic to the kernel of α, while H is isomorphic to the cokernel G/β_*K of β. Conversely, any normal subgroup $N \lhd G$ yields a short exact sequence $1 \to N \to G \to G/N \to 1$ with morphisms the insertion $N \to G$ and the projection $G \to G/N$.

Given a short exact sequence (3), we say that the middle group G is an *extension* of the group K by the group H. This extension is said to *split* when there is a morphism $\gamma: H \to G$ of groups with $\alpha \circ \gamma = 1_H$. Two extensions G and G' (that is, two short exact sequences with the same end terms K and H and with middle terms G and G') are called "equivalent" when there is an isomorphism $\psi: G \cong G'$ such that the diagram

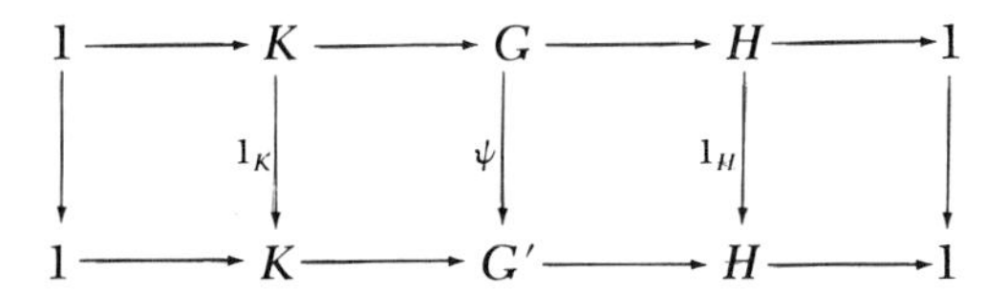

commutes. For example, the direct product $G \times H$ with the morphisms $g \mapsto (g, 1)$ and $(g, h) \mapsto h$ is an extension

$$1 \to G \to G \times H \to H \to 1$$

split by $h \mapsto (1, h)$. For each natural number $n \geqslant 3$, the dihedral group Δ_n of all symmetries of the regular n-gon (§II.5) contains a cyclic subgroup C_n of order n generated by the rotation of the n-gon through $360°/n$; with this, Δ_n is an extension

$$1 \to C_n \to \Delta_n \to \Delta_n / C_n \to 1$$

of a cyclic group of order n by the group Δ_n / C_n of order 2. This extension splits (map the generator of Δ_n / C_n of order 2 to any reflection of the n-gon).

We recall (§II.1) that the automorphisms of any group B form under composition a group Aut B.

The product $B \times H$ of two groups B and H may be generalized to the construction of a "semidirect" product $B \times_\theta H$ depending on B, on H, and on a morphism $\theta : H \to \operatorname{Aut} B$. Write each automorphism $\theta(h) : B \to B$ as $\theta(h)b = h * b$, for $h \in H$ and $b \in B$. For all $g, h \in H$ and all $b, c \in B$ the identities

$$h * (bc) = (h * b)(h * c), \qquad (gh) * b = g * (h * b) \tag{4}$$

then hold, the first because $b \mapsto h * b$ is an endomorphism of B, the second because $h \mapsto \theta h$ is a morphism $H \to \operatorname{Aut} B$ of groups. In the set $B \times H$ of all pairs (b, g) we now define a product by

$$(b, g)(c, h) = (b(g * c), gh). \tag{5}$$

(*Motivation:* To move g past c, write $gc = (gcg^{-1})g$ and regard gcg^{-1} as $g * c$.)

A calculation with the rules (4) shows that this product is associative. The pair (1, 1) of the identity elements for B and for H is an identity for this multiplication. Also,

$$(b, g)(g^{-1} * b^{-1}, g^{-1}) = (b(g * (g^{-1} * b^{-1})), gg^{-1}) = (bb^{-1}, 1) = (1, 1),$$

so each element (b, g) has a right inverse $(g^{-1} * b^{-1}, g^{-1})$. This element is also a left inverse. Hence, the set of all pairs (b, g) is a group $B \times_\theta H$ under this multiplication. It is called the *semidirect product* of B and H, relative to θ.

There are many examples. When $\theta : H \to \operatorname{Aut} B$ is the trivial morphism which takes every $h \in H$ to the identity automorphism 1_B, the semidirect product $B \times_\theta H$ is just the usual product $B \times H$. Again, for any group B, take $H = \operatorname{Aut} B$ and θ the identity $1 : \operatorname{Aut} B \to \operatorname{Aut} B$. The resulting semidirect product is called the "holomorph" of the given group B, written Hol (B).

Let C_n be any cyclic group of order n with generator a. If k is any integer prime to n, a^k also generates C_n, so the assignment $a^i \mapsto a^{ki}$ for all $i \in \mathbf{n}$ defines an automorphism α of C_n; this automorphism has $\alpha^m = 1$ for any integer m with $k^m \equiv 1 \pmod{n}$. Thus, if C_m is a cyclic group of order m and generator d, $\theta(d) = \alpha$ defines a morphism $\theta: C_m \to \operatorname{Aut} C_n$. The corresponding semidirect product $C_n \times_\theta C_m$ consists of the pairs (a^i, d^j) for $0 \leqslant i < n$ and $0 \leqslant j < m$. If we write a for $(a, 1)$ and d for $(1, d)$, the definition (5) gives $(a^i, d^j) = a^i d^j$ and $da = a^k d$. Hence, $C_n \times_\theta C_m$ is generated by two elements a and d satisfying the relations

$$a^n = 1, \qquad d^m = 1, \qquad dad^{-1} = a^k. \tag{6}$$

These relations suffice to calculate any product $(a^i d^j)(a^h d^l)$, so they are defining relations for this group, which is known as a *metacyclic* group. This group $C_n \times_\theta C_m$ depends on the three positive integers m, n, k with $k^m \equiv 1 \pmod{n}$. In particular, if $k = n - 1$ and $m = 2$, these equations describe the dihedral group Δ_n of (II.17).

Any semidirect product $B \times_\theta H$ may be described by an exact sequence. By the definition (5) of the multiplication, $(b, 1)(c, 1) = (bc, 1)$ so $b \mapsto (b, 1)$ is a morphism $\beta: B \to B \times_\theta H$. Similarly, $(b, g) \mapsto g$ is a morphism α. These morphisms give a sequence

$$1 \to B \xrightarrow{\beta} B \times_\theta H \xrightarrow{\alpha} H \to 1 \tag{7}$$

which is exact, for β is a monomorphism, $\operatorname{Im} \beta = \operatorname{Ker} \alpha$, and α is an epimorphism. Also $h \mapsto (1, h)$ is a morphism $\gamma: H \to B \times_\theta H$ with $\alpha \circ \gamma = 1$, so this sequence splits. If $B_0 = \beta_* B$ and $H_0 = \gamma_* H$, the subgroups B_0 and H_0 satisfy $B_0 \cap H_0 = 1$ and $B_0 \vee H_0 = B \times_\theta H$. This observation has the following converse.

Proposition 5. *Any group G with subgroups $N \lhd G$ and S such that both $N \cap S = 1$ and $N \vee S = G$ is a semidirect product. Explicitly, if $\theta: S \to \operatorname{Aut} N$ assigns to each $s \in S$ the automorphism $n \mapsto sns^{-1}$ of N, an isomorphism $\phi: N \times_\theta S \cong G$ is given by $\phi(n, s) = ns$.*

Proof: Since $N \lhd G$, conjugation by s is an automorphism of N; as above, we write this automorphism as $n \mapsto s * n = sns^{-1}$. Since $N \vee S = G$, every element in G is a product ns of $n \in N$ and $s \in S$; as $N \cap S = 1$, this representation is unique. Moreover, $(ns)(n't) = n(sn's^{-1})st = n(s * n')st$. Comparison with the multiplication rule (5) for the semidirect product shows that $(n, s) \mapsto ns$ is a morphism of groups, hence an isomorphism, as desired.

This proposition may be summarized by the diagram

$$\begin{array}{ccccccccc}
1 & \longrightarrow & N & \xrightarrow{\beta} & N \times_\theta S & \xrightarrow{\alpha} & S & \longrightarrow & 1 \\
 & & \downarrow{\scriptstyle 1_N} & & \downarrow{\scriptstyle \phi} & & \downarrow{\scriptstyle \psi} & & \\
1 & \longrightarrow & N & \longrightarrow & G & \xrightarrow{p} & G/N & \longrightarrow & 1.
\end{array} \tag{8}$$

In this diagram, the top row is the split short exact sequence (7) for the semidirect product $N \times_\theta S$, and the bottom row is the short exact sequence (also split) given by the projection of G on its quotient group. The vertical map ϕ in the middle is the isomorphism of the theorem, and the vertical map ψ at the right is the isomorphism $\psi: S \cong G/N$ which is a special case of the diamond isomorphism theorem with $G = N \vee S$ and $1 = N \cap S$. Since $\phi\beta(n) = \phi(n, 1) = n$ and $\psi\alpha(n, s) = \psi(s) = Ns = p\phi(n, s)$, the diagram is commutative.

Corollary. *Any split short exact sequence*

$$1 \to K \to G \to H \to 1$$

yields an isomorphism $G \cong K \times_\theta H$ *for a suitable* $\theta: H \to \operatorname{Aut} K$.

Proof: Let the given sequence be split by a morphism $\gamma: H \to G$. Then the image N of K and $S = \gamma_* H$ satisfy the hypotheses of the proposition.

Appendix: *Character Groups of Abelian Groups.* Let T denote the additive group $T = \mathbf{R}/\mathbf{Z}$; it is isomorphic to the multiplicative group of complex numbers of absolute value 1. The *character group* of an abelian group A is defined to be the group

$$\operatorname{Ch}(A) = \operatorname{Hom}_z(A, T). \tag{9}$$

If $\alpha: A \to B$ is a morphism of abelian groups, then $\operatorname{Ch}(\alpha): \operatorname{Ch}(B) \to \operatorname{Ch}(A)$ is the morphism which sends each character $g: B \to T$ of B into the character $g \circ \alpha: A \to T$ of A. For each group A, a morphism $\chi = \chi_A: A \to \operatorname{Ch}(\operatorname{Ch} A)$ is defined by $(\chi a)f = f(a)$, for $a \in A$ and f a character $f: A \to T$. Some of the basic properties of these character groups are set down in Exercises 6–10 below.

EXERCISES

1. For the holomorph, prove $\operatorname{Hol}(\mathbf{Z}_3) \cong S_3$ and $\operatorname{Hol}(\mathbf{Z}_2 \times \mathbf{Z}_2) \cong S_4$. (*Hint:* Construct a split exact sequence with S_4 as middle term.)

*2. Define a "holomorphism" h of a group B to be a bijection $h: B \to B$ such that $h(ab^{-1}c) = (ha)(hb)^{-1}(hc)$ for all $a, b, c \in B$. Show that the set of all holomorphisms of B under composition form a group isomorphic to $\operatorname{Hol}(B)$.

3. If $(h, b) \mapsto h * b$ is any function $H \times B \to B$ satisfying the identities (4), show that there is a unique morphism $\theta: H \to \operatorname{Aut} B$ with $(\theta h)b = h * b$.

4. Exhibit four nonisomorphic semidirect products $C_7 \times_\theta C_6$; for k as in (6), show that $k = 1$ yields an abelian group, $k = -1$ the group $C_3 \times \Delta_7$, $k = 2$ a group $C_2 \times H_{21}$, where H_{21} is a nonabelian group generated by d^2 and a, and $k = 3$ the group $\operatorname{Hol}(C_7)$.

5. (*The Short Five Lemma.*) Consider a commutative diagram

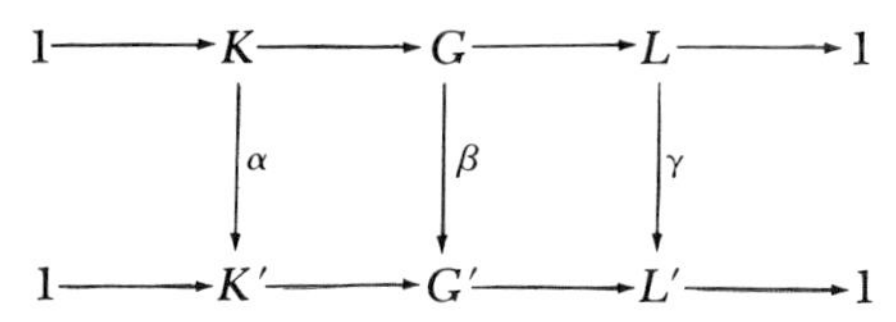

of morphisms of groups, with each row a short exact sequence. Prove:

α and γ monomorphisms imply β a monomorphism;

α and γ epimorphisms imply β an epimorphism.

(*Hint:* See the corresponding lemma of §IX.9 for modules.)

6. For each positive integer n, show that the group $T = \mathbf{R}/\mathbf{Z}$ has exactly one cyclic subgroup of order n.

7. (a) For any positive n, prove $\mathrm{Ch}\,(\mathbf{Z}_n) \cong \mathbf{Z}_n$.

(b) For A finite abelian, prove $\mathrm{Ch}\,(A) \cong A$.

(c) For A finite abelian, deduce that $\chi_A : A \to \mathrm{Ch}\,(\mathrm{Ch}\,A)$ is an isomorphism.

8. (a) Show that $A \mapsto \mathrm{Ch}\,(A)$ and $\alpha \mapsto \mathrm{Ch}\,(\alpha)$ define a contravariant functor on the category of all abelian groups to itself.

(b) For any $\alpha : A \to B$, prove that $\mathrm{Ch}\,(\mathrm{Ch}\,\alpha) \circ \chi_A = \chi_B \circ \alpha$, and draw the diagram.

(c) Given A, $\chi_{\mathrm{Ch}\,(A)} : \mathrm{Ch}\,(A) \to \mathrm{Ch}\,\mathrm{Ch}\,\mathrm{Ch}\,(A)$ and

$$\mathrm{Ch}\,(\chi_A) : \mathrm{Ch}\,\mathrm{Ch}\,\mathrm{Ch}\,(A) \to \mathrm{Ch}\,(A),$$

prove that $\mathrm{Ch}\,(\chi_A) \circ \chi_{\mathrm{Ch}\,(A)} = 1$.

9. (a) If $0 \to A \xrightarrow{\alpha} B \xrightarrow{\beta} C \to 0$ is a short exact sequence of abelian groups, prove $0 \to \mathrm{Ch}\,(C) \to \mathrm{Ch}\,(B) \to \mathrm{Ch}\,(A)$ exact, with maps $\mathrm{Ch}\,\beta$, $\mathrm{Ch}\,\alpha$.

(b) If B is finite, show also that $\mathrm{Ch}\,(B) \to \mathrm{Ch}\,(A) \to 0$ is exact.

10. If $S \subset A$ and $j : S \to A$ is the insertion, the kernel of $\mathrm{Ch}\,(j)$: $\mathrm{Ch}\,(A) \to \mathrm{Ch}\,(S)$ is called the "annihilator" of S in $\mathrm{Ch}\,(A)$, and written $\mathrm{An}\,(S)$.

(a) For A finite abelian, prove that $S \mapsto \mathrm{An}\,(S)$ is a bijection from the lattice of all subgroups of A to that of all subgroups of $\mathrm{Ch}\,(A)$, and that this bijection carries meets to joins and joins to meets (a "dual automorphism").

(b) For A finite abelian, prove that $\mathrm{An}\,(S) \cong \mathrm{Ch}\,(A/S)$ and also that $\mathrm{Ch}\,(S) \cong (\mathrm{Ch}\,A)/(\mathrm{An}\,S)$.

3. Characteristic Subgroups

A subgroup C of a group G is called a *characteristic subgroup* of G when $\alpha_* C = C$ for every automorphism α of G. Because the inverse of each α is also an automorphism, it is enough to require that $\alpha_* C \subset C$ for all α. Since the automorphisms include the inner automorphisms $g \mapsto aga^{-1}$ (for a fixed in G), every characteristic subgroup of G is automatically normal in G. Conversely, however, not every normal subgroup is characteristic; for example, the four-group $\mathbf{Z}_2 \times \mathbf{Z}_2$ is abelian, so all its subgroups are normal but the only characteristic subgroups are the improper subgroups 1 and $\mathbf{Z}_2 \times \mathbf{Z}_2$.

For an additive abelian group A we have already used in Chapter XI several characteristic subgroups: The torsion subgroup (all elements of finite order in A) and, for each positive integer n, the subgroups An (all multiples na) and the subgroups ${}^{(n)}A$ (all elements a with $na = 0$).

Characteristic subgroups of nonabelian groups G may be constructed similarly, by using properties invariant under all automorphisms of G. For example, the *center* Z of G, defined as

$$Z(G) = \{a \mid a \in G \text{ and } ag = ga \text{ for all } g \in G\}, \tag{10}$$

is an abelian subgroup of G. If $a \in Z(G)$ and β is any automorphism of G, then $(\beta a)g = \beta(a(\beta^{-1}g)) = \beta((\beta^{-1}g)a) = g(\beta a)$ for all $g \in G$, and so $\beta a \in Z(G)$. Hence, the center is a characteristic subgroup.

Evidently, a group G is abelian if and only if it coincides with its center. For other groups, the notion of the *commutator*

$$[a, b] = aba^{-1}b^{-1} \tag{11}$$

of two elements a and b is of interest. Note that $ab = [a, b]ba$; in particular, the commutator $[a, b]$ is 1 precisely when the elements a and b commute. Since $(aba^{-1}b^{-1})^{-1} = bab^{-1}a^{-1}$, the inverse of each commutator $[a, b]$ is the commutator $[b, a]$, and the identity $1 = [1, 1]$ is a commutator. However, the product of two commutators need not be a commutator. Therefore, the set of all products $[a_1, b_1] \cdots [a_k, b_k]$ of commutators is a subgroup $[G, G]$, called the *commutator subgroup* of G.

Theorem 6. *In any group G, the commutator subgroup $[G, G]$ is a characteristic subgroup, the quotient group $G/[G, G]$ is abelian, and the projection $p: G \to G/[G, G]$ is universal among morphisms from G to abelian groups.*

Proof: Since any automorphism α of G carries a commutator $[a, b]$ to a commutator $[\alpha a, \alpha b]$, the subgroup $[G, G]$ is characteristic and hence normal. In the quotient group $G/[G, G]$ all commutators are 1, so this quotient group is abelian. If $\alpha: G \to A$ is any morphism to an abelian group A, $\alpha(aba^{-1}b^{-1}) = (\alpha a)(\alpha b)(\alpha a)^{-1}(\alpha b)^{-1} = 1$ because A is abelian, so the kernel of α includes all commutators. By the main theorem on the projection $p: G \to G/[G, G]$ to a quotient group, $\alpha = \alpha' \circ p$ for a unique α'. This means exactly that p is universal, as stated.

Here is an equivalent way of formulating this universality:

COROLLARY. *A normal subgroup $N \lhd G$ has an abelian quotient group G/N if and only if $N \supset [G, G]$.*

EXERCISES

1. In an abelian group A, show for each prime p that the set $T_p(A)$ of all those elements of A with order some power of p is a characteristic subgroup of A.

2. For subgroups S and T of G, let $[S, T]$ be the subgroup of G generated by all commutators $[s, t]$ for $s \in S$ and $t \in T$. If S and T are characteristic in G, prove that $[S, T]$ is also a characteristic subgroup of G.

3. In any group, prove the identity

$$[ab, c] = (a[b, c]a^{-1})[a, c] = [b, c][[c, b], a][a, c].$$

4. Consider the dihedral group Δ_n with generators a, d and defining relations $a^n = d^2 = 1$ and $dad^{-1} = a^{-1}$.
   (a) Prove that $a^2 \in [\Delta_n, \Delta_n]$.
   (b) If n is odd, prove $[\Delta_n, \Delta_n]$ cyclic of order n.
   (c) If n is even, prove $[\Delta_n, \Delta_n]$ cyclic of order $n/2$.

*5. If G is the multiplicative group of all 3×3 matrices, over a field F, of the form

$$\begin{bmatrix} 1 & a & b \\ 0 & 1 & c \\ 0 & 0 & 1 \end{bmatrix},$$

with $a, b, c \in \mathbf{Z}$, prove that the commutator group of G equals its center.

6. If $N \lhd G$, show that any characteristic subgroup C of N is a normal subgroup of G.

4. Conjugate Classes

To examine the possible orders of elements and of subgroups in a finite group we will use the operation of conjugation in a group. *Conjugation* by an element a in a group G is the function $\gamma_a : G \to G$ defined for each $g \in G$ by $\gamma_a(g) = aga^{-1}$, just as in §II.7. Since

$$(\gamma_a g)(\gamma_a h) = (aga^{-1})(aha^{-1}) = a(gh)a^{-1} = \gamma_a(gh),$$

this function is an endomorphism of G. Successive conjugation, first by b and then by a, is

$$(\gamma_a \gamma_b)g = \gamma_a(\gamma_b g) = a(bgb^{-1})a^{-1} = (ab)g(ab)^{-1} = \gamma_{ab} g \qquad (12)$$

for all $g \in G$; hence, $\gamma_a \circ \gamma_b = \gamma_{ab}$. In particular, $\gamma_a \circ \gamma_{a^{-1}} = \gamma_1 = 1_G$, so each endomorphism γ_a has $\gamma_{(a^{-1})}$ as two-sided inverse and is therefore an automorphism of G. Any automorphism of G of the form γ_a is called an *inner automorphism* of G.

The equation $\gamma_a\gamma_b = \gamma_{ab}$ states more: That $a \mapsto \gamma_a$ is a morphism $\gamma: G \to \operatorname{Aut}(G)$ on G to the group $\operatorname{Aut}(G)$ of all automorphisms of G. The kernel of γ consists of all elements a with $aga^{-1} = g$ for every g, hence is the center $Z(G)$ of G. The image of γ is a subgroup of $\operatorname{Aut}(G)$ called the *group* $\operatorname{In}(G)$ *of inner automorphisms*. Since γ has kernel $Z(G)$ and image $\operatorname{In}(G)$, one has $\operatorname{In}(G) \cong G/Z(G)$.

Proposition 7. *For any group G, $\operatorname{In}(G)$ is a normal subgroup of* $\operatorname{Aut}(G)$.

Proof: For $\alpha: G \to G$ any automorphism of G and γ_a an inner automorphism of G one must prove $\alpha\gamma_a\alpha^{-1}$ an inner automorphism. But for any $g \in G$, as α is a morphism,

$$(\alpha\gamma_a\alpha^{-1})g = \alpha\left[a(\alpha^{-1}g)a^{-1}\right] = (\alpha a)(\alpha\alpha^{-1}g)(\alpha a^{-1}) = (\alpha a)g(\alpha a)^{-1};$$

this states that $\alpha\gamma_a\alpha^{-1} = \gamma_{\alpha a}$ is inner, as required.

Since $\operatorname{In}(G) \lhd \operatorname{Aut}(G)$ one may form the corresponding quotient group $\operatorname{Aut}(G)/\operatorname{In}(G)$ and with it the exact sequence

$$1 \to Z(G) \to G \xrightarrow{\gamma} \operatorname{Aut}(G) \to \operatorname{Aut}(G)/\operatorname{In}(G) \to 1. \tag{13}$$

The exactness of this sequence states that γ has kernel $Z(G)$ and cokernel $\operatorname{Aut}(G)/\operatorname{In}(G)$.

Each group G acts on itself by conjugation, as in §II.7. Explicitly, $a \in G$ acts on $g \in G$ by $(a, g) \mapsto \gamma_a g = aga^{-1}$. Under this action the subgroup of all elements a leaving g fixed is the subgroup

$$C_G(g) = \{a \mid aga^{-1} = g\},$$

called the "centralizer" of g in G, while the orbit of g is the set

$$D = D_g = \{aga^{-1} \mid a \in G\}$$

or "conjugate class" which consists of all conjugates of g.

Now suppose G finite. If g is fixed, Theorem II.21 states that the number of conjugates of g is simply the index, written $[G:C_G(g)]$, of the centralizer of g in G. Every element $g \in G$ belongs to exactly one conjugate class. If Δ is the set of all conjugate classes, and if we choose one element g_D from each conjugate class D, the total number of elements in G (the index of 1 in G) is

$$[G:1] = \sum_{D \in \Delta} [G : C_G(g_D)], \qquad g_D \in D. \tag{14}$$

This is called the *class equation* for the finite group G. Now the conjugate class of g consists of g alone if and only if $aga^{-1} = g$ for all $a \in G$; that is, if and only if g is in the center $Z(G)$ of G. In other words, each element in the center is in a conjugate class by itself, and these one-element classes together make up $[Z(G):1]$ elements. Therefore, for a finite group G, the class equation (14) may be rewritten as

$$[G:1] = [Z(G):1] + \sum_{D \in \Delta'} [G : C_G(g_D)], \qquad g_D \in D, \quad (15)$$

where the sum is taken over the set Δ' of all those conjugate classes D having more than one element.

We now draw some consequences. The first, due to Cauchy, is a partial converse of Lagrange's theorem. (The order of every element divides the order of G.)

Theorem 8 (*Cauchy*). *If a prime number p divides the order of a finite group G, then G has an element of order p.*

Proof: First suppose that G is abelian. By the Corollary XI.6 on the structure of finite abelian groups, G is a product of cyclic groups, and the order of G is the product of the orders of these factors. Hence, any prime factor p of the order of G must divide the order of at least one of these cyclic factors, so that factor (and hence G) does contain an element of order p.

With this special case in hand, we now prove the theorem by induction on the order of G. If G has any proper subgroup S with order divisible by p, then by the induction assumption this subgroup S (and hence G) contains the desired element of order p. Otherwise, the order of every proper subgroup S is prime to p. Since p divides the order of G, which is $[G:1] = [G:S][S:1]$, it follows that p divides the index $[G:S]$ of every proper subgroup S. Now consider the class equation (15). The prime p divides $[G:1]$ and each $[G:C_G] > 1$; hence, it must also divide the one remaining term $[Z(G):1]$. But this term is the order of the center $Z(G)$, an abelian group. Hence, by the previous special case, $Z(G)$ contains the desired element of order p.

Let p be any prime number. A group G is called a *p-group* when the order of every element in G is some power of p.

Corollary. *A finite group is a p-group if and only if its order is some power p^e of the prime p.*

Proof: If $[G:1] = p^e$, then by Lagrange's theorem the order of every element in G is divisor of p^e, hence is some power of p. Conversely, suppose that G is a p-group, and write its order as a product $n = p^e m$, where the second factor m is relatively prime to p. If $m \neq 1$, the order of G is divisible by some prime $q \neq p$, so by the Cauchy theorem G contains an element of order q, a contradiction to the assumption that every element had order a power of p.

Many groups have a trivial center (for example, every symmetric group S_n with $n \neq 2$). But the class equation shows that p-groups must have nontrivial centers:

THEOREM 9 (*Burnside*). *Any finite p-group $G > 1$ has center $Z(G) > 1$.*

Proof: If not, the class equation (15) expresses the order p^e of G as a sum of 1 plus a number of indices $[G:C]$ which are all positive powers of p. This is a contradiction.

COROLLARY. *Any group of order p^2 (p a prime) is abelian.*

For, otherwise, the nontrivial center would be of order p, hence cyclic with a generator c such that $c^p = 1$. If b is an element not in the center, then b and c together must generate the whole group. But $bc = cb$ because c is in the center, hence the group is indeed abelian.

EXERCISES

1. Find the number of conjugates of the cycle $(1\ 2\ 3)$ in S_4 and in S_5.
2. Find the number of conjugates of $(1\ 2)(3\ 4)$ in A_4 and in S_4.
3. If $G/Z(G)$ is cyclic, prove that G must be abelian (and hence equal to $Z(G)$).
4. Prove that every non-abelian group of order p^3 has center of order p.

*5. Give a direct proof of Theorem 8 for an additive abelian group A by listing all elements $a_1, \cdots, a_n$ of A, taking the cyclic subgroup C_{a_i} generated by each a_i, constructing an epimorphism $C_{a_1} \times \cdots \times C_{a_n} \to A$, and arguing thence that p divides the order of some a_i.

*6. If p is a prime number and a group G of order $p + 1$ has an automorphism α of order p, prove that G is abelian and that there is a prime number q with $g^q = 1$ for every $g \in G$.

5. The Sylow Theorems

The Sylow theorems study *all* the possible p-subgroups of a finite group G. Let the order of G be $n = p^e m$, with second factor m prime to p. Since any finite p-group has order some power p^f, any p-group which is a subgroup of G must have order some divisor of p^e. A p-subgroup $P \subset G$ with the maximal possible order p^e is called a *Sylow* p-subgroup of G. This definition thus implies that $S \subset G$ is a Sylow p-subgroup of G if and only if the order of S is a power of p and the index $[G:S]$ is relatively prime to p.

First we study these subgroups for an abelian group A. For each prime p, as in §XI.5, the set $T_p(A)$ of *all* elements in A of order some power of p is a subgroup of A. The primary decomposition theorem (Theorem XI.9) states that the finite abelian group A is a biproduct $T_{p_1}(A) \oplus \cdots \oplus T_{p_k}(A)$ of a finite list of these subgroups $T_p(A)$. Each factor T_{p_i} is a p_i-group, so has order some power $p_i^{e_i}$. Therefore, the order of A is $n = p_1^{e_1} \cdots p_k^{e_k}$. This shows that the list of primes $p_1, \cdots, p_k$ contains all the prime divisors of n. Therefore, each subgroup $T_{p_i}(A)$ is a Sylow p_i-subgroup of A. Now this subgroup was defined to contain all elements in A of order some power $p_i^{f_i}$, and therefore must contain every p_i-subgroup of A. In particular, $T_{p_i}(A)$ itself is the only p_i-subgroup of order $p_i^{e_i}$, so is the only Sylow p_i-subgroup of A.

These observations may be summarized as follows:

Proposition 10. *A finite abelian group A contains for each prime factor p of its order exactly one Sylow p-subgroup. This subgroup contains all p-subgroups of A, and A is the biproduct of its Sylow p-subgroups.*

For nonabelian groups the situation is less simple. Just for example, the symmetric group S_3 of order 6 contains one Sylow 3-subgroup (the cyclic group generated by the cycle (1 2 3)) and three Sylow 2-subgroups (the cyclic subgroups of order 2 generated by each of (12), (13), and (23)). In general, we prove first that a Sylow subgroup always exists.

Theorem 11 (*The First Sylow Theorem*). *Any finite group G contains to each prime p at least one Sylow p-subgroup.*

The proof is by induction on the order n of G. If n is relatively prime to p or if $n = p$ the result is immediate. Hence, let G have order $n = p^e m$ with m prime to p, and suppose that the theorem is true for all groups of order smaller than n. If the center $Z(G)$ has order divisible by p, then $Z(G)$ must contain an element a of order p. The cyclic subgroup C generated by a, as part of the center, is surely normal in G, so we may construct the factor group G/C, with order $p^{e-1}m$. By the induction assumption, G/C contains a Sylow p-subgroup P' of order p^{e-1}. By Corollary 1.1, this subgroup P' of G/C has the form $P' = P/C$, where the subgroup P of G must have order $p^{e-1}p = p^e$. Hence, the subgroup P is the desired Sylow p-subgroup of G.

There remains the case when the order of $Z(G)$ is relatively prime to p. We take the class equation in the second form (15). Since $[G:1]$ but not $[Z(G):1]$ is divisible by p, at least one of the remaining indices $[G:C_G(g_D)]$ in this equation must be relatively prime to p. This means that at least one of the centralizers $C_G(g_D)$ has order divisible by p^e, hence, by the induction assumption, contains a subgroup Q of order p^e. This Q is also a Sylow p-subgroup in the whole group G. The proof is complete.

Now let $\mathbb{S}$ be the set of all subgroups S of G. Then the assignment $(g, S) \mapsto gSg^{-1}$ defines an action of G on $\mathbb{S}$. The orbit of a subgroup S under this action is the set of all its conjugate subgroups. The isotropy group of S (the subgroup of G leaving S fixed in this action) is

$$N_G(S) = \{\text{all } g \mid g \in G \text{ and } gSg^{-1} = S\}. \tag{16}$$

This subgroup of G is called the *normalizer* of the subgroup S in G. It is a subgroup of G which contains S and has S as normal subgroup—indeed, it contains all those subgroups H of G in which S is normal. Moreover, Theorem II.21 states that the number of distinct conjugates of S in G is the index $[G:N_G(S)]$ of its normalizer.

Using this action of G on the set of its subgroups we can establish some further Sylow theorems. First, a preliminary result:

Lemma. *If P is a Sylow p-subgroup of the finite group G, any p-subgroup of the normalizer $N_G(P)$ of P is contained in P.*

Proof: Let R be a p-group contained in $N_G(P)$. Since $P \lhd N_G(P)$ and $R \subset N_G(P)$, the diamond isomorphism theorem for $N_G(P)$ shows that the join $P \vee R$ is the product PR and that $RP/P \cong R/(P \cap R)$; hence the order of RP/P is a power of p. The same is then true for the whole group RP. But the Sylow subgroup P of G has order the maximal possible power of p. Hence, $RP = P$ and therefore $R \subset P$, as required.

Theorem 12 (*The Second Sylow Theorem*). *Any two Sylow p-subgroups of a finite group G are conjugate in G. The number k of distinct Sylow p-subgroups of G is a divisor of $[G:1]$ and satisfies $k \equiv 1 \pmod p$.*

Proof: Let G act (by conjugation) on the set $\mathbb{S}$ of all its subgroups. Take a Sylow p-subgroup $P = P_1$; its orbit under this action is the set

$$\mathfrak{U} = \{P = P_1, P_2, P_3, \cdots, P_t\}$$

of all subgroups of G conjugate to P. Each of these subgroups is isomorphic (via conjugation) to P, hence has the same order as P, hence is a Sylow p-subgroup of G. Moreover, by Theorem II.21, the number t of these conjugates is the index in G of the normalizer of P, hence is a divisor of $[G:1]$.

Now let the given subgroup P act—again by conjugation—on this set $\mathfrak{U}$ of subgroups of G. Conjugation by any element of P leaves P fixed, so $P \in \mathfrak{U}$ is a one-point orbit in $\mathfrak{U}$ under this action. If P_i is some group in $\mathfrak{U}$ left fixed by this action, then P lies in the normalizer of P_i, hence $P \subset P_i$ by the lemma, and hence $P = P_i$ (since both these Sylow subgroups have the same order). Therefore, P is the *only* one-point orbit in $\mathfrak{A}$. Thus the number t of elements in $\mathfrak{A}$ is $t = 1 + k_2 + \cdots + k_r$, where each k_i is the number of points in an orbit and so is an index $[P:C]$ of some proper subgroup $C \subset P$, hence is a positive power of p. Therefore, $t \equiv 1 \pmod p$.

Now suppose that G had some Sylow p-subgroup Q not a conjugate of P, hence not in the set $\mathfrak{A}$. Let this subgroup Q act on the set $\mathfrak{A}$ by conjugation. By the lemma again, Q cannot normalize any P_i, so there are no one-point orbits under the action of Q. Thus the number of elements in $\mathfrak{A}$ is $t = l_1 + \cdots + l_s$, where each l_i is an index $[Q:C'] > 1$, hence a positive power of p. This gives $t \equiv 0 \pmod p$, in contradiction to the previous result that $t \equiv 1 \pmod p$. This contradiction shows that every Sylow p-subgroup of G is one of the conjugates P_i of P in the set $\mathfrak{A}$, and hence that the number of these subgroups is as stated.

Corollary (*The Frattini Argument*). *If M is a normal subgroup of G and P a Sylow p-subgroup of M with normalizer $N_G(P)$ in G, then the product $MN_G(P)$ is G.*

Proof: For any $g \in G$ the conjugate subgroup gPg^{-1} is still contained in the normal subgroup $M = gMg^{-1}$, so both P and gPg^{-1} are Sylow p-subgroups of M. By the second Sylow theorem, they are then conjugate by an element $h \in M$, so $hPh^{-1} = gPg^{-1}$, and therefore $g^{-1}hP(g^{-1}h)^{-1} = P$. Thus gh^{-1} is in $N_G(P)$ and $g = h(gh)^{-1}$ is in the product $MN_G(P)$ as desired.

The Sylow p-subgroups of G are the p-subgroups of maximal possible order. We can say something about the other p-subgroups.

Theorem 13 (*The Third Sylow Theorem*). *Any p-subgroup R of a finite group G is contained in some Sylow p-subgroup.*

Proof: Take once again the set $\mathfrak{A}$ of all Sylow p-subgroups of G, as above, and let the given p-subgroup R act on the set $\mathfrak{A}$, by conjugation. Each orbit has 1 or $p^f > 1$ elements; since the total number of points P in $\mathfrak{A}$ is $t \equiv 1 \pmod p$, there must be at least one orbit with just one element; that is, there must be some P_i such that R lies in its normalizer. By the lemma, this means that $R \subset P_i$. Hence, R is contained in some Sylow subgroup, just as asserted.

As an application, we prove that any group of order pq, with p and q distinct primes, must be metacyclic. Since p and q are distinct, assume that $p > q$. Every Sylow p-subgroup is cyclic of order p, and the number k of such subgroups is either 1 or at least $p + 1$, by Theorem 12. Now two distinct such subgroups meet only in 1, so $p + 1$ such subgroups would contain together $(p - 1)(p + 1) + 1 = p^2$ different elements, too many for the order pq of G. Therefore there is just one Sylow p-subgroup P, cyclic of order p and with generator (say) a. There is also at least one Sylow q-subgroup, cyclic with generator d of order q. Since P equals each of its conjugates, $P \lhd G$, and $a \mapsto dad^{-1}$ gives an automorphism of P, of order 1 or q. Therefore, $dad^{-1} = a^k$ for some exponent k with $k^q \equiv 1 \pmod{p}$. We have thus derived the defining equations (6) for a metacyclic group of order pq.

For the next section we need one more Sylow theorem.

Theorem 14. *If P is a Sylow p-subgroup of the finite group G, $N_G(N_G(P)) = N_G(P)$.*

In other words, the normalizer $N_G(P)$ of any Sylow subgroup is its own normalizer.

Proof: Take any $t \in N_G(N_G(P))$. Then $tPt^{-1} \subset N_G(P)$, so by the lemma above, $tPt^{-1} \subset P$ and hence $tPt^{-1} = P$. Thus $t \in N_G(P)$, as asserted.

EXERCISES

1. Find, up to isomorphism, all groups of orders 15 and 21.

2. The quaternion group M of Exercise II.5.8 is the group with generators a and b and defining relations $a^4 = 1$, $b^2 = a^2$, $bab^{-1} = a^{-1}$; by Exercise VIII.8.4, M has order 8 and is isomorphic to the group of 8 elements $\{\pm 1, \pm i, \pm j, \pm k\}$ with the multiplication rules $i^2 = j^2 = k^2 = -1$, $ij = k = -ji$, $jk = i = -kj$, $ki = j = -ik$.

(a) Prove that M has center Z of order 2 and that $M/Z \cong \mathbf{Z}_2 \times \mathbf{Z}_2$.

(b) Show that M is not isomorphic to the dihedral group Δ_4.

***3.** Prove that a nonabelian group of order 8 is isomorphic to M or to Δ_4.

4. If a p-subgroup S of a finite group G is not a Sylow p-subgroup, prove that S is a proper subgroup of $N_G(S)$. (*Hint:* Let S act on the set of all conjugates of S in G.)

5. Let N be the normalizer in G of a Sylow p-subgroup P of the finite group G, while S and T are subgroups with $N \subset S \subset T \subset G$.

(a) Prove that $N_G(S) = S$.

(b) Prove that $[T:S] \equiv 1 \pmod{p}$. (*Hint:* Let P act by multiplication on the right cosets of S in T.)

6. Nilpotent Groups

The p-groups, although not necessarily abelian, all have nontrivial centers. They are examples of a larger class of groups, the nilpotent groups, which we now describe. The process of taking the center $Z(G)$ of a group G may be iterated, to define for each $k \in \mathbf{N}$ a normal subgroup $Z_k(G) \triangleleft G$, the kth center of G. For $k = 0$, $Z_0(G) = 1$, while for $k = 1$, $Z_1(G) = Z(G)$ is the center. Given $Z_k(G)$ normal in G, the quotient group $G/Z_k(G)$ has a center which, like all subgroups of G/Z_k, must have the form S/Z_k for a unique $S \subset Z_k$. This subgroup S is the $(k + 1)$-st *center*; in symbols,

$$Z_{k+1}(G)/Z_k(G) = Z(G/Z_k(G)). \tag{17}$$

Since $Z(G/Z_k(G))$ is normal in G/Z_k, Theorem 1 shows that Z_{k+1} is normal in G. These successive centers give a (possibly) increasing series

$$1 \subset Z_1(G) \subset Z_2(G) \subset \cdots \subset Z_k(G) \subset Z_{k+1}(G) \subset \cdots \tag{18}$$

of normal subgroups of G. This series is called the "ascending central series" or the "upper central series" of the group G. In general, it may never reach G; for example, when $G = \mathbf{Z}_2 \times S_3$, $Z_1(\mathbf{Z}_2 \times S_3) = \mathbf{Z}_2 \times 1$ and also every $Z_k(\mathbf{Z}_2 \times S_3)$ is $\mathbf{Z}_2 \times 1$. However, a finite group G is defined to be *nilpotent* when there is some index c with $Z_c(G) = G$ (and the first such index c is called the "class of nilpotency" of G). The Burnside theorem (Theorem 9) implies that *every* finite p-group is nilpotent. One may also show that a product of two nilpotent groups is nilpotent (Exercise 2). Here is a converse statement, which describes all (finite) nilpotent groups in terms of p-groups, and is an analog of the primary decomposition theorem for finite abelian groups as summarized in Proposition 10 above:

THEOREM 15. *Any finite nilpotent group is the direct product of its Sylow subgroups.*

The proof rests on the following lemma.

LEMMA. *Any proper subgroup S of a nilpotent group G is a proper subgroup of its normalizer $N_G(S)$ in G.*

Proof: Given $S \neq G$, find in the ascending central series the last index k with $S \supset Z_k(G)$. By the very definition of Z_{k+1}, $g \in Z_{k+1}$ and $s \in S$ have $gsg^{-1} \equiv s \pmod{Z_k}$. Since $Z_k \subset S$, this proves that $gSg^{-1} \subset S$ and hence that every element $g \in Z_{k+1}$ is in $N_G(S)$. As S does not contain Z_{k+1}, S is a proper subgroup of $N_G(S)$, Q.E.D.

Now return to the theorem. Let P be a Sylow p-subgroup of a nilpotent group G. If $P = G$, P is the only Sylow subgroup of G, and the theorem follows. Otherwise, by the lemma, P is a proper subgroup of $N_G(P)$. Since $N_G(P)$ is its own normalizer (Theorem 14), the Lemma then shows that $N_G(P)$ must be all of G. In other words, P is a normal subgroup of G. Hence, there is just one Sylow p-subgroup P for each prime divisor p of $[G:1]$. Take all of these Sylow subgroups $P_1, \cdots, P_k$, one for each prime factor p of

$[G:1]$. Since their orders are relatively prime, $P_i \cap P_j = 1$ if $i \neq j$. If $g \in P_1$ and $h \in P_2$, then $(ghg^{-1})h^{-1} = g(hg^{-1}h^{-1})$ is both in P_1 and in P_2, hence is 1. Therefore by Proposition II.13, the join $P_1 \vee P_2$ is the direct product of P_1 and P_2. By induction on k, G is the direct product of the subgroups $P_1, \cdots, P_k$.

EXERCISES

1. Show in detail that every finite p-group is nilpotent.
2. Prove that the product of two nilpotent groups is nilpotent.
3. (a) If G has order p^e with p prime, while $0 < f < e$, prove that G has a subgroup of order p^f.

   (b) If a nilpotent group G has order n while m divides n, prove that G has a subgroup of order m.
4. Prove that each $Z_k(G)$ is a characteristic subgroup of G.
5. Prove: A finite group is nilpotent if and only if it is a product of p-groups.
6. Using the commutator subgroup $[S, T]$ defined in Exercise 3.2, define a series $\Gamma_0(G) = G \supset \Gamma_1(G) \supset \cdots \supset \Gamma_k(G) \supset \cdots$ (the "descending central series" or the "lower central series" of G) by recursion on k as $\Gamma_0(G) = G$, $\Gamma_{k+1}(G) = [\Gamma_k(G), G]$.

   (a) Prove that each $\Gamma_k(G)$ is a characteristic subgroup of G.

   *(b) Show that $Z_c(G) = G$ if and only if $\Gamma_c(G) = 1$; when this is the case, show that $\Gamma_k(G) \subset Z_{c-k}(G)$ for all $k = 0, \cdots, c$.
7. If G is nilpotent of class 1 or 2, prove the identities

$$[ab, c] = [a, c][b, c], \qquad [a, bc] = [a, b][a, c].$$

7. Solvable Groups

Just as the class of nilpotent groups is defined by examining iterated centers, so the larger class of solvable groups is defined by examining iterated commutator subgroups.

The characteristic subgroup of G generated by all the commutators $aba^{-1}b^{-1}$, as in (11), is called the *commutator subgroup* $[G, G]$ or the "derived subgroup" $G' = [G, G]$ of G. Any morphism $\alpha: G \to H$ carries the commutator $aba^{-1}b^{-1}$ in G to the commutator $(\alpha a)(\alpha b)(\alpha a)^{-1}(\alpha b)^{-1}$ in H, hence α restricts to a morphism $\alpha': G' \to H'$ for the derived subgroups. If α is a monomorphism, so is α'; if α is an epimorphism, so is α'. The assignments $G \mapsto G'$ and $\alpha \mapsto \alpha'$ define a functor from groups to groups. This functor may be iterated, as $G^{(0)} = G$ and $G^{(k+1)} = (G^{(k)})'$ for any $k \in \mathbf{N}$; we call $G^{(k)}$ the kth *derived subgroup* of G. This defines a (possibly) decreasing series

$$G \supset G' \supset G^{(2)} \supset \cdots \supset G^{(k)} \supset G^{(k+1)} \supset \cdots$$

of characteristic subgroups of G. Finally, a group G is defined to be *solvable* when $G^{(k)} = 1$ for some k. (The word "solvable" comes from an application to the solution of equations by radicals, to be discussed in Chapter XIII.) Every abelian group A has $A' = 1$, so is automatically solvable. In §9 below we will show that the symmetric groups S_n for $n > 4$ are not solvable.

THEOREM 16. *All subgroups and all quotient groups of a solvable group are solvable. Any extension of a solvable group by a solvable group is solvable.*

Proof: If $S \subset G$, the insertion $j: S \to G$ is a monomorphism, and so is each $j^{(k)}: S^{(k)} \to G^{(k)}$. Hence, $G^{(k)} = 1$ for some k implies $S^{(k)} = 1$ for the same k, so S is solvable. The argument for a quotient group is similar. Finally, if G is an extension of a solvable group K by a solvable group H, as in the short exact sequence

$$1 \to K \xrightarrow{\beta} G \xrightarrow{\alpha} H \to 1,$$

then $H^{(k)} = 1$ shows that α restricts to $\alpha^{(k)}: G^{(k)} \to H^{(k)} = 1$, hence that $G^{(k)}$ is contained in the kernel $\beta_* K$ of α. But $K^{(m)} = 1$ implies $1 = (G^{(k)})^{(m)} = G^{(k+m)}$, so G is solvable, as claimed.

COROLLARY. *Any product of solvable groups is solvable.*

PROPOSITION 17. *Any nilpotent group is solvable.*

Proof: By the definition of the kth center, each $Z_k(G)/Z_{k-1}(G)$ is abelian, so any commutator of two elements of $Z_k(G)$ must lie in $Z_{k-1}(G)$. Hence, if G is nilpotent of class c, so that $Z_c(G) = G$, then $G' \subset Z_{c-1}$, $G'' \subset Z_{c-2}, \cdots, G^{(c)} = 1$, and therefore G is solvable.

Conversely, however, a solvable group need not be nilpotent; for example, S_3 is not nilpotent.

EXERCISES

1. Prove that each derived subgroup $G^{(k)}$ is a characteristic subgroup of G.
2. Show that S_3 and S_4 is solvable, but not nilpotent.
3. Show that the holomorph of any cyclic group is solvable.
4. If $M \lhd G$ and $N \lhd G$ with G/M and G/N solvable, prove that $G/M \cap N$ is solvable.
5. If G is a group of order pqr, where p, q, and r are three distinct prime numbers, prove that G is solvable.

8. The Jordan–Hölder Theorem

The utility of the ascending series of iterated centers and the descending series of derived groups suggests an examination of other possible series of subgroups in any group G. A series S of subgroups

$$G \supset N_1 \supset N_2 \supset \cdots \supset N_k = 1, \qquad G = N_0, \tag{19}$$

reaching from $G = N_0$ to the identity subgroup $1 = N_k$ is called a *subnormal series* (the customary term is "normal series") when each N_i is a normal subgroup of the preceding group of the series: $N_i \lhd N_{i-1}$, for $i = 1, \cdots, k$. A subgroup of G is said to be "subnormal" in G precisely when it is a term in such a series. A subnormal series for G of length k, as shown, determines a list $G/N_1, N_1/N_2, \cdots, N_{k-1}/N_k$ of k quotient groups, called the "factors" of the series. A second subnormal series $G \supset M_1 \supset M_2 \supset \cdots \supset M_k = 1$ of the same length for the same group G is said to have *factors isomorphic* to those of the first series when, after a permutation of factors, each factor of the second series is isomorphic to the corresponding factor of the first series.

A group S is called *simple* when it has no proper normal subgroups; for example, every cyclic group of prime order is simple; in the next section we shall meet examples of simple permutation groups. Other typical examples of simple groups can be constructed from matrices over finite fields F. One such is the projective special linear group $PSL(n, F)$, to be described at the end of §9. Recently, a number of other "sporadic" examples of large finite simple groups have been discovered.

A *composition series* for G is a subnormal series in which all the factors are simple, and $\neq 1$; hence, a composition series is a subnormal series with no refinements except itself. (Here a "refinement" of a series S is any series obtained from S by successive insertions of an additional subgroup properly between two successive subgroups of the series). Every finite group has at least one composition series, for we may take N_1 to be any maximal proper normal subgroup of G, N_2 to be any maximal proper normal subgroup of this N_1, and so on. However, the infinite group $\mathbf{Z}$ has no composition series.

For example, a cyclic group C of prime power order p^e with generator a has the composition series $C \supset C_1 \supset \cdots \supset C_e = 1$ of length e, in which each C_i is the cyclic subgroup generated by a^{p^i}. This is (Exercise 1) the only composition series for C. Next consider products $G \times G'$. By Proposition II.28, $N \lhd G$ and $N' \lhd G'$ imply $(N \times N') \lhd (G \times G')$ and also that $(G \times G')/(N \times N') \cong (G/N) \times (G'/N')$; hence subnormal series

$$G \rhd N \supset 1 \qquad \text{and} \qquad G' \rhd N' \supset 1$$

give two subnormal series for $G \times G'$:

$$G \times G' \rhd G \times N' \rhd G \times 1 \rhd N \times 1 \rhd 1 \times 1,$$
$$G \times G' \rhd N \times G' \rhd 1 \times G' \rhd 1 \times N' \rhd 1 \times 1.$$

In these two series the factors (suitably permuted) are isomorphic. Since any finite abelian group A of order $n = p_1^{e_1} \cdots p_k^{e_k}$ is a product of cyclic groups of prime power orders, it follows that A has (at least) one composition series in which the factors are all cyclic of prime order; indeed, every composition series so obtained for A will have exactly e_i cyclic factors of order p_i, for each $i \in \mathbf{k}$. These examples (and many others, see Exercise 2) suggest the following general theorem:

Theorem 18 (*Jordan–Hölder*). *Any two composition series for a finite group G have the same length and isomorphic factors.*

By induction on k we shall prove that every group G with a composition series $G \supset N_1 \supset \cdots \supset N_k = 1$ of length k has all its other composition series $G \supset M_1 \supset \cdots \supset M_l = 1$ of the same length $l = k$ and with isomorphic factors. For $k = 1$, this means that $G \supset 1$ is a composition series for G, hence that G is simple, which implies that $G \supset 1$ is the *only* composition series for G. By induction, suppose the stated result holds for all smaller k, and consider the two composition series $\{N_i\}$ and $\{M_j\}$ for G. If $N_1 = M_1$, the induction assumption applied to the group N_1 with composition series $N_1 \supset N_2 \supset \cdots \supset N_k = 1$ of length $k - 1$ establishes the result for G. If $N_1 \neq M_1$, we form the intersection $N_1 \cap M_1$, a normal subgroup of G, and the corresponding inclusion diagram, as shown below.

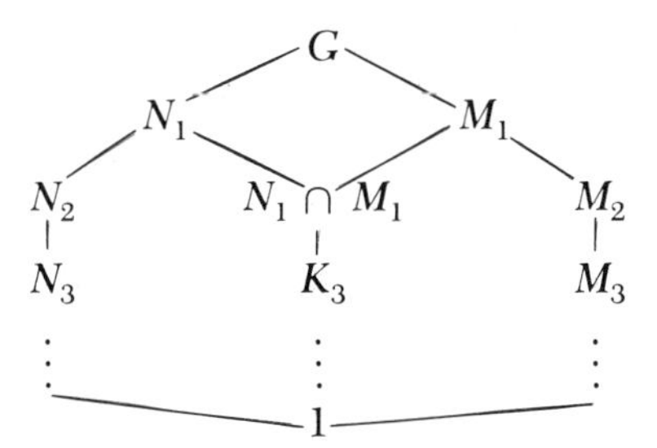

Since $G = N_1 \vee M_1$, the diamond isomorphism theorem gives $G/N_1 \cong M_1/N_1 \cap M_1$ and $G/M_1 \cong N_1/N_1 \cap M_1$. Continue down from the intersection $N_1 \cap M_1$ with some composition series $N_1 \cap M_1 \supset K_3 \supset \cdots \supset K_t = 1$. This gives four composition series for G, along the four possible descending paths visible in the diagram. The first two series start through N_1, so they have the same length and isomorphic factors by the induction assumption (applied to $N_1 \supset \cdots \supset N_k = 1$). The second and third series pass over opposite sides of the diamond and continue as $K_3 \supset \cdots \supset K_t = 1$, so they have the same length t and—by the diamond isomorphism theorem—isomorphic factors. The third and fourth series compare by the induction assumption again, this time for $M_1 \supset \cdots$. The three successive comparisons show that the N and M series for G do have the same length and isomorphic factors as well.

This theorem shows that each finite group G determines its list of "composition factors" $G/N_1, N_1/N_2, \cdots, N_{k-1}/N_k$ uniquely, up to order and isomorphism.

Corollary. *A finite group G is solvable if and only if its composition factors are cyclic of prime order and also if and only if it has a subnormal series with abelian factors.*

Proof: Suppose first that G is solvable; then the derived groups $G \supset G' \supset \cdots \supset G^{(k)} = 1$ are a subnormal series (and more; a series with each $G^{(i)}$ normal in G) with factors $G^{(i)}/G^{(i+1)}$ abelian. Suppose now that G has a subnormal series $G \supset N_1 \supset \cdots \supset N_m = 1$ with abelian factors. Each abelian factor group N_i/N_{i+1} has, as noted above, a composition series in which each factor is cyclic of prime order. Inserting the corresponding subgroups of N_i between N_i and N_{i+1} refines the given subnormal series of G to a composition series with all factors cyclic of prime order. Finally, suppose $G \supset K_1 \supset \cdots \supset K_t = 1$ is any composition series for G with cyclic factors. To prove G solvable it will suffice to prove for each derived group that $G^{(i)} \subset K_i$, for then $G^{(t)} \subset K_t = 1$, and G is solvable by definition. But we may prove $G^{(i)} \subset K_i$ by induction on i, as follows. Since the factor K_i/K_{i+1} is given abelian, the derived group K_i' has $K_i' \subset K_{i+1}$, while by the induction assumption, $G^{(i)} \subset K_i$. Hence, $G^{(i+1)} = (G^{(i)})' \subset K_i' \subset K_{i+1}$. This completes the induction and thus the proof of the corollary.

EXERCISES

1. Prove that a finite cyclic group of prime power order has only one composition series.

2. Exhibit all composition series for each of the following groups: (a) $\mathbf{Z}_6 \times \mathbf{Z}_5$; (b) Δ_4; (c) Δ_5; (d) A_4.

3. Determine the composition factors of S_4 and Δ_6.

4. A "chief series" for a group G is a series $G \supset N_1 \supset \cdots \supset N_k = 1$ with each N_i normal in the whole group G and which cannot be further refined so as to preserve this property. Prove that any two chief series for a finite group G have the same length and isomorphic factors. (*Hint:* Proof like that for Jordan–Hölder, with the diamond at the *bottom* so as to get normality in all of G.)

***5.** (a) (*The Zassenhaus Lemma.*) If $G \supset S \triangleright S' \supset 1$ and $G \supset T \triangleright T' \supset 1$, use the diamond isomorphism theorem to establish isomorphisms

$$\frac{S'(S \cap T)}{S'(S \cap T')} \cong \frac{S \cap T}{(S \cap T')(S' \cap T)} \cong \frac{T'(S \cap T)}{T'(S' \cap T)}.$$

(b) (*The Schreier–Zassenhaus Refinement Theorem.*) Prove: Any two subnormal series for the same group G (*not* necessarily finite) have refinements of the same length with isomorphic factors. (*Hint:* Given series $\{N_i\}$

and $\{M_j\}$, set $N_{ij} = N_{i+1}(N_i \cap M_j)$ and $M_{ij} = M_{j+1}(M_j \cap N_i)$ and use the Zassenhaus Lemma to compare $N_{ij}/N_{i,j+1}$ with a suitable M-quotient.)

(c) Deduce the Jordan–Hölder theorem from (b).

***6.** If G is solvable of order $p^r q^s$ (p, q primes), and if the factors of the composition series of G can occur in any order, show that G is nilpotent. (*Hint:* Any maximal normal subgroup N of G must have unique Sylow subgroups, by induction on order.)

7. Show by an example that nonisomorphic groups can have isomorphic composition factors (thus the length and the factors of a composition series do not determine the group).

9. Simplicity of A_n

The following result is the source of the fact that a polynomial equation of degree 5 or more cannot in general be solved "by radicals".

THEOREM 19. *For $n \geqslant 5$, the alternating group A_n is simple.*

Recall that A_n is the group of all even permutations of $\mathbf{n} = \{1, \cdots, n\}$; in particular, any 3-cycle (cycle of length 3) is a product $(1\ 2\ 3) = (1\ 2)(2\ 3)$ of two transpositions, hence is even and in A_n. First an observation on the generators of A_n:

LEMMA 1. *For $n \geqslant 3$, A_n is generated by the set of all its 3-cycles.*

Proof: Since any permutation is a product of transpositions, by Proposition II.16, an even permutation is a product of an even number of transpositions. Thus A_n is generated by all products of two transpositions. However, any product of two transpositions can be written as a product of 3-cycles, as indicated by the two typical cases $(1\ 2)(2\ 3) = (1\ 2\ 3)$ and $(1\ 2)(3\ 4) = (1\ 2\ 3)(2\ 3\ 4)$. This proves the Lemma.

LEMMA 2. *For $n > 1$, A_n is a characteristic subgroup of S_n.*

Proof: The result is immediate for $n = 2$, so assume that $n \geqslant 3$. Every square of a permutation of S_n is even, hence in A_n. Since $(1\ 2\ 3)^2 = (1\ 3\ 2)$, Lemma 1 thus implies that A_n is the subgroup of S_n generated by all squares of elements of S_n. This description is invariant under all automorphisms of S_n.

COROLLARY. *Any automorphism $\alpha: S_n \to S_n$ carries normal subgroups of A_n to normal subgroups of A_n.*

Lemma 3. *If $n > 4$ and $1 \neq N \lhd S_n$, then N contains A_n. (In other words, A_n is the only proper normal subgroup of S_n for $n > 4$.)*

Proof: Take any permutation $\mu \neq 1$ in N. Since the center of S_n is 1, there is some permutation of S_n which does not commute with μ, and hence some transposition τ which does not commute with μ. The commutator $[\mu, \tau] = \mu\tau\mu^{-1}\tau^{-1}$, as the product of $\mu \in N$ by one of its conjugates, is then in the normal subgroup N. On the other hand, $\mu\tau\mu^{-1}$ is a transposition, different from τ. Now $(\mu\tau\mu^{-1})\tau^{-1}$, as the product of two different transpositions, is, as in the proof of Lemma 1, either a 3-cycle (ijk) or the product $(ij)(kl)$ of two disjoint 2-cycles.

In the first case, the normal subgroup N contains all conjugates of (ijk). By the rule of Proposition II.15 for calculating conjugates, these conjugates are all of the 3-cycles of S_n. Thus N contains all of A_n.

In the second case, by the same rule, N contains with $(ij)(kl)$ all products of two disjoint transpositions. Since $n > 4$, we can choose a fifth digit m not one of i, j, k, or l. Thus N contains $(ij)(kl)$, $(jm)(kl)$ and their product (ijm). Consequently, N again contains all 3-cycles and so all of A_n, Q.E.D.

Lemma 4. *If $\mu \in S_n$ commutes with all its conjugates $\tau\mu\tau^{-1}$ for τ a transposition and if $n > 3$, then μ has order 2.*

Proof: We write μ as a product of disjoint cycles. If all these cycles are transpositions, then μ has order 2, as asserted. Otherwise, at least one cycle has length 3 or more, so μ has the form $\mu = (ijk \cdots)\cdots$. Let τ be the transposition $\tau = (kl)$ with $l \neq i, j, k$, so that $\tau\mu\tau^{-1} = (ijl \cdots)\cdots$. Then $\mu(\tau\mu\tau^{-1})(i) = k$ while $(\tau\mu\tau^{-1})\mu(i) = l$, so that μ and $\tau\mu\tau^{-1}$ do not commute, contrary to hypothesis.

Now we can prove the theorem. Let M be a maximal proper normal subgroup of A_n. By Lemma 3, $M \lhd S_n$ is impossible; hence the normalizer of M in S_n is precisely A_n. No transposition τ is in A_n, so $\tau M\tau^{-1} \neq M$, and any two transpositions τ, τ' give the same $\tau M\tau^{-1} = \tau' M\tau'^{-1}$. By the corollary to Lemma 2, $\tau M\tau^{-1} \lhd A_n$. Since $M \cap (\tau M\tau^{-1}) \lhd S_n$, it follows by Lemma 3 again that $M \cap (\tau M\tau^{-1}) = 1$. As $M \vee (\tau M\tau^{-1}) \lhd A_n$ and M is maximal, $M \vee (\tau M\tau^{-1}) = A_n$. Therefore, A_n is the direct product $M \times \tau M\tau^{-1}$, and this for *every* transposition τ. In particular, therefore, every $\mu \in M$ commutes with all its conjugates $\tau\mu\tau^{-1}$ by transpositions τ, so μ has order 2 by Lemma 4. Since order is preserved by inner automorphisms, the same is true for all elements of $\tau M\tau^{-1}$ and hence for all elements of the direct product $A_n = M \times \tau M\tau^{-1}$, in contradiction to the fact that A_n contains every 3-cycle.

Corollary. *For $n > 4$, S_n is not solvable.*

Proof: For $n > 4$, the composition series $S_n \supset A_n \supset 1$ does not have all cyclic factors.

There are other examples of finite groups which are simple. For instance, the center Z of the special linear group $SL(n, F)$ over the field F is the subgroup of all those scalar matrices κI with $\kappa^n = 1$. The quotient group $SL(n, F)/Z$ is the "projective special linear group" $PSL(n, F)$. It can be proved to be simple for every field F and $n \geqslant 3$. In particular, if F is a finite field, this defines a large class of finite simple groups.

EXERCISES

1. (a) Using the group V of Exercise II.10.5, exhibit for S_4 a composition series of the form $S_4 \supset A_4 \supset V \supset C \supset 1$, for a suitable C.

(b) Find all possible composition series for S_4.

2. Describe all the conjugate classes in A_6.

3. If a simple group G has a proper subgroup of index n, prove that the order of G is at most $n!$

4. Let G be the group of all rotations of the three-dimensional inner product space $\mathbf{R}^3$, while t and t' in G are two rotations through the same angle.

(a) Prove t and t' conjugate in G.

(b) If the composite tt' is a rotation through an angle β, show β a continuous function of the angle between the axes of t and t'.

(c) Deduce that G is a simple group.

5. (a) Show that the order of $SL(n, \mathbf{Z}_p)$ is (see Exercise VII.6.16)

$$p^{n(n-1)/2}(p^n - 1)(p^{n-1} - 1) \cdots (p^2 - 1).$$

(b) Show that its Sylow p-subgroups are the conjugates of the subgroup S of matrices of the form $I + U$, where U is strictly triangular.

***6.** (a) Show that $PSL(3, \mathbf{Z}_2)$ has order 168. (For this reason, it is often designated G_{168}.)

(b) Show that the Sylow 2-subgroup of G_{168} is isomorphic to the octic group Δ_4. (*Hint:* Use Exercise 5b.)

(c) Show that in G_{168}, the matrices A that satisfy $A\epsilon_3 = \epsilon_3$ form a subgroup of order 24 and index 7.

(d) Prove that the matrix B shown below generates a Sylow 7-subgroup of G_{168}.

$$\begin{pmatrix} 1 & 1 & 0 \\ 0 & 0 & 1 \\ 1 & 0 & 0 \end{pmatrix}.$$

(e) Show that G_{168} is simple. (*Hint:* Use Sylow's theorems, and show that it contains no *normal* subgroup of order 84, 56, or 24.)

CHAPTER XIII

Galois Theory

1. Quadratic and Cubic Equations

Galois theory arose from the problem of solving polynomial equations

$$a_n x^n + a_{n-1}x^{n-1} + \cdots + a_1 x + a_0 = 0$$

in one "unknown" x and with real or complex coefficients $a_n, a_{n-1}, \cdots, a_0$. The familiar formula for the solutions of a quadratic equation ($n = 2$) in terms of square roots long ago suggested the problem of getting similar formulas in terms of cube and higher roots for equations of any degree. This proved possible for cubic ($n = 3$) and quartic ($n = 4$) equations, but the realization that such formulas were "in general" not possible for degrees $n > 4$ depended on a more penetrating analysis of the behavior and structure of roots of such polynomial equations. As we will see, this analysis involves the study of a group associated with each polynomial and applies to polynomials with coefficients in any field F.

Recall that a quadratic equation

$$ax^2 + bx + c = 0$$

is solved by "completing the square" so as to remove the linear term. This amounts to making the transformation $y = 2ax + b$. Then

$$y^2 = 4a^2x^2 + 4abx + b^2 = 4a(ax^2 + bx) + b^2 = b^2 - 4ac.$$

The quadratic in x is thus replaced by a new quadratic $y^2 - c' = 0$ (with $c' = b^2 - 4ac$) which can be solved by extracting a square root. Then, provided that F has characteristic $\neq 2$, the solutions of the original quadratic equation are

$$x = \left(-b \pm \sqrt{b^2 - 4ac}\,\right)/2a.$$

The solutions are thus obtained from the coefficients by rational operations (addition, subtraction, multiplication, and division) and one square root. Any field is closed under rational operations, so if F is a field containing the coefficients a, b, and c of the given quadratic, then the (possibly) larger field $K = F(\sqrt{b^2 - 4ac}\,)$ generated by F and the square root $\sqrt{b^2 - 4ac}$ contains *both* the roots of the given quadratic. This field K can also be described as the field generated by the elements of F and the two solutions $(-b \pm \sqrt{b^2 - 4ac})/2a$. In other words, our attention has shifted from the particular solutions of an equation to the whole field which they generate. The same shift happens with equations of higher degrees.

Given a cubic equation $ax^3 + bx^2 + cx + d = 0$ over a field of characteristic $\neq 3$ a similar transformation by $y = 3ax + b$ will remove the quadratic term. Hence, it suffices to consider a cubic equation of the form

$$y^3 + py + q = 0 \tag{1}$$

with coefficients p and q and no quadratic term. In this equation try the substitution $y = z + w$; it changes the equation to

$$z^3 + w^3 + q + (3zw + p)(z + w) = 0.$$

The last term will vanish if $3zw + p = 0$; that is, if $w = -p/3z$, so that the substitution $y = z + w$ is $y = z - p/3z$. The equation for z is then

$$z^3 - p^3/(27z^3) + q = 0.$$

This, cleared of fractions, is a quadratic in z^3; its solution

$$z^3 = -q/2 + (q^2/4 + p^3/27)^{1/2}$$

will give a value γ for z which is a cube root of an expression involving a square root,

$$\gamma = \left[-q/2 + (q^2/4 + p^3/27)^{1/2}\right]^{1/3}. \tag{2}$$

Hence, a root $y = \beta$ for the original cubic is

$$\beta = \gamma - p/3\gamma.$$

The cubic has thus been solved "by radicals". The solution involves division by 2 and 3, so applies to any field of characteristic not 2 or 3.

To get all three roots of the given cubic, we may use the cube roots of 1; that is, the roots of $x^3 - 1$. Since $x^3 - 1 = (x - 1)(x^2 + x + 1)$, we take ω to be a root of $x^2 + x + 1$, so that

$$0 = 1 + \omega + \omega^2. \tag{3}$$

It follows that $x^3 - 1 = (x - 1)(x - \omega)(x - \omega^2)$, so that 1, ω, and ω^2 are the three cube roots of 1. The equation for z now has three roots

$$\gamma_1 = \gamma, \qquad \gamma_2 = \omega\gamma, \qquad \gamma_3 = \omega^2\gamma.$$

Hence we may write three expressions for roots of the cubic (1) as

$$\beta_1 = \gamma - p/3\gamma, \qquad \beta_2 = \omega\gamma - p/3\omega\gamma, \qquad \beta_3 = \omega^2\gamma - p/3\omega^2\gamma. \tag{4}$$

To show that these are all the roots of this cubic we need the identity

$$y^3 + py + q = (y - \beta_1)(y - \beta_2)(y - \beta_3). \tag{5}$$

This follows by comparing coefficients of each power of y. The coefficient of y^2 on the right is

$$-(\beta_1 + \beta_2 + \beta_3) = -(1 + \omega + \omega^2)\gamma + (1 + \omega + \omega^2)p/3\gamma = 0,$$

which is 0, by (3), while that of y on the right is

$$\beta_1\beta_2 + \beta_1\beta_3 + \beta_2\beta_3 = (1 + \omega + \omega^2)(\gamma^2 + p^2/9\gamma^2) - 3(\omega + \omega^2)p/3 = p,$$

and the constant term on the right is

$$\begin{aligned} -\beta_1\beta_2\beta_3 &= -\gamma^3 + (1 + \omega + \omega^2)(p\gamma/3 - p^2/9\gamma) + p^3/27\gamma^3 \\ &= -\gamma^3 + p^3/27\gamma^3 = q. \end{aligned}$$

In other words, *all* the roots β_1, β_2, and β_3 of the cubic are expressed by radicals, as in (2).

The cube root ω of 1 used in this solution can be any root $\omega \neq 1$ of $x^3 - 1$; that is, either root of $x^2 + x + 1$. By the formula for a quadratic equation, $\omega = (-1 + \sqrt{-3}/2)$—in agreement with a familiar formula for ω as a complex number with argument $2\pi/3$ and absolute value 1.

The *discriminant* of the cubic (5) is defined to be

$$D = [(\beta_1 - \beta_2)(\beta_1 - \beta_3)(\beta_2 - \beta_3)]^2.$$

Thus $D = 0$ if and only if the cubic has a multiple root. Moreover, D can be expressed rationally in terms of the coefficients. For, expanding the formula for D and using the relations

$$\beta_1 + \beta_2 + \beta_3 = 0, \qquad \beta_1\beta_2 + \beta_1\beta_3 + \beta_2\beta_3 = p, \qquad \beta_1\beta_2\beta_3 = -q$$

between the roots and the coefficients one verifies that

$$D = -4p^3 - 27q^2.$$

Therefore the square root involved in the formula (1) is

$$(q^2/4 + p^3/27)^{1/2} = \sqrt{-3D}/18.$$

Initially, the problem of solving a polynomial equation appeared to be that of finding "formulas" for its roots. The process of finding those formulas, however, shifts our attention to the field generated by those roots. In the quadratic case this is the field $F(\sqrt{b^2 - 4ac})$ generated by a root of the transformed quadratic $y^2 - b^2 + 4ac = 0$. For a cube root $\sqrt[3]{a}$, any field containing all three roots of $z^3 - a$ must contain the three cube roots 1, ω, and ω^2 of 1. Thus for a general cubic $y^3 + py + q$ with coefficients in a field F we are led to consider the field $F(\omega, \beta_1, \beta_2, \beta_3)$ generated from F by the cube root ω of 1 and all three roots β_1, β_2, and β_3 of the cubic. The auxiliary element γ used above turns out to be one of the elements in this field. In general, our study of polynomial equations will be focused not just on the roots, but on all the elements in the fields which these roots generate.

EXERCISES

1. For the discriminant D, verify that $D = -4p^3 - 27q^2$.

2. For the auxiliary quantity γ for the cubic, express γ in terms of $\beta_1, \beta_2, \beta_3$ and ω, thus proving that $\gamma \in F(\omega, \beta_1, \beta_2, \beta_3)$.

3. (Solutions of a quartic.) (a) By translation, show that any quartic equation can be put in the form $y^4 + py^2 + qy + r = 0$.

(b) For all z, show this equation equivalent to

$$(y^2 + z/2)^2 - [(z-p)y^2 - qy + (z^2/4 - r)] = 0.$$

(c) Let γ be a root of the cubic $(z^2/4 - r)(z - p) = q^2/4$, while $\alpha^2 = \gamma - p$. Show that the equation of (b) becomes

$$(y^2 + \gamma/2 + \delta)(y^2 + \gamma/2 - \delta) = 0,$$

where $\delta = \alpha y - q/2\alpha$. Conclude that the quartic can be solved by radicals.

2. Algebraic and Transcendental Elements

In a field K, the intersection of any collection of subfields is itself a subfield of K. In particular, if $F \subset K$ is a subfield of K and $u_1, \cdots, u_m$ are elements of K, the intersection of all those subfields which contain F and all u_i is a subfield, called the subfield *generated* by F and $u_1, \cdots, u_n$, and written as

$$F(u_1, \cdots, u_n) \tag{6}$$

with round brackets. For example, the field $\mathbf{Q}(\sqrt{2})$ discussed in §III.13 is the subfield of the field $\mathbf{R}$ of real numbers generated by $\sqrt{2}$ (and $\mathbf{Q}$). Similarly, the expression

$$F[u_1, \cdots, u_n], \tag{7}$$

with square brackets, designates the sub*ring* of K generated by F and the elements $u_1, \cdots, u_n$. An example is the ring $F[x]$ of polynomials in x with coefficients in a given field F.

An element $u \in K$ is said to be *algebraic* over a subfield $F \subset K$ when there is a non-zero polynomial f in $F[x]$ with $f(u) = 0$; that is, when u is a root of some polynomial $f = \Sigma a_i x^i$ with coefficients a_i in F and not all zero. Now the set D of all those polynomials f in $F[x]$ with root u is clearly an ideal in $F[x]$; indeed, it is a prime ideal, for $f(u)g(u) = 0$ implies that either $f(u) = 0$ or $g(u) = 0$. Like any ideal in $F[x]$, D is a principal ideal, in that it consists of all the multiples of some one polynomial h; moreover, we may take h to be monic (that is, with leading coefficient 1). Since the ideal D is prime, h is irreducible. This polynomial h is called the *minimal polynomial* of u over F; it may be described directly as the monic polynomial h of least degree for which $h(u) = 0$.

The *degree* of an algebraic element (over F) is the degree of its minimal polynomial. Classically, an element of $\mathbf{C}$ algebraic over $\mathbf{Q}$ is called an *algebraic number*. For example, the minimal polynomial of $\sqrt[5]{2}$ over $\mathbf{Q}$ is $x^5 - 2$, while if ω is a complex cube root of unity, its minimal polynomial over $\mathbf{Q}$ is $x^2 + x + 1$.

An element $u \in K$ is said to be *transcendental* over a subfield $F \subset K$ when it is not algebraic over F; that is, when $f(u) = 0$ for $f \in F[x]$ implies that $f = 0$ in the polynomial ring $F[x]$. It follows that the homomorphism of rings defined for each f by $f \mapsto f(u)$ is a monomorphism $F[x] \to K$ and therefore an isomorphism $F[x] \cong F[u]$; indeed, an isomorphism of rings which is the identity on the common subfield F, and which sends x to u. Hence any two elements u and $u' \in K$ both transcendental over a subfield F generate isomorphic rings $F[u] \cong F[u']$ and thus isomorphic fields $F(u) \cong F(u')$, for these fields are just the fields of quotients of $F[u]$ and $F[u']$, in the sense of §III.5.

Up to this point, the "base field" F has been considered as a subfield of a given larger field K, so that $F(u)$ has been described as the subfield of the given K generated by F and u. The typical example is $\mathbf{Q}$ as a subfield of $\mathbf{C}$, where $\sqrt{2}$ and i in $\mathbf{C}$ are algebraic over $\mathbf{Q}$, while real numbers such as e and π can be shown to be transcendental over $\mathbf{Q}$. Alternatively, starting *only* with a field F we may wish to construct a larger field. An example is the construction (§III.6) of the polynomial ring $F[x]$ in the indeterminate x over F; its field of quotients is then a field $F(x)$ generated by F and one element x transcendental over F. In the same way, given only F, one can construct a field $F(u)$ generated by an element u algebraic over F with a specified irreducible polynomial as its minimal polynomial.

THEOREM 1. *If h is a monic irreducible polynomial in $F[x]$, for F any field, there exists a field $K = F(u)$ generated by F and an element u with minimal polynomial h. If $K' = F(u')$ is any other such field generated by F and some u', with the same minimal polynomial h, there is a unique isomorphism $\phi: K \cong K'$, which is the identity on F and has $\phi u = u'$.*

Proof: In the polynomial ring $F[x]$ the principal ideal (h) generated by h is a maximal ideal, so the quotient ring $F[x]/(h)$ is a field (Proposition III.

30) with elements the cosets $(h) + f$ for $f \in F[x]$. This field contains F as a subfield (identify each $b \in F$ with the coset of b), and is generated by these constants and the coset $u = (h) + x$. Hence, $F[x]/(h) = F(u)$. Moreover (exactly as in Proposition III.31 with $u = \zeta$), this coset u is a root of h, so $F(u)$ is the required field. If $K' = F(u')$ is any other field generated from F by an element u' with the same minimal polynomial h, substitution of u' for x in the polynomial ring gives a homomorphism $F[x] \to F(u')$ with kernel (h) and hence (by universality properties of the quotient ring) an isomorphism $F(u) = F[x]/(h) \cong F(u')$. Since $F(u)$ is generated by F and u, this map is clearly the only possible homomorphism $\phi: F(u) \cong F(u')$ with $\phi(a) = a$ for $a \in F$ and $\phi(u) = u'$.

The formula $F(u) \cong F[x]/(h)$ shows that every element v of the field $F(u)$ can be written as a polynomial expression $\Sigma c_i u^i$ in u, with coefficients $c_i \in F$. If n is the degree of the minimal polynomial h, the equation $h(u) = 0$ can be used to replace any powers u^i with $i \geqslant n$. Hence, every $v \in F(u)$ can be written uniquely in the form

$$v = a_0 + a_1 u + \cdots + a_{n-1} u^{n-1}, \tag{8}$$

with coefficients $a_i \in F$; this linear expression is unique because h is minimal. We may write (8) as $v = f(u)$, where f is the polynomial $\Sigma a_i x^i$. If $v \neq 0$, its inverse v^{-1} in the field $F(u)$ must *also* be expressible as such a polynomial. To find this expression, note that f in $v = f(u)$ has degree less than that of the minimal polynomial h for u. Since h is irreducible, f and h must be relatively prime in the ring $F[x]$, so that their g.c.d. 1 can be written in the familiar form (Theorem III.23)

$$1 = sf + th \tag{9}$$

for suitable polynomials $s, t \in F[x]$. Substituting $x = u$ and using $h(u) = 0$ gives $1 = s(u)f(u)$, which expresses v^{-1} as the polynomial $s(u)$.

The Euclidean algorithm of §III.12 will explicitly construct s and t.

The isomorphism $\phi: F(u) \cong F(u')$ of Theorem 1 sends v of (8) in $F(u)$ to $a_0 + a_1 u' + \cdots + a_{n-1}(u')^{n-1}$ in $F(u')$. We will need a similar isomorphism in a more general case when an isomorphism $\theta: F \cong F'$ of the base fields is given. Such a θ extends to an isomorphism $\theta_{\#}: F[x] \cong F'[x]$ of polynomial rings in the evident way, with $\theta_{\#}(\Sigma a_i x^i) = \Sigma(\theta a_i) x^i$, as in Proposition III.16.

Proposition 2. *Let $\theta: F \cong F'$ be an isomorphism of fields. If the fields $K = F(u)$ and $K' = F'(u')$ are fields generated by elements u and u' with the minimal polynomials h and h' over F and F', respectively, and if θ maps h into h', as $\theta_{\#} h = h'$, then there is a unique way of extending θ to an isomorphism $\phi: F(u) \cong F'(u')$ with $\phi u = u'$.*

Proof: Set $\phi(\Sigma a_i u^i) = \Sigma(\theta a_i)u'^i$. This is the only choice for ϕ, since ϕa_i must be θa_i and $\phi u = u'$.

If F is a subfield of K, we also call K an *extension* of F; it is a *simple extension* if $K = F(u)$ for some one element $u \in K$. In general, for elements $u_1, u_2, \cdots$ of K we may form a sequence of subfields

$$F \subset F(u_1) \subset F(u_1, u_2) \subset F(u_1, u_2, u_3) \subset \cdots .$$

In this sequence each successive simple extension is either transcendental or algebraic.

We say that u_1 and u_2 are *algebraically independent* over F when u_1 is transcendental over F and u_2 is transcendental over $F(u_1)$. This property is actually symmetric in u_1 and u_2, and $(F(u_1))(u_2) = F(u_1, u_2) = (F(u_2))(u_1)$.

EXERCISES

1. Show that if u is algebraic (over F), then so are u^2 and $u + 3$, and conversely.

2. Find five polynomial equations of different degrees for $\sqrt{3}$ and show explicitly that they are all multiples of the monic irreducible equation for $\sqrt{3}$ (over the field $\mathbf{Q}$).

3. In the simple extension $\mathbf{Q}(u)$ generated by a root u of the irreducible equation $u^3 - 6u^2 + 9u + 3 = 0$, express each of the following elements in terms of the elements $1, u, u^2$, as in (8): u^4, u^5, $3u^5 - u^4 + 2$, $1/(u + 1)$, $1/(u^2 - 6u + 8)$.

4. Exhibit an automorphism, not the identity, of each of the following fields: $\mathbf{Q}(\sqrt{2})$, $\mathbf{Q}(\sqrt{-3})$, $\mathbf{Q}(i)$.

5. Exhibit nonreal fields of complex numbers isomorphic to each of the real fields $\mathbf{Q}(\sqrt[3]{5})$, $\mathbf{Q}(\sqrt[4]{2})$.

6. If $K \supset F$ and $u \in K$ is a root of some monic polynomial $f \in F[x]$ which is irreducible over F, show that f is the minimal polynomial of u over F.

7. If $K = F(u_1, \cdots, u_k)$ show that, after permutation of the u_i, there is an integer r, $1 \leqslant r \leqslant k$, such that u_i is transcendental or algebraic over $F(u_1, \cdots, u_{i-1})$ according as $i \leqslant r$ or $i > r$ (that is, the transcendental extensions can be made first).

3. Degrees

In an extension $F \subset K$ of fields, the field K is a vector space over the base field F, with addition of vectors the addition in K and (scalar) multiplication of $k \in K$ by a scalar $a \in F$ the product ak in K ("forget" the rest of the multiplication in K). The dimension of this vector space is called the *degree* of the field K over F, in symbols $[K:F]$. Thus $[K:F] = n$ means that there are n elements $u_1, \cdots, u_n$ (a basis of the vector space K over F) such that every element $v \in K$ can be written uniquely as a linear combination

$$v = a_1 u_1 + \cdots + a_n u_n, \qquad a_i \in F. \tag{10}$$

In particular, if $K = F(u)$ where u is algebraic over F with minimal polynomial h, the degree $[F(u):F]$ of the field extension is exactly the degree n of the polynomial h—because $1, u, \cdots, u^{n-1}$ form a basis as in (8). Therefore, $u \in K$ is algebraic or transcendental over F according as the degree $[F(u):F]$ is finite or infinite.

If the degree $[K:F]$ is finite, then *every* element v of K is algebraic over F. Specifically, if $[K:F] = m$ and $v \in K$, then the $m + 1$ elements $1, v, \cdots, v^m$ of the vector space K must be linearly dependent over F, and this dependence $\sum a_i v^i = 0$ gives a polynomial in $F[x]$ with root v, making v algebraic over F.

Degrees of successive finite extensions multiply, as follows:

Theorem 3. *For extensions $F \subset K \subset L$ of finite degree,*

$$[L : F] = [L : K][K : F]. \tag{11}$$

This result is included in the following more explicit fact: If $u_1, \cdots, u_n$ is a basis of K (as a vector space over F) and $v_1, \cdots, v_m$ is a basis for L over K, then the nm products $u_i v_j$ for $i = 1, \cdots, n$ and $j = 1, \cdots, m$ form a basis for L over F.

Proof: Any element $w \in L$ can be written as a linear combination $w = \sum_j r_j v_j$ with coefficients $r_j \in K$. Each coefficient r_j is in turn a linear combination $r_j = \sum_i a_{ji} u_i$ with coefficients a_{ji} in F. Substituting these expressions into that for w gives

$$w = \sum_j \sum_i a_{ji} u_i v_j,$$

so w is indeed a linear combination with coefficients in F of the nm products $u_i v_j$. The same type of argument proves that these nm elements are linearly independent over F. Hence, they do constitute the required basis.

This theorem has several immediate and useful corollaries. Recall that the degree of an algebraic element is the degree of its minimal polynomial.

Corollary 1. *If $F \subset K$ is an extension of finite degree, then every element $u \in K$ is algebraic over F and has degree a divisor of $[K:F]$.*

Proof: By the theorem, $[K:F] = [K:F(u)][F(u):F]$.

Corollary 2. *If $F \subset K$ has finite degree, then an element $u \in K$ generates all of K (from F) if and only if its degree over F is $[K:F]$.*

Corollary 3. *All algebraic numbers form a field.*

Proof: If u and v are algebraic numbers (that is, elements of $\mathbf{C}$ algebraic over $\mathbf{Q}$), then $\mathbf{Q}(u)$ and $\mathbf{Q}(v)$ are finite extensions of $\mathbf{Q}$, as is $\mathbf{Q}(u, v)$. Now $u + v$, uv, and u/v for $v \neq 0$, all lie in $Q(u, v)$, hence are also algebraic, Q.E.D.

Corollary 4. *Given field extensions $F \subset K \subset L$ with all elements of K algebraic over F, each $w \in L$ algebraic over K is also algebraic over F.*

Proof: Each element w algebraic over K has a minimal polynomial $h = \Sigma u_i x^i$ with coefficients $u_i \in K$. Therefore, w is still algebraic (with the *same* minimal polynomial) over the field $F(u_0, \cdots, u_n)$ generated by F and all these coefficients; moreover, this field has finite degree over F because each u_i is algebraic over F. Therefore, w is contained in the field $F(u_0, \cdots, u_n, w)$ of finite degree over F, hence is algebraic over F.

The determination of the degrees of specific elements requires essentially some explicit proof that the apparent minimal polynomial is indeed irreducible; for example, over $\mathbf{Q}$ we may use the Eisenstein criterion (Exercise III.12.8) and Exercise 10 below. For each prime p, the polynomial $x^3 - p$ is irreducible over $\mathbf{Q}$ because the unique factorization of integers shows that $x^3 - p$ has no rational roots, hence no linear factors. As a further example, $x^2 + 1$ is irreducible over $\mathbf{R}$ because $\mathbf{R}$ is an ordered field (cf. Proposition VIII.1). Another example is the following calculation of a degree:

$$\left[\mathbf{Q}(\sqrt[3]{7}, i) : \mathbf{Q}\right] = \left[\mathbf{Q}(\sqrt[3]{7}, i) : \mathbf{Q}(\sqrt[3]{7})\right]\left[\mathbf{Q}(\sqrt[3]{7}) : \mathbf{Q}\right] = 6. \quad (12)$$

The second factor on the right is 3, by the irreducibility of $x^3 - 7$, while the first factor on the right is 2 because by Theorem 3 an extension of degree 3 cannot contain an element i of degree 2 (or, because a subfield $\mathbf{Q}(\sqrt[3]{7})$ of the real numbers cannot contain the nonreal complex number i).

EXERCISES

1. Each of the following numbers is algebraic over $\mathbf{Q}$. In each case, find the monic irreducible equation satisfied by the number. (a) $2 + \sqrt{3}$; (b) $\sqrt[4]{5} + \sqrt{5}$; (c) $\sqrt[3]{2} + \sqrt[3]{4}$; (d) $u^2 - 1$, where u satisfies $u^3 = 2u + 2$; (e) $u^2 + u$ where u satisfies $u^3 = -3u^2 + 3$.

2. (a) If K is an extension of degree 2 of the field $\mathbf{Q}$ of rationals, prove that $K = \mathbf{Q}(\sqrt{d})$, where d is an integer which is not a square and which has no factors that are squares of integers.

(b) How much of this result remains true if $\mathbf{Q}$ is replaced by an arbitrary field F of characteristic ∞? By a field F of prime characteristic?

3. In Theorem 3 prove in detail that the nm elements $u_i v_j$ are linearly independent over F.

4. If p is a polynomial over F of degree m and irreducible over F and if K is a finite extension of F of degree relatively prime to m, prove that p is irreducible over K.

5. Determine the degree of each of the following extensions of the field $\mathbf{Q}$ of rational numbers. Give reasons.

(a) $\mathbf{Q}(\sqrt{3}, i)$. (b) $\mathbf{Q}(\sqrt[3]{5}, \sqrt{-2})$. (c) $\mathbf{Q}(\sqrt{18}, \sqrt[4]{2})$.
(d) $\mathbf{Q}(\sqrt{8}, 3 + \sqrt{50})$. (e) $\mathbf{Q}(\sqrt[3]{2}, u)$, where $u^4 + 6u + 2 = 0$.
(f) $\mathbf{Q}(\sqrt{3}, \sqrt{-5}, \sqrt{7})$. (g) $\mathbf{Q}(\sqrt{3}, \sqrt{2})$.

6. Give a basis over $\mathbf{Q}$ for each field of Exercise 5.

7. Determine whether the polynomial given is irreducible over the field indicated. Give reasons.

(a) $x^2 + 3$, over $\mathbf{Q}(\sqrt{7})$. (b) $x^2 + 1$, over $\mathbf{Q}(\sqrt{-2})$.
(c) $x^3 + 8x - 2$, over $\mathbf{Q}(\sqrt{2})$.
(d) $x^5 + 3x^3 - 9x - 6$, over $\mathbf{Q}(\sqrt{7}, \sqrt{5}, 1 + i)$.

8. If K is an extension of F of prime degree, prove that any element in K but not in F generates all of K over F.

9. Let F be any field contained in an integral domain D. Prove:

(a) D is a vector space over F.

(b) If, as a vector space, D has finite dimension over F, then D is a field.

10. (a) A polynomial f in $\mathbf{Z}[x]$ is said to be *primitive* when the greatest common divisor of its coefficients is 1. Prove the Gauss lemma: The product of two primitive polynomials is primitive.

(b) Using the Gauss lemma, prove that a polynomial f irreducible in $\mathbf{Z}[x]$ is also irreducible in $\mathbf{Q}[x]$.

4. Ruler and Compass

The classical problem of "duplicating the cube" is this: Given a line segment in the plane which is the edge of a cube, find the edge of a cube of double the volume, using only straight edge and compass constructions in the plane. It turns out that this is impossible: If the given cube has edge 1, the required cube has edge x where $x^3 = 2$, and this cubic polynomial $x^3 - 2$ is irreducible over $\mathbf{Q}$.

In detail, consider any figure S in the Euclidean plane. Such a figure consists of a finite number of points, lines, and circles; we will assume that it contains at least two perpendicular lines. Take these lines as coordinate axes, choose a unit distance, and refer all the items in the figure to these coordinates. The figure then determines a field $F(S)$ as follows: $F(S)$ is the field generated from the field $\mathbf{Q}$ of rational numbers by (a) the coordinates of all (the finite number of) points in S; (b) the intercepts (on the axes) of all lines in S; (c) the numbers A, B, C, for each circle with equation $x^2 + y^2 + Ax + By + C = 0$ in S.

Now each step in a construction by straight edge and compass enlarges the figure in one of the following ways: (i) adjoin the point of intersection of two lines; (ii) adjoin the line through two given points; (iii) adjoin the circle with center at a given point and passing through another given point; (iv) adjoin the two points of intersection of a line and a circle; and (v) adjoin the two points of intersection of two circles.

Each of these steps causes a (possible) extension of the field of the figure. By familiar facts about linear equations, steps (i), (ii), and (iii) do not enlarge the field, while steps (iv) and (v) may enlarge it by the adjunction of the solution of one quadratic equation. Hence, if a figure T is obtained from a figure S by straight edge and compass, the field extension $F(S) \subset F(T)$ consists of successive quadratic extensions. Hence, by Theorem 3, $[F(T):F(S)]$ is a power of 2. But we may assume that $F(S) = \mathbf{Q}$, while the desired equation $x^3 - 2 = 0$ has no rational roots, so is irreducible over $\mathbf{Q}$. By Corollary 1 of Theorem 3, it has no root in the field $F(T)$ of degree 2^n over $\mathbf{Q}$.

Sometimes the Euclidean constructions are understood to include one more possible step: (vi) adjoin a point chosen "at random", in the sense that the coordinates u, v of the new point satisfy no "special" relation to given coordinates; that is, that u and v are algebraically independent (satisfy no non-zero algebraic identities) over the field $F(S)$ of the given figure. This step (vi) will not duplicate the cube, because one can prove that an extension $F \subset F(u)$ with u transcendental over F adds to F no new elements algebraic over F.

A similar argument will show that "in general" it is impossible to trisect an angle by straight edge and compass: Writing the trigonometric equation for the cosine of 3θ in terms of the cosine of θ gives an equation of degree 3 for $\cos\theta$ (in terms of $\cos 3\theta$), which for most angles θ will be irreducible.

EXERCISES

1. If u is transcendental over F, prove that every element $y \in F(u)$ which is algebraic over F actually lies in F.

2. (a) If $b = \cos 3\theta$, show that $\cos\theta$ is a root of $4x^3 - 3x - b = 0$.

(b) In case $\theta = 20^0$, show that this cubic polynomial is irreducible over $\mathbf{Q}$.

5. Splitting Fields

A polynomial f in an indeterminate x *splits* in a field L when it has a factorization

$$f = a(x - u_1) \cdots (x - u_n)$$

for constants $a \neq 0, u_1, \cdots, u_n$ in L. Here n is the degree of f, while $a \in L$ is its leading coefficient, and $u_1, \cdots, u_n$ are its n roots, not necessarily all different. A *splitting field* for a polynomial $f \in F[x]$ is a field $L \supset F$ such that f splits in L as $f = a(x - u_1) \cdots (x - u_n)$ *and* such that $L = F(u_1, \cdots, u_n)$. For example, if ω is a cube root of 1, then $x^3 - 2$ has the roots $\sqrt[3]{2}$, $\omega\sqrt[3]{2}$, and $\omega^2\sqrt[3]{2}$; hence, $\mathbf{Q}(\sqrt[3]{2}, \omega) \supset \mathbf{Q}$ is a splitting field for $x^3 - 2$ over $\mathbf{Q}$.

THEOREM 4. *Any polynomial $f \in F[x]$ has a splitting field L over F; if $L' \supset F$ is a second splitting field for f over F, there is an isomorphism $\phi: L \to L'$, which is the identity on F.*

Proof: The existence of L will be proved simultaneously for *all* fields F, by induction on the degree of f. The existence is immediate when this degree is 0 or 1. In general, for a polynomial f of degree $n > 1$, take a factor h of f which is irreducible over F and construct as in Theorem 1 a field $K = F(u)$ generated by F and one root u of this irreducible h. Then u is a root of f, so that f factors over K as $f = (x - u)g$, for some polynomial g in $K[x]$ of degree $n - 1$. By the induction assumption, the polynomial g has a splitting field over K, say a field $L = K(u_2, \cdots, u_n)$ generated by roots $u_2, \cdots, u_n$ of g. Since $K = F(u)$, this field $L = F(u, u_2, \cdots, u_n)$ is generated from F by roots $u, u_2, \cdots, u_n$, and therefore is the required splitting field for f over F.

It remains to show that this splitting field L is unique up to an isomorphism ϕ which is the identity on F. To prove this we will replace the identity on F by an arbitrary θ isomorphism and prove instead the following slightly more general result: If $\theta: F \cong F'$ is an isomorphism of fields which carries a polynomial $f \in F[x]$ to $\theta_{\#} f = f' \in F'[x]$ and if L and L' are splitting fields for f and f' over F and F', respectively, then there is an isomorphism $\phi: L \cong L'$ which extends θ. The proof is by induction on the degree n of f, as in the diagram

$$\begin{array}{ccccc} F & \subset & F(u) & \subset & F(u, u_2, \cdots, u_n) = L \\ \downarrow \theta & & \vdots\ \theta_1 & & \vdots\ \phi \\ F' & \subset & F'(u') & \subset & F'(u', u_2', \cdots, u_n') = L'. \end{array}$$

If f has degree 0 or 1, the result is immediate. Otherwise, consider f of degree $n > 1$ and take a factor h of f which is irreducible over F; its image h' under the isomorphism θ is then irreducible over F'. Now L splits h and L' splits h', so we may choose a root u of h in L and a root u' of h' in L'. Then by Proposition 2 the given isomorphism $\theta: F \to F'$ extends to an isomorphism $\theta_1: F(u) \cong F'(u')$. But now f and f' factor as $f = (x - u)g$ and

$f' = (x - u')g'$, so that $L \supset F(u)$ and $L' \supset F'(u')$ are splitting fields for corresponding polynomials g and g'. By the induction assumption, θ_1 then extends to an isomorphism $L \cong L'$, as desired.

This isomorphism shows that for a given polynomial we get essentially the same splitting field, no matter how it is constructed. Thus $x^3 - 2$ has over $\mathbf{Q}$ a splitting field $\mathbf{Q}(\sqrt[3]{2}, \omega)$ constructed as a subfield of the field $\mathbf{C}$ of complex numbers, where $\sqrt[3]{2}$ is a real cube root of 2 and ω a complex cube root of 1. As in the calculation (12) above, this shows that $\mathbf{Q}(\sqrt[3]{2}, \omega)$ has degree 6 over $\mathbf{Q}$. It follows that the minimal polynomial $x^2 + x + 1$ for ω over $\mathbf{Q}$ is still irreducible over $\mathbf{Q}(\sqrt[3]{2})$, and hence is still the minimal polynomial for ω over $\mathbf{Q}(\sqrt[3]{2})$. We can construct a different splitting field for $x^3 - 2$ by taking successive quotient rings

$$K = \mathbf{Q}[x]/(x^3 - 2), \qquad L = K[x]/(x^2 + x + 1),$$

so that K has a root α (the coset of x) of the polynomial $x^3 - 2$ and there is an isomorphism $K \cong \mathbf{Q}(\sqrt[3]{2})$ sending α to $\sqrt[3]{2}$. Since $x^2 + x + 1 \in \mathbf{Q}[x]$ was still irreducible over $\mathbf{Q}(\sqrt[3]{2})$, it is also irreducible over the isomorphic field K. Hence, the quotient ring L above is a field; it contains a root ω' of $x^2 + x + 1$ (ω' is the coset of the new indeterminate x). Therefore, $L = \mathbf{Q}(\alpha, \omega')$ is a splitting field for $x^3 - 2$. For that matter, one can construct still another formally different splitting field for $x^3 - 2$ by taking successive quotient rings in the opposite order:

$$K' = \mathbf{Q}[x]/(x^2 + x + 1), \qquad L' = K'[x]/(h),$$

where h is a monic irreducible factor of $x^3 - 2$ over K'. Then L' is a field containing a root $\omega \neq 1$ of $x^3 - 1$ and a root β of $x^3 - 2$. By the theorem, L' must therefore have degree 6 over $\mathbf{Q}$, so the polynomial $x^3 - 2$ must be irreducible over K', so h above is actually $x^3 - 2$.

Now return to the theorem. It states that any two splitting fields of f are isomorphic, and this means that the roots of f in the first splitting field F must be "like" those in the second splitting field. In more detail, the isomorphism $\phi: L \cong L'$ of splitting fields which is the identity on the subfield F must map the polynomial f (of degree n) to itself and hence must map the factorization $f = a(x - u_1) \cdots (x - u_n)$ of f in L to the factorization $f = a(x - u_1') \cdots (x - u_n')$ in L'. Therefore, ϕ must carry the roots $u_1, \cdots, u_n$ of f in L in some order into the roots $u_1', \cdots, u_n'$ of f in the second splitting field L'. In particular, if the n roots are all different in L, they are all different in L'. We can therefore define: A polynomial $f \in F[x]$ of degree n is *separable* when it has n distinct roots in *some* (and hence in every) splitting field.

To reformulate this notion, we define the multiplicity of a root. A root u of a polynomial $f \in F[x]$ is said to have *multiplicity* m when f factors in $F(u)$ as $f = (x - u)^m g$ where $g(u) \neq 0$. Roots of multiplicities $1, 2, 3, \cdots$ are called simple, double, triple, $\cdots$ roots; if $u_1, \cdots, u_k$ are all the distinct roots of f in some splitting field L and $m_1, \cdots, m_k$ their multiplicities, then

$$f = a(x - u_1)^{m_1}(x - u_2)^{m_2} \cdots (x - u_k)^{m_k}, \qquad a \neq 0. \qquad (13)$$

If L' is a second splitting field, then the isomorphism $\phi: L \cong L'$ of the theorem must map this factorization (13) of f in $L[x]$ into a factorization in $L'[x]$ with k distinct roots $u'_1, \cdots, u'_k$ with the *same* set of multiplicities $m_1, \cdots, m_k$. In particular, a polynomial is separable when it has no root of multiplicity $m > 1$.

In the factorization (13) the sum $m_1 + \cdots + m_k$ of the multiplicities of the roots is the degree n of the polynomial f. In this sense, a polynomial of degree n always has n roots "counted according to their multiplicity". This idea, like its generalizations, is much used in algebraic geometry, as for example in Bezout's theorem: Two algebraic curves of degrees m and n in the (complex projective) plane always intersect in mn points "counted according to their multiplicities".

For a separable polynomial f we can also count the number of isomorphisms $L \cong L'$ of splitting fields in the preceding theorem.

THEOREM 5. *If a separable polynomial $f \in F[x]$ has a splitting field $L \supset F$, while $\theta: F \cong F'$ is an isomorphism of fields sending f to a polynomial $f' \in F'[x]$ with splitting field $L' \supset F'$, then θ can be extended in exactly $[L:F]$ ways to an isomorphism $\phi: L \cong L'$.*

The proof follows from the previous uniqueness proof, viewed more quantitatively. In detail, we proceed by induction on the degree of f, the case of degree 0 or 1 being trivial. For f of degree $n > 1$ and u one of its roots in L, the minimal polynomial h of u is a factor of f, say of degree d. The given isomorphism θ maps h to a polynomial $h' \in F'[x]$ of the same degree; since f is separable, so are h and h', so h' has exactly d roots in L'. Now every isomorphism ϕ of L to L' which extends θ must map the chosen root u to one of the d roots of h' in L'. In other words, $\theta: F \cong F'$ extends in exactly $d = [F(u):F]$ ways to an isomorphism

$$\theta_i : F(u) \cong F'(u'_i), \qquad u'_i \text{ a root of } h'.$$

But $f = (x - u)g$ over $F(u)$, and L is a splitting field for g over $F(u)$, while by the induction assumption each θ_i extends in exactly $[L:F(u)]$ ways to $L \cong L'$. All told this gives precisely $[L:F(u)][F(u):F] = [L:F]$ extensions of the original isomorphism θ, Q.E.D.

This theorem may be applied in particular when $L = L'$:

Corollary. *The splitting field $L \supset F$ of a separable polynomial $f \in F[x]$ has exactly $[L:F]$ automorphisms which are the identity on F.*

EXERCISES

1. Find the degrees of the splitting fields of the following polynomials over $\mathbf{Q}$:

(a) $x^3 - x^2 - x - 2$. (b) $x^4 - 7$. (c) $(x^2 - 2)(x^2 - 5)$.

2. Prove: The splitting field of a polynomial of degree n over a field F has degree at most $n!$ over F.

3. (a) If ζ is a primitive nth root of unity in the field $\mathbf{C}$, prove that $\mathbf{Q}(\zeta)$ is the splitting field of $x^n - 1 = 0$ over $\mathbf{Q}$.

(b) Compute its degree for $n = 3, 4, 5, 6$.

4. If the polynomial $f \in F[x]$ is separable but not necessarily irreducible over F, show that the quotient ring $F[x]/(f)$ is a product of fields (cf. Exercises 5 and 6 of §III.13).

6. Galois Groups of Polynomials

The automorphisms of a splitting field form a group called the Galois group.

DEFINITION. The Galois group *of a polynomial $f \in F[x]$ over the base field F, for any splitting field $L \supset F$ of f, is the group Γ of all those automorphisms of the splitting field L which leave every element of F fixed. Because the splitting field L is unique, up to an isomorphism, so is this group Γ (cf. Exercise 6).*

For example, a splitting field of $x^2 + 1$ over $\mathbf{R}$ is the field $\mathbf{C} = \mathbf{R}(i)$ of complex numbers; there are exactly two automorphisms, the identity and complex conjugation ($a + bi \mapsto a - bi$), so the Galois group is cyclic of order 2, with generator the automorphism "conjugation". Put differently, the Galois group is (isomorphic to) the group of all permutations of the two roots $\{i, -i\}$ of the polynomial $x^2 + 1$. Again, a splitting field of $x^3 - 2$ over $\mathbf{Q}$ is the extension $\mathbf{Q}(\sqrt[3]{2}, \omega)$ of degree 6; by the corollary of Theorem 5 it has six automorphisms, which must be given by the $6 = 3!$ permutations of the three roots $\sqrt[3]{2}$, $\omega\sqrt[3]{2}$, and $\omega^2\sqrt[3]{2}$ of $x^3 - 2$. An alternative construction of a splitting field for $x^3 - 2$, as described after Theorem 4 above, would still lead to a Galois group consisting of all six permutations of the three roots of $x^3 - 2$ in the splitting field.

These examples suggest that the Galois group of f is isomorphic to a group of permutations of the roots of f (we will soon see that it need not consist of *all* permutations!):

Proposition 6. *Each automorphism θ in the Galois group of a separable polynomial f of degree n over F permutes the roots of f in the splitting field L. The function sending θ to the permutation induced on the n roots of f is a monomorphism from the Galois group Γ to the symmetric group S_n of all permutations on n letters.*

Proof: Let $L = F(u_1, \cdots, u_n)$ be the splitting field of f, with $u_1, \cdots, u_n$ the n (distinct) roots. Each root u_i satisfies the polynomial equation $f(u_i) = 0$, with coefficients in F. Each automorphism θ in the Galois group Γ must leave F pointwise fixed, and so maps this equation to an equation $f(\theta u_i) = 0$. Therefore, the image θu_i is also one of the roots of f. Since θ is an automorphism and hence a bijection, it must carry distinct roots to distinct roots, hence it determines a permutation θ^*

$$u_1 \mapsto \theta u_1, \quad u_2 \mapsto \theta u_2, \quad \cdots, \quad u_n \mapsto \theta u_n \tag{14}$$

on the set $\{u_1, \cdots, u_n\}$. Since L is generated as a field by F and $u_1, \cdots, u_n$, the automorphism θ is completely determined by the permutation θ^*. Therefore, $\theta \mapsto \theta^*$ is an injection. Since $(\theta\phi)^* = \theta^*\phi^*$, it is a monomorphism of groups, $\Gamma \to S_n$, as asserted.

The Galois group can also be described as this permutation group; more exactly, as the group of those permutations of the roots which preserve all "rational relations" between the roots. Here a permutation preserves all rational relations precisely when it preserves all sums and all products of rational combinations of the roots; that is, is an automorphism.

As an example, we construct the Galois group of the (separable) polynomial $x^4 - 7$ over $\mathbf{Q}$. Since 7 is prime, this polynomial is irreducible over $\mathbf{Q}$ by the Eisenstein criterion. One of its roots is the real fourth root $u = \sqrt[4]{7}$ of 7; the others are iu, $-u$, and $-iu$, where $i^2 = -1$. A splitting field is thus $\mathbf{Q}(u, i)$; since i is complex, this field has degree 2 over $\mathbf{Q}(u)$ and hence degree 8 over $\mathbf{Q}$. Since it is also the splitting field of $x^2 + 1$ over the intermediate field $\mathbf{Q}(u)$, the field $\mathbf{Q}(u, i)$ has an automorphism θ transposing the roots $i, -i$; this automorphism θ is determined by

$$\theta i = -i, \qquad \theta u = u, \qquad \theta^2 = 1.$$

Since $\mathbf{Q}(u, i)$ is the splitting field of $x^4 - 7$ over $\mathbf{Q}(i)$, it has an automorphism ϕ sending u to the root iu; thus ϕ is determined by

$$\phi i = i, \qquad \phi u = iu, \qquad \phi^4 = 1.$$

These equations determine θ and ϕ on the whole splitting field; for example, they show that ϕ is the following cyclic permutation

$$u \mapsto iu \mapsto -u \mapsto -iu \mapsto u$$

of the roots of $x^4 - 7$. Using these equations, one can calculate the composites $\theta\phi$, $\phi\theta$, and so on. This gives $(\theta\phi\theta)i = i$ and $(\theta\phi\theta)u = -iu$, so that $\theta\phi\theta = \phi^3$. Thus θ and ϕ satisfy the relations

$$\theta^2 = 1, \qquad \phi^4 = 1, \qquad \theta\phi\theta = \phi^3.$$

The group Δ of automorphisms which they generate contains the cyclic subgroup Σ of order 4 generated by ϕ. By the third relation above this subgroup is normal in Δ, so the whole group consists of the cosets Σ and $\theta\Sigma$, and has order 8. Since this is the degree of the splitting field in question, it is the whole Galois group of $x^4 - 7$ over $\mathbf{Q}$. This group of order 8 is the dihedral group Δ_4 (the group of the square) of §II.5. It is a proper subgroup of index 3 in the whole symmetric group S_4 of all permutations on the roots.

The groups and fields involved may be displayed as follows:

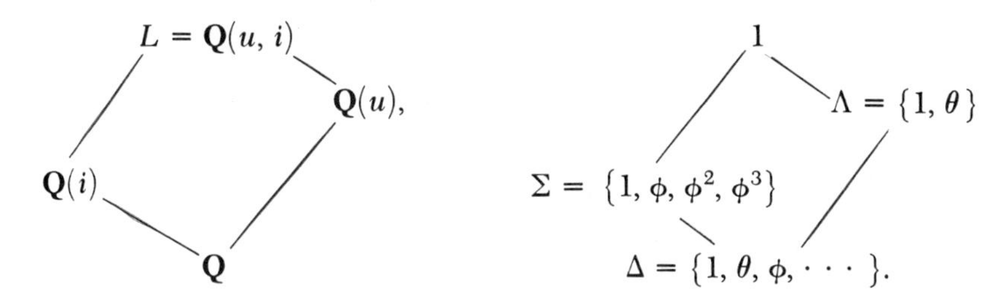

Here the whole group Δ is the Galois group of $x^4 - 7$ over $\mathbf{Q}$, the normal subgroup Σ is the Galois group of $x^4 - 7$, with splitting field L, over $\mathbf{Q}(i)$ and the subgroup Λ of index 4 in Δ is the Galois group of $x^2 + 1$ over $\mathbf{Q}(u)$. Note that each subgroup corresponds to an intermediate field, and is the Galois group of the polynomial $x^4 - 7$ over that intermediate field. We will see in §9 that this is a general fact.

The following result will be useful:

Theorem 7. *Any finite multiplicative group in a field F is cyclic.*

Proof: The group G in question is a subgroup of the group F^* of all non-zero elements of F. Hence, G is a finite abelian group, so the Corollary of Theorem XI.6 shows that G is a direct product $\mathbf{Z}_{m_1} \times \cdots \times \mathbf{Z}_{m_k}$ of cyclic groups, where $n = m_1 \cdots m_k$ is the order of G and $m_1 = m$ is the least common multiple of the orders of all the elements of G. Thus every element

b of G satisfies the equation $b^m = 1$. But the polynomial $x^m - 1$ of degree m can have at most m roots b in the field F, while this set G has $n = mm_2 \cdots m_k$ elements b. Hence, $n = m$, and $G \cong \mathbf{Z}_{m_1}$ is cyclic.

EXERCISES

1. Represent each element of the Galois group Δ of $x^4 - 7$ over $\mathbf{Q}$ as a permutation on the roots of $x^4 - 7$.

2. (a) In the group Δ of the text, find the subgroup which is the Galois group of $x^2 + 1$ over $\mathbf{Q}(iu)$.

(b) The same for $x^4 - 7$ over $\mathbf{Q}(u^2)$.

3. Show that the Galois group of $(x^2 - 2)(x^2 + 2)$ over $\mathbf{Q}$ is the four group, and represent it as a group of permutations on the roots.

4. Show that the Galois group of $x^3 - 1$ over $\mathbf{Q}$ is cyclic of order 2.

5. Show that the Galois group of $x^6 - 7$ over $\mathbf{Q}$ is a dihedral group Δ_6, using the fact that $-\omega$ is a primitive 6th root of unity.

6. If $\phi: L \cong L'$ is an isomorphism between two splitting fields of a polynomial f over a field F, show that the Galois groups Γ and Γ' of automorphisms of L and L' over F are isomorphic under $\theta \mapsto \phi\theta\phi^{-1}$ for all $\theta \in \Gamma$.

The following sequence of exercises determines the Galois group of $x^p - a$ over $\mathbf{Q}$ for p and a both prime.

7. Show that the multiplicative group of integers modulo p is cyclic (a generator of this cyclic group is called a *primitive root* modulo p).

8. Show that the splitting field of $x^p - 1$ over $\mathbf{Q}$ is $\mathbf{Q}(\eta)$, of degree $p - 1$, where η is a primitive pth root of 1. (*Hint:* The irreducibility of the minimal polynomial $x^{p-1} + \cdots + x + 1$ follows by the Eisenstein irreducibility criterion after the substitution $x = y + 1$.)

9. Show that the Galois group of $x^p - 1$ over $\mathbf{Q}$ is cyclic of order $p - 1$, generated by an automorphism θ with $\theta(\eta) = \eta^b$, where b is a primitive root modulo p.

10. Show that the Galois group of $x^p - a$ over $\mathbf{Q}(\eta)$, for a and p prime and η as in Exercise 8, is cyclic of order p, generator an automorphism ϕ with $\phi(u) = \eta u$, where $u^p = a$.

11. Show for a and p prime that the Galois group of $x^p - a$ over $\mathbf{Q}$ has order $p(p - 1)$ and is generated by θ, ϕ with

$$\theta^{p-1} = 1, \qquad \phi^p = 1, \qquad \theta\phi\theta^{-1} = \phi^b,$$

where b is a primitive root modulo p.

7. Separable Polynomials

It now behooves us to examine when an irreducible polynomial is separable. To do this we use formal derivatives to recognize multiple roots of polynomials.

The formal derivative Dx^k of the monomial x^k (kth power) is defined to be

$$Dx^k = kx^{k-1} = x^{k-1} + \cdots + x^{k-1} \qquad (k \text{ summands}),$$

just as in the calculus. More generally, for any polynomial $f = \Sigma a_k x^k$ in $F[x]$ its *formal derivative* Df is defined to be the polynomial

$$Df = \Sigma k a_k x^{k-1}, \qquad k a_k = a_k + \cdots + a_k \qquad (k \text{ summands}). \quad (15)$$

From this definition it follows that $D(af) = aDf$, $D(f + g) = Df + Dg$, and

$$D(fg) = (Df)g + fDg; \quad (16)$$

in other words, the formal derivative D is a linear transformation (of F-modules) $D{:}F[x] \to F[x]$ which satisfies the "Leibniz rule" (16) for the differentiation of products.

PROPOSITION 8. *A polynomial $f \in F[x]$ is separable if and only if f and Df have greatest common divisor* 1 *in the ring $F[x]$.*

Proof: The greatest common divisor (g.c.d.) of f and Df can be computed in the polynomial ring $F[x]$ by the Euclidean algorithm (§III.12). For any extension field $L \supset F$, the same computation holds in the extended polynomial ring $L[x]$, so f and Df have the same g.c.d. in $F[x]$ and in $L[x]$. But we can take $L \supset F$ to be a splitting field for f, and find this g.c.d. directly in $L[x]$, as follows.

If f is inseparable, it has in L a root u of multiplicity $m > 1$, so that $f = (x - u)^m g$ for some polynomial g. Then by the Leibniz rule (16) its formal derivative is

$$Df = m(x - u)^{m-1}g + (x - u)^m Dg = (x - u)^{m-1}(mg + (x - u)Dg).$$

Therefore, f and Df have a nonconstant factor $(x - u)^{m-1}$ in common. On the other hand, if f is a separable polynomial, each root u has multiplicity $m = 1$, and the preceding calculation shows that u is not a root of Df. Hence, no linear factor f is a factor of Df, and therefore f and Df have greatest common divisor 1. These results prove the Proposition.

PROPOSITION 9. *If the field F has characteristic ∞, every polynomial f irreducible over F is separable.*

Proof: Recall that F has characteristic ∞ when, for every positive integer k, $k1 = 1 + \cdots + 1$ (k summands) is non-zero. Hence, $a_k \neq 0$ implies $ka_k = (k1)a_k \neq 0$. Thus any f of positive degree has a formal derivative Df as in (15) which is non-zero and of smaller degree. If f is irreducible, it has no proper factors and in particular can have no common factors with Df

except constants. Therefore, by the previous proposition, f is separable, Q.E.D.

The characteristic of a field is either ∞ or a prime p (Theorem III.26); in the latter case $pa = a + \cdots + a = 0$ for every a in the field. This means that a polynomial of the form $f = x^p - b$ over a field of characteristic p has a formal derivative $Df = px^{p-1} = 0$, which is identically zero. In such a case, the proof of the proposition above could not apply; it will indeed turn out that such polynomials can be both irreducible and inseparable for characteristic p. This will use the

LEMMA. *For all elements a and b in a field F of characteristic p,*

$$(a + b)^p = a^p + b^p. \tag{17}$$

Proof: The binomial expansion (§III.1, Exercise 9) is

$$(a + b)^p = a^p + pa^{p-1}b + \cdots + (k, p - k)a^k b^{p-k} + \cdots + b^p,$$

where the binomial coefficients $(k, p - k) = p!/[k!(p - k)!]$ are

$$(k, p - k) = p(p - 1) \cdots (p - k + 1)/[1 \cdot 2 \cdots k];$$

they are integers and for $0 < k < p$ the prime factor p appears in the numerator and not the denominator, so these intermediate coefficients are all zero in characteristic p. This leaves just $a^p + b^p$, as required.

This equation plus the equation $(ab)^p = a^p b^p$, valid in any field, shows that $a \mapsto a^p$ preserves both sums and products in characteristic p, hence is an endomorphism $F \to F$ for any field F of characteristic p.

It follows also that any polynomial of the form $f = x^p - b$ is inseparable in characteristic p. For if u is one root of f, then $u^p = b$, and $(x - u)^p = x^p - b = f$, so u is a p-fold root. Thus we can exhibit an irreducible *and* inseparable polynomial of degree p, thus showing that the hypothesis of "characteristic ∞" is essential to Proposition 9 above. Take the field F to be the extension $\mathbf{Z}_p(t)$ of the field $\mathbf{Z}_p$ of p elements by one transcendental element t. The polynomial $f = x^p - t$ over this field is inseparable by the remark above, and is also irreducible. Indeed, it is *linear* in t, hence manifestly irreducible as a polynomial over $\mathbf{Z}_p$ in x and t. It is thus irreducible as a polynomial in x over the integral domain $D = \mathbf{Z}_p[t]$. This suffices as in Exercise 3.10 to prove f irreducible over F, the field of quotients of D.

The inseparable polynomial $x^p - b$ has formal derivative zero. More generally, if some polynomial f has the form

$$f = b_m x^{mp} + b_{m-1}x^{(m-1)p} + \cdots + b_1 x^p + b_0, \qquad b_i \in F, \tag{18}$$

with the degree of each non-zero term some multiple kp of p, then the formula (15) shows that the formal derivative Df is identically zero. Conversely, $Df = 0$ holds precisely when f has the form (18); this amounts to the statement that f has the form $g(x^p)$, where g is a polynomial

$$g = b_m y^m + b_{m-1} y^{m-1} + \cdots + b_1 y + b_0.$$

It turns out that all such polynomials f are indeed inseparable (Exercise 4).

An element $u \in K$ algebraic over a subfield F is said to be *separable* over F when its minimal polynomial over F is separable.

An extension $K \supset F$ of fields is called *algebraic* when every element u of K is algebraic over F and *separable algebraic* when every $u \in K$ is algebraic and separable over F. In characteristic ∞, every algebraic extension is automatically separable, by Proposition 9.

EXERCISES

1. For the field $\mathbf{Z}_p(t)$ considered in the text, show that

$$[\mathbf{Z}_p(t) : \mathbf{Z}_p(t^p)] = p, \qquad [\mathbf{Z}_p(t) : \mathbf{Z}_p(t^{p^2})] = p^2.$$

2. Show that the Galois group of $x^p - t$ over $\mathbf{Z}_p(t)$ is the identity.

3. Exhibit an irreducible polynomial of degree p^e (e a natural number, p prime) with all roots equal.

4. Show in characteristic p that every polynomial f of the form $f = g(x^p)$ of (18) is inseparable, with every root of multiplicity some multiple of p.

8. Finite Fields

A field F with a finite number of elements must have characteristic some prime p. By Theorem III.28 it is therefore isomorphic to an extension of the field $\mathbf{Z}_p$ of integers modulo p. If the degree $[F:\mathbf{Z}_p]$ is n, one can choose a basis $u_1, \cdots, u_n$ of n elements for F as a vector space over $\mathbf{Z}_p$. Then each element $v \in F$ has a unique expression $v = \Sigma a_i u_i$ with coefficients a_i which are integers modulo p. There are p choices for each coefficient a_i, so F has exactly p^n elements.

THEOREM 10. *Every finite field has p^n elements, where p is the characteristic and n some positive integer. Conversely, for each prime p and each such n there is, up to isomorphism, exactly one field of p^n elements; namely, the splitting field of $x^{p^n} - x$ over $\mathbf{Z}_p$.*

It remains only to prove the converse assertion. To prove the existence of the required field we start from the field $\mathbf{Z}_p$ of integers mod p; by Theorem

7 its multiplicative group is cyclic of order $p - 1$, so $a^p = a$ for every a in $\mathbf{Z}_p$. (This is just the little Fermat theorem: $a^p \equiv a \pmod p$ for integers a.) Now construct the splitting field $L \supset \mathbf{Z}_p$ of the polynomial $x^{p^n} - x$. For elements a and b in this field L, we have

$$(a + b)^{p^n} = a^{p^n} + b^{p^n}, \qquad (ab)^{p^n} = a^{p^n}b^{p^n}, \qquad (a/b)^{p^n} = a^{p^n}/b^{p^n},$$

the first by (17) and the last for $b \neq 0$. Hence, the subset S of L which consists just of all the roots u of $x^{p^n} - x$ contains $\mathbf{Z}_p$ and is closed under addition, subtraction, multiplication, and division (except by zero). This set S is therefore itself a field, and this field S contains all roots of $f = x^{p^n} - x$, so it must be the whole of the splitting field L. Moreover, the formal derivative $Df = p^n x^{p^n - 1} - 1 = -1 \neq 0$ has no factors in common with f, so f is separable and therefore has p^n distinct roots. Thus $L = S$ is a field with exactly p^n elements, as required. This field is often called the "Galois field" $GF(p^n)$.

Finally, let L' be any other field with p^n elements. Its multiplicative group has order $p^n - 1$, so all non-zero elements of L' are roots of $x^{p^n - 1} - 1$ and *all* elements of L' are roots of $x^{p^n} - x$. Since L' contains p^n such roots, it must be a splitting field for this polynomial. By the uniqueness of splitting fields, $L' \cong L$.

For any field F of characteristic p, the correspondence $a \mapsto a^p$ is by (17) an endomorphism $F \to F$ of fields, hence a monomorphism. For a finite field L, this endomorphism must be an automorphism $\theta : L \cong L$, with $\theta a = a^p$ for all $a \in L$. Moreover, $\theta^n a = a^{p^n} = a$, so this automorphism θ has order n. Thus θ generates a cyclic group of n automorphisms of L. But L is a splitting field over $\mathbf{Z}_p$, so by the corollary of Theorem 5 it has exactly $[L:\mathbf{Z}_p] = n$ automorphisms. Therefore, the group of all automorphisms of L is the cyclic group of order n generated by θ. This group Γ is the Galois group of the polynomial $x^{p^n} - x$ over $\mathbf{Z}_p$.

Proposition 11. *In the field L of p^n elements, there is for each integral divisor m of n exactly one subfield of p^m elements, and every subfield of L has this form for some m.*

Proof: If m divides n, then $a^{p^m} = a$ implies $a^{p^n} = a$. Since L contains all roots of the latter equation, it contains all roots of the former, hence contains the field with p^m elements. On the other hand, by Theorem 3 any subfield M of L has a degree m over $\mathbf{Z}_p$, which is a divisor of $n = [L:\mathbf{Z}_p]$, hence M must be the subfield of p^m elements, Q.E.D.

The lattice of all subfields of L is thus isomorphic to the lattice of all positive divisors of n.

If $\theta: L \cong L$ is the automorphism $a \mapsto a^p$ of L, then $\theta^m a = a^{p^m}$. Hence, the subfield M of p^m elements is exactly the subfield of all those elements of L which are left fixed by the automorphism θ^m; that is, by the subgroup of Γ consisting of all the powers of θ^m. But every subgroup of the cyclic group Γ is cyclic and generated by θ^m for some divisor m of n. Hence, the assignment which sends each subgroup Σ of Γ to the subfield of all those elements of L which are left fixed by the automorphisms of Σ is a bijection from subgroups of Γ to subfields of L. This is another example of the general property already illustrated in §6 for the subfields of the splitting field of $x^4 - 7$.

EXERCISES

1. In any field of characteristic p, show that an element has at most one pth root.

2. Show that every element in a finite field of characteristic p has a (unique) pth root.

3. (a) Prove that any finite field L of characteristic p is a simple extension of its prime field $\mathbf{Z}_p$.

(b) For each n, prove that there is an irreducible polynomial of degree n over $\mathbf{Z}_p$.

4. If $L \supset \mathbf{Z}_p$ is a finite field, prove that an element of L left fixed by all automorphisms of L is necessarily an element of $\mathbf{Z}_p$.

5. If f is an irreducible polynomial of degree n over $\mathbf{Z}_p$, prove that its Galois group is cyclic of order n.

9. Normal Extensions

Let $K \supset F$ be an extension of a field. The set

$$\Gamma = \Gamma(K/F) = \{\theta : K \to K \text{ and } \theta a = a \text{ for all } a \in F\},$$

consisting of all field automorphisms $\theta: K \cong K$ which leave every element of F fixed, is a group under composition. It is called the *Galois group* of the extension $K \supset F$; for example, if $f \in F[x]$ is a polynomial and L its splitting field over F, the Galois group $\Gamma(L/F)$ of this extension is just the Galois group of the polynomial f, as previously defined in §6. However, for some other extensions, the Galois group may be quite different. For example, the extension $\mathbf{Q}(\sqrt[3]{2}) \supset \mathbf{Q}$ has a Galois group which consists only of the identity. Indeed, any nonidentity automorphism of this field must be the identity on $\mathbf{Q}$ and must map $\sqrt[3]{2}$ to some other root of its minimal polynomial $x^3 - 2$, but the other roots are complex, so not in this field. Similarly, the extension $\mathbf{Q}(\sqrt[3]{2}, \sqrt{7}) \supset \mathbf{Q}$ has only two automorphisms: The identity and an automorphism θ with $\theta(\sqrt{7}) = -\sqrt{7}$ and $\theta(\sqrt[3]{2}) = \sqrt[3]{2}$.

To pick out extensions with enough automorphisms we make a

DEFINITION. An extension $K \supset F$ of fields is normal *(or, K is normal over F) when for each u in K but not in F there is an automorphism θ in $\Gamma(K/F)$ with $\theta u \neq u$. (In other words, each element of K not in F is moved by some automorphism of K over F.)*

For example, the extensions $\mathbf{Q}(\sqrt[3]{2})$ and $\mathbf{Q}(\sqrt[3]{2}, \sqrt{7})$ of $\mathbf{Q}$ discussed above are not normal.

For later purposes it is useful to think of the set of all elements of K unmoved under automorphisms of K over F. This set

$$\Gamma^{\#} = \{u \mid u \in K \text{ and } \theta u = u \text{ for all } \theta \in \Gamma\}, \tag{19}$$

consisting of all elements of K left fixed by every automorphism of Γ, is clearly a subfield of K, and $K \supset \Gamma^{\#} \supset F$. Also, the extension $K \supset F$ is *normal* if and only if $\Gamma^{\#} = F$.

We now show how to recognize normal extensions of finite degree.

THEOREM 12. *For an extension $K \supset F$ of fields of finite degree the following properties are equivalent:*

(i) K is the splitting field over F of a separable polynomial.
(ii) $K \supset F$ is normal.
(iii) K is a separable extension of F, and every polynomial q irreducible in $F[x]$ and with one root in K splits in K.
(iv) The degree $[K:F]$ is the order $[\Gamma:1]$ of the Galois group $\Gamma(K/F)$.

Proof: First, (i) implies (ii). For let K be the splitting field over F of some separable polynomial f; then by the corollary to Theorem 5, the Galois group $\Gamma = \Gamma(K/F)$ has exactly $[K:F]$ elements (so that (iv) holds). Now form the subfield $\Gamma^{\#}$ of (19) consisting of all elements left fixed by Γ. Then K is still the splitting field of the same polynomial f over the field $\Gamma^{\#}$, and has over $\Gamma^{\#}$ the same group of automorphisms. By the corollary of Theorem 5 again, Γ has $[K:\Gamma^{\#}]$ elements. Therefore, $[K:\Gamma^{\#}] = [K:F]$. Since $\Gamma^{\#} \supset F$, this implies that $\Gamma^{\#} = F$, which states that $K \supset F$ is normal.†

Next, (ii) implies (iii). For $K \supset F$ normal, consider any element $u \in K$. Since K is a finite extension of F, u is algebraic over F with some minimal polynomial g; therefore, any automorphism $\theta \in \Gamma(K/F)$ must map u into some one of the roots of this polynomial. Let $u_1 = u, u_2, \cdots, u_m$ be all the

†Some authors (e.g., Birkhoff and Mac Lane) define $K \supset F$ to be normal when every polynomial q in $F[x]$ and with one root in K splits in K. Normal in this sense *plus* separable gives normal as used above.

(different) images of u under automorphisms of Γ. Then each $\theta \in \Gamma$ must permute the m different elements $u_1, \cdots, u_m$. Consider the polynomial

$$h = (x - u_1)(x - u_2) \cdots (x - u_m) \tag{20}$$

of degree m. Each $\theta \in \Gamma$ permutes the factors of this polynomial, hence must leave the polynomial itself and all its coefficients fixed. Since $K \supset F$ is given to be normal, these coefficients must all lie in F, so that h is a polynomial $h \in F[x]$ with root u. It must then be a multiple of the minimal polynomial g for u over F. This minimal polynomial g therefore splits in K, with linear factors some of the (all different) linear factors $x - u_i$. Therefore, each $u \in K$ is separable over F, and its minimal polynomial splits, as required for (iii).

The polynomial h used in this proof expands as

$$h = x^m - (u_1 + \cdots + u_m)x^{m-1} + \cdots + (-1)^m u_1 u_2 \cdots u_m.$$

Its coefficients (which may be formed for any elements u_i) are polynomials in the u_i known as the *elementary symmetric functions* of $u_1, \cdots, u_m$.

Next, (iii) implies (i). Since K is an extension of F of finite degree, we can generate K as $K = F(w_1, \cdots, w_k)$. By (iii) each generator w_i is separable over F, with a minimal polynomial f_i which splits in K. The product polynomial $f = f_1 \cdots f_k$ is then separable unless some f_i and f_j have a common factor over K; since they are irreducible over F, this happens only when $f_i = f_j$. Then f with duplicate factors omitted is a separable polynomial f and K is its splitting field over F, as required in (i).

Finally, that (i) implies (iv) was already shown on the way to proving that (i) implies (ii). Conversely, we prove that (iv) implies (i). Given only that $[K:F] = [\Gamma:1]$, construct the subfield $\Gamma^{\#}$ of K consisting of all elements of K left fixed by all the automorphisms of Γ. The extension $K \supset \Gamma^{\#}$ has Galois group Γ; by construction it is a normal extension. Hence (since we know that (ii) implies (i) and thus (iv)), $[K:\Gamma^{\#}] = [\Gamma:1] = [K:F]$. But $[K:\Gamma^{\#}] = [K:F]$ with $\Gamma^{\#} \supset F$ implies that $\Gamma^{\#} = F$. This states that $K \supset F$ is normal, which is (ii), and which implies (i). The proof is complete.

Reciprocally, we consider the field fixed by any finite group of automorphisms.

Lemma. (*Artin*). *If Δ is any finite group of automorphisms of a field K while $\Delta^{\#}$ is the subfield of all those elements of K which are left fixed by every automorphism in Δ, then the degree $[K:\Delta^{\#}]$ is at most the order $[\Delta:1]$ of Δ:*

$$[K : \Delta^{\#}] \leqslant [\Delta : 1].$$

Proof: Let Δ have order n and suppose to the contrary that some $n + 1$ elements $u_1, \cdots, u_{n+1}$ of K were linearly independent over $\Delta^{\#}$. Form the n equations

$$x_1(\theta u_1) + \cdots + x_{n+1}(\theta u_{n+1}) = 0$$

in $n + 1$ unknowns, one for each of the n automorphisms θ in Δ. These simultaneous homogeneous linear equations have more unknowns than equations, hence (by Theorem VI.20) have a non-zero solution in the field K. Among the solutions, choose one with the smallest number r of non-zero x's, and rearrange so that just $x_1, \cdots, x_r$ are non-zero. Since the equations are homogeneous, we can also take $x_1 = 1$. Then $r > 1$ and at least one of the x_i is not in the fixed field $\Delta^{\#}$, because otherwise the first equation, with θ the identity automorphism, would make $u_1, \cdots, u_{n+1}$ linearly dependent over $\Delta^{\#}$. Let us assume $x_2 \notin \Delta^{\#}$. By definition, this means that there is some automorphism $\theta_0 \in \Delta$ with $\theta_0 x_2 \neq x_2$. Now apply this θ_0 to all n equations, to get n equations ($\theta \in \Delta$)

$$(\theta_0 x_1)(\theta_0\theta u_1) + \cdots + (\theta_0 x_{n+1})(\theta_0\theta u_{n+1}) = 0.$$

Since $\theta_0\theta$ runs over all of Δ with θ, we get the same equations as before, but with x_i replaced by $\theta_0 x_i$, and thus we get a second solution $x_1 = 1, \theta_0 x_2, \cdots, \theta_0 x_r, 0, \cdots, 0$. Subtracting this solution from the first solution $x_1, \cdots, x_r, 0, \cdots, 0$, where $x_1 = 1$, gives a new solution of the form $0, x_2 - \theta_0 x_2, \cdots, x_r - \theta_0 x_r, 0, \cdots, 0$ with a smaller number of non-zero unknowns. This is a contradiction.

In this lemma, every element of K not in the subfield $\Delta^{\#}$ is by definition moved by one of the automorphisms of Δ. Therefore, the extension $K \supset \Delta^{\#}$ is normal, by definition of normality. Also, the Galois group $\Gamma(K/\Delta^{\#})$ contains δ. Hence, by part (iv) of Theorem 12 and the Artin lemma, $[\Gamma:1] = [K:\Delta^{\#}] \leqslant [\Delta:1] \leqslant [\Gamma:1]$. In other words, $\Gamma = \Delta$ and $[K:\Delta^{\#}] = [\Delta:1]$. Therefore, starting with a group Δ of automorphisms of K, the degree of K over the corresponding field of fixed elements is exactly the order of Δ.

Theorem 12 above gives a similar comparison of sizes: Starting with a normal extension $K \supset F$, the order of $\Gamma(K/F)$ is exactly the degree $[K:F]$. These two "size" comparisons are the essential part of the "fundamental theorem" of the next section.

EXERCISES

1. If $M \supset F$ is a separable algebraic extension of finite degree, prove that there is a field $N \supset M$ with $N \supset F$ normal. Describe the smallest such field.
2. Exhibit the elementary symmetric functions of u_1, u_2, u_3, u_4.
3. How much of Theorem 12 remains true when $K \supset F$ is an algebraic extension of infinite degree?

10. The Fundamental Theorem

For a normal extension $N \supset F$ with Galois group Γ we now construct two functions between subsets of N and subsets of Γ. For each subset Σ of Γ the subset

$$\Sigma^{\#} = \{u \mid u \in N \text{ and } \sigma u = u \text{ for all } \sigma \in \Sigma\} \subset N$$

is clearly a subfield of N, called the "fixed field" of Σ. Moreover,

$$\Sigma_1 \subset \Sigma_2 \quad \Rightarrow \quad \Sigma_1^{\#} \supset \Sigma_2^{\#}. \tag{21}$$

To each subset S of N the subset

$$S^* = \{\theta \mid \theta \in \Gamma \text{ and } \theta s = s \text{ for all } s \in S\} \subset \Gamma$$

is clearly a subgroup of Γ, called the "fixing subgroup" for S. Moreover,

$$S_1 \subset S_2 \quad \Rightarrow \quad S_1^* \supset S_2^* \tag{22}$$

(if more is to be fixed, fewer automorphisms will do the job). Furthermore, given any $\Sigma \subset \Gamma$ and any $S \subset N$,

$$S \subset \Sigma^{\#} \quad \Leftrightarrow \quad \Sigma \subset S^* \tag{23}$$

(S is among the things fixed by Σ, if and only if Σ is among the things fixing S).

Theorem 13 (*Fundamental Theorem of Galois Theory*). *If $N \supset F$ is a normal extension of fields of finite degree with Galois group Γ, there is a bijection between subgroups Δ of Γ and fields L with $N \supset L \supset F$ ("intermediate" fields). This bijection $\Delta \mapsto \Delta^{\#}$ assigns to each subgroup Δ the intermediate field $\Delta^{\#}$ consisting of all elements of N left fixed by every automorphism of Δ. Its inverse $L \mapsto L^*$ assigns to each intermediate field L the subgroup of all those automorphisms of N which leave every element of L fixed. Each $N \supset L$ is a normal extension and L^* is its Galois group. Moreover, the degrees and orders satisfy*

$$[N : L] = [L^* : 1], \qquad [L : F] = [\Gamma : L^*]. \tag{24}$$

We picture the bijection as

$$\begin{array}{ccc} N & \longmapsto & N^* = 1 \\ \cup & & \cap \\ \Delta^{\#} = L & \longmapsto & L^* = \Delta \\ \cup & & \cap \\ F & \longmapsto & F^* = \Gamma. \end{array} \tag{25}$$

Observe that the degrees of the field extensions (on the left) correspond to indices of the subgroups (on the right).

Proof: Start first with an intermediate field L. Since $N \supset F$ is normal, it is the splitting field over F of some separable polynomial, so is also the splitting field over L of the same separable polynomial. Therefore, by Theorem 12, $N \supset L$ is normal. Its Galois group is by definition the group of all those automorphisms of N which leave fixed all elements of L; this is just the group L^*. Since $N \supset L$ is normal, the definition of normality shows that $L^{*\#} = L$ and Theorem 12 gives $[N:L] = [L^*:1]$. This is the first equation of (24). It includes the case $[N:F] = [\Gamma:1]$ since $F^* = \Gamma$; this gives

$$[N : L][L : F] = [N : F] = [\Gamma : 1] = [\Gamma : L^*][L^* : 1].$$

Using the first equation of (24), we now obtain the second.

Next we start with any subgroup $\Delta \subset \Gamma$. Then

$$[N : \Delta^{\#}] \leqslant [\Delta : 1] \leqslant [\Delta^{\#*} : 1], \tag{26}$$

the first inequality holds by the Artin lemma of §9, and the second because of the trivial inclusion $\Delta \subset \Delta^{\#*}$. But $\Delta^{\#}$ is a field L, so $[N:\Delta^{\#}] = [\Delta^{\#*}:1]$ by (24). Thus both inequalities of (26) are equalities; in particular, $\Delta = \Delta^{\#*}$.

Since $\Delta^{\#*} = \Delta$ and $L = L^{*\#}$, the functions * and # are inverses of each other, so the desired bijection is established.

Since * and # both reverse inclusions, it can be proved (Exercise 6) that the bijection is a lattice anti-isomorphism between the lattice of intermediate fields L and the lattice of subgroups of Γ (here "anti" means that the morphism sends join to meet and meet to join; thus $\vee$ to $\cap$ and $\cap$ to $\vee$).

The index equalities (24) may also be written in terms of subgroups Δ as

$$[\Gamma : \Delta] = [\Delta^{\#} : F], \qquad [\Delta : 1] = [N : \Delta^{\#}]. \tag{27}$$

Since the Galois group Γ is finite, it has only a finite number of subgroups. We conclude that a normal extension $N \supset F$ of finite degree has only a finite number of intermediate fields—a fact not obvious just by inspection of the fields alone.

Theorem 14. *For $N \supset L \supset F$ as in Theorem 12, the intermediate field L is a normal extension of F if and only if L^* is a normal subgroup of Γ. In this case, the quotient group Γ/L^* is isomorphic to the Galois group of $L \supset F$.*

Proof: If L is a normal extension of F, then every $u \in L$ has over F a minimal polynomial which splits in L. This means that each automorphism $\theta \in \Gamma$ must map each $u \in L$ to some θu which is a root of the same minimal polynomial, hence also in L. In other words, $\theta : N \to N$ with domain and codomain restricted to L is still an automorphism $\theta|L{:}L \to L$; and the assignment

$$\theta \mapsto \theta|L, \qquad \Gamma \to \Gamma(L/F)$$

is a morphism of groups, from Γ to the Galois group of $L \supset F$. The kernel of this morphism is the set of all those θ with $\theta|L$ the identity; that is, the set of all those θ in L^*. Hence, $L^* \lhd \Gamma$. The image of the homomorphism is therefore isomorphic to the quotient group Γ/L^*. By the equation (24) on indices, the order of the group is $[L{:}F] = [\Gamma(L/F){:}1]$. Therefore, the image is the whole Galois group: $\Gamma/L^* \cong \Gamma(L/F)$.

Conversely, let Δ be a normal subgroup of Γ. For $\delta \in \Delta$ each conjugate $\delta_1 = \theta^{-1}\delta\theta$ by any θ in Γ is then also an element of Δ. For each element $u \in \Delta^\#$, $\delta_1 u = u$, so

$$\delta(\theta u) = \theta\delta_1 u = \theta u,$$

and θu is again fixed by all elements of Δ; that is, $\theta u \in \Delta^\#$. Therefore, each θ in Γ restricted to the field $\Delta^\# = L$ is an automorphism $\theta|L{:}L \cong L$. Moreover, $\theta|L = \theta'|L$ if and only if $\theta\theta'^{-1} \in L^* = \Delta$; that is, if and only if θ and θ' lie in the same coset of Δ. Thus $L \supset F$ has $[\Gamma{:}\Delta]$ automorphisms, hence by Theorem 12 is normal.

Correspondences $S \mapsto S^*$ and $\Sigma \mapsto \Sigma^\#$ between posets (here the poset of all subsets S of N and that of all subsets Σ of Γ) are called *Galois correspondences* (or "polarities") when they enjoy the properties (21), (22), and (23). (See Exercise 9.) The same properties will reappear in more general guise in the study of adjoint functors (§XV.8).

EXERCISES

1. Illustrate the fundamental theorem by exhibiting the complete correspondence between intermediate fields and subgroups in the case of a finite field, regarded as an extension of its prime field.

2. Exhibit all subfields and the corresponding subgroups for the splitting field $x^4 - 7$, as discussed in §6.

3. In the fundamental theorem, let L_1 and L_2 be intermediate fields with $L_1 \supset L_2$. Show that $[L_1{:}L_2] = [L_2^*{:}L_1^*]$, and that L_1 is a normal extension of L_2 if and only if L_1^* is a normal subgroup of L_2^*.

4. If $K \supset F$ is any separable extension of finite degree, prove that there is only a finite number of intermediate fields L, with $K \supset L \supset F$.

5. In the fundamental theorem, two intermediate fields L_1 and L_2 are said to be *conjugate* when there exists a $\theta \in \Gamma(N/F)$ with $\theta(L_1) = L_2$. Prove that this is the case if and only if L_1^* and L_2^* are conjugate subgroups of Γ.

6. For the lattice of all intermediate fields L, with $N \supset L \supset F$, and the lattice of all subgroups of Γ prove that the bijection of the Fundamental

Theorem satisfies

$$(L \cap M)^* = L^* \vee M^*, \qquad (L \vee M)^* = L^* \cap M^*.$$

7. If $K \supset F$ is an extension of finite degree with only a finite number of intermediate fields L, prove that there is an element $w \in K$ with $K = F(w)$. (*Hint:* If F is infinite, and $u, v \in K$, the infinitely many $a \in F$ yield an infinite number of intermediate fields $F(u + av)$.)

8. Use Exercises 4 and 7 to prove that a finite separable extension $K \supset F$ of fields always has an element w with $K = F(w)$.

9. From equations (21), (22), and (23) deduce that

$$S \subset S^{*\#}, \qquad \Sigma \subset \Sigma^{\#*}, \qquad S^* = S^{*\#*}, \qquad \Sigma^\# = \Sigma^{\#*\#}.$$

10. For vector spaces W and V and $\psi : W \times V \to F$ bilinear show that there is a Galois correspondence between subsets S of W and T of V given by $S \mapsto S^\perp$, $T \mapsto T^\perp$, where $S^\perp$ is the annihilator of S (all $v \in V$ with $\psi(w, v) = 0$ for every $w \in S$). Find conditions (cf. §VI.4) when this correspondence is a bijection between subspaces.

11. The Solution of Equations by Radicals

A polynomial equation $f = 0$ is said to have a solution by radicals when there is a formula for a root of the equation in terms of rational operations and radicals such as $\sqrt[n]{a}$. If the exponent n of such a radical factors as $n = hk$, then $\sqrt[n]{a} = \sqrt[h]{(\sqrt[k]{a})}$, so it suffices to consider radicals with a prime exponent $n = p$. Such a radical $\sqrt[p]{a}$ for a in a given field F is just a root of the polynomial $x^p - a$, and this polynomial has a splitting field $L \supset F$ which can be described explicitly as follows. If the prime exponent p is the characteristic of the field F, then $x^p - a$ is inseparable, so has just one p-fold root $u = \sqrt[p]{a}$; the splitting field is therefore $F(u)$. If the exponent p is not the characteristic of F, and if $a \neq 0$, then the polynomial $x^p - a$ is separable and so has p distinct roots $u_1, \cdots, u_p$. The ratios $1, u_2/u_1, \cdots, u_p/u_1$ are then p distinct pth roots of 1, so the splitting field of $x^p - a$ contains that of $x^p - 1$. Now these pth roots of 1 lie in the splitting field and so form a multiplicative group of order p there which must be cyclic by Theorem 7. Let ζ be a generator of this group; call it a *primitive* pth root of unity. (If the field F is a subfield of the field $\mathbf{C}$ of complex numbers, ζ can be taken to be $\zeta = e^{2\pi i/p}$ in the complex plane.) Then if $u = \sqrt[p]{a}$ is any one root of $x^p - a$, all the roots are $u, \zeta u, \cdots, \zeta^{p-1}u$ and the splitting field is $F(u, \zeta)$. Its degree over F is at most $p(p - 1)$, since it is generated by elements u and ζ satisfying equations of degrees p and $p - 1$ over F.

Lemma. *Over any field F, the polynomial $x^p - a$ with p prime either has a root in F or is irreducible over F.*

Proof: Assume first that p is not the characteristic of F. Then in a splitting field, $x^p - a = (x - u)(x - \zeta u) \cdots (x - \zeta^{p-1}u)$. If $x^p - a$ were reducible in $F[x]$, any proper factor g of degree $m < p$ would be a product of m of the linear factors $x - \zeta^i u$, so the constant term b of this factor g is a product of m roots $\zeta^i u$, and has the form $b = \zeta^j u^m$ for some j. Then $b^p = u^{mp} = a^m$. Now m and p are relatively prime integers, so by Theorem III.23 there are integers s and t with $1 = sm + tp$. Therefore,

$$a = a^1 = a^{sm}a^{tp} = b^{ps}a^{tp} = (b^s a^t)^p,$$

and a has a pth root $b^s a^t$ in F.

If p is the characteristic of F, then $x^p - a$ is inseparable, but the same argument applies after replacing all ζ's by 1's.

For p not the characteristic of F we examine the Galois group of $x^p - a$. Adjoining ζ and then u gives a chain of subfields (upside down) of the splitting field,

$$\begin{array}{ll} F & \Gamma \to \Gamma/\Delta = \Sigma \\ \cap & \cup \qquad \cup \\ F(\zeta) & \Delta \to 1 \\ \cap & \cup \\ F(\zeta, u) & 1 \end{array}$$

with corresponding subgroups, as displayed. The field $F(\zeta)$ generated by all pth roots of unity is the splitting field of $x^p - 1$ so, as a normal extension of F, corresponds to a normal subgroup $\Delta \lhd \Gamma$, and has a Galois group Σ, as displayed above. This group Σ consists of the automorphisms θ of the field $F(\zeta)$ over F. Each such automorphism θ is determined by its action on ζ. But $\theta\zeta = \zeta^i$ for some integer i relatively prime to p. Therefore, θ is also an automorphism of the cyclic multiplicative group $\{1, \zeta, \cdots, \zeta^{p-1}\}$. Now the group A of *all* automorphisms of this cyclic group is just the multiplicative group of all integers i prime to p, taken modulo p, and this group is cyclic of order $p - 1$. Therefore, Σ, as a subgroup of the cyclic group A, is also cyclic (of order some divisor of $p - 1$).

The group Δ of all those automorphisms of $F(\zeta, u)$ which leave ζ fixed is also the Galois group of $F(\zeta, u)$ over $F(\zeta)$. This extension is generated by one root u of the polynomial $x^p - a$, which by the lemma is either irreducible over $F(\zeta)$ or has one (and hence in this case all) of its roots in $F(\zeta)$. In the first case, Δ is a group of order p, hence is cyclic. In the second case, Δ is the identity. In either case, the Galois group Γ of $x^p - a$ over F is an extension of the cyclic normal subgroup Δ by the cyclic group $\Gamma/\Delta \cong \Sigma$, so is solvable. We have proved

Theorem 15. *If the prime p is not the characteristic of the field F, then for each $a \in F$ the Galois group of $x^p - a$ is solvable.*

A field K is said to be a *radical extension* of F if there is some finite chain of intermediate fields of the form

$$F = K_0 \subset K_1 \subset \cdots \subset K_{t-1} \subset K_t = K, \qquad K_i = K_{i-1}(u_i), \qquad u_i^{n_i} = a_i,$$

in which each step arises by adjoining an n_ith root u_i of some element a_i in K_{i-1}.

Proposition 16. *If the field F has characteristic ∞, then any radical extension $K \supset F$ can be embedded in a radical extension $N \supset F$ which is normal over F and has a solvable Galois group over F.*

Proof: Since we can evidently make an induction on the number t of radicals used (the number t of intermediate fields in the chain just above) it will be enough to prove that if $M \supset F$ is a normal and radical extension of F with a solvable Galois group and $a \in M$, then $M(u)$, where $u^n = a$, can be embedded in another such extension N. We can also assume that the exponent n is a prime p. Take Σ to be the Galois group of $M \supset F$. The polynomial $f = \prod_\sigma (x^p - \sigma a)$, with the product taken over all $\sigma \in \Sigma$, has coefficients left fixed by all $\sigma \in \Sigma$, hence has coefficients in F. The splitting field N of this polynomial over M then clearly contains the given radical extension $M(u)$. On the other hand, the splitting field can be constructed by successive steps as a chain of intermediate fields (with corresponding subgroups of $\Gamma(N \supset F)$):

$$F \subset M \subset M(\zeta) \subset M(\zeta, \sqrt[p]{a}) \subset \cdots \subset N,$$

$$\Gamma \supset \Delta \supset \Delta_0 \supset \Delta_1 \supset \cdots \supset 1.$$

Here we have adjoined to M first a primitive pth root ζ of 1, then the pth root of a, then the pth root of some other σa, and so on. Hence, each step beyond M is a normal extension of the previous field, so each corresponding subgroup Δ_i is a normal subgroup of the previous Δ_{i-1}. Moreover, each quotient Δ_i/Δ_{i+1} is a cyclic group—either the cyclic Galois group of $x^p - 1$ over M, or the cyclic group (of order p or order 1) of some polynomial $x^p - \sigma a$. By hypothesis, the normal field M has a solvable Galois group, and this group is the quotient group $\Gamma/\Delta \cong \Sigma$. The whole group Γ is therefore solvable, since any extension of a solvable group by a solvable group is itself solvable, by Theorem XII.16. Moreover, the field N, like M, is a radical extension, Q.E.D.

We can now prove the main result about the solution of equations by radicals.

Theorem 17. *If a polynomial f irreducible over a field F of characteristic ∞ has any one root in a radical extension of F, then the Galois group of f is solvable.*

Proof: We can take the root w of f to lie in a normal radical extension $N \supset F$ constructed as in Proposition 16. Since the extension is normal and one root w of the irreducible f lies in it, all roots w_i of f are there. Hence, the splitting field $L = F(w_1, \cdots, w_n)$ of f over F is contained in N, with corresponding normal subgroup Λ, as in the diagram

$$\begin{array}{ccc} N & \longmapsto & 1 \\ \cup & & \cap \\ L & \longmapsto & \Lambda \\ \cup & & \cap \\ F & \longmapsto & \Gamma. \end{array}$$

By the fundamental theorem, the Galois group $\Gamma(N/F)$ is then the quotient Γ/Λ of the solvable group Γ, hence is solvable.

It follows that an equation whose Galois group is not solvable cannot be solved by radicals. For example, we know by Theorem XII.19 that the alternating group A_5 is simple, and hence that the symmetric group S_5 is not solvable ($S_5 \triangleright A_5 \triangleright 1$ is its only composition series).

Next, we construct an equation of fifth degree whose Galois group is the symmetric group S_5. Take a field M of characteristic ∞, form the polynomial ring $M[x_1, \cdots, x_5]$ in five simultaneous indeterminates $x_1, \cdots, x_5$ and its field of quotients $K = M(x_1, \cdots, x_5)$; this field K can also be described as the result of five successive simple transcendental extensions $M \subset M(x_1) \subset M(x_1, x_2) \subset \cdots \subset K$ of the given field M. Now form the polynomial f in a new indeterminate t with the five roots $x_1, \cdots, x_5$:

$$f = (t - x_1)(t - x_2) \cdots (t - x_5) = t^5 - \sigma_1 t^4 + \sigma_2 t^3 - \sigma_3 t^2 + \sigma_4 t - \sigma_5.$$

Its coefficients σ_i are the "elementary symmetric functions" of the five given x's; they are

$$\begin{array}{ll} \sigma_1 = x_1 + x_2 + x_3 + x_4 + x_5, & \sigma_2 = x_1x_2 + x_1x_3 + \cdots + x_4x_5, \\ \sigma_3 = x_1x_2x_3 + \cdots + x_3x_4x_5, & \sigma_4 = x_1x_2x_3x_4 + \cdots + x_2x_3x_4x_5, \\ \sigma_5 = x_1x_2x_3x_4x_5. & \end{array}$$

The polynomial f with coefficients in the field $F = M(\sigma_1, \cdots, \sigma_5)$ thus has $K = M(x_1, \cdots, x_5)$ as its splitting field over F, so its Galois group is given by certain permutations of the x_i. But *every* permutation of the x_i gives an automorphism of K, and each such automorphism leaves fixed all the symmetric functions $\sigma_1, \cdots, \sigma_5$ and hence the whole field F. Therefore, the Galois group of f is the whole symmetric group S_5, and the polynomial equation $f = 0$ cannot be solved over F by radicals.

This polynomial equation is, in fact, the "general" (monic) polynomial of degree 5. (This is because the $\sigma_1, \cdots, \sigma_5$ are algebraically independent over M, in the sense defined in §2.) Hence, the general equation of degree 5 cannot be solved by radicals. One can also construct polynomials of degree 5 with coefficients in $\mathbf{Q}$ which cannot be solved by radicals.

Theorem 17 has a converse (which we will not prove here): If f irreducible over F of characteristic ∞ has a solvable Galois group, then its splitting field can be embedded in a radical extension of F.

EXERCISES

1. If ζ is a primitive 13th root of 1, list all subfields of $\mathbf{Q}(\zeta)$.

2. Prove that any finite group G is the Galois group of a normal extension of a suitable field.

3. (a) For n indeterminates $x_1, \cdots, x_n$, define the elementary symmetric functions $\sigma_1, \cdots, \sigma_n$.

(b) Show that any polynomial in $x_1, \cdots, x_n$ which is "symmetric" (that is, invariant under the action of S_n) can be written as a rational function of the elementary symmetric functions $\sigma_1, \cdots, \sigma_n$.

*(c) Show that any symmetric polynomial in $x_1, \cdots, x_n$ can be written as a *polynomial* in $\sigma_1, \cdots, \sigma_n$.

4. If F has characteristic p, $a \in F$ and $x^p - a$ has no root u in F, show that its splitting field $F(u)$ is *inseparable* over F, in the sense that every element not in F has its minimal polynomial inseparable over F.

CHAPTER XIV

Lattices

The present chapter develops the notions of posets and lattices, introduced in §IV.6. Various special kinds of lattices will be distinguished, including modular lattices, distributive lattices, and Boolean algebras. Additional basic properties enjoyed by such special kinds of lattices will be proved, and all finite distributive lattices enumerated.

1. Posets: Duality Principle

We recall that a *poset* is a set P with a binary relation $x \leqslant y$, for $x, y \in P$, which satisfies:

P1. For all x, $x \leqslant x$ (Reflexivity).
P2. If $x \leqslant y$ and $y \leqslant x$, then $x = y$ (Antisymmetry).
P3. If $x \leqslant y$ and $y \leqslant z$, then $x \leqslant z$ (Transitivity).

If $x \leqslant y$ and $x \neq y$, one writes $x < y$, and says that x is "less" than or "properly contained" in y. The relation $x \leqslant y$ is also written $y \geqslant x$, and the relation $x < y$ as $y > x$. A poset is called a *chain* (or a "simply ordered" set) when its elements satisfy

P4. Given x and y, either $x \leqslant y$ or $y \leqslant x$.

Familiar examples of posets include the following (see also §IV.6). The power set $P(I)$ of all subsets $X, Y, \cdots$ of a given set I is a poset when $X \leqslant Y$ means that $X \subset Y$. The set of all subgroups $S, T, \cdots$ of a given group G is a poset when $S \leqslant T$ means that S is a subgroup of T. The set $\mathbf{P}$ of all positive integers $m, n, \cdots$ is a poset when $m \leqslant n$ means that m divides n (that is, that $m|n$). The set $\mathbf{Q}^n$ of all lists ξ of n rational numbers is a poset if $\xi \leqslant \eta$ means that $\xi_i \leqslant \eta_i$ for each $i \in \mathbf{n}$.

DEFINITION. By a least *element of a subset X of a poset P, we mean an element $a \in X$ such that $a \leqslant x$ for all $x \in X$. By a* greatest *element of X, we mean an element $b \in X$ such that $b \geqslant x$ for all $x \in X$. The (unique) least and greatest elements of the whole poset P, when they exist, are called the* universal bounds *of P, and are denoted by O and I, respectively.*

Thus a poset P has universal bounds O and I, if and only if $O \leqslant x \leqslant I$ for every $x \in P$.

These concepts of "least" and "greatest" elements are different from the concepts of "minimal" and "maximal" elements. A *minimal* element of a subset X of a poset P is an element a such that $x < a$ for no $x \in X$; a *maximal* element is defined dually. Clearly, a least element must be minimal and a greatest element maximal, but the converse is not true. Whereas X can have many minimal elements, it can have only one least element.

PROPOSITION 1. *Any finite non-empty subset X of a poset has minimal and maximal elements.*

Proof: Let X be the set $\{x_1, \cdots, x_n\}$. Define $m_1 = x_1$, and m_k as x_k if $x_k < m_{k-1}$ and m_{k-1} otherwise. Then m_n will be minimal. Dually, X has a maximal element.

LEMMA. *For a chain, the notions of minimal and least (maximal and greatest) element are equivalent. Hence, any finite chain has a least (first) and greatest (last) element.*

Proof: If $x < a$ for no $x \in X$, then, by P4, $x \geqslant a$ for every $x \in X$.

PROPOSITION 2. *Every finite chain of n elements is isomorphic to the ordered set* $\mathbf{n}$.

In other words, there is a bijection ϕ from the chain X to $\mathbf{n} = \{1, 2, \cdots, n\}$ such that $x_1 \leqslant x_2$ if and only if $\phi(x_1) \leqslant \phi(x_2)$.

Proof: Let ϕ map the least $x \in X$ into 1, and the least of the remaining $x \in X$ into 2, and so on.

The converse of a binary relation R is, as in §I.3, the relation $R^{\smile}$ such that $xR^{\smile}y$ (read, "x is in the relation $R^{\smile}$ to y") if and only if yRx. Thus the converse of the relation "includes" is the relation "is included in". It is obvious from inspection of conditions P1–P3 that

THEOREM 3 (*Duality Principle*). *The converse of any partial ordering is itself a partial ordering.*

DEFINITION. The dual *of a poset X is that poset X^{op} defined by the converse partial ordering relation on the same elements. A function $f{:}P \to Q$ between posets is a* dual morphism *of order when $x \leqslant y$ in P implies $fx \geqslant fy$ in Q.*

Clearly, the relation of duality is symmetric; the composite of any two dual morphisms of order is a morphism of order, and $X^{\text{op op}} \cong X$ (any poset is isomorphic to the dual of its dual). Moreover, every theorem or proof about posets has a "dual", in which every property or relation is replaced by its dual (for example, $>$ by $<$, O by I, maximal by minimal, etc.).

The idea of an inclusion diagram introduced in §IV.6 can be defined more precisely, and generalized to any finite poset, using the following "covering" relation, which is analogous to the relation of "immediate superior" in a hierarchy.

DEFINITION. By a covers *b in a poset P, it is meant that $a > b$, but that $a > x > b$ for no $x \in P$.*

Using this covering relation, one can construct a *diagram* of any finite poset P, as follows. Draw a small circle to represent each element a of P, placing a higher than b whenever $a > b$. Draw a straight segment joining a to b whenever a covers b.

Since $a > b$ if and only if one can move from a to b downward along some broken line, it is clear that any finite poset is defined up to isomorphism by its diagram. It is also clear that a diagram of the dual P^{op} of a poset P is obtained from one of P by turning the latter upside down.

The length of a finite chain isomorphic to $\mathbf{n}$ is defined to be $n - 1$, and the *length* $l[P]$ of a poset P is defined as the least upper bound of the lengths of the chains in P. When $l[P]$ is finite, P is said to be of *finite length*. Any poset of finite length is defined up to isomorphism by its covering relation: $a > b$ in P if and only if a finite sequence $x_0, x_1, \cdots, x_n$ exists, such that $a = x_0$, $b = x_n$, and x_{i-1} covers x_i for $i = 1, \cdots, n$.

The isomorphism or nonisomorphism of two finite posets can often be tested most simply by comparing their diagrams, as in Figure XIV-1.

In a poset P of finite length with a universal lower bound O, the *height* or "dimension" $h[x]$ of an element $x \in P$ is, by definition, the l.u.b. of the lengths of the chains $O = x_0 < x_1 < \cdots < x_l = x$ between O and x. If P has a universal upper bound I, then clearly $h[I] = l[P]$. Clearly also, $h[x] = 1$ if and only if x covers O; such elements x are called the "atoms" or "points" of the poset P.

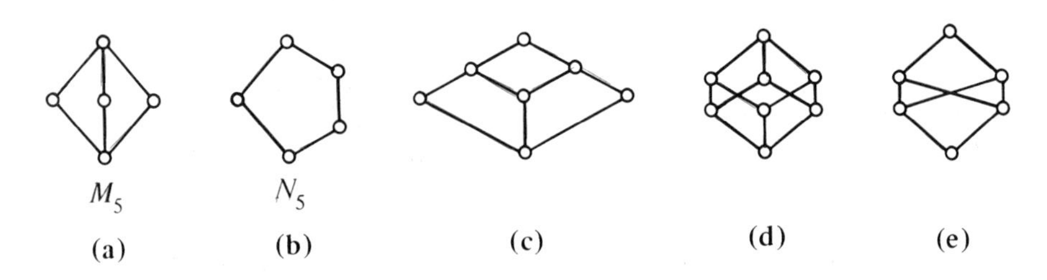

FIGURE XIV-1

EXERCISES

1. Prove Theorem 3 in detail.

2. Prove that, in any poset P, $x < x$ for no x, while $x < y$ and $y < z$ imply $x < z$.

3. Prove that, if a relation $<$ on a set S satisfies the two conditions of Exercise 3, then the relation $x \leqslant y$ defined to mean that $x < y$ or $x = y$ partially orders S.

4. Prove that, in any poset P, $x_1 \leqslant x_2 \leqslant \cdots \leqslant x_n \leqslant x_1$ implies $x_1 = x_2 = \cdots = x_n$.

5. Draw a diagram for each of the following posets:

(a) The subgroups of a cyclic group of order 72.

(b) The subgroups of the quaternion group.

(c) The positive integral divisors of 24 under the partial ordering defined by divisibility.

(d) The subgroups of a cyclic group of order 54.

(e) The ideals of the ring $\mathbf{Z}_{40}$.

(f) The subsets of a set of four points, under inclusion.

6. Show that the posets of Exercises 5c–5e are isomorphic.

7. Let P be a poset of finite length. Show that any two elements of P have an upper bound if and only if P has a greatest element I ("universal upper bound").

2. Lattice Identities

A *lattice*, as defined in §IV.6, is a poset† P in which any two elements x and y have a g.l.b. or "meet" $x \wedge y$ and a l.u.b. or "join" $x \vee y$. We have seen many types of algebraic systems in which the subsystems of a given system form a lattice under inclusion—as for the subfields of a field, the subgroups of a group, and the submodules of a module. We now consider in more detail the algebraic properties of the binary operations $\wedge$ and $\vee$ in any lattice. First, one easily verifies:

LEMMA 1. *In any poset P, the operations of meet and join satisfy the following laws, whenever the expressions referred to exist:*

L1. $x \wedge x = x, \quad x \vee x = x$ (*idempotent*).

L2. $x \wedge y = y \wedge x, \quad x \vee y = y \vee x,$ (*commutative*).

L3. $x \wedge (y \wedge z) = (x \wedge y) \wedge z,$ (*associative*).
$x \vee (y \vee z) = (x \vee y) \vee z,$

L4. $x \wedge (x \vee y) = x \vee (x \wedge y) = x$ (*absorption*).

Moreover, $x \leqslant y$ is equivalent to each of the conditions

$$x \wedge y = x \text{ and } x \vee y = y \qquad (\textit{consistency}).$$

†Note that the void set is also a lattice, vacuously. The void lattice $\varnothing$ is exceptional in various, mostly trivial, respects.

Proof: The idempotent and commutative laws are evident. The dual associative laws L3 are also evident; for example, $x \wedge (y \wedge z)$ and $(x \wedge y) \wedge z$ are both equal to the g.l.b. of x, y, and z whenever all expressions referred to exist. The equivalence between $x \leqslant y$, $x \wedge y = x$, and $x \vee y = y$ is easily verified, and implies L4, Q.E.D.

One also easily verifies that, if a poset P has a O, then

$$O \wedge x = O \qquad \text{and} \qquad O \vee x = x \qquad\qquad \text{for all } x \in P.$$

Dually, if P has a universal upper bound I, then

$$x \wedge I = x \qquad \text{and} \qquad x \vee I = I \qquad\qquad \text{for all } x \in P.$$

The proofs will be left to the reader.

LEMMA 2. *In any lattice, the operations of join and meet are* isotone*:*

$$\text{If } y \leqslant z, \qquad \text{then } x \wedge y \leqslant x \wedge z \qquad \text{and} \qquad x \vee y \leqslant x \vee z. \tag{1}$$

Proof: By L1–L3 and consistency, $y \leqslant z$ implies

$$x \wedge y = (x \wedge x) \wedge (y \wedge z) = (x \wedge y) \wedge (x \wedge z),$$

whence $x \wedge y \leqslant x \wedge z$ by consistency. The second inequality of this lemma can be proved dually (Duality Principle).

LEMMA 3. *Any lattice satisfies the distributive inequalities*

$$x \wedge (y \vee z) \geqslant (x \wedge y) \vee (x \wedge z), \tag{2}$$

$$x \vee (y \wedge z) \leqslant (x \vee y) \wedge (x \vee z). \tag{2'}$$

Proof: First, $x \wedge y \leqslant x$, and $x \wedge y \leqslant y \leqslant y \vee z$; hence, $x \wedge y \leqslant x \wedge (y \vee z)$. Also, $x \wedge z \leqslant x$, $x \wedge z \leqslant z \leqslant y \vee z$; hence, $x \wedge z \leqslant x \wedge (y \vee z)$. Now (2) follows from the definition of $\vee$.

The second distributive inequality (2′) follows from (2) by duality.

LEMMA. 4. The elements of any lattice satisfy the modular inequality*:*

$$x \leqslant z \text{ implies } x \vee (y \wedge z) \leqslant (x \vee y) \wedge z. \tag{3}$$

Proof: Clearly, $x \leqslant x \vee y$ and $x \leqslant z$. Hence, $x \leqslant (x \vee y) \wedge z$. Also, $y \wedge z \leqslant y \leqslant x \vee y$ and $y \wedge z \leqslant z$, whence $y \wedge z \leqslant (x \vee y) \wedge z$. Combining these results with the definition of $\vee$, $x \vee (y \wedge z) \leqslant (x \vee y) \wedge z$, Q.E.D.

We now prove that, conversely, the identities L1–L4 completely characterize lattices.

DEFINITION. A set with a single binary operation which is idempotent, commutative, and associative is called a semilattice.

Lemma 1 has the following immediate corollary; a dual corollary for joins also holds.

COROLLARY. *Let P be any poset in which any two elements have a meet. Then P is a semilattice with respect to the binary operation $\wedge$.*

Such semilattices are called *meet*-semilattices. Conversely:

LEMMA 5. *If S is a semilattice under the binary operation $\square$, the definition*

$$x \leqslant y \quad \Leftrightarrow \quad x \square y = x \tag{4}$$

makes S a poset in which $x \square y = \text{g.l.b.}\ \{x, y\}$.

Proof: The idempotent law $x \square x = x$ implies the reflexive law $x \leqslant x$. By the commutative law $x \square y = y \square x$, $x \leqslant y$ (that is, $x \square y = x$) and $y \leqslant x$ (that is, $y \square x = y$) imply $x = x \square y = y \square x = y$. This proves the antisymmetric law P2. The associative law for $\square$ makes $x \leqslant y$ and $y \leqslant z$ imply $x = x \square y = x \square (y \square z) = (x \square y) \square z = x \square z$, whence $x \leqslant z$, proving the transitive law P3. We leave it to the reader to prove that $x \square y \leqslant x$ and $x \square y \leqslant y$. Finally, if $z \leqslant x$ and $z \leqslant y$, then by associativity and the definition (4), $z \square (x \square y) = (z \square x) \square y = z \square y = z$, whence $z \leqslant x \square y$, proving that $x \square y = \text{g.l.b.}\ \{x, y\}$.

THEOREM 4. *Any set L with two binary operations which satisfy L1–L4 is a lattice, and conversely.*

Proof: First, by Lemma 5, any L which satisfies L1–L3 is a poset in which $x \wedge y = \text{g.l.b.}\ \{x, y\}$, so that $x \leqslant y$ means $x \wedge y = x$. Second, by L4, $x \wedge y = x$ implies $x \vee y = (x \wedge y) \vee y = y$, and (by duality) conversely. Hence, $x \leqslant y$ is also equivalent to $x \vee y = y$. By duality, it follows that $x \vee y = \text{l.u.b.}\ \{x, y\}$; hence, L is a lattice. The converse was shown in Lemma 1, completing the proof.

EXERCISES

1. Show that any chain is a lattice. What are $x \wedge y$ and $x \vee y$ in this lattice?

2. Show that, in a lattice of finite length, there always exist universal bounds O and I.

3. Show that any meet-semilattice L is a commutative semigroup. When is it a monoid?

4. Prove that $(a \wedge b) \vee (c \wedge d) \leqslant (a \vee c) \wedge (b \vee d)$ for any four elements a, b, c, d in a lattice.

5. (a) Draw diagrams of five nonisomorphic lattices of five elements, and show that three are self-dual.

(b) Show that every lattice of five elements is isomorphic to one of these.

6. (a) Show that there are just four nonisomorphic nonvoid lattices of less than five elements.

*(b) Show that there are just 15 nonisomorphic lattices of six elements, of which seven are self-dual. (*Hint:* Consider the posets of four elements not O or I.)

7. Show that any poset of five or fewer elements with universal bounds O and I is a lattice.

8. Draw a diagram of the lattice of all partitions π of the set **4**, letting $\pi \leqslant \pi'$ mean that π is a "subpartition" of π' (that is, that every set in π is a subset of one in π').

3. Sublattices and Products of Lattices

A *morphism* of *lattices*, as defined in §IV.6, is a function $f:L \to M$ on a lattice L to a lattice M such that

$$f(x \wedge y) = f(x) \wedge f(y) \qquad \text{and} \qquad f(x \vee y) = f(x) \vee f(y) \qquad (5)$$

for all $x, y \in L$. Thus, for any element x of a set U, there is an epimorphism f_x from the lattice $P(U)$ of all subsets of U to the lattice $\{O, I\} = \mathbf{2}$, which maps the subsets containing x onto I and the others onto O; here $P(U)$ is a Boolean algebra, as defined in §8.

A lattice S is a *sublattice* of a lattice L when S is a subset of L and the insertion $S \to L$ is a morphism of lattices. Equivalently, a subset S (possibly empty) of a lattice L is a sublattice of L if and only if S is a lattice under the restrictions to S of the binary operations $\wedge$ and $\vee$ of L. Since a lattice is described by the identities L1–L4 on these operations, it follows (much as in the case of subgroups and subrings) that a subset S of a lattice L is a sublattice if and only if S is closed under $\wedge$ and $\vee$.

If $f:L \to M$ is a morphism of lattices, one easily shows that the image f_*S of any sublattice S of L is a sublattice of M, and that the inverse image (counter image) f^*T of any sublattice T of M is a sublattice of L (but note that f^*T may be empty even if $T \neq \varnothing$).

In any lattice L the empty subset and any one-element subset are sublattices. More generally, for any two elements $a \leqslant b$ of a lattice L, the "interval" $[a, b]$ of all x such that $a \leqslant x \leqslant b$ is a sublattice of L. It is important to recognize that a subset of a lattice L can be a lattice under the partial ordering of L without being a sublattice. For example, this is true if L is the lattice of all *subsets* of a finite group G, and S is the set of all *subgroups* of G. As in §IV.6, $S \subset L$ and S is a lattice, yet S is not a sublattice of L because the set-union of two subgroups is not a subgroup.

Among sublattices of a given lattice, ideals are especially noteworthy. An *ideal* of a lattice L is a nonvoid subset J of L such that: (i) $a \in J$ and $x \leqslant a$ implies $x \in J$, and (ii) $a \in J$ and $b \in J$ imply $a \vee b \in J$. Ideals in lattices are somewhat analogous to ideals in rings; thus the inverse image of O under any morphism of lattices is an ideal (although the converse is not true, in general). For any $a \in L$, the set $J(a) = a \wedge L$ of all $x \leqslant a$ in L is an ideal; such ideals are called "principal" ideals.

Lattices L in which *every* subset X has a g.l.b. inf X and a l.u.b. sup X are called *complete lattices*. Setting $X = L$, we see that any nonvoid complete lattice L contains a least element O and a greatest element I. Evidently, the dual of any lattice is a lattice, and the dual of any complete lattice is a complete lattice, with meets and joins interchanged. Any finite lattice or any lattice of finite length is complete.

The lattice of all subgroups of a (possibly infinite) group is a typical example of a complete lattice, because being a subgroup of a given group is a closure property in the following sense.

DEFINITION. A property of subsets of a set I is a closure *property when (i) I has the property, and (ii) any intersection of subsets having the given property itself has this property.*

Theorem 5. *Let L be any complete lattice, and let S be any subset of L such that (i) $I \in S$, and (ii) $T \neq \varnothing$ and $T \subset S$ implies* inf $T \in S$. *Then S is a complete lattice.*

Proof: For any (nonvoid) subset T of S, evidently inf T (in L) is a member of S by (ii), and is the greatest lower bound of T in S. Dually, let $U \subset S$ be the set of all upper bounds of T in S; it is nonvoid since $I \in S$. Then inf $U \in S$ is also an upper bound of T, but is a *least* upper bound since inf $U \leqslant u$ for all $u \in U$. This proves that S is a complete lattice.

Corollary. *Those subsets of any set which have a given closure property form a complete lattice, in which the lattice meet of any family of subsets $\{S, S', \cdots\}$ is their intersection, and their lattice join is the intersection of all subsets T which contain every $S, S', \cdots$.*

Besides occurring naturally in other branches of mathematics, new lattices can also be constructed from old ones by various processes. One such process consists in forming products.

DEFINITION. The product $P \times Q$ *of two posets* P *and* Q *is the set of all pairs* (x, y) *with* $x \in P$, $y \in Q$, *partially ordered by the rule that* $(x_1, y_1) \leqslant (x_2, y_2)$ *if and only if* $x_1 \leqslant x_2$ *in* P *and* $y_1 \leqslant y_2$ *in* Q.

THEOREM 6. *The product* $L \times M$ *of any two lattices is a lattice.*

Proof: For any two elements (x_i, y_i) in $L \times M$ $(i = 1, 2)$, the element $(x_1 \vee x_2, y_1 \vee y_2)$ contains both of the (x_i, y_i)—hence is an upper bound for them. Moreover, for every other upper bound (u, v) of the (x_i, y_i), $u \geqslant x_i$ $(i = 1, 2)$ and hence (by definition of l.u.b.) $u \geqslant x_1 \vee x_2$; likewise, $v \geqslant y_1 \vee y_2$, and so $(u, v) \geqslant (x_1 \vee x_2, y_1 \vee y_2)$. This shows that

$$(x_1 \vee x_2, y_1 \vee y_2) = (x_1, y_1) \vee (x_2, y_2), \tag{6}$$

whence the latter exists. Dually,

$$(x_1 \wedge x_2, y_1 \wedge y_2) = (x_1, y_1) \wedge (x_2, y_2), \tag{7}$$

which proves that L is a lattice.

EXERCISES

1. Show that a lattice L is a chain if and only if every subset of L is a sublattice.
2. Show that any lattice of finite length is complete.
3. Show that the sublattices of any lattice L form a complete lattice under set-inclusion.
4. (a) Show that the ideals of any lattice L with O form a complete lattice $\hat{L}$ under set-inclusion.

   (b) Show that if L is finite, then L and $\hat{L}$ are isomorphic.
5. In any lattice L, show that the function $a \mapsto a \wedge L$ is a morphism of order which carries meets into set-intersections:

$$(a \wedge b \wedge L) = (a \wedge L) \cap (b \wedge L).$$

6. Show that the eight-element lattice of all subsets of a set of three points contains no seven-element sublattice.

*7. Show that any lattice of $n > 6$ elements contains a sublattice of exactly six elements.

8. If $L \times M$ is the product of the lattices L and M, show that the projections $L \times M \to L$ and $L \times M \to M$ are morphisms of lattices with the expected universal properties (§IV.7).

4. Modular Lattices

Just as many of the most important groups are commutative, and still more are solvable, so many important lattices enjoy additional special properties not possessed by all lattices. Among these special classes of lattices, modular and distributive lattices are the most important; we shall discuss them next, treating modular lattices first.

DEFINITION. A lattice is modular *when it satisfies the following modular identity:*

L5. *If* $x \leqslant z$, *then* $x \vee (y \wedge z) = (x \vee y) \wedge z$.

Modular lattices arise naturally in the theories of groups and modules, as the following result shows.

Theorem 7. *The normal subgroups of any group G form a modular lattice.*

Proof: The normal subgroups $M, N, \cdots$ of G certainly form a lattice, in which $M \wedge N = M \cap N$ is the intersection of M and N and $M \vee N = MN$ is the set of products xy with $x \in M$, $y \in N$. To prove that the lattice is modular, it suffices by the modular inequality (Lemma 4) to show that $L \subset N$ implies $(L \vee M) \cap N \subset L \vee (M \cap N)$. To this end, let a belong to $(L \vee M) \cap N$. Then $a \in L \vee M = LM$ so that $a = bc$, where $b \in L$, $c \in M$. Now if we prove that $c \in M \cap N$, it will follow that $a \in L(M \cap N) = L \vee (M \cap N)$, as desired.

But $c = b^{-1}a$ and $b^{-1} \in L \subset N$, and $a \in N$ because $a \in (L \vee M) \cap N$. Thus $c \in N$. But by choice of c, we have $c \in M$. Therefore, $c \in M \cap N$; this completes the proof.

Not every lattice is modular. For example, the lattice N_5 whose diagram is shown in Figure XIV-2(b) is easily verified to be nonmodular. We now show that the lattice of Figure XIV-2(b) is the smallest nonmodular lattice. In fact, one can prove a sharper result.

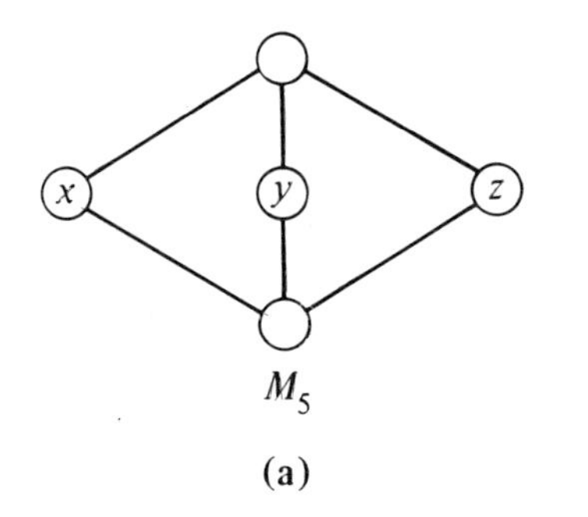

(a)

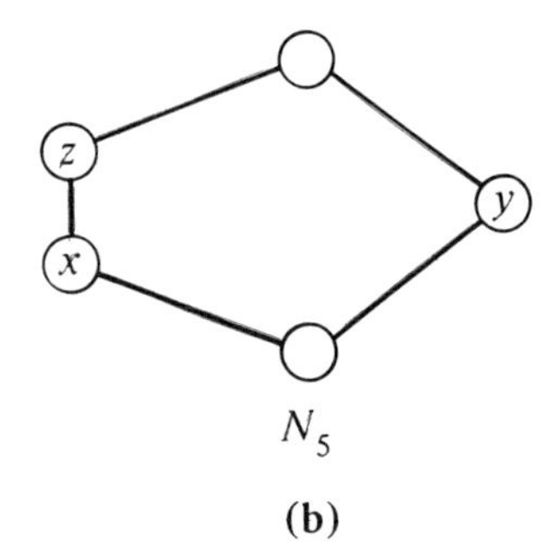

(b)

Figure XIV-2

Theorem 8. *Any nonmodular lattice L contains the lattice N_5 of Figure XIV-2(b) as a sublattice.*

Proof: By definition and the modular inequality, L contains elements u, y, and w such that $u < w$ and $u \vee (y \wedge w) < (u \vee y) \wedge w$. Then the elements $x = u \vee (y \wedge w)$, $z = (u \vee y) \wedge w$, y, together with

$$x \vee y = [u \vee (y \wedge w)] \vee y = u \vee [(y \wedge w) \vee y] = u \vee y = z \vee y$$

and (dually) $x \wedge y = y \wedge z$ form a sublattice which is isomorphic to the lattice of Figure XIV-2(b), with least element $y \wedge z$ and greatest element $u \vee y$.

On the other hand, the five-element lattice M_5 whose diagram is shown in Figure XIV-2(a) is isomorphic with the lattice of all subgroups of the (abelian) four-group $\mathbf{Z}_2 \times \mathbf{Z}_2$; hence it is modular. Furthermore, any sublattice S of a modular lattice M is modular, since then L5 holds a fortiori in S.

Theorem 9. *For any ring R, the R-submodules of any R-module A form a modular lattice.*

Proof: By the preceding remark, it suffices to prove that the R-submodules of A form a sublattice of the modular lattice of all subgroups of A, considered as an abelian group under addition. (This lattice is modular by Theorem 7, since every subgroup of an abelian group is normal.) Therefore, it suffices to prove that if S and T are both R-submodules of A, then so are $S \cap T$ and the set $S + T$ of all sums $s + t$ for $s \in S$, $t \in T$. This was shown, however, in (V.15).

5. Jordan–Hölder–Dedekind Theorem

We now establish for modular lattices an analog of the Diamond Isomorphism Theorem for modules (Theorem XII.4). First, we associate with any comparable pair $c \leqslant d$ of elements of a lattice L the *interval* $[c, d]$ consisting of all $x \in L$ such that $c \leqslant x \leqslant d$. It is obvious that any interval is a sublattice. The following result is less obvious.

Theorem 10. *If a and b are elements of a modular lattice M, then the map $x \mapsto x \wedge a$ is a lattice isomorphism $\phi_a:[b, a \vee b] \to [a \wedge b, a]$ with inverse ψ_b given by $y \mapsto y \vee b$.*

Proof: If $b \leqslant x \leqslant a \vee b$, then the isotone (by Lemma 2 of §2) mappings of ϕ_a and ψ_b satisfy

$$\psi_b(\phi_a(x)) = (x \wedge a) \vee b = x \wedge (a \vee b) \quad \text{by L5,} \qquad \text{since } x \geqslant b,$$
$$= x, \qquad \text{since } x \leqslant a \vee b.$$

Hence, $\psi_b\phi_a$ is the identity function of $[b, a \vee b]$; dually, $\phi_a\psi_b$ is the identity function of $[a \wedge b, a]$. Since any isotone bijection of lattices preserves l.u.b. and g.l.b., hence is a lattice isomorphism, the proof is complete.

COROLLARY. *Let $a \neq b$ in a modular lattice L. Then the following covering conditions hold:*

(ξ) *If a and b both cover c, then $a \vee b$ covers both a and b.*
(ξ') *Dually, if c covers both a and b, then both a and b cover $a \wedge b$.*

Proof: As $a \neq b$, and a and b cover c, $a \wedge b = c$, $a \vee b > a$ ($a \vee b = a$ is impossible), and $a \vee b > b$. But by Theorem 10, $[a, a \vee b]$ is isomorphic to $[a \wedge b, b]$; since b covers $a \wedge b$, it follows that $a \vee b$ covers a. Similarly, $a \vee b$ covers b, proving (ξ). The proof of (ξ') is dual.

DEFINITION. In a modular lattice, two intervals of the form $[a \wedge b, a]$ and $[b, a \vee b]$ will be called transposes; *two intervals $[c, d]$ and $[c^*, d^*]$ will be called* projective *when there exists a sequence of intervals $[c_k, d_k]$, $k = 0, \cdots, n$, such that: (i) $[c_0, d_0] = [c, d]$, (ii) $[c_n, d_n] = [c^*, d^*]$, and (iii) $[c_{k-1}, d_{k-1}]$ and $[c_k, d_k]$ are transposes, for $k = 1, \cdots, n$.*

Since transposed intervals are isomorphic, by Theorem 10, and since isomorphism is transitive, it follows in any modular lattice that *any two projective intervals are isomorphic sublattices*. Using the Diamond Isomorphism Theorem for modules, we have, similarly:

LEMMA. *Let L(A) be the modular lattice of all R-submodules of any R-module A. In L(A), projective intervals $[S, T]$ and $[S^*, T^*]$ correspond to isomorphic quotient-modules T/S and T^*/S^*.*

Proof: By the Diamond Isomorphism Theorem, transposed intervals correspond to isomorphic quotient-modules. Since isomorphism is transitive, the lemma follows.

Now define a *connected chain* of length m in a poset P to be a list of elements $x_0, x_1, \cdots, x_m$ of P in which each x_i covers x_{i-1} ($i = 1, \cdots, m$). Let M be any lattice of finite length in which the covering condition (ξ') holds. For each positive integer m, let $P(m)$ be the following proposition: If one connected chain $\gamma: a = x_0 < x_1 < \cdots < x_m = b$ from a to b in M has length m, then every such connected chain has length m.

The truth of $P(1)$ is trivial in any lattice; we shall now show that $P(m-1)$ implies $P(m)$ in any lattice of finite length in which (ξ') holds. By duality, the same result will then follow from (ξ) (lattices in which (ξ) or (ξ') holds are called "semimodular").

Let $\gamma': a = y_0 < y_1 < \cdots < y_n = b$ be any other connected chain from a to b in L; set $z = x_{m-1} \wedge y_{n-1}$. In case $x_{m-1} = y_{n-1} = z$, $P(m-1)$ trivially implies $m - 1 = n - 1$, and so $m = n$. In the contrary case that $x_{m-1} \neq y_{n-1}$, the assumed covering condition (ξ') shows that both x_{m-1} and y_{n-1} cover $x_{m-1} \wedge y_{n-1} = z$; by the induction hypothesis $P(m-1)$, any connected chain $\delta: a < \cdots < z < x_{m-1}$ from a to x_{m-1} will have length $m - 1$. Hence (see Figure XIV-3), the connected chain $\delta': a < \cdots < z < y_{n-1}$, which coincides with δ between a and z, will also have length $m - 1$. Therefore, by $P(m-1)$ again, the chain $a < y_1 < \cdots < y_{n-1}$ will have length $m - 1$, whence $m = n$. We have proved the first statement of the following *Jordan–Hölder–Dedekind Theorem* for modular lattices.

Theorem 11. *Let M be any lattice of finite length in which either covering condition (ξ) or (ξ') holds. Then any two connected chains in M which have the same ends have the same length. If M is modular, then the intervals $[x_{i-1}, x_i]$ and $[y_{j-1}, y_j]$ of any two connected chains joining O and I can be paired, so that paired intervals are projective.*

To prove the second statement, it suffices to observe that the interval-pairs $[x_{m-1}, b]$, $[z, y_{n-1}]$ and $[y_{n-1}, b]$, $[z, x_{m-1}]$ are projective in Figure XIV-3, and to use an induction argument similar to that used in proving the first statement.

Corollary. *Let the modular lattice $L(A) = M$ of all R-submodules of an R-module A have finite length, and let $O = B_0 < B_1 < \cdots < B_m = A$*

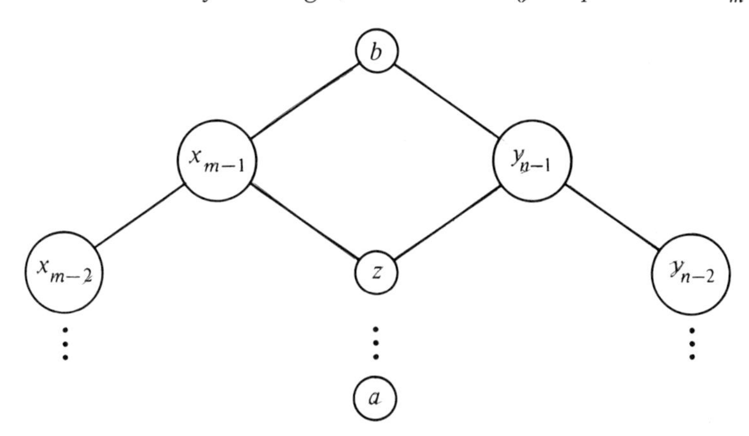

Figure XIV-3

and $O = C_0 < C_1 < \cdots < C_m = A$ *be any two connected chains from* O *to* A *in* $L(A)$. *Then the* B_j/B_{j-1} *and* C_k/C_{k-1} *can be so paired that paired quotient-modules are isomorphic.*

EXERCISES

1. Show that the lattice N_5 of Figure XIV-2(b) is not modular.
2. Show that the dual of any modular lattice is modular.
3. Show that any lattice of length two is modular.
4. Show that any sublattice of a modular lattice is again modular.
5. Prove that, in any lattice, L5 is equivalent to the identity $x \wedge (y \vee z) = x \wedge [[y \wedge (x \vee z)] \vee z]$.
6. Show that the product of any two modular lattices is a modular lattice.
7. In a modular lattice, let $a \wedge b < c < a$ and $a \wedge b < d < b$. Show that the set $\{a, b, c, d\}$ generates a sublattice isomorphic with the direct product of two chains of length two.
8. Prove that the partitions of any finite set form a lattice which satisfies the covering condition (ξ).
9. (a) Show that the subgroups of any p-group form a lattice that satisfies (ξ').

   (b) Prove that, nevertheless, the lattice of all subgroups of the octic group (group of the square) is not modular.

6. Distributive Lattices

Many important lattices are distributive, in the (self-dual) sense that their elements satisfy both distributive laws:

$$\left.\begin{array}{l} \text{L6}'.\ x \wedge (y \vee z) = (x \wedge y) \vee (x \wedge z), \\ \text{L6}''.\ x \vee (y \wedge z) = (x \vee y) \wedge (x \vee z), \end{array}\right\} \quad \text{all } x, y, z \in L.$$

Thus, three of the examples of the posets of §1 are distributive lattices, as will be shown below. The demonstration will be considerably simplified by the following surprising result.

THEOREM 12. *In any lattice,* L6′ *and* L6″ *are equivalent.*

Proof: We shall prove that L6′ implies L6″. We have, for any x, y, and z,

$$\begin{aligned} (x \vee y) \wedge (x \vee z) &= [(x \vee y) \wedge x] \vee [(x \vee y) \wedge z] && \text{by L6}', \\ &= x \vee [z \wedge (x \vee y)] && \text{by L4, L2}, \\ &= x \vee [(z \wedge x) \vee (z \wedge y)] && \text{by L6}', \\ &= [x \vee (z \wedge x)] \vee (z \wedge y) && \text{by L3}, \\ &= x \vee (z \wedge y) && \text{by L4}. \end{aligned}$$

The converse implication follows by duality, completing the proof.

LEMMA. *Any chain is a distributive lattice.*

In fact, $x \wedge y$ is the lesser of x and y; $x \vee y$ is the greater of x and y; $x \wedge (y \vee z)$ and $(x \wedge y) \vee (x \wedge z)$ are both equal to x, if x is smaller than y or z; and both equal to $y \vee z$ in the alternative case that x is bigger than y and z.

Likewise, any sublattice of a product of chains (or of other distributive lattices) is a distributive lattice. We shall now prove a special case of this result.

Consider the set D^S of all functions $f:S \to D$ from an arbitrary set S to a fixed distributive lattice D. Define $f \leqslant g$ to mean that $f(x) \leqslant g(x)$ for all $x \in S$. For any $f, g \in D^S$, define functions $M, m:S \to D$ by

$$M(x) = f(x) \vee g(x), \quad m(x) = f(x) \wedge g(x) \qquad \text{for all } x \in S. \tag{8}$$

Clearly, $M = f \vee g$ and $m = f \wedge g$ in the poset defined by the relation $\leqslant$; hence, D^S is a lattice. Moreover, for all $x \in S$, $u = f \wedge (g \vee h)$ and $v = (f \wedge g) \vee (f \wedge h)$ satisfy

$$\begin{aligned} u(x) = f(x) \wedge (g(x) \vee h(x)) &= [f(x) \wedge g(x)] \vee [f(x) \wedge h(x)] \\ &= [(f \wedge g)(x)] \vee [(f \wedge h)(x)] = v(x). \end{aligned} \tag{8'}$$

Hence (by Theorem 12), the lattice D^S is distributive. We have thus proved

THEOREM 13. *The functions from any set S to any distributive lattice D form a distributive lattice, if $f \leqslant g$ is defined to mean that $f(x) \leqslant g(x)$ for all $x \in S$.*

Now consider the lattice $\mathbf{Z}^+$ of positive integers ordered by divisibility, as in §1. Here $m \wedge n$ is the g.c.d. of m and n, while $m \vee n$ is their l.c.m. Again, each $m \in \mathbf{Z}^+$ can be uniquely written as a prime-power product

$$m = 2^{e(1)}3^{e(2)}5^{e(3)} \cdots p_k^{e(k)} \cdots,$$

where p_k is the kth prime and $e(k)$ is the highest power of p_k dividing m. This decomposition assigns to each positive integer m a function $e:\mathbf{Z}^+ \to \mathbf{N}$ from the set of positive integers to $\mathbf{N}$. Moreover, g.c.d. (m, n) goes into $e \wedge f$ and l.c.m. (m, n) into $e \vee f$ under this correspondence—while evidently mn goes into $e + f$.

Hence, the positive integers under the partial ordering $m|n$ form a *distributive lattice*—specifically, an ideal of $\mathbf{N}^{\mathbf{Z}^+}$ as ordered in Theorem 13.

Distributive lattices have another distinguishing property, which will be wanted below: The unicity of "relative complements".

THEOREM 14. *In a distributive lattice, if $c \wedge x = c \wedge y$ and $c \vee x = c \vee y$, then $x = y$.*

Proof: Using repeatedly the hypotheses, L4, L2, and L6, we have

$$x = x \wedge (c \vee x) = x \wedge (c \vee y) = (x \wedge c) \vee (x \wedge y)$$
$$= (c \wedge y) \vee (x \wedge y) = (c \vee x) \wedge y = (c \vee y) \wedge y = y.$$

This completes the proof.

7. Rings of Sets

Two of the most basic concepts of set theory are those of a "ring" and of a "field" of sets, defined as follows.

DEFINITION. A ring of sets *is a family* Φ *of subsets of a set* I *which contains with any two sets* S *and* T *also their (set-theoretic) intersection* $S \cap T$ *and their union* $S \cup T$. A field of sets *is a ring of sets which contains with any* S *also its set-complement* S' *in* I.

Any ring of sets is a distributive lattice under the natural ordering $S \subset T$. Thus the open sets of the real line form a distributive lattice; the closed sets form a dual distributive lattice. We now construct a wide class of *finite* distributive lattices.

LEMMA 1. *Let* P *be any poset. Call a subset* $A \subset P$ *"closed below" when it contains with any* a *all* $x \in P$ *with* $x \leqslant a$. *Then the family of all subsets of* P *which are closed below is a ring of sets.*

We leave the proof as an exercise.

We can order the elements of any finite poset P in a sequence, so that if $a < b$ in P, then a precedes b in the sequence. We choose for a_1 any minimal element of P, in the sense of §1; then a_2 as any minimal one of the remaining elements, and so on. By construction, each of the sets $A_k = \{a_1, \cdots, a_k\} \subset P$ is closed below; moreover,

$$\varnothing = A_0 < A_1 < \cdots < A_n = P,$$

where n is the number of elements of P. Further, each A_k covers A_{k-1}, since it contains just one point not in A_{k-1}. Hence, the chain

$$\varnothing = A_0 < A_1 < \cdots < A_n = P$$

is connected. But by Theorem 11, all such chains in the ring of the subsets closed below in P have the same length. We conclude:

LEMMA 2. *In any finite poset* P, *the ring of all subsets of* P *which are closed below is a distributive lattice whose length equals the number of elements of* P.

The rest of this section will be devoted to showing that, conversely, every distributive lattice L of finite length n is isomorphic to the ring of all subsets closed below in the poset P of all "join-irreducible" elements of L, defined as follows.

DEFINITION. An element $a \neq O$ of a lattice L is join-irreducible *when $b \vee c = a$ implies $b = a$ or $c = a$.*

LEMMA 3. *If p is join-irreducible in a distributive lattice L, then*

$$p \leqslant \bigvee_{i=1}^{k} x_i \qquad \textit{implies} \qquad p \leqslant x_i \qquad\qquad \textit{for some } i.$$

(Here $\bigvee_{i=1}^{k} x_i$ stands for $x_1 \vee \cdots \vee x_k$.)

Proof: By $p \leqslant \bigvee_{i=1}^{k} x_i$, we have $p = p \wedge \bigvee_{i=1}^{k} x_i = \bigvee_{i=1}^{k} (p \wedge x_i)$, by distributivity. Since p is join-irreducible, $p = p \wedge x_i$ for some i. Hence, $p \leqslant x_i$ for some i.

COROLLARY. *In a distributive lattice of finite length, each element a has a strictly unique representation as the join of the* maximal *join-irreducible elements which it contains.*

Proof: By induction on height, every a in L is the join of a finite set of join-irreducible elements. Hence, it is the join of the subset of *maximal* such elements.

LEMMA 4. *If a distributive lattice L of finite length contains exactly n join-irreducible elements $p_1, \cdots, p_n$, then $l[L] = n$.*

Proof: Arrange the p_i so that p_1 is minimal, p_2 minimal among the remaining p_i, and so on. Then the chain

$$O < p_1 < p_1 \vee p_2 < \cdots < \bigvee_{i=1}^{k} p_i = q_k$$

has length k, because $p_1 \vee p_2 \cdots \vee p_j = p_1 \vee p_2 \cdots \vee p_j \vee p_{j+1}$ would imply $p_{j+1} \leqslant p_1 \vee p_2 \cdots \vee p_j$, which by Lemma 3 implies $p_{j+1} \leqslant p_i$ for some $i < j + 1$. This is impossible since p_i was minimal among $p_i, \cdots, p_n$. Therefore, the q_k form a chain of length n in L.

Conversely, since every element a of L is a join of all $p_i \leqslant a$, each chain $a_1 < \cdots < a_m$ in L determines a chain of subsets $A_1 < \cdots < A_m$ of the set of all join-irreducible elements p_i of L, and so no chain in L can be longer than n.

Theorem 15. *Let L be any distributive lattice of length n. Then the poset X of all join-irreducible elements $p_i > O$ of L has n elements. Moreover L is isomorphic with the ring of all subsets of X which are closed below.*

Proof: As before, every a in L is the join $\bigvee_A p_i$ of the finite set A of the join-irreducible elements $p_i > O$ which it contains. Also, if $p_i \leqslant a$ and $p_j \leqslant p_i$, then $p_j \leqslant a$; that is, every set A is closed below in X. But conversely, by Lemma 3, if A is closed below, then $\bigvee_A p_i$ contains no p_k not in A. Hence, the assignment $a \mapsto A$, which is obviously isotone, is bijective, and so an isomorphism of L to the ring of all those subsets of X which are closed below.

Corollary. *The number of (nonisomorphic) distributive lattices of length n is equal to the number of (nonisomorphic) posets of n elements.*

Thus there are 2 nonisomorphic distributive lattices of length two, 5 of length three, 16 of length four, and 63 of length five.

EXERCISES

1. Exhibit diagrams of the 5 nonisomorphic distributive lattices of length three, shading their join-irreducible elements.
2. In the poset of positive integers ordered by divisibility, show that the join-irreducible elements are the powers of primes.
3. Show that, if $n = p_1^{e_1} \cdots p_r^{e_r}$ $(p_1 < \cdots < p_r)$ is the representation of a positive integer as a product of prime-powers, then the divisors of n partially ordered by divisibility form the distributive lattice $E_1 \times \cdots \times E_r$, where E_i is a chain of length e_i.
4. Show that the two lattices of Figures XIV-2(a) and XIV-2(b) are the only nondistributive lattices of five elements.
5. Show that L6′ and L1–L3 imply $x \vee (x \wedge y) = x \wedge (x \vee y)$.
6. Show that if we add to a distributive lattice L two new elements O, I satisfying $O < x < I$ for all $x \in L$, the result is still a distributive lattice.

*7. Show that the poset of all join-irreducible elements in any finite distributive lattice L is isomorphic to the poset of all meet-irreducible elements of L.

8. Boolean Algebras

By a *complement* of an element x in a lattice L with O and I is meant an element $y \in L$ such that $x \wedge y = O$ and $x \vee y = I$. The lattice L is called *complemented* if all its elements have complements. In the lattice $P(I)$ of all subsets S of a set I, the usual set-theoretic complement S' of S in I does satisfy $S \wedge S' = O$ and $S \vee S' = I$; hence, the lattice $P(I)$ is complemented (in fact, S' is the only complement of S in this lattice). The modular lattice of all subspaces of a finite-dimensional vector space V is complemented, by Corollary VI.8.4. The case of all subspaces of $(\mathbf{Z}_2)^2$ gives the modular lattice M_5 of Figure XIV-1(a).

Theorem 16. *In a complemented distributive lattice, each element x has one and only one complement. Moreover, with O and I the complements satisfy*

$$\begin{array}{lll}
\text{L7.} & x \wedge O = O, & x \vee O = x, \\
 & x \wedge I = x, & x \vee I = I. \\
\text{L8.} & x \wedge x' = O, & x \vee x' = I. \\
\text{L9.} & (x')' = x. & \\
\text{L10.} & (x \wedge y)' = x' \vee y', & (x \vee y)' = x' \wedge y'.
\end{array}$$

Proof: In a distributive lattice, complements are unique when they exist, by Theorem 14. Hence, the unique complement x' of each x in the given lattice L satisfies L8, and defines a unary operation $x \mapsto x'$ on L. By the symmetry of the definition of a complement, x is then a complement of x', so $(x')' = x$ as asserted in L9.

Next we prove that

$$x \wedge a = O \quad \Leftrightarrow \quad x \leqslant a'. \tag{9}$$

Indeed, if $x \leqslant a'$, then $x \wedge a \leqslant a' \wedge a = O$, so $x \wedge a = O$. Conversely, if $x \wedge a = O$, then

$$x = x \wedge I = x \wedge (a \vee a') = (x \wedge a) \vee (x \wedge a') = O \vee (x \wedge a') = x \wedge a',$$

and so $x \leqslant a'$.

Now $x \leqslant y$ in L implies that $y' \wedge x \leqslant y' \wedge y = O$ and hence by (9) that $y' \leqslant x'$. In other words, the bijection $x \mapsto x'$ of L to L inverts order; it must, therefore, carry meets to joins and joins to meets, as asserted in L10.

DEFINITION. A Boolean algebra *A is a distributive lattice, under binary operations $\wedge$ and $\vee$, which also has universal bounds O and I and a unary operation $'$ satisfying* L8. *If A and B are Boolean algebras, a function $f: A \to B$ is a morphism of Boolean algebras when it is a morphism of $\vee$, of $\wedge$, and of $'$.*

Theorem 16 asserts that any complemented distributive lattice is a Boolean algebra, and also that the involution law L9 and De Morgan's law L10 hold in every Boolean algebra. Any field of sets is a Boolean algebra; in particular, the lattice of all subsets of a set is a Boolean algebra.

By L8, any morphism f of Boolean algebras has $f(O) = O$ and $f(I) = I$, and so is a morphism of the nullary operations "select O" and "select I". There is a partial converse as follows.

Proposition 17. *If A and B are Boolean algebras, then any function $f{:}A \to B$ with $f(O) = O$, $f(I) = I$ which is a morphism of $\vee$ and $\wedge$ is a morphism of Boolean algebras.*

Proof: For each $x \in A$, the function f applied to L8 gives

$$(fx) \wedge (fx') = O, \qquad (fx) \vee (fx') = I.$$

These equations in B state that $f(x')$ is the unique complement $(fx)'$ of $f(x)$, so $f(x') = (fx)'$, and f is a morphism of Boolean algebras.

A Boolean algebra S is a (Boolean) *subalgebra* of another Boolean algebra A when S is a subset of A and the insertion $S \to A$ is a morphism of Boolean algebras. This implies that the operations $\vee$, $\wedge$, and $'$ in S are restrictions to S of these operations in A *and* that the universal bounds of S are those of A. As in previous cases, a subset S of a Boolean algebra A is a Boolean subalgebra if and only if it contains O and I and is closed under the three operations $\vee$, $\wedge$, and $'$. Note in particular that a proper interval sublattice $[a, b]$ of A, although (Exercise 5) a Boolean sublattice, is not closed under complement and so is not a Boolean subalgebra of A. The intersection of any family of Boolean subalgebras of A is again a subalgebra. In particular, the intersection of all Boolean subalgebras of A containing n given elements $a_1, \cdots, a_n$ is a Boolean subalgebra of A, called the subalgebra "generated" by these elements a_i.

We now show that any distributive lattice with O and I has a largest "Boolean subalgebra".

Theorem 18. *The complemented elements of any distributive lattice with universal bounds form a sublattice, and hence a Boolean algebra.*

Proof: If x and y are complemented, then

$$(x \wedge y) \wedge (x' \vee y') = (x \wedge y \wedge x') \vee (x \wedge y \wedge y') = O \vee O = O$$

and dually. Hence, $x \wedge y$ has the complement $x' \vee y'$, as in L8. Dually $x \vee y$ has the complement $x' \wedge y'$, completing the proof.

It is easy to determine all Boolean algebras of finite length. For, unless an element a covers O, we can find p such that $O < p < a$, whence $q = p' \wedge a$ satisfies $q < a$ and

$$p \vee q = p \vee (p' \wedge a) = (p \vee p') \wedge (p \vee a) = I \wedge a = a;$$

hence, only elements a which cover O ("points") can be join-irreducible. On the other hand, all points are join-irreducible, and the set X of all points is totally unordered: Every subset of X is closed below. By Theorem 15, we conclude

Theorem 19. *Every complemented distributive lattice of finite length n is isomorphic to the field of all subsets of its n points.*

Corollary. *Every Boolean algebra of finite length is a product of n copies of a chain of two elements.*

Remark. Theorem 4 and the first part of Theorem 16 show that conditions L2, L3, L4, L6–L6′, L7, and L8 together suffice to define a Boolean algebra. Actually, even these postulates are redundant; thus, since

$$x \vee (x \wedge y) \overset{(L7)}{=} (x \wedge I) \vee (x \wedge y) \overset{(L6)}{=} x \wedge (I \vee y) \overset{(L2)}{=} x \wedge (y \vee I) \overset{(L7)}{=} x \wedge I = x$$

and dually, L4 is implied by L8, L2, and L6–L6′. With more work, one can show that conditions L2, L3, L6–L6′, L7, and L8 form a set of postulates for Boolean algebra. See Exercises 9–11 below.

EXERCISES

1. Prove that $x = y$ in a Boolean algebra if and only if $(x \wedge y') \vee (x' \wedge y) = 0$.

2. Prove Poretzky's law: For given x and t, $x = 0$ if and only if $t = (x \wedge t') \vee (x' \wedge t)$.

3. Prove that: (a) $y \leqslant x'$ if and only if $x \wedge y = 0$, and (b) $y \geqslant x'$ if and only if $x \vee y = I$.

4. (a) Show that the dual of any Boolean algebra is a Boolean algebra.
   (b) Show that the product of any two Boolean algebras is a Boolean algebra.

5. Prove that every interval $[a, b]$ of a Boolean algebra L is a Boolean algebra under the given partial order, in which the complement of any x is $(a \vee x') \wedge b$.

*6. Generalize Exercise 5 to complemented modular lattices. (*Caution:* Complements need not be unique.)

7. Find a modular lattice of seven elements in which the complemented elements do not form a sublattice.

8. Let S be a subalgebra of the Boolean algebra A of order 2^n with points (minimal proper elements) $p_1, \cdots, p_n$.
   (a) Define the binary relation $p_i S p_j$ to mean that, for all $s \in S$, $p_i \leqslant s$ implies $p_j \leqslant s$. Show that S is an equivalence relation on the points of A.
   (b) Conversely, prove that for any equivalence relation σ on the points of A, the set of all $a \in A$ such that $p_i \leqslant a$ and $p_i \sigma p_j$ imply $p_j \leqslant a$ constitutes a subalgebra $S(\sigma)$ of A.

9. Prove in detail that L2, L3, L6–L6′, and L7 imply L4.

***10.** (a) In an algebraic system $[A; \wedge, \vee]$ satisfying L2, L3, L6, and L7, show that for each $x \in A$ there is at most one x' satisfying L8.

(b) Show that L4 is redundant in the definition of a Boolean algebra. (*Hint:* Use Exercises 9 and 10a.)

11. Show that the laws L2, L3, L6–L6′, L8, and the identities $O \vee x = x = x \wedge I$ imply that $O \wedge x = O$ and $x \vee I = I$.

9. Free Boolean Algebras

We will prove in this section the surprising fact that any Boolean algebra generated by a finite number of elements is necessarily finite. For instance, let A be a Boolean algebra generated by two elements x and y. A contains x' and y' and hence the four meets

$$x \wedge y, \qquad x \wedge y', \qquad x' \wedge y, \qquad x' \wedge y'. \tag{10}$$

Now we consider joins of one or more of these four elements, such as

$$(x \wedge y') \vee (x' \wedge y), \qquad (x \wedge y) \vee (x' \wedge y) \vee (x' \wedge y').$$

Counting O, there are 2^4 such expressions, corresponding to the 2^4 subsets of the set (10) of four elements. Any one of these 2^4 expressions is called a *disjunctive canonical* expression (a "disjunctive Boolean polynomial") in x and y; here "disjunctive" means "join".

Let S be the set of all elements of A which can be written as disjunctive canonical expressions in x and y; we will show that S is a subalgebra. First, since any element of S is a join of elements (10), S is closed under join. It is also closed under meet, for the meet of any two different elements (10) is zero (for example, $(x \wedge y) \wedge (x \wedge y') = x \wedge y \wedge y' = x \wedge O = O$). Since

$$x = x \wedge (y \vee y') = (x \wedge y) \vee (x \wedge y'), \qquad x' = (x' \wedge y) \vee (x' \wedge y')$$

the set S also contains x, y, x', and y'. Now any element of S is a join of meets of elements x, y, x', y', so by L10 any complement of an element of S is a meet of joins of elements x, y, x', y' in S. Since S is closed under meets and joins, it is therefore closed under complement. Hence, S, as a subalgebra containing x and y, is the algebra generated by these two elements. It has at most 2^4 different elements.

A similar argument for a Boolean algebra with three generators x, y, and z show that each of its elements can be written as a join of some of the following eight elements (like the four elements of (10)): $x^e \wedge y^{e'} \wedge z^{e''}$, where each x^e is either x or x', each $y^{e'}$ is either y or y', etc. Hence, such an algebra has at most 2^8 elements.

For a Boolean algebra A with n generators $x_1, \cdots, x_n$ there will be 2^n meets like (10), of the form $x_1^{e_1} \wedge x_2^{e_2} \wedge \cdots \wedge x_n^{e_n}$, where each $x_i^{e_i}$ is either x_i or its complement x_i'. Let $\{-, '\}$ be the set with two elements, the "exponents" $-$ and $'$; then each e_i is an element $e_i \in \{-, '\}$, with the understanding that $x_i^- = x_i$ and that x_i' is the complement of x_i. The function set $\{-, '\}^{\mathbf{n}}$ is the set of all 2^n lists $\mathbf{e}:\mathbf{n} \to \{-, '\}$ of n such exponents e_i, and each such list $\mathbf{e}$ determines a meet $x_1^{e_1} \wedge \cdots \wedge x_n^{e_n} \in A$. If E is any subset of the function set $\{-, '\}^{\mathbf{n}}$, let $h(E)$,

$$h(E) = \bigvee_{\mathbf{e} \in E} (x_1^{e_1} \wedge x_2^{e_2} \wedge \cdots \wedge x_n^{e_n}), \tag{11}$$

be the join of all these meets, for $\mathbf{e} \in E$, with the convention that $h(\varnothing) = O$.

Theorem 20. *If $x_1, \cdots, x_n$ are n elements of a Boolean algebra A while $B = P(\{-, '\}^{\mathbf{n}})$ is the Boolean algebra of all subsets E of the function set $\{-, '\}^{\mathbf{n}}$, then* (11) *defines a morphism $h:B \to A$ of Boolean algebras whose image is the subalgebra of A generated by $x_1, \cdots, x_n$.*

Proof: Let S and T be two subsets of $\{-, '\}^{\mathbf{n}}$. By the idempotent law $a \vee a = a$, $h(S) \vee h(T)$ is $h(S \cup T)$. As for the intersection, $e_i \neq e_i'$ means that $x_i^{e_i} \wedge x_i^{e_i'} = x_i \wedge x_i' = O$; hence, $h(S) \wedge h(T) = h(S \cap T)$. By convention, $h(\varnothing) = O$ for the empty subset $S = \varnothing$. On the other hand, $x_i \vee x_i' = I$ in A, so

$$I = (x_1 \vee x_1') \wedge (x_2 \vee x_2') \wedge \cdots \wedge (x_n \vee x_n').$$

The right side expanded by the distributive law gives 2^n terms which are exactly all the terms in $h(I_B)$, where $I_B = \{-, '\}^{\mathbf{n}}$ is the universal upper bound for B. We have shown that h is a morphism of join, meet, and universal bounds. By Proposition 17, it is therefore a morphism of Boolean algebras.

The image of h is a Boolean subalgebra of A. Each x_i is

$$x_i = \cdots (x_{i-1} \vee x_{i-1}') \wedge x_i \wedge (x_{i+1} \vee x_{i+1}') \cdots; \tag{12}$$

expanded by the distributive law, this is an $h(S)$; hence, the image of h contains every x_i and is thus the subalgebra generated by these elements x_i, as required.

Corollary. *A Boolean algebra generated by n elements $x_1, \cdots, x_n$ can contain at most $2^{(2^n)}$ elements.*

Theorem 21. *There is a Boolean algebra B and a list $S_1, \cdots, S_n$ of n elements of B with the following property: For any list $x_1, \cdots, x_n$ of n elements of a Boolean algebra A there is exactly one morphism $h:B \to A$ of Boolean algebras with $h(S_i) = x_i$ for all $i \in \mathbf{n}$.*

A Boolean algebra B with a list $\mathbf{S}$ with this property is called a "free" Boolean algebra on the set $\{S_1, \cdots, S_n\}$ of free generators, in exact analogy to the free modules of Theorem V.8. Again, the theorem states that the list $\mathbf{S} \in B^{\mathbf{n}}$ is a universal among lists $\mathbf{x}$ of n elements in a Boolean algebra.

Proof: In the Boolean algebra $B = P(\{-, '\}^{\mathbf{n}})$ used above, introduce for each $i \in \mathbf{n}$ the subset $S_i \subset \{-, '\}^{\mathbf{n}}$ with

$$S_i = \{\mathbf{e} | \mathbf{e} \in \{-, '\}^{\mathbf{n}} \text{ and } e_i = -\}.$$

Thus $\mathbf{e} \in S_i$ if and only if $e_i = -$, so the formula (11) for the morphism $h: B \to A$ above shows that $h(S_i) = x_i$ for all i. Now the definition of S_i shows that $\{\mathbf{e}\} = S_1^{e_1} \cap \cdots \cap S_n^{e_n}$ for each list $\mathbf{e}$ and hence that each S in B is

$$S = \bigcup_{\mathbf{e} \in S} (S_1^{e_1} \cap S_2^{e_2} \cap \cdots \cap S_n^{e_n}).$$

By this formula, any morphism $B \to A$ of Boolean algebras with $S_i \mapsto x_i$ must be given by the formula (11) for h, so h is unique, as required.

From this theorem it follows that two Boolean expressions in n letters $x_1, \cdots, x_n$ determine the same function in every Boolean algebra if and only if they reduce to the same disjunctive canonical expression.

EXERCISES

1. (a) In any distributive lattice show that

$$(x \wedge y) \vee (y \wedge z) \vee (z \wedge x) = (x \vee y) \wedge (y \vee z) \wedge (z \vee x).$$

(b) In a Boolean algebra, compute the disjunctive canonical expression for the preceding formula.

2. In a Boolean algebra test each of the following proposed equalities, by reducing each side to its disjunctive canonical form:

(a) $[x \wedge (y \vee z)']' = (x \wedge y)' \vee (x \wedge z)$.

(b) $x = (x' \vee y')' \vee [z \vee (x \vee y)']$.

3. (a) In a Boolean algebra A generated by n elements $x_1, \cdots, x_n$ show that every element has a "conjunctive canonical expression" as a meet of terms of the form $x_1^{e_1} \vee \cdots \vee x_n^{e_n}$.

(b) If A is free on n generators, show that this conjunctive canonical expression is unique.

4. (a) Draw a diagram of the free Boolean algebra with two generators.

(b) Show that this algebra has exactly twelve pairs (and twenty-four ordered pairs) of two generators.

(c) Show that the group of its automorphisms is solvable.

***5.** Let A be a set with elements O, I, binary operations $\wedge$, $\vee$ and a unary operation $'$ satisfying L2, L6–L6′, and L8. Show that if O and I satisfy $O \vee x = x$ and $x \wedge I = x$ for all x, then: (i) L1 holds, (ii) $O \wedge x = O$ and $x \vee I = I$ for all x, and (iii) L4 holds.

***6.** (a) Assuming the results of Exercise 5, show that its hypotheses also imply: (iv) If $x \wedge y = O$ and $x \vee y = I$, then $y = x'$, (v) L9, and (vi) L3. (*Hint:* For (vi), let $a = x \wedge (y \wedge z)$ and $b = (x \wedge y) \wedge z$. Prove that $x \vee a = x \vee b$ and $x' \vee a = x' \vee b$; then use $(x \vee a) \wedge (x' \vee a) = (x \vee b) \wedge (x' \vee b)$ to show $a = b$.)

(b) Conclude that if $[A; \wedge, \vee, ', O, I]$ satisfies the hypotheses of Exercise 5, it is a Boolean algebra. (F. Gerrish.)

CHAPTER XV

Categories and Adjoint Functors

THE NOTION of a universal element has played a central role in many of the constructions of the previous chapters. Each such element is "universal" for a suitable functor, and each such functor is defined on an appropriate category. In this chapter we will make a more systematic study of the notions of "category" and "functor", leading up to the concept of a pair of "adjoint" functors between two categories. This concept, like that of universal elements, is of wide applicability.

1. Categories

The morphisms of groups, or of rings, or of modules played a central role in our study of each of these types of algebraic systems. For each type of algebraic system, the class of all systems $X, Y, \cdots$ of that type and of all their morphisms $f: X \to Y$ form a concrete category (§IV.5). The basic operation in such a category is the composition $g \circ f$ of morphisms f and g, defined when domain (g) = codomain (f), as in the diagram

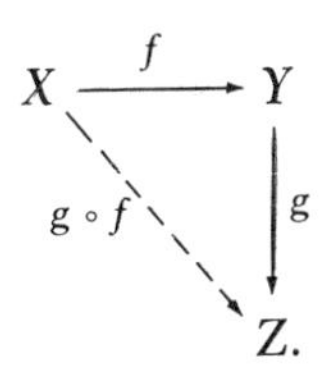

In a concrete category, each object X was associated to an actual underlying set $\mathfrak{U}(X)$ and each morphism $f: X \to Y$ was a function $\mathfrak{U}(f): \mathfrak{U}(X) \to \mathfrak{U}(Y)$ defined on the underlying sets, so the composition of morphisms was the actual composition of functions. More generally, we now wish to suitably compose arrows (morphisms) in a category even when there are no underlying sets. For this, we must specify the necessary properties of composition by suitable axioms, now to follow. These axioms describe composition in an "abstract" category in just the way in which the axioms for multiplication in an "abstract" group reflect the properties of multiplication in transformation groups. The essential feature of these axioms on composition must use the objects of the category to specify when the composite $g \circ f$ is defined.

DEFINITION. Let **X** *be a class of "objects", $X, Y, Z, \cdots$ together with two functions, as follows:*

(i) A function assigning to each pair (X, Y) of objects of **X** *a set* $\hom_{\mathbf{X}}(X, Y) = \hom(X, Y)$. *An element $f \in \hom(X, Y)$ in this set is called an* arrow *$f: X \to Y$ of* **X**, *with* domain *X and* codomain *Y. Each such arrow has a unique domain and a unique codomain.*

(ii) A function assigning to each triple (X, Y, Z) of objects of **X** *a function*

$$\hom(Y, Z) \times \hom(X, Y) \to \hom(X, Z). \tag{1}$$

For arrows $g: Y \to Z$ and $f: X \to Y$, this function is written as $(g, f) \mapsto g \circ f$, and the arrow $g \circ f: X \to Z$ is called the composite *of g with f.*

The class **X** *with these two functions is called a* category *when the following two axioms hold:*

Associativity: *If $h: Z \to W$, $g: Y \to Z$ and $f: X \to Y$ are arrows of* **X** *with the indicated domains and codomains, then*

$$h \circ (g \circ f) = (h \circ g) \circ f. \tag{2}$$

Identity: *For each object Y of* **X** *there exists an arrow $1_Y: Y \to Y$ such that*

$$f: X \to Y \quad \Rightarrow \quad 1_Y \circ f = f; \qquad g: Y \to Z \quad \Rightarrow \quad g \circ 1_Y = g. \tag{3}$$

Properties (3) state that 1_Y acts both as a left and as a right identity for composition whenever this composition is defined; they clearly determine 1_Y uniquely. We call 1_Y the *identity arrow* of the object Y.

The axioms for a category are much like the axioms for a multiplicative monoid (§I.11), except that the product (= composite) $g \circ f$ of morphisms g and f in a category is defined only when the domain of g is the codomain of f. Indeed, for each object X in a category, the set $\hom(X, X)$ is a monoid under composition, so that a category with just one object X is essentially just a monoid $\hom(X, X)$.

An example is the category used in §VII.1 to analyze matrix algebra over any ring R. The objects are the natural numbers n, with $\hom(n, m)$ the set of all $m \times n$ matrices A over R and composition the usual matrix product. This is a category in the sense of the definition above. It is not a concrete category because its objects n are not (regarded as) sets, and its arrows A are correspondingly not functions. It could be altered to a concrete category by replacing each object n by the free module R^n and each A by the corresponding linear transformation $t_A: R^n \to R^m$. This replaces the formal composition of matrices by the actual composition of the transformations t_A.

Many examples of categories are provided by the various concrete categories we have already discussed, in virtue of the following result (parallel to the fact that every transformation group is a group).

Theorem 1. *Every concrete category is a category.*

The proof is straightforward. Each concrete category, as defined in §IV.5, is a class **P** of objects in which each object P has an underlying set $\mathfrak{U}(P)$ and each pair of objects P, Q a set hom (P, Q) of functions on $\mathfrak{U}(P)$ to $\mathfrak{U}(Q)$. Since the second axiom for a concrete category states that $f \in \text{hom}(P, Q)$ and $g \in \text{hom}(Q, R)$ have their composite function $g \circ f$ in hom (P, R), the actual composition of functions provides the composition of arrows required for (1) above. The associativity axiom (2) then holds because the composition of functions is always associative, while the identity axiom (3) holds because each set hom (P, P) in a concrete category contains the identity function of the set $\mathfrak{U}(P)$. This completes the proof.

Examples of (concrete) categories; including some from §IV.5, are

S, the category of all sets, is the class of all sets $X, Y, \cdots,$ with each hom (X, Y) the set Y^X of all functions on X to Y, and with the usual composition.

$\mathbf{Mod}_R$, the category of all right R-modules for a fixed ring R, is the class of all these modules $A, B, \cdots,$ with each hom (A, B) the set of all morphisms $A \to B$ of modules, under the usual composition.

Mon, the category of all monoids M.

Grp, the category of all groups.

Rng, the category of all rings.

Each type of algebraic system yields in the same way a corresponding (concrete) category. Thus we may speak of the category of all lattices, of all ordered rings, or of all Boolean algebras, or again of the category whose objects are sets X equipped with one binary and one ternary operation, and whose morphisms $X \to Y$ are functions which are morphisms both for the binary and for the ternary operation. A useful case is the *category* $\mathbf{S}_*$ of *pointed sets*. A pointed set (X, x_0) is a set X together with a selected element $x_0 \in X$; a morphism $f:(X, x_0) \to (Y, y_0)$ of pointed sets is a function $f: X \to Y$ with $f(x_0) = y_0$. Since giving a nullary operation $\mathbf{1} \to X$ on the set X amounts to giving an element $x_0 \in X$ (as the image of 1), a pointed set may be regarded as a (quite rudimentary) type of algebraic system: A set X equipped with a single nullary operation.

Many categories arise as subcategories of given ones. Thus let **W** be a subclass (of the class of objects) of a category **X**. To each pair U, V of objects in this subclass **W**, take the set $\text{hom}_{\mathbf{W}}(U, V)$ of morphism in **W** to be just the set $\text{hom}_{\mathbf{X}}(U, V)$ of morphisms in **X**, while composition in **W** is the restriction to these sets of composition in **X**. This clearly makes **W** a category; it is called a *full subcategory* of **X**.

For example, in the category of groups the subclass of all abelian groups determines in this way a full subcategory, called the category **Abgrp** of abelian groups. Full subcategories of the category of rings include the category of commutative rings, the category of commutative noetherian rings, the category of integral domains, the category of principal ideal domains, and the category of fields. For a given ring R, the category $\mathbf{Mod}_R$ of all R-modules has as full subcategories the category of all R-modules of finite type, the category of all free R-modules, and the category of all cyclic R-modules.

More generally, a category $\mathbf{U}$ is a subcategory of $\mathbf{X}$ when (i) the class $\mathbf{U}$ of objects of $\mathbf{U}$ is a subclass of $\mathbf{X}$; (ii) for $U, V \in \mathbf{U}$ the set $\hom_{\mathbf{U}}(U, V)$ of arrows $f: U \to V$ in $\mathbf{U}$ is a subset of $\hom_{\mathbf{X}}(U, V)$; and (iii) composition of two arrows in $\mathbf{U}$, when defined, is their composite in $\mathbf{X}$. It follows that, for each U, the identity arrow 1_U in $\mathbf{U}$ is the identity arrow of U in $\mathbf{X}$. For example, the category whose objects are all integral domains D and whose arrows are all *mono*morphisms $D \to D'$ of domains is a subcategory of $\mathbf{Rng}$. It is not a full subcategory.

There are many other examples of categories not initially given as concrete ones, as in the following cases.

Each poset P determines a category $\mathbf{P}$; the objects are the elements $x \in P$, while each hom (x, y) is empty unless $x \leqslant y$, in which case it contains exactly one morphism $f: x \to y$. Since $x \leqslant y$ and $y \leqslant z$ imply $x \leqslant z$ in a poset, the composite of two morphisms $x \to y$ and $y \to z$ can then be defined (in exactly one way), and is associative. Since $x \leqslant x$ in a poset, each object x has an identity morphism. As $x \leqslant y$ and $y \leqslant x$ imply $x = y$ in a poset, each morphism of $\mathbf{P}$ with a two-sided inverse is an identity. Conversely, any category $\mathbf{X}$ in which each set hom (X, Y) has at most one element and every morphism with a two-sided inverse is an identity arises in this way from a poset.

To a poset P regarded as a category $\mathbf{P}$ one may apply the definition (§IV.7) of a product object. A product $x \times y$ of two elements $x, y \in P$ is thus an element $x \times y$ with arrows (projections) $x \times y \to x$ and $x \times y \to y$ such that any z with arrows $z \to x$ and $z \to y$ has a suitable arrow $z \to x \times y$. In other words, translating to the inclusion relation for the poset, a product has $x \times y \leqslant x$, $x \times y \leqslant y$, so that any z with $z \leqslant x$ and $z \leqslant y$ has $z \leqslant x \times y$. In other words, a product of x and y in P is just a g.l.b. of x and y in P.

The "arrow category" $\downarrow$ is the category with just two objects 1, 2 and one nonidentity arrow $1 \to 2$. There are several useful concrete categories with objects all the finite sets $\mathbf{n} = \{1, \cdots, n\}$ and morphisms either all functions $f: \mathbf{n} \to \mathbf{m}$, all injections f, or all functions f which are *monotonic*, in the sense that $i \leqslant j$ implies $fi \leqslant fj$.

One may construct a category from any commutative diagram. For instance, the diagram below

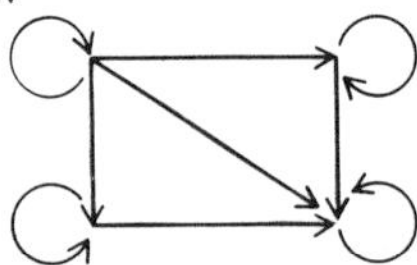

denotes a category with four objects (the vertices of the square) and with arrows as indicated (the loops are the identity arrows), composition being defined so that the diagram commutes.

Certain notation and terminology (in any category **X**) will be convenient. An arrow $u:X \to Y$ is *invertible* in **X** if there is an arrow $u':Y \to X$ in **X** with both $u' \circ u = 1_X$ and $u \circ u' = 1_Y$. A familiar argument shows that if such a u' exists, it is unique; hence, it is usually written as $u' = u^{-1}$, and called the inverse of u. If the composite $v \circ u$ of two invertible arrows v and u is defined, it is invertible and has $(v \circ u)^{-1} = u^{-1} \circ v^{-1}$; also, every identity arrow is invertible, with itself as inverse. Two objects X and Y are *equivalent* (that is, "isomorphic") in **X** if there is an invertible arrow $u:X \to Y$. This relation of equivalence between objects of **X** is reflexive, symmetric, and transitive.

An arrow $m:Y \to Z$ is said to be *monic* in **X** when $m \circ f = m \circ f'$ for two arrows $f, f':X \to Y$ always implies an equality $f = f'$; thus m monic means that m can be "canceled on the left" from any equation between arrows of **X**. If m has a left inverse in **X**, it is monic; the converse need not hold. In the category **S** of sets, m is monic if and only if m (as a function) is an injection. Indeed, suppose first that $m:Y \to Z$ is monic and consider two elements y, y' in the set Y with $m(y) = m(y')$. Now take the set **1** with just one element 1; the two functions $f, f':\mathbf{1} \to Y$ with $f(1) = y$ and $f'(1) = y'$ have composites $m \circ f = m \circ f'$; since m is monic, $f = f'$ and hence $y = y'$; thus m is an injection. Conversely, suppose m an injection, and consider two functions $f, f':X \to Y$ with $m \circ f = m \circ f'$. For each element $x \in X$, $mf(x) = mf'(x)$; since m is injective, $f(x) = f'(x)$. This equality for all x gives $f = f'$; hence, m is monic. In any concrete category **P**, an arrow $g:P \to Q$ is always a function $g:\mathfrak{U}(P) \to \mathfrak{U}(Q)$ on the underlying sets. Hence, the proof just given shows that g injective as a function always means g monic as an arrow in **P**. The converse need not hold, but does hold in each of the concrete categories listed above after Theorem 1.

An arrow $e:X \to Y$ is said to be *epic* in **X** when an equality $g \circ e = g' \circ e$ for two arrows $g, g':Y \to Z$ always implies $g = g'$; in brief, e epic means that e is right-cancellable. In the category of sets, epic means surjective (see Exercise 3). In a concrete category **P**, an arrow $f:P \to Q$ is always a function $f:\mathfrak{U}(P) \to \mathfrak{U}(Q)$ on the underlying sets, and it is immediate that f surjective as a function makes f epic as an arrow of **P**. The converse assertion need not hold in a concrete category. For instance, in the category of all integral domains the insertion $j:\mathbf{Z} \to \mathbf{Q}$ is not a surjection but is epic, because two different morphisms $g, g':\mathbf{Q} \to D$ of domains must differ on some integer m, hence must have $g \circ j \neq g' \circ j$.

An object I is said to be *initial* in a category **X** if to each object X in the category there is exactly one morphism $I \to X$ with domain I and codomain X. An object T is *terminal* in **X** if there is to each object $X \in \mathbf{X}$ exactly one morphism $X \to T$. Any two terminal objects T and T' in the same category are equivalent there. Indeed, since T is terminal there is an arrow $u: T' \to T$; since T' is terminal, there is $u': T \to T'$. The composite $u' \circ u: T' \to T'$ must be the unique arrow from T' to the terminal T', but the identity $1_{T'}: T' \to T'$ is another such; hence $u' \circ u = 1_{T'}$. Since T is also terminal a corresponding argument shows $u \circ u' = 1_T$. Thus each of u, u' is the inverse of the other, so $u: T \to T'$ is invertible, and the two terminal objects T, T' are equivalent, as asserted. In the category **Grp** of groups, any group $\{1\}$ with just one element is both initial and terminal. In the category **S** of sets, any set such as $\mathbf{1} = \{1\}$ with just one element is terminal, for there is to each set X exactly one function $X \to \mathbf{1}$; on the other hand, the empty set $\varnothing$ is initial, for there is exactly one function $\varnothing \to X$; namely, the function with graph the empty set (see §I.3).

The definition of an initial object I ($I \to X$ unique) is just like that of a terminal object T ($T \leftarrow X$ unique) except that *all arrows are reversed*. Such pairs of concepts are said to be "dual" to each other. Thus monic (= left cancellable) is dual to epic (= right cancellable). In general, the *dual* of any statement about a category is obtained as in §IV.7 by reversing all arrows, inverting the order of all composites, and interchanging "domain" with "codomain". Now observe in the definition of a category that the dual of each axiom is an axiom (for example, each 1_Y is required to be both a left identity and a right identity). Hence, any proof using only those axioms, when dualized, is still a proof, so the dual of any demonstrable theorem about categories is also demonstrable. This result is called the *duality principle*; it is "metamathematical", in the sense that it is a theorem about theorems. For example, we proved above that any two terminal objects in a category are equivalent. Now, "terminal" is dual to "initial", while the notion of equivalence of objects is clearly self-dual; hence, the duality principle allows us to conclude that any two initial objects in a category are equivalent.

EXERCISES

1. In each of the following concrete categories, show that an arrow is monic if and only if it is injective as a function:
   (a) **Grp**. (b) **Rng**. (c) $\mathbf{Mod}_R$.
2. Show that the l.u.b. of two elements in a poset P is a coproduct of those elements in the corresponding category **P**.
3. In each of the following concrete categories show that an arrow is epic if and only if it is surjective as a function:
   (a) **S**. (b) **Abgrp**. (c) $\mathbf{Mod}_R$.
4. In the category of commutative additive monoids with cancellation ($a + b = a + c$ implies $b = c$), show that the insertion $\mathbf{N} \to \mathbf{Z}$ is a morphism which is epic but not a surjection.
5. In each of the following categories, determine whether there are initial and terminal objects; if there are such, describe them:
   (a) $\mathbf{S}_*$ (pointed sets). (b) $\mathbf{Mod}_R$. (c) **Rng**.

6. Show that a group is the same thing as a category with exactly one object and every arrow invertible.

7. Determine (up to isomorphism) all categories with exactly three arrows.

8. Let an "induction algebra" be a set S with one nullary and one unary operation. Show that $(\mathbf{N}, 0, \sigma)$ is an initial object in the category of all induction algebras.

9. Let $\mathbf{\Delta}$ be the (concrete) category with objects the sets $\mathbf{n}$ and arrows $f:\mathbf{n} \to \mathbf{m}$ the monotonic functions. Show that there are exactly $n+1$ monic arrows $d_i:\mathbf{n} \to \mathbf{n}+1$, $i = 1, \cdots, n+1$, and that any monic arrow $f:\mathbf{n} \to \mathbf{m}$ with $m \geqslant n$ is a composite of such d_i's. Find a similar description of those arrows in $\mathbf{\Delta}$ which are epic.

2. Functors

Functors are defined for arbitrary categories just as for concrete categories (§IV.2, §IV.5).

DEFINITION. If $\mathbf{X}$ *and* $\mathbf{X}'$ *are two categories, a* functor $\mathfrak{T}:\mathbf{X} \to \mathbf{X}'$ *is a pair of functions, an* object function *and a* mapping function. *The object function assigns to each object* X *of the first category* $\mathbf{X}$ *an object* $\mathfrak{T}(X)$ *of* $\mathbf{X}'$*; the mapping function assigns to each arrow* $f:X \to Y$ *of the first category an arrow* $\mathfrak{T}(f):\mathfrak{T}(X) \to \mathfrak{T}(Y)$ *of the second category* $\mathbf{X}'$. *These functions must satisfy two requirements:*

$$\mathfrak{T}(1_X) = 1_{\mathfrak{T}(X)} \qquad \textit{for each identity } 1_X \textit{ of } \mathbf{X}; \tag{4}$$

$$\mathfrak{T}(g \circ f) = (\mathfrak{T}g) \circ (\mathfrak{T}f) \qquad \textit{for each composite } g \circ f \textit{ defined in } \mathbf{X}. \tag{5}$$

These requirements amount to the statement that a functor is a *morphism of categories*; in particular, an *isomorphism* $\mathfrak{T}:\mathbf{X} \cong \mathbf{X}'$ of categories is a functor $\mathfrak{T}$ for which both the object and the mapping functions are bijections. For each category $\mathbf{X}$, the identity $1:\mathbf{X} \to \mathbf{X}$ is a functor; if $\mathfrak{T}':\mathbf{X}' \to \mathbf{X}''$ and $\mathfrak{T}:\mathbf{X} \to \mathbf{X}'$ are functors, their composite $\mathfrak{T}' \circ \mathfrak{T}$, defined in the evident way, is also a functor $\mathfrak{T}' \circ \mathfrak{T}:\mathbf{X} \to \mathbf{X}''$. These statements suggest that we may consider the "category of all categories"; the objects are the categories, the morphisms are functors.

There is an "underlying set functor" $\mathfrak{U}$ for any concrete category $\mathbf{P}$. Indeed, for each object P in the concrete category $\mathbf{P}$ we are given an underlying set $\mathfrak{U}(P)$, while each morphism $f:P \to Q$ of the category is a function $f:\mathfrak{U}(P) \to \mathfrak{U}(Q)$. Clearly, $P \mapsto \mathfrak{U}(P)$ is the object function and $f \mapsto f$ the mapping function of a functor $\mathfrak{U}:\mathbf{P} \to \mathbf{S}$. This functor is also called a "forgetful" functor.

There are other types of forgetful functors. Thus a ring R has been defined to be an abelian group with a binary operation of multiplication satisfying certain axioms, while a morphism $R \to R'$ of rings is a morphism of abelian groups which is also a morphism of multiplication. The functions assigning to each ring R the same set R, regarded just as an abelian group, and to each morphism f of rings the same function f, regarded just as a morphism of abelian groups, together give a functor on the category of rings to the category of abelian groups. There are many other such examples (R-modules to abelian groups, inner product spaces to real vector spaces, etc.).

Let A be a fixed object in any category $\mathbf{X}$. The set

$$\text{hom}\,(A, X) = \{\text{all } t : A \to X \,|\, t \text{ an arrow of } \mathbf{X}\} \tag{6}$$

is the object function $h_A(X) = \text{hom}\,(A, X)$ of a functor $h_A : \mathbf{X} \to \mathbf{S}$. The corresponding mapping function assigns to each arrow $f : X \to Y$ of the category the function

$$h_A(f) = f_* : \text{hom}\,(A, X) \to \text{hom}\,(A, Y) \tag{7}$$

defined for each $t : A \to X$ to be the composite $f_*(t) = f \circ t : A \to Y$. Since $1_X = 1$ and $(g \circ f)_*(t) = g \circ (f \circ t) = g_*(f_* t)$, this is a functor. As in §IV.2, it is called the (covariant) *hom functor* $h_A = \text{hom}\,(A, -) : \mathbf{X} \to \mathbf{S}$.

Many functors arise as subfunctors of such hom-functors. For instance, in the category $\mathbf{Mod}_R$, with D a submodule of A, the projection $p : A \to A/D$ is a universal element for the functor Ann_D, where $\text{Ann}_D\,(B)$ is the set of those morphisms $t : A \to B$ with $t_* D = 0$, while $\text{Ann}_D\,(f)$ for $f : B \to C$ is the function $t \mapsto f \circ t$. Here each set $\text{Ann}_D\,(B)$ is a subset of $h_A(B)$ and each function $\text{Ann}_D\,(f)$ is a restriction of $h_A(f)$, so Ann_D is a "subfunctor" of the hom-functor h_A.

In general, if $\mathfrak{F} : \mathbf{X} \to \mathbf{S}$ is a functor to sets, a subfunctor $\mathfrak{G} : \mathbf{X} \to \mathbf{S}$ is one with each $\mathfrak{G}(X)$ a subset of $\mathfrak{F}(X)$ and each $\mathfrak{G}(f)$ a restriction of $\mathfrak{F}(f)$. Each subfunctor $\mathfrak{G}$ of $\mathfrak{F}$ is determined by its object function.

Let $\mathbf{G}$ be a category with one object 1 and every arrow invertible. Then the arrows of $\mathbf{G}$ form a group G under composition, and every group can be realized in this way as the arrows of a one-object category. A functor $\mathfrak{T} : \mathbf{G} \to \mathbf{S}$ sends the single object 1 of $\mathbf{G}$ to a set $X = \mathfrak{T}(1)$ and each arrow $g : 1 \to 1$ in $\mathbf{G}$ to a function $\mathfrak{T}(g) : X \to X$ satisfies (4) and (5). If we write $\mathfrak{T}(g)x$ as gx for each $x \in X$, these two conditions (4) and (5) become $1x = X$ and $(gf)x = g(fx)$. In other words, the functor $\mathfrak{T}$ is just an *action* of the group G on the set X, as defined in (II.23). Conversely, any action of a group G on a set X arises in this way from a functor $\mathfrak{T} : \mathbf{G} \to \mathbf{S}$ with $\mathfrak{T}(1) = X$. Put differently, the notion of a functor on a category to sets is the natural generalization of the notion of an action of a group on a set.

Let F be a field, so that $\mathbf{Mod}_F$ is the category of vector spaces V over F. A functor $\mathfrak{F} : \mathbf{G} \to \mathbf{Mod}_F$ then picks out a vector space $\mathfrak{F}(1) = V$ and assigns to each g in the group G an invertible linear transformation $\mathfrak{F}(g) : V \to V$ with

$$\mathfrak{F}(1) = 1, \qquad \mathfrak{F}(gh) = \mathfrak{F}(g)\mathfrak{F}(h).$$

Such an $\mathfrak{F}$ is a *representation* of the group G by linear transformations.

In order to consider functors of several arguments, we introduce for each pair of categories $\mathbf{X}$, $\mathbf{X}'$ a *product category* $\mathbf{X} \times \mathbf{X}'$. An object of this product is an ordered pair (X, X') of objects of $\mathbf{X}$ and $\mathbf{X}'$, respectively; an arrow $(X, X') \to (Y, Y')$ with the indicated domain and codomain is an ordered pair (f, f') of arrows $f:X \to Y, f':X' \to Y'$. The composition of arrows is defined termwise; thus (f, f') as just above and a second such ordered pair $(g, g'):(Y, Y') \to (Z, Z')$ have the composite

$$(g, g') \circ (f, f') = (g \circ f, g' \circ f') : (X, X') \to (Z, Z'). \tag{8}$$

The axioms for a category are readily verified. The evident projections are functors $\mathbf{X} \times \mathbf{X}' \to \mathbf{X}$ and $\mathbf{X} \times \mathbf{X}' \to \mathbf{X}'$. This gives a product diagram (§IV.7) in the category of all categories.

A functor $\mathfrak{B} : \mathbf{X} \times \mathbf{X}' \to \mathbf{W}$ on a product category $\mathbf{X} \times \mathbf{X}'$ to a category $\mathbf{W}$ is called a *bifunctor* on $\mathbf{X}$ and $\mathbf{X}'$ to $\mathbf{W}$. For example, the cartesian product of sets is a bifunctor $\mathbf{S} \times \mathbf{S} \to \mathbf{S}$. Its object function assigns to each pair (X, X') of sets their cartesian product $X \times X'$; its mapping function assigns to each pair of functions $f:X \to Y, f':X' \to Y'$ their product $f \times f':X \times X' \to Y \times Y'$, defined as a function in the usual way (§I.3).

In view of the definition of a product category, a bifunctor $\mathfrak{B}$ must assign to each pair of objects X and X' of $\mathbf{X}$ and $\mathbf{X}'$ an object $\mathfrak{B}(X, X')$ of $\mathbf{W}$ and to each pair of arrows $f:X \to Y$ and $f':X' \to Y'$ an arrow

$$\mathfrak{B}(f, f') : \mathfrak{B}(X, X') \to \mathfrak{B}(Y, Y') \tag{9}$$

so that the analogs of conditions (4) and (5) hold; that is, so that

$$\mathfrak{B}(1_X, 1_{X'}) = 1_{\mathfrak{B}(X, X')} \qquad \text{for all objects } X \in \mathbf{X}, X' \in \mathbf{X}'; \tag{10}$$

$$\mathfrak{B}(g \circ f, g' \circ f') = \mathfrak{B}(g, g') \circ \mathfrak{B}(f, f') \tag{11}$$

for $g \circ f$ and $g' \circ f'$ defined. The last condition may be replaced by three simpler conditions:

$$\mathfrak{B}(g \circ f, 1) = \mathfrak{B}(g, 1) \circ \mathfrak{B}(f, 1), \quad \mathfrak{B}(1, g' \circ f') = \mathfrak{B}(1, g') \circ \mathfrak{B}(1, f'), \tag{12}$$

$$\mathfrak{B}(f, 1) \circ \mathfrak{B}(1, f') = \mathfrak{B}(f, f') = \mathfrak{B}(1, f') \circ \mathfrak{B}(f, 1); \tag{13}$$

here each "1" designates a suitable identity arrow. The first of these conditions (12) simply says that $\mathfrak{B}(-, X')$ is a functor in the first variable when the second variable X' is held constant; the second is the corresponding statement about $\mathfrak{B}(X, -)$.

EXERCISES

1. Let **Abgrp** be the category of all abelian groups, and R a fixed ring:

(a) If A is a fixed right R-module, construct a functor $H_A:\mathbf{Mod}_R \to \mathbf{Abgrp}$ with object function $H_A(B) = \mathrm{Hom}_R(A, B)$ for each right R-module B (cf. Theorem V.1).

(b) If $\mathfrak{U}:\mathbf{Abgrp} \to \mathbf{S}$ is the underlying set functor, show that the composite functor $\mathfrak{U} \circ H_A$ is the functor h_A of the text.

2. Construct a functor $\mathfrak{S}:\mathbf{Grp} \to \mathbf{S}$ whose object function assigns to each group $\mathbf{G}$ the set of all subgroups of G.

3. Construct a functor $\mathfrak{T}$ on the category of commutative rings K to itself with each $\mathfrak{T}(K)$ the polynomial ring $K[x]$ in one indeterminate x.

4. Show that the product of two groups is the object function of a bifunctor $\mathbf{Grp} \times \mathbf{Grp} \to \mathbf{Grp}$.

5. For any ring R, show that the biproduct $\oplus$ of R-modules gives a bifunctor $\mathbf{Mod}_R \times \mathbf{Mod}_R \to \mathbf{Mod}_R$.

6. Represent Equation (11) by a commutative diagram.

7. Show that (11) implies (12) and (13).

8. Show that (12) and (13) together imply (11).

9. Establish the following isomorphisms of categories:

$$\mathbf{X} \times \mathbf{Y} \cong \mathbf{Y} \times \mathbf{X}; \qquad (\mathbf{X} \times \mathbf{Y}) \times \mathbf{Z} \cong \mathbf{X} \times (\mathbf{Y} \times \mathbf{Z}).$$

10. If $\mathfrak{F}:\mathbf{X} \to \mathbf{S}$ is a functor and for each $X \in \mathbf{X}$, $\mathfrak{G}_0(X)$ is a subset of $\mathfrak{F}(X)$ such that $x \in \mathfrak{G}_0(X)$ and $f:X \to Y$ imply $\mathfrak{F}(f)x \in \mathfrak{G}_0(Y)$, prove that $\mathfrak{G}_0$ is the object function of a unique subfunctor of $\mathfrak{F}$.

11. To each field F construct the category $\mathbf{E}$ whose objects are the pairs (V, t) of vector spaces V over F and endomorphisms $t: V \to V$ and whose arrows $f:(V, t) \to (V', t')$ are the linear transformations $f: V \to V'$ with $t' \circ f = f \circ t$. Show this category isomorphic to the category of all $F[x]$-modules (compare Proposition XI.7).

3. Contravariant Functors

If R is a ring, each right R-module A determines a dual left R-module $A^* = \mathrm{Hom}(A, R)$ and each morphism $f:A \to B$ of right R-modules a dual $f^*:B^* \to A^*$, in the opposite direction, which is a morphism of left R-modules. Moreover, each $(1_A)^*$ is 1_{A^*}, and each $(g \circ f)^*$ is $f^* \circ g^*$, as in Proposition V.20. These conditions are just like those defining a functor, except that composition is inverted and arrows are reversed. For this reason, we call the dual a "contravariant" functor. In this terminology, the functors previously defined are said to be *covariant*.

In general, a *contravariant functor* $\mathfrak{C}$ on a category $\mathbf{X}$ to a category $\mathbf{X}'$ is a pair of functions which assign to each object X in $\mathbf{X}$ an object $\mathfrak{C}(X)$ in $\mathbf{X}'$, and to each arrow $f:X \to Y$ in $\mathbf{X}$ a morphism $\mathfrak{C}(f):\mathfrak{C}(Y) \to \mathfrak{C}(X)$ in $\mathbf{X}'$, assigning to each identity arrow 1_X the identity of $\mathfrak{C}(X)$ and to each composite $g \circ f$ of arrows of X the composite $\mathfrak{C}(g \circ f) = \mathfrak{C}(f) \circ \mathfrak{C}(g)$, as suggested in the diagram:

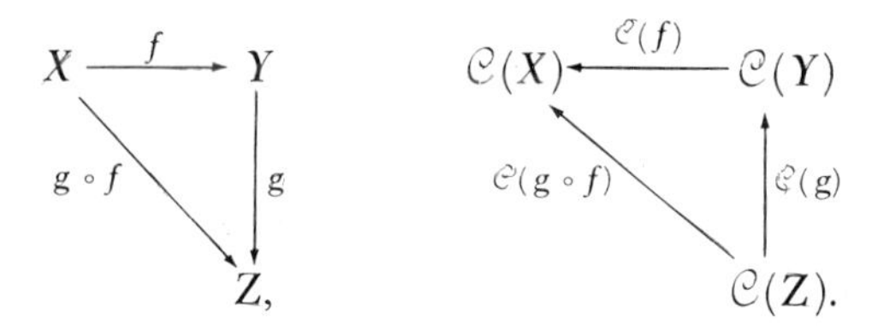

Clearly, the composite of two contravariant functors is a covariant one.

There is another description of contravariance. We recall that each multiplicative group G determines an "opposite" group (§II.2, rule 6) with product $(a, b) \mapsto ba$. Similarly, to each category $\mathbf{X}$ we can construct an *opposite category* $\mathbf{X}^{\mathrm{op}}$, as follows:

Objects of $\mathbf{X}^{\mathrm{op}}$ are all the objects of $\mathbf{X}$.

Arrows of $\mathbf{X}^{\mathrm{op}}$ are $f^{\mathrm{op}}: Y \to X$, one for each arrow $f: X \to Y$ in $\mathbf{X}$.

Composites $f^{\mathrm{op}} \circ g^{\mathrm{op}} = (g \circ f)^{\mathrm{op}}$ in $\mathbf{X}^{\mathrm{op}}$ are defined wherever $g \circ f$ is defined.

This description shows that $\mathbf{X}^{\mathrm{op}}$ is a category and that the two functions $X \mapsto X$, $f \mapsto f^{\mathrm{op}}$ yield a contravariant functor on $\mathbf{X}$ to $\mathbf{X}^{\mathrm{op}}$ which we call $\mathcal{J}$; its inverse is also a contravariant functor $\mathcal{J}^{-1}$ on $\mathbf{X}^{\mathrm{op}}$ to $\mathbf{X}$.

Each contravariant functor $\mathcal{C}$ on $\mathbf{X}$ to $\mathbf{X}'$ determines a covariant functor $\bar{\mathcal{C}}: \mathbf{X}^{\mathrm{op}} \to \mathbf{X}'$ on the opposite category as the composite $\bar{\mathcal{C}} = \mathcal{C} \circ \mathcal{J}^{-1}$. Moreover, each covariant $\mathcal{D}: \mathbf{X}^{\mathrm{op}} \to \mathbf{X}'$ on the opposite category arises in this way from a unique contravariant functor $\mathcal{C} = \mathcal{D} \circ \mathcal{J}$. Hence, instead of introducing a contravariant functor $\mathcal{C}$ on $\mathbf{X}$ to $\mathbf{X}'$ it is possible (and usually is more convenient) to introduce the corresponding covariant functor $\mathcal{D}: \mathbf{X}^{\mathrm{op}} \to \mathbf{X}'$.

As in this case, an arrow $\to$ from domain category to codomain will always designate a *co*variant functor.

A standard example is the contravariant hom-functor defined, much as in §IV.7, for any category $\mathbf{X}$. Fix an object B in $\mathbf{X}$; each object Y then determines the set hom (Y, B) of all arrows $t: Y \to B$ in $\mathbf{X}$, while each arrow $f: X \to Y$ gives by composition $t \mapsto t \circ f = f^*(t)$ a function f^* from hom (Y, B) to hom (X, B). More formally, for each B in $\mathbf{X}$ the *contravariant hom-functor* is the functor hom $(-, B) = h^B: \mathbf{X}^{\mathrm{op}} \to \mathbf{S}$ defined by

$$h^B(Y) = \mathrm{hom}\,(Y, B) = \{\text{all } t : Y \to B \text{ in } \mathbf{X}\},$$
$$h^B(f^{\mathrm{op}}) = f^* : h^B(Y) \to h^B(X) \qquad \text{for each } f : X \to Y \text{ in } \mathbf{X}.$$

For example, if $\mathbf{X}$ is the category of right R-modules, then $h^R(X) =$ hom (X, R) is just the set of elements in the dual module $X^* = \mathrm{Hom}\,(X, R)$, so the contravariant hom-functor h^R is just the composite $\mathcal{U} \circ \mathcal{D}$, where $\mathcal{D}$ is the (contravariant) functor assigning to each right module its dual left module, and $\mathcal{U}$ the "forgetful" functor assigning to each left module its underlying set.

The set hom (A, B) for A fixed gives the covariant hom functor h_A and for B fixed the contravariant h^B. If neither is fixed, this set gives the object function of a bifunctor

$$\hom : \mathbf{X}^{\mathrm{op}} \times \mathbf{X} \to \mathbf{S}. \tag{14}$$

Its mapping function assigns to each pair of arrows $f:A \to B$ and $g:X \to Y$ the function

$$\hom(f^{\mathrm{op}}, g) : \hom(B, X) \to \hom(A, Y)$$

defined for each $t:B \to X$ as $[\hom(f^{\mathrm{op}}, g)](t) = g \circ t \circ f$, as shown below:

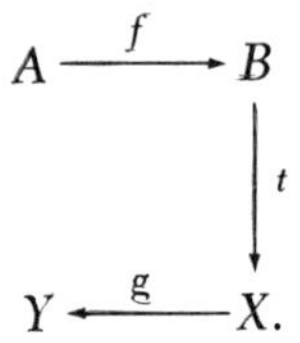

$$\begin{array}{ccc} A & \xrightarrow{f} & B \\ & & \downarrow t \\ Y & \xleftarrow{g} & X. \end{array}$$

(More briefly, $\hom(f^{\mathrm{op}}, g)$ is the composite $f^* \circ g_* = g_* \circ f^*$, in either order, of the maps f^* and g_* induced as in (7) and (15) by f and g.)

Other examples of contravariance include the contravariant power set functor $\mathcal{P}^*:\mathbf{S}^{\mathrm{op}} \to \mathbf{S}$, as described in §IV.8, and the contravariant functor (on the category of abelian groups to itself) assigning to each abelian group its character group (§XII.2, Appendix).

EXERCISES

1. Show that there is a bijection from the set of all contravariant functors on $\mathbf{X}$ to $\mathbf{X}'$ to the set of all covariant functors $\mathbf{X} \to \mathbf{X}'^{\mathrm{op}}$.

2. If a category $\mathbf{X}$ has a terminal object, show that $\mathbf{X}^{\mathrm{op}}$ has an initial object; similarly, show that if any statement about categories is true for a category $\mathbf{X}$, then the dual statement is true for $\mathbf{X}^{\mathrm{op}}$.

3. Establish the following isomorphisms of categories:

$$(\mathbf{X}^{\mathrm{op}})^{\mathrm{op}} \cong \mathbf{X}; \qquad (\mathbf{X} \times \mathbf{Y})^{\mathrm{op}} \cong \mathbf{X}^{\mathrm{op}} \times \mathbf{Y}^{\mathrm{op}}.$$

4. Natural Transformations

For a ring R, the double dual $A \mapsto A^{**}$ of a right R-module A provides a covariant functor $\mathbf{Mod}_R \to \mathbf{Mod}_R$. Moreover, we constructed in (V.48) in a "natural" way for each module A a morphism $\omega_A:A \to A^{**}$. The fact that those morphisms ω are "natural" and independent of choices (for example, of bases) is related to the fact that the diagram (below)

$$\begin{array}{ccc} B & \xrightarrow{\omega_B} & (B^*)^* \\ \downarrow s & & \downarrow (s^*)^* \\ A & \xrightarrow{\omega_A} & (A^*)^* \end{array}$$

of (V.49) commutes for every choice of the morphism $s: B \to A$. Because of the latter property, we will call ω a "natural transformation" from the identity functor to the double dual functor $A \mapsto A^{**}$. The general definition is as follows.

DEFINITION. If $\mathfrak{S}, \mathfrak{T}: \mathbf{X} \to \mathbf{X}'$ *are functors, a* natural transformation $\tau: \mathfrak{S} \to \mathfrak{T}$ *from* $\mathfrak{S}$ *to* $\mathfrak{T}$ *is a function which assigns to each object* X *of* $\mathbf{X}$ *an arrow* $\tau_X: \mathfrak{S}(X) \to \mathfrak{T}(X)$ *of* $\mathbf{X}'$ *in such a way that every arrow* $f: X \to Y$ *of* $\mathbf{X}$ *yields a commutative diagram*

$$\begin{array}{ccc} \mathfrak{S}(X) & \xrightarrow{\tau_X} & \mathfrak{T}(X) \\ \downarrow \mathfrak{S}(f) & & \downarrow \mathfrak{T}(f) \\ \mathfrak{S}(Y) & \xrightarrow{\tau_Y} & \mathfrak{T}(Y). \end{array} \tag{15}$$

A natural transformation $\tau: \mathfrak{S} \to \mathfrak{T}$ is also called a "morphism of functors".

In case each τ_X is invertible in $\mathbf{X}'$, we call $\tau: \mathfrak{S} \to \mathfrak{T}$ a *natural isomorphism* or a *natural equivalence*, and note that the inverses $(\tau_X)^{-1}$ then constitute a natural isomorphism $\tau^{-1}: \mathfrak{T} \to \mathfrak{S}$.

For example, characteristic functions yield a natural transformation from the contravariant power-set functor $\mathfrak{P}^*$ to the contravariant hom-functor h^2. First recall (§IV.8) that each subset $S \subset Y$ is determined by its characteristic function $t_S: Y \to \mathbf{2} = \{1,2\}$, which is defined from S by setting

$$t_S y = 1 \quad \text{if } y \in S, \qquad t_S y = 2 \quad \text{if } y \notin S.$$

Moreover, the assignment $S \mapsto t_S$ is a function ψ_Y from the power set $P(Y)$ to the function set $\mathbf{2}^Y$. Since every function $t: Y \to \mathbf{2}$ is the characteristic function of some subset S of Y (namely, the set S of all y with $ty = 1$), this assignment is a bijection

$$\psi_Y: P^*(Y) \cong \mathbf{2}^Y.$$

This bijection is defined for every set Y, so can be viewed as a "transformation" from the contravariant power set functor $P^*:\mathbf{S}^{\mathrm{op}} \to \mathbf{S}$ on the left to the contravariant hom-functor $h^2:\mathbf{S}^{\mathrm{op}} \to \mathbf{S}$ on the right. We assert that this Y is natural. By the definition (15) this assertion means for every $f:Y \to X$ that the diagram

$$\begin{array}{ccc} T \in P^*(X) & \xrightarrow{\psi_X} & \mathbf{2}^X \\ {\scriptstyle f^*}\downarrow & & \downarrow{\scriptstyle \mathbf{2}^f} \\ P^*(Y) & \xrightarrow{\psi_Y} & \mathbf{2}^Y \end{array}$$

commutes. To prove this, start with any subset $T \subset X$. The function $(\psi_Y \circ f^*)T$ is the characteristic function of the inverse image set $f^*T \subset Y$, while $(\mathbf{2}^f \circ \psi_T)T$ is the composite $t_T \circ f: Y \to \mathbf{2}$. But $t_T f y = 1$ precisely when $fy \in T$; that is, precisely when $y \in f^*T$. Hence, $t_T f = \psi_Y(f^*T)$, so the diagram above is commutative. Therefore, ψ is natural.

Next we give an example of naturality for bifunctors. For each ring R the biproduct $A \oplus B$ of two R-modules A and B gives the object function of a bifunctor $\mathbf{Mod}_R \times \mathbf{Mod}_R \to \mathbf{Mod}_R$; the mapping function assigns to two morphisms $s:A \to A'$ and $t:B \to B'$ the morphism $s \oplus t:A \oplus B \to A' \oplus B'$ given by $(a, b) \mapsto (sa, tb)$. The commutative law for biproducts is the isomorphism $\phi:A \oplus B \cong B \oplus A$, given on elements as $\phi(a, b) = (b, a)$. This isomorphism $\phi = \phi_{A, B}$ is defined for each pair of R-modules A and B; the diagram shown commutes for all s and t, so the isomorphism ϕ is natural.

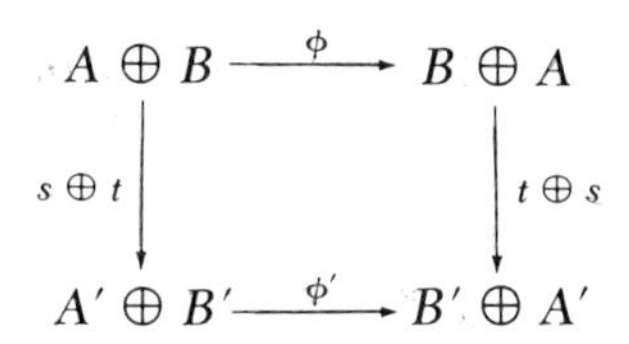

Incidentally, to prove that this square diagram commutes we can prove instead the commutativity of the two simpler squares indicated below:

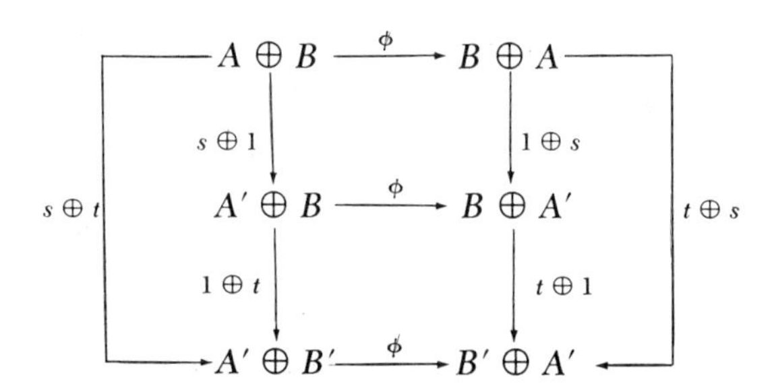

The same simplification works for arbitrary bifunctors, as follows:

Proposition 2. *If* $\mathfrak{S}$ *and* $\mathfrak{T}:\mathbf{X}\times\mathbf{X}'\to\mathbf{Y}$ *are bifunctors, while* ϕ *is a function assigning to each pair of objects* X, X' *a morphism* $\phi_{X,X'}$ *from* $\mathfrak{S}(X, X')$ *to* $\mathfrak{T}(X, X')$, *then* ϕ *is a natural transformation,* $\phi:\mathfrak{S}\to\mathfrak{T}$, *if and only if* $\phi_{X,X'}$ *is natural in* X *for each fixed* X' *and natural in* X' *for each fixed* X.

The proof is left to the reader (Exercise 6).

There are also natural transformations involving functors of more than two variables. For example, given a cartesian product $S \times T$ of two sets, with its projections p and q, and given a third set X, the function $\phi = \phi_{S,T,X}$ assigning to each function $h:X\to S\times T$ the pair (ph, qh) is, as in IV(22) a bijection

$$\phi_{S,T,X} : (S \times T)^X \cong S^X \times T^X.$$

Each side here is the object function of a functor $\mathbf{S}\times\mathbf{S}\times\mathbf{S}^{\mathrm{op}}\to\mathbf{S}$ of three variables, covariant in the first two (S and T) and contravariant in the third variable X. The transformation ϕ is readily proved to be a natural one between trifunctors. As in the case of two variables (Proposition 2) it suffices to verify naturality in each of the three variables separately.

If $\sigma:\mathfrak{R}\to\mathfrak{S}$ and $\tau:\mathfrak{S}\to\mathfrak{T}$ are natural transformations between functors $\mathfrak{R}$, $\mathfrak{S}$, and $\mathfrak{T}:\mathbf{X}\to\mathbf{X}'$, one may define a *composite transformation* $\tau\circ\sigma$ by $(\tau\circ\sigma)_X = \tau_X\circ\sigma_X:\mathfrak{R}(X)\to\mathfrak{T}(X)$. One sees readily (juxtaposing two commutative squares like (15)) that $\tau\circ\sigma$ is a natural transformation $\tau\circ\sigma:\mathfrak{R}\to\mathfrak{T}$. This suggests that one might form a "functor category" with functors as the objects and natural transformations as the arrows.

More formally, let $\mathbf{Y}$ be any category and $\mathbf{X}$ a (sufficiently) small category. Define the *functor category* $\mathbf{Y}^{\mathbf{X}}$ to be the class of all functors $\mathfrak{S}:\mathbf{X}\to\mathbf{Y}$, with each hom $(\mathfrak{S}, \mathfrak{T})$ the *set* of all natural transformations $\tau:\mathfrak{S}\to\mathfrak{T}$. This collection hom $(\mathfrak{S}, \mathfrak{T})$ is a set when the category $\mathbf{X}$ is sufficiently small—for example, not as large as the category of *all* sets or *all* groups.

As a first example of a functor category, take $\mathbf{X}$ to be the arrow category $\downarrow$, which has just two objects 1, 2 and one nonidentity arrow $\alpha:1\to 2$. Each functor $\mathfrak{F}:\downarrow\to\mathbf{Y}$ is thus determined by two objects $\mathfrak{F}(1)$ and $\mathfrak{F}(2)$ in Y and one arrow $f = F(\alpha):\mathfrak{F}(1)\to\mathfrak{F}(2)$. Each natural transformation $\tau:\mathfrak{F}\to\mathfrak{G}$ is determined by the pair of arrows $\tau(1)$ and $\tau(2)$ for the two objects 1, 2 of $\downarrow$, which by (15) form a commutative diagram

$$\begin{array}{ccc} \mathfrak{F}(1) & \xrightarrow{f} & \mathfrak{F}(2) \\ \downarrow{\scriptstyle \tau(1)} & & \downarrow{\scriptstyle \tau(2)} \\ \mathfrak{G}(1) & \xrightarrow{g} & \mathfrak{G}(2). \end{array}$$

In other words, the functor category $\mathbf{Y}^{\downarrow}$ is the category with objects the arrows f of $\mathbf{Y}$ and having as arrows $f \to g$ the commutative squares in $\mathbf{Y}$ with top f and bottom g.

Similarly, if D is any commutative arrow diagram, regarded as a (small) category, the functor category $\mathbf{Y}^D$ is the category with objects all diagrams in $\mathbf{Y}$ of the form D.

Let $\mathbf{G}$ be a category with one object 1 and all arrows invertible, so that $\mathbf{G}$ is actually just the group G with multiplication the composition of all the arrows $g:1 \to 1$. A functor $\mathbf{G} \to \mathbf{S}$ is then, as observed in §2, just a set $\mathbf{X}$ with an action of G on X. Consequently, the functor category $\mathbf{S}^{\mathbf{G}}$ is just the category with objects all actions of G on a set X. Here an arrow τ between an action on X and an action on Y is just a function $\tau:X \to Y$ which commutes with the action of each $g \in G$, in the sense that $g(\tau x) = \tau(gx)$ for every $x \in X$.

Appendix: When is the functor category legitimate?

The set of all sets is a troublesome object, because of the Russell paradox. This paradox considers the set R whose elements are exactly those sets X not members of themselves:

$$R = \{X | X \notin X\};$$

Then $R \in R$ means that R is one of these X's, so $R \notin R$. On the other hand, $R \notin R$ means that R is one of these X's, so $R \in R$—a contradiction.

For this reason we have described the category $\mathbf{S}$ of sets not as the *set* of all sets, but as the *class* of all sets. Similarly, **Grp** is the *class* of all groups. This usage appeals to the Gödel–Bernays form of the axioms for set theory. This form takes set theory to deal with two kinds of collections, *sets* and *classes*. A set is a collection whose elements are sets, a class is a collection whose elements are sets, and every set is also a class. However, a class which is not a set cannot be an element of anything else, class or set, so there is no such thing as a class of classes. However, there is a class U of *all* sets, called the "universal class", and this class is the desired category $\mathbf{S}$ of all sets. Moreover, given any property of sets, the axioms on classes imply that one can construct the *class* of all the sets with that property—for example, the class **Grp** of all groups. (The Gödel–Bernays axioms may be found in Gödel's book on the continuum hypothesis, cited in the bibliography.)

The axioms provide for the construction of many classes which *are* sets. We have defined a category as a *class* $\mathbf{X}$ of objects with each hom (X, Y) a *set*. When the set $\mathbf{X}$ of all objects is itself a set, we say that $\mathbf{X}$ is a *small* category. Our construction of the functor category $\mathbf{Y}^{\mathbf{X}}$ applies for any category $\mathbf{Y}$, provided the exponent $\mathbf{X}$ is a small category in this sense. The proof depends on a suitable use of the axioms on classes.

Similarly, one may form the category **Cat** whose objects are all *small* categories, with hom $(\mathbf{X}, \mathbf{X}')$ the *set* of all functors $\mathbf{X} \to \mathbf{X}'$.

There are alternative approaches to this problem in foundations. For example, the Zermelo–Fraenkel axioms for set theory specify all the necessary properties of sets without making reference to any "classes". To these axioms one may add an axiom requiring the existence of a set U, called the "universe", such that the sets which are elements of U themselves satisfy the Zermelo–Fraenkel axioms. (This says that U is so large that any legitimate operation of set-theory on the sets in U yields a set in U.) On this foundation, a category $\mathbf{X}$ can be described as a *set* of objects, with hom-sets as in our definition. A set $\mathbf{X}$ or a category is called *small* when it is an element of the fixed universe U. One then takes $\mathbf{S}$ to be the category of all *small* sets, **Grp** to be the category of all small groups, and so on. The functor category $\mathbf{Y}^{\mathbf{X}}$ can always be constructed, but it need not be small.

Alternatively, one may replace the axioms on sets and their elements by axioms on sets and the functions between them. This amounts to using axioms for the category of all sets, or more generally for the categories which are called "elementary topoi".

EXERCISES

1. For sets X, Y, and Z establish natural equivalences
$$X \times Y \cong Y \times X, \qquad X \times (Y \times Z) \cong (X \times Y) \times Z.$$

2. Show that the factor-commutator group $G/[G, G]$ is a functor on **Grp** to **Abgrp** and that the projection $p: G \to G/[G, G]$ is a natural transformation of functors.

3. Show that the disjoint union $S \dot{\cup} T$ of two sets S and T, defined as in §IV.7, is the object function of a bifunctor $\mathbf{S} \times \mathbf{S} \to \mathbf{S}$, and establish for it natural commutative and associative equivalences, like those of Exercise 1.

4. For $S \dot{\cup} T$ as in Exercise 3, show $X^{S \dot{\cup} T} \cong X^S \times X^T$ natural in all three variables.

5. Show the equivalence $S^{(X \times Y)} \cong (S^X)^Y$ of (IV.24) to be natural in all three variables.

6. Prove Proposition 2.

7. Show that the isomorphisms of (V.35) and (V.36) are natural.

8. (a) Given functors $\mathcal{S}, \mathcal{T}: \mathbf{X} \to \mathbf{X}'$ and $\mathcal{S}', \mathcal{T}': \mathbf{X}' \to \mathbf{X}''$ and natural transformations $\tau: \mathcal{S} \to \mathcal{T}$ and $\tau': \mathcal{S}' \to \mathcal{T}'$ show that
$$(\tau * \tau')_X = \mathcal{T}'(\tau_X) \circ \tau'_{\mathcal{S}X} = \tau'_{\mathcal{T}X} \circ \mathcal{S}'(\tau_X)$$
defines a natural transformation $\tau * \tau': \mathcal{S}' \circ \mathcal{S} \to \mathcal{T}' \circ \mathcal{T}$.

(b) Describe $\tau * \tau'$ when $\mathcal{S} = \mathcal{T}$ and τ is the identity.

(c) If $\mathbf{X}$ and $\mathbf{X}'$ are both small categories, deduce that composition of functors $(\mathcal{S}', \mathcal{S}) \mapsto \mathcal{S}' \circ \mathcal{S}$ defines a bifunctor
$$\mathbf{X}''^{\mathbf{X}'} \times \mathbf{X}'^{\mathbf{X}} \to \mathbf{X}''^{\mathbf{X}}.$$

5. Representable Functors and Universal Elements

In previous chapters we have seen many examples of universal elements for functors from some concrete category to the category of sets. We recall the definition (§IV.3).

DEFINITION. Let $\mathcal{S}: \mathbf{X} \to \mathbf{S}$ *be a functor to the category of sets. A universal element for* $\mathcal{S}$ *is a pair* (u, R) *consisting of an object* R *of* $\mathbf{X}$ *and* *an element $u \in \mathbb{S}(R)$ with the following property: To any object X of* $\mathbf{X}$ *and any element $s \in \mathbb{S}(X)$, there is exactly one arrow $f:R \to X$ with $\mathbb{S}(f)u = s$.*

If (u, R) is universal, then for each object X the assignment $f \mapsto \mathbb{S}(f)u$ is a bijection $\phi_X:\hom(R, X) \cong \mathbb{S}(X)$ of sets, as stated in the representation theorem (Theorem IV.2). Now we are in a position to state more about this bijection: It is a natural equivalence. A bijection with this property is called a "representation" of the functor $\mathbb{S}$. This concept turns out to be equivalent to that of a universal element.

DEFINITION. A representation *of a functor $\mathbb{S}:\mathbf{X} \to \mathbf{S}$ to the category of sets is a pair (R, ϕ) consisting of an object R of* $\mathbf{X}$ *and a family of bijections*

$$\phi_X : \hom_{\mathbf{X}}(R, X) \cong \mathbb{S}(X) \tag{16}$$

natural in X. A functor $\mathbb{S}$ with such a representation is said to be representable, *and to be* represented *by R.*

In brief, a representation of $\mathbb{S}$ is a natural isomorphism $\phi:h_R \cong \mathbb{S}$.

THEOREM 3. *For each functor $\mathbb{S}:\mathbf{X} \to \mathbf{S}$ the formulas*

$$u = \phi_R(1_R), \qquad \phi_X f = (\mathbb{S}f)u \tag{17}$$

for $1_R:R \to R$ the identity and $f:R \to X$ any arrow, establish a bijection from representations (R, ϕ) of $\mathbb{S}$ to universal elements (u, R) for $\mathbb{S}$.

Proof: First we take a representation (R, ϕ) of the functor $\mathbb{S}$. Since ϕ is natural, the diagram below commutes for each $f:R \to X$. Now observe that the set $\hom(R, R)$ at the upper left contains a distinguished element, the identity arrow $1_R:R \to R$. Take u to be the image of this distinguished element under ϕ_R. The commutativity of the diagram

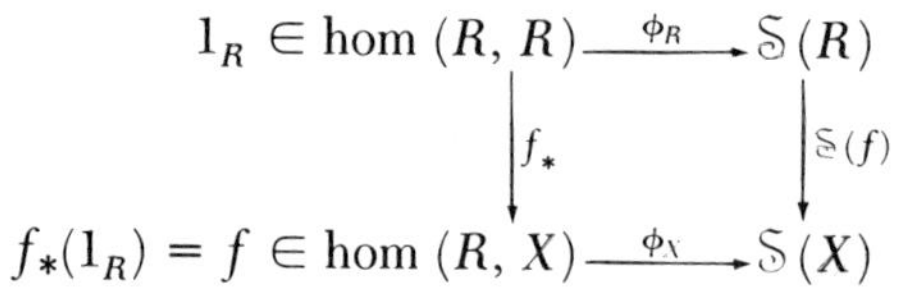

$$\begin{array}{ccc} 1_R \in \hom(R, R) & \xrightarrow{\phi_R} & \mathbb{S}(R) \\ \downarrow f_* & & \downarrow \mathbb{S}(f) \\ f_*(1_R) = f \in \hom(R, X) & \xrightarrow{\phi_X} & \mathbb{S}(X) \end{array}$$

then reads $\phi_X(f) = (\mathbb{S}f)u$. But ϕ_X is a bijection, so each element $a \in \mathbb{S}(X)$ is $\phi_X(f)$ for a unique f, and so each a is $(\mathbb{S}f)u$ for a unique f. This states that the element u is a universal element.

Conversely, let (u, R) be a universal element for $\mathbb{S}$. For each object X, define $\phi_X:\hom(R, X) \to \mathbb{S}(X)$ by $\phi_X(f) = (\mathbb{S}f)u$. The universality of u then states that ϕ_X is a bijection. It is natural, for any $g:X \to Y$ gives $\mathbb{S}(g)\phi_X f = \mathbb{S}(g)\mathbb{S}(f)u = \mathbb{S}(gf)u = \phi_Y(gf) = \phi_Y g_* f$, as required for naturality (draw a diagram!). Hence, ϕ is a representation of $\mathbb{S}$.

One checks at once that the functions $(R, \phi) \mapsto (u, R)$ and $(u, R) \mapsto (R, \phi)$ so constructed are inverses of each other. In any representation, this theorem allows us to speak of the *representing object* R, the *natural bijection* ϕ, and the *universal element* u for the representation.

For example, let N be a normal subgroup of G. For any group H, let $\mathcal{F}(H)$ be the set of all morphisms $f: G \to H$ of groups with $f_* N = 1$. Then $\mathcal{F}$ is a subfunctor of hom $(G, -)$, and the main theorem on quotient groups (Theorem II.26) states that the projection $p: G \to G/N$ is a universal element for this functor. By the theorem above, the corresponding representation $\mathcal{F}(H) \cong \text{hom}\,(G/N, H)$ is a bijection of sets, natural in H. In the same way, each of the universal elements which we have constructed for other functors to $\mathbf{S}$ leads to a representation.

The uniqueness theorem for universal elements (Theorem IV.1) may be restated as follows to include the representations.

THEOREM 4. *If (R, ϕ, u) and (R', ϕ', u') are two representations of the same functor $\mathcal{S}: \mathbf{X} \to \mathbf{S}$, there is an invertible morphism $\theta: R \to R'$ of $\mathbf{X}$ for which $\mathcal{S}(\theta)u = u'$ and for which ϕ' is the composite $\phi \circ \theta^*$,*

$$\text{hom}\,(R', X) \xrightarrow{\theta^*} \text{hom}\,(R, X) \xrightarrow{\phi} \mathcal{S}(X).$$

In particular, any two representing objects R and R' for the same functor $\mathcal{S}$ are equivalent in $\mathbf{X}$ by $\theta: R \cong R'$.

Proof: As $u \in \mathcal{S}(R)$ is a universal element and $u' \in \mathcal{S}(R')$ an element of $\mathcal{S}$, there is a unique arrow $\theta: R \to R'$ of $\mathbf{X}$ with $\mathcal{S}(\theta)u = u'$. Since u' is also universal, there is a unique arrow $\theta': R' \to R$ with $\mathcal{S}(\theta')u' = u$. As $\mathcal{S}$ is a functor, $u = \mathcal{S}(\theta')\mathcal{S}(\theta)u = \mathcal{S}(\theta'\theta)u$. Hence, $\theta'\theta: R \to R$ is the unique arrow of $\mathbf{X}$ with $u = \mathcal{S}(\theta'\theta)u$; since also $u = \mathcal{S}(1_R)u$, this unique arrow $\theta'\theta$ is the identity arrow 1_R. By a similar argument, $\theta\theta' = 1_{R'}$. Therefore, θ' is a two-sided inverse of θ, so θ is invertible, as asserted.

Now we consider any morphism $f': R' \to X$. By the description (17) of the representation $\phi': \text{hom}\,(R', X) \to \mathcal{S}(X)$, we have $\phi' f' = \mathcal{S}(f')u' = \mathcal{S}(f')\mathcal{S}(\theta)u = \mathcal{S}(f'\theta)u$. On the other hand, $\theta^* f' = f' \circ \theta: R \to X$, so by (17) again, this time for ϕ, we have $(\phi \circ \theta^*)f' = \phi(f' \circ \theta) = \mathcal{S}(f'\theta)u$. This proves that $\phi \circ \theta^* = \phi'$, as required.

This theorem may also be proved by using initial objects in a suitably constructed category. Given a functor $\mathcal{S}: \mathbf{X} \to \mathbf{S}$, define an *$\mathcal{S}$-pointed object* of the category $\mathbf{X}$ to be a pair (a, X) consisting of an object X and an element $a \in \mathcal{S}(X)$. Define an arrow $f: (a, X) \to (b, Y)$ of *$\mathcal{S}$-pointed objects* to be an arrow $f: X \to Y$ of $\mathbf{X}$ for which $\mathcal{S}(f)a = b$. The $\mathcal{S}$-pointed objects with these arrows form a category, say $\mathbf{X}_{\mathcal{S}*}$. For example, if $\mathcal{S} = I$ is the identity functor on sets, an $\mathcal{S}$-pointed set is just a pointed set in the sense described in §1. Again, if $\mathcal{U}: \mathbf{Rng} \to \mathbf{S}$ is the underlying set functor on rings, then a $\mathfrak{U}$-pointed ring is just a pointed ring in the sense used in our discussion (IV.6) of the universal properties of polynomial rings.

With this definition, it is clear that a universal element (u, R) for $\mathfrak{S}$ is the same thing as an initial object (u, R) in the category $\mathbf{X}_{\mathfrak{S}*}$ of all $\mathfrak{S}$-pointed objects of $\mathbf{X}$. Since any two initial objects in a category are equivalent, the theorem follows.

If $\mathbf{D}$ is the category of integral domains D, with arrows the monomorphisms of domains, the quotient field $Q(D)$ was characterized (§IV.1) by the fact that the injection $j:D \to Q(D)$ was universal among monomorphisms $\alpha:D \to F$ to a field F. Each such α is then uniquely a composite $\alpha = \alpha' \circ j$, and the correspondence $\alpha' \mapsto \alpha$ is a representation

$$\hom_{\mathbf{D}}(Q(D), F) \cong \hom_{\mathbf{D}}(D, F).$$

Here the functor hom $(D, -)$ which is represented depends on a "parameter" D, so the representing object $Q(D)$ also depends on D. Moreover, $Q(D)$ is a functor of D. This is a general fact, as follows, about representations with a "parameter" M.

Theorem 5 (*The Parameter Theorem*). *Let* $\mathbf{M}$ *and* $\mathbf{X}$ *be categories and*

$$\mathfrak{T} : \mathbf{M}^{\mathrm{op}} \times \mathbf{X} \to \mathbf{S}$$

a bifunctor such that for each object M in the category $\mathbf{M}$ the functor $\mathfrak{T}(M, -)$ of one variable has a representation (R_M, ϕ_M). Then there is a unique functor $\mathfrak{R} : \mathbf{M} \to \mathbf{X}$ with object function $\mathfrak{R}(M) = R_M$ such that the equivalence

$$\phi_M : \hom(R_M, X) \cong \mathfrak{T}(M, X) \tag{18}$$

is natural in M as well as in X. The mapping function of the functor $\mathfrak{R}$ is described as follows: Given any arrow $k:M \to N$ in $\mathbf{M}$, $\mathfrak{R}(k)$ is the unique arrow taking the universal element $u_M \in \mathfrak{T}(M, R_M)$ to $\mathfrak{T}(k, 1)u_N$.

Proof: To each arrow $k:M \to N$ in the category $\mathbf{M}$ we wish to choose an arrow $R_k:R_M \to R_N$ in the category $\mathbf{X}$ so that ϕ_M will be natural in M, as in the diagram

$$\begin{array}{rcl} \phi_N : \hom(R_N, X) \cong & \mathfrak{T}(N, X) & \\ & \downarrow \mathfrak{T}(k, 1) & \\ \phi_M : \hom(R_M, X) \cong & \mathfrak{T}(M, X) & \end{array}$$

So set $X = R_N$ here and start with the identity $1:R_N \to R_N$ upper left. There is then a unique $R_k:R_M \to R_N$ with

$$\phi_M(R_k) = \mathfrak{T}(k, 1)\phi_N(1). \tag{19}$$

With this choice, one readily verifies that $\mathfrak{R}$ becomes a functor $M \mapsto R_M$, $k \mapsto R_k$ on **M** to **X**.

It remains to show that this choice makes ϕ_M natural in M; this amounts to showing for each k that the diagram

$$\begin{array}{ccc} \hom(R_N, X) & \xrightarrow{\phi_N} & \mathfrak{T}(N, X) \\ \downarrow {\scriptstyle (R_k)^*} & & \downarrow {\scriptstyle \mathfrak{T}(k,1)} \\ \hom(R_M, X) & \xrightarrow{\phi_M} & \mathfrak{T}(M, X), \end{array} \qquad \begin{array}{l} \phi_N(g) = \mathfrak{T}(1, g)u_N, \\ \\ \phi_M(f) = \mathfrak{T}(1, f)u_M \end{array}$$

is commutative. But, by naturality, $\phi_N g = \mathfrak{T}(1, g)\phi_N 1$, thus stating that $\phi_N(1)$ is the universal element. Hence,

$$\mathfrak{T}(k, 1)\phi_N(g) = \mathfrak{T}(k, 1)\mathfrak{T}(1, g)\phi_N(1) = \mathfrak{T}(1, g)\mathfrak{T}(k, 1)\phi_N(1),$$

the latter because $\mathfrak{T}$ is a bifunctor as in (13). Now put in (19), to continue as

$$= \mathfrak{T}(1, g)\phi_M(R_k) = \phi_M(gR_k) = \phi_M[(R_k)^*g].$$

This gives the desired commutativity, and so completes the proof.

The whole discussion of representable functors (like any set of theorems about categories) can be dualized by replacing the category **X** by its opposite, and so the covariant hom-functor by the contravariant one. The results thus obtained are automatic; for convenience of reference we may summarize them as follows.

Let $\mathfrak{K}$ be a contravariant functor on **X** to **S**. A *corepresentation* of $\mathfrak{K}$ is a pair (R, ψ) where R is an object of **X** and ψ_X a family of bijections

$$\psi_X : \hom(X, R) \cong \mathfrak{K}(X) \tag{20}$$

natural in X. Thus ψ is a natural isomorphism $\psi : h^R \cong \mathfrak{K}$. A *universal element* (v, R) for the contravariant functor $\mathfrak{K}$, as defined in §IV.7, consists of an object R of **X** and an element $v \in \mathfrak{K}(R)$ such that for each element $c \in \mathfrak{K}(X)$ there is exactly one morphism $f : X \to R$ with $\mathfrak{K}(f)v = c$ [since $\mathfrak{K}$ is contravariant, $\mathfrak{K}(f)$ is a function $\mathfrak{K}(f) : \mathfrak{K}(R) \to \mathfrak{K}(X)$]. Alternatively, we may define a $\mathfrak{K}$*-copointed object* of X to be a pair (c, X) with $c \in \mathfrak{K}(X)$, while a morphism $f : (c, X) \to (d, Y)$ of such objects is a morphism $f : X \to Y$ of **X** with $\mathfrak{K}(f)d = c$. A universal element for the functor $\mathfrak{K}$ may now be described as an object (v, R) terminal in the category of all $\mathfrak{K}$-copointed objects of **X**. The formulas

$$v = \psi_R(1_R), \qquad \psi(g) = (\mathfrak{K}g)v \qquad \text{for } g : X \to R \tag{21}$$

provide a bijection between corepresentations (R, ψ) of $\mathfrak{K}$ and universal elements (v, R) for $\mathfrak{K}$. Since each such universal element is terminal in a category, it is unique up to equivalence. In particular, the corepresenting object R is unique up to an equivalence in **X**.

A product diagram $G \overset{p}{\leftarrow} G \times G' \overset{p'}{\rightarrow} G'$ in a category $\mathbf{X}$ was defined in (IV.22) by the requirement that each pair of arrows $k:L \rightarrow G$, $k':L \rightarrow G'$ can be written as $k = ph$, $k' = p'h$ for a unique $h:L \rightarrow G \times G'$. This states that $h \mapsto (ph, p'h)$ is a bijection

$$\hom(L, G \times G') \cong \hom(L, G) \times \hom(L, G'). \tag{22}$$

It is natural in L, so is a representation of the contravariant functor of L given by the right-hand side. If we introduce the product category $\mathbf{X} \times \mathbf{X}$ and the diagonal functor $\Delta:\mathbf{X} \rightarrow \mathbf{X} \times \mathbf{X}$ with $\Delta(L) = (L, L)$ and $\Delta(f) = (f, f)$, this right-hand side can be rewritten to give

$$\hom(L, G \times G') \cong \hom_{\mathbf{X} \times \mathbf{X}}(\Delta L, (G, G')). \tag{23}$$

EXERCISES

1. Show that the underlying set functor $\mathcal{U}:\mathbf{Mod}_R \rightarrow \mathbf{S}$ on the category of right R-modules is representable (and give a representation).

2. Show that the underlying set functor $\mathcal{U}:\mathbf{Rng} \rightarrow \mathbf{S}$ on the category of rings is representable.

3. Let two functors $\mathcal{S}, \mathcal{S}':\mathbf{X} \rightarrow \mathbf{S}$ have representations (R, ϕ) and (R', ϕ'), respectively. Show that there is for each natural transformation $\tau:\mathcal{S} \rightarrow \mathcal{S}'$ a unique morphism $f:R' \rightarrow R$ of $\mathbf{X}$ such that $\tau\phi = \phi' f^*:\hom(R, X) \rightarrow \mathcal{S}'(X)$.

4. (The Yoneda Lemma) Let $\mathcal{S}:\mathbf{X} \rightarrow \mathbf{S}$ be any functor. Show that there is for each object X a bijection from the set $\mathcal{S}(X)$ to the set of all natural transformation $h_X \rightarrow \mathcal{S}$.

***5.** Show that the bijection of Exercise 4 is "natural" in X and in $\mathcal{S}$—explaining what this means.

6. State and prove the analog of Theorem 5 for corepresentations.

7. Describe the coproduct of two objects in a category as the representing object for a suitable functor.

8. (a) Show that products exist in the category of $\mathbf{S}_*$ of pointed sets, and describe these products.

(b) Show that coproducts (often called "wedge products") exist in the category $\mathbf{S}_*$, and describe them.

9. Show that products do not exist in the category of all inner product spaces, with morphisms all orthogonal transformations, as defined in §X.7.

10. Show that the category of fields does not have products (*Hint:* Every morphism of fields is a monomorphism).

11. Show that a poset P is a lattice if and only if the corresponding category $\mathbf{P}$ (as defined in §1) has both products and coproducts.

6. Adjoint Functors

Consider the description of "free objects" in a concrete category such as the category $\mathbf{Mod}_R$ of all right modules over a fixed ring R. Each set X determines a corresponding free R-module $\mathfrak{F}(X) = R^{(X)}$, as in §V.5, and for each function $g:X \to Y$ there is a corresponding morphism $\mathfrak{F}(g):\mathfrak{F}(X) \to \mathfrak{F}(Y)$ of free modules, so $\mathfrak{F}$ is a functor on $\mathbf{S}$ to $\mathbf{Mod}_R$. Compare this functor with the underlying set functor $\mathfrak{U}$ which assigns to each R-module A the set $\mathfrak{U}(A)$ of its elements. These functors are opposite in direction:

$$\mathfrak{F} : \mathbf{S} \to \mathbf{Mod}_R, \qquad \mathfrak{U} : \mathbf{Mod}_R \to \mathbf{S}.$$

The statement that the module $\mathfrak{F}(X)$ is free on its subset X means that the function $j = j_X:X \to \mathfrak{U}(\mathfrak{F}(X))$ inserting the set X in the underlying set of its free module has the following universal property (as in the definition of a free module in §V.5): Any function $f:X \to \mathfrak{U}(A)$ from X to the underlying set of an R-module A factors as $f = \mathfrak{U}(t) \circ j$ for a unique morphism $t:\mathfrak{F}(X) \to A$ of modules, as in the commutative diagram below:

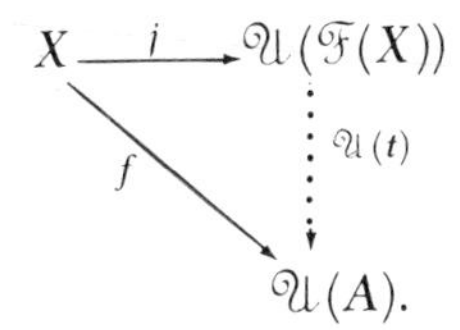

The corresponding representation

$$\hom_{\mathbf{Mod}_R} (\mathfrak{F}(X), A) \cong \hom_{\mathbf{S}} (X, \mathfrak{U}(A)) \tag{24}$$

is the bijection $t \mapsto f = \mathfrak{U}(t) \circ j$ carrying morphisms t of modules to functions f between sets; as with any representation with a parameter, this bijection is natural in X and in A. Because of this representation, we call the free module functor $\mathfrak{F}$ "left adjoint" to the underlying set functor $\mathfrak{U}$.

More generally, consider any two categories $\mathbf{T}$ and $\mathbf{X}$ and two functors

$$\mathfrak{F} : \mathbf{X} \to \mathbf{T}, \qquad \mathfrak{G} : \mathbf{T} \to \mathbf{X} \tag{25}$$

in opposite directions between them. For an object X in $\mathbf{X}$ and an object T in $\mathbf{T}$ we may then compare the set $\hom(\mathfrak{F}X, T)$ of all arrows in $\mathbf{T}$ from $\mathfrak{F}(X)$ to T with the set $\hom(X, \mathfrak{G}T)$ of all arrows in $\mathbf{X}$ from X to $\mathfrak{G}(T)$. This leads to the following definition.

DEFINITION. *An* adjunction *of the functor* $\mathfrak{F}$ *to the functor* $\mathfrak{G}$ *of* (25) *is a bijection*

$$\phi = \phi_{T,X} : \hom_{\mathbf{T}} (\mathfrak{F}X, T) \cong \hom_{\mathbf{X}} (X, \mathfrak{G}T) \tag{26}$$

of sets, defined for all objects T and X and natural in these arguments T and X. Given such an adjunction, the functor $\mathfrak{F}$ is called a left adjoint of $\mathfrak{G}$, while $\mathfrak{G}$ is called a right adjoint of $\mathfrak{F}$.

This definition is closely related to the notion of representability:

THEOREM 6. *A functor $\mathfrak{G}:\mathbf{T} \to \mathbf{X}$ has a left adjoint if and only if each functor* $\hom(X, \mathfrak{G}-) = h_X \circ \mathfrak{G}$ *is representable, for each object X in $\mathbf{X}$. When this is the case, a functor $\mathfrak{F}$ left adjoint to $\mathfrak{G}$ assigns to each X an object $\mathfrak{F}(X)$ representing $h_X \circ \mathfrak{G}$. Any other functor left adjoint to $\mathfrak{G}$ is naturally equivalent to $\mathfrak{F}$.*

Proof: First, let $\mathfrak{G}$ have a left adjoint $\mathfrak{F}$. For fixed X, the adjunction (26) above is then a representation of $\hom_{\mathbf{X}}(X, \mathfrak{G}T)$, as a functor of T, by the object $\mathfrak{F}X$ in $\mathbf{T}$. If $\mathfrak{F}':\mathbf{X} \to \mathbf{T}$ is any other functor left adjoint to $\mathfrak{G}$, then for each X the objects $\mathfrak{F}X$ and $\mathfrak{F}'X$ both represent the same functor, hence must be equivalent. This equivalence may be shown to be natural.

Conversely, suppose for each X that the functor $T \mapsto \hom(X, \mathfrak{G}T)$ is representable, and let $F(X)$ be a representing object. This means that there is for each T and X a bijection $\phi_{T,X}$ like (26), natural in T for fixed X. But the parameter theorem (Theorem 5) states then that $F(X)$ is the object function of a functor $\mathfrak{F}$ such that (26) is natural in both arguments, and hence an adjunction, as required.

For $\mathfrak{G}$ to have a left adjoint, each composite functor $h_X \circ \mathfrak{G}:\mathbf{T} \to \mathbf{S}$ to sets must be representable. Now we proved in Theorem 3 that a representable functor can be described in terms of a universal element and that this universal element may be obtained from the representation as follows. In (26), set T equal to the representing object $\mathfrak{F}X$ and take the image under ϕ of the identity map $\mathfrak{F}X \to \mathfrak{F}X$. This image is an arrow $u = u_X:X \to \mathfrak{G}(\mathfrak{F}X)$ of $\mathbf{X}$. This proves the

COROLLARY. *A functor $\mathfrak{G}:\mathbf{T} \to \mathbf{X}$ has a left adjoint if and only if there is for each object X an object $\mathfrak{F}X$ of $\mathbf{T}$ and an arrow $u = u_X:X \to \mathfrak{G}(\mathfrak{F}X)$ such that the pair $(u_X, \mathfrak{F}(X))$ is a universal element for the functor $h_X \circ \mathfrak{G}:\mathbf{T} \to \mathbf{S}$.*

In other words, $\mathfrak{G}$ has a left adjoint if and only if there is to each object $X \in \mathbf{X}$ an object FX and an arrow $u:X \to \mathfrak{G}FX$, universal among arrows from X to $\mathfrak{G}(-)$.

As usual, this universality means: Given an arrow $f:X \to \mathfrak{G}(T)$ in $\mathbf{X}$, there is a unique arrow $h:\mathfrak{F}(X) \to T$ with $f = \mathfrak{G}(h) \circ u$, as displayed in the

diagrams

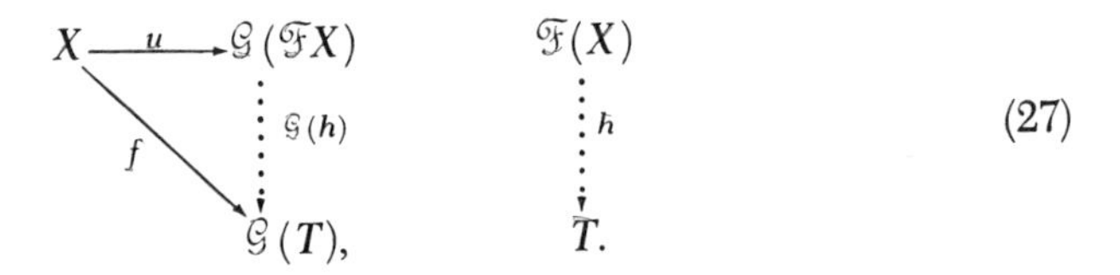

In case **T** is a concrete category, while $\mathcal{G} = \mathcal{U} : \mathbf{T} \to \mathbf{S}$ is its underlying set functor, an adjoint $\mathcal{F}$ of $\mathcal{U}$, when it exists, is called a *free-object functor* for **T**, while the object $\mathcal{F}(X)$ for each set X is the *free* **T**-*object* on the set X of generators. This includes in particular the case of the free-module functor discussed at the start of this section. There are many other such examples. If **T** is the (concrete) category of all commutative rings, a free-object functor exists; it assigns to each set X the polynomial ring $\mathbf{Z}[X]$ in the set X of indeterminates. In particular, if $X = \{x_1, \cdots, x_n\}$ is a finite set, the polynomial ring $\mathbf{Z}[x_1, \cdots, x_n]$ is a free commutative ring on these n generators x_i, in keeping with the universality properties of this polynomial ring, as described in Exercise IV.4.2. There is a free-group functor for the category of all groups. The category of all abelian groups also has a free-abelian-group functor, which is a special case (with $R = \mathbf{Z}$) of the functor already described for R-modules. Exercises 5, 6, and 8 give other examples of free-object functors.

Another important adjunction is involved in the process of regarding a function $f(z, y) \in X$ of two variables as a function $F(z)$ of one variable, defined by $F(z)(y) = f(z, y)$, as in Theorem IV.7. Here the values of $F(z)$ are in the function set X^Y, so the assignment $f \mapsto F$ is, as in (IV.24), a bijection

$$\phi : X^{Z \times Y} \cong (X^Y)^Z. \tag{28}$$

In this bijection, both sides represent trifunctors on **S** to **S**, contravariant in Z and Y and covariant in X, and the bijection ϕ is natural in all three arguments. Since the function set X^Y is just the set hom (Y, X) of all arrows from X to Y in **S**, this natural bijection can be rewritten as

$$\phi : \hom_{\mathbf{S}}(Z \times Y, X) \cong \hom_{\mathbf{S}}(Z, \hom_{\mathbf{S}}(Y, X)). \tag{29}$$

Now hold Y constant. This bijection is then an adjunction of the functor $- \times Y : \mathbf{S} \to \mathbf{S}$ to the functor hom $(Y, -) : \mathbf{S} \to \mathbf{S}$. Put differently, the functor "cartesian product with Y" has a right adjoint which is the covariant hom-functor h_Y.

A closely related example is that of tensor product and Hom in the category of all modules over a commutative ring K (Exercise 4).

EXERCISES

1. Show that the underlying set functor $\mathcal{G}:\mathbf{S}_* \to \mathbf{S}$ from pointed sets to sets has as left adjoint the functor $\mathcal{A}$ which assigns to each set X the disjoint union $X \dot{\cup} \{*\}$ of X with a new point $*$.

2. If $\mathbf{M}$ is the category of monoids, show that the forgetful functor $\mathcal{U}:\mathbf{Rng} \to \mathbf{M}$ which assigns to each ring R the same set with the same multiplication (forget the addition in the ring) has a left adjoint (cf. Exercise III.6.6).

3. Let $\mathbf{F}$ be the category of fields, $\mathbf{D}$ that of integral domains, with arrows the monomorphisms of domains and $\mathcal{U}:\mathbf{F} \to \mathbf{D}$ the forgetful functor which regards each field F just as an integral domain $\mathcal{U}(F)$. Show that the construction of the field of quotients of an integral domain (§III.5) provides a functor $\mathcal{Q}:\mathbf{D} \to \mathbf{F}$ left adjoint to $\mathcal{U}$.

4. Let K be a commutative ring and B a fixed K-module. Using Corollary IX.20, prove that $-\otimes B:\mathbf{Mod}_K \to \mathbf{Mod}_K$ is left adjoint to the functor $\mathrm{Hom}_K(B, -):\mathbf{Mod}_K \to \mathbf{Mod}_K$.

5. (a) Show that the forgetful functor $\mathcal{U}:\mathbf{Mod}_K \to \mathbf{Mod}_{\mathbf{Z}}$ which regards each K-module just as an additive abelian group has a left adjoint which assigns to each abelian group C the (right) K-module $C \otimes_{\mathbf{Z}} K$.

(b) For K a subring of the commutative ring K', show that Theorem IX.21 on change of rings provides a pair of adjoint functors between the categories $\mathbf{Mod}_K$ and $\mathbf{Mod}_{K'}$.

6. Construct a left adjoint for each covariant hom functor $\mathbf{S}_* \to \mathbf{S}$ on the category $\mathbf{S}_*$ of pointed sets.

7. Show that the category of monoids has a free-object functor.

***8.** Show that the category of groups has a free-object functor.

9. State and prove the dual of Theorem 6.

10. In the corollary to Theorem 6, show that u is a natural transformation $u:I \to \mathcal{G} \circ \mathcal{F}$, where I is the identity functor $I:\mathbf{X} \to \mathbf{X}$.

11. Show that the dual of the corollary to Theorem 6 gives a natural transformation $\mu:\mathcal{F} \circ \mathcal{G} \to I'$, where I' is the identity functor on $\mathbf{T}$.

12. (a) If $\mathbf{C}$ is a category with products, as defined in §IV.7, show that the product functor $(C, C') \mapsto C \times C'$ is right adjoint to the diagonal functor $\Delta:\mathbf{C} \to \mathbf{C} \times \mathbf{C}$ from $\mathbf{C}$ to the product category $\mathbf{C} \times \mathbf{C}$ defined by $\Delta(C) = (C, C)$ and $\Delta(f) = (f, f)$.

(b) Show that a coproduct functor is similarly left adjoint to Δ.

13. Prove that $\mathcal{F}:\mathbf{X} \to \mathbf{T}$ is left adjoint to $\mathcal{G}:\mathbf{T} \to \mathbf{X}$ if and only if there are natural transformations $u:I \to \mathcal{G} \circ \mathcal{F}$ and $\mu:\mathcal{F} \circ \mathcal{G} \to I$ which satisfy

$$\mathcal{G}(\mu_T)u_{\mathcal{G}T} = 1_{\mathcal{G}T} \qquad \text{and} \qquad \mu_{\mathcal{F}X}\mathcal{F}(u_X) = 1_{\mathcal{F}X}$$

for all objects X in $\mathbf{X}$ and T in $\mathbf{T}$. (*Hint:* cf. u and μ in Exercise 10, 11.)

14. For adjoint functors $\mathcal{F}$ and $\mathcal{G}$ as, for example, in Exercise 13, let $\mathcal{M} = \mathcal{G} \circ \mathcal{F}:\mathbf{X} \to \mathbf{X}$ be the composite, I the identity functor.

(a) For u and μ as in Exercise 13 (in the notation * of Exercise 4.8, with 1 an identity natural transformation) construct natural transformations

$$u * 1 : \mathcal{M} = \mathcal{M} \circ I \to \mathcal{M} \circ \mathcal{M}, \qquad 1 * u : \mathcal{M} = I \circ \mathcal{M} \to \mathcal{M} \circ \mathcal{M},$$

$$m = 1 * \mu * 1 : \mathcal{M} \circ \mathcal{M} \to \mathcal{M}.$$

(b) Prove that m has the property

$$m \circ (m * 1) = m \circ (1 * m) : \mathfrak{M} \circ \mathfrak{M} \circ \mathfrak{M} \to \mathfrak{M}.$$

(c) Prove that m and u satisfy

$$m \circ (u * 1) = 1 = m \circ (1 * u) : \mathfrak{M} \to \mathfrak{M}.$$

(d) Compare the results of (b) and (c) with the definition of an algebra, (IX.49), (IX.50).

Note: The object consisting of a functor $\mathfrak{M} : \mathbf{X} \to \mathbf{X}$ with natural transformations $m: \mathfrak{M} \circ \mathfrak{M} \to \mathfrak{M}$ and $u: I \to \mathfrak{M}$ satisfying (b) and (c) occurs frequently and is called a *monad* or a *triple* in $\mathbf{X}$. In 1965 Kleisli and Eilenberg–Moore proved that every monad arises, as in this exercise, from a pair of adjoint functors.

CHAPTER XVI

Multilinear Algebra

AMONG multilinear functions the tensor product, as defined in Chapter IX, is universal. Given a vector space V over a field F, the iterated tensor products of V with itself (and with its dual V^*) are called spaces of "tensors" over V. From the properties of the tensor product one may derive bases for these tensor spaces and formulas for the changes in their coordinates. All sums of tensors over V form an algebra (that is, a ring which is also a vector space) when the product of two tensors is defined to be their tensor product. Explicitly, if we call an element in the m-fold tensor product $V \otimes \cdots \otimes V$ a tensor of degree m, then tensors of degrees m and n, respectively, have a product which is a tensor of degree $m + n$. Because of this property of the degrees, this tensor algebra is called a "graded" algebra.

The alternating multilinear functions on $V \times V$, $V \times V \times V$, and so on, play a special role, as for example in the description of a determinant as an alternating function. In this chapter we shall define a universal alternating multilinear function, called the exterior product and written $V \times V \times V \to V \wedge V \wedge V$. The resulting sequence of the exterior powers, $F, V, V \wedge V, V \wedge V \wedge V, \cdots$ of V can again be made into a graded algebra, called the exterior algebra $\Lambda(V)$ of the vector space V. The formal properties of this algebra provide efficient descriptions of properties of determinants and of subspaces of V.

These developments of tensor and exterior algebras apply not just to vector spaces, but also to free modules of finite rank over any commutative ring K. Since K is commutative, these modules may be regarded either as right K-modules or as left K-modules, as may be convenient.

1. Iterated Tensor Products

To any two modules A and B over the same commutative ring K, we have constructed a tensor product, which is a module $A \otimes B = A \otimes_K B$, and a bilinear function, $(a, b) \mapsto a \otimes b$ on $A \times B$ to $A \otimes B$, which is universal (Theorem IX.15) for bilinear functions on $A \times B$ to a variable third module.

PROPOSITION 1. *If A, B, and C are three K-modules, the function $A \times B \times C \to A \otimes (B \otimes C)$ given by $(a, b, c) \mapsto a \otimes (b \otimes c)$ is a universal trilinear function on $A \times B \times C$ to K-modules.*

Proof: Since $a \otimes b$ is bilinear, this function $a \otimes (b \otimes c)$ is trilinear (linear in each of a, b, and c). We must show it universal; that is, that for each trilinear function $k:A \times B \times C \to D$ there is exactly one linear transformation $t:A \otimes (B \otimes C) \to D$ with $t(a \otimes (b \otimes c)) = k(a, b, c)$.

First, fix a. Then $k(a, -, -)$ is a bilinear function $B \times C \to D$, so the universality of $B \otimes C$ shows that there is exactly one linear map $s_a:B \otimes C \to D$ with $s_a(b \otimes c) = k(a, b, c)$ for all elements $b \in B$ and $c \in C$. Next we consider a linear combination $\kappa_1 a_1 + \kappa_2 a_2$ in the module A. It determines a map $s_{\kappa_1 a_1 + \kappa_2 a_2}$ and also a map $\kappa_1 s_{a_1} + \kappa_2 s_{a_2}$, each in Hom $(B \otimes C, D)$ and each sending $b \otimes c$ to $k(\kappa_1 a_1 + \kappa_2 a_2, b, c)$. The uniqueness part of the universality of $B \otimes C$ asserts that these maps are equal:

$$s_{\kappa_1 a_1 + \kappa_2 a_2} = \kappa_1 s_{a_1} + \kappa_2 s_{a_2} : B \otimes C \to D.$$

This means that $a \mapsto s_a$ is a linear function, put differently, that s is a bilinear function $s:A \times (B \otimes C) \to D$, as in the diagram

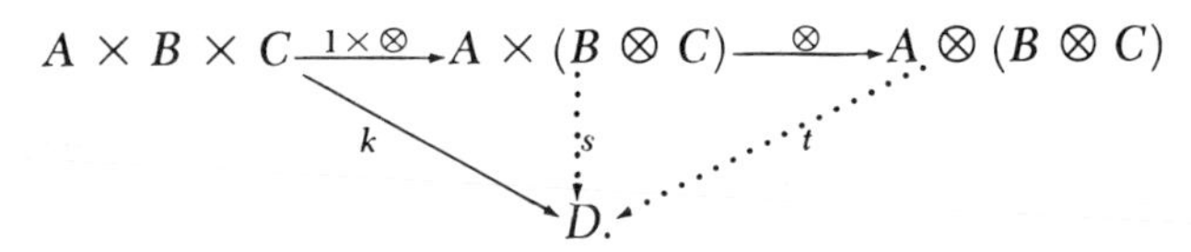

The universality of $\otimes:A \times (B \otimes C) \to A \otimes (B \otimes C)$ now asserts that there is exactly one linear transformation $t:A \otimes (B \otimes C) \to D$ for which $t(a \otimes (b \otimes c)) = s_a(b \otimes c)$ for all a, b, and c. Since $s_a(b \otimes c) = k(a, b, c)$, this means that there is exactly one t with $t(a \otimes (b \otimes c)) = k(a, b, c)$, and this is the required universality of $(a, b, c) \mapsto (a \otimes (b \otimes c))$.

Corollary. *The assignment $a \otimes (b \otimes c) \mapsto (a \otimes b) \otimes c$ defines an isomorphism*

$$A \otimes (B \otimes C) \cong (A \otimes B) \otimes C. \tag{1}$$

Proof: The proposition showed $a \otimes (b \otimes c)$ a universal trilinear function. A similar argument shows that $(a \otimes b) \otimes c$ is also a universal trilinear function. But the uniqueness of universals gives an isomorphism from the codomain $A \otimes (B \otimes C)$ of the first universal function to that of the second, carrying each $a \otimes (b \otimes c)$ onto $(a \otimes b) \otimes c$, as required.

Now we can write a triple tensor product $A \otimes B \otimes C$ *without* bothering to put in parentheses, for this corollary gives a standard or "canonical" isomorphism of one parenthesized product $A \otimes (B \otimes C)$ to the other one $(A \otimes B) \otimes C$. We shall identify the element $a \otimes (b \otimes c)$ with $(a \otimes b) \otimes c$ by this isomorphism, and write $a \otimes b \otimes c$ for this element. Similarly, given four modules A, B, C, and D, we may define a universal quadrilinear function on $A \times B \times C \times D$ may be defined by assigning to (a, b, c, d) any one of the elements

$$a \otimes (b \otimes (c \otimes d)), \cdots, (a \otimes b) \otimes (c \otimes d), ((a \otimes b) \otimes c) \otimes d.$$

We write $a \otimes b \otimes c \otimes d$ for any one (not specified) of these expressions.

For any list of $k \geqslant 2$ modules $A_1, \cdots, A_k$ one can now define by recursion a k-fold "tensor product" module $A_1 \otimes \cdots \otimes A_k$ and a function $(a_1, \cdots, a_k) \mapsto a_1 \otimes \cdots \otimes a_k$ and prove by induction on k that this function $A_1 \times \cdots \times A_k \to A_1 \otimes \cdots \otimes A_k$ is a universal multilinear function for the given k-fold product domain $A_1 \times \cdots \times A_k$. This universality means: To each multilinear function $h: A_1 \times \cdots \times A_k \to B$ there is exactly one linear transformation $t: A_1 \otimes \cdots \otimes A_k \to B$ with

$$t(a_1 \otimes \cdots \otimes a_k) = h(a_1, \cdots, a_k)$$

for all $a_i \in A_i$, $i \in \mathbf{n}$. This fact will be used freely below to construct such linear maps t. We emphasize that this construction works with *any* distribution of parentheses in $a_1 \otimes \cdots \otimes a_k$ *and* that the function t is completely determined by its values on those elements of the iterated tensor product which can be written as $a_1 \otimes \cdots \otimes a_k$. The general element of the module $A_1 \otimes \cdots \otimes A_k$ is a sum of such special elements, but we need not write down the value of t on these sums.

EXERCISES

1. Give an explicit proof that $(a, b, c) \mapsto (a \otimes b) \otimes c$ is a universal trilinear function.
2. Prove in detail that $(a, b, c, d) \mapsto (a \otimes b) \otimes (c \otimes d)$ is a universal quadrilinear function for four modules.
3. Show that there are five different ways of putting parentheses on $A \otimes B \otimes C \otimes D$, and exhibit the five corresponding universal quadrilinear functions on $A \times B \times C \times D$.
4. For any two modules A and B, derive by universality an isomorphism $A \otimes B \cong B \otimes A$.
5. Express the universality of $a \otimes b \otimes c$ as a representation of a suitable functor.
6. Show that any iterated tensor product of free modules is free.
7. Show that the module $A \otimes B \otimes C$ is spanned by all of the elements $a \otimes b \otimes c$ for $a \in A$, $b \in B$, and $c \in C$.
8. Prove the isomorphism $A \otimes (B \otimes C) \cong (A \otimes B) \otimes C$ of the Corollary natural in A, B, and C.

2. Spaces of Tensors

From a given finite-dimensional vector space V over a field F, one may construct iterated tensor product spaces such as $V \otimes V$, $V \otimes V \otimes V$, as well as tensor products such as $V \otimes V^*$, $V^* \otimes V \otimes V^*$, involving the space V^* dual to V. The elements of these spaces are known as "tensors"; they are used extensively in differential geometry and in its applications. This section will describe these tensors, their coordinates, and the formalism for change of coordinates.

The elements of the space V will be called *contravariant vectors* or *tensors of type* (0, 1). If $\mathbf{b}$ is a basis of n vectors for V, the dual space V^* (the space of *covariant vectors*) has a dual basis consisting of n linear forms $x_i: V \to F$. These forms will now be written as $b^i = x_i$, with the index i as a superscript. Thus $b^i: V \to F$ is defined by $b^i(b_j) = \delta_j^i$ for all $i, j \in \mathbf{n}$, where δ is the Kronecker delta written with one index as a superscript, so that δ_j^i is 1 or 0 according as $i = j$ or $i \neq j$. The scalars $\xi^i = b^i(v)$, also written with a superscript, are the coordinates of the vector v relative to the basis $\mathbf{b}$, so the vector v can be written as the linear combination $v = \Sigma_i b_i \xi^i$. In such a formula, the symbol Σ_i for summation over an index i which appears once as subscript index and once as superscript index is often omitted (this is the "summation convention") so that the *same* formula is written $v = b_i \xi^i$, with summation over $i = 1, \cdots, n$ understood.

The elements of the tensor product space $V \otimes V$ are the finite sums $r = v_1 \otimes w_1 + \cdots + v_k \otimes w_k$, where the v_i and w_i are vectors of V; these sums are tensors of type (0, 2) or tensors in two contravariant indices. If $\mathbf{b}$ is a basis of n vectors for V, then by Proposition IX.16 the n^2 tensors $b_i \otimes b_j$ for $i, j \in \mathbf{n}$ form a basis for $V \otimes V$. Each tensor $r \in V \otimes V$ thus has a unique representation as a linear combination

$$r = (b_i \otimes b_j)\rho^{ij} \tag{2}$$

(summation understood over i and j) with n^2 scalar coefficients ρ^{ij}. We call these ρ^{ij} the *coordinates* (or the "components") of the tensor r, relative to the basis $\mathbf{b}$; they are written with the two contravariant indices i and j as superscripts.

Now we consider the effect of changing from the dual basis pair (b^i, b_j) for V to another dual basis pair (c^h, c_k), where we have used $h, k \in \mathbf{n}$ for the indices of the second pair. The change matrix P from $\mathbf{b}$ to $\mathbf{c}$ is the $n \times n$ matrix with the scalar entries $P_i^h = c^h b_i$—just as in (VII.34), except that the indices are now written P_i^h and not P_{hi}. The corresponding formulas (VII.35) and (VII.36) for the new coordinates c^h in terms of the old coordinates and the old basis vectors in terms of the new ones, both written with the summation convention, are

$$c^h = P_i^h b^i, \qquad b_j = c_k P_j^k. \tag{3}$$

In $V \otimes V$, this implies that $b_i \otimes b_j = (c_h \otimes c_k) P_i^h P_j^k$, so the expression (2) for the tensor r becomes $r = (c_h \otimes c_k) P_i^h P_j^k \rho^{ij}$ (with summation over all four indices). This is an expression $r = (c_h \otimes c_k)\bar{\rho}^{hk}$ for r in terms of the new basis

$c_h \otimes c_k$ with scalar coordinates $\bar{\rho}^{hk}$ given by

$$\bar{\rho}^{hk} = P_i^h P_j^k \rho^{ij}, \qquad h, k = 1, \cdots, n. \tag{4}$$

To summarize: Each basis **b** of n vectors in V determines a basis of n^2 tensors $b_i \otimes b_j$ in $V \otimes V$, so that every tensor r in $V \otimes V$ has n^2 coordinates ρ^{ij} relative to this basis. A change of basis in V by a change matrix P replaces the coordinates ρ^{ij} by the $\bar{\rho}^{hk}$ of (4). This is the source of the classical description: A tensor of type (0, 2) is something with n^2 coordinates ρ^{ij} which change with change of basis according to formula (4).

In the tensor product space $V \otimes \cdots \otimes V$ with p factors the n^p tensors $b_{i_1} \otimes \cdots \otimes b_{i_p}$, where **i** is any list of p elements of **n**, form a basis, so any tensor $r \in V \otimes \cdots \otimes V$ can be expressed as $r = (b_{i_1} \otimes \cdots \otimes b_{i_p})\rho^{i_1 \cdots i_p}$, with an array of n^p scalar coordinate $\rho^{i_1 \cdots i_p}$. The change of basis from **b** to **c**, as above, yields new coordinates $\bar{\rho}$ expressed in terms of the old as

$$\bar{\rho}^{h_1 \cdots h_p} = P_{i_1}^{h_1} \cdots P_{i_p}^{h_p} \rho^{i_1 \cdots i_p}. \tag{5}$$

The tensor product space $V^* \otimes V$ of type (1, 1) is the space of "mixed tensors" in two indices, one contravariant and one covariant. Given a basis **b** of V, the n^2 mixed tensors $b^i \otimes b_j$, with $i, j \in \mathbf{n}$, form a basis of $V^* \otimes V$, by Proposition IX.16. Hence any mixed tensor t in this space has a unique representation as a linear combination $t = (b^i \otimes b_j)\tau_i^j$ with n^2 scalar coefficients τ_i^j, called the *coordinates* of t (rel **b**). These coordinates involve one "contravariant" index j (written as a superscript) and one "covariant" index i (a subscript). We shall write these coordinates here as τ_i^j, though the more explicit notation $\tau_i{}^j$, with covariant index before the contravariant one, would indicate that t is in the tensor product space $V^* \otimes V$ with the factors V^* and V, in that order.

We now calculate the effect on these bases and these coordinates of a change from the basis **b** to a new basis **c** for V. The corresponding change matrix P is invertible. If we write its inverse as $P^{-1} = Q$, with n^2 scalar entries Q_h^i such that $Q_h^i P_j^h = \delta_j^i$ (with summation over h, by convention), then the Equation (3) for the new dual basis c^h in terms of the old one implies that $b^i = \delta_j^i b^j = \gamma Q_h^i P_j^h b^j = Q_h^i c^h$. With the equation $b_j = c_k P_j^k$, the old basis elements of $V^* \otimes V$ become $b^i \otimes b_j = (c^h \otimes c_k)Q_h^i P_j^k$, so that the tensor $t \in V^* \otimes V$ can be written as

$$t = (b^i \otimes b_j)\tau_i^j = (c^h \otimes c_k)Q_h^i P_j^k \tau_i^j.$$

This is an expression of the form $t = (c^h \otimes c_k)\bar{\tau}_h^k$; therefore, the new coordinates $\bar{\tau}$ of the tensor t are expressed in terms of the old coordinates as

$$\bar{\tau}_h^k = Q_h^i P_j^k \tau_i^j, \qquad h, k \in \mathbf{n}. \tag{6}$$

In other words, a tensor t of type (1, 1) is something with n^2 coordinates τ_i^j

which change with change of basis according to the formula (6) involving both the change matrix P and its inverse Q. More formally, this (classical) description of a tensor replaces the element $t \in V^* \otimes V$ by a function T on the set of all bases of V to $n \times n$ matrices which assigns to a basis $\mathbf{b}$ the $n \times n$ matrix $T(\mathbf{b}) = \tau$ with entries τ_i^j, to $\mathbf{c}$ the $n \times n$ matrix $T(\mathbf{c}) = \bar{\tau}$, and so on:

Proposition 2. *Each tensor $t \in V^* \otimes V$ determines a function T which assigns to every basis* $\mathbf{b}$ *of V an $n \times n$ matrix τ of scalars in such a way that change from* $\mathbf{b}$ *to any other basis* $\mathbf{c}$ *replaces the matrix $\tau = T(\mathbf{b})$ by the matrix $\bar{\tau} = T(\mathbf{c})$ given by* (6). *Conversely, every function T with this property arises in this way from exactly one tensor $t \in V^* \otimes V$.*

Proof: The argument above showed that the coordinates of each mixed tensor t do constitute such a function T. Conversely, given T, take any one basis $\mathbf{b}$ of V and define t as $(b^i \otimes b_j)\tau_i^j$ where $\tau = T(\mathbf{b})$; then in every basis this tensor t has the coordinates assigned by the given function T.

Our definition of a mixed tensor as an element of $V^* \otimes V$ has the advantage of great conceptual simplicity; the alternative description of t by the function T giving all its coordinates is frequently used. One example is the observation that a Jacobian matrix defines a tensor of type (1, 1). In detail, let V be an n-dimensional real vector space and $f: V \to V$ a function which is smooth but not necessarily linear. If $\mathbf{b}$ is any basis of V, then each value $f(\sum b_i \xi^i)$ of the function f can be written as $\sum b_j \mu^j$ for n scalars μ^j; this defines n functions $(\xi^1, \cdots, \xi^n) \mapsto \mu^j$ on $\mathbf{R}^n$ to $\mathbf{R}$; these are just the usual functions used to describe f by coordinates. We assume all these functions differentiable. Now define $T(\mathbf{b}) = \tau$, where τ_i^j is the partial derivative $\partial \mu^j / \partial \xi^i$ evaluated at the origin (in other words, τ is the "Jacobian matrix" of the function f relative to the basis $\mathbf{b}$). We assert that these matrices τ consist of the coordinates of a tensor $t \in V^* \otimes V$. For, change $\mathbf{b}$ to $\mathbf{c}$; then $\sum b_i \xi^i = \sum c_h \bar{\xi}^h$ with $\xi^i = Q_h^i \bar{\xi}^h$, while $\sum b_j \mu^j = \sum c_k \bar{\mu}^k$ with $\bar{\mu}^k = P_j^k \mu^j$. The usual formula for the partial derivative of a composite function then shows that

$$\bar{\tau}_h^k = \frac{\partial \bar{\mu}^k}{\partial \bar{\xi}^h} = \sum_{j=1}^{n} \sum_{i=1}^{n} \frac{\partial \bar{\mu}^k}{\partial \mu^j} \frac{\partial \mu^j}{\partial \xi^i} \frac{\partial \xi^i}{\partial \bar{\xi}^h} = P_j^k \tau_i^j Q_h^i,$$

with all partial derivatives evaluated at the origin $\xi^i = \bar{\xi}^h = 0$. This is exactly the formula (6), so the function T has the property in Proposition 2. Therefore, T defines a unique tensor $t \in V^* \otimes V$, the Jacobian of f.

Next we define the operation of "contraction" on a mixed tensor. Since V^* is the vector space dual to V, each linear form $f \in V^*$ may be evaluated at each vector $v \in V$ to give a scalar $f(v)$, and this assignment $(f, v) \mapsto f(v)$ is, as in (V.46) and (V.47), a bilinear form $V^* \times V \to F$. Since the tensor product $\otimes : V^* \times V \to V^* \otimes V$ is a universal bilinear function on $V^* \times V$, there is a unique linear transformation

$$\gamma : V^* \otimes V \to F \qquad \text{with} \qquad \gamma(f \otimes v) = f(v) \tag{7}$$

for all $f \in V^*$ and all $v \in V$. This linear transformation γ is called *contraction*. By the definition of a dual basis, $\gamma(b^i \otimes b_j) = b^i(b_j) = \delta_j^i$ and hence, for any tensor t, $\gamma[(b^i \otimes b_j)\tau_i^j] = \delta_j^i \tau_i^j = \tau_i^i$. In words: The contraction $\gamma(t)$ of a mixed tensor $t \in V^* \otimes V$ is found from any one matrix τ_i^j of coordinates of t as the sum $\gamma(t) = \tau_i^i$.

Mixed tensors with more than two indices may be described in a similar way. Given natural numbers p and q, the space $T_q^p(V)$ of all p-times covariant and q-times contravariant tensors [the space of *tensors of type* (p, q)] is defined to be the iterated tensor product $V^* \otimes \cdots \otimes V^* \otimes V \otimes \cdots \otimes V$, with p factors V^* and q factors V. Given a basis $\mathbf{b}$ for V and the corresponding dual basis for V^*, Proposition IX.16 again shows that this tensor space $T_q^p(V)$ has a basis consisting of all the tensor products of the given basis elements; explicitly, all the tensor products $b^{i_1} \otimes \cdots \otimes b^{i_p} \otimes b_{j_1} \otimes \cdots \otimes b_{j_q}$, where $\mathbf{i}:\mathbf{p} \to \mathbf{n}$ is a list of p indices in $\mathbf{n}$ and $\mathbf{j}:\mathbf{q} \to \mathbf{n}$ a list of q indices; the basis then has n^{p+q} elements. Relative to this basis, each tensor $t \in T_q^p(V)$ has n^{p+q} scalars $\tau_{i_1 \cdots i_p}^{j_1 \cdots j_q}$ as coordinates. If the basis $\mathbf{b}$ is changed to a basis $\mathbf{c}$ in V with change matrix P and inverse $P^{-1} = Q$, the formulas $b_j = c_k P_j^k$ and $b^i = Q_h^i c^h$, applied as in the previous case, show that the new coordinates $\bar{\tau}$ of t are expressed in terms of the old coordinates by the equations

$$\bar{\tau}_{h_1 \cdots h_p}^{k_1 \cdots k_q} = Q_{h_1}^{i_1} \cdots Q_{h_p}^{i_p} P_{j_1}^{k_1} \cdots P_{j_q}^{k_q} \tau_{i_1 \cdots i_p}^{j_1 \cdots j_q}. \tag{8}$$

(Note that the summation convention here calls for summation over all the indices $i_1, \cdots, i_p, j_1, \cdots, j_q$.)

Contraction can be applied to such a mixed tensor t, if we first select one contravariant index j_l and one covariant index i_m (so that $l \in \mathbf{q}$ and $m \in \mathbf{p}$). For example, selecting $l = q$ and $m = p$, the corresponding contraction is the unique linear transformation $\gamma : T_q^p(V) \to T_{q-1}^{p-1}(V)$ with

$$\begin{aligned}\gamma(f^1 \otimes \cdots \otimes f^p \otimes v_1 \otimes \cdots \otimes v_q) \\ = (f^1 \otimes \cdots \otimes f^{p-1}) \otimes (v_1 \otimes \cdots \otimes v_{q-1}) f^p(v_q)\end{aligned}$$

for all linear forms $f^i : V \to F$ and all vectors $v_j \in V$. One may prove that the coordinates of the contracted tensor $\gamma(t)$ may be obtained from the coordinates τ of t (relative to any basis $\mathbf{b}$) by setting $j_l = i_m = i$ in the indices of τ and applying the summation convention—which amounts to adding the resulting values of τ for all $i = 1, \cdots, n$.

In physics and in differential geometry one uses not only tensor spaces over the fields $\mathbf{R}$ and $\mathbf{C}$ but also "tensor fields": Differentiable functions from a portion of Euclidean space or of a manifold to one of the spaces of tensors which we have here described.

The dual of any tensor space over V is another tensor space over V, as one may prove (Exercise 7) from the following proposition:

Proposition 3. *For any two finite-dimensional vector spaces V and W over the same field F there is an isomorphism*

$$\theta : V^* \otimes W^* \cong (V \otimes W)^* \tag{9}$$

with $[\theta(f \otimes g)](v \otimes w) = (fv)(gw)$ for all linear forms $f \in V^*$, $g \in W^*$ and all vectors $v \in V$, $w \in W$. If $(\mathbf{x}, \mathbf{b})$ and $(\mathbf{y}, \mathbf{c})$ are, respectively, dual basis pairs for the spaces V and W, then the forms $\theta(x_i \otimes y_j)$ constitute a basis of $(V \otimes W)^*$ dual to the basis of all vectors $b_i \otimes c_j$ of $V \otimes W$.

Proof: First, fix f and g. Then $(v, w) \mapsto (fv)(gw) \in F$ is a bilinear form $V \times W \to F$. By the universality of the tensor product, there is a unique linear transformation $\eta(f, g): V \otimes W \to F$ with $\eta(f, g)[v \otimes w] = (fv)(gw)$. Thus $\eta(f, g)$ is an element of the dual space $(V \otimes W)^*$; moreover, $(f, g) \mapsto \eta(f, g)$ is a bilinear function $V^* \times W^* \to (V \otimes W)^*$. By the universality of tensor product again, there is a unique linear transformation $\theta: V^* \otimes W^* \to (V \otimes W)^*$ with $\theta(f \otimes g) = \eta(f, g)$.

To show this θ an isomorphism, as asserted, let V have dimension n and W dimension m, and choose dual basis pairs as in the statement of the proposition. The vectors $x_i \otimes y_h$ for $i \in \mathbf{n}$ and $h \in \mathbf{m}$ then form a basis for the tensor product $V^* \otimes W^*$, as do $b_j \otimes c_k$ with $j \in \mathbf{n}$ and $k \in \mathbf{m}$ for $V \otimes W$. By the definition of θ,

$$[\theta(x_i \otimes y_h)](b_j \otimes c_k) = (x_i b_j)(y_h c_k) = \delta_{ij}\delta_{hk}$$

for all $i, j \in \mathbf{n}$ and all $h, k \in \mathbf{m}$. This states that the $\theta(x_i \otimes y_h)$ form the basis of $(V \otimes W)^*$ dual to $b_j \otimes c_k$. Since θ is linear and carries a basis to a basis, it must be an isomorphism, as asserted.

For example, since $V^{**} \cong V$ for V finite-dimensional, this proposition gives

$$(V^* \otimes V \otimes V)^* \cong V^{**} \otimes V^* \otimes V^* \cong V \otimes V^* \otimes V^*.$$

In an inner product space U over the field of real numbers, the inner product gives an explicit isomorphism $U^* \cong U$. Hence, in this case the distinction between covariant and contravariant tensors can be ignored.

EXERCISES

1. Derive in detail the formula (5) for the change of coordinates of a contravariant tensor.

2. Derive in detail the formula (8) for the change of coordinates of a mixed tensor.

3. Describe three different contractions $V^* \otimes V \otimes V \otimes V \to V \otimes V$.

4. Prove the rule stated in the text for calculating the coordinates of a tensor $\gamma(t)$ for any contraction $\gamma: T_q^p(V) \to T_{q-1}^{p-1}(V)$.

5. From a basis for V construct a basis for the kernel of the contraction $V^* \otimes V \to F$.

6. Show how iterated contractions yield exactly two different linear transformations $V^* \otimes V^* \otimes V \otimes V \to F$.

7. For V finite-dimensional, prove that $[T_q^p(V)]^* \cong T_p^q(V)$.

8. Show that the isomorphism θ of Proposition 3 is natural (as a morphism of functors of V and W in the sense of Chapter XV).

9. Let K be a commutative ring:

(a) Use the formulas of Proposition 3 to define for K-modules A and B a K-linear map $\theta: A^* \otimes B^* \to (A \otimes B)^*$. If $A = K^n$ is free, prove this map θ an isomorphism.

(b) Construct a natural morphism

$$\text{Hom}\,(A, A') \otimes \text{Hom}\,(B, B') \to \text{Hom}\,(A \otimes B, A' \otimes B')$$

which will include the map θ of (a) as the special case with $A' = K = B'$.

3. Graded Modules

Throughout this section, K is a fixed commutative ring and "module" means "K-module". Given a module C, we frequently have occasion to consider sequences of modules, such as the sequences $(K, C, C, C, \cdots)$, $(K, C^*, (C \otimes C)^*, (C \otimes C \otimes C)^*, \cdots)$ or finally $(K, C, C \otimes C, C \otimes C \otimes C, \cdots)$. The sequence last mentioned is the sequence of tensor powers $T_p(C)$; it may be defined by recursion on p as

$$T_0(C) = K, \qquad T_{p+1}(C) = T_p(C) \otimes C. \tag{10}$$

Any sequence of modules will be called a "graded module"; in other words, a *graded module* is a sequence

$$G = (G_0, G_1, G_2, G_3, \cdots) \tag{11}$$

of modules G_p, one for each index $p \in \mathbf{N}$. Certain of the concepts already developed for modules apply almost automatically to graded modules, as we shall now indicate. First, a *morphism* $t: G \to G'$ of graded modules is a sequence $t_0, t_1, t_2, \cdots$ of morphisms $t_p: G_p \to G'_p$ of modules, one for each $p \in \mathbf{N}$. If $s: G' \to G''$ is another such morphism of graded modules, the composite $s \circ t: G \to G''$ is the morphism of graded modules defined for each p by $(s \circ t)_p = s_p \circ t_p$. With this composition, the graded modules and their morphisms form a category.

In a graded module G, the elements of G_p are said to have degree p. By an *element* g of G, we mean an element $g \in G_p$ of some specified degree, so that an "element" is given together with its degree. Thus two elements of G have a sum in G only when they have the same degree.

A *graded submodule* of a graded module G is defined to be a graded module S with each S_p a submodule of G_p. A graded submodule so defined is *not* a subset of G. However, for any two graded submodules S and S' of the same graded module G, we may define $S \subset S'$ (S is "contained" in S' or "included" in S') to mean that $S_p \subset S'_p$ for each $p \in \mathbf{N}$. Then $S \subset S'$ and $S' \subset S$ together imply that S and S' are identical, so that the graded submodules of G are partly ordered by inclusion.

If $t: G \to G'$ is a morphism of graded modules, the *image* of t is a graded submodule of G', and is the sequence consisting of the images of the morphisms $t_p: G_p \to G'_p$, while the *kernel* of t is the sequence of the kernels of the t_p, hence is a graded submodule of G. For any graded submodule $S \subset G$, the *quotient module* G/S is defined to be that graded module G/S with $(G/S)_p = G_p/S_p$ for each p, while the *projection* $\pi: G \to G/S$ of G on the quotient is the sequence $\pi_p: G_p \to G_p/S_p$ of the usual projections. This projection $\pi: G \to G/S$ has kernel S. Given G and S, π is a universal element for the functor which assigns to each graded module H the set of all those morphisms $G \to H$ of graded modules whose kernels contain S.

For graded modules, the most important construction is the formation of the (graded) *tensor product* of two graded modules, G and H. It is defined to be the graded module $G \otimes H$ whose nth term is the following $(n + 1)$-fold biproduct

$$\begin{aligned}(G \otimes H)_n &= \bigoplus_{p+q=n} (G_p \otimes H_q) \\ &= (G_n \otimes H_0) \oplus \cdots \oplus (G_0 \otimes H_n), \qquad n \in \mathbf{N}. \qquad (12)\end{aligned}$$

Thus, given elements $g \in G_p$ and $h \in H_q$, their tensor product $g \otimes h$ is an element of $G_p \otimes H_q$ which we also shall identify (by the insertion of $G_p \otimes H_q$ in the biproduct) with an element of $(G \otimes H)_{p+q}$. This makes the map $G_p \times H_q \to (G \otimes H)_{p+q}$ given by $(g, h) \mapsto g \otimes h$ bilinear for all p and q. It is a universal such family of bilinear functions, in the following sense.

Proposition 4. *Let G, H, and M be graded modules. If for all p and q there are given bilinear functions $t_{p,q}: G_p \times H_q \to M_{p+q}$, there is exactly one morphism $t: G \times H \to M$ of graded modules with $t(g \otimes h) = t_{p,q}(g, h)$ for all $g \in G_p$, $h \in H_q$.*

Proof: By the universality of the tensor product of modules, there is for each p and q a unique morphism $s_{p,q}: G_p \otimes H_q \to M_{p+q}$ of modules with

$s_{p,q}(g \otimes h) = t_{p,q}(g, h)$ for all g and h. By the universality of the insertions of the factors $G_p \otimes H_q$ in the biproduct $(G \otimes H)_{p+q}$, there is for each n a unique morphism $t_n:(G \otimes H)_n \to M_n$ with $t_n(g \otimes h) = s_{p,q}(g \otimes h)$ for all p and q with $p + q = n$. The sequence of these morphisms t_n is the desired morphism t of graded modules.

The chief use of the tensor product of graded modules is this result, which describes the family of bilinear functions $t_{p,q}$ by a single morphism t of graded modules. For example, given morphisms $u:G \to G'$ and $v:H \to H'$ of graded modules, the family of tensor product functions $u_p \otimes v_q:G_p \otimes H_q \to G'_p \otimes H'_q$ combine in this way to determine a morphism $G \otimes H \to G' \otimes H'$ of graded modules which we call $u \otimes v$. With this construction, the tensor product becomes a bifunctor on graded modules to graded modules.

For three graded modules G, H, and M the tensor product is associative: The assignment $g \otimes (h \otimes m) \mapsto (g \otimes h) \otimes m$ yields an isomorphism $G \otimes (H \otimes M) \cong (G \otimes H) \otimes M$ of graded modules. We use this isomorphism to "identify" $G \otimes (H \otimes M)$ with $(G \otimes H) \otimes M$.

A graded module H is said to be "concentrated in degree 0" when $H_n = 0$ for $n > 0$. Then H is the sequence $(H_0, 0, 0, \cdots)$; it is convenient to identify this graded module with the (ordinary) module H_0. In particular, the letter K will denote the ring of scalars regarded as a graded module concentrated in degree 0. For any graded module G, the graded tensor product $G \otimes K$ thus has $(G \otimes K)_n = G_n \otimes K \cong G_n$. This isomorphism (and the similar one for the tensor product $K \otimes G$) is used to make the following identifications:

$$G \otimes K = G, \qquad K \otimes G = G. \tag{13}$$

Appendix: *Internally Graded Modules*. Each graded module G determines an (ordinary) module ΣG in the following way. Consider all sequences $\mathbf{f} = (f_0, f_1, f_2, \cdots)$ of elements $f_p \in G_p$; let the "support" of such a sequence $\mathbf{f}$ be the set of all those indices p with $f_p \neq 0$. Take ΣG to be the set of all such sequences $\mathbf{f}$ with finite support. With module operations defined pointwise, as in $(\mathbf{f} + \mathbf{g})_p = \mathbf{f}_p + \mathbf{g}_p$ and $(\mathbf{f}\kappa)_p = f_p\kappa$ for each $\kappa \in K$, this set ΣG is a K-module. If $t:G \to G'$ is a morphism of graded modules, then the assignment $\mathbf{f} \mapsto t \circ \mathbf{f}$ is a morphism $\Sigma t:\Sigma G \to \Sigma G'$ of (ordinary) modules. The assignments $G \mapsto \Sigma G$, $t \mapsto \Sigma t$ define a functor Σ on graded modules to modules; the module ΣG is often written as $\Sigma_{p=0}^{\infty} G_p$.

Which (ordinary) modules have the form ΣG for some graded module G? Call a module S a *sum* of its submodules D_p, $p = 0, 1, 2, \cdots$ when every element $s \neq 0$ in S can be written in one and only one way as a finite sum $s = d_0 + \cdots + d_p$ with each $d_i \in D_i$ and $d_p \neq 0$. Then $\{D_p\}$ is a graded module D and S is isomorphic to ΣD. Indeed, many authors define an (internally) "graded module" to be a module S which is such a sum of submodules. In general, $S = \Sigma D$ has many more elements than the modules $\{D_p\}$, regarded as a graded module.

EXERCISES

1. Show that the submodules of a given graded module, when partly ordered by inclusion (as defined in the text), form a modular lattice. Given two submodules in this lattice, describe explicitly the elements belonging to their join and to their meet.

2. If $t:G \to G'$ is a morphism of graded modules, define the direct and the inverse image functions t_* and t^* on submodules, and show that they have all properties valid for the analogous image functions for morphisms of ordinary modules (see (V.17) and (V.18)).

3. Show that the Short Five Lemma (§IX.9) holds for exact sequences of graded modules.

4. (a) Describe the biproduct $G \oplus H$ of two graded modules G and H.

(b) State and prove the universal property of the pair of injections $G, H \to G \oplus H$.

(c) Do the same for the pair of projections $G \oplus H \to G, H$.

5. State and prove the "diamond" isomorphism theorem for graded modules.

6. For a graded module G, show that ΣG may have more submodules than G, in the sense that there can be (ordinary) submodules T of ΣG which do not have the form $T = \Sigma S$ for a graded submodule S of G.

7. (The "Universal" description of the direct sum Σ as an "infinite coproduct".) For G a fixed graded module, consider the functor (on modules to sets) which assigns to each module C the set of all sequences $\mathbf{s}$ of morphisms $s_p:G_p \to C$ of modules. Construct a universal element $\mathbf{u}$ for this functor, where u_p assigns to each $a \in G_p$ the element $u_p a$ of ΣG with $(u_p a)_p = a$ and $(u_p a)_q = 0$ for $p \neq q$.

8. Construct a functor $\mathfrak{T}$ on modules C to graded modules with $(\mathfrak{T}C)_p = C$ for all p. Use the results of Exercise 7 to show that Σ is the left adjoint of $\mathfrak{T}$.

9. (Infinite products of modules.) For a graded module G, let ΠG be the module of all sequences $\mathbf{f}$ with $f_p \in G_p$ for each $p \in \mathbf{N}$, under pointwise module operations.

(a) For each p show that $\mathbf{f} \mapsto f_p$ is a morphism $\rho_p:\Pi G \to G_p$ of modules.

(b) If A is a module and $\mathbf{s}$ a sequence of morphisms $s_p:A \to G_p$ of modules, all with domain A, prove that there is a unique morphism $t:A \to \Pi G$ of modules with $\rho_p \circ t = s_p$ for every p.

(c) Deduce that a product of infinitely many modules need not be isomorphic to the coproduct of the same modules.

4. Graded Algebras

An algebra (over the fixed ring K) was defined in §IX.12 to be a module A which is also a ring in such a way that the product $(a, b) \mapsto ab$ in the ring satisfies $(ab)\kappa = a(b\kappa) = (a\kappa)b$ for all scalars κ. This condition, with the distributive law for multiplication, states that the product $(a, b) \mapsto ab$ is a bilinear function $A \times A \to A$. By the universality of the tensor product, this means that the product determines a unique morphism $\pi: A \otimes A \to A$ of modules with $\pi(a \otimes b) = ab$ for all $a, b \in A$. The unit element 1 of the ring also determines a morphism $u: K \to A$ of modules, by $u(\kappa) = 1\kappa \in A$. The remaining ring axioms, requiring that multiplication be associative and that 1 be both a left identity and a right identity for multiplication, are expressed exactly by the commutativity of three diagrams

$$\begin{array}{ccc} A \otimes A \otimes A & \xrightarrow{1 \otimes \pi} & A \otimes A \\ {\scriptstyle \pi \otimes 1}\downarrow & & \downarrow{\scriptstyle \pi} \\ A \otimes A & \xrightarrow{\pi} & A, \end{array} \qquad \begin{array}{ccccc} K \otimes A & = & A & = & A \otimes K \\ {\scriptstyle u \otimes 1}\downarrow & & \downarrow{\scriptstyle 1} & & \downarrow{\scriptstyle 1 \otimes u} \\ A \otimes A & \xrightarrow{\pi} & A & \xleftarrow{\pi} & A \otimes A. \end{array} \tag{14}$$

(The equalities at the top in the right-hand diagram are the identifications of (13).) In brief (Theorem IX.23), an algebra A is a module together with two morphisms $\pi: A \otimes A \to A$ and $u: K \to A$ of modules such that the diagrams (14) commute.

A graded algebra is now defined in exactly the same way: A *graded algebra* A is a graded module with two morphisms $\pi: A \otimes A \to A$ and $u: K \to A$ of graded modules such that the diagrams (14) commute.

This is a definition by diagrams, in which no elements but only arrows appear. It may be recast as a more conventional description in terms of elements:

Proposition 5. *Let A be a graded module with a selected element $1 \in A_0$ and a given function assigning to each pair of elements a, b of A an element $ab \in A$, called their "product", so as to satisfy the conditions:*

(i)	*degree* $(ab) =$ *degree* $a +$ *degree* b,	*for all* $a, b \in A$;
(ii)	$a(b_1 + b_2) = ab_1 + ab_2$,	
	$(b_1 + b_2)a = b_1a + b_2a$,	*if* $\deg b_1 = \deg b_2$;
(iii)	$(ab)\kappa = a(b\kappa) = (a\kappa)b$,	*for all* $a, b \in A, \kappa \in K$;
(iv)	$a(bc) = (ab)c$,	*for all* $a, b, c \in A$;
(v)	$1a = a = a1$,	*for all* $a \in A$.

Then there are unique morphisms $\pi: A \otimes A \to A$ and $u: K \to A$ of graded modules such that $\pi(a \otimes b) = ab$ for all $a, b \in A$ and $u(\kappa) = 1\kappa$ for all $\kappa \in K$. With these morphisms, A is a graded algebra, and any pair of morphisms π, u making A a graded algebra arises in this way.

Proof: Suppose first that A is a graded algebra, as defined by the diagrams (14). For any two elements $a, b \in A$, take the product ab to be the image $\pi(a \otimes b)$ of $a \otimes b$ under the morphism π; this product does satisfy the conditions (i)–(iv). Next let 1 be the image of the unit of K under the given morphism u. As K is concentrated in degree zero, 1 is an element in A_0, while by the commutativity of the diagram (14) it satisfies condition (v).

Conversely, let (i)–(v) hold. The first three of these conditions together state that $(a, b) \mapsto ab$, for each pair of degrees p and q, is a bilinear function $\pi_{p,q}: A_p \times A_q \to A_{p+q}$. These bilinear functions then combine by Proposition 4 to determine a unique morphism $\pi: A \otimes A \to A$ with $\pi(a \otimes b) = ab$. Similarly, $1 \in A_0$ determines a morphism $u: K \to A$ with $u(\kappa) = 1\kappa$. The final conditions (iv) and (v) assert exactly that the diagrams (14) commute for these morphisms π and u, so that A is a graded algebra, as required.

A *graded subalgebra* of a graded algebra A is a graded submodule S of A which contains $1 \in A_0$ and is closed under multiplication. For example, A_0 is a graded subalgebra, one which is concentrated in degree 0.

To construct examples we shall write Ke for the free K-module on one generator e; in particular, K is (as usual) the free K-module generated by the unit $1 \in K$.

The *Grassmann algebra* $G^{(1)}$ on one generator is the graded module

$$G^{(1)} = (K, Ke, 0, 0, \cdots), \tag{15}$$

with unit 1 and product given by the product in K, bilinearity, and the rules $1e = e1 = e$ and $ee = 0$. For example, this means that the product mapping $K \times Ke \to Ke$ is the bilinear function $(\kappa, \lambda e) \mapsto \kappa\lambda e$. This product is manifestly associative and makes G a graded algebra.

The *Grassmann algebra* $G^{(2)}$ on two generators e and f is the graded module

$$G^{(2)} = (K, Ke \oplus Kf, Kef, 0, 0, \cdots), \tag{16}$$

with unit 1 and product given by the product in K, bilinearity, and the rules

$$1e = e1 = e, \qquad 1f = f1 = f, \qquad e^2 = f^2 = 0, \qquad fe = -ef.$$

For example, these rules determine the product of any two elements of degree 1 as $(\kappa e + \lambda f)(\mu e + \nu f) = (\kappa\nu - \lambda\mu)ef$. In particular, this shows that the square of any element of degree 1 (like e^2 and f^2) is zero. One readily sees that this product is associative and makes $G^{(2)}$ a graded algebra.

The *graded polynomial algebra* $P = P_K{}^{(1)}$ on one generator x of degree 2 is the graded module with $P_{2n+1} = 0$ and $P_{2n} = Kx_n$ for all n, so that

$$P = (Kx_0, 0, Kx_1, 0, Kx_2, 0, \cdots), \tag{17}$$

and with product given by the bilinear functions $Kx_p \times Kx_q \to Kx_{p+q}$ defined by $(\kappa x_p, \lambda x_q) \mapsto \kappa\lambda x_{p+q}$. This product is clearly associative and has $x_0 = 1$ as unit, while, for $p > 0$, each x_p is just the pth power x^p. Hence, setting $x = x_1$ and $x_p = x^p$, P is the graded algebra consisting of all the monomials κx^p in the usual polynomial ring $K[x]$ of §III.6.

The *graded polynomial algebra* $P = P_K^{(2)}$ on two generators, each of degree 2, starts with the graded module P for which each P_{2n+1} is zero and each P_{2n} the free module on $n+1$ free generators $e_{n,0}, e_{n-1,1}, \cdots, e_{0,n}$. The product $e_{p,q}e_{r,s} = e_{p+r,q+s}$ thus defines a bilinear function $P_{2n} \times P_{2m} \to P_{2(n+m)}$; this product is associative, makes $e_{0,0}$ the identity, and so makes P a graded algebra. If we set $x = e_{1,0}$ and $y = e_{0,1}$, then each $e_{p,q}$ is the product $x^p y^q$; in this notation, with scalars κ_i, the elements of P_{2n} are

$$\kappa_n x^n + \kappa_{n-1}x^{n-1}y + \kappa_{n-2}x^{n-2}y^2 + \cdots + \kappa_0 y^n;$$

these combinations of x and y are usually called the "homogeneous" polynomials of degree n with coefficients κ_i in K. Moreover, the product in P of two such homogeneous polynomials is exactly the usual product of such polynomials in the polynomial ring $K[x, y]$ of §IV.4.

For two elements a and b of odd degree in a graded algebra, it is often the case that $ab = -ba$, as for instance with $ef = -fe$ in the Grassmann algebra above. For this reason, a graded algebra A is said to be *commutative* (sometimes "skew commutative") when for all elements a and b of the algebra

$$a_p b_q = (-1)^{pq} b_q a_p, \qquad a_p \in A_p, \quad b_q \in A_q. \tag{18}$$

In words: To interchange the order of two elements in a commutative graded algebra, affix the sign -1 if both degrees are odd, otherwise the sign $+1$. All the Grassmann and polynomial algebras described above are commutative in this sense.

The *graded algebra of multilinear forms* on a given module C provides an example of a graded algebra which is not commutative. Indeed, for $p > 0$ let C^p be the p-fold product set $C \times \cdots \times C$. The set $M_p(C)$ of all multilinear forms $h: C^p \to K$ is then a module under the usual pointwise operations; explicitly, each linear combination $h_1\kappa_1 + h_2\kappa_2$ is defined by

$$(h_1\kappa_1 + h_2\kappa_2)(c_1, \cdots, c_p) = h_1(c_1, \cdots, c_p)\kappa_1 + h_2(c_1, \cdots, c_p)\kappa_2. \tag{19}$$

We extend this definition to the case $p = 0$ by agreeing that a function $h: C^0 \to K$ (that is, a function of $p = 0$ variables in C) is just a function $h: 1 \to K$; that is, an element of K. We take *all* such h to be "multilinear", so that $M_0(C) \cong K$. The sequence $(M_0(C), M_1(C), \cdots, M_p(C), \cdots)$ is thus a graded module. Given two elements $h \in M_p(C)$ and $k \in M_q(C)$ of this module, define their product to be the function hk on C^{p+q} for which

$$(hk)(c_1, \cdots, c_{p+q}) = h(c_1, \cdots, c_p)k(c_{p+1}, \cdots, c_{p+q})$$

for all $c_i \in C$. Then $hk: C^{p+q} \to K$ is multilinear, hence an element of $M_{p+q}(C)$ (even if $p = 0$ or $q = 0$), while the assignment $(h, k) \mapsto hk$ is a bilinear function $M_p \times M_q \to M_{p+q}$. As this product hk is manifestly associative and has the identity element of K as its unit, the sequence $M(C)$ is a graded algebra, called the algebra of all multilinear forms on the module C.

The graded algebras are the objects of a category in which the morphisms are defined in the expected way: A morphism $t: A \to B$ of graded algebras is a morphism t of graded modules such that both the diagrams

$$\begin{array}{ccc} A \otimes A & \xrightarrow{\pi} & A \\ {\scriptstyle t \otimes t}\downarrow & & \downarrow{\scriptstyle t} \\ B \otimes B & \xrightarrow{\pi} & B, \end{array} \qquad \begin{array}{ccc} K & \xrightarrow{u} & A \\ {\scriptstyle 1}\downarrow & & \downarrow{\scriptstyle t} \\ K & \xrightarrow{u} & B \end{array} \tag{20}$$

commute. In terms of elements, a morphism $t: A \to B$ of graded algebras is thus a sequence of morphisms $t_p: A_p \to B_p$ of modules with $t_{p+q}(ab) = (t_p a)(t_q b)$ for all p, all q, and all $a \in A_p$ and all $b \in A_q$, and with $t_0(1_A) = 1_B$, where 1_A and 1_B are the unit elements of the graded algebras A and B, respectively.

The kernel of any morphism $t: A \to B$ of graded algebras is an ideal in A. Here a (two-sided) *ideal* in a graded algebra A is defined to be a graded submodule D of A such that

$$d \in D \text{ and } a \in A \quad \Rightarrow \quad ad \in D \text{ and } da \in D.$$

If D is an ideal in the graded algebra A and $\rho: A \to A/D$ the projection on the graded quotient module, there is exactly one way in which to define a product on A/D so that A/D becomes a graded algebra and ρ a morphism of such algebras. Namely, given elements a, b in A, take the product $(\rho a)(\rho b)$ to be $\rho(ab)$. The usual calculation shows the definition unambiguous: If $\rho a = \rho a'$ and $\rho b = \rho b'$, then $a - a' = d \in D$ and $b - b' = e \in D$, so $ab - a'b' = a'e + db' + de$ is in D because D is an ideal; hence, $\rho(ab) = \rho(a'b')$. The essential fact about this *graded quotient algebra* A/D is that the projection $\rho: A \to A/D$ is universal for the functor which assigns to each graded algebra B the set of all those morphisms $t: A \to B$ of graded algebras with kernel containing D. The proof of this fact is straightforward (Exercise 7).

This construction of graded quotient algebras does include the quotient rings defined in §III.3; this is just the special case when $K = \mathbf{Z}$ and the algebra A is concentrated in degree zero.

For example, let n be a fixed positive integer. Then in any graded algebra A the sequence $D^{(n)} = (0, \cdots, 0, A_n, A_{n+1}, A_{n+2}, \cdots)$ formed by taking all elements in A of degree n or greater is clearly an ideal in A. The corresponding quotient algebra $A/D^{(n)}$ may be called the algebra A "truncated" at degree n.

The set of all ideals D in a given graded algebra A is partly ordered by inclusion, and the intersection $D \cap D'$ of two ideals (in more detail, the sequence $p \mapsto D_p \cap D'_p$ of intersections of modules) can be shown to be an ideal in A. Similarly, the intersection of any set of ideals of A is an ideal of A. In particular, if X is any subset (of the set of elements) of A, the intersection of all those ideals of A which contain X is an ideal of A, called the ideal *generated* by the subset X.

Lemma. *If X is any set of elements of a graded algebra A, then, in the ideal $D = D(X)$ of A generated by X, each D_p is the set of all finite sums*

$$a_1x_1b_1 + a_2x_2b_2 + \cdots + a_nx_nb_n \tag{21}$$

with $a_i, b_i \in A$, each $x_i \in X$, and each $a_ix_ib_i$ of degree p.

Proof: Each set D_p, as described in (21), is a submodule of A_p. If one of the elements (21) of D_p is multiplied, on the left or on the right, by any element $c \in A_q$, the product is clearly an element of D_{p+q}. Therefore, the sequence D is an ideal of A. This ideal contains all the elements $x \in X$; on the other hand, any ideal containing X must contain all products axb, all their sums, and hence all elements of D. Therefore D is, as required, the ideal of A generated by X.

This description is convenient for the explicit construction of ideals. If $D(X)$ is the ideal generated by X, the projection $A \to A/D(X)$ is universal for those morphisms $t:A \to B$ of graded algebras with $t(x) = 0$ for every $x \in X$.

Appendix: *Internally Graded Algebras.* At the end of §3 we constructed a functor Σ taking each graded module G to an "internally graded" module ΣG. If $G = A$ is a graded algebra, the corresponding module ΣA can be made into an (ordinary) algebra. Indeed, each element $\mathbf{f} \in \Sigma A$ is a sequence $\mathbf{f}$ with finite support and with $f_p \in A_p$. Then ΣA does become an algebra when the product of two sequences $\mathbf{f}$ and $\mathbf{g}$, each with finite support, is defined from the given products $A_p \otimes A_q \to A_{p+q}$ to be the sequence $\mathbf{fg}$ with nth term

$$(\mathbf{fg})_n = f_0g_n + f_1g_{n-1} + \cdots + f_ng_0.$$

In particular, for the graded polynomial algebra $P^{(1)}$, $\Sigma P^{(1)}$ then is the ordinary $K[x]$, for this product definition is precisely the convolution product used in our original definition of the polynomial ring $K[x]$ (III.28).

EXERCISES

1. Describe the Grassmann algebra on three generators, giving a detailed proof of the associative law.

2. (a) Construct a graded algebra H with each $H_p = Ku_p$ free on one generator u_p and with product $u_pu_q = u_{p+q}$ if one at least of p or q is even and $u_pu_q = 0$ when both p and q are odd.

(b) Show that this algebra H is commutative and contains both the algebras $G^{(1)}$ and $P^{(1)}$ as subalgebras.

3. Let D be the graded $\mathbf{Z}$-module with $D_{2p+1} = 0$ for every p and with each D_{2p} a free $\mathbf{Z}$-module on one generator t_p.

(a) Show that $t_p t_q = (p, q)t_{p+q}$, with (p, q) the binomial coefficient, makes D a commutative graded $\mathbf{Z}$-algebra. (For notation, see Exercise III.1.9.)

(b) Show that $t_p \mapsto x^p/p!$ defines a monomorphism $D \to P_{\mathbf{Q}}{}^{(1)}$ of graded $\mathbf{Z}$-algebras (regarding the $\mathbf{Q}$-algebra $P_{\mathbf{Q}}{}^{(i)}$ of §4 as a $\mathbf{Z}$-algebra). (*Note:* The algebra D, called the polynomial algebra on one generator with "divided powers", arises in topological studies.)

4. If $C = K$ is the free K-module on one generator, show that the algebra $M(C)$ of multilinear forms has each $M_p(C)$ free on one generator u_p, with product given by $u_p u_q = u_{p+q}$.

5. Show that $C \mapsto M(C)$ is the object function of a contravariant functor on modules to graded algebras.

6. If the ring K of scalars is a field F, show that every ideal in the graded polynomial algebra $P_F{}^{(1)}$ is generated by some power x^n.

7. For D an ideal in a graded algebra A, prove the universality, as stated in the text, of the projection $A \to A/D$.

8. For X a set of elements in a graded algebra A, prove the universality, as stated in the text, of the projection $A \to A/D(X)$.

9. Prove that the ideals D in a graded algebra form a modular lattice under inclusion.

10. Define a product $A \times A'$ of two graded algebras A and A' and prove that the projections $A \times A' \to A$ and $A \times A' \to A'$ are universal.

11. Show in detail that Σ is a functor on graded algebras to algebras.

12. Let A and A' be commutative graded algebras. Construct a commutative graded algebra C and morphisms $j: A \to C$, $j': A' \to C$ of graded algebras with the following universal property: To every pair $s: A \to B$, $s': A' \to B$ of morphisms of graded algebras to a commutative graded algebra B there is a unique morphism $t: C \to B$ of graded algebras with $t \circ j = s$ and $t \circ j' = s'$. (*Hint:* As a graded module, C is $A \otimes A'$.)

13. The algebra C of Exercise 12, unique up to isomorphism by the uniqueness of universals, is called the "tensor product" $C = A \otimes A'$ of graded algebras:

(a) For Grassmann algebras, prove $G^{(1)} \otimes G^{(1)} \cong G^{(2)}$.

(b) For polynomial algebras, prove $P^{(1)} \otimes P^{(1)} \cong P^{(2)}$.

(c) For any three commutative graded algebras A, A', and A'', show that there is an isomorphism $A \otimes (A' \otimes A'') \cong (A \otimes A') \otimes A''$ of graded algebras.

5. The Graded Tensor Algebra

The graded tensor module $T(C)$ of the module C, as defined in (10), has $T_0(C) = K$ and $T_p(C) = C \otimes \cdots \otimes C$, with p factors C. Since p factors C

followed by q factors C makes $p + q$ factors C, there is an isomorphism

$$\phi_{p,q} : T_p(C) \otimes T_q(C) \cong T_{p+q}(C) \tag{22}$$

of modules. In detail, for $p = 0$, $T_0(C) = K$ and $\phi_{0,q}$ is the isomorphism $K \otimes T_q \cong T_q$ of (13). If $p > 0$ and $q > 0$, we take lists $c_1, \cdots, c_p$ and $c'_1, \cdots, c'_q$ of elements of C and observe that

$$(c_1, \cdots, c_p, c'_1, \cdots, c'_q) \mapsto (c_1 \otimes \cdots \otimes c_p) \otimes (c'_1 \otimes \cdots \otimes c'_q) \in T_p \otimes T_q,$$
$$(c_1, \cdots, c_p, c'_1, \cdots, c'_q) \mapsto c_1 \otimes \cdots \otimes c_p \otimes c'_1 \otimes \cdots \otimes c'_q \in T_{p+q}$$

are both universal $(p + q)$-linear functions. Hence, by the uniqueness of universals, there is an isomorphism $\phi_{p,q}: T_p \otimes T_q \cong T_{p+q}$ taking the first function to the second one, so that

$$\phi_{p,q}((c_1 \otimes \cdots \otimes c_p) \otimes (c'_1 \otimes \cdots \otimes c'_q)) = c_1 \otimes \cdots \otimes c_p \otimes c'_1 \otimes \cdots \otimes c'_q.$$

This isomorphism is used to turn the graded module $T(C)$ into a graded algebra, defining the product of $u \in T_p$ and $v \in T_q$ to be $uv = \phi_{p,q}(u \otimes v)$. Thus, if $u = c_1 \otimes \cdots \otimes c_p$ and $v = c'_1 \otimes \cdots \otimes c'_q$, while $\kappa, \lambda \in K$,

$$(c_1 \otimes \cdots \otimes c_p)(c'_1 \otimes \cdots \otimes c'_q) = c_1 \otimes \cdots \otimes c_p \otimes c'_1 \otimes \cdots \otimes c'_q, \tag{23}$$

$$(c_1 \otimes \cdots \otimes c_p)\kappa = c_1\kappa \otimes \cdots \otimes c_p\kappa, \tag{24}$$
$$\lambda(c'_1 \otimes \cdots \otimes c'_q) = \lambda c'_1 \otimes \cdots \otimes c'_q.$$

In brief, to *multiply two elements* u and v in $T(C)$, simply *place* the *symbol* $\otimes$ *between them*; this multiplication is associative and has the unit of $K = T_0$ as its unit. Hence, $T(C)$ is a graded algebra which we call the *graded tensor algebra* (in brief, the "tensor algebra") of the module C. We note that T is a functor on modules to graded algebras.

In this tensor algebra $T(C)$, $T_1(C) = C$ is the given module C, and every element $c_1 \otimes \cdots \otimes c_p$ is a p-fold product $c_1 \cdots c_p$ of elements c_i of C. Thus any element of $T_p(C)$ is a sum of products of elements of C, so $T(C)$ is generated as a graded algebra by the elements of C.

To illustrate this tensor algebra, take $C = K$ to be the free module on one generator 1; then $T_p(K) = K \otimes \cdots \otimes K \cong K$ is also free on one generator $1 \otimes \cdots \otimes 1$. If we call this generator y_p, the product of two generators is $y_p y_q = y_{p+q}$, so each $y_p = y_1^p$, and $T(K)$ is the familiar graded algebra $(K, Ky_1, Ky_1^2, Ky_1^3, \cdots)$.

If $C = Ka \oplus Kb$ is free on two generators a and b, then $T_2(C)$ is free on four generators $a^2 = a \otimes a$, $ab = a \otimes b$, $ba = b \otimes a$, and $b^2 = b \otimes b$; in particular $ab \neq \pm ba$. Similarly, $T_3(C)$ is free on eight generators a^3, a^2b, aba, ba^2, ab^2, bab, b^2a, and b^3, while each $T_p(C)$ is the free module with generators all the 2^p formally different products of p letters, each letter being either a or b. Thus, $T(C)$ is the "universal" (noncommutative) graded algebra with two generators a and b of degree 1.

Let C be the free module with basis $b_1, \cdots, b_n$. Then, by Proposition IX.16, $T_2(C) = C \otimes C$ is the free module with basis the n^2 elements $b_{i_1} \otimes b_{i_2}$ and each $T_p(C)$ is the free module with basis the n^p elements $b_{i_1} \otimes b_{i_2} \otimes \cdots \otimes b_{i_p}$, one for each list $\mathbf{i}:\mathbf{p} \to \mathbf{n}$. The product (23) of two of these basis elements is

$$(b_{i_1} \otimes \cdots \otimes b_{i_p})(b_{j_1} \otimes \cdots \otimes b_{j_q}) = b_{i_1} \otimes \cdots \otimes b_{i_p} \otimes b_{j_1} \otimes \cdots \otimes b_{j_q}.$$

To find the product of any two elements $r \in T_p(C)$ and $s \in T_q(C)$ take their coordinates ρ and σ relative to these bases and write r and s, using the summation convention of §2, as

$$r = (b_{i_1} \otimes \cdots \otimes b_{i_p})\rho^{i_1 \cdots i_p}, \qquad s = (b_{j_1} \otimes \cdots \otimes b_{j_q})\sigma^{j_1 \cdots j_q}.$$

The product tensor rs is therefore the element with coordinates

$$\rho^{i_1 \cdots i_p} \sigma^{j_1 \cdots j_q}, \tag{25}$$

rel $\mathbf{b}$. This applies in particular when C is a vector space. It shows that two contravariant tensors of types $(0, p)$ and $(0, q)$, respectively, have as product a contravariant tensor of type $(0, p + q)$. This product is defined by (23) and bilinearity, independent of any choice of coordinates; relative to any basis of C, the coordinates of the product tensor are the products of coordinates as in (25).

This tensor algebra $T(C)$ is generated by the n basis elements $b_1, \cdots, b_n$; since all formally different products $b_{i_1} \otimes \cdots \otimes b_{i_p}$ of these generators are linearly independent, we might describe $T(C)$ as the "most general" (noncommutative) graded algebra generated by the n elements b_i; that is, by the module C. This observation may be stated as a universality property as follows.

Theorem 6. *Let A be a graded algebra, so that A_1 denotes the module of all elements of degree 1 in A. Then to any morphism $s:C \to A_1$ of modules there is a unique morphism $t:T(C) \to A$ of graded algebras with $t_1 = s$.*

The graded algebras $T(C)$ and A, displayed as sequences of modules, form the diagram

$$\begin{array}{ccccccc} T(C) = (K, & C, & T_2(C), & \cdots, & T_p(C) = C \otimes \cdots \otimes C, & \cdots) \\ \vdots t & \downarrow u & \downarrow s = t_1 & \vdots t_2 & \vdots t_p & \\ A \;\; = (A_0, & A_1, & A_2, & \cdots, & A_p, & \cdots) \end{array}$$

with given morphisms u (on the units) and s; we are required to fill in the morphisms $t_2, \cdots, t_p, \cdots$ so as to get a morphism of graded algebras.

Proof: The unit of $T(C)$ is that of $K = T_0(C)$ and the product in $T(C)$ is the tensor product. Hence, any such morphism t of algebras must have $t_0(\kappa) = \kappa 1 = u(\kappa)$ for each scalar κ and

$$t_p(c_1 \otimes \cdots \otimes c_p) = (sc_1) \cdots (sc_p), \qquad c_i \in C, \tag{26}$$

for each $p > 0$. By this formula t, if it exists, is unique. Conversely, $(c_1, \cdots, c_p) \mapsto (sc_1) \cdots (sc_p)$ is a multilinear function while $(c_1, \cdots, c_p) \mapsto c_1 \otimes \cdots \otimes c_p$ is a universal multilinear function. Therefore the formula (26) above does define for each p a unique morphism $t_p : T_p(C) \to A_p$ of modules. Then the formula (23) for the product in $T(C)$ shows that

$$t_{p+q}((c_1 \otimes \cdots \otimes c_p)(c_1' \otimes \cdots \otimes c_q')) = t_p(c_1 \otimes \cdots \otimes c_p) t_q(c_1' \otimes \cdots \otimes c_q').$$

Therefore, t is a morphism of graded algebras, as required.

This theorem is a universality statement. For a fixed module C, let $\mathfrak{F}$ be the functor on graded algebras A to sets which assigns to each graded algebra A the set hom (C, A_1) of all morphisms $C \to A_1$ of modules. The theorem asserts that the pair $(1_C, T(C))$ is a universal element for this functor $\mathfrak{F}$. As we now know, by Theorem XV.3 this amounts to stating that the functor $\mathfrak{F}$ is representable; that is, $t \mapsto t_1$ gives a bijection

$$\hom_{\mathbf{Gr\ alg}}(T(C), A) \cong \hom_{\mathbf{Mod}}(C, A_1),$$

natural in A; on the left, "hom" stands for morphisms of graded algebras; on the right, for morphisms of modules. This is a representation with C as a parameter, while $A \mapsto A_1$ is a functor $\mathfrak{U}_1$ from graded algebras to modules (the functor "take the module of all elements of degree 1"). Thus our theorem can also be read as follows: The functor T from modules to graded algebras is adjoint to the functor $\mathfrak{U}_1$ from graded algebras to modules.

EXERCISES

1. Show that the Grassmann algebra on one generator is a quotient of the tensor algebra $T(Kb)$ on a free module with one generator b.
2. Show that the Grassmann algebra on two generators is a quotient of the tensor algebra $T(Ka \oplus Kb)$. What is the corresponding ideal of the tensor algebra?

For Exercises 3–7, we construct to each module C the graded module $U(C) = T(C) \otimes T(C^*)$ with $U_0(C) = K$ and

$$U_n(C) = [T_n(C) \otimes T_0(C^*)] \oplus [T_{n-1}(C^*) \otimes T_1(C)] \oplus \cdots \oplus [T_0(C) \otimes T_n(C^*)].$$

3. For all natural numbers p, q, p', q' establish an isomorphism

$$[T_p(C) \otimes T_q(C^*)] \otimes [T_{p'}(C) \otimes T_{q'}(C^*)] \cong T_{p+p'}(C) \otimes T_{q+q'}(C^*).$$

4. Use the isomorphism of Exercise 3 to define a product in $U(C)$ and show that this product makes $U(C)$ a graded algebra.

5. Show that the graded algebra $U(C)$ of Exercise 4 contains both the graded algebras $T(C)$ and $T(C^*)$ as graded subalgebras.

6. Show that U, as in Exercise 4, may be regarded as a functor to the category of graded algebras from the category whose objects are all modules C and whose morphisms are all isomorphisms of modules.

7. If K is a field F and $C = V$ a finite-dimensional vector space over F, then $U(V)$ is the algebra of all tensors (of all types (p, q)) over V. Given two tensors r and s, describe the coordinates of their product rs in $U(V)$ in terms of the coordinates of r and s, all relative to some one basis of V.

6. The Exterior Algebra of a Module

We next study graded algebras A over K with the property

$$a \in A_1 \quad \Rightarrow \quad a^2 = 0: \tag{27}$$

All squares of elements of degree 1 are zero. The two Grassmann algebras described in §4 are typical examples of such algebras. Any commutative graded algebra A over a field $K = F$ of characteristic not 2 has this property (27), for degree $a = 1$ implies $aa = -aa$ by commutativity, hence $2a^2 = 0$ and so $a^2 = 0$ because the scalar 2 has an inverse in F.

For each K-module C we shall now construct a graded algebra A with $A_1 = C$ and the property (27). In the tensor algebra $T(C)$, let $D(C^2)$ be the ideal generated by all squares c^2 of elements of C, and define the *exterior algebra* of the module C to be the graded quotient algebra

$$\Lambda(C) = T(C)/D(C^2). \tag{28}$$

It is customary to write the binary operation of multiplication for two elements $a, b \in \Lambda(C)$ as $(a, b) \mapsto a \wedge b$, and to call this operation the *exterior product*.

Proposition 7. *For any K-module C, the (graded) exterior algebra $\Lambda(C)$ has $\Lambda_0(C) = K$, is generated by the set $\Lambda_1(C) = C$ of its elements of degree 1, is commutative, and has the square $a \wedge a$ of every element a of odd degree zero.*

Proof: Since the ideal $D(C^2)$ is generated by elements c^2 of degree 2, it contains no non-zero elements of degree less than 2. Hence, $\Lambda_1(C) =$

$T_1(C)/D_1 = T_1(C) = C$. In the tensor algebra, every element is a sum of products of elements of degree one. Therefore, this is also the case in the quotient algebra $\Lambda(C)$, so an element of degree p in $\Lambda(C)$ is a sum of products of the form $c_1 \wedge \cdots \wedge c_p$ for $c_i \in C = \Lambda_1(C)$. Hence, $\Lambda(C)$ is generated as a graded algebra by C. Since $c^2 = c \wedge c = 0$ by definition, any two elements $c_1, c_2 \in C$ satisfy

$$\begin{aligned} 0 = (c_1 + c_2)^2 &= c_1{}^2 + c_1 \wedge c_2 + c_2 \wedge c_1 + c_2{}^2 \\ &= c_1 \wedge c_2 + c_2 \wedge c_1. \end{aligned}$$

Therefore, $c_1 \wedge c_2 = -c_2 \wedge c_1$; in other words, any two elements of degree 1 commute, with the correct sign -1, since both degrees are odd. By induction on p,

$$(c_1 \wedge \cdots \wedge c_p) \wedge c = (-1)^p c \wedge (c_1 \wedge \cdots \wedge c_p),$$

because c is interchanged with p terms c_i of degree 1; by another induction

$$\begin{aligned} &(c_1 \wedge \cdots \wedge c_p) \wedge (c_1' \wedge \cdots \wedge c_q') \\ &\qquad = (-1)^{pq}(c_1' \wedge \cdots \wedge c_q') \wedge (c_1 \wedge \cdots \wedge c_p). \end{aligned}$$

As any element of degree p is a sum of such elements $c_1 \wedge \cdots \wedge c_p$, this shows that the algebra $\Lambda(C)$ is commutative.

It remains to show that the square of any element a of odd degree is zero. By definition, this is true for any element c of degree 1. If a is a product $c_1 \wedge \cdots \wedge c_p$, then $a^2 = a \wedge a$ has a factor $c_1 \wedge c_1 = 0$, hence is zero. If a is a sum of such products, as $a = a_1 + \cdots + a_k$, then

$$a^2 = \sum a_i^2 + \sum_{i<j} (a_i \wedge a_j + a_j \wedge a_i) = \sum_{i<j} (a_i \wedge a_j - a_i \wedge a_j) = 0$$

becasue a_i and a_j both have odd degree. This completes the proof.

As an example, let C be the free module Kb on one generator b. Then any element $c_1 \wedge c_2$ in $\Lambda(C)$ is zero, for each c_i is $c_i = \kappa_i b$ for some scalar κ_i, and $\kappa_1 b \wedge \kappa_2 b = \kappa_1\kappa_2(b \wedge b) = 0$. Thus $\Lambda(C)$ is the graded module

$$\Lambda(Kb) = (K, Kb, 0, 0, \cdots)$$

with product given by $b^2 = 0$; it is isomorphic to the Grassmann algebra in one generator, as described in (15).

Next let the module $C = Kb_1 \oplus Kb_2$ be free on two generators b_1 and b_2. Since each $\Lambda_p(C)$ is spanned by p-fold products $c_1 \wedge \cdots \wedge c_p$ of elements of C and each c_i is a linear combination of b_1 and b_2, $\Lambda_p(C)$ is spanned by the products of p factors, with each factor b_1 or b_2. But $b_1{}^2 = 0 = b_2{}^2$, so any product of three or more such factors must be zero. Hence, $\Lambda_p(C) = 0$ for $p > 2$, while $\Lambda_2(C)$ is spanned by one product $b_1 \wedge b_2$. This element

$b_1 \wedge b_2$ is *not* zero. To see this, note that $\Lambda_2(C)$ by definition is the quotient module $T_2(C)/D_2$, where D_2 is the module of elements of degree 2 in the ideal $D(C^2)$ generated by all squares c^2 in $T(C)$. This means that D_2 is the set of all sums of such squares c^2. But $T_2(C)$ is free on four generators $b_1^2, b_1b_2, b_2b_1, b_2^2$, and hence also free on the four generators, b_1^2, $b_1b_2 + b_2b_1, b_2^2, b_1b_2$, while any square c^2 in $T_2(C)$ is a linear combination of the first three of these generators. Therefore, $b_1b_2 \notin D_2$, so that $b_1 \wedge b_2$ is non-zero in $T_2(C)/D_2$. Thus $\Lambda(C)$ is the graded module

$$\Lambda(Kb_1 \oplus Kb_2) = (K, Kb_1 \oplus Kb_2, K(b_1 \wedge b_2), 0, 0, \cdots)$$

with product given by $b_1^2 = b_2^2 = 0$ and $b_1 \wedge b_2 = -\, b_2 \wedge b_1$; it is isomorphic to the Grassmann algebra on two generators, as described in (16).

Before making a similar analysis of free modules with more generators it will be convenient to formulate a universal property for the exterior algebra $\Lambda(C)$.

Theorem 8. *If S is a graded algebra with $s^2 = 0$ for every element $s \in S_1$ of degree 1 while $h: C \to S_1$ is a morphism of modules, there is a unique morphism $t: \Lambda(C) \to S$ of graded algebras with $t_1 = h$.*

Proof: If we think of the ordinary module C as a graded module concentrated in degree 1, we can regard the given morphism $h: C \to S_1$ as a morphism of graded modules, zero in degrees not one. Similarly, the insertion $j: C \to T(C)$ is a morphism of graded modules. So is the projection $\pi: T(C) \to T(C)/D(C^2) = \Lambda(C)$ of the tensor algebra onto its quotient algebra $\Lambda(C)$. Thus we are given the solid arrows j, h, and π in the diagram shown below. By the universal property of the tensor algebra $T(C)$, there is exactly one morphism k of

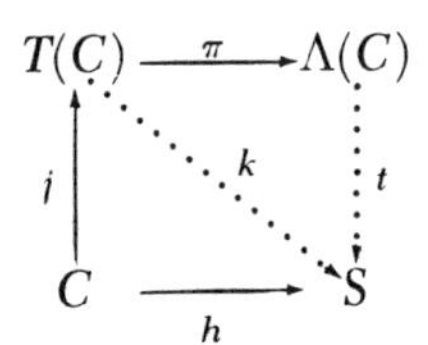

graded algebras (dotted arrow) so that the lower triangle commutes. But $c \in C$ implies that $k(c) \in S_1$; since $S_1^2 = 0$, $k(c^2) = 0$. Therefore, the kernel of k contains the ideal $D(C^2)$ of $T(C)$ generated by all squares c^2. By the universal property of the projection π with this kernel, there is a unique morphism t of graded algebras (dotted vertical arrow) which makes the second triangle commute. This t is the morphism required.

This theorem, in effect, describes a pair of adjoint functors. Assign to each graded algebra S the module S_1; this gives a functor from graded algebras S

with $s^2 = 0$ for $s \in S_1$ to modules. The theorem states that its adjoint is the functor Λ which assigns to each module C the graded algebra $\Lambda(C)$.

For fixed p, the module $\Lambda_p(C) = C \wedge \cdots \wedge C$ of all elements of degree p in the exterior algebra $\Lambda(C)$ also has a universal property. We recall that a multilinear function $h: C^p \to E$ on the p-fold product $C^p = C \times \cdots \times C$ of a module C to another module E is called "alternating" (§IX.1) when

$$i \neq j \text{ and } c_i = c_j \quad \Rightarrow \quad h(c_1, \cdots, c_p) = 0.$$

Since $c \wedge c = 0$, $(c_1, \cdots, c_p) \mapsto c_1 \wedge \cdots \wedge c_p$ is an alternating multilinear function on C^p to $\Lambda_p(C)$. It is the universal such function on C^p:

Theorem 9. *For modules C and E, there is to any alternating multilinear function $h: C^p \to E$ a unique morphism $t: \Lambda_p(C) \to E$ of modules with*

$$t(c_1 \wedge \cdots \wedge c_p) = h(c_1, \cdots, c_p) \tag{29}$$

for all $c_1, \cdots, c_p \in C$.

Proof: We first describe Λ_p directly. By its construction, $\Lambda_p(C)$ is a quotient module of the p-fold tensor product $T_p(C) = C \otimes \cdots \otimes C$. The kernel of the projection $T_p \to \Lambda_p$ is the module of all elements of degree p in the ideal $D(C^2)$ generated by all squares c^2. By the Lemma of §4, this kernel consists of sums of terms ac^2b of degree p, with $a, b \in T(C)$. Expressing a and b as sums of products of elements of C, it follows that the kernel of $T_p \to \Lambda_p$ is the submodule of T_p spanned by all products $c_1 \otimes \cdots \otimes c_p$ of p factors with two successive factors equal.

Now let $h: C^p \to E$ be any alternating multilinear function. Since h is multilinear, it can be written as $h(c_1, \cdots, c_p) = s(c_1 \otimes \cdots \otimes c_p)$ for a unique linear map $s: T_p(C) \to E$, while h alternating means that s vanishes whenever two successive arguments are equal; that is, on the kernel of the projection $T_p(C) \to \Lambda_p(C)$. Therefore, s factors through this projection to give the desired morphism t. (This may be visualized by a diagram like that for the previous theorem.)

Thus $\Lambda_p(C)$ is the codomain of a universal alternating multilinear function $(c_1, \cdots, c_p) \mapsto c_1 \wedge \cdots \wedge c_p$. Since such a universal function is unique, up to isomorphism, the module $\Lambda_p(C)$ could have been constructed directly as the codomain of such a function, without reference to the rest of the exterior algebra $\Lambda(C)$.

Appendix: The *Symmetric Algebra*. The "symmetric" algebra is to be a free commutative graded algebra generated by a given module C of elements of *even* degree 2 (in the exterior algebra, the generators C had *odd* degree 1). To define it, first modify the tensor algebra $T(C)$ by multiplying all degrees by 2; call the resulting graded algebra

$$\hat{T}(C) = (K, O, C, O, C \otimes C, O, C \otimes C \otimes C, \cdots).$$

Now define the "symmetric algebra" of the module C to be the graded quotient algebra $\hat{T}(C)/D$, where D is the ideal generated by all differences $cc' - c'c$ *for* $c, c' \in C$. Some of the basic properties of this algebra $S(C)$ are set forth in the exercises below.

EXERCISES

1. (a) If C is a free module on one generator, show that the symmetric algebra $S(C)$ is isomorphic to the graded polynomial algebra $P^{(1)}$ of (17).

(b) If C is free on two generators, prove that $S(C)$ is isomorphic to the graded polynomial algebra $P^{(2)}$ of §4.

2. For any module C, prove $S(C)$ commutative.

3. If C is a module, A is a graded algebra, and $s: C \to A_2$ is a morphism of modules, show that there exists a unique morphism $t: \hat{T}(C) \to A$ of graded algebras with $t_2 = s$.

4. (a) If A is a commutative graded algebra and $s: C \to A_2$ a morphism of modules, show that there exists a unique morphism $t: S(C) \to A$ of graded algebras with $t_2 = s$.

(b) Deduce that the functor S (on modules to commutative graded algebras) is the adjoint of a suitable functor.

(c) Make the corresponding deduction for Exercise 3.

7. Determinants by Exterior Algebra

The exterior algebra $\Lambda(C)$ on a free module C with basis $b_1, \cdots, b_n$ has an explicit basis like that already described in the special cases when $n = 1$ or $n = 2$. In each degree p, $\Lambda_p(C)$ is spanned by products of p elements of C and hence by products of p of the basis elements b_i. Since $b_i \wedge b_j = - b_j \wedge b_i$ for $i \neq j$, it is enough to consider just products in which the list $\mathbf{k}$ of subscripts is increasing, as in

$$b_{\mathbf{k}} = b_{k_1} \wedge \cdots \wedge b_{k_p}, \qquad 1 \leqslant k_1 < \cdots < k_p \leqslant n.$$

Since the list $\mathbf{k}: \mathbf{p} \to \mathbf{n}$ is "increasing" ($k_i < k_j$ for all $i < j$) it is determined by its image, which is a subset $\{k_1, \cdots, k_p\}$ of p elements in $\mathbf{n}$. The number of such subsets is the binomial coefficient $(p, n - p) = n!/[p!(n - p)!]$.

THEOREM **10.** *If C is a free module with basis $b_1, \cdots, b_n$, its exterior algebra $\Lambda(C)$ is zero in degrees exceeding n, while for degrees p with $0 \leqslant p \leqslant n$, $\Lambda_p(C)$ is a free module with basis the $(p, n - p)$ elements $b_{\mathbf{k}}$, one for each increasing list $\mathbf{k}: \mathbf{p} \to \mathbf{n}$.*

Proof: We shall prove first that $b_1 \wedge \cdots \wedge b_n \neq 0$ in $\Lambda_n(C)$. Indeed, since C is a free module on n generators, we know that there is a non-zero alternating multilinear form $d: C^n \to K$; by Theorem IX.2 the determinant provides such a form. Since $c_1 \wedge \cdots \wedge c_n$ is a universal alternating multilinear function, there is a linear transformation $t: \Lambda_n(C^n) \to K$ with $t(c_1 \wedge \cdots \wedge c_n) = d(c_1, \cdots, c_n)$. In particular,

$$t(b_1 \wedge \cdots \wedge b_n) = d(b_1, \cdots, b_n) = 1 \neq 0,$$

so $b_1 \wedge \cdots \wedge b_n$ must be non-zero.

For any degree p, we suppose now that the elements $b_{\mathbf{k}}$ spanning $\Lambda_p(C)$ are linearly dependent, so that $\Sigma_{\mathbf{k}} b_{\mathbf{k}} \lambda_{\mathbf{k}} = 0$, where the sum is taken over all increasing lists $\mathbf{k}$ of p indices and has some coefficient $\lambda_{\mathbf{h}} \neq 0$. The indices not in the list $\mathbf{h}$ form an increasing list $h_1', \cdots, h_q'$ of $q = n - p$ indices. Then $b_{\mathbf{h}} \wedge b_{\mathbf{h}'} = \pm\, b_1 \wedge \cdots \wedge b_n$, while if $\mathbf{k} \neq \mathbf{h}$, then $b_{\mathbf{k}}$ and $b_{\mathbf{h}'}$ have at least one index in common, so that $b_{\mathbf{k}} \wedge b_{\mathbf{h}'} = 0$. Therefore,

$$0 = b_{\mathbf{h}'} \wedge (\Sigma b_{\mathbf{k}} \lambda_{\mathbf{k}}) = (b_{\mathbf{h}'} \wedge b_{\mathbf{h}}) \lambda_{\mathbf{h}} = \pm\, (b_1 \wedge \cdots \wedge b_n) \lambda_{\mathbf{h}}.$$

But $b_1 \wedge \cdots \wedge b_n \neq 0$, so $\lambda_{\mathbf{h}} = 0$, a contradiction to the supposed linear dependence.

We now arrange the $(p, n - p)$ elements $b_{\mathbf{k}} \in \Lambda_p(C)$ in any way in a list. We have proved that this list spans $\Lambda_p(C)$ and is linearly independent. Hence, it is a basis, as required.

THEOREM 11. *If C is a free module of rank n, then the determinant $|t|$ of each endomorphism $t: C \to C$ satisfies*

$$tc_1 \wedge \cdots \wedge tc_n = (c_1 \wedge \cdots \wedge c_n)|t| \tag{30}$$

for all lists $c_1, \cdots, c_n$ of n elements of C.

Proof: The determinant of an endomorphism t was defined in §IX.4 to be that scalar $|t|$ with

$$d(tc_1, \cdots, tc_n) = d(c_1, \cdots, c_n)|t|$$

for all $c_i \in C$, whenever d is an alternating n-form on C and hence also whenever d is an alternating n-linear function on C to a free module of rank 1. If we take d to be the alternating function $(c_1, \cdots, c_n) \mapsto c_1 \wedge \cdots \wedge c_n$ to the free module $\Lambda_n(C)$, for all $c_i \in C$ and all alternating n-forms d on C, we get the desired formula (30); since $b_1 \wedge \cdots \wedge b_n \neq 0$, this formula does determine $|t|$.

Put differently: $\Lambda_n(C)$ is a functor of C, so each endomorphism t of C induces an endomorphism $\Lambda_n(t)$ of $\Lambda_n(C)$. But $\Lambda_n(C)$ is a free module of rank 1, so this endomorphism $\Lambda_n(t)$ must be just multiplication by some scalar. That scalar is $|t|$.

This theorem will seem more striking when we show that it provides an explicit means of calculating out determinants. Thus let t have matrix A relative to the basis $\mathbf{b}$; this means that $tb_j = \Sigma_i b_i A_{ij}$. The formula (30) for $|t|$ with $\mathbf{c} = \mathbf{b}$ then reads

$$\left(\sum_i b_i A_{i1}\right) \wedge \cdots \wedge \left(\sum_i b_i A_{in}\right) = (b_1 \wedge \cdots \wedge b_n)|A|. \qquad (31)$$

For example, when $n = 2$, multilinearity gives

$$\begin{aligned}(b_1 A_{11} + b_2 A_{21}) &\wedge (b_1 A_{12} + b_2 A_{22}) \\ &= b_1{}^2 A_{11} A_{12} + (b_1 \wedge b_2) A_{11} A_{22} + (b_2 \wedge b_1) A_{21} A_{12} + b_2{}^2 A_{21} A_{22} \\ &= (b_1 \wedge b_2)(A_{11} A_{22} - A_{21} A_{12}).\end{aligned}$$

That is (as it ought to be!) just the usual formula for the determinant of the 2×2 matrix A. The same sort of calculation works for $n \times n$ matrices. For, (31) expanded by multilinearity gives n^n terms, one for each function $f:\mathbf{n} \to \mathbf{n}$, where f selects from the jth factor $\Sigma_i b_i A_{ij}$ the summand $b_i A_{ij}$ with $i = f_j$. Then

$$\sum_f (b_{f_1} \wedge \cdots \wedge b_{f_n})(A_{f_1 1} \cdots A_{f_n n}) = (b_1 \wedge \cdots \wedge b_n)|A|,$$

with sum over all functions $f:\mathbf{n} \to \mathbf{n}$. But any exterior product of b's in which some one b_i appears twice is zero, so there remain just those terms with the function f a permutation $f = \sigma$ of $\mathbf{n}$. Now interchanging two b's changes the sign:

$$b_{\sigma_1} \wedge \cdots \wedge b_{\sigma_n} = (-1)^{\operatorname{sgn} \sigma} b_1 \wedge \cdots \wedge b_n.$$

Thus we have recovered the conventional explicit formula for $|A|$ as

$$\sum_{\sigma \in S_n} (-1)^{\operatorname{sgn} \sigma} A_{\sigma_1 1} \cdots A_{\sigma_n n} = |A|.$$

We consider next the minors of an $n \times m$ matrix $A:\mathbf{n} \times \mathbf{m} \to K$. Let $\mathbf{h}:\mathbf{p} \to \mathbf{n}$ and $\mathbf{k}:\mathbf{p} \to \mathbf{m}$ be two increasing lists of p indices each. Then $\mathbf{h} \times \mathbf{k}:\mathbf{p} \times \mathbf{p} \to \mathbf{n} \times \mathbf{m}$, so the composite $A \circ (\mathbf{h} \times \mathbf{k}):\mathbf{p} \times \mathbf{p} \to K$ is a $p \times p$ matrix. Its entries are $A_{h(i),\, k(j)}$ for $i, j \in \mathbf{p}$, so it may be described as that matrix obtained from A by using only the entries in the rows $h_1 < \cdots < h_p$ and in the columns $k_1 < \cdots < k_p$. A $p \times p$ *minor* of A is by definition the determinant $|A \circ (\mathbf{h} \times \mathbf{k})|$ of such a matrix. These minors arise in the calculation of exterior products as follows:

Lemma. *Let C be a free module of rank n with basis* $\mathbf{b}$, *while* $\mathbf{a}$ *is a list of m elements of C, and A the matrix rel* $\mathbf{b}$ *of this list, in the sense that*

$a_j = \sum_i b_i A_{ij}$ *for* $j \in \mathbf{m}$. *Then, for each increasing list* $\mathbf{k}: \mathbf{p} \to \mathbf{m}$,

$$a_{k_1} \wedge \cdots \wedge a_{k_p} = \sum_{\mathbf{h}} b_{\mathbf{h}} |A \circ (\mathbf{h} \times \mathbf{k})|, \tag{32}$$

with the sum at the right taken over all increasing lists $\mathbf{h}: \mathbf{p} \to \mathbf{n}$.

Proof: Since $\Lambda_p(C)$ has all the products $b_{\mathbf{h}}$ as basis, any exterior product of p elements of C must be a linear combination of these $b_{\mathbf{h}}$. To find the coefficients in this combination, we expand

$$a_{k_1} \wedge \cdots \wedge a_{k_p} = \Big(\sum_i b_i A_{ik_1}\Big) \wedge \cdots \wedge \Big(\sum_i b_i A_{ik_p}\Big).$$

All terms on the right involving $b_{\mathbf{h}} = b_{h_1} \wedge \cdots \wedge b_{h_p}$ will be obtained if we restrict each sum on i to the range $i = h_1, \cdots, h_p$. By the basic formula (31), this is just $b_{\mathbf{h}}|A \circ (\mathbf{h} \times \mathbf{k})|$, Q.E.D.

If $\mathbf{h}: \mathbf{p} \to \mathbf{n}$ is an increasing list of p indices, the $n - p$ indices remaining form another increasing list, the *complement* $\mathbf{h}': \mathbf{n} - \mathbf{p} \to \mathbf{n}$ of $\mathbf{h}$: here $\mathbf{n} - \mathbf{p}$ is $\{1, \cdots, n - p\}$. Together, the product of two so-indexed exterior products is

$$(u_{h_1} \wedge \cdots \wedge u_{h_p}) \wedge (u_{h'_1} \wedge \cdots \wedge u_{h'_{n-p}}) = (-1)^{\operatorname{sgn} \mathbf{h}}\, u_1 \wedge \cdots \wedge u_n, \tag{33}$$

where sgn $\mathbf{h}$ denotes the parity of the permutation

$$\begin{pmatrix} 1 & \cdots & p & p+1 & \cdots & n \\ h_1 & \cdots & h_p & h'_1 & \cdots & h'_{n-p} \end{pmatrix}.$$

Theorem 12 (*The Laplace Expansion of a Determinant*). *Let A be an $n \times n$ matrix over a commutative ring K and* $\mathbf{k}: \mathbf{p} \to \mathbf{n}$ *an increasing list of p indices. Then*

$$|A| = (-1)^{\operatorname{sgn} \mathbf{k}} \sum_{\mathbf{h}} (-1)^{\operatorname{sgn} \mathbf{h}} |A \circ (\mathbf{h} \times \mathbf{k})|\, |A \circ (\mathbf{h}' \times \mathbf{k}')|, \tag{34}$$

where the sum is taken over all increasing lists $\mathbf{h}: \mathbf{p} \to \mathbf{n}$ *of p indices, while* $\mathbf{h}'$, $\mathbf{k}'$ *are the increasing lists complementary to* $\mathbf{h}$ *and* $\mathbf{k}$, *respectively.*

If $p = 1$, this formula is the familiar expansion (IX.9) of $|A|$ by cofactors of the column k_1. For any p, it is a similar expansion, expressing $|A|$ as a sum of products of minors $|A \circ (\mathbf{h} \times \mathbf{k})|$ from the given p columns $k_1, \cdots, k_p$ by (complementary) minors from the remaining $n - p$ columns of A.

Proof: If we regard A as the matrix, rel $\mathbf{b}$, of some list $a_1, \cdots, a_n$ of n

elements of C, the basic formula (31) for $|A|$ reads

$$a_1 \wedge \cdots \wedge a_n = (b_1 \wedge \cdots \wedge b_n)|A|.$$

Reorder the exterior product on the left in the order indicated by the given $\mathbf{k}$ as

$$a_1 \wedge \cdots \wedge a_n = (-1)^{\operatorname{sgn} \mathbf{k}} (a_{k_1} \wedge \cdots \wedge a_{k_p}) \wedge (a_{k'_1} \wedge \cdots \wedge a_{k'_{n-p}}).$$

Expand each product on the right by the lemma, so that

$$a_1 \wedge \cdots \wedge a_n = (-1)^{\operatorname{sgn} \mathbf{k}} \Big(\sum_{\mathbf{h}} b_{\mathbf{h}} |A \circ (\mathbf{h} \times \mathbf{k})| \Big) \wedge \Big(\sum_{\mathbf{l}} b_{\mathbf{l}} |A \circ (\mathbf{l} \times \mathbf{k}')| \Big)$$

with sums over all increasing lists $\mathbf{h}:\mathbf{p} \to \mathbf{n}$ and $\mathbf{l}:\mathbf{n} - \mathbf{p} \to \mathbf{n}$. But $b_{\mathbf{h}} = b_{h_1} \wedge \cdots \wedge b_{h_p}$ and $b_{\mathbf{l}}$ have a factor in common and hence have product zero unless $\mathbf{l}$ is the complement $\mathbf{l} = \mathbf{h}'$, while in that case $b_{\mathbf{h}} \wedge b_{\mathbf{h}'} = (-1)^{\operatorname{sgn} \mathbf{h}}\, b_1 \wedge \cdots \wedge b_n$, just as in (33) above. Thus

$$\begin{aligned} &a_1 \wedge \cdots \wedge a_n \\ &\quad = (-1)^{\operatorname{sgn} \mathbf{k}} (b_1 \wedge \cdots \wedge b_n) \sum_{\mathbf{h}} (-1)^{\operatorname{sgn} \mathbf{h}} |A \circ (\mathbf{h} \times \mathbf{k})|\, |A \circ (\mathbf{h}' \times \mathbf{k}')|, \end{aligned}$$

which gives the result required.

EXERCISES

1. (a) Let c be a vector and $a_1, \cdots, a_n$ be n linearly independent vectors in an n-dimensional vector space. Show that the solutions ξ_i of the vector equation $\Sigma_i a_i \xi_i = c$ are the scalars ξ_i determined for each i by

$$(a_1 \wedge \cdots \wedge a_n)\xi_i = a_1 \wedge \cdots \wedge a_{i-1} \wedge c \wedge a_{i+1} \wedge \cdots \wedge a_n.$$

(b) Show how this result implies Cramer's rule (IX.15).

(c) Show how this result gives the coordinates of a vector relative to a given basis.

2. If A is an $n \times n$ matrix, while $\mathbf{k}:\mathbf{p} \to \mathbf{n}$ and $\mathbf{l}:\mathbf{n} - \mathbf{p} \to \mathbf{n}$ are two increasing lists, with $\mathbf{l}$ not the complement of $\mathbf{k}$, prove that

$$0 = \sum_{\mathbf{h}} (-1)^{\operatorname{sgn} \mathbf{h}} |A \circ (\mathbf{h} \times \mathbf{k})|\, |A \circ (\mathbf{h}' \times \mathbf{l})|,$$

where the sum is over all increasing lists $\mathbf{h}:\mathbf{p} \to \mathbf{n}$.

3. For a free module C with the usual basis show that a multiplication table for the basis elements $b_{\mathbf{h}}$ of $\Lambda(C)$ is given in terms of the image sets $\mathbf{h}_*\mathbf{p}$ and $\mathbf{k}_*\mathbf{q}$ of the increasing lists $\mathbf{h}:\mathbf{p} \to \mathbf{n}$ and $\mathbf{k}:\mathbf{q} \to \mathbf{n}$ as follows:

$$\begin{aligned} b_{\mathbf{h}} \wedge b_{\mathbf{k}} &= 0, && \text{if } (\mathbf{h}_*\mathbf{p}) \cap (\mathbf{k}_*\mathbf{q}) \neq \varnothing, \\ &= (-1)^{\alpha} b_{\mathbf{l}}, && \text{if } (\mathbf{h}_*\mathbf{p}) \cap (\mathbf{k}_*\mathbf{q}) = \varnothing, \end{aligned}$$

where $\mathbf{l}:\mathbf{p}+\mathbf{q}\to\mathbf{n}$ is the increasing list combining the lists $\mathbf{h}$ and $\mathbf{k}$, while α is the number of ordered pairs (i, j) with $i \in \mathbf{h}_*\mathbf{p}$, $j \in \mathbf{k}_*\mathbf{q}$, and $i > j$.

4. Let C be a free module of rank n and basis $\mathbf{b}$, while, for each p, $f:\Lambda_p(C)\to\Lambda_{p+1}(C)$ is the morphism "exterior product with b_n". Prove the following sequence exact:

$$0 \to \Lambda_0(C) \xrightarrow{f} \Lambda_1(C) \xrightarrow{f} \Lambda_2(C) \to \cdots \to \Lambda_{n-1}(C) \xrightarrow{f} \Lambda_n(C) \to 0.$$

8. Subspaces by Exterior Algebra

In this section we shall consider an n-dimensional vector space V over a field F and its exterior algebra $\Lambda(V)$. An element of $\Lambda_p(V)$ is called a *p-vector*, while those p-vectors which are exterior products $v_1 \wedge \cdots \wedge v_p$ of p elements of V are called *decomposable p-vectors*.

THEOREM 13. *A list $v_1, \cdots, v_p$ of vectors in V is linearly independent if and only if $v_1 \wedge \cdots \wedge v_p \neq 0$ in $\Lambda_p(V)$.*

Proof: If the given list is independent, it is part of a basis $v_1, \cdots, v_n$. Then the exterior product $v_1 \wedge \cdots \wedge v_p$ is one of the elements in the corresponding basis of $\Lambda_p(V)$, as described in Theorem 10. Hence, $v_1 \wedge \cdots \wedge v_p$ is surely non-zero.

On the other hand, if the given list is linearly dependent, one of its vectors, say v_p, is a linear combination $v_p = v_1\kappa_1 + \cdots + v_{p-1}\kappa_{p-1}$ of the remaining ones. If we expand the exterior product

$$v_1 \wedge \cdots \wedge v_p = v_1 \wedge \cdots \wedge v_{p-1} \wedge (v_1\kappa_1 + \cdots + v_{p-1}\kappa_{p-1}),$$

every term has a repeated factor, so is zero.

This theorem gives an effective way of calculating when p given vectors are independent. Thus, if the p vectors, say $a_1, \cdots, a_p$, have an $n \times p$ matrix A relative to some given basis $\mathbf{b}$ of V, the formula (32) of the Lemma of §7 shows that the calculation of the product $a_1 \wedge \cdots \wedge a_p$ amounts exactly to the calculation of all the $p \times p$ minors of the matrix A. The vectors are linearly independent if and only if at least one of these $p \times p$ minors is non-zero.

This suggests the following related result:

COROLLARY. *The rank of an $n \times m$ matrix A (with entries in a field) is the largest integer r such that A has a non-zero $r \times r$ minor.*

Proof: Take an n-dimensional vector space V with basis $\mathbf{b}$; then A is the matrix, rel $\mathbf{b}$, of a list $a_1, \cdots, a_m$ of vectors of V, and the rank of A is the dimension of the subspace spanned by this list of vectors. By the theorem, the rank r is thus the largest integer p for which some exterior product $a_{h_1} \wedge \cdots \wedge a_{h_p} \neq 0$. Express this product in terms of the standard basis of $\Lambda_p(V)$ as in (32). The coefficients in this expression are $p \times p$ minors of A, so the rank is the largest integer p with some such minor non-zero, Q.E.D.

The rows and the columns of A enter symmetrically in this description of the rank of A. Hence, using the identity $|A| = |A^T|$ for the minors of A, we have another proof that row rank equals column rank.

Lemma. *For each p-vector $s \in \Lambda_p(V)$ the subset*

$$S = \{w | w \in V \quad \text{and} \quad w \wedge s = 0\}, \tag{35}$$

of V is a subspace of V of dimension at most p. If $b_1, \cdots, b_q$ is a basis of this subspace S, there is a $(p - q)$-vector u with $s = b_1 \wedge \cdots \wedge b_q \wedge u$.

Proof: Since the exterior product $(w, s) \mapsto w \wedge s$ is bilinear, the set S is a subspace. We may take a basis $b_1, \cdots, b_q$ for S, extend it to a basis of V, take the corresponding basis of p-vectors $b_{\mathbf{h}}$ in $\Lambda_p(V)$, and write the given p-vector s in terms of this basis as $s = \sum b_{\mathbf{h}}\kappa_{\mathbf{h}}$ for suitable scalars $\kappa_{\mathbf{h}}$, one for each increasing list $\mathbf{h}:\mathbf{p} \to \mathbf{n}$. Now $w \in S$ is defined to mean that $w \wedge s = 0$; in particular, that $b_1 \wedge s = \cdots = b_q \wedge s = 0$. But, $b_i \wedge s = \sum_{\mathbf{h}}(b_i \wedge b_{\mathbf{h}})\kappa_{\mathbf{h}}$, and in this sum the factors $b_i \wedge b_{\mathbf{h}}$ are zero when the list $\mathbf{h}$ contains the index i, while the remaining factors $b_i \wedge b_{\mathbf{h}}$ are linearly independent. Thus, if a coefficient $\kappa_{\mathbf{h}}$ in s is non-zero, the corresponding basis element $b_{\mathbf{h}} = b_{h_1} \wedge \cdots \wedge b_{h_p}$ must have each of $b_1, \cdots, b_q$ as one of its p factors. Therefore, $q \leqslant p$. In the formula for s, taking out the common factor $b_1 \wedge \cdots \wedge b_q$ gives $s = (b_1 \wedge \cdots \wedge b_q) \wedge (\sum b_{\mathbf{h}^{(0)}}\kappa_{\mathbf{h}})$, where $\mathbf{h}^{(0)}$ is the list $\mathbf{h}$ with $1, \cdots, q$ deleted. Then $\sum b_{\mathbf{h}^{(0)}}\kappa_{\mathbf{h}}$ is the $(p - q)$-vector u required.

Theorem 14. *Two linearly independent lists $v_1, \cdots, v_p$ and $w_1, \cdots, w_p$ of p vectors, each in a finite-dimensional vector space V, span the same subspace if and only if $v_1 \wedge \cdots \wedge v_p$ is a non-zero scalar multiple of $w_1 \wedge \cdots \wedge w_p$.*

Proof: Suppose first that they span the same subspace. Then each w_i can be expressed as a linear combination of $v_1, \cdots, v_p$. If we put these combinations in the exterior product $w_1 \wedge \cdots \wedge w_p$ and expand, the only non-zero terms are multiples of $v_1 \wedge \cdots \wedge v_p$, so $w_1 \wedge \cdots \wedge w_p = (v_1 \wedge \cdots \wedge v_p)\kappa$.

Conversely, suppose that $w_1 \wedge \cdots \wedge w_p = (v_1 \wedge \cdots \wedge v_p)\kappa$ for some scalar $\kappa \neq 0$. The p-vector $s = w_1 \wedge \cdots \wedge w_p$ defines as in (35) the subspace S of all those vectors w with $w \wedge s = 0$. This subspace S contains every v_i and every w_i, while by the lemma its dimension is at most p.

Therefore, the independent lists $v_1, \cdots, v_p$ and $w_1, \cdots, w_p$ are both bases for S, hence do span the same subspace.

We call two non-zero p-vectors s and s' of V "equivalent" if there is a non-zero scalar κ with $s' = s\kappa$.

Corollary. *In a finite-dimensional vector space V, assign to each decomposable non-zero p-vector s the set S of all vectors $w \in V$ with $w \wedge s = 0$. This assignment*

$$s \mapsto S = \{w | w \in V \text{ and } w \wedge s = 0\}$$

induces a bijection from the set of equivalence classes of non-zero decomposable p-vectors to the set of all p-dimensional subspaces of V.

Proof: If $s \neq 0$ is decomposable, say as $s = v_1 \wedge \cdots \wedge v_p$, the corresponding subspace S contains the p linearly independent vectors $v_1, \cdots, v_p$, hence is indeed of dimension p. Each subspace arises so, say from the exterior product $v_1 \wedge \cdots \wedge v_p$ of the vectors of any one of its bases.

This corollary raises the general question: In an n-dimensional space, which p-vectors are decomposable? If $p = 1$, a p-vector is just a vector, and is decomposable by definition. If $p = n$, any n-vector is a multiple of $b_1 \wedge \cdots \wedge b_n$, so is also decomposable. The next case is $p = n - 1$:

Proposition 15. *In an n-dimensional vector space every $(n-1)$-vector is decomposable.*

Proof: Take a basis **b** for V. The corresponding basis for $\Lambda_{n-1}(V)$ consists of the $(n-1)$-vectors

$$s_i = b_1 \wedge \cdots \wedge b_{i-1} \wedge b_{i+1} \wedge \cdots \wedge b_n \quad (\text{omit } b_i), \qquad i = 1, \cdots, n.$$

Any $(n-1)$-vector s then is $s = \Sigma s_i \kappa_i$ for suitable scalars κ_i. A vector $v = \Sigma b_i \xi_i$ has $v \wedge s = (b_1 \wedge \cdots \wedge b_n)\Sigma(-1)^{i-1}\xi_i \kappa_i$. Hence, the set S of all vectors w with $w \wedge s = 0$ is given by the set of all solutions ξ of one homogeneous linear equation $\Sigma(-1)^{i-1}\xi_i\kappa_i = 0$ (with coefficients κ_i not all zero when $s \neq 0$). Therefore, S is an $(n-1)$-dimensional subspace, so by the corollary above s is decomposable.

However, p-vectors are not usually decomposable. In a four-dimensional vector space V with basis **b** over **Q** take the two-vector $s = b_1 \wedge b_2 + b_3 \wedge b_4$. The corresponding subspace S consists of all vectors $\Sigma b_i \xi_i$ with $(\Sigma b_i \xi_i) \wedge s = 0$. But $(\Sigma b_i \xi_i) \wedge s$ is $(b_1 \wedge b_2 \wedge b_3)\xi_3 + (b_1 \wedge b_2 \wedge b_4)\xi_4 +$

$(b_1 \wedge b_3 \wedge b_4)\xi_1 + (b_2 \wedge b_3 \wedge b_4)\xi_2$; this is zero only when $\xi_1 = \xi_2 = \xi_3 = \xi_4 = 0$. Therefore, $S = 0$; by the corollary, s cannot be decomposable.

EXERCISES

1. In a finite-dimensional vector space V, let $s \neq 0$ be a decomposable p-vector, $t \neq 0$ a decomposable q-vector, and S, T the corresponding subspaces of V, as in (35).

(a) If $p \leqslant q$, prove that $S \subset T$ if and only if there is a $(q - p)$-vector r such that $t = r \wedge s$.

(b) Show that $S \cap T = 0$ if and only if $s \wedge t \neq 0$, and, when this is the case, show that the join $S + T$ is the subspace corresponding to the decomposable $(p + q)$-vector $s \wedge t$.

(c) If $p = q$, show that $s + t$ is decomposable if and only if $S \cap T$ has dimension at least $p - 1$.

2. In a four-dimensional space V with basis $\mathbf{b}$, show that the two-vector

$$b_1 \wedge b_2 + b_1 \wedge b_3 + b_1 \wedge b_4 + b_2 \wedge b_3 + b_2 \wedge b_4 + b_3 \wedge b_4$$

is not decomposable.

3. (The quadratic condition on the Plücker coordinates of two-spaces in four-space.) In a four-dimensional vector space V with basis b_1, b_2, b_3, b_4 write each two-vector s as $s = \Sigma(b_i \wedge b_j)\xi_{ij}$, with sum over all pairs (i, j) with $1 \leqslant i < j \leqslant 4$ (i.e., over all increasing lists $\mathbf{2} \to \mathbf{4}$). Call the scalars ξ_{ij} the (Plücker) coordinates of the two-vector s, rel $\mathbf{b}$.

(a) Prove that the coordinates of any decomposable two-vector satisfy

$$\xi_{12}\xi_{34} - \xi_{13}\xi_{24} + \xi_{14}\xi_{23} = 0;$$

*(b) Show that any two-vector satisfying this condition is decomposable.

9. Duality in Exterior Algebra

Each K-module C has a dual K-module $C^* = \text{Hom}(C, K)$, so the identity is an isomorphism $C^* \to \text{Hom}(C, K)$. For exterior algebra we will construct a corresponding morphism

$$\Psi_p : \Lambda_p(C^*) \to \text{Hom}\left(\Lambda_p(C), K\right) = \left[\Lambda_p(C)\right]^*.$$

We will show that this is an isomorphism when C is free of finite rank.

Given any C and given p linear forms $f_1, \cdots, f_p \in C^*$ and p elements $c_1, \cdots, c_p \in C$, we first form the $p \times p$ matrix A with entries $A_{ij} = f_i(c_j)$. Now any determinant is an alternating multilinear function of its rows and of its columns. Hence, for a fixed list $f_1, \cdots, f_p$ of linear forms, $|A|$ is an alternating p-linear function of $c_1, \cdots, c_p$. Since $(c_1, \cdots, c_p) \mapsto c_1 \wedge \cdots \wedge c_p$ is a universal such function, there is a linear transformation

$t(f_1, \cdots, f_p):\Lambda_p(C) \to K$ with

$$[t(f_1, \cdots, f_p)](c_1 \wedge \cdots \wedge c_p) = |A|, \qquad A_{ij} = f_i(c_j).$$

Each $t(f_1, \cdots, f_p)$ is thus an element of the dual module $[\Lambda_p(C)]^*$. Since the determinant $|A|$ is an alternating multilinear function of its rows, the assignment $(f_1, \cdots, f_p) \mapsto t(f_1, \cdots, f_p)$ is alternating and multilinear. Because $(f_1, \cdots, f_p) \mapsto f_1 \wedge \cdots \wedge f_p \in \Lambda_p(C^*)$ is universal, there is a linear transformation $\Psi:\Lambda_p(C^*) \to [\Lambda_p(C)]^*$ with $\Psi(f_1 \wedge \cdots \wedge f_p) = t(f_1, \cdots, f_p)$; that is, with

$$[\Psi(f_1 \wedge \cdots \wedge f_p)](c_1 \wedge \cdots \wedge c_p) = |A|, \tag{36}$$

where A has the entries $A_{ij} = f_i(c_j)$ for $i, j \in \mathbf{p}$.

Theorem 16. *If C is a free module of finite rank over a commutative ring, then for each natural number p the linear transformation Ψ of (36) is an isomorphism*

$$\Psi : \Lambda_p(C^*) \cong [\Lambda_p(C)]^*. \tag{37}$$

Proof: When C is free of finite rank n, each basis $b_1, \cdots, b_n$ for C gives a dual basis $x_1, \cdots, x_n$ for C^* defined just as for finite-dimensional vector spaces, so that $x_i(b_j) = \delta_{ij}$ for $i, j \in \mathbf{n}$. Then $\Lambda_p(C^*)$ has basis all $x_{\mathbf{h}} = x_{h_1} \wedge \cdots \wedge x_{h_p}$, for $\mathbf{h}$ an increasing list $\mathbf{h}:\mathbf{p} \to \mathbf{n}$, while $\Lambda_p(C)$ has basis all $b_{\mathbf{k}} = b_{k_1} \wedge \cdots \wedge b_{k_p}$ for $\mathbf{k}:\mathbf{p} \to \mathbf{n}$ also increasing. Now $[\Psi x_{\mathbf{h}}](b_{\mathbf{k}}) = |A|$, where A is the $p \times p$ matrix with entries $A_{ij} = x_{h_i}(b_{k_j})$. Given a row index i in this matrix, $x_{h_i}(b_{k_j})$ is zero unless $k_j = h_i$. Therefore some row of A vanishes unless each h_i equals some k_j; this can happen only when the two increasing lists are equal: $\mathbf{k} = \mathbf{h}$. In that case A is the $p \times p$ identity matrix, with determinant 1. Hence, $(\Psi x_{\mathbf{h}})b_{\mathbf{k}}$ is zero or one according as $\mathbf{h} \neq \mathbf{k}$ or $\mathbf{h} = \mathbf{k}$. This states that the elements $\Psi x_{\mathbf{h}}$ constitute a basis of $[\Lambda_p(C)]^*$ dual to the basis $b_{\mathbf{k}}$ of $\Lambda_p(C)$. In other words, Ψ carries a basis of $\Lambda_p(C^*)$ into one of $[\Lambda_p(C)]^*$, hence must be an isomorphism, as asserted.

The relation between elements $c \in C$ and $f \in C^*$ is often written symmetrically, as $\langle f, c\rangle = f(c) \in K$, just as in (V.46). In this notation, Theorem 16 becomes

Corollary. *For a free module C of finite rank and for each p,*

$$\langle \Psi(f_1 \wedge \cdots \wedge f_p), c_1 \wedge \cdots \wedge c_p\rangle = |A|, \qquad A_{ij} = \langle f_i, c_j\rangle,$$

for all $f_i \in C^*$ and $c_j \in C$. For a basis $\mathbf{b}$ of C, $\langle \psi b^h, b_k\rangle = \delta_{h,k}$.

Curiously enough, there is another way of expressing the dual of $\Lambda_p(C)$, subject to one choice. For C free of rank n, the module $\Lambda_n(C)$ is isomorphic to the ring K of scalars. Let us choose an isomorphism $\eta : \Lambda_n(C) \cong K$; this amounts to choosing one element as basis for $\Lambda_n(C)$, say the usual element $b_1 \wedge \cdots \wedge b_n$.

Theorem 17. *If C is a free module of finite rank with a chosen isomorphism $\eta : \Lambda_n(C) \cong K$, then the bilinear map $\Lambda_{n-p}(C) \times \Lambda_p(C) \to K$ given for $s \in \Lambda_{n-p}$ and $t \in \Lambda_p$ by $(s, t) \mapsto \eta(s \wedge t)$ induces an isomorphism*

$$\theta : \Lambda_{n-p}(C) \cong [\Lambda_p(C)]^*, \qquad (\theta s)t = \eta(s \wedge t).$$

Proof: Let us take the usual basis b_i; then the $b_{\mathbf{h}}$ with $\mathbf{h}:\mathbf{n} - \mathbf{p} \to \mathbf{n}$ an increasing list form a basis for Λ_{n-p}, while the $b_{\mathbf{k}}$ with $\mathbf{k}:\mathbf{p} \to \mathbf{n}$ an increasing list form one for Λ_p. By the usual rules for exterior products, $b_{\mathbf{h}} \wedge b_{\mathbf{k}} = 0$ unless $\mathbf{h}$ is the complement $\mathbf{k}'$ of $\mathbf{k}$; when this is the case, $b_{\mathbf{h}} \wedge b_{\mathbf{k}} = \pm b_1 \wedge \cdots \wedge b_n$. Now suppose that the chosen isomorphism η has $\eta(b_1 \wedge \cdots \wedge b_n) = 1$. By the definition of θ, this means that $(\theta b_{\mathbf{h}})(b_{\mathbf{k}})$ is 0 or ± 1 according as $\mathbf{h} \neq \mathbf{k}'$ or $\mathbf{h} = \mathbf{k}'$. Therefore, $\theta b_{\mathbf{h}}$ is, except for signs, the basis of $[\Lambda_p(C)]^*$ dual to the basis $b_{\mathbf{k}}$ of $\Lambda_p(C)$. Hence, θ is an isomorphism.

These duality theorems can be applied to an inner product space U over the field $\mathbf{R}$ of real numbers.

Theorem 18. *If U is a finite-dimensional inner product space over the field $\mathbf{R}$ of real numbers, then, for each p, $\Lambda_p(U)$ is an inner product space, with an inner product given, for two lists v and w of p vectors each, as*

$$\langle v_1 \wedge \cdots \wedge v_p, w_1 \wedge \cdots \wedge w_p \rangle = |A|, \tag{38}$$

where A is the $p \times p$ matrix of real numbers defined in terms of the inner product in V as $A_{ij} = \langle v_i, w_j \rangle$, for $i, j \in \mathbf{p}$.

Proof: The inner product $\langle\ ,\ \rangle$ in U yields as in Theorem X.9 an isomorphism $\theta : U \cong U^*$ with $(\theta v)(w) = \langle v, w \rangle$ for all vectors $v, w \in U$. For each p, there is thus an isomorphism $\Lambda_p(\theta) : \Lambda_p(U) \cong \Lambda_p(U)^*$ on p-vectors. As the inner product of two p-vectors r and $s \in \Lambda_p(U)$ we propose, using (37),

$$\langle r, s \rangle = \langle \Psi \Lambda_p(\theta) r, s \rangle.$$

This is a bilinear map $\langle\ ,\ \rangle : \Lambda_p(U) \times \Lambda_p(U) \to \mathbf{R}$. For decomposable p-vectors $r = v_1 \wedge \cdots \wedge v_p$ and $s = w_1 \wedge \cdots \wedge w_p$, the Corollary of Theorem 16 converts this definition to the form (38). But all p-vectors are sums of decomposables, so (38) determines the inner product and proves that it is symmetric. For an orthonormal basis $\mathbf{b}$ of U, the p-vectors $b_{\mathbf{h}}$ for $\mathbf{h}:\mathbf{p} \to \mathbf{n}$ increasing form a basis of $\Lambda_p(U)$. By the formula of the corollary, $\langle b_{\mathbf{h}}, b_{\mathbf{k}} \rangle = \delta_{\mathbf{hk}}$. This states that the matrix of the inner product relative to the basis $b_{\mathbf{h}}$ is the identity matrix. Hence, the proposed inner product (38) is positive definite, so that $\Lambda_p(U)$ is indeed an inner product space.

For this inner product in $\Lambda_p(U)$, the length of a p-vector $v_1 \wedge \cdots \wedge v_p$ can be interpreted as the p-dimensional volume of the parallelopiped whose edges are the vectors $v_1, \cdots, v_p$.

EXERCISES

1. Let U be an inner product space over $\mathbf{R}$; use the corresponding inner product of Theorem 18 in each $\Lambda_p(U)$.

(a) If $\mathbf{b}$ is an orthonormal basis for U, show that the p-vectors $b_{\mathbf{h}}$ for all increasing lists $\mathbf{h}:\mathbf{p} \to \mathbf{n}$ form an orthonormal basis for $\Lambda_p(U)$.

(b) If $t:U \to U$ is orthogonal, prove that $\Lambda_p(t)$ is also orthogonal.

2. If U is a unitary space over $\mathbf{C}$, show that there is a natural way to make each $\Lambda_p(U)$ a unitary space.

10. Alternating Forms and Skew-Symmetric Tensors

The modules $\Lambda_p(V)$ occurring in the exterior algebra of a finite-dimensional vector space V can be described in two additional ways.

The first way uses alternating forms. For each K-module C, the set $\text{Alt}_p\,(C)$ of all alternating p-linear forms $h:C^p \to K$ is itself a K-module under the usual pointwise module operations of (19). This module is the dual of $\Lambda_p(C)$:

Proposition 19. *For any module C, each linear form $t:\Lambda_p(C) \to K$ determines an alternating p-linear form $h:C^p \to K$ by $h(c_1, \cdots, c_p) = t(c_1 \wedge \cdots \wedge c_p)$. The assignment $t \mapsto h$ is an isomorphism*

$$[\Lambda_p(C)]^* \cong \text{Alt}_p\,(C)$$

of modules.

Proof: The assignment $t \mapsto h$ is a morphism of (pointwise) module operations. By the universality of $c_1 \wedge \cdots \wedge c_p$ (Theorem 9), this assignment is an isomorphism.

In case K is a field and $C = V$ a finite-dimensional vector space, this implies that $\Lambda_p(V) \cong [\Lambda_p(V)]^{**} \cong [\text{Alt}_p\,(V)]^*$.

The second way of describing Λ_p uses skew-symmetric tensors over a vector space V. First we note that the symmetric group S_p acts on the space $T_p(V)$ of all tensors of type $(0, p)$ by $\sigma(v_1 \otimes \cdots \otimes v_p) = v_{\sigma_1} \otimes \cdots \otimes v_{\sigma_p}$. A tensor $r \in T_p(V)$ is said to be *skew-symmetric* when $\sigma(r) = (-1)^{\text{sgn}\,\sigma} r$ for all permutations $\sigma \in S_p$; in other words, r is skew-symmetric when the application of any odd permutation to r changes the sign of r. The set of all skew-symmetric tensors in $T_p(V)$ is clearly a subspace of $T_p(V)$.

Now suppose that the field F of scalars has characteristic ∞. This assumption allows us to define an endomorphism Sk of $T_p(V)$ with image exactly the subspace of skew-symmetric tensors.

Lemma. *For each natural number p and each tensor $r \in T_p(V)$, the formula*

$$\mathrm{Sk}\,(r) = (1/p!) \sum (-1)^{\mathrm{sgn}\,\sigma} \sigma r, \tag{39}$$

summed over all $\sigma \in S_p$, defines an endomorphism $\mathrm{Sk}: T_p(V) \to T_p(V)$, with image the skew-symmetric tensors, such that $\mathrm{Sk}\,(r) = r$ when r is skew-symmetric. For any permutation ρ and any $r \in T_p$,

$$\mathrm{Sk}\,(\rho r) = (-1)^{\mathrm{sgn}\,\rho}\,\mathrm{Sk}\,(r). \tag{40}$$

Proof: If r is skew-symmetric, each σr is $(-1)^{\mathrm{sgn}\,\sigma} r$, so the sum in the definition (39) of Sk has $p!$ terms all equal to r, and hence $\mathrm{Sk}\,(r) = r$; (note that characteristic ∞ is used to form the reciprocal of $p!$).

Now let r be any tensor and ρ any permutation of $\mathbf{p}$. Then

$$\rho\,\mathrm{Sk}\,(r) = (1/p!) \sum_\sigma (-1)^{\mathrm{sgn}\,\sigma} \rho\sigma r = (-1)^{\mathrm{sgn}\,\rho} (1/p!) \sum_\sigma (-1)^{\mathrm{sgn}\,\rho\sigma} \rho\sigma r,$$

because $\sigma \mapsto (-1)^{\mathrm{sgn}\,\sigma}$ is a morphism $S_p \to \{\pm 1\}$ of groups. But for ρ fixed, $\sigma \mapsto \rho\sigma$ is a bijection $S_p \to S_p$, so the right-hand summation equals $\Sigma_\sigma (-1)^{\mathrm{sgn}\,\sigma} \sigma r$, and the formula becomes

$$\rho\,\mathrm{Sk}\,(r) = (-1)^{\mathrm{sgn}\,\rho}\,\mathrm{Sk}\,(r).$$

This proves that the image of Sk consists of skew-symmetric tensors.

Finally, for r and ρ as before, a similar argument yields (40).

Theorem 20. *If V is a finite-dimensional vector space over a field of characteristic ∞, there is a monomorphism $t: \Lambda_p(V) \to T_p(V)$ of vector spaces with image the subspace of all skew-symmetric tensors. Explicitly, if $v_1, \cdots, v_p$ are any p vectors of V, this monomorphism is defined by*

$$t(v_1 \wedge \cdots \wedge v_p) = \mathrm{Sk}\,(v_1 \otimes \cdots \otimes v_p). \tag{41}$$

Proof: The result is immediate for $p = 1$, so assume that $p \geqslant 2$. Let $\tau = (ij)$ be any transposition in the symmetric group S_p. By equation (40), $\mathrm{Sk}\,(v_1 \otimes \cdots \otimes v_p) = -\,\mathrm{Sk}\,(\tau(v_1 \otimes \cdots \otimes v_p))$. In particular, if $v_i = v_j$, then $\tau(v_1 \otimes \cdots \otimes v_p) = v_1 \otimes \cdots \otimes v_p$, and hence

$$2\,\mathrm{Sk}\,(v_1 \otimes \cdots \otimes v_p) = 0.$$

Since $2 \neq 0$ in a field of characteristic ∞, $\operatorname{Sk}(v_1 \otimes \cdots \otimes v_p) = 0$ whenever $v_i = v_j$ for $i \neq j$. This shows that the p-linear function $(v_1, \cdots, v_p) \mapsto \operatorname{Sk}(v_1 \otimes \cdots \otimes v_p)$ is alternating. Since $(v_1, \cdots, v_p) \mapsto v_1 \wedge \cdots \wedge v_p$ is the universal alternating p-linear function, there is a (unique) morphism t which has the property (41).

To show that t is a monomorphism, take any basis $\mathbf{b}$ of V. The corresponding basis of $\Lambda_p(V)$ consists of the p-vectors $b_{h_1} \wedge \cdots \wedge b_{h_p}$ for all *increasing* lists $\mathbf{h}:\mathbf{p} \to \mathbf{n}$ of p indices, while the basis of $T_p(V)$ consists of all products $b_{i_1} \otimes \cdots \otimes b_{i_p}$ for *all* lists $\mathbf{i}:\mathbf{p} \to \mathbf{n}$ of p indices. Since $t(b_{h_1} \wedge \cdots \wedge b_{h_p})$ is (except for the factor $1/p!$) just a signed sum of all permutations of $b_{h_1} \otimes \cdots \otimes b_{h_p}$, the image of the basis $b_{\mathbf{h}}$ of $\Lambda_p(V)$ is manifestly linearly independent in $T_p(V)$, and hence t is a monomorphism. Since any skew-symmetric tensor lies in the image of Sk, it is also in the image of t.

On the one hand, this result provides an explicit basis for the space of skew-symmetric tensors. On the other hand, it makes it possible to describe the exterior algebra $\Lambda(V)$ as the graded algebra of skew-symmetric tensors.

EXERCISES

1. Show that the skew-symmetric tensors over V of dimension $n > 1$ do not form a graded subalgebra of $T(V)$.

2. For a p-vector v and a q-vector w, show that

$$t_{p+q}(v \wedge w) = \operatorname{Sk}(t_p v \otimes t_q w).$$

3. Use the result of Exercise 2 to make the skew-symmetric tensors into a graded algebra, isomorphic by t to $\Lambda(V)$.

4. A tensor r is *symmetric* when $\sigma(r) = r$ for all $\sigma \in S_p$. Using the symmetric algebra $S(V)$ of the exercises of §6, obtain an analog of Theorem 20 for symmetric tensors.

APPENDIX

Affine and Projective Spaces

In a vector space the origin plays a special role; for example, it is fixed under all automorphisms of the space. To make a vector space homogeneous, so that all its vectors are equivalent under the group of the space, one must adjoin to the general linear group the translations of the space. Together these generate the affine group, which acts on the vectors, regarded as the "points" of the corresponding "affine space". By adjoining translations to an orthogonal group one similarly gets the Euclidean group acting on an Euclidean space. Finally, each vector space determines a corresponding projective space with a projective group which contains an affine group as proper subgroup.

These constructions will be illustrated first in the case of the affine line.

1. The Affine Line

The real *affine line* $L = \mathbf{R}$ is defined to be the set $\mathbf{R}$ of real numbers, regarded just as a *set* of "points" $l \in L$; its geometry will be described by certain "affine transformations" $L \to L$. These transformations include scalar multiplication $l \mapsto \kappa l$ of a point l by a fixed real scalar κ and translations $l \mapsto \mu + l$ of l by a real scalar μ. The most general affine transformation $l \mapsto \kappa l \mapsto \kappa l + \mu$ is a scalar multiplication followed by a translation; in other words, two scalars κ and μ in $\mathbf{R}$ determine an affine transformation $a = a(\mu, \kappa): L \to L$ by $a(l) = \kappa l + \mu$. The composite of two affine transformations is $l \mapsto \kappa l + \mu \mapsto \kappa'\kappa l + \kappa'\mu + \mu'$, so is the affine transformation

$$a(\mu', \kappa') \circ a(\mu, \kappa) = a(\mu' + \kappa'\mu, \kappa'\kappa). \tag{1}$$

An affine transformation $a(\mu, \kappa)$ is a bijection (an "affine automorphism") if and only if $\kappa \neq 0$; by (1), its inverse is then the affine transformation $a(-\mu/\kappa, 1/\kappa)$. The set of all affine automorphisms is the one-dimensional *real affine group* A_1. By the formula (1), the assignment $a(\mu, \kappa) \mapsto \kappa$ is an epimorphism $A_1 \to \mathbf{R}^*$ from the affine group to the multiplicative group $\mathbf{R}^*$ of non-zero real numbers. The kernel of this epimorphism consists of all the translations $a(\mu, 1)$. Since $a(\mu', 1) \circ a(\mu, 1) = a(\mu' + \mu, 1)$, this kernel may be described as the image of the monomorphism $\mu \mapsto a(\mu, 1)$ from the additive group $\mathbf{R}^+$ of real numbers to A_1. These observations about the structure of the affine group A_1 may be summarized by the statement that there is a sequence of morphisms of groups,

$$0 \to \mathbf{R}^+ \to A_1 \to \mathbf{R}^* \to 1, \tag{2}$$

which is "exact". The exactness of this sequence has the same meaning as for sequences of morphisms of modules (§ IX.9): At each group G in the sequence, the image of the arrow *to* G is the kernel of the arrow *from* G. In the present sequence (2), exactness at $\mathbf{R}^+$ (with 0 the trivial additive group) means that $\mathbf{R}^+ \to A_1$ is a monomorphism, exactness at A_1 that the kernel of $A_1 \to \mathbf{R}^*$ is the image of $\mathbf{R}^+ \to A_1$ (that is, the normal subgroup of translations), and exactness at $\mathbf{R}^*$ (with 1 the trivial multiplicative group) that $A_1 \to \mathbf{R}^*$ is an epimorphism.

The real affine group A_1 is isomorphic to the group of all pairs (μ, κ) of real scalars, with $\kappa \neq 0$, multiplied by the rule

$$(\mu', \kappa')(\mu, \kappa) = (\mu' + \kappa'\mu, \kappa'\kappa). \tag{3}$$

A property P of two or more points l, l' on the line L is called an *affine property* when it is invariant under the action of the affine group; that is, when $P(l, l')$ holds if and only if $P(a(l), a(l'))$ holds for all transformations $a \in A_1$. For example, the property "l is two units distant from l'" is not an affine property, because scalar multiplication can alter the distance between points. On the other hand, "l is the midpoint of l_1 and l_2" is an affine property (of three points); indeed, the equation $l = (l_1 + l_2)/2$ expresses the fact that l is the midpoint of l_1 and l_2; applying any affine transformation $a = a(\mu, \kappa)$,

$$(a(l_1) + a(l_2))/2 = (\kappa l_1 + \mu + \kappa l_2 + \mu)/2 = \kappa[(l_1 + l_2)/2] + \mu,$$

so $a(l)$ is the midpoint of $a(l_1)$ and $a(l_2)$. More generally, for any two real numbers ω_1 and ω_2 with sum $\omega_1 + \omega_2 = 1$, the *average* of the two points l_1 and l_2 with the *weights* ω_1 and ω_2 is the point $l = l_1\omega_1 + l_2\omega_2$. A calculation like that above, for any affine transformation a, shows that $a(l_1\omega_1 + l_2\omega_2) = a(l_1)\omega_1 + a(l_2)\omega_2$. Therefore an affine transformation preserves all (weighted) averages. Conversely, it can be shown that any function $f: L \to L$ which does preserve all averages is an affine transformation (Exercise 5).

The real affine line L may be described *just* in terms of its translations $l \mapsto \mu + l$. Explicitly, $(\mu, l) \mapsto \mu + l$ defines an action of the additive group $\mathbf{R}^+$ on the set L; for each point l this action gives a bijection $\mu \mapsto \mu + l$ of $\mathbf{R}$ to L. For scalars ω_1 and ω_2 with $\omega_1 + \omega_2 = 1$, the average $l\omega_1 + (\mu + l)\omega_2$ is $\mu\omega_2 + l$ and so can be obtained by translating the point l by the amount $\mu\omega_2$; hence averages can be defined via translations. Similarly, affine transformations can then be described as functions $a: L \to L$ which preserve all averages. This approach to affine geometry will be used systematically in the next section.

A typical affine problem is that of classifying real quadratic functions $c: L \to L$ for equivalence under the affine group. By definition, a quadratic function c is a function on L to L of the form $x \mapsto \alpha x^2 + \beta x + \gamma$, for $x \in L$ and real scalars α, β, and γ. If $\alpha \neq 0$, the affine transformation $a(0, \sqrt{\pm\alpha})$ will reduce the coefficient of x^2 to ± 1. Completion of the square and a translation then removes the coefficient β, so the function c becomes $x \mapsto \pm x^2 + \gamma'$ for some real γ'. On the other hand, if $\alpha = 0$ and $\beta \neq 0$, the function c can similarly be reduced to the identity function. Hence, any quadratic function on the real line L is equivalent under the affine group to one of the functions $\pm x^2 + \gamma'$, x, or γ. One of the main objectives of this chapter is to find similar reductions of quadratic functions of several variables—and to give a coordinate-free description of what is meant by a quadratic function. This will lead to a classification of the corresponding figures (the hyperquadrics) under both the affine and the Euclidean groups.

The real affine plane is like the affine line. An affine transformation of the plane is the composite of a linear transformation and a translation, and all these transformations preserve averages. Two figures (that is, two sets) in the plane are affine equivalent when there exists an affine automorphism of the plane carrying the first figure to the second; for example, one may readily see that any two non-degenerate triangles are affine equivalent. The fact that affine transformations preserve averages makes them useful for elementary geometric arguments. For example, one may prove the familiar theorem that the medians of any triangle T meet in a point by noting that the theorem holds (say) for an equilateral triangle, that there is an affine transformation a carrying this equilateral triangle to T, and that a carries lines to lines, midpoints to midpoints, and hence medians to medians, Q.E.D. Going further, one can also establish that the point of intersection of the medians is $\frac{1}{3}$ of the way along each median from base to vertex. Indeed, this is evidently the case for the equilateral triangle, the point $\frac{1}{3}$ of the way from l to l' is just the weighted average $l(\frac{1}{3}) + l'(\frac{2}{3})$, and affine transformations of the plane preserve such averages.

EXERCISES

1. Show that the morphism $A_1 \to \mathbf{R}^*$ of (2) has a right inverse which is a morphism of groups.

2. Prove that the affine automorphisms of the line L leaving any one point of L fixed form a group isomorphic to $\mathbf{R}^*$.

3. If $l_1 \neq l_2$ and $l_1' \neq l_2'$ are two pairs of points of L, show that there is exactly one affine transformation a with $a(l_1) = l_1'$ and $a(l_2) = l_2'$.

4. If $l_1 \neq l_2$ are two distinct points of L, show that any point $l \in L$ can be represented in exactly one way as an average $l_1\omega_1 + l_2\omega_2$, with $\omega_1 + \omega_2 = 1$.

5. Prove that a function $f: L \to L$ is an affine transformation of the line L if and only if it preserves all averages.

6. (a) Show that every parallelogram is affine equivalent to a square.

(b) Prove by affine methods that the diagonals of a parallelogram meet in a point which is the midpoint of both diagonals.

7. Prove by affine methods that in a trapezoid the two diagonals and the line joining the midpoints of the parallel sides all meet in a point.

8. An affine automorphism $a = a(v, t)$ of the real plane $\mathbf{R}^2$ is the composite of a linear automorphism $t: \mathbf{R}^2 \to \mathbf{R}^2$ and the translation $\xi \mapsto v + \xi$ by a vector $v \in \mathbf{R}^2$.

(a) Derive a formula like (1) for the composite of two such automorphisms.

(b) Derive an exact sequence like (2) for the group A_2 of all these automorphisms.

9. (a) Prove that any parallelepiped is affine equivalent to a cube.

(b) Prove that the four diagonals of any parallelepiped have a common midpoint (it is their center of gravity).

10. Show that the affine group A_1 is not abelian, and hence not isomorphic to the product $\mathbf{R}^+ \times \mathbf{R}^*$.

2. Affine Spaces

For the remainder of this chapter, F will be an arbitrary *field of characteristic not* 2. This assumption implies that $\frac{1}{2}$ exists in F, and hence that midpoints may be defined by the familiar formula $l(\frac{1}{2}) + l'(\frac{1}{2})$. An affine space P over the field F will now be defined in terms of the action on P of a vector space V over F, just as the real affine line was described in § 1 as a set with an action (of "translation") by real numbers.

DEFINITION. An affine space P over F is a non-void set for which there exists a finite-dimensional vector space V over F and a function $V \times P \to P$, written $(v, p) \mapsto v + p$, such that

(*i*) *For all vectors $v, w \in V$ and all points $p \in P$,*

$$0 + p = p, \qquad (v + w) + p = v + (w + p); \tag{4}$$

(*ii*) *For any two points $p, q \in P$ there is exactly one vector $v \in V$ with*

$$v + q = p. \tag{5}$$

The dimension *of P is the vector space dimension of V.*

Note that the same symbol + is used both for the sum of two vectors and for the action of a vector on a point.

Axiom (*ii*) states for each point $q \in P$ that $v \mapsto v + q$ is a bijection $V \cong P$ of sets.

Axiom (*i*) states that the additive group of the vector space V acts on the set P. The action $p \mapsto v + p$ of each vector $v \in V$ is a bijection $v' : P \cong P$ which we call a *translation* of P. If we define the linear combination $v'\kappa + w'\lambda$ of two translations v' and w' to be $(v\kappa + w\lambda)'$, the set of all these translations is a vector space $P^{\#}$, isomorphic via $v' \mapsto v$ to the given vector space V. We call $P^{\#}$ the *space of translations* of the affine space P. Since the space V in the definition of P may be replaced by any isomorphic vector space, we will normally replace it by $P^{\#}$, identifying each v with v'.

For example, the affine line L discussed in § 1 is an affine space with the one-dimensional real vector space $\mathbf{R}$ as its translation space, provided each scalar $\mu \in \mathbf{R}$ is interpreted as the translation $l \mapsto \mu + l$.

If V is any vector space over F and $P = S + u_0$ any coset of a finite-dimensional subspace S of V, then P is an affine space over F with translation space $P^{\#} = S$, when we interpret each vector $t \in S$ as the bijection $s + u_0 \mapsto t + s + u_0$ for all $s \in S$ (that is, as the left action of $t \in S$ on the left coset $S + u_0$).

The special case when S is all of V (and hence the coset is just V itself) is worthy of note. In detail, for any finite-dimensional vector space V, let $V^{\flat}$ denote the set V, so a point $p \in V^{\flat}$ is just a vector u of V. This point $p = u \in V^{\flat}$ and a vector $v \in V$ have a sum $v + p$ which we regard as a point in $V^{\flat}$, and the axioms (*i*) and (*ii*) evidently hold. Thus the vector space V has been turned into an affine space $V^{\flat}$ with translation space $(V^{\flat})^{\#} \cong V$. The affine space $V^{\flat}$ is the same *set* as the vector space V, but with a different algebraic structure.

Soon we will show that every affine space is affine-isomorphic to a space of the form $V^{\flat}$ for some vector space V.

We now examine again the axioms for an affine space. In axiom (*ii*), the unique vector v with $v + q = p$ will be denoted as $v = p - q$ and called the "vector from q to p". This defines, by $(p, q) \mapsto p - q$, a function $P \times P \to V$ with the following evident properties, for all points p, q, and r:

$$(p - q) + p = p, \qquad (p - q) + (q - r) = p - r, \tag{6}$$

$$p - q = 0 \quad \Leftrightarrow \quad p = q. \tag{7}$$

Note that the *difference* $p - q$ of two points is a vector, while a sum, vector plus point, is a point. Together these operations enjoy the properties that

$$v + (p - q) = (v + p) - q, \; (v + p) - (w + q) = (v - w) + (p - q). \tag{8}$$

From these axioms we may now define "averages" in P. A list $\boldsymbol{\omega}$ of n *weights* is a list of n scalars ω_i with sum $\omega_1 + \cdots + \omega_n = 1$. If $\mathbf{p}$ is any list of n points of P, then, for any two points p_0, q_0,

$$\left(\sum_i (p_i - p_0)\omega_i\right) + p_0 = \left(\sum_i (p_i - q_0)\omega_i\right) + q_0.$$

Indeed, set $q_0 - p_0 = v$; then the left-hand side is

$$\begin{aligned}\left(\sum_i (p_i - q_0 + v)\omega_i\right) + p_0 &= \left(\sum_i (p_i - q_0)\omega_i + \sum_i v\omega_i\right) + p_0\\ &= \left(\sum_i (p_i - q_0)\omega_i\right) + (v + p_0)\\ &= \left(\sum_i (p_i - q_0)\omega_i\right) + q_0,\end{aligned}$$

because $\sum \omega_i = 1$. Thus the point $q = \sum (p_i - p_0)\omega_i + p_0$ of P does not depend on the choice of p_0, but only on the list $\mathbf{p}$ and the weights $\boldsymbol{\omega}$. We call this point the *average* of the list $\mathbf{p}$ with *weights* $\boldsymbol{\omega}$, and we write this average as $q = \sum p_i \cdot \omega_i$, so that

$$\sum p_i \cdot \omega_i = \left(\sum (p_i - p_0)\omega_i\right) + p_0, \qquad \text{for any } p_0 \in P. \tag{9}$$

For each list $\boldsymbol{\omega}$ of weights, $\mathbf{p} \to \sum p_i \cdot \omega_i$ is thus an n-ary operation $P^n \to P$. The result of this operation is written in the notation $\sum p_i \cdot \omega_i$, "as if" it were an actual linear combination in some vector space. But notice that the "scalar multiple" $p_i \cdot \omega_i$ (written with a dot between, as a reminder) has no meaning by itself; only the *whole* expression $\sum p_i \cdot \omega_i$ has meaning (and is a point), and this only when $\sum \omega_i = 1$.

By this definition, these averages may be manipulated just as if they were actual linear combinations, provided always that the sum of the weights used is 1. For instance, for any permutation $\sigma : \mathbf{n} \to \mathbf{n}$, one has $\sum p_i \cdot \omega_i = \sum p_{\sigma i} \cdot \omega_{\sigma i}$. The general distributive law

$$\sum_{j=1}^{n}\left(\sum_{i=1}^{m_j} p_{ij} \cdot \eta_{ij}\right) \cdot \omega_j = \sum_{j=1}^{n}\sum_{i=1}^{m_j} p_{ij} \cdot (\eta_{ij}\omega_j) \tag{10}$$

is correct as soon as it makes sense (n and each of $m_1, \ldots, m_n$ a positive integer, $\omega_1 + \cdots + \omega_n = 1$ and $\eta_{1j} + \cdots + \eta_{m_j,j} = 1$ for each $j \in \mathbf{n}$); the right-hand side is a single average over $\sum m_j$ points. For *any* list $\boldsymbol{\kappa}$ of n scalars and any list $\mathbf{w}$ of n vectors,

$$\left(\sum w_i \kappa_i\right) + p = p \cdot \left(1 - \sum \kappa_i\right) + \sum (w_i + p) \cdot \kappa_i, \tag{11}$$

where the right-hand side is an $(n + 1)$-fold average—as follows at once from the definition (9) of this average with the choice $p_0 = p$. In particular, this formula (11) includes the cases

$$(v + w) + p = p \cdot (-1) + (v + p) \cdot 1 + (w + p) \cdot 1, \tag{12}$$

$$w\kappa + p = p \cdot (1 - \kappa) + (w + p) \cdot \kappa, \tag{13}$$

which state how the sum and the scalar multiples of translations v and w can be expressed in terms of averages of points. An especially useful average is

$$p \cdot 1 + q \cdot (-1) + r \cdot 1 = (p - q) + r;$$

it gives the fourth vertex of the parallelogram with vertices p, q, and r.

Lemma 1. *Any n-fold average, for $n > 2$, can be written as a composite of 2-fold averages.*

Proof: We first derive two identities for n-fold averages. For $n > 2$ and $\omega_1 \neq 1$, set $\lambda = (1 - \omega_1)^{-1}$. Then, by the distributive law (10),

$$p_1 \cdot \omega_1 + \cdots + p_n \cdot \omega_n = p_1 \cdot \omega_1 + [p_2 \cdot (\omega_2\lambda) + \cdots + p_n \cdot (\omega_n\lambda)] \cdot (1 - \omega_1)$$

is an $(n - 1)$-fold average followed by a 2-fold one. For $n > 3$ and $\omega_1 + \omega_2 \neq 1$, set $\mu = (1 - \omega_1 - \omega_2)^{-1}$. Then, by the distributive law again,

$$p_1 \cdot \omega_1 + \cdots + p_n \cdot \omega_n$$
$$= p_1 \cdot \omega_1 + p_2 \cdot \omega_2 + [p_3 \cdot (\omega_3\mu) + \cdots + p_n \cdot (\omega_n\mu)] \cdot (1 - \omega_1 - \omega_2)$$

an $(n - 2)$-fold average followed by a 3-fold one. Now for $n = 3$, $\omega_1 + \omega_2 + \omega_3 = 1$ implies (since $2 \neq 0$) that at least one ω_i is not 1; hence any 3-fold average may be reduced by the first of these identitie For $n > 3$, at least one of ω_1, ω_2, or $\omega_1 + \omega_2$ is not 1, so any n-fold average reduces by one or the other of these two identities. With induction on n, the lemma is proved.

We will now regard the affine space P as an algebraic system with many n-ary operations, namely the operations of n-fold average with weights ω for *every* n and *every* ω. In particular, an affine transformation (a "morphism of affine spaces") will be defined to be a morphism of all these operations:

DEFINITION. *If P and P' are affine spaces over the same field, an* affine transformation $a: P \to P'$ *is a function on P to P' with*

$$a(p_1 \cdot \omega_1 + \cdots + p_n \cdot \omega_n) = a(p_1) \cdot \omega_1 + \cdots + a(p_n) \cdot \omega_n \tag{14}$$

for all n, all points $p_i \in P$ and all lists ω with $\omega_1 + \cdots + \omega_n = 1$.

By Lemma 1, it suffices to assume this property for $n = 2$.

Lemma 2. *Every translation is an affine transformation.*

Proof: We consider translation by a vector v. By the definition of an average,

$$\begin{aligned} v + \sum p_i \cdot \omega_i &= [\, v + \sum (p_i - p_0)\omega_i \,] + p_0 \\ &= [\, \sum v\omega_i + \sum (p_i - p_0)\omega_i \,] + p_0 \\ &= [\, \sum (v + (p_i - p_0))\omega_i \,] + p_0 = \sum (v + p_i) \cdot \omega_i; \end{aligned}$$

this shows that $p \mapsto v + p$ preserves averages, hence is affine.

Lemma 3. *A bijection $f: P \to P$ is a translation if and only if*

$$f(q) = q \cdot 1 + f(p) \cdot 1 + p \cdot (-1) \tag{15}$$

for all points p, $q \in P$.

This formula has an evident geometric meaning; a translation taking p to p' and q to q' gives a parallelogram with parallel sides $pp' \parallel qq'$, and

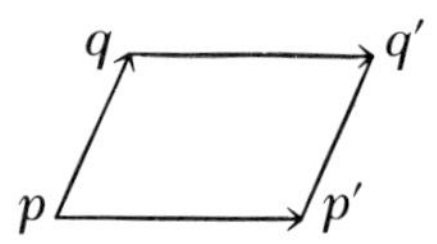

this formula gives the fourth vertex q' of the parallelogram as the evident average of the other three vertices.

Proof: A translation f has $f(p) = v + p$ for some vector v. By the definition of an average (with $p_0 = p$)

$$\begin{aligned} q \cdot 1 + (v + p) \cdot 1 + p \cdot (-1) &= (q - p) + ((v + p) - p) \cdot 1 + 0(-1) + p \\ &= (q - p) + v + p = v + q = f(q), \end{aligned}$$

which proves (15). Conversely, let the bijection f satisfy (15). Choose any point p and define a vector v by $v = fp - p$. Then, by (15) and the definition of an average,

$$\begin{aligned} f(q) &= q \cdot 1 + (fp) \cdot 1 + p \cdot (-1) \\ &= [(q - p) + (fp - p) + (p - p)(-1)] + p \\ &= [(q - p) + v] + p = v + q \end{aligned}$$

for any point q, so f is indeed translation by the vector v.

Theorem 1. *Every affine space P over F is affine-isomorphic to the affine space $V^\flat$ defined from some finite-dimensional vector space V over F. Explicitly, if p_0 is any point in P, the assignment $v \mapsto v + p_0$ is an isomorphism $(P^{\#})^\flat \cong P$ of affine spaces.*

Proof: By the axioms, we know that $v \mapsto v + p_0$ is a bijection from the *set* $P^{\#\flat}$ of translations of P to the set P of points. Now the formula (11) with $\sum \kappa_i = 1$ and $p = p_0$ shows that this bijection takes averages $\sum w_i \kappa_i$ of vectors to averages of points, hence is an affine transformation $(P^{\#})^{\flat} \to P$, as required.

Proposition 2. *If V and V' are finite-dimensional vector spaces over F with $V^{\flat}$ and $V'^{\flat}$ the corresponding affine spaces, then a function $f: V \to V'$ is a linear transformation if and only if $f(0) = 0$ and $f: V^{\flat} \to V'^{\flat}$ is an affine transformation.*

Proof: Any linear transformation f preserves linear combinations, hence is surely an affine transformation. Conversely, suppose f affine, and let λ_1, λ_2 be any two scalars. Any linear combination $v_1\lambda_1 + v_2\lambda_2$ of vectors of V may then be written in $V^{\flat}$ as a 3-fold average

$$v_1\lambda_1 + v_2\lambda_2 = 0 \cdot (1 - \lambda_1 - \lambda_2) + v_1 \cdot \lambda_1 + v_2 \cdot \lambda_2.$$

Since f preserves averages *and* 0, it preserves linear combinations, hence is linear.

Appendix on affine axioms: We have defined an affine space in terms of a given vector space of translations. However, an affine space P over F can be described directly as an algebraic system: A set P equipped with one n-ary operation $\mathbf{p} \mapsto \sum p_i \cdot \omega_i$ for each n and each list ω of n weights of F, subject to the following axioms on these operations:

(i) If $\mathbf{p}$ is a list (or row) of m points, M an $m \times n$ matrix of scalars with the sum of each column 1, and ω a list of n weights, then

$$(\mathbf{p} \cdot M) \cdot \omega = \mathbf{p} \cdot (M \cdot \omega), \tag{16}$$

where the matrix product $\mathbf{p} \cdot M$ is the list of the n points $\sum_i p_i \cdot M_{ij}$, well defined since $\sum_i M_{ij} = 1$;

(ii) The general distributive law (10);

(iii) For every point p, $p \cdot 1 = p$;

(iv) For all points p and q, $p \cdot 1 + q \cdot 0 = p$.

These axioms suffice to show that the averages have all the algebraic properties of actual linear combinations with coefficient sum always 1; from them one may define translations and derive the previous definition of an affine space, as stated in Exercises 5–7 below.

EXERCISES

1. Prove the general distributive law (10) from the definition of an average.

2. Establish the following alternative description of an affine space (W. F. Eberlein): Let P be a set, V a finite-dimensional vector space over F, and

$(p, q) \mapsto p - q$ a function $P \times P \to V$ satisfying the following axioms: (i) $(p - q) + (q - r) = p - r$; (ii) (7); (iii) To each $q \in P$ and each $v \in V$ there exists $p \in P$ with $p - q = v$. Prove that P is an affine space with translation space $P^{\#}$ isomorphic by $\theta: V \cong P^{\#}$ to V in such a way that $\theta(p - q)$ is the difference of points as defined in the text.

3. If F has characteristic 2, show that Lemma 1 holds with "2-fold" replaced by "3-fold".

4. Define a product $P \times P'$ of two affine spaces over F, and prove it a product in the categorical sense (§ III.11).

Exercises on the Appendix (Using the Axioms Stated There)

5. (a) For any permutation σ of $\mathbf{n}$, prove that $\sum p_i \cdot \omega_i = \sum p_{\sigma i} \cdot \omega_{\sigma i}$ (*Hint:* Use (16) of axiom (i), with M a permutation matrix.)

(b) For all $\omega_1 + \omega_2 = 1$, prove $p \cdot \omega_1 + p \cdot \omega_2 = p$.

(c) For all $\eta_1 + \eta_2 = 0$, prove $p \cdot 1 + q \cdot \eta_1 + q \cdot \eta_2 = p$.

6. Prove Lemma 1 for these averages.

7. (a) Define a "translation" of P by the parallelogram formula, show that every translation is an affine transformation and that the translations of P form an abelian group under composition.

(b) Show that the set $P^{\#}$ of all translations of P is a vector space over F under the vector operations defined by (12) and (13).

(c) If $P^{\#}$ is finite-dimensional, deduce that P and $V = P^{\#}$ satisfy the definition of the text.

8. (a) If P is an affine space of dimension at least two and $\square: P \times P \to P$ a binary operation such that every affine transformation $a: P \to P$ is a morphism of $\square$, prove that $\square$ is an average.

(b) Establish a corresponding result for n-ary operations.

3. The Affine Group

We have already defined an affine transformation $P \to P'$ as a function preserving averages; from this definition it is clear that the identity function $P \to P$ is an affine transformation and that the composite of two affine transformations, when defined, is affine. Hence the class of all affine spaces, with these affine transformations as morphisms, constitutes a concrete category, the *category of affine spaces* over F. The assignment $P \mapsto P^{\#}$ sending each affine space to its vector space of translations becomes a functor on this category to the category of vector spaces over F, when combined with a suitable assignment $a \mapsto a^{\#}$ for affine transformations a.

Theorem 3. *Let P and P' be affine spaces and $P^{\#}$, $P'^{\#}$ the corresponding vector spaces of translations. Then to each affine transformation $a: P \to P'$ there is a unique linear transformation $a^{\#}: P^{\#} \to P'^{\#}$ such that*

$$a(w + p) = a^{\#}(w) + a(p) \tag{17}$$

for all translations w of P and all points $p \in P$. If this equation holds for all translations w and one point $p = p_0$, it holds for all translations w and all points p.

Proof: For fixed p_0, Theorem 1 asserts that $w \mapsto w + p_0$ is an affine isomorphism $P^{\#\flat} \to P$. Now for each affine transformation a the composite

$$\begin{array}{ccccccc} P^{\#\flat} & \to & P & \xrightarrow{a} & P' & \to & P'^{\#\flat} \\ w & \mapsto & w + p_0 & \mapsto & a(w + p_0) & \mapsto & a(w + p_0) - a(p_0), \end{array}$$

is affine and takes 0 to 0, hence is a linear transformation $a^{\#}: P^{\#} \to P'^{\#}$ by Proposition 2. Its definition as a composite reads $a^{\#}(w) = a(w + p_0) - a(p_0)$. This is the desired Equation (17) for $p = p_0$. Any other point p_1 can be written as $p_1 = w_1 + p_0$ for some translation w_1, so

$$\begin{aligned} a(w + p_1) &= a(w + w_1 + p_0) = a^{\#}(w + w_1) + a(p_0) \\ &= a^{\#}(w) + a^{\#}(w_1) + a(p_0) = a^{\#}(w) + a(p_1). \end{aligned}$$

This proves the desired Equation (17) for all $p = p_1$.

Each affine map a thus determines a linear map $a^{\#}$, called the *trace* of a, by the condition (17) or the equivalent condition

$$a^{\#}(q - p) = a(q) - a(p), \qquad \text{all } p, q \in P. \tag{18}$$

Clearly the trace of a composite of affine maps is the composite of their traces, in the same order. Hence $P \mapsto P^{\#}$ and $a \mapsto a^{\#}$ is a functor.

If p_0' is a fixed point in P', the *constant* function $P \to P'$ sending every point of P to this one point $p_0' \in P'$ clearly preserves all averages, hence is an affine transformation, of trace zero. Every transformation of trace zero must by (17) have this form. Similarly, we may construct all the affine maps with a given trace:

Theorem 4. *In the affine spaces P and P' choose points $p_0 \in P$ and $p_0' \in P'$. Then to each linear transformation $t: P^{\#} \to P'^{\#}$ there is exactly one affine transformation $a: P \to P'$ with $a(p_0) = p_0'$ and with trace $a^{\#} = t$. This transformation a is defined by $a(w + p_0) = t(w) + p_0'$ for all $w \in P^{\#}$.*

Proof: The composite $p \mapsto p - p_0 \mapsto t(p - p_0) \mapsto t(p - p_0) + p_0'$, as in

$$P \to P^{\#\flat} \xrightarrow{t} P'^{\#\flat} \to P',$$

is an affine transformation a, has $a(p_0) = p_0'$, and $a(w + p_0) = t(w) + a(p_0)$. This last equation determines a uniquely from t and p_0 and shows $a^{\#} = t$, as required.

This theorem describes all affine transformations $P \to P'$, in terms of given "origins" p_0 and p_0'. There is one such transformation a for each pair (u', t) consisting of a vector $u' \in P'^{\#}$ and a linear transformation $t: P^{\#} \to P'^{\#}$; namely, the transformation $a(u', t)$ carrying p_0 to $u' + p_0'$ with trace t. In other words, $a(u', t)$ is defined for all $w \in P^{\#}$ by

$$[a(u', t)](w + p_0) = tw + u' + p_0'. \tag{19}$$

Composing two such transformations $P \to P'$ and $P' \to P''$ gives

$$a(u'', t') \circ a(u', t) = a(u'' + t'u', t' \circ t) \tag{20}$$

—parallel to the formula (1) for the affine line.

This construction may also be formulated in vector space terms, as follows. To vector spaces V and V' construct the corresponding affine spaces $V^{\flat}$ and $V'^{\flat}$; an affine map $V^{\flat} \to V'^{\flat}$ is then just a linear map $V \to V'$ followed by a translation of V' (thus, $w \mapsto tw \mapsto tw + u'$). Briefly: Affine maps are composites of linear maps and translations.

By (19) for $P = P'$, an affine endomorphism $a = a(u', t)$ of P is invertible if and only if $t: P^{\#} \to P^{\#}$ is invertible, while by (20) its inverse then is $a(-t^{-1}(u'), t^{-1})$.

The *affine group* of an affine space P is the group $A(P)$ of all invertible affine transformations $a: P \to P$. We have already described in § 1 the group of the affine line; the group $A(P)$ has a similar structure:

Theorem 5. *For the group $A(P)$ of an affine space P there is a short exact sequence of morphisms of groups,*

$$0 \to (P^{\#})^+ \to A(P) \xrightarrow{\tau} GL(P^{\#}) \to 1. \tag{21}$$

Here $P^{\#}$ is the vector space of all translations of P, $(P^{\#})^+$ its additive group, and $GL(P^{\#})$ its general linear group; moreover, the monomorphism $(P^{\#})^+ \to A(P)$ is the inclusion (every translation is an affine map), while $\tau: A(P) \to GL(P^{\#})$ assigns to each affine transformation a its trace $a^{\#}$. This morphism τ has a right inverse which is a morphism of groups.

Proof: The exactness of the sequence means that the image of each morphism is the kernel of the next. At $(P^{\#})^+$, this is immediate. At $A(P)$, the kernel of τ consists of those affine transformations $a: P \to P$ with trace the identity. By (17), such a transformation has $a(w + p) = w + a(p)$ and hence $a(w + p) - (w + p) = a(p) - p$; this states that the vector $u = a(p) - p$ is independent of p and that the transformation a is translation by $u \in (P^{\#})^+$. Hence the kernel of τ is the image of $(P^{\#})^+$. Finally, exactness at $GL(P^{\#})$ states that τ is an epimorphism; this will follow from the construction of the right inverse of τ. To construct this, choose an origin p_0. Then, in the notation (19), $t \mapsto a(0, t)$ is a morphism $GL(P^{\#}) \to A(P)$ of groups which is a right inverse for τ.

Relative to the origin p_0, the group $A(P)$ can be described as the set $(P^{\#})^{+} \times GL(P^{\#})$ of all pairs (u, t), with the multiplication given by (20).

Affine geometry may be described as the study of equivalence under the action of the affine group $A(P)$ on P and on the set of subsets of P. For example, a "figure" in P is any subset X of P. Two figures X and X' are *affine equivalent,* if and only if there is an invertible affine transformation $a: P \to P$ with $a(X) = X'$. Also, X is *parallel* to X' when there is a translation $w: P \to P$ with $w_*(X) = X'$. (This is equivalence under the action of the group of translations.)

To describe coordinates, let P have dimension n. A *frame* $\mathbf{f}$ in P is a list $f_0, \ldots, f_n$ of $n + 1$ points of P such that the vectors $f_1 - f_0, \ldots, f_n - f_0$ form a basis of $P^{\#}$. Relative to a frame f, each point p may therefore be written as

$$p = [(f_1 - f_0)\xi_1 + \cdots + (f_n - f_0)\xi_n] + f_0 \tag{22}$$

for a unique list $\boldsymbol{\xi}$ of n scalars, called the *linear coordinates* of the point p relative to the frame f. This expression may also be written as an average

$$p = f_0 \cdot (1 - \xi_1 - \cdots - \xi_n) + f_1 \cdot \xi_1 + \cdots + f_n \cdot \xi_n. \tag{23}$$

Note that choosing a frame involves choosing an "origin" f_0 in P and a basis $\mathbf{f} - f_0 = (f_1 - f_0, \ldots, f_n - f_0)$ in $P^{\#}$.

By Theorem 4, an affine transformation $a: P \to P'$ is determined by giving $a(f_0)$ and the trace $a^{\#}$, and $a^{\#}$ in turn is determined by giving the images $a^{\#}(f_i - f_0)$ of the basis vectors of $P^{\#}$. This proves

Proposition 6. *If* $\mathbf{f}$ *is a frame in an affine space* P *of dimension* n *and* $\mathbf{p}'$ *is a list of* $n + 1$ *points in a second affine space* P', *there is exactly one affine transformation* $a: P \to P'$ *with* $a(f_i) = p_i'$ *for* $i = 0, \ldots, n$.

It follows that any two frames in P are affine equivalent.

For a vector space V over F, a "form" is a function on V to the field F of scalars, as in the case of linear "forms" and quadratic "forms". For an affine space P over F, functions on P to F (i.e., to $F^{\flat}$) will be termed "functionals". For example, an *affine functional* k on P is defined to be an affine transformation $k: P \to F^{\flat}$ on P to the one-dimensional affine space $F^{\flat}$ (the vector space F regarded as an affine space). Let P have dimension n, and take a frame $\mathbf{f}$ in P. By Proposition 6, there is for each $i \in \mathbf{n}$ exactly one affine functional $x_i: P \to F^{\flat}$ with $x_i(f_i) = 1$ and $x_i(f_j) = 0$ for $i \neq j$. Since this functional preserves averages, it takes the point p of (23) to its ith coordinate ξ_i. We call this affine functional $x_i: P \to F^{\flat}$ the ith *coordinate functional* of P, relative to the frame $\mathbf{f}$. An arbitrary affine functional

$k: P \to F^\flat$ is determined by its value $k(f_0) = \nu$ at the origin and by the n values $k^\#(f_i - f_0) = \mu_i$ of its trace $k^\#$. By (22) one then has

$$k(p) = \sum \mu_i \xi_i + \nu = \sum \mu_i x_i(p) + \nu$$

for any point $p \in P$. This result may be recorded as an equation between functionals

$$k = \mu_1 x_1 + \cdots + \mu_n x_n + \nu \tag{24}$$

(ν is the constant functional with value $\nu \in F$ at every point of P). In other words, an affine functional k is one which can be expressed as a non-homogeneous linear combination of the coordinates (relative to any frame).

Next let $a: P \to P'$ be any affine transformation and choose frames f and f' in the affine spaces P and P' of dimensions n and m, respectively. Since each coordinate functional $x_i': P' \to F^\flat$ is an affine transformation, so is the composite $x_i' \circ a$; it may then be expressed as in (24) as a linear combination

$$x_i' \circ a = \sum_{j=1}^{n} A_{ij} x_j + B_i, \qquad i = 1, \ldots, m, \tag{25}$$

where A is an $m \times n$ matrix and B an $m \times 1$ matrix, both over F. These two matrices A and B determine $a: P \to P'$, relative to the frames $\mathbf{f}$ and $\mathbf{f}'$. We may also describe the matrices A and B for the affine transformation a by using (19). With $p_0 = f_0$ and $p_0' = f_0'$ as origins, a has the form $a = a(u', t)$ for some vector $u' \in P'^\#$ and some linear transformation $t: P^\# \to P'^\#$, and A is the matrix of t, rel $\mathbf{f} - f_0$ and $\mathbf{f}' - f_0'$; B is the list of $(\mathbf{f}' - f_0')$-coordinates of u'. If $P' = P$ and $f' = f$, the Equations (25) with $x_i' = x_i$ represent a transformation of P to itself (an alibi); if A is invertible, they can also be interpreted as the equations describing a change of coordinates (an alias).

Proposition 7. *Any two non-constant affine functionals on an affine space P are affine equivalent.*

Proof: If the functional $k = \sum \mu_i x_i + \nu$ of (24) is not constant, some $\mu_i \neq 0$, so $\mu_1, \ldots, \mu_n$ is the first column of an $n \times n$ invertible matrix A. This matrix A in (25), with $B_1 = \nu$, determines an affine automorphism $a: P \to P$ with $x_1 \circ a = k$. Thus any *non constant affine functional k is equivalent to a coordinate functional x_j*; in particular, any two such k are equivalent to each other.

Appendix on barycentric coordinates: The expression (23) for p in terms of a frame may be written more symmetrically as an average

$$p = f_0 \cdot \zeta_0 + f_1 \cdot \zeta_1 + \cdots + f_n \cdot \zeta_n \tag{26}$$

with $n + 1$ scalars ζ_i of sum 1. This (unique) list of scalars is called the list of *barycentric coordinates* of p, relative to the frame $\mathbf{f}$. For example, in the real affine plane, a frame consists of the vertices of a (non-degenerate) triangle, as in the figure below.

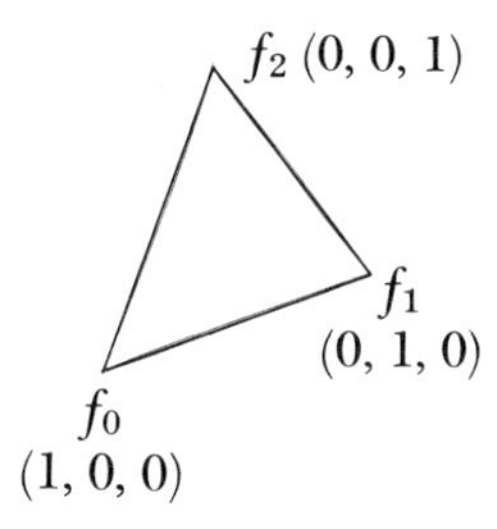

The vertices of the triangle have the barycentric coordinates (1, 0, 0), (0, 1, 0) and (0, 0, 1), while $(\frac{1}{3}, \frac{1}{3}, \frac{1}{3})$ are the barycentric coordinates of the center of gravity of the triangle. Also, the points inside the triangle are the points with all barycentric coordinates positive, a point on the side of the line f_1f_2 opposite f_0 has $\zeta_0 < 0$, and so on.

EXERCISES

1. Show that an affine transformation is an injection (a surjection, or a bijection, respectively) if and only if its trace is a monomorphism (an epimorphism or an isomorphism, respectively).

2. Show that the subgroup of all invertible affine transformations $P \to P$ leaving a given point p_0 fixed is isomorphic to the general linear group of $P^{\#}$.

3. Show that the affine transformations $a: P \to P'$ can be represented, relative to frames $\mathbf{f}$ and $\mathbf{f}'$ in P and P' as in (25), by $(m + 1) \times (n + 1)$ matrices of the form

$$a \mapsto \begin{bmatrix} A & B \\ 0 & 1 \end{bmatrix},$$

in such a way that the composition of transformations corresponds to the product of matrices.

4. Show that the set hom (P, P') of all affine transformations $P \to P'$ is itself an affine space when the averages of transformations are defined pointwise.

5. (a) In an affine space P show that the following three properties of a list $p_0, p_1, \ldots, p_k$ of $k + 1$ points are logically equivalent:

(i) The vectors $p_1 - p_0, \ldots, p_k - p_0$ are linearly independent in $P^{\#}$;

(ii) No two averages $\sum_i p_i \cdot \omega_i$ with distinct lists of weights $\omega_0, \ldots, \omega_k$ are equal;

(iii) No p_j is an average of the remaining points of the list.

(b) A list of points $p_0, \ldots, p_k$ with one (and hence all) of these properties is said to be *affine independent*. Show that a frame in P is a maximal affine independent list.

6. (a) For given n, show that the set of all lists ζ of $n + 1$ scalars with $\zeta_0 + \cdots + \zeta_n = 1$ is an affine space when the averages of such lists are defined termwise. (Thus a two-fold average of lists ζ and ζ', with weights $\omega_1 + \omega_2 = 1$, is to be

$$\zeta \cdot \omega_1 + \zeta' \cdot \omega_2 = (\zeta_0\omega_1 + \zeta_0'\omega_2, \ldots, \zeta_n\omega_1 + \zeta_n'\omega_2);$$

(b) Prove that the space so constructed is isomorphic to any n-dimensional affine space P over the same field in such a way that the ζ's become barycentric coordinates in P.

4. Affine Subspaces

In plane analytic geometry the line through two distinct points (x_0, y_0) and (x_1, y_1) in the plane consists of the points (x, y) given by the equations

$$x = x_0(1 - t) + x_1 t, \qquad y = y_0(1 - t) + y_1 t,$$

in terms of the real "parameter" t. If we write p for (x, y), these equations express the point p as an average $p = p_0 \cdot (1 - t) + p_1 \cdot t$. A line L in any real affine plane P may be described as a proper subset of P with at least two points and such that *any* average of points of L is still a point of L. As in § 1, this line is itself a (one-dimensional) affine space when its translations are taken to be the restrictions to L of those translations u of P for which $u_*(L) \subset L$ (those translations which slide L along itself).

For any affine space P, an affine space S is defined to be an *affine subspace* of P when the set S is a subset of P and the insertion $j: S \to P$ of this subset is an affine transformation. As j, like any affine transformation, must carry averages to averages, this subset S must be closed under all averages (i.e., any n-fold average in P of points in S is in S). Conversely, this closure condition on a non-void subset of P suffices to make that subset an affine subspace. The proof of this fact is more delicate than in previous cases (Theorem III.10 for groups or Theorem IV.4 for rings) because an affine space P has been defined not in terms of operations on P itself, but in terms of vector space operations on a set $P^\#$ of bijections of P.

Theorem 8. *Let P be an affine space. Any affine subspace of P is closed under all averages. Conversely, if S is a non-void subset of P closed under all averages in P, then S is an affine space and a subspace of P for exactly one vector space $S^\#$ of translations of S.*

Proof: First we show that there is only one choice for $S^\#$. If the insertion $j: S \to P$ is to be an affine transformation, its trace $j^\#$ must be a linear transformation, with $u + s = j(u + s) = j^\#(u) + s$ for all points $s \in S$ and all translations u of S. This equation states that the translation $s \mapsto u + s$

of S by u is the restriction to S of the translation $j^{\#}(u)$ of P. Since $u \neq u'$ implies $u + s \neq u' + s$, $j^{\#}: S^{\#} \to P^{\#}$ is a monomorphism. Since $S^{\#}$ contains a translation $s' - s$ for each pair of its points s', s, the image of $j^{\#}$ must be all the translations $js' - js \in P^{\#}$. These results completely specify $S^{\#}$.

Now suppose S a subset of P closed under all averages; we wish to construct these translations of S. Let $D(S) = \{w \mid w \in P^{\#} \text{ and } w_* S \subset S\}$; the formulas (12) and (13) for sums and scalar multiples of translations in terms of averages in P show that $D(S)$ is a vector subspace of $P^{\#}$; we call it the "direction" of S. This subspace $D(S)$ contains all differences $w = s' - s$ of points s', $s \in S$, for, if r is any other point of S, the "parallelogram" formula (15) for the translation w shows that

$$w + r = r \cdot 1 + (w + s) \cdot 1 + s \cdot (-1) = r \cdot 1 + s' \cdot 1 + s \cdot (-1);$$

this is a 3-fold average of points in S, hence in S by closure. These conclusions show that the function $P^{\#} \times P \to P$ given as $(v, p) \mapsto v + p$ restricts to a function $D(S) \times S \to S$ which satisfies both the axioms (i) and (ii) in the definition of an affine space. The translations of this space S are then the restrictions to S of the translations $w \in D(S)$ of P. The definition of averages in terms of translations shows that the insertion $j: S \to P$ is affine, completing the proof.

From this theorem it follows that the intersection of two affine subspaces, if not empty, is an affine subspace. Here is another way of generating subspaces:

Corollary. *If $k: P \to F^{\flat}$ is an affine functional on P, then for each $\mu \in F$ the inverse image $k^{-1}\{\mu\}$ is empty or an affine subspace of P.*

Proof: An average of any number of equal μ's is μ, so the inverse image set in question is closed under all averages.

EXERCISES

1. If V is a finite-dimensional vector space and $V^{\flat}$ the corresponding affine space, show that a subset of $V^{\flat}$ is an affine subspace if and only if it is a coset of some vector subspace of V.

2. Given points $p_1, \ldots, p_k$ of P, show that the set of all k-fold averages of these points is the smallest affine subspace of P containing them all (the "affine span" of P).

3. Let P be an affine space and $D(S) \subset P^{\#}$ the direction of any affine subspace S of P, as described in the text. Prove:

(a) S is parallel to S′ if and only if $D(S) = D(S')$.

(b) $S = S'$ if and only if $D(S) = D(S')$ and $S \cap S' \neq \varnothing$.

(c) Given a vector subspace T of $P^{\#}$ and a point $p_0 \in P$, there is exactly one affine subspace of P containing the point p_0 and with direction T.

4. Prove that two affine subspaces of P are equivalent (under the action of the affine group of P on subsets of P) if and only if they have the same dimension.

5. (a) Show that the set of all solutions of a system of m non-homogeneous linear equations $\sum_j A_{ij}x_j = b_i$, $i = 1, \ldots, m$, in n unknowns $x_1, \ldots, x_n$ and with coefficients A_{ij} and b_i in a field F is an affine subspace of $(F^n)^\flat$ (as usual, the characteristic of F is not 2).

*(b) Prove that every affine subspace of $(F^n)^\flat$ can be represented in this form.

5. Biaffine and Quadratic Functionals

If P and P' are affine spaces over F, a function $b: P \times P' \to F^\flat$ is called a *biaffine functional* when for each point $p' \in P$ the partial function $p \mapsto b(p, p')$ is an affine transformation $P \to F^\flat$ and for each p the partial function $p' \mapsto b(p, p')$ is an affine transformation $P' \to F^\flat$. If $P = P'$, a biaffine functional $b: P \times P \to F^\flat$ is called *symmetric* when $b(p, q) = b(q, p)$, for all points $p, q \in P$.

The product of two affine functionals is biaffine. Explicitly, if $k: P \to F^\flat$ and $k': P' \to F^\flat$ are both affine functionals, their product (taken pointwise on F) is the functional $kk': P \times P' \to F^\flat$ given by $(p, p') \mapsto (kp)(k'p')$; it is biaffine.

Biaffine functionals may be constructed from bilinear forms on the corresponding translation spaces $P^\#$ and $P'^\#$. Given points $p_0 \in P$ and $p_0' \in P'$ (to serve as "origins" in P and P'), a scalar γ, and functions

$$h: P^\# \times P'^\# \to F, \qquad f: P^\# \to F, \qquad f': P'^\# \to F,$$

with h bilinear and f and f' both linear, the formula

$$b(w + p_0, w' + p_0') = h(w, w') + f(w) + f'(w') + \gamma, \qquad w \in P^\#, \; w' \in P'^\#, \tag{27}$$

defines a function $b: P \times P' \to F^\flat$. For w' fixed, this function b is the sum of a linear function of w and a scalar $f'(w') + \gamma$, hence it is an affine functional of $w + p_0$. Symmetrically, b is affine in $p' = w' + p_0'$ when w is fixed. Hence b is a biaffine functional. The scalar γ is the value $\gamma = b(p_0, p_0')$ of b at the origins. In case $P = P'$ and $p_0 = p_0'$, b is symmetric if and only if the bilinear form h is symmetric and the linear forms f and f' are equal.

We will soon prove that every biaffine functional can be expressed in the form (27), but first we rewrite the formula (27) in terms of coordinates. Choose any (linear) coordinates x_i in P and x_j' in P' for $i \in \mathbf{n}$ and $j \in \mathbf{m}$. Then $x_i: P \to F^\flat$ and $x_j': P' \to F^\flat$ are affine functionals, and so the product $x_i x_j': P \times P' \to F^\flat$ is biaffine, as is any constant functional κ or any product

$\kappa x_j'$ of x_j' by a constant $\kappa : P \to F^\flat$. In (27), h is expressed by an $n \times n$ matrix A, f by a $1 \times n$ matrix B, f' by B'; all told, b has the form

$$b(w + p_0, w' + p_0') = \sum_{i,j} x_i A_{ij} x_j' + \sum_i B_i x_i + \sum_j B_j' x_j' + \gamma; \quad (28)$$

in other words, the biaffine functional b is expressed as a polynomial in the coordinate functionals, of degree at most 1 in each of the lists $x_1, \ldots, x_n$ and $x_1', \ldots, x_m'$. The bilinear part $h = \sum x_i A_{ij} x_j'$ can be described directly from b without reference to coordinates or choice of origin. To do this, take two points $p = w + p_0$ and $q = v + p_0$ in P. Then, (27), the difference $b(p, p') - b(q, p') = h(w - v, w') + f(w - v)$ involves neither f' nor γ; taking still another difference will eliminate f. This observation motivates the following theorem.

Theorem 9. *To each biaffine functional $b : P \times P' \to F^\flat$ there is a unique bilinear form $h : P^\# \times P'^\# \to F$ on the corresponding translation spaces $P^\#$ and $P'^\#$ such that*

$$b(p, p') - b(q, p') - b(p, q') + b(q, q') = h(p - q, p' - q') \quad (29)$$

for all points p, $q \in P$ and p', $q' \in P'$. If $P = P'$, b symmetric implies h symmetric.

As in previous cases, we call h the (bilinear) *trace* of b, and write $h = b^\#$.

Proof: For p' fixed, the partial function $b(-, p') : P \to F^\flat$ is affine, hence has trace a function $d : P^\# \times P' \to F$ with the usual description

$$d(p - q, p') = b(p, p') - b(q, p')$$

of a trace. For each p', $w \mapsto d(w, p')$ is a trace, hence linear in w. For each $w = p - q$, $p' \mapsto d(w, p')$ is the difference of two affine functionals $b(p, -)$ and $b(q, -)$, hence is affine. If we take the trace of this affine functional we obtain a function $h : P^\# \times P'^\# \to F$ with the usual description

$$h(w, p' - q') = d(w, p') - d(w, q')$$

of a trace (in the second argument). As a trace, $h(w, w')$ is linear in w'; as a difference of two linear functions of w, it is also linear in w. This makes h bilinear. Setting $w = p - q$ in the last equation and using the previous equation gives the desired conclusion (29). This formula (29) clearly determines h uniquely. Finally, when $P = P'$ the symmetries $b(p, p') = b(p', p)$, etc., imply that h is symmetric (as a bilinear form).

Proposition 10. *Let $b : P \times P' \to F^\flat$ be a biaffine functional with trace $b^\# : P^\# \times P'^\# \to F$. Given points $p_0 \in P$ and $p_0' \in P'$, there are* *unique linear forms $f:P^{\#} \to F$ and $f':P'^{\#} \to F$ such that*

$$b(w + p_0, w' + p_0') = b^{\#}(w, w') + f(w) + f'(w') + b(p_0, p_0') \qquad (30)$$

holds for all vectors $w \in P^{\#}$ and $w' \in P'^{\#}$. If $P = P'$, $p_0 = p_0'$, and b is symmetric, then $f = f'$ and $b^{\#}$ is symmetric as a bilinear form.

Once proved, this will establish that every biaffine functional is indeed of the sort cited as an example in (27) above.

The linear forms f and f' in (30) depend not only on the given functional b, but also on the choice of "origins" p_0 and p_0'. Indeed, set $w' = 0$ in the conclusion (30); it reads

$$f(w) = b(w + p_0, p_0') - b(p_0, p_0'),$$

and so states that f is the trace of the affine transformation $b(-, p_0'):P \to F^{\flat}$; this functional, and its trace, depends on the choice of the origin p_0' in P'.

Proof: Let f be the trace just described, and define $f':P'^{\#} \to F$ analogously as a trace by $f'(w') = b(p_0, w' + p_0') - b(p_0, p_0')$. In (29) set $p = w + p_0$, $q = p_0$, $p' = w' + p_0'$, and $q' = p_0'$ and substitute the expressions for f and f'. The result is the desired formula (30).

Next we examine quadratic functions, using the fact that the characteristic of F is not 2, so that F contains a scalar $\frac{1}{2}$.

DEFINITION. A function $c:P \to F^{\flat}$ on an affine space P over F is a quadratic functional *when the formula*

$$b(p, q) = c[p \cdot (\tfrac{1}{2}) + q \cdot (\tfrac{1}{2})] \cdot 2 + (cp) \cdot (-\tfrac{1}{2}) + (cq) \cdot (-\tfrac{1}{2}) \qquad (31)$$

defines a (necessarily symmetric) biaffine functional $b:P \times P \to F^{\flat}$, called the functional obtained by polarizing *c.*

This is an "affine" definition because the argument of the first c in (31) is an average (with weights $(\frac{1}{2}, \frac{1}{2})$), as is the expression for b in terms of c (with weights $(2, -\frac{1}{2}, -\frac{1}{2})$). To show that this is a reasonable definition, we observe that a quadratic form g on $P^{\#}$ (in the sense of our previous definition, but now written g and not q as in Chapter XI) does define a quadratic functional in this sense:

Lemma. *If $g:P^{\#} \to F$ is a quadratic form and p_0 a point of P, then $c(w + p_0) = g(w)$ for all $w \in P^{\#}$ defines a quadratic functional $c:P \to F^{\flat}$.*

Proof: For this c, the polarized functional b of (31) is given by

$$b(v + p_0, w + p_0) = g[v(\tfrac{1}{2}) + w(\tfrac{1}{2})]2 - (gv)(\tfrac{1}{2}) - (gw)(\tfrac{1}{2}).$$

Since $g(2v) = 4g(v)$ for g quadratic, this becomes

$$2b(v + p_0, w + p_0) = g(v + w) - g(v) - g(w).$$

But the definition of a quadratic form g states that the expression on the right is bilinear in v and w, and hence that b is biaffine, as required.

More trivially, any linear form $f: P^\# \to F$ on the translation space $P^\#$ defines a quadratic functional $c_f: P \to F^\flat$ by $c_f(w + p_0) = f(w)$; in this case the polarized biaffine functional of c_f is given by $b(v + p_0, w + p_0) = [f(v) + f(w)](\frac{1}{2})$.

In the vector space case, every symmetric bilinear form could be obtained by polarizing a quadratic form. The analogous result holds here:

Theorem 11. *Every symmetric biaffine functional $b: P \times P \to F^\flat$ on an affine space P over F may be obtained by polarizing exactly one quadratic functional; namely, the functional $c: P \to F^\flat$ with $c(p) = b(p, p)$ for all points p.*

Proof: Given c, the definition (31) for $p = q$ states that $b(p, p) = c(p)$. Conversely, given b symmetric and biaffine, we define c by $c(p) = b(p, p)$, and check that the biaffine property of b makes the definition (31) an identity.

To express a quadratic functional in terms of coordinate functionals, choose a frame; that is, choose an origin $p_0 \in P$ and coordinates $\mathbf{x}$ in $P^\#$. Then $c(p) = b(p, p)$ and the coordinate formula (28) for b, with $B_i' = B_i$ because of symmetry, expresses the functional c as

$$c = \sum_{i,j=1}^{n} x_i A_{ij} x_j + 2 \sum_{i=1}^{n} B_i x_i + \gamma. \tag{32}$$

Here n is the dimension of P, A is an $n \times n$ symmetric matrix, B a list of n scalars, and γ a scalar, while each term $x_i A_{ij} x_j$ denotes the quadratic functional $p \mapsto x_i(p) A_{ij} x_i(p)$. In other words, *a quadratic functional on P is simply a function on P to F which can be expressed in one coordinate system* (and hence in every such system) *as a polynomial of degree at most 2 in the coordinates.*

The evident question is: How much can the expression be simplified by suitable choice of the frame? First choose a basis in the translation space $P^\#$ which diagonalizes the quadratic terms; for these coordinates $\mathbf{y}$, the functional c is

$$c(w + p_0) = \alpha_1 y_1^2 + \cdots + \alpha_r y_r^2 + 2B_1 y_1 + \cdots + 2B_n y_n + \gamma,$$

with r the rank of $b^\#$ and with all the scalars $\alpha_i \neq 0$. A translation of the origin will replace y_j by coordinates $z_j = y_j - (B_j/\alpha_j)$ for $j = 1, \ldots, r$. This makes the new coefficients $B_1, \ldots, B_r$ all zero, and changes γ. There remains a linear term $2B_{r+1} y_{r+1} + \cdots + 2B_n y_n$; it is a linear form on the subspace of $P^\#$ with equations $z_1 = \cdots = z_r = 0$. If it is not zero, choose this linear form to be twice the first new coordinate z_{r+1} in the subspace;

a translation will then eliminate γ. For these new coordinates (and new origin) c has the form

$$c(w + p_0) = \alpha_1 z_1^2 + \cdots + \alpha_r z_r^2 + 2z_{r+1}, \qquad z_i = z_i(w). \tag{33}$$

Otherwise, the linear terms are zero, and c becomes (with γ changed to δ)

$$c(w + p_0) = \alpha_1 z_1^2 + \cdots + \alpha_r z_r^2 + \delta. \tag{34}$$

One problem remains: Are these two quadratic functionals (33) and (34) really different? If so, can we describe the difference intrinsically, without referring to a choice of coordinates (an "alias"); that is, by invariants of c under the action of the affine group (an "alibi").

Hence we consider invariants of a quadratic functional $c: P \to F^\flat$. The polarized symmetric bilinear functional $b: P \times P \to F^\flat$ has as trace a symmetric bilinear form $b^\#: P^\# \times P^\# \to F$, which in turn can be obtained (Theorem XI.2) as the polarized bilinear form of a quadratic form $g: P^\# \to F$. We call this quadratic form g the *trace* $g = c^\#$ of c and its rank the *rank* of c. In terms of coordinates, this trace is just the function given by the quadratic terms $\sum x_i A_{ij} x_j$ in the formula (32) for c. One can also see from this formula that a translation of the origin p_0, replacing each x_i by $x_i + \kappa_i$, will not alter these quadratic terms. Any linear invariant of the trace of c will be an affine invariant of c; in particular, the rank r of $c^\#$ is such an invariant, and this rank is the number of nonzero quadratic terms in the diagonal forms (33) and (34). We will also use the radical U of $c^\#$, defined as in § XI.3 to be the set of all those vectors $u \in P^\#$ for which $c^\#(v + u) = c^\#(v)$ for every $v \in P^\#$. Finally, we define the *center* C of the quadratic functional c to be the subset of P consisting of all those points $q \in P$ such that

$$c(w + q) = c^\#(w) + c(q) \qquad \text{for all } w \in P^\#. \tag{35}$$

From this formula it follows at once that $q \in C$ and $u \in U$ imply $u + q \in C$.

In terms of the radical and the center we can now describe the two types (33) and (34) of quadratic functionals:

Theorem 12. *Let $c: P \to F^\flat$ be a quadratic functional on an n-dimensional affine space P over F. Then the center of c is either empty or an affine subspace of P. In the first case, the rank r of c is less than n, while for a suitable origin p_0 and suitable coordinates $z_1, \ldots, z_n$ in $P^\#$, c is*

$$c(w + p_0) = \alpha_1 z_1^2 + \cdots + \alpha_r z_r^2 + 2z_{r+1}, \qquad z_i = z_i(w), \tag{36}$$

for scalars $\alpha_1, \ldots, \alpha_r$, none zero; moreover, the radical of $c^\#$ is the subspace $z_1 = \cdots = z_r = 0$. If the center C of c is an affine subspace of P, its direction $D(C)$ is the radical of $c^\#$, and c has the same value $c(q) = \delta$

for all points q in the center. For an origin p_0 in the center and suitable coordinates $z_1, \ldots, z_n$ in $P^\#$, c is

$$c(w + p_0) = \alpha_1 z_1^2 + \cdots + \alpha_r z_r^2 + \delta, \qquad z_i = z_i(w), \tag{37}$$

for scalars $\alpha_1, \ldots, \alpha_r$, none zero, and the center of c is the subspace of all points $w + p_0$ with $z_1(w) = \cdots = z_r(w) = 0$.

Proof: Take c in one of the forms (33) and (34) above. Its quadratic trace is then $c^\#(w) = \alpha_1 z_1^2 + \cdots + \alpha_r z_r^2$, where the z_i are the coordinates of w, so its radical is the $n - r$ dimensional subspace $z_1 = \cdots = z_r = 0$ of $P^\#$. Suppose that $q = u + p_0$ is a point in the center, and write $y_1, \ldots, y_n$ for the coordinates of q. Then $w + q = w + u + p_0$. In the two cases (33) and (34), calculation gives

$$\begin{aligned} c(w + q) = c(w + u + p_0) &= c^\#(w) + c(q) + 2z_{r+1} + 2\sum_{i=1}^{r} \alpha_i y_i z_i, \\ &= c^\#(w) + c(q) + 2(\alpha_1 y_1 z_1 + \cdots + \alpha_r y_r z_r), \end{aligned}$$

respectively. For q to be in the center, the right side must reduce to $c^\#(w) + c(q)$ for all choices of z. For the first formula, this is never possible (choose z_{r+1} suitably!), so the center in this case is empty. This gives the first form (36). For the second formula, this is possible exactly when $y_1 = \cdots = y_r = 0$, so that u must be in the radical U of $c^\#$. Hence, the center is the affine subspace $U + p_0$; it is the subspace passing through p_0, and with direction U. For any point q in this center, the formula (34) shows that $c(q) = \delta$ is constant. This gives the second form (37).

In these formulas, the terms of degree at most 1 can be described directly in the following way, which will be useful in the treatment of Euclidean quadrics in § 7.

Proposition 13. *Let $U \subset P^\#$ be the radical of the trace $c^\#$ of a quadratic functional $c: P \to F^\flat$. Then there is a unique linear form $k: U \to F$ such that, for all points $p \in P$ and all vectors $u \in U$*

$$c(u + p) = 2k(u) + c(p). \tag{38}$$

We will call k the *linear trace* of c on its radical U.

Proof: For each choice of an origin p_0, Equation (30) expresses the polarized biaffine functional b for c in terms of its trace $b^\#$ as

$$b(w + p_0, w' + p_0) = b^\#(w, w') + f(w) + f(w') + b(p_0, p_0);$$

here $f: P^\# \to F$ is a linear form which depends on the choice of origin. Now $c(p) = b(p, p)$ and $c^\#(w) = b^\#(w, w)$, so the formula becomes

$$c(w + p_0) = c^\#(w) + 2f(w) + c(p_0). \tag{39}$$

Now we define $k: U \to F$ to be the restriction of f to $U \subset P^{\#}$. If w is in the radical U of $c^{\#}$, then $c^{\#}(w) = 0$, so this equation (39) becomes the desired Equation (38) for the particular point $p = p_0$. It remains only to prove (38) for all points $p = v + p_0$ in P. But, by (39),

$$c(u + p) = c(u + v + p_0) = c^{\#}(u + v) + 2f(u + v) + c(p_0).$$

But u in the radical means that $c^{\#}(u + v) = c^{\#}(v)$, so

$$c(u + p) = c^{\#}(v) + 2f(u) + 2f(v) + c(p_0) = c(v + p_0) + 2k(u)$$

which is $c(p) + 2k(u)$, just as in (38). This equation also shows that the linear form k on U is uniquely determined by c, as required.

In case F is the field $\mathbf{R}$ of real numbers, the coefficients of the quadratic terms of Theorem 12 can be chosen as ± 1, and in this case there is an additional invariant, the signature s (number of terms $+\ 1$ in the diagonal form). The same theorem now holds, with the two forms (36) and (37) replaced by

$$x_1^2 + \cdots + x_s^2 - x_{s+1}^2 - \cdots - x_r^2 + 2x_{r+1}, \tag{40}$$

$$x_1^2 + \cdots + x_s^2 - x_{s+1}^2 - \cdots - x_r^2 + \delta. \tag{41}$$

The theorem states that these forms (40) and (41), for all suitable s, r, n, and δ, are canonical: By a suitable choice of linear coordinates $\mathbf{x}$, each real quadratic functional can be expressed in exactly one of these forms. This result is an "alias" (a choice of new coordinates), it may be restated as an "alibi". With the usual agreement that two quadratic functionals c and c' on the same real affine space P are affine equivalent when there is an affine automorphism $a: P \to P$ such that $c' = c \circ a$, this "alibi" reads as follows:

Corollary. *Let P be an affine space over $\mathbf{R}$. Two quadratic functionals $c, c': P \to \mathbf{R}^\flat$ are affine equivalent if and only if they have the same rank and the same signature and are either both centerless or both centered, with the same value on all points of their respective centers.*

Each quadratic functional $c: P \to F^\flat$ determines a figure, called a *quadric*, defined as the set (or "locus") of all points $p \in P$ with $c(p) = 0$. The above classification of real quadratic functionals gives a classification of real quadrics. Since $c(p) = 0$ and $\kappa c(p) = 0$ for $\kappa \neq 0$ give the same locus, the scalar δ in (41) may be replaced by 0 or 1. Hence, any quadric in real affine n-space is affine equivalent to one of the loci

$$x_1^2 + \cdots + x_s^2 - x_{s+1}^2 - \cdots - x_r^2 = \begin{cases} -2x_{r+1}, \\ 1, \\ 0. \end{cases}$$

If the quadratic functional $c: P \to \mathbf{R}^\flat$ has a center, then the associated quadric is symmetrical about any point q in the center, in the sense that whenever a point $w + q$ is on the quadric, so is the point $-w + q$. (The point $-w + q$ may be described as the reflection of $w + q$ in q.) To prove this, note that $w + q$ on the quadric means that $c(w + q) = 0$ and hence by (35) that $c^\#(w) = -c(q)$. Since $c^\#(-w) = c^\#(w)$, this implies that $c^\#(-w) = -c(q)$ and hence $c(-w + q) = 0$, as required.

In the real affine plane ($n = 2$) a quadric is called a *conic section*. The possible such affine conics of positive rank $r = 2$ and $r = 1$ are as follows (where in case $r = 2$ the center is at the origin $x = 0$, $y = 0$):

$r = 2$		$r = 1$	
$x^2 + y^2 - 1 = 0,$	circle	$x^2 + 2y = 0,$	parabola
$x^2 - y^2 - 1 = 0,$	hyperbola	$-x^2 + 2y = 0,$	parabola
$-x^2 - y^2 - 1 = 0,$	no locus	$x^2 - 1 = 0,$	two parallel lines
$x^2 + y^2 = 0,$	one point	$-x^2 - 1 = 0,$	no locus
$x^2 - y^2 = 0,$	two intersecting lines	$x^2 = 0,$	one line
$-x^2 - y^2 = 0,$	one point		

For the circle and the hyperbola, the center has its usual meaning, while the parabola has no center and the two parallel lines $x^2 = 1$ have center the line $x = 0$ half way between them. The invariance of the signature implies that a circle is *not* affine equivalent to a hyperbola; the invariance of rank implies that neither is affine equivalent to a parabola.

EXERCISES

1. Determine the canonical form of each of the following real quadratic functionals:

(a) $x^2 + 4y^2 + 9z^2 + 4xy + 6xz + 12yz + 8x + 16y + 24z + 15.$
(b) $x^2 - 6xy + 10y^2 + 2xz - 20z^2 - 10yz - 40z - 17.$
(c) $x^2 + 4z^2 + 4xz + 4x + 4z - 6y + 6.$
(d) $-2x^2 - 3y^2 - 7z^2 + 2xy - 8yz - 6xz - 4x - 6y - 14z - 6.$

2. Exhibit the linear trace on the radical of each of the quadratic functionals c of (36) and (37), and prove that this linear trace is zero if and only if c has no center.

3. (a) Give a complete set of canonical forms for quadrics in real affine three-space.

(b) Give a brief geometric description of each canonical form, specifying in particular its center (if any) and its radical.

4. Show that $q: V \to F$ is a quadratic form (in the sense defined in Chap. XI) if and only if $2q((u + v)/2) - (qu)/2 - (qv)/2$ is bilinear.

6. Euclidean Spaces

Next we consider affine spaces in which "distance" and "angle" are defined.

DEFINITION. A Euclidean space *is an affine space P over the field* $\mathbf{R}$ *of real numbers together with an inner product $P^{\#} \times P^{\#} \to \mathbf{R}$ which makes the translation space $P^{\#}$ of P into an inner product space.*

With the usual notations $\langle v, w \rangle$ for the inner product of two vectors, $|w| = \langle w, w \rangle^{1/2}$ for the length of a vector $w \in P^{\#}$, and $q - p \in P^{\#}$ for the difference of two points, the *distance* between two points p and q of a Euclidean space P is defined to be

$$\rho(p, q) = |q - p|. \tag{42}$$

With this distance function $\rho : P \times P \to \mathbf{R}$ there is a more geometric way of describing a Euclidean space:

Theorem 14. *A Euclidean space is a real affine space P together with a function $\rho : P \times P \to \mathbf{R}$ assigning to each pair of points p, q a real number $\rho(p, q)$ as distance in such a way that*

(*i*) *$\rho(p, q) \geqq 0$ and $\rho(p, q) = 0$ if and only if $p = q$;*
(*ii*) *$\rho(p, q) = \rho(q, p)$ for all points p and q;*
(*iii*) *The distance ρ is invariant under translation, in that*

$$\rho(w + p, w + q) = \rho(p, q)$$

for all points p, q and all translations w of P;
(*iv*) *For any three points p, q, and r, $\rho(q, r)^2 - \rho(q, p)^2 - \rho(r, p)^2$ is a bilinear function of the translations $q - p$ and $r - p$.*

Proof: The distance as defined above certainly has these four properties. Conversely, suppose $\rho : P \times P \to \mathbf{R}$ a function with these four properties. The property (iii) implies that the distance $\rho(p, q)$ depends only on the translation $v = q - p$ which carries p to q. For, suppose $q' - p' = q - p$, write v for $q - p$ and u for $p' - p$, so $q = v + p$, $p' = u + p$, and $q' = (v + u) + p$. Then, by (iii),

$$\rho(p', q') = \rho(u + p, v + u + p) = \rho(p, v + p) = \rho(p, q).$$

Thus $\rho(p, q)$ depends on $v = q - p$; in other words, $|v| = \rho(p, v + p)$ defines a length function $P^{\#} \to \mathbf{R}$ satisfying (42).

If the pythagorean theorem were true for an arbitrary triangle

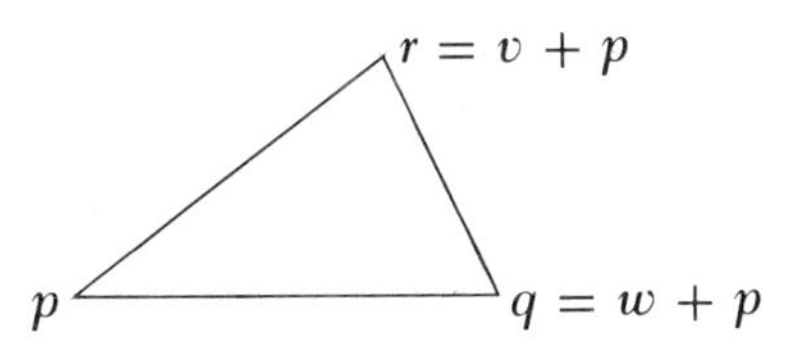

with vertices p, $q = w + p$ and $r = v + p$ the expression

$$\rho(q, r)^2 - \rho(q, p)^2 - \rho(r, p)^2$$

would be zero. In general, it is not zero, but by (iii) it depends only on the translations v and w; property (iv) states that this number, call it

$$\begin{aligned}\langle v, w\rangle &= \rho(w + p, v + p)^2 - \rho(w + p, p)^2 - \rho(v + p, p)^2\\ &= |w - v|^2 - |w|^2 - |v|^2,\end{aligned}$$

is a bilinear function of w and v. By the symmetry property (ii), $|-w|^2 = |w|^2$. These two properties are exactly the ones which state that $w \mapsto |w|^2$ is a quadratic form on $P^\#$. Property (i) states that this quadratic form is positive definite. Hence $P^\#$ is indeed an inner product space.

In § XI.5, we defined a metric space to be a set X with a distance function $\rho: X \times X \to \mathbf{R}$ satisfying three axioms: (i) and (ii) of the theorem above, and the triangle axiom $\rho(q, r) \leqq \rho(q, p) + \rho(p, r)$. In the notations v and w above, the triangle axiom becomes $|w - v| \leqq |w| + |v|$; it is a consequence of the Schwarz inequality. This proves

Corollary 1. *A Euclidean space is a metric space under its distance function.*

This corollary is just a restatement of the fact that an inner product space can be regarded as a metric space. To be explicit, just as every affine space is isomorphic to the affine space $V^\flat$ obtained from some vector space V, so every Euclidean space P is isomorphic to the space $U^\flat$ obtained from some real inner product space U (say, with U the translation space of P).

Corollary 2. *Every affine subspace R of a Euclidean space P is Euclidean, under the distance function obtained by restricting the distance function of the whole space P.*

This follows at once, since the inner product in an inner product space U may be restricted to any vector subspace of U.

Any average of points may now be described in terms of distance, just as in elementary geometry. For example, an average $r = p \cdot \kappa + q \cdot (1 - \kappa)$ can be shown (Exercise 2) to be the unique point r which satisfies both

$$\rho(r, p) = |1 - \kappa|\rho(q, p) \qquad \text{and} \qquad \rho(r, q) = |\kappa|\rho(p, q). \tag{43}$$

A *rigid map* (or, an "isometry") $f: P \to P'$ between Euclidean spaces P and P' is defined to be a function on P to P' which preserves distance, in the sense that $\rho(fp, fq) = \rho(p, q)$ for all $p, q \in P$. In particular, a rigid map $f: P \to P$ of a Euclidean space into itself is called a *rigid motion* of P. The definition of a rigid map does not assume that the map is affine, but this fact can be deduced.

Theorem 15. *A function $f: P \to P'$ between Euclidean spaces P and P' is a rigid map if and only if it is an affine transformation with an orthogonal trace.*

Proof: In one direction the result is almost evident: Any affine transformation with orthogonal trace is an orthogonal transformation followed by a translation, and both orthogonal transformations and translations are rigid. More formally, if $a: P \to P'$ is affine with trace $a^{\#}$ orthogonal, then $a(w + p) - a(p) = a^{\#}(w)$; hence, for $w = q - p$,

$$\rho(ap, aq) = |a(w + p) - a(p)| = |a^{\#}(w)| = |w| = \rho(p, q);$$

this equation states that a is a rigid map.

The interest lies in the converse proof: Assuming only that f is a rigid map, to prove f affine. Pick some point p_0 in P and define a function $s: P^{\#} \to P'^{\#}$ by $s(w) = f(w + p_0) - f(p_0)$; thus s would be the trace of f if f had a trace. For vectors $v, w \in P^{\#}$,

$$|v - w| = \rho(v + p_0, w + p_0) = \rho(f(v + p_0), f(w + p_0)),$$

because f is rigid. By the definition of s this is

$$|v - w| = \rho(s(v) + f(p_0), s(w) + f(p_0)) = |s(v) - s(w)|.$$

This states that s preserves lengths of differences. Since inner products are defined as $\langle v, w \rangle = |v - w|^2 - |v|^2 - |w|^2$, it follows that s preserves inner products. Take an orthonormal basis $b_1, \ldots, b_n$ of $P^{\#}$; since s preserves inner products, $sb_1, \ldots, sb_n$ is orthonormal in $P'^{\#}$, and hence is an orthonormal basis $s\mathbf{b}$ for some subspace U' of $P'^{\#}$ (namely, for $U' = s_*(P^{\#})$). Any vector $w = \sum b_j \xi_j$ in $P^{\#}$ has coordinates ξ_j, rel $\mathbf{b}$, given as the inner products $\xi_j = \langle b_j, w \rangle$. The image vector $s(w)$ has coordinates relative to $s\mathbf{b}$ the inner products $\langle sb_j, sw \rangle$. Since s preserves inner products, these coordinates are the same scalars ξ_j. In other words, each vector $w = \sum b_j \xi_j$ has image $s(w) = \sum (sb_j)\xi_j$. This states that s is a linear transformation (indeed, that linear transformation taking **b** to s**b**). Since we already know that s preserves inner products, $s: P^\# \to P'^\#$ is even orthogonal. But the original rigid map f satisfies the equation $f(w + p) = s(w) + f(p)$ for all vectors w and one point $p = p_0$. Hence by Theorem 4 this function f is an affine map, and by Theorem 3 it satisfies this equation for all p, so has the linear transformation s as trace, Q.E.D.

COROLLARY. *Any rigid motion f (of P to itself) is a bijection, with inverse a rigid motion. Hence all rigid motions of P form a group under composition.*

Proof: The trace $f^\#$ is orthogonal, hence a bijection $P^\# \cong P^\#$. Therefore f has an inverse; since f preserves distances, so does its inverse, so this inverse is also a rigid motion.

The group $E(P)$ of all rigid motions of the Euclidean space P is known as the *Euclidean group* of the space. Since two Euclidean spaces of the same dimension are isomorphic, there is (up to isomorphism) just one Euclidean group for each dimension. The Euclidean group $E(P)$ is a subgroup of the affine group $A(P)$ and the morphism $\tau: A(P) \to GL(P^\#)$ given by the trace has by Theorem 15 the property that $\tau(a) = a^\#$ is orthogonal if and only if a is a rigid map. Hence τ restricts to an epimorphism $\tau': E(P) \to O(P^\#)$, where $O(P^\#)$ denotes the orthogonal group of the translation-space $P^\#$, and the exact sequence (21) yields a new exact sequence

$$0 \to (P^\#)^+ \to E(P) \xrightarrow{\tau'} O(P^\#) \to 1 \tag{44}$$

of morphisms of groups.

The elements of $E(P)$ can also be described as pairs (u, t) of vectors $u \in P^\#$ and orthogonal maps $t \in O(P^\#)$ with the product

$$(u', t')(u, t) = (u' + t'u, t' \circ t). \tag{45}$$

In an n-dimensional Euclidean space P, an *orthonormal frame* $r_0, r_1, \ldots, r_n$ is a list of $n + 1$ points in P such that the differences $r_1 - r_0, \ldots, r_n - r_0$ form an orthonormal basis for $P^\#$. Each point $p \in P$ has n (linear) coordinates $x_i(p) = \langle r_i - r_0, p - r_0 \rangle$ relative to such a frame. A rigid motion $f: P \to P$ is given relative to such a frame **r** by the n equations

$$x_i' = \sum_{j=1}^{n} A_{ij}x_j + B_j, \qquad \text{where } x_i' = x_i(fp),\ x_i = x_i(p), \tag{46}$$

$i = 1, \ldots, n$, where A is an $n \times n$ orthogonal matrix and B a $1 \times n$ matrix. Exactly the same equations describe a change of basis, this time with $x_i = x_i(p)$ the linear coordinates of p rel **r** and x_i' the linear coordinates of p relative to some other orthonormal frame **r**$'$.

EXERCISES

1. If p, q, and r are three points in a Euclidean space, show that the equation $\rho(p, r) = \rho(p, q) + \rho(q, r)$ implies that the vector $p - q$ is a scalar multiple of the vector $p - r$ (and hence that p, q, and r are collinear).

2. (a) Show that the weighted average $r = p \cdot \kappa + q \cdot (1 - \kappa)$ does satisfy the conditions (43).

(b) Prove that any solution r of these equations must be $p \cdot \kappa + q \cdot (1 - \kappa)$. (*Hint:* Use Exercise 1.)

3. If L and L' are lines in a Euclidean plane P, show that there is a rigid motion of P taking L to L'.

4. Describe in detail the equations for change of orthonormal frame, both with frames directly and with the corresponding coordinates.

5. Establish the description (45) of the group $E(P)$.

6. Derive in detail the Equations (46) for a rigid motion.

7. In the exact sequence (44), show that τ' has a right inverse which is a morphism of groups.

7. Euclidean Quadrics

Throughout this section P is a Euclidean space of dimension n over the field $\mathbf{R}$ and $P^\#$ is its translation space. We shall study equivalence under action of the Euclidean group on figures and on functionals. Two figures X and X' are equivalent under the Euclidean group when there is rigid motion $f: P \to P$ carrying the set $X \subset P$ onto the set X'. Two quadratic functionals $c, c': P \to \mathbf{R}$ are equivalent under this group when there is a rigid motion f with $c \circ f = c'$. Since the trace of a rigid motion is an orthogonal map $t: P^\# \to P^\#$, while the trace of the functional c is a quadratic form $c^\#$ on $P^\#$, $c \circ f = c'$ implies $c^\# \circ t = c'^\#$; hence any orthogonal invariant of the form $c^\#$ is a Euclidean invariant of the functional c. Such invariants include the eigenvalues of $c^\#$. Similarly, c determines a linear form, its trace on the radical of $c^\#$; the norm of this form is a Euclidean invariant of c, and we know by Proposition XI.16 that the norm of a linear form is a complete orthogonal invariant of that form.

Given c, we take an orthonormal frame and express c in the linear coordinates relative to this frame. We can diagonalize the quadratic terms by the principal axis theorem and then simplify the linear terms, just as in the affine case. This argument proves

Theorem 16. *A quadratic functional $c: P \to \mathbf{R}$ of rank r on a Euclidean space P has a list of r non-zero real eigenvalues $\lambda_1, \ldots, \lambda_r$. Let M be the norm of the (linear) trace of c on its radical. If $M \neq 0$, then $r < n$ and the center of c is empty; for a suitable orthonormal frame with an origin*

$f_0 = p_0$, c has the form

$$c(w + p_0) = \lambda_1 z_1^2 + \cdots + \lambda_r z_r^2 + 2Mz_{r+1},$$

where the $z_i = z_i(w)$ are the n (linear) coordinates of w relative to this frame. If $M = 0$, the center of c is not empty; for a suitable orthonormal frame with an origin p_0 in the center, c has the form

$$c(w + p_0) = \lambda_1 z_1^2 + \cdots + \lambda_r z_r^2 + \delta,$$

where δ is the (constant) value of c on points of its center.

Changing this "alias" result to "alibi" gives the

Corollary 1. *Two quadratic functionals $c, c' : P \to \mathbf{R}^\flat$ are Euclidean equivalent if and only if they agree in the following invariants: rank, eigenvalues, norm of linear trace, value in center.*

The figures we consider are the *quadric hypersurfaces*, given as the loci $c(p) = 0$; for $n = 2$, they are the *conic sections*, for $n = 3$ the *quadric surfaces*. For their classification, we can multiply the function c by any non-zero scalar without altering the locus; hence we can arrange to have $M = -1$ or $M = 0$; if $M = 0$, to have $\delta = -1$ or $\delta = 0$; if $M = \delta = 0$, to have $\lambda_1 = 1$ or $r = 0$.

For example, consider the conics ($n = 2$). For rank 2 the canonical forms are

$$\lambda_1 x^2 + \lambda_2 y^2 - 1 = 0 \quad \text{or} \quad \lambda_1 x^2 + \lambda_2 y^2 = 0, \qquad \lambda_1 \neq 0, \lambda_2 \neq 0.$$

If both eigenvalues λ_i are positive, the first equation gives an ellipse; the center of the quadratic functional is the usual center of the ellipse; while if one compares this equation with the conventional equation $x^2/a^2 + y^2/b^2 = 1$, it appears that the eigenvalues determine the length of the semi-axes of the ellipse. Our corollary thus includes, in particular, the fact that two ellipses are equivalent under the Euclidean group if and only if their axes have the same length (and likewise for a hyperbola, with $\lambda_1 > 0, \lambda_2 < 0$). For rank 1 the conics are

$$\lambda_1 x^2 - 2y = 0, \qquad \lambda_1 x^2 - 1 = 0, \qquad x^2 = 0$$

(respectively a parabola, two parallel lines, "two coincident" lines). This conclusion may be formulated as a theorem in the elementary analytic geometry of the plane, as follows:

Corollary 2. *A polynomial equation of degree 2 in cartesian coordinates in the Euclidean plane has as locus one of the following: The empty set, one point, one line, two parallel lines, two intersecting lines, a parabola, an ellipse, or a hyperbola.*

EXERCISES

1. Classify under the Euclidean group the following quadratic functionals:
(a) $4xz + 4y^2 + 8y + 8$.
(b) $9x^2 - 4xy + 6y^2 + 3z^2 + 25x + 45y + 12z + 16$.
2. List all canonical forms of quadric surfaces in three-space, describing in each case the geometric meaning of the eigenvalues.

8. Projective Spaces

In the affine plane over a field F, any two points lie on a unique line, and any two *non-parallel* lines "intersect" in a unique point. We shall now construct, over the same field, a *projective* plane, in which

(i) *Any* two distinct points lie on a unique line;
(ii) *Any* two distinct lines intersect in a unique point.

The incidence properties (i) and (ii) are clearly dual to each other, in the sense that the interchange of the words "point" and "line," plus a minor change in terminology, changes property (i) into property (ii) and vice versa.

To construct a projective plane P, start with any three-dimensional vector space V over F and define $P = P(V)$ by the statements:

A *point* $p \in P(V)$ is a one-dimensional vector subspace of V,
A *line* L of $P(V)$ is a two-dimensional vector subspace of V,

while the point p is *on* the line L (or "incident" with L) when the subspace p is contained in the subspace L. For these definitions, the incidence property (i) holds: If $p \neq q$ are points, they are different one-dimensional subspaces of V, so their sum is a two-dimensional subspace L, hence a line of $P(V)$—and the only line containing p and q. The incidence property (ii) also holds: Two distinct lines L and L' are distinct two-dimensional subspaces, hence their vector-space sum $L + L'$ must be the whole three-dimensional space V. The dimension of their intersection is therefore, by (VII.14),

$$\dim (L \cap L') = \dim L + \dim L' - \dim (L + L') = 2 + 2 - 3 = 1;$$

hence, $L \cap L'$ is a one-dimensional subspace, thus is a point p of $P(V)$ and the unique point in which L and L' intersect.

This may be visualized as follows. The points of P are the lines through the origin of V and the lines of P are the planes through the origin of V; two such lines span a unique plane through the origin: This is (i); two distinct planes through the origin meet in a unique line: This is (ii).

To examine the relation to the affine plane when F is the real field $\mathbf{R}$ a different description is useful. In $V = \mathbf{R}^3$, the unit sphere S^2 is the locus

$x_1^2 + x_2^2 + x_3^2 = 1$. Each line through the origin of $\mathbf{R}^3$ meets this unit sphere in two diametrically opposite points, and each plane through the origin meets the sphere in a great circle. (See Figure XII.1.) Hence one can describe $P(\mathbf{R}^3)$ thus:

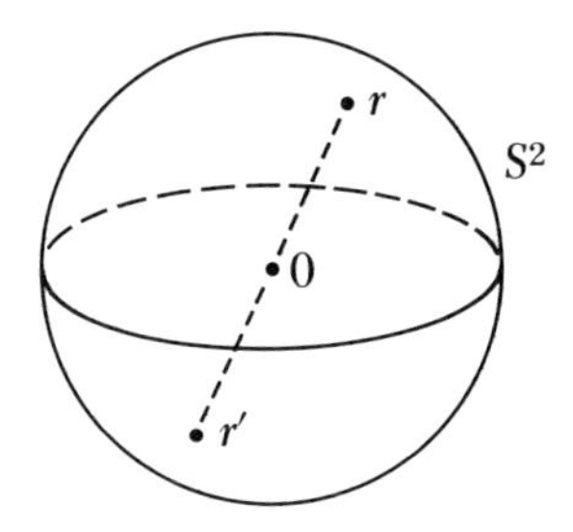

FIGURE XII.1

A *point* $p \in P(\mathbf{R}^3)$ is a pair (r, r') of diametrically opposite points (r, r') of S^2;

A *line* L of $P(\mathbf{R}^3)$ is a great circle on S^2;

while the point $p = (r, r')$ lies *on* the line L when the great circle L passes through r and r'. Again, it is evident that two distinct great circles on the sphere S^2 meet in exactly one pair (r, r') of diametrically opposite points. This is the incidence property (ii) of the projective plane.

There is a variant of this description. Let D^2 be the lower hemisphere of S^2, consisting of all points (x_1, x_2, x_3) in $\mathbf{R}^3$ with $x_1^2 + x_2^2 + x_3^2 = 1$ and $x_3 \leqq 0$. Its boundary is the equator S^1 of S^2, with $x_1^2 + x_2^2 = 1$ and $x_3 = 0$. For each pair (r, r') of diametrically opposite points of S^2, one of r or r' must lie in this lower hemisphere D^2, and both lie there only when they are diametrically opposite points of the equator S^1. Thus one can describe $P(\mathbf{R}^3)$ as follows:

A *point* p of $P(\mathbf{R}^3)$ is a point on the hemisphere D^2, with the proviso that diametrically opposite points of its equator S^1 are to be identified.

A *line* L of $P(\mathbf{R}^3)$ is the intersection with D^2 of any great circle of S^2. In particular, the equator itself (with diametrically opposite points identified) is a line L of $P(\mathbf{R}^3)$.

Now take a plane π tangent to the hemisphere at its south pole $(0, 0, -1)$, and project the hemisphere radically from the origin 0 to the plane π. To project a point r of the hemisphere not on its equator S^1, take the line segment $0r$ and prolong till it meets π in r^*; the projection is $r \mapsto r^*$. (See Figure XII.2.) This projection f is thus a bijection from the complement of S^1 in D^2 to all of π; moreover, the f-image of each line of $P(\mathbf{R}^3)$—each

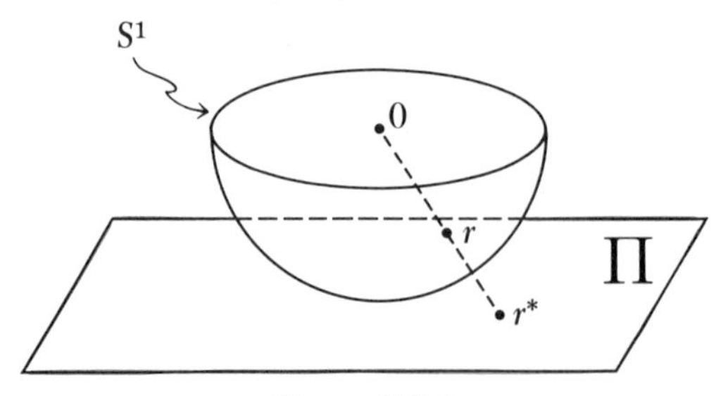

FIGURE XII.2

great circle of S^2—is exactly a line in the plane π. Now regard f^{-1} as a monomorphism from the affine plane π to the projective plane $P(\mathbf{R}^3)$. It carries points to points, preserving incidence. Its image is the set of all points of $P(\mathbf{R}^3)$ except the points on the line L_0 represented by the equator S^1 of D^2. Thus the projective plane can be obtained from the affine plane by adjoining all the points on one new (projective) line L_0, called the "line at infinity". Any set of parallel lines in the affine plane is mapped by f^{-1} to a set of great circles on D^2 passing through the ends of a diameter of S^1, hence meeting in a pair (r, r') of diametrically opposite points of the equator S^1. Such a pair is, by agreement, a single point of the projective plane $P(\mathbf{R}^3)$. In this way, adding the line at infinity to the affine plane has the precise effect of adding points of intersection for all those pairs of parallel lines which did not meet in π. Observe that the line L_0 at infinity is not an affine line but a projective line: An affine line plus one "point at infinity".

We also consider projective spaces of higher dimension and over any field. Such a projective space is to be a set of points together with certain distinguished subsets, called its projective subspaces (or, its "hyperplanes"). Rather than describing such a space axiomatically, we will construct such a projective space $P(V)$ of dimension n over a field F from any vector space V of positive dimension $n + 1$ over the same field. Define the points of $P(V)$ to be the one-dimensional vector subspaces of V; for each vector subspace S of V, take $P(S)$ to be the subset of all those points p of $P(V)$ (i.e., all those one-dimensional vector subspaces p of V) for which $p \subset S$; in particular, $P(\{0\}) = \varnothing$. This set $P(V)$ is a *projective space* with distinguished subsets all the sets $P(S)$; if S has vector-space dimension $k + 1$, the distinguished subset $P(S)$ is called a *projective subspace* of dimension k in $P(V)$. In particular, a two-dimensional vector subspace S is a *projective line* $P(S)$ in V, a three-dimensional vector subspace a *projective plane* in $P(V)$, and so on, while the trivial vector subspace $\{0\}$ determines the empty projective subspace of $P(V)$. If $S \subset T$; that is, if S is a vector subspace of another vector subspace T, one has $P(S) \subset P(T)$, for each one-dimensional subspace of V contained in S is also contained in T. When

$P(S) \subset P(T)$, we say that the projective subspace $P(S)$ *lies on* or is "incident" with $P(T)$. Under this inclusion relation $P(S) \subset P(T)$, the projective subspaces of $P(V)$ form a lattice, in which the meet of $P(S)$ and $P(T)$ is the intersection $P(S) \cap P(T) = P(S \cap T)$, while the join is the projective subspace $P(S + T)$ corresponding to the vector space sum $S + T$. The assignment $S \mapsto P(S)$ is clearly an isomorphism of the lattice of vector subspaces of V to the lattice of projective subspaces of $P(V)$. Each subspace $P(T)$ may be regarded as a projective space in its own right, with distinguished subsets the $P(S)$ for $S \subset T$.

Let $t: V \cong V'$ be an isomorphism of V to another vector space V'. The image t_*S of any vector subspace of V is a subspace of V', of the same dimension. Thus $P(S) \mapsto P(t_*S)$ is a bijection which preserves all incidence relations, and hence is an isomorphism of lattices. In particular, for points, $p \mapsto t_*p$ defines a bijection $P(t): P(V) \cong P(V')$ which we call a *projective isomorphism.*

Under composition, the set of all projective automorphisms of $P(V)$ is a group, the *projective group* $PGL(V)$. Since $P(t_1t_2) = P(t_1)P(t_2)$, $t \mapsto P(t)$ is an epimorphism P from the general linear group to the projective group.

Theorem 17. *For each finite-dimensional vector space V over the field F, with F^* the multiplicative group of all nonzero scalars of F, there is a short exact sequence*

$$1 \to F^* \xrightarrow{M} GL(V) \xrightarrow{P} PGL(V) \to 1$$

of morphisms of groups, where the morphism M assigns to each scalar $\kappa \neq 0$ the linear transformation $v \mapsto v\kappa$ given by scalar multiplication by κ.

Proof: We need only show that the image of M is the kernel of P. Trivially, each automorphism $v \mapsto v\kappa$ maps each one-dimensional subspace of V to itself, so the image of M is contained in the kernel of P. Conversely, suppose $t: V \to V$ has $t_*p = p$ for every one-dimensional subspace p of V. Let v_1 and v_2 be any two non-zero vectors of V. Since t_* maps the one-dimensional subspace spanned by each v_i to itself, one must have $tv_i = v_i\kappa_i$ for some scalars κ_1 and κ_2, and $t(v_1 + v_2)$ must be $(v_1 + v_2)\mu$ for some μ. If v_1 and v_2 are linearly dependent, $\kappa_1 = \kappa_2$. If they are linearly independent,

$$v_1\mu + v_2\mu = t(v_1 + v_2) = tv_1 + tv_2 = v_1\kappa_1 + v_2\kappa_2,$$

so $\mu - \kappa_1 = 0 = \mu - \kappa_2$ and therefore $\kappa_1 = \kappa_2$. Thus t multiplies every vector v by the same scalar κ, so t is in the image of the morphism M, as required.

For the standard vector space F^{n+1} with elements the lists

$$\xi = (\xi_1, \ldots, \xi_{n+1}),$$

the corresponding projective space $P(F^{n+1})$ has as points all the one-dimensional vector subspaces of F^{n+1}, and such a subspace consists of all the multiples $\xi\kappa$ of one list $\xi \neq 0$ by a variable $\kappa \neq 0$ in F. One often says that a point of $P(F^{n+1})$ *is* a list $\xi \neq 0$, with the understanding that each such list ξ is to be identified with all lists $\xi\kappa$ for $\kappa \neq 0$. The projective group of $P(F^{n+1})$ is called $PGL(n, F)$, the *projective group* in dimension n for the field F. By the Theorem above and the fundamental theorem on quotient groups, this projective group is isomorphic to the quotient group $GL(n + 1, F)/(M_*F^*)$: The group of all $(n + 1) \times (n + 1)$ invertible matrices over F modulo the normal subgroup of all non-zero scalar multiples of the $(n + 1) \times (n + 1)$ identity matrix.

Two projective spaces of the same dimension over F, like two vector spaces of the same dimension, are isomorphic. In particular, each basis **b** of $n + 1$ vectors of V gives as in (VII.7) an isomorphism $L_{\mathbf{b}}: F^{n+1} \cong V$ of vector spaces and hence a projective isomorphism $P(L_{\mathbf{b}}): P(F^{n+1}) \cong P(V)$. The inverse of $L_{\mathbf{b}}$ assigns to each vector $v \in V$ the list ξ of its coordinates, rel **b**. Similarly, $P(L_{\mathbf{b}}^{-1}): P(V) \cong P(F^{n+1})$ assigns to each point p of the projective space $P(V)$ a point ξ. This list $\xi_1, \ldots, \xi_{n+1}$ is called the list of *homogeneous coordinates* of p, relative to the basis **b**. We note that the homogeneous coordinates of a point are never all zero, and are determined only up to a non-zero scalar multiple κ.

EXERCISES

1. In the projective three-space over a field F, prove
   (a) Any two distinct points lie on one and only one line;
   (b) Any three points not on a line lie on one and only one plane.
2. Generalize Exercise 1 to projective n-space.
3. List all points and lines and the points on each line in the projective plane over the field $\mathbf{Z}_2$.
4. In the projective plane over a finite field with n elements, show that there are $n^2 + n + 1$ points and $n^2 + n + 1$ lines, as well as $n + 1$ points on each line.

9. Projective Quadrics

Let $q: V \to F$ be a quadratic form on a finite-dimensional vector space V over F. From the definition of a quadratic form, it follows directly that $q(v\kappa) = q(v)\kappa^2$; this property is also evident from the expression of q as a homogeneous quadratic polynomial in any one list of coordinates **x** of V. From it we see that the set of all vectors v with $q(v) = 0$ contains with each v all scalar multiples $v\kappa$, and thus contains that point p of the projective space $P(V)$ which is the one-dimensional subspace of all $v\kappa$ $(v \neq 0)$.

The set of all these projective points p with $q(p) = 0$ is called a *projective quadric*. If $P(V)$ has dimension 2, it is called a *projective conic*. If the quadric form q has rank r, the quadric $q = 0$ is said to have rank $r - 1$.

To illustrate, we now examine these conics when F is $\mathbf{R}$.

Theorem 18. *In the real projective plane, any two non-empty projective conics of rank 2 are equivalent under the projective group.*

Proof: The conic is the locus $q = 0$, where q is a quadratic form of rank 3. We may describe q, relative to suitable coordinates, by one of the following canonical forms:

$$\begin{aligned}
\text{signature } 3: &\quad x_1^2 + x_2^2 + x_3^2 = 0\\
2: &\quad x_1^2 + x_2^2 - x_3^2 = 0\\
1: &\quad x_1^2 - x_2^2 - x_3^2 = 0\\
0: &\quad -x_1^2 - x_2^2 - x_3^2 = 0.
\end{aligned}$$

In the real field, the first and the fourth equations have no real solutions except (0, 0, 0), which does not represent a point of the projective plane. Hence these two projective conics are empty. The second equation may be turned into the third by multiplying by -1 and permuting the variables by $1 \mapsto 2$, $2 \mapsto 3$ and $3 \mapsto 1$. Hence these two conics are projectively equivalent, as asserted.

Now remove from the real projective plane the points of the projective line $x_3 = 0$; as in § 8, the points remaining form the real affine plane. The homogeneous coordinates of its points have $x_3 \neq 0$; since homogeneous coordinates may be multiplied by any non-zero scalars, we may write these coordinates as $(x_1, x_2, 1)$ where $x_3 = 1$ and x_1, x_2 are linear coordinates in the affine plane. Now take the intersection of each conic above with the affine plane $x_3 = 1$. The two non-void conics above become the affine loci $x_1^2 + x_2^2 - 1 = 0$ (signature 2, the ellipse) and $x_1^2 - x_2^2 - 1 = 0$ (signature 1, the hyperbola). In other words, on restriction of the projective plane to the affine plane, the ellipse is a conic which does not meet the line $x_3 = 0$ at infinity, while the hyperbola is a conic which does meet the line at infinity in the two points with the homogeneous coordinates $(1, \pm 1, 0)$. Moreover, under the projective group (with these points "at infinity" added to the hyperbola) the ellipse and the hyperbola are equivalent.

We might have started with the projective conic $x_1^2 - 2x_2x_3 = 0$. The corresponding quadratic form still has rank 3, for in the coordinates $y_1 = x_1$, $y_2 = (x_2 + x_3)/2$, $y_3 = (x_2 - x_3)/2$ it is $y_1^2 - y_2^2 + y_3^2 = 0$. In the x-coordinates it meets the line $x_3 = 0$ at infinity in the single point with homogeneous coordinates $(0, 1, 0)$. Its intersection with the affine plane $x_3 = 1$ is the parabola given by $x_1^2 - 2x_2 = 0$. Thus the projective conic meeting the line at infinity in a *single* point, on restriction to the affine plane, becomes a parabola. Moreover, under the projective group (and with this point "at infinity" added) the parabola is equivalent to the ellipse and the hyperbola.

10. Affine and Projective Spaces

We have already indicated in § 8 how the real projective plane can be obtained by adjoining to the real affine plane the points of one new (projective) line, to be the "line at infinity". This process is possible more generally: Over any field F, the projective n-space can be obtained by adjoining to the affine n-space the points of one new projective $(n - 1)$-space "at infinity". This situation is better described in the opposite direction: From a projective n-space remove all the points of a projective subspace of dimension $n - 1$; what remains is an affine n-space with affine transformations specified in the following way by projective ones.

Theorem 19. *Let V be an $(n + 1)$-dimensional vector space over F and W a subspace of V of dimension n, so that $P(W)$ is a projective subspace of dimension $n - 1$ in the projective n-space $P(V)$. Let G be the group of all those projective automorphisms $P(t)$ of $P(V)$ for which $[P(t)]_* P(W) = P(W)$. Then the points of $P(V)$ not in $P(W)$ form an affine space Q of dimension n over F with translation space $Q^\# \cong W$, in such a way that the restrictions to Q of the automorphisms $P(t) \in G$ are the affine automorphisms of Q, while the translations of Q are the restrictions to Q of those $P(t) \in G$ which restrict to the identity on $P(W)$.*

The last conclusion is geometrically plausible for $n = 2$: Each translation of an affine plane carries a set S of (all) mutually parallel lines to itself, hence does not move the "point at infinity" in which this set of parallel lines meet. Here $P(W)$ is the set of points "at infinity" and Q the set of all other points of $P(V)$.

Proof: Choose a vector $v_0 \in V$ not in W and construct the coset $W + v_0$ of W in V. Each point $q \in Q$ is a one-dimensional vector subspace of V not contained in W, so it intersects this coset $W + v_0$ in exactly one point $w + v_0$; conversely, each $w + v_0$ for $w \in W$ determines a point $q \in Q$ consisting of all the multiples $(w + v_0)\kappa$ for $\kappa \in F$. Thus we can and will identify the set Q of the theorem with the coset $W + v_0$. As a coset of W, Q is thus (§ 2) an affine n-space with translation space isomorphic to W.

Now let the projective automorphism $P(t)$ lie in the group G described in the theorem, so that $t(w) \in W$ for each $w \in W$, while tv_0 is not in W, so is $tv_0 = v_0\kappa + u$ for some $\kappa \neq 0$ and some $u \in W$. Multiplying t by κ^{-1} does not change the projective automorphism $P(t)$, and gives a new t with $tv_0 - v_0 \in W$. Thus we can and will identify the group G with the group of all those vector space automorphisms $t: V \to V$ with $tv_0 - v_0 \in W$ and $t_* W = W$. Now each vector $u \in W$ and each automorphism $s \in GL(W)$ together determine an automorphism $t = (u, s) \in G$ by the equations

$$(u, s)v_0 = u + v_0, \qquad (u, s)w = s(w) \qquad \text{for all } w \in W,$$

for these equations will determine the effect of the linear transformation (u, s) on every vector of V. Every element t of the group G has the form $t = (u, s)$, for u the vector $u = tv_0 - v_0 \in W$ and s the restriction of t to W. The elements $(u, 1)$ for $u \in W$ form a subgroup H of G consisting exactly of all those $t \in G$ which restrict to the identity on W. Now when any $t = (u, s)$ in G is restricted to the set $Q = P(V) - P(W)$ its action on any point $w + v_0 \in Q$ is $(u, s)(w + v_0) = s(w) + u + v_0$. If $s = 1$, this equation states that the subgroup H is the (additive) group of all translations of the affine space $Q = W + v_0$. For any $s \in GL(W)$, this equation states that (u, s) is that affine automorphism of Q which has trace s and which carries the "origin" v_0 to the point $u + v_0 \in Q$. Since this includes all possible automorphisms of the affine space Q, the proof is complete.

This theorem is often used to derive affine geometry from projective geometry.

Bibliography

GENERAL

BIRKHOFF, GARRETT, and BARTEE, T. C. *Modern Applied Algebra.* New York: McGraw-Hill, 1970.

BIRKHOFF, GARRETT, and MAC LANE, SAUNDERS. *A Survey of Modern Algebra* (4th ed.). New York: Macmillan, 1977.

COHEN, P.M. *Algebra.* 2 vols. New York: Wiley, 1974, 1977.

GODEMENT, ROGER. *Algebra* (translated from the French). Houghton Mifflin, 1968.

HERSTEIN, I. N. *Topics in Algebra* (2nd ed.). New York: Wiley, 1975.

HUNGERFORD, THOMAS W. *Algebra.* New York: Holt, Rinehart and Winston, 1974. Reprinted, New York: Springer-Verlag.

JACOBSON, N. *Basic Algebra I; Basic Algebra II.* San Francisco: W. H. Freeman, 1974, 1976; 2nd ed. 1985, 1980.

LANG, SERGE, *Algebra.* Reading, Mass.: Addison-Wesley, 1965; 2nd ed. 1984.

VAN DER WAERDEN, B. L. *Algebra* (translation of *Moderne Algebra*). New York: Ungar, 1970.

SET THEORY

GODEL, KURT. *The Consistency of the Axiom of Choice and of the Generalized Continuum-Hypothesis with the Axioms of Set-Theory* (Annals of Mathematics Studies No.2). Princeton, N. J.: Princeton University Press, 1940.

HALMOS, PAUL R. *Naive Set Theory* (2nd ed.). New York: Springer-Verlag, 1974.

JECH, THOMAS. *Set Theory.* New York: Academic Press, 1978.

LINEAR ALGEBRA

BOURBAKI, NIKOLAS. *Elements of Modern Algebra I* (translated from the French). Reading, Mass.: Addison-Wesley, 1974.

CURTIS, CHARLES W. *Linear Algebra, An Introductory Approach.* New York: Springer-Verlag, 1981.

GREUB, WERNER. *Linear Algebra* (4th ed.). New York: Springer-Verlag, 1974.

HALMOS, PAUL. *Finite-Dimensional Vector Spaces* (2nd ed.). New York: Springer-Verlag, 1974.

SOLIAN, ALEXANDRU. *Theory of Modules* (translated from the Romanian). New York: Wiley, 1977.

STRANG, GILBERT. *Linear Algebra and Its Applications* (2nd ed.). New York: Academic Press, 1980.

GROUP THEORY

FUCHS, LASZO. *Abelian Groups.* New York: Academic Press. Vol. 1, 1970; vol. 2, 1973.

GORENSTEIN, DANIEL. *Finite Groups.* (2nd ed.). New York: Chelsea Pub. Co., 1980.

ROTMAN, J. J. *The Theory of Groups* (3rd ed.). Boston: Allyn and Bacon, 1984.

GALOIS THEORY

ARTIN, E. *Galois Theory* (Notre Dame Mathematical Lectures No. 2; 2nd ed.). Notre Dame, Ind.: Notre Dame University Press, 1944.

GAAL, LISL. *Classical Galois Theory with Examples* (3rd ed.). New York: Chelsea Pub. Co., 1979.

GARLING, D. J. H. *A Course in Galois Theory*. Cambridge, England: Cambridge University Press, 1986.

KAPLANSKY, IRVING. *Fields and Rings* (2nd ed.). Chicago: University of Chicago Press, 1972.

STEWART, IAN. *Galois Theory. New York:* Wiley, 1973.

NUMBER THEORY

HARDY, G. H., and WRIGHT, E. M. *An Introduction to the Theory of Numbers* (5th ed.). New York: Oxford University Press, 1979.

NIVEN, IVAN, and ZUCKERMAN, H. S. *An Introduction to the Theory of Numbers* (3rd ed.). New York: Wiley, 1972.

O'MEARA, O. T. *Introduction to Quadratic Forms.* New York: Springer-Verlag, 1973.

ALGEBRAIC NUMBER THEORY

LANG, SERGE. *Algebraic Number Theory* (2nd ed.). New York: Springer-Verlag, 1986.

SAMUEL, PIERRE. *Algebraic Theory of Numbers* (translated from the French). London: Kersham Publ. Co. Ltd., 1971.

STEWART, I. N., and TOLL, D. O. *Algebraic Number Theory.* New York: Wiley, 1979.

WEIL, ANDRÉ. *Basic Number Theory.* New York: Springer-Verlag, 1967.

WEISS, E. *Algebraic Number Theory.* (2nd ed.). New York: Chelsea.

COMMUTATIVE ALGEBRA

KAPLANSKY, IRVING. *Commutative Rings* (2nd ed.). Chicago: University of Chicago Press, 1974.

ZARISKI, OSCAR and SAMUEL, PIERRE. *Commutative Algebra I and II* (Graduate Texts in Mathematics, Vols. 28 and 29; 2nd ed.). New York: Springer-Verlag, 1975.

RING THEORY

AUSLANDER, MAURICE, and BUCHSBAUM, DAVID A. *Groups, Rings, Modules.* New York: Harper and Row, 1974.

HERSTEIN, I. N. *Topics in Ring Theory.* Chicago: University of Chicago Press, 1965.

LATTICE THEORY AND BOOLEAN ALGEBRA

ABBOTT, J. C. *Sets, Lattices and Boolean Algebras.* Boston: Allyn and Bacon, 1969.

BIRKHOFF, GARRETT. *Lattice Theory* (3rd ed.). Providence, R.I.: American Mathematical Society, 1966.

SIKORSKI, R. *Boolean Algebra* (Ergebnisse de Mathematik und ihrer Grenzgebiete, neue Folge, Vol. 25; 2nd ed.). New York: Springer-Verlag, 1984.

HOMOLOGICAL ALGEBRA

HILTON, P. J. and STAMMBACH, U. A. A Course in Homological Algebra. New York: Springer-Verlag, 1971.

MAC LANE, SAUNDERS, *Homology*. New York: Springer-Verlag, 1963.

CATEGORY THEORY

BARR, MICHAEL, and WELLS, CHARLES. *Toposes, Triples and Theories*. New York: Springer-Verlag, 1985.

JOHNSTONE, PETER. *Topos Theory*. New York: Academic Press, 1978.

MAC LANE, SAUNDERS. *Categories for the Working Mathematician*. New York: Springer-Verlag, 1972.

PAREIGIS, BODO. *Categories and Functors*. New York: Academic Press, 1970.

UNIVERSAL ALGEBRA

BURRIS, STANLEY, and SANKAPPANAVAR, H. P. *A Course in Universal Algebra*. New York: Springer-Verlag, 1980.

COHN, P. M. *Universal Algebra* (Rev. Ed.). Hingham, Mass.: Kluwer Academic Publishers, 1981.

GRÄTZER, GEORGE. *Universal Algebra* (2nd ed.). New York: Springer-Verlag, 1979.

SCHUBERT, HORST. *Categories* (translated from the German). New York: Springer-Verlag, 1972.

MULTILINEAR ALGEBRA AND DIFFERENTIAL GEOMETRY

FLANDERS, H. *Differential Forms with Applications to the Physical Sciences*. New York: Academic Press, 1963.

GREUB, WERNER. *Multilinear Algebra* (2nd ed.). New York: Springer-Verlag, 1978.

KOBAYASHI, S. and NOMIZU, K. *Foundations of Differential Geometry*, Vol. I. New York: Wiley, 1963.

STERNBERG, S. *Lectures on Differential Geometry* (2nd ed.). New York: Chelsea Pub. Co., 1983.

THORPE, JOHN A. *Elementary Topics in Differential Geometry*. New York: Springer-Verlag, 1979.

GROUP REPRESENTATIONS

CURTIS, CHARLES W., and REINER, IRVING. *Methods of Representation Theory*. New York: Wiley, Vol. 1, 1981; Vol. 2, 1987.

SHAW, RONALD. *Linear Algebra and Group Representation*. Orlando, Fla: Academic Press, Vol. 1, 1982; Vol. 2, 1983.

Index

To the Appendix

Index